Biological Regulation and Development

Volume 2

Molecular Organization
and Cell Function

Biological Regulation and Development

Series Editor
ROBERT F. GOLDBERGER, *Laboratory of Biochemistry, National Cancer Institute*

Editorial Board
PAUL BERG, *Department of Biochemistry, Stanford University Medical Center*
ROBERT T. SCHIMKE, *Department of Biological Sciences, Stanford University*
KIVIE MOLDAVE, *Department of Biological Chemistry, California College of Medicine, University of California, Irvine*
PHILIP LEDER, *Laboratory of Molecular Genetics, National Institute of Child Health and Human Development*
LEROY E. HOOD, *Biology Department, California Institute of Technology*

A Continuation Order Plan is available for this series. A continuation order will bring delivery of each new volume immediately upon publication. Volumes are billed only upon actual shipment. For further information please contact the publisher.

Biological Regulation and Development

Volume 2

Molecular Organization and Cell Function

Edited by
Robert F. Goldberger
National Cancer Institute
Bethesda, Maryland

PLENUM PRESS · NEW YORK AND LONDON

Library of Congress Cataloging in Publication Data

Main entry under title:

Molecular organization and cell function.

   (Biological regulation and development; v. 2)
   Includes bibliographical references and index.
   1. Cell physiology. 2. Molecular biology. 3. Biological control systems. I. Goldberger,
Robert F. II. Series.
QH631.M57   574.87'6   80-19935
ISBN 0-306-40486-9

Contributors

Merton R. Bernfield
  Department of Pediatrics
  Stanford University School of
    Medicine
  Stanford, California

Jay A. Berzofsky
  The Metabolism Branch
  National Cancer Institute
  National Institutes of Health
  Bethesda, Maryland

Max D. Cooper
  The Cellular Immunobiology Unit of
    the Tumor Institute, Departments
    of Pediatrics and Microbiology, and
    the Comprehensive Cancer Center
  University of Alabama in Birmingham
  Birmingham, Alabama

J. C. Gerhart
  Department of Molecular Biology
  University of California
  Berkeley, California

John F. Kearney
  The Cellular Immunobiology Unit of
    the Tumor Institute, Departments
    of Pediatrics and Microbiology, and
    the Comprehensive Cancer Center
  University of Alabama in Birmingham
  Birmingham, Alabama

Jonathan King
  Department of Biology
  Massachusetts Institute of Technology
  Cambridge, Massachusetts

Alexander R. Lawton
  The Cellular Immunobiology Unit of
    the Tumor Institute, Departments
    of Pediatrics and Microbiology, and
    the Comprehensive Cancer Center
  University of Alabama in Birmingham
  Birmingham, Alabama

Paul C. Letourneau
  Department of Anatomy
  University of Minnesota
  Minneapolis, Minnesota

Robert M. Macnab
  Department of Molecular Biophysics
    and Biochemistry
  Yale University
  New Haven, Connecticut

Dale L. Oxender
  Department of Biological Chemistry
  The University of Michigan Medical
    School
  Ann Arbor, Michigan

Steven C. Quay
    Department of Pathology
    Harvard Medical School and
        Massachusetts General Hospital
    Boston, Massachusetts
    Present Address: Department of
        Pathology
    Stanford University School of
        Medicine
    Stanford, California

Peter N. Ray
    Department of Medical Genetics
    University of Toronto
    Toronto, Canada

B. D. Sanwal
    Department of Biochemistry
    University of Western Ontario
    London, Canada

Alan N. Schechter
    Laboratory of Chemical Biology
    National Institute of Arthritis,
        Metabolism, and Digestive Diseases
    National Institutes of Health
    Bethesda, Maryland

James A. Spudich
    Department of Structural Biology
    Sherman Fairchild Center
    Stanford University School of
        Medicine
    Stanford, California

Kenneth A. Thomas
    Merck, Sharp and Dohme Research
        Laboratories
    Rahway, New Jersey

Dennis G. Uyemura
    Department of Biochemistry
    State University of New York
    Stony Brook, New York

Preface

The motivation for us to conceive this work on regulation was mainly our belief that it would be fun, and at the same time productive, to approach the subject in a way that differs from that of other treatises. We thought it might be interesting and instructive—for both author and reader—to examine a particular area of investigation in a framework of many different problems. Cutting across the traditional boundaries that have separated the subjects in past volumes on regulation is not an easy thing to do—not because it is difficult to think of what interesting topics should replace the old ones, but because it is difficult to find authors who are willing to write about areas outside those pursued in their own laboratories. Anyone who takes on the task of reviewing a broad area of interest must weave together its various parts by picking up the threads from many different laboratories, and attempt to produce a fabric with a meaningful design. Finding persons who are likely to succeed in such tasks was the most difficult part of our job.

In the first volume of this treatise, most of the chapters dealt with the mechanisms of regulation of gene expression in microorganisms. This second volume involves a somewhat broader area, spanning the prokaryotic–eukaryotic border. It begins with a discussion of the various mechanisms by which the catalytic activities of enzymes are regulated, the ways in which altering the rate of protein turnover affects the life of the cell, and the general principles that apply to the process by which nascent polypeptide chains fold into the specific three-dimensional structures of the corresponding native proteins. The volume goes on to discuss regulation and development as reflected in phage morphogenesis and in cell motility, shape, and adhesion. Regulation of transport processes, the role of gradients in development, and how cells sense their environment are considered next. The final chapters concern current concepts of cellular regulation in the field of immunology.

For many of the chapters in this volume, publication comes rather a long time after their original submission. We hope that the extensive editing and rewriting that took so much of this time will prove to have been worthwhile. It was our conviction that the greatest value of these chapters is as organized groups of concepts rather than as a source of details of the latest experiments. But we also must apologize to the authors who may have found that they were in for a lot more work—and a longer delay—than they had anticipated.

Robert F. Goldberger   Philip Leder
Paul Berg   Kivie Moldave
Leroy E. Hood   Robert T. Schimke

vii

Contents

3   Regulation of Structural Protein Interactions as Revealed in Phage Morphogenesis

JONATHAN KING

4   Mechanisms Regulating Pattern Formation in the Amphibian Egg and Early Embryo

J. C. GERHART

5   Biochemistry and Regulation of Nonmuscle Actins: Toward
    an Understanding of Cell Motility and Shape
    Determination

DENNIS G. UYEMURA AND JAMES A. SPUDICH

6   The Regulation of Cell Behavior by Cell Adhesion

PAUL C. LETOURNEAU, PETER N. RAY, AND MERTON R.
BERNFIELD

7   Sensing the Environment: Bacterial Chemotaxis

ROBERT M. MACNAB

8   Regulation of Membrane Transport

STEVEN C. QUAY AND DALE L. OXENDER

9   Immunoglobulin Diversity: Regulation of Expression of Immunoglobulin Genes during Primary Development of B Cells

ALEXANDER R. LAWTON, JOHN F. KEARNEY, AND MAX D. COOPER

10 Immune Response Genes in the Regulation of Mammalian Immunity

JAY A. BERZOFSKY

Regulation of Enzyme Activity

B. D. SANWAL

1 Introduction

A glance at any map showing the scope, theory, principles, and techniques of metabolic pathways is sufficient for one to realize the need for regulation of the myriad enzymes that catalyze the individual reactions in these pathways. A controlled and coordinated flow of metabolites in the interconnected catabolic, anabolic, and amphibolic routes is absolutely essential to support the ever-changing energy and growth demands of a cell (Davis, 1961; Sanwal, 1970). This is achieved in all living systems principally by the regulation of critical, strategically placed enzymes of individual pathways. (See Goldberger's definition of "regulation" in Volume 1, Chapter 1.) Control can be exerted by a change in the quantities or catalytic efficiencies of the enzymes. The quantity of an enzyme can be regulated by an alteration of either its rate of synthesis or its rate of degradation. The former encompasses regulatory mechanisms that act at the genetic level, such as induction and repression, and are discussed in Volume 1 of this treatise; the latter encompasses regulatory mechanisms that act at the level of protein turnover, and are discussed in Chapter 2 of this volume.

We will primarily be concerned here with regulatory mechanisms that *reversibly* alter the catalytic properties of enzymes, deliberately omitting discussion of mechanisms that lead to irreversible alterations of enzymic proteins. Examples of irreversible changes are proenzyme–enzyme, prohormone–hormone, or other conversions resulting from limited proteolysis, such as those that occur in the formation of collagen from procollagen (Bellamy and Bornstein, 1971), the conversion of fibrinogen to fibrin (Davie and Fujikawa, 1975), the formation of serine proteases from their corresponding zymogens (Neurath *et al.*, 1973), and conversion of neutral fructose diphosphatase of mammalian organs to the alkaline form (Horecker *et al.*, 1975). Although much less well understood than the above-mentioned

B. D. SANWAL • Department of Biochemistry, University of Western Ontario, London, Canada, N6A 5C1

systems, certain other phenomena, such as the modifications of peptide acceptors by NH_2-terminal addition of aminoacyl residues (Soffer, 1974; Deutch and Soffer, 1975) or tyrosylations (Raybin and Flavin, 1975), are also examples of irreversible modifications. There are indications (Deutch and Soffer, 1975) that the purpose of some of these alterations may be to prepare the proteins for proteolytic degradation, rather than to regulate their activites. The proenzyme–enzyme conversions, on the other hand, are terminal events in the pathway of the biosynthesis of the enzymes and may have regulatory significance in the initiation of physiological functions (Neurath et al., 1973; Horecker et al., 1975). The regulation in these cases is functionally, although not operationally equivalent to induction of enzyme synthesis and is not involved in the control of enzyme activity per se.

In the following account the basic concepts of regulation of enzyme activity have been emphasized, without too much concern over the details of individual enzymes and their regulatory effectors. This approach has necessitated a careful and perhaps a somewhat biased selection of published references for discussion. Detailed descriptions of metabolic regulatory circuits, however, have already appeared (Sanwal, 1970; Stadtman, 1970; Sanwal et al., 1971) and are reviewed in various journals.

2 Categories of Enzymic Controls

In order to systematize the immense amount of knowledge in this area, the following categories of controls may be considered: (1) active site controls, (2) allosteric controls, and (3) enzyme modifications. This classification is somewhat artificial but helps in focusing the principal features of control in each category. Physiologically, the purpose of each of these control mechanisms is the same, namely, to maintain metabolic fluxes through pathways commensurate with the growth and energy demands of a cell.

3 Active Site Controls

All enzyme-catalyzed reactions are subject to the laws of mass action, and the rate of catalysis depends on the amounts of substrate and products available and their affinities for the enzyme. Examination of the Michaelis–Menten substrate–velocity curves makes this point immediately apparent. Two properties of the Michaelian hyperbola should be emphasized. At values below or around the K_m of the substrate any increase in substrate level increases the velocity of the enzyme in a sensitive manner, whereas at concentrations above the K_m, increases in substrate level have only slight effects on the velocity. High substrate concentrations produce a "buffering effect" on the velocity of the reaction. This property is of obvious regulatory significance when large amounts of substrates (such as a carbohydrate energy source) become suddenly available to a cell (Stebbing, 1974).

Regulation of enzymes by substrate availability is a generalized mechanism and must operate with all enzymes irrespective of whether those enzymes are also controlled by other means, such as feedback inhibition. To the extent that it operates on all enzymes, active site control may be considered as an example of "primitive" control—that is, one that depends on the nature of the active site of the enzyme alone. As we will see in Section 4, much more sophisticated regulatory devices are built around the presence of allosteric sites that are distinct from active sites and apparently are a more recent acquisition of certain crucial enzymes of given metabolic pathways (Engel, 1973). This is not to suggest that active site or primitive controls are of lesser significance than are allosteric controls. Indeed, it has been proposed that catalytic sites in some cases may have evolved to fulfill regulatory roles (Atkinson, 1970). This may be so particularly in the case of active sites that bind coenzymes and adenine nucleotides (Atkinson, 1970). There are several coenzymes that exist in the cell in an oxidized and a reduced form. Examples include the nicotinamide adenine nucleotides, NAD^+ and NADH, and $NADP^+$ and NADPH. The total intracellular pool of these coenzymes (reduced + oxidized) and the reduced coenzyme/oxidized coenzyme ratios, averaged over a short time scale, tend not to fluctuate widely because of the activities of several dehydrogenases that catalyze equilibrium reactions (Krebs, 1969). Because of the uncoordinated activity of these dehydrogenases in the cell, however, the concentration of NADH or NAD^+ may vary momentarily without significant alterations in its total intracellular concentration. The NAD^+ mole fraction ($NAD^+/NAD^+ + NADH$) may then determine or regulate the activity of the dehydrogenases that bind NAD^+ and NADH with different affinities at the active coenzyme site (Atkinson, 1970; Shen and Atkinson, 1970). Contingent on these affinities the velocity response curves can, at least theoretically, vary widely (from a sigmoidal to a hyperbolic shape) with a change in NAD^+ mole fraction (Atkinson, 1970).

Like the coenzymes discussed above another class of compounds, the adenine nucleotides (ATP, ADP), are interconverted by various kinases during metabolic transformations. In biosynthetic reactions ATP is utilized and ADP is generated, whereas the reverse is true of biodegradative reactions. It has been suggested that the activities of the kinases are modulated by the varying proportions of ATP and ADP in the adenylate pool via competition at the active adenylate sites of enzymes (Atkinson, 1968; Atkinson, 1970). Perhaps the active sites of the kinases have evolved to provide adjustment for an optimal ratio of binding energies for the two nucleotides. For the kinases involved in biosynthesis (those utilizing ATP) the velocity response curves with changing ATP mole fractions *in vitro* (Atkinson, 1968) are similar to those obtained with the coenzyme pair $NAD^+ : NADH$. The adenylates, however, introduce some added complications; essentially, unlike the two-component system of the coenzymes, adenylates consist of a three-component system (ATP, ADP, and AMP). This is because of the presence of the ubiquitous adenylate kinase that catalyzes the reaction

$$ATP + AMP \rightleftharpoons 2ADP \tag{1}$$

The mole fraction of ATP and ADP in the adenylate pool varies with the so-called *energy charge* (Atkinson and Walton, 1967):

$$\text{Energy charge} = \frac{ATP + \frac{1}{2}ADP}{ATP + ADP + AMP} \tag{2}$$

The important point for our discussion, however, is that purely competitive effects between ATP and ADP at catalytic sites may lead to regulation of enzyme activity and the velocity response of the enzyme may be related to the inherent properties of the active center. [Adenylate pairs (ATP–AMP and ATP–ADP) also regulate enzyme activities by competition at allosteric sites, but regulation of this kind (see Section 4) is completely different from the active site controls.] Lest the impression be obtained that the controls discussed above have actually been proved to occur *in vivo*, it should be pointed out that the possibility of active site controls has been suggested only on teleological grounds and on the basis of work done with purified enzymes *in vitro*. Two main objections may be raised against these studies. First, the concentrations of enzymes *in vivo* are several orders of magnitude higher than those ordinarily utilized in *in vitro* experiments (Srere, 1967), and the steady-state assumptions usually applied to kinetic measurements, in which the concentration of substrate is much greater than is the concentration of free enzyme, may not hold under physiological conditions. The second objection, valid for enzyme systems with two substrates and two or more products (kinases and dehydrogenases, for instance), is that the rate–concentration plots with varying energy charge can be markedly altered depending on whether or not the enzyme being assayed is saturated with the nonnucleotide substrate (Purich and Fromm, 1972). Although these objections are theoretically valid for all enzyme systems, including allosteric ones, they are rather minor ones. More recent studies on several enzymes, especially those of prokaryotes and lower eukaryotes, have shown that *in situ* (in permeabilized cells) the enzymes do not show any marked deviations from results obtained with the same enzymes *in vitro* (Reeves and Sols, 1973; Weitzman and Hewson, 1973; Serrano *et al.,* 1973; Reitzer and Neet, 1974). In addition Sols and Marco (1970) have pointed out that the assumption that free substrate concentration is higher than is enzyme concentration probably remains valid for most of the enzymes and metabolites *in vivo*.

4 Allosteric Regulation

4.1 General Description

While all enzymes are regulated by catalytic site controls, a few enzymes, strategically placed at the beginnings or branch points of metabolic sequences, are subject to regulation by end products or other metabolites. The existence of this "feedback" regulation of enzymes was first clearly pointed out by Umbarger (1956) and by Yates and Pardee (1956). In a now classic paper, Umbarger (1956) demonstrated that in *Escherichia coli* isoleucine inhibits the first enzyme of its pathway, threonine deaminase. Similarly, cytidine triphosphate (CTP), the end product of the pathway for pyrimidine biosynthesis in *E. coli,* inhibited the activity of the initial enzyme of the pathway, aspartate transcarbamylase (Yates and Pardee, 1956; Pardee, 1971). Following on the heels of these two discoveries, several biosynthetic pathways in prokaryotes were shown to have initiating enzymes, the activities of which were inhibited by the specific end products (Monod *et al.,* 1963). Two generalizations emerged early during the course of these studies. First, the susceptible, or target, enzyme invariably catalyzes a nonequilibrium reaction; and second, there is no steric resemblance between the end-product inhibitor and the substrate with which that inhibitor seemed to compete at the enzyme surface. With both threonine deaminase (Changeux, 1961) and aspartate transcarbamylase (Gerhart and Pardee, 1961, 1962), it was further found that the use of various treatments (time-dependent aging or heating) could alter the feedback sensitivity of the enzymes without appreciably changing the catalytic activities. All these findings led Monod *et al.* (1963) to suggest that regulatory enzymes must possess a site distinct from the active site, which they labeled *allosteric* (*allo* = other). Allosteric regulation—that is, control of the catalytic site of an enzyme by binding an allosteric ligand (effector or modulator) at the allosteric site, is either negative (decreasing the activity of the enzyme) or positive (increasing the activity of the enzyme). One of the earliest examples of allosteric regulation that was discovered, though not conceived of in this way, was positive regulation. Cori *et al.* (1938) found that the activity of phosphorylase *b* was almost totally dependent on the positive modulator, AMP. Several enzymes in the amphibolic and degradative pathways have now been found to be regulated by positive feed-forward mechanisms, or precursor activation (Sanwal, 1970; Sanwal *et al.*, 1971).

Seldom is an allosteric enzyme exclusively controlled by either positive or negative means. The majority of allosteric enzymes are affected by both positive and negative modulators. The two modulators may bind either at the same allosteric site or at two dinstinct allosteric sites. Depending on the number and nature of the interactions between the allosteric sites, regulatory systems of several kinds (feedback inhibition as well as activation) are now recognized. The important ones are cooperative feedback modulation, concerted or multivalent inhibition, and cumulative inhibition.

Cooperative feedback was first described by Caskey *et al.* (1964) in the pathway for purine biosynthesis in animal cells. The first reaction of the purine pathway, catalyzed by glutamine phosphoribosylpyrophosphate amidotransferase, was found to be inhibited separately by the 6-hydroxypurine ribonucleotides, GMP and IMP, and by the 6-aminopurine ribonucleotides, AMP and ADP. In the presence of a mixture of these two groups of nucleotides (GMP and AMP or IMP and ADP), however, the total amount of inhibition was more than additive. Similarly, cooperative activation occurs in the control of phosphoenolpyruvate carboxylase of *Salmonella*. This enzyme catalyzes the formation of oxalacetate from phosphoenolpyruvate (Maeba and Sanwal, 1969), and is activated by both fructose 1,6-diphosphate and GTP separately, but when these activators are added together the activation is much higher than the sum of the individual activations. Cooperative modulation has recently been shown to be due to the presence of a separate allosteric site for each of the effectors on the enzyme surface; binding of one effector at its site increases the affinity of the other site for the second effector (Smando *et al.*, 1974).

In the case of concerted inhibition, the two or more feedback inhibitors of an enzyme do not significantly inhibit the enzyme singly, but when present together they cause considerable inhibition. Aspartokinases of some prokaryotes are affected in this way by the two end-products of the branched aspartate pathway, threonine and lysine (Datta and Gest, 1964; Paulus and Gray, 1964). In eukaryotes, an example of concerted inhibition is provided by beef-liver glutamate dehydrogenase, which requires two effectors, NADH and GTP, for inhibition (Goldin and Frieden, 1971). The molecular basis of concerted inhibition is that the binding of one effector at its allosteric site is necessary for significant binding of another effector at a separate site, and only when both effectors are bound is inhibition of the enzyme produced.

Finally, in the case of cumulative inhibition there is no cooperation or antagonism between several inhibitors of an enzyme. Each of the inhibitors brings about some degree of inhibition, irrespective of whether other inhibitors are present. If two inhibitors are simultaneously present in saturating concentrations, the total residual activity of the enzyme is equal to the product of the residual activities obtained in the presence of each of the inhibitors tested separately. This remarkable regulatory mechanism was discovered by Woolfolk and Stadtman (1964) for control of the glutamine synthase of *E. coli,* and so far this example remains unique for control of this kind. The physical basis for cumulative inhibition has not been elucidated with any degree of certainty. Equilibrium binding studies (Stadtman and Ginsburg, 1974) suggest distinct binding sites for each of the effectors on the enzyme surface without significant conformational interaction between those sites.

However, electron spin resonance measurements indicate not only that several effectors share common sites at which they compete but also that some effectors may compete at or very near the active site (Dahlquist and Purich, 1975). Clearly, the molecular basis for cumulative inhibition is quite complex.

Almost all the effects described above occur in the initiating allosteric enzymes of unidirectional pathways. The unidirectionality is conferred on the pathways mostly by the physiologically irreversible nature of the reaction catalyzed by the allosteric enzyme. Since this occurs in several pathways some workers have concluded that most allosteric enzymes catalyze irreversible reactions. This is not necessarily so. In recent years several enzymes catalyzing highly reversible reactions have been found to be allosteric. In at least three cases the allosteric effectors are capable of modulating the activities of the enzymes only, or predominantly, in *one* direction. The first case of this kind was discovered by LeJohn (1968): the NAD^+-dependent glutamate dehydrogenase of an aquatic mold, *Blastocladiella*, that catalyzes the oxidative deamination of glutamate in the forward direction and the reductive amination of α-ketoglutarate in the reverse direction.

$$\text{L-Glutamate} + NAD^+ + H_2O \rightleftharpoons NH_4^+ + NADH + \alpha\text{-ketoglutarate} \qquad (3)$$

The oxidative deamination is specifically inhibited by several effectors that have no effect on the reductive amination. Perhaps only the binding of NAD^+ to the enzyme is inhibited by the effectors, but this is indeed remarkable because NAD^+ and $NADH$ must obviously bind to the same catalytic site.

The second case of this kind (Klemme, 1976) is the enzyme phosphoenolpyruvate carboxykinase of *Rhodospirillum rubrum* that catalyzes the freely reversible reaction

$$\text{Phosphoenolpyruvate} + CO_2 + ADP \rightleftharpoons \text{oxalacetate} + ATP \qquad (4)$$

The allosteric inhibitor of this enzyme, ATP, inhibits the enzyme only in the direction of carboxylation. In a third case, that of the multifunctional glucose-6-phosphatase of mammalian tissues, discriminant control on the synthetic activity (formation of glucose-6-phosphate) and hydrolytic activity (production of glucose) is exerted by various effectors. For example, long-chain acyl-CoA esters and cupric ions activate the phosphate transferase activity while inhibiting hydrolytic activity (Nordlie, 1976). The molecular mechanisms of these unidirectional effects are completely unknown.

4.2   Properties of Allosteric Systems

4.2.1   The Basic Allosteric Concept

As was mentioned earlier, Monod *et al.* (1963, 1965) proposed the allosteric concept mainly to emphasize the fact that regulatory enzymes have sites that bind regulatory ligands and are separate from the catalytic centers. All the effects on the activity of the enzyme, both positive and negative, were considered to be due entirely to the indirect changes of the active site produced as a consequence of effector binding at the allosteric site. Monod *et al.* (1965) further pointed out that, in addition to possessing regulatory sites, several allosteric enzymes also display sigmoidal rate–concentration plots rather than the conventional hyperbolic ones. Since the publication of these classical papers (Monod *et al.*, 1963, 1965), a vast amount of literature has accumulated on the kinetics of several enzyme systems, many of which have been labeled "allosteric" purely on the grounds that the substrate yields cooperative, or sigmoidal, initial velocity plots. While demonstration of a regulatory, or allosteric, site would be sufficient for an enzyme to be classified as allosteric, this classification cannot be made on the basis of kinetic data alone—that is, the presence of a sigmoidal rate–concentration plot.

The presence of a specific allosteric site has been demonstrated directly in several enzymes by thermodynamic binding studies, but such observations are few and far between (Goldin and Frieden, 1971; Ginsburg, 1969; Levitzki and Koshland, 1972; Smando *et al.*, 1974; Waygood *et al.*, 1976). This is largely attributable to the difficulty in getting sufficiently large amounts of enzyme to be able to investigate equilibrium binding of ligands. Fortunately, however, demonstration of an allosteric site is not entirely contingent on the application of physicochemical techniques. Early in the study of allosteric enzymes it was discovered that mild treatments of the enzymes, such as aging (storage at temperatures slightly above freezing), treatment with mercurials (Monod *et al.*, 1963; Gerhart and Pardee, 1962), treatment with urea (Gerhart and Pardee, 1962), high ionic strength (Murphy and Wyatt, 1965), freezing (Caskey *et al.*, 1964), moderate heating (Gerhart and Pardee, 1962), and diverse other mild chemical and physical treatments (Monod *et al.*, 1963) led to a partial or complete loss of the sensitivity of the enzyme to all allosteric effectors, positive or negative, without any significant change in catalytic properties. This phenomenon was termed *desensitization* by Monod *et al.* (1963), and its existence demonstrated that effector and substrate binding areas on the enzyme surface are topographically distinct. With suitable manipulations most of the allosteric enzymes can be desensitized. Along with the physicochemical agents cited above, these manipulations may involve genetic manipulations. For example, treatment of bacteria with an inhibitor analogue that binds at the allosteric site of a feedback sensitive enzyme may result in growth of mutant cells that have a desensitized enzyme (Cohen, 1965). Revertants of catalytic site mutants of allosteric enzymes also frequently produce desensitized enzymes (Cohen, 1965; Monod *et al.*, 1963).

The desensitized forms of enzymes usually resemble the native forms in most physical characteristics, such as molecular weight and subunit composition. Genetic evidence suggests that desensitization is not due to local denaturation of the allosteric site, but probably is due to a change in one critical amino acid residue that is required to maintain the conformational integrity of the site (Monod *et al.*, 1963, 1965). In a so far unique case, namely, aspartate transcarbamylase, desensitization by any of the physical and chemical agents leads to a dissociation of the enzyme into constituent regulatory and catalytic subunits (Gerhart and Schachman, 1965). The phenomenon of desensitization has figured prominently in the formulation of the allosteric transition theory of Monod and co-workers (Monod *et al.*, 1965).

4.2.2 Nature of Regulatory Enzymes

As discussed in Section 4.2.1 an allosteric site that serves as the receptor for chemical signals distinguishes a regulatory enzyme from others that are merely susceptible to catalytic site controls. In addition to the relatively simple question of whether or not an enzyme has a regulatory site, two other mutually interdependent aspects of allosteric enzymes have been intensively studied: subunit structure and kinetics. Studies on both these aspects of allosteric enzymes have helped to elucidate the molecular mechanisms involved in regulation of enzyme activity.

4.2.2a   Subunit Structure.   There are two questions of crucial interest in this area. First, is the subunit* that binds the regulatory effectors distinct from the catalytic subunit? Second, is a polymeric structure absolutely required for the control of enzymes activity?

From among the several allosteric enzymes that have been purified to homogeneity and studied by physicochemical means (a handy catalog is provided in Darnall and Klotz, 1975), only one is definitely known to have distinct regulatory and catalytic subunits. This is the aspartate transcarbamylase of *E. coli* alluded to in Section 4.2.1 (Gerhart and Schachman, 1965). This enzyme can be easily dissociated by treatment with *p*-hydroxymercuribenzoate into two subunits (Gerhart and Schachman, 1968). One of these, the regulatory subunit, binds only the allosteric effectors, CTP and ATP, while the other, the catalytic subunit, binds only the substrate, aspartic acid (or its analogues) in the presence of the second substrate, carbamyl phosphate. Six molecules of substrate are bound to each molecule of enzyme, corresponding to the actual number of monomers per catalytic subunit (Weber, 1968; Wiley and Lipscomb, 1968). In the native form, the enzyme shows positive cooperativity in the binding of substrate—that is, the binding of one molecule of substrate facilitates binding of a second molecule of substrate. This property is lost on dissociation— that is, the enzyme becomes desensitized—although the catalytic subunit remains active.

For some time it seemed that the ribonucleoside diphosphate reductase of *E. coli* might turn out to function like aspartate transcarbamylase, with distinct regulatory and catalytic subunits (Brown *et al.*, 1967). Ribonucleoside diphosphate reductase, which converts ribonucleoside diphosphates to the corresponding deoxyribonucleotides, is a complex of two dissimilar subunits, B1 and B2, each of which by itself is inactive. The B1 subunit has recently been found to contain both the substrate-binding site and the allosteric ligand-binding site, whereas the B2 protein contains nonheme iron and participates in the formation of the catalytic site (Döbeln and Reichard, 1976). Thus it is clear that the molecular arrangement of ribonucleoside diphosphate reductase is completely different from that of aspartate transcarbamylase.

Like ribonucleoside diphosphate reductase, several other enzymes are now known that have subunits of two different kinds. Among these are both allosteric and nonallosteric enzymes. Some examples of such *heteromeric* enzymes are carbamyl phosphate synthase (Trotta *et al.*, 1974), ribulose diphosphate carboxylase (Buchanan and Schürmann, 1973), and glutamate synthase (Miller and Stadtman, 1972).

The data available on the amino acid compositions of regulatory enzymes have shown nothing to distinguish these enzymes from other enzymes. The same lack of distinction is observed in studies on other parameters, such as cold lability, association–dissociation of subunits, and heat stability in the presence of substrates. The association of some regulatory enzymes into bifunctional complexes, such as aspartokinase I–homoserine dehydrogenase I of *E. coli* K12 (Cohen, 1969), also finds its counterpart in nonallosteric enzymes (Ginsburg and Stadtman, 1970).

It is clear from the above discussion, that, apart from a few exceptions, such as aspartate transcarbamylase, there are no basic and remarkable differences between the subunit structures and compositions of regulatory enzymes and those of nonregulatory enzymes.

---

*The terminology proposed by Monod *et al.* (1965) has been used in this chapter. An enzyme containing a finite number of identical subunits is referred to as an *oligomer*. Identical subunits of an oligomer are called *protomers*. The term *monomer* is used to describe a single polypeptide chain of a fully dissociated oligomer. The term subunit is left undefined and is used to refer to any chemically or physically identifiable submolecular entity within an oligomer (Monod *et al.*, 1965).

The question may now be asked whether or not allosteric phenomena are *always* associated with the oligomeric state of proteins. In other words, can strictly monomeric enzymes (those possessing only a single polypeptide chain) be allosteric? The idea that regulatory enzymes must be composed of more than one polypeptide chain has been firmly entrenched in biochemical thought. The origins of this dogma can be traced to the concerted transition theory formulated by Monod *et al.* (1965). Recently, several monomeric enzymes have been found to display cooperative rate–concentration plots: bovine pancreatic ribonuclease I (Rübsamen *et al.*, 1974), wheat germ hexokinase (Meunier *et al.*, 1974), and yeast hexokinase (Shill and Neet, 1975). These need not be considered allosteric enzymes because, as noted above, the presence of a distinct regulatory site, not kinetic behavior, is the only sufficient condition to categorize an enzyme as allosteric. Even within the framework of this narrow definition, at least three allosteric enzymes have now been shown to be monomeric *in vitro*. These are the ribonucleoside diphosphate reductase of *Lactobacillus leichmannii* (Panagou *et al.*, 1972; Goulian and Beck, 1966; Chen *et al.*, 1974), homoserine transacetylase of *Bacillus polymyxa* (Wyman and Paulus, 1975), and phosphoenolypyruvate carboxykinase of *Escherichia coli* (Wright and Sanwal, 1969; Goldie and Sanwal, 1980). The activity of ribonucleotide reductase is modulated by several nucleotide effectors (Panagou *et al.*, 1972; Goulian and Beck, 1966; Chen *et al.*, 1974) much as is the polymeric reductase from *E. coli*. Homoserine transacetylase, the first enzyme of methionine biosynthesis, is subject to concerted feedback inhibition by the end products, L-methionine and S-adenosylmethionine. Inhibitor specificity studies have suggested that there are separate sites on the enzyme for these inhibitors, distinct from the active site (Wyman and Paulus, 1975). Phosphoenolpyruvate carboxykinase is activated by calcium and an allosteric site for this ligand has been shown to occur (Goldie and Sanwal, 1980). The question of interest with all of these enzymes is whether the monomeric state is an artifact of purification. It is conceivable that these allosteric enzymes function as oligomers *in vivo*, a possibility that has been tested with only one enzyme, homoserine transacetylase, in permeabilized cells. In contrast to the hyperbolic rate–concentration curves *in vitro*, the enzymes assayed *in situ* yields sigmoidal plots for the inhibitors (Wyman et al., 1975). This behavior does not necessarily mean that the enzyme is oligomeric *in vivo*. Such studies remain to be done with other monomeric enzymes. It must be emphasized, however, that there is no compelling reason to believe that allosteric enzymes must always be polymeric.

4.2.2b   Kinetics and Related Properties.   While studies of subunit structure and physicochemical properties of regulatory enzymes have been few and far between, a vast amount of kinetic data has been accumulated that is difficult to analyze. It has become increasingly clear, however, that the original classification of allosteric enzymes into K and V systems, proposed by Monod *et al.* (1965), does not conform to reality. It may be recalled that according to this classification, the enzymes belonging to the K system yield sigmoidal, or cooperative (positive), initial velocity plots for both substrates and inhibitors *(homotropic interactions)*. The cooperativity is considerably increased in the presence of an allosteric inhibitor, but it disappears in the presence of an activator at saturating concentration—that is, the rate–concentration plot becomes hyperbolic (Fig. 1a). The allosteric effectors change not the V_{max} but only the K_m of the substrate (hence the name K system). In enzymes belonging to the V system, on the contrary, the rate–concentration plots of the substrate are hyperbolic (Fig. 1d), both in the absence and presence of the allosteric effectors. The effectors change only the V_{max}, not the K_m of the substrate. Plots of activity against inhibitor concentrations are, however, sigmoidal *(heterotropic interactions—that is, interactions between substrate and inhibitor sites)*. Frequently cited examples of K and V systems are phosphofructokinase of *E. coli* (Blangy *et al.,* 1968) and fructose diphosphatase from mammalian sources (Taketa and Pogell, 1965), respectively. It must be emphasized that "perfect" examples of V and K systems are rarities, if they exist at all, and the classification of Monod *et al.* (1965) is at present of purely historical interest.

Operationally, most of the regulatory enzymes may be placed into three major categories (Sanwal, 1970): (1) modulator-independent cooperative systems (MIC), (2) modulator-dependent cooperative systems (MDC), and (3) allosteric noncooperative systems (ANC). The MIC category includes enzymes in which one or both substrates give a cooperative response in the absence of effectors (Fig. 1a,c). Enzymes of this class may be further subdivided on the basis of whether the rate–concentration, or binding, curve for substrate shows a positive cooperativity [which is true of a majority of allosteric enzymes (Fig. 1a)], or negative cooperativity (Levitzki and Koshland, 1976; Fig. 1c). The MDC category includes enzymes that yield hyperbolic rate–concentration plots for substrate(s) in the absence of allosteric inhibitors, but cooperative plots in their presence (Fig. 1b). A well-known example of this class is the threonine deaminase of some bacteria (Maeba and Sanwal, 1966). Theoretically, the cooperative plots in MDC systems may be positive or negative, but in almost all the cases studied so far only positive cooperative plots have been obtained (Sanwal, 1970). Enzymes of the ANC system yield hyperbolic rate–concentration plots in the presence and in the absence of modulators (Fig. 1d): examples of such enzymes are ribonucleoside diphosphate reductase (Brown *et al.,* 1967; Brown and Reichard, 1969) and fructose diphosphatase from rabbit liver (Taketa and Pogell, 1965).

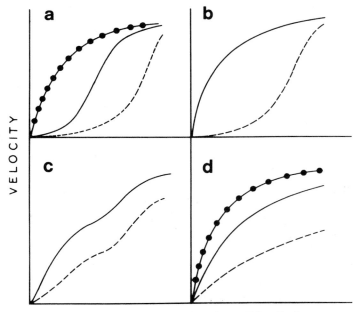

SUBSTRATE CONCENTRATION

Figure 1. The general shape of rate–concentration plots for allosteric enzymes of various kinds. The solid lines (———) denote the shapes of the curves in the presence of substrate alone. The broken lines (---) are plots in the presence of less than saturating amounts of an inhibitor. The beaded lines (•-•-•) are plots in the presence of saturating amounts of an activator. (a) Modulator-independent positive cooperative system, (b) modulator-independent negative cooperative system, (c) modulator-independent cooperative system, and (d) allosteric noncooperative system. Additional details are given in the text.

The categorization of allosteric enzymes on the basis of initial rate–concentration plots is purely for the sake of systematization of knowledge; it has no physical basis at all. For instance, the cooperativity of substrate in different enzymes of the MIC or MDC systems may occur on the basis of completely different molecular mechanisms. Furthermore, it is difficult to judge from velocity data alone (on which the classification is based) whether in ANC systems, as an example, noncooperativity is only apparent or indeed exists. For instance, in initial velocity studies fructose diphosphatase invariably yields hyperbolic plots with the substrate (Taketa and Pogell, 1965), but equilibrium binding studies show that it binds cooperatively (Pontremoli *et al.,* 1968). Conversely, examples are also known in which cooperativity of the substrate is seen only in initial velocity studies—that is, under conditions in which dilute solutions of enzymes are employed. Muscle phosphofructokinase (Mansour, 1963) and lactate dehydrogenase from *Aerobacter* (Sawula and Suzuki, 1970), for instance, yield positive cooperative plots in velocity measurements, but when high concentrations of the proteins are employed in equilibrium binding studies (Kemp and Krebs, 1967), or stopped-flow measurements (Sawula and Suzuki, 1970), the cooperativity completely, or nearly, disappears. This situation can very easily arise if ligand binding causes the enzyme to undergo aggregation or dissociation, a situation by no means uncommon in allosteric systems (Levitzki and Schlessinger, 1974; Levitzki, 1975). These examples do point to the hazards of considering all initial velocity data as equilvalent to thermodynamic binding at equilibrium. Binding data are preferable for making comparisons among different enzyme systems and are a necessity if comparisons are to be made between molecular models proposed for allosteric phenomena (Levitzki, 1975).

The classification of regulatory enzymes into the categories discussed above is of importance also from a physiological point of view. The substrate may be viewed not only as a substrate but also as a controlling element, negative or positive, in the MIC and MDC systems but not in ANC systems. The physiological advantages that accrue to the organism in having enzymes of one or the other of these categories at different points in various pathways is not known. It is clear, however, that positive cooperativity, because it inherently creates threshold effects, may be advantageous in controlling metabolic flow in an extremely sensitive manner. Negative cooperativity, on the other hand, would tend to insulate the enzyme from changes in metabolite concentration (Conway and Koshland, 1968); this may be necessary for some enzymes of major linear pathways that are required to have constant activity despite wide fluctuations in substrate concentrations. Somewhat similar considerations may apply to the control of initial enzymes of highly branched pathways or to those utilizing common intermediates. Let us take, as an example, two different enzymic pathways in which a common substrate (A) is converted to two end products (X and Y). X and Y are allosteric regulators of their respective initiating enzyme. A sudden increase in X might cause severe inhibition of the initial enzyme of its own pathway. This would result in an unnecessary diversion of A to Y, unless a mechanism existed to buffer the initial enzyme of pathway A $\rightarrow$ X against inhibition, at least in the short term. Frieden (1970) has pointed out that there are several allosteric enzymes that respond slowly (seconds to minutes) to sudden changes in ligand concentration. This response has been termed *hysteresis,* and enzymes displaying this response have been named *hysteretic* enzymes. These enzymes can be found in all MIC, MDC, and ANC categories. Several molecular mechanisms can cause hysteresis, such as ligand-induced isomerization of the enzyme, displacement of tightly bound ligands by other ligands, and association–dissociation reactions (Frieden, 1970).

In recent years considerable effort has been made to understand the molecular mechanisms underlying cooperativity in proteins and the kinetics of regulatory systems, but no consensus has emerged. Part of the confusion has arisen because of the overt or covert tendency of investigators to either generalize from very narrow studies made on a single enzyme, or to seek unifying concepts (where none may exist) from a study of few enzymes. For example, negative cooperativity (Fig. 1b) can arise in an oligomeric enzyme by any of a number of mechanisms: (1) by destabilizing interactions among protomers—that is, by successively weaker binding of ligand to subunits—if all subunits carry equivalent active sites; (2) by the binding of the substrate or other ligands to nonequivalent sites; or (3) by the binding of a ligand with different affinities to an associating–dissociating system. Two of these mechanisms, at least, have been documented in the literature for two different enzymes. Conway and Koshland (1968) demonstrated that, in rabbit-muscle glyceralde-hyde-3-phosphate dehydrogenase, NAD^+ binds with successively weaker affinities to the four monomers of the enzyme. In *E. coli* citrate synthase (Wright and Sanwal, 1971), however, negative cooperativity in the binding of α-ketoglutarate, an inhibitor, emanates from the differential binding of the ligand to an equilibrium mixture of tetramer and octamer. Thus there is no reason to believe that the same fundamental mechanism must determine negative cooperativity in all systems. Atkinson and Walton (1967) have correctly pointed out that each individual case must have evolved with a considerable degree of independence, and it need not be assumed that the mechanism of cooperativity (negative or positive) is the same for all regulatory enzymes.

Like negative cooperativity, positive cooperativity can *theoretically* invlove several different mechanisms. In fact, several theories have been put forward to explain sigmoidicity of rate–concentration plots. Theories based on alternate pathway mechanisms (Sweeny and Fisher, 1968; Ferdinand, 1966) and those that postulate "activating" sites to explain positive homotropic effects of substrates have largely remained unproven and need not be considered further here. Indeed, some common features of interaction of substrate and modifier binding to regulatory enzymes that have begun to emerge recently are not easily explicable on the basis of alternate pathway mechanisms. For instance, in several allosteric systems it is now known that the binding of a negative effector is either considerably enhanced by, or obligatorily requires, that the substrate be bound to the enzyme. Examples are bovine glutamate dehydrogenase (Frieden, 1967), citrate synthase of *E. coli* (Wright and Sanwal, 1971), ATP-phosphoribosyl pyrophosphate phosphorylase of *E. coli* (Klingsøyr and Atkinson, 1970), and fructose diphosphatase of rabbit liver (Sarngadharan *et al.*, 1969).

It is fair to state that the two leading theories that are capable of explaining cooperative phenomena and other general properties of regulatory enzymes in molecular terms are those of Monod *et al.* (1965) and Koshland *et al.* (1966). For technical and detailed evaluations of these theories, the reader is referred to two excellent reviews on this subject (Koshland, 1970; Levitzki, 1975). Very briefly, the model of Monod *et al.* (1965), the *concerted transition theory,* proposes that all allosteric enzymes are oligomers, and can exist in two conformations, or states (Fig. 2a), termed T (tight) and R (relaxed). The T and R states are in dynamic equilibrium defined by a dimensionless allosteric constant, L. For enzymes of the so-called K system (see Section 4.2.2b), the T state has a high affinity for allosteric inhibitors (I) and has a low affinity for substrates (S) and allosteric activators (A); the opposite is true for the R state. For enzymes of the V system, however, S has the same affinity for both T and R states. In keeping with the allosteric concept (Monod *et al.*, 1963), the binding sites for S, I, and A are considered to be quite distinct from each other. In a perfect enzyme of the K system, where the value of the constant L is very large (that is, almost all of the enzyme exists in the T state), binding of S or A is supposed to cause a conversion of the enzyme from the T to the R state. The most important assumption of the model is that when the conversion occurs there is a concerted transition of all of the subunits of the oligomer. In other words, there cannot be a hybrid enzyme molecule in which some subunits are in the T state and other subunits of the same molecule are in the R state (Fig. 2a). It is apparent that on the basis of this mechanism, S and A will yield sigmoidal rate–concentration or binding plots [homotropic interactions (Fig. 1a)] and I will yield only binding isotherms because most of the enzyme is assumed to be in the T state. In the presence of saturating A most of the enzyme will be in the R form, S will give hyperbolic binding curves, and I will give sigmoidal binding curves (heterotropic interactions). The phenomenon of desensitization is easily understood on the basis of the theory. Since transition of the oligomer from one state to another involves the realignment of a few critical

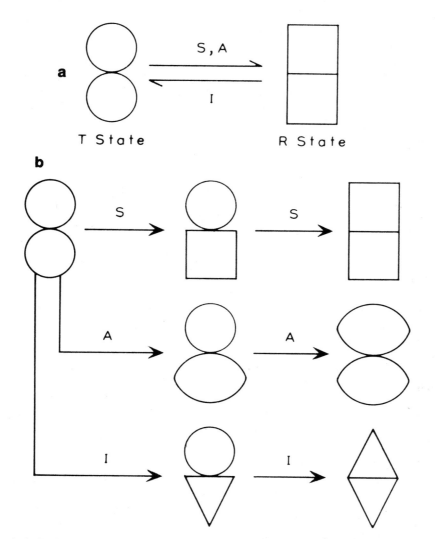

Figure 2. Stylized representation of the concerted model (a) and sequential model (b) for a dimeric enzyme. The various shapes are meant to convey conformational states diagrammatically. The symbols represent: T, tight; R, relaxed; S, substrate; A, activator; and I, inhibitor. Details are discussed in the text.

bonds, any treatment that modifies these bonds (genetic mutation, chemical or physical treatment, etc.) or prevents realignment will result in the "locking" of the oligomer into one or the other state. An enzyme of the K system locked in the R state, for instance, will be insensitive to inhibition by I because I binds preferentially to the enzyme in the T state, which is not available, and will be insensitive to activation by A because the enzyme in the R state is already in the high-affinity state for S. Simultaneously all homotropic and heterotropic interactions of the various ligands will be lost. With appropriate assumptions it is possible to explain the kinetic behavior and peculiarities of most regulatory enzymes on the basis of the theory of Monod *et al.* (1965).

An alternative model to explain the kinetic behavior of allosteric enzymes has been put forward by Koshland *et al.* (1966). This model, termed the *sequential model*, was developed from the "induced fit" theory of enzyme specificity proposed by Koshland earlier (1970). The major assumptions of the sequential model are that the oligomeric protein exists in only *one* conformation in the absence of a ligand (substrate or effector), and when a molecule of the ligand is bound to one subunit, a conformational change occurs not only in the occupied subunit but also in the neighboring vacant subunit. This latter (vacant) subunit is then able to bind substrate with altered affinity. The binding of the second molecule not only stabilizes the altered conformation but influences the conformation of the third (vacant) subunit, and so on. Contrary to the hypothesis of Monod *et al.* (1965), the subunits of an oligomeric protein can exist in more than two forms, and several subunit hybrid states are also possible (Fig. 2b). Using simplified situations, one could easily explain the kinetic behavior of enzymes of the K system by assuming three conformations of the subunits: the naked (vacant) subunit, the active subunit with allosteric activator and/ or substrate bound, and the inactive subunit with allosteric inhibitor bound (Fig. 2b).

The general nature of this review precludes consideration of the experiments that may distinguish between these two models for the kinetics of particular allosteric enzymes. Suffice it to say that in actual practice it is quite difficult to refute or confirm one or the other of these models on the basis of traditional initial velocity data. Indeed, almost all of the peculiarities of allosteric systems (except negative cooperativity and enzymes with half-of-the sites reactivity) can be explained, at least qualitatively, on the basis of either of the two theories by using proper, and in some cases justifiable, assumptions. Consider, for instance, the occurrence of association–dissociation phenomena evoked in some regulatory proteins by the presence of various allosteric ligands. Several examples of this kind are now known (Gerlt *et al.*, 1973). Acetyl-CoA carboxylase polymerizes in the presence of the allosteric activator, citrate (Vagelos *et al.*, 1963), as does the catabolic threonine deaminase in the presence of AMP (Hirata *et al.*, 1965; Gerlt *et al.*, 1973). The substrate (ATP and UTP) dimerizes CTP synthase (Levitzki, 1975) and $NADP^+$ promotes dimer formation in glucose-6-phosphate dehydrogenase (Levy *et al.*, 1966). The allosteric inhibitor threonine favors an aggregated state of homoserine dehydrogenase, whereas the activators methionine and isoleucine favor a dissociated state (Datta *et al.*, 1964). The question of primary interest here is whether the allosteric effects in the associating–dissociating systems are caused by the binding of a ligand preferentially to one species, thereby causing displacement of the equilibrium, or whether a direct binding to one species causes conversion into another? Again, the available data can be interpreted on the basis of both the concerted and the sequential models, although it would be fair to say that the weight of evidence favors the latter model. Frieden (1963), for example, has demonstrated that association–dissociation of glutamate dehydrogenase in the presence of effectors and substrates is probably due to ligand-induced conformational changes. If conformational changes that result from ligand binding can be communicated to the neighboring subunits, it is easy to understand how, in some cases, distortion of the subunit may lead to a sufficient degree of weakening of the interprotomeric bonds to induce dissociation. Similarly, polymerization can be caused by a ligand-induced conformational change in a monomer that exposes certain amino acid residues required for noncovalent interactions among subunits or monomers. If conformational changes do precede association–dissociation in some enzymes, the question may be raised whether the associated and dissociated states in other systems could be equated with the R and T states in the concerted model. It has already been mentioned that in some allosteric enzymes, such as homoserine dehydrogenase (Datta *et al.*, 1964), the inhibitors stabilize an associated state (in which the enzyme has a low affinity for substrate), and the activators stabilize a dissociated state (in which the enzyme has a high affinity for substrate). This is the opposite of what happens in the case of acetyl-CoA carboxylase, in which the activator promotes an associated (high-affinity) state and the inhibitors promote a dissociated (low-affinity) state (for reviews, see Numa, 1974; and Lane *et al.*, 1975). In still other cases, such as deoxycytidylate deaminase (Duncan *et al.*, 1972), both the allosteric inhibitor dTTP and the activator dCTP stabilize a polymerized state. Thus it seems that there are also enzymes in which association–dissociation may be caused by conformational effects of ligand binding that are *not* related to equilibrium isomeric states of discrete activity.

Some other recent observations in various allosteric systems are incompatible with the concerted transition model. It may be recalled that quaternary structural hybrids containing monomers of both the T and the R states in a single molecule are not permitted in the concerted model. Recent resonance spectroscopy studies of hemoglobin suggest that just such hybrids do occur (Huestis and Raftery, 1972*a,b*). Multiple conformational states, rather than just two, have also been shown by indirect measurements in CTP synthase (Levitzki, 1975). Reference was made earlier to the existence of negative cooperativity in some proteins, such as glyceraldehyde-3-phosphate dehydrogenase (Conway and Koshland, 1968), beef-liver glutamate dehydrogenase (Engel and Dalziel, 1969), and several others (a list is provided in Levitzki and Koshland, 1976). Negative cooperativity (Fig. 1b) has been shown to arise by a progressive decrease in the value of the binding constant as each successive molecule of the substrate becomes bound at an additional active site in the oligomeric enzyme (Conway and Koshland, 1968). In other words, the binding of the first substrate molecule to one subunit makes it more difficult for the second and subsequent substrate molecules to bind to the remaining sites of the same molecule. Thus negative cooperativity is permitted in the sequential model, but not in the concerted model. In the latter case, increasing substrate concentration is supposed to shift the T–R equilibrium only toward that state which binds the substrate preferentially (R form). This mechanism can therefore give rise only to positive cooperativity. Similarly, the phenomenon referred to as *half-of-the-sites reactivity* (Levitzki and Koshland, 1976) is inexplicable on the basis of the concerted model. According to the sequential theory, however, half-of-the-sites reactivity is merely viewed as an extreme, and limiting, case of negative cooperativity. Reactivity of this type, in which a ligand binds to only half its potential binding sites on the enzyme surface, is now known to be of wide occurrence in allosteric and nonallosteric systems (Levitzki and Koshland, 1976). On the basis of the sequential model, binding of a molecule of ligand to one subunit of the enzyme induces a structural change in the neighboring subunit, thereby reducing the affinity of the neighboring subunit for the ligand to zero. In all fairness it should be mentioned, however, that half-of-the-sites reactivity is also easily accounted for by a preexisting asymmetry model (Matthews and Bernhard, 1973), that assumes that the identical monomers of an oligomer are assembled in such a way as to produce subunits of two types: those with high affinity and those with low affinity. In such an oligomer, reactivity of half of the sites can be explained without postulating any conformational change resulting from ligand binding.

Before leaving the topic of this section, the basis of the cooperativity (positive or negative) of monomeric proteins should be examined. Well-documented examples of monomeric allosteric proteins were discussed in Section 4.2.2a. Both the concerted and sequential models definitely require an oligomer structure for the genesis of cooperativity of any kind. What, then, is the plausible kinetic basis of cooperativity in monomeric proteins? Ricard *et al.* (1974) have proposed the so-called *mnemonical* concept for monomeric enzymes following ordered reaction mechanisms—that is, where there is ordered binding of two or more substrates on the enzyme surface. The free enzyme is considered to exist in two conformational states in equilibrium, and the nonhyperbolic behavior is considered to result from cooperation of these forms in the overall reaction process.

Similarly, it has been suggested (Shill and Neet, 1975; Neet and Ainslie, 1976) that cooperativity in monomeric enzymes is caused by a ligand-induced slow transition, or isomerization, between different functional or conformational states of the enzyme. A general prediction of these "tautomeric state" theories is that cooperativity would not appear in equilibrium-binding studies but would appear in kinetic experiments.

5 Enzyme Modifications

Complementing, and in addition to, the catalytic site and allosteric controls of enzymes is a regulatory mechanism that involves the modification of the catalytic efficiencies and other properties of enzymes. These modifications are brought about by (1) noncovalent and reversible interactions with other polypeptides, and (2) enzyme-catalyzed covalent changes. The noncovalent interactions are referred to here as *protein–protein interactions* and the covalent changes as *covalent conversion*.

5.1 Protein–Protein Interactions

Protein–protein interactions are of widespread occurrence in various biological processes (Frieden, 1971). Viewed broadly, the association of subunits of oligomeric proteins or polymerization of these proteins into higher aggregate forms can be considered as protein–protein interactions. Although enzyme activity is regulated if such interactions are reversible (homomeric protein interactions), those interactions that occur between an enzyme whose activity is to be regulated and an unrelated polypeptide (heteromeric protein interactions) that may or may not have enzyme activity by itself will be our principal concern here. Homomeric protein interactions occur in self-associating systems. Some of these, such as acetyl-CoA carboxylase, have been discussed in Section 4 because most often the self-association is triggered by the binding of an allosteric effector or substrate. The heteromeric protein interactions are of two kinds. First, and most interesting, is the mechanism of regulation referred to as *molecular conversion* (Monod and Jacob, 1961). In this case the substrate *specificity* of an enzyme is altered by association of the enzyme with another protein. In the interactions of the second kind, referred to simply as *noncovalent interactions*, the activity of an enzyme is reversibly affected (inhibited or activated) by its association with another protein.

A prototype of molecular conversions is bovine glutamate dehydrogenase, which consists of six identical subunits (Goldin and Frieden, 1971). The enzyme undergoes a concentration-dependent polymerization–depolymerization, and in dilute solutions (such as those utilized for enzyme assays) exists primarily in the dissociated form. This protomeric form of the enzyme was shown by Tomkins *et al.* (1965) to exhibit remarkable variations in enzyme specificity, depending on the presence of various regulatory ligands. In the presence of ADP, an allosteric effector, the enzyme is more or less specific for L-glutamate. In the presence of CTP, steroids, and some other regulatory ligands, however, the enzyme shows markedly enhanced capacity to oxidize L-alanine and various other monocarboxylic acids. The capacity for L-alanine oxidation does not, however, exceed more than 1–2% of that for L-glutamate oxidation. Despite this fact, and notwithstanding the homomeric interactions in the change of substrate specificity, the findings with glutamate dehydrogenase did suggest for the first time the possibility that similar regulation may occur in other enzyme systems. Several instances of molecular conversions are now known in which heteromers, and in some cases allosteric effectors, participate in changing the substrate specificities of catalytic sites. These systems are discussed in the following sections.

5.1.1a   Ribonucleotide Diphosphate Reductase.   The enzyme ribonucleotide diphosphate reductase from *E. coli* converts ribonucleoside diphosphates to the corresponding deoxyribonucleotides. It seems to be the only enzyme that provides all the precursors necessary for DNA synthesis—the four deoxynucleoside diphosphates. *In vitro* the enzyme can utilize a small protein, thioredoxin, as a hydrogen donor, but the *in vivo* donor is not known with certainty (Mark and Richardson, 1976). The enzyme consists of two nonidentical subunits, B1 (a dimer of nonidentical monomers) and B2 (a dimer of identical monomers), each of which is inactive by itself (Thelander, 1973; Thelander and Reichard, 1979). A fully functional enzyme is formed when B1 and B2 are mixed together in the presence of magnesium ions. Protein B2 participates in the formation of the catalytic site and contains nonheme iron and an organic free radical essential for enzymic activity. The B1 component contains dithiols that participate in the oxidation reactions of electron transport, and this protein also has both the substrate-binding site for various ribose diphosphates and the allosteric sites (Döbeln and Reichard, 1976). There are two classes of allosteric sites (Brown and Reichard, 1969). Binding of allosteric effectors (ATP, dATP, dGTP, and dTTP) to sites of one class regulates the substrate specificity of the enzyme. These sites are named *h* sites and bind dATP with highest affinity but also have the capacity to bind the other effectors. Depending on the nature of the nucleoside triphosphates occupying the *h* sites, the enzyme will reduce either pyrimidine or purine ribonucleoside diphosphates, or both. The reduction of CDP or UDP, for instance, is facilitated by the presence of ATP, and the reduction of ADP and GDP is enhanced in the presence of dGTP. Sites of the second class, termed *l* sites, bind dATP and ATP with low affinity (compared to the *h* site) but have no detectable capacity to bind dTTP or dGTP. Binding of the allosteric effectors at the *l* sites determines the overall activity of the enzyme.

5.1.1b   RNA Polymerase of Prokaryotes.   The role of certain polypeptides in altering the substrate specificity of the enzyme is crucial. The DNA-dependent RNA polymerase of bacteria, in particular *E. coli*, has been the subject of several recent reviews (e.g., Chamberlin, 1974) that consider the molecular, catalytic, and regulatory properties of the enzyme. The simplest form of RNA polymerase that can catalyze DNA-dependent synthesis of RNA is known as the *core enzyme*, which has the subunit structure $\alpha_2\beta\beta'$. Burgess and co-workers (1969) found that core polymerase has a rather low ability to transcribe the DNA of certain phages, such as T4, but this ability is enhanced some 30- to 40-fold when a small polypeptide referred to as *sigma* is added *in vitro*. The sigma factor actually forms a part of the polymerase holoenzyme, but it can be separated from the core enzyme by chromatography on phosphocellulose (Burgess *et al.*, 1969). The role of sigma in directing the specificity of the core enzymes for particular base sequences (promoter regions) on DNA templates has been carefully studied with phage T7 DNA (Chamberlin, 1974). Either core polymerase or the holoenzyme can bind to T7 DNA in thousands of places, termed class B sites, that probably form a continuum throughout the length of the DNA. Binding to class B sites by core enzyme is rather strong, but when sigma is associated with the enzyme it binds only weakly to these sites and dissociates quickly. Sites of a second class, class A, a maximum of about 8 in one molecule of T7 DNA, bind only the holoenzyme, and binding is extremely tight. The available evidence indicates that these sites are probably the promoter regions of DNA, where RNA chain initiation begins. Once initiation has occurred, and the first few bases have been transcribed, the extremely specific, high affinity of holopolymerase for the promoter regions (class A sites) of the DNA becomes a hindrance to elongation of the RNA chain, which cannot occur unless the polymerase is able to move away from its binding site, transcribing the DNA as it goes. To solve this *impasse*, the sigma factor dissociates from the DNA–polymerase complex. The core polymerase, because of its rather general, nonspecific affinity for all DNA, now moves on to elongate the initiated RNA chain. The role of sigma, then, is to ensure that RNA core polymerase binds specifically to promoter regions of DNA in such a way that some melting of the DNA to single-strand regions can occur, and a few bases can be transcribed into RNA. Sigma is not needed for the elongation of the initiated chain and is probably an impediment to it; therefore, it dissociates and takes no further part in the synthesis of the particular RNA chains it helped to initiate correctly.

Apart from the sigma factor, several other polypeptides are probably able to modify the specificity of the core polymerase. During phage T4 infection of *E. coli*, for example, two phage-directed polypeptides are found associated with RNA polymerase (Horvitz, 1973; Ratner, 1974). The same is also observed with phage SPO1 growing in *Bacillus subtilis* (Duffy *et al.*, 1975; Fox, 1976; Talkington and Pero, 1978). There are reasons to believe that these proteins are able to switch the promoter specificities of the core enzyme to produce late mRNA after the early genes (needed for phage-specific RNA metabolism) have been transcribed. There is some evidence that even during normal sporulation in *B. subtilis* the RNA polymerase of the vegetative cells is modified by association of a protein of molecular weight 37,000 with the core polymerase (Haldenwang and Losick, 1979). There is little information about the chemical properties of the phage-directed polypeptides, or even of the sigma factor itself, that could account for the effects of these phage peptides on RNA polymerase.

5.1.1c The P_{II} Regulatory Protein in the Glutamine Synthase Adenylylation System. Glutamine synthase of *E. coli* catalyzes the ATP-dependent formation of L-glutamine from ammonia and α-ketoglutarate. The system for regulating the activity of this enzyme has been extensively studied by Stadtman and co-workers (for review, see Ginsburg and Stadtman, 1975).

$$GS + 12ATP \xrightarrow[\text{glutamine}]{\text{ATase, } P_{II}A} GS(AMP)_{12} + 12PP_i \qquad (5)$$

$$GS(AMP)_{12} + 12P_i \xrightarrow[\alpha\text{-ketoglutarate, ATP}]{\text{ATase, } P_{II}D} GS + 12ADP \qquad (6)$$

These covalent modifications are in addition to the elaborate system of allosteric inhibitions that have also been found in this enzyme (Ginsburg and Stadtman, 1975). In reactions (5) and (6), GS is glutamine synthase, a dodecamer of identical subunits and $GS(AMP)_{12}$ is the adenylylated, or less active, form of the enzyme. ATase is an adenylyltransferase catalyzing the interconversion of GS and $GS(AMP)_{12}$. The direction in which the interconversion takes place is regulated by a protein, designated P_{II}, that also exists in two forms: $P_{II}A$, promoting adenylylation [reaction (5) above] and $P_{II}D$, promoting deadenylylation [reaction (6) above]. The interconversion of $P_{II}A$ and $P_{II}D$ is also under enzymic control (discussed in Section 5.2.2).

P_{II} is clearly one of the more interesting of the regulatory proteins currently under study. When it exists in one form, it *specifies* and promotes a net reaction in one direction; when it is in the other form the direction specified and stimulated is the reverse. This alternating function is especially remarkable in view of the fact that P_{II} has very short polypeptide chains.

5.1.1d Fatty Acid Synthetase from Mammary Gland. The series of enzymes involved in the synthesis of fatty acids from acetyl-CoA and malonyl-CoA (acetyl and malonyl transacylases, β-ketothioester synthetase, β-ketothioester reductase, β-hydroxy-acylthioester dehydrase, enoylthioester reductase and acylthioester hydrolase) in various mammalian tissues are associated in a multienzyme complex (for review, see Bloch and Vance, 1977). The fatty acids undergoing condensation and reduction remain bound by thioester linkages to an acyl carrier protein. The multienzyme fatty acid synthetase complex predominantly synthesizes palmitoyl-acyl carrier protein. Palmitic acid is cleaved from the thioester linkage by a long-chain acylthioester hydrolase (thioesterase I) which forms an integral part of the enzyme complex. While palmitic acid is the major end product of fatty acid synthetase, the lactating mammary gland also produces medium chain fatty acids characteristic of milk fat (Smith and Abraham, 1975). In search of the mechanism by which fatty acid synthetase in the mammary gland switches from the predominant production of long-chain fatty acids to the formation of medium-chain fatty acids during lactation, several workers have found "factors" in the cytosol of mammary glands of rabbits (Strong *et al.*, 1973) and rats (Smith and Abraham, 1975) which modify the product specificity of the fatty acid synthetase. The "factor" has recently been identified as a medium-chain acyl-CoA hydrolase (thioesterase II) which is different from the long-chain acyl-CoA hydrolase (thioesterase I) normally associated with the fatty acid synthetase complex (Knudsen *et al.*, 1976; Libertini and Smith, 1978) in nonlactating tissues. Thioesterase II seems to be unique to the mammary gland (Libertini and Smith, 1978) and is a monomer of molecular weight 29,000 to 33,000 (depending upon the tissue of origin). It is capable of shifting the product specificity of rat mammary gland fatty acid synthetase from predominantly C_{14}, C_{16}, and C_{18} fatty acids (long chain) to mainly C_8, C_{10}, and C_{12} (medium chain) fatty acids. The actual molecular mechanism of the alteration in product specificity of the multienzyme complex in the presence of thioestase II is not known but there are indications that it is connected with the ability of thioesterase II to hydrolyze acylthioester bonds (Libertini and Smith, 1978).

5.1.2 Noncovalent Interactions

Regulation of enzyme activity by noncovalent interactions is of special interest because it illustrates one of the most subtle consequences of protein–protein interactions. In noncovalent regulation, the enzyme polypeptide has an activity of its own, but a second, unrelated peptide, often acting in conjunction with one or more small molecules, alters some parameter of that activity. The simplest alteration is to turn the enzyme on or off as in the case of enzymes composed of *regulatory subunits* and *catalytic subunits*. In such enzymes the interactions between subunits of the two types are sufficiently strong that the aggregate can be purified as one protein species using conventional procedures. More complex noncovalent interactions are to be found in cases in which the activity of an oligomeric enzyme is enhanced or inhibited by interaction with another protein that may or may not function by itself as an enzyme. A fascinating infrastructure has been evolved by organisms whereby the regulatory protein is produced in some cases only in response to the presence in the growth medium of end products of the pathway in which the regulated enzyme has a functional role. Superficially, the process therefore resembles the well-known end product repression, where the rate of synthesis of one or more enzymes of a pathway is repressed by the presence of the end product during growth. In the cases to be discussed here, however, unlike the case of end-product repression, the metabolite *induces* the synthesis of a protein that inhibits the activity of the relevant enzyme(s) of the pathway by protein–protein interactions. It has been proposed that the name "antizyme" be applied to such protein inhibitors (Heller *et al.*, 1976).

Several cases have been reported in the literature in which the catalytic efficiencies of enzymes are increased or decreased by interaction with other enzymes or with other proteins *in vitro*. Examples include inhibition of glucose-6-phosphate dehydrogenase (Bonsignore *et al.*, 1968), ribonuclease (Blobel and Potter, 1966), and hepatic 3-hydroxy-3-methylglutaryl conenzyme A reductase (McNamara and Rodwell, 1975). However, it is unlikely that such interactions occur *in vivo*, or that they have any regulatory significance. This conclusion is based on the fact that enzyme activity is easily affected by conformational changes of catalytic sites brought about by electrostatic or hydrophobic interactions with both small molecules and macromolecules. Several unnatural polycations (such as polylysine) and polyanions activate a number of enzymes (Sanwal *et al.*, 1966, and references therein). Similarly, several naturally occurring proteins, having proper density and distribution of charged or hydrophobic groups on their surfaces, may incidentally affect *in vitro* the activity of enzymes whose conformations are susceptible to change by electrostatic or hydrophobic interactions. Such protein–protein interactions, demonstrable only *in vitro*, may have absolutely no physiological relevance. Despite this uncertainty, however, there are two examples that deserve mention. One is the activation of rabbit-liver fructose diphosphatase by phosphofructokinase, both enzymes being oligomeric and allosteric (Pogell *et al.*, 1968); the other example is the activation or inhibition of a large number of enzymes by a calcium-dependent regulatory protein (Cheung, 1969; Kakiuchi *et al.*, 1973; Brostrom *et al.*, 1976; Stevens *et al.*, 1976) referred to as calmodulin (Wang and Waisman, 1979; Cheung, 1980). In the former case, not only is the activity of fructose diphosphatase enhanced by interaction with phosphofructokinase, but also its allosteric properties are altered. A potent allosteric inhibitor of fructose diphosphatase, AMP, no longer shows any effect whatsoever when the enzyme is complexed with phosphofructokinase. Of particular interest in this system is the fact that both enzymes catalyze the same sugar phosphate interconversions, phosphofructokinase in one direction and fructose diphosphatase in the other. What the physiological significance of this protein–protein interaction could be, if indeed such interaction occurs *in vivo* at all, is difficult to ascertain.

In the second example, calmodulin, an ubiquitous protein present in all tissues so far investigated (Cheung, 1980) activates a large number of enzymes (Cheung, 1980). Among the enzymes affected are cyclic AMP phosphodiesterase, adenylate cyclase, myosin light chain kinase, phosphorylase kinase, NAD kinase, and Ca^{2+}-Mg^{2+}–ATPase. The mode of action of calmodulin was first studied extensively with the cyclic AMP phosphodiesterase system (Kakiuchi *et al.,* 1973; Wang and Waisman, 1979; Cheung, 1980). Calmodulin requires Ca^{2+} before it affects the enzyme. The calmodulin–Ca^{2+} complex binds reversibly to the phosphodiesterase resulting in its activation. A similar mechanism also seems to operate in the activation of other enzymes listed above, except perhaps phosphorylase kinase (Cohen *et al.,* 1978). In this case, calmodulin exists in the kinase as a subunit of the oligomeric enzyme and serves to bind Ca^{2+}.

The mechanism of activation by calmodulin has many similarities with the regulation of actomyosin ATPase by another calcium-binding protein, troponin (for review, see Weber and Murray, 1973). Indeed, both calmodulin and muscle troponin C have very similar amino acid compositions, molecular weights, and other physical and chemical properties (Stevens *et al.,* 1976; Cheung, 1980), including extensive segments of amino acid homology.

The effect of calmodulin on diverse enzyme systems and processes in cells has raised the question of the physiological significance of this regulation. Although in the current state of our knowledge all hypotheses in this area are bound to be highly speculative, the consensus appears to be emerging that calmodulin probably serves as the mediator of most of the effects of calcium on cellular metabolism.

Well studied systems in which noncovalent interactions do play a definite part in the regulation of enzyme activity *in vivo* are discussed in the following section.

5.1.2a   Lactose Synthase. An enzyme originally shown to be present in mammary gland, galactosyltransferase (also called protein A), catalyzes the following reaction (Brew *et al.,* 1968):

$$\text{UDP-galactose} + N\text{-acetylglucosamine} \xrightarrow{Mn^{2+}} N\text{-acetyllactosamine} + \text{UDP} \quad (7)$$

This enzyme is now known to be also present in particulate fractions prepared from tissues other than mammary gland (reviewed in Ebner and Magee, 1975; and Hill and Brew, 1975). Although galactosyltransferase from all sources will catalyze reaction (7), it is likely that *in vivo* the enzyme plays a role in glycoprotein biosynthesis by catalyzing the formation of a β-galactose(1→4)β-N-acetylglucosamine linkage (Ebner and Magee, 1975).

The activity and apparent specificity of galactosyltransferase changes when it is combined with an entirely unrelated nonenzymic protein, α-lactalbumin (also called protein B), a plentiful constituent of whey. The protein A–B complex catalyzes the following reaction:

$$\text{UDP-galactose} + \text{glucose} \xrightarrow{Mn^{2+}} \text{lactose} + \text{UDP} \quad (8)$$

The protein complex is referred to as lactose synthase. Formation of the complex causes a marked suppression of the ability to catalyze synthesis of N-acetyllactosamine [Eq. (7)]. One molecule of transferase combines with one of α-lactalbumin to form lactose synthase (Klee and Klee, 1972; Challand and Rosemeyer, 1974). The complex forms only weakly unless one of the substrates, UDP-galactose or N-acetylglucosamine, is present.

Since the apparent specificity of galactosyltransferase changes on combination with α-lactalbumin, the latter was called a *specifier* protein to distinguish it from other regulatory proteins and regulatory associations (Brew *et al.,* 1968). However, it has now been shown that the galactosyltransferase itself catalyzes the transfer of galactose to glucose to form lactose in the complete absence of α-lactalbumin, provided the glucose concentration in the assay medium is very high (K_m for glucose = 1.4 M) (Fitzerald *et al.,* 1970). Furthermore, the V_{max} of the reaction (lactose synthesis) is the same with or without α-lactalbumin. Thus, it is amply clear that what α-lactalbumin really does is not to cause a change in the substrate specificity of the transferase, as was thought earlier (Brew *et al.,* 1968), but simply to lower the K_m values of the substrates for the active sites—that is, to increase the affinity for the glucose site and decrease the affinity for the *N*-acetylglucosamine site. In addition to glucose, other mono- and disaccharides, as well as glycerides, which are marginal substrates for the A protein by itself, exhibit reasonably high affinities when the A protein is combined with α-lactalbumin. It should be pointed out, however, that α-lactalbumin does not appreciably inhibit the transfer of galactose to glycoproteins, although, as mentioned earlier, the transfer of galactose to *N*-acetylglucosamine is markedly inhibited. It would appear that *in vivo* lactose synthase, once formed, is responsible for the synthesis of both lactose and glycoproteins. The organ specificity and control of lactose synthesis (mammary gland) is related to the concentration of α-lactalbumin produced in the lactating gland under hormonal stimulation.

5.1.2b Cyclic AMP-Dependent Protein Kinase. The transfer of the terminal phosphoryl group of nucleoside triphosphates to serine or threonine residues of diverse protein substrates is catalyzed by cyclic AMP-dependent kinase (Krebs, 1972; Rubin and Rosen, 1975). It has been shown to consist of two different polypeptides, one regulatory (R) and the other catalytic (C) (Krebs, 1972). The enzyme from bovine heart is actually a tetramer, with two each of the catalytic and regulatory chains (Rubin and Rosen, 1975). The enzyme (RC complex) is inactive, but is activated in the presence of cyclic AMP. The enzyme undergoes dissociation by the selective binding of cyclic AMP to the R subunit (Gill and Garren, 1969; Krebs, 1972).

$$\text{RC} + \text{cyclic AMP} \rightleftharpoons \text{R cyclic AMP} + \text{C} \tag{9}$$

The C subunit is the activated form of the enzyme and the R subunit acts as an inhibitor. Cyclic AMP is not entirely specific in the dissociation of the RC complex. Histones, for example, also dissociate the complex (Miyamoto *et al.,* 1971). This is probably due to the fact that the R polypeptide is acidic, and histones, being basic, bind selectively to this subunit and displace the C subunit (Krebs, 1972). In addition, protein kinase from rabbit muscle is dissociated into R and C subunits by treatment with *p*-hydroxymercuribenzoate, a finding that suggests involvement of sulfhydryl groups in the interaction between the subunit (Murray *et al.,* 1974).

Two types of protein kinases are known in mammalian cells, referred to as type I and type II. Their relative proportions are species- as well as tissue-specific. Both of the isoenzymes are tetrameric and probably share the same catalytic subunit of molecular weight 40,000. The R subunits of the two isoenzymes, however, differ from each other; type I subunit having a molecular weight of 49,000 and type II a molecular weight of 56,000 (Beavo *et al.,* 1975; Rosen *et al.,* 1975). In addition, the R subunit of type II kinase is capable of being phosphorylated by its catalytic subunit in the presence of ATP. The phosphorylated kinase requires lower concentrations of cyclic AMP for dissociation of the R subunit and activation of the enzyme. In addition, the phosphorylated subunit does not readily associate with the C subunit in the absence of cyclic AMP (for review, see Hoppe and Wagner, 1979).

5.1.2c Yeast Ornithine Transcarbamylase–Arginase Interactions. Ornithine transcarbamylase in yeast, as in other organisms, is concerned with the biosynthesis of arginine. It catalyzes the reaction of ornithine with carbamoyl phosphate:

$$\text{Ornithine} + \text{carbamoyl-P} \rightarrow \text{citrulline} + \text{P}_i \tag{10}$$

The product, citrulline, is converted to arginine by two additional steps of the Krebs urea cycle.

When arginine is available in the medium as a source of nitrogen, it induces the formation of arginase, an enzyme that hydrolyzes arginine to urea and ornithine. The urea can be further degraded to yield ammonia and CO_2 (the urease reaction), and the ornithine can be converted to glutamate, whose α-amino group is widely available to other biosynthetic processes via transamination. It is important, however, that the ornithine transcarbamylase activity be suppressed under these conditions, so that the ornithine produced by arginase not be returned to the urea cycle. If arginase and the transcarbamylase were both to function, a futile cycle would result in which ornithine would repeatedly be converted to arginine and arginine converted back to ornithine, with the expenditure of five energy-rich bonds per cycle—two at the argininosuccinate synthase step of the cycle (which hydrolyzes ATP to AMP plus pyrophosphate), and three for the synthesis of carbamoyl phosphate (Wiame, 1971). Such a waste of energy is prevented as follows. Ornithine itself, at higher concentrations, begins to inhibit ornithine transcarbamylase. This inhibition is reinforced by the enzyme arginase, and the addition of arginine increases the inhibition still more. In the presence of ornithine and arginine together, arginase and ornithine transcarbamylase combine into a stoichiometric complex, apparently containing one molecule of each enzyme (Wiame, 1971). Treatment of ornithine transcarbamylase with the acetylating agent, acetylimidazole, leads to a loss of ornithine inhibition, while transcarbamylase activity remains intact; such a desensitized enzyme is no longer inhibited by arginase plus arginine and cannot form a complex with it. Heating the ornithine transcarbamylase has the same effects. Mutant strains of yeast are known in which the ornithine transcarbamylase is insensitive to arginase, and other strains have been isolated in which the arginase is inactive but still able to inhibit the transcarbamylase (Wiame, 1971). The regulatory interaction between the ornithine transcarbamylase and arginase of yeast has been demonstrated also in *Bacillus subtilis* (Issaly and Issaly, 1974).

Both ornithine transcarbamylase and the arginase of yeast are trimers (Penninckx *et al.*, 1974). It is interesting to note that this unusual quaternary structure is also exhibited by the catalytic subunit of aspartate transcarbamylase, a functionally related enzyme.

Regulation of the type described above is a scarce example of what Wiame (1971) has called *epienzymatic control*. In this particular case, arginase acts as a regulatory subunit for ornithine transcarbamylase, although it is also an enzyme in its own right.

5.2 Covalent Conversions

The concept that enzyme activity can be controlled by chemical modification originated with the discovery of Krebs and Fischer (1956) that mammalian glycogen phosphorylase can exist in two interconvertible forms, phosphorylated *a* (active) and dephosphorylated *b* (inactive). Since that time several enzyme systems have been found to undergo enzyme-catalyzed covalent conversions. So far, three principal types of conversions have been found. In one, the enzyme is phosphorylated by a specific kinase that transfers the γ-phosphoryl group of ATP to certain seryl residues in the protein. The modified enzyme is dephosphorylated by a protein phosphatase. In the second, the enzyme undergoes reversible adenylylation and deadenylylation, cataylzed by specific enzyme systems. Regulation of this kind is known so far in glutamine synthase (reviewed in Ginsburg and Stadtman, 1975) and possibly in the lysine-sensitive aspartokinase of *E. coli* (Niles and Westhead, 1973). In covalent conversions of the third kind, the enzyme is covalently modified by the transfer of the ADP-ribose moiety of NAD^+ to acceptor groups. There is considerable evidence that yet another kind of protein modification, viz., methylation–demethylation, participates in the regulation of certain behavioral mechanisms, such as bacterial chemotaxis (for review, see, Springer *et al.*, 1979) and human monocyte chemotaxis (Pike *et al.*, 1978). In these processes the carboxyl groups of certain proteins involved in signal transduction in membranes are methylated by methyltransferases utilizing the universal methyl donor, S-adenosyl-L-methionine, and demethylated by methylesterases. However, the structure of the methyl-accepting proteins is far from clear and intimate details of the methylation–demethylation reactions have not yet been clarified.

It should be pointed out that covalent conversions of all types lead to a change not only in the catalytic activity of the enzyme but also in the allosteric and other control properties of the enzyme. Also, within the stimulus–response time scale required by various regulatory devices, covalent conversions are intermediate between purely allosteric regulation on the one hand, and genetic regulation (repression or induction) on the other. While the latter process requires any time from minutes to hours, allosteric regulation (inhibition or activation) operates within the microsecond–millisecond range (Kirschner *et al.*, 1966), unless hysteresis is superimposed on the allosteric control (Frieden, 1970). Covalent conversion by a cascade system (see below), takes about 2 sec for 50% conversion in the one case in which the temporal parameter has been well studied—namely, the conversion of muscle phosphorylase *b* to phosphorylase *a*. While an effector has to be continuously present for allosteric regulation to occur, covalent conversion may be triggered by even a transient stimulus, and the altered state of the enzyme persists long after the stimulus has completely disappeared. This mode of regulation may have peculiar advantages when hormones constitute the primary stimulus that triggers enzyme modification.

In most systems in which covalent conversions occur, a series of steps intervene between the receptor of a primary stimulus and the final conversion. This is particularly true of phosphorylation–dephosphorylation interconversions. The primary step is the activation of a protein kinase by cyclic AMP that is itself generated by adenylate cyclase in response to activation by hormones. Cascading systems of this kind provide a mechanism for the amplification of regulatory signals.

5.2.1 Regulation by Phosphorylation–Dephosphorylation

Change in enzyme activity by phosphorylation–dephosphorylation is frequently utilized for control in nature. Detailed information, however, is available only in a few enzyme systems. Selected systems and the concepts that have emerged from studying them are discussed below. To date nearly twenty enzymes of intermediary metabolism have been shown to undergo phosphorylation–dephosphorylation (Krebs and Beavo, 1979) and this number will doubtless increase in future.

5.2.1a Regulation of Enzymes for Synthesis and Degradation of Glycogen. Glycogen is synthesized in all organisms by the enzyme glycogen synthase and is degraded to glucose-1-phosphate by glycogen phosphorylase. The essentially cyclical nature of synthesis and degradation demands control systems that are coordinated in such a way that conditions that suppress glycogen synthesis, promote glycogen degradation, and vice versa. This is achieved by phosphorylation–dephosphorylation mechanisms such that when phosphorylase *b* is phosphorylated to phosphorylase *a* it becomes activated (Krebs and Fischer, 1956). Conversely, when glycogen synthase *a* is phosphorylated, it becomes inactivated

(Villar-Palasi and Larner, 1960). Recent evidence indicates that coordination may occur at the dephosphorylation step catalyzed by a single protein phosphatase that utilizes both phosphorylase *a* and glycogen synthase *b* as substrates (Killilea *et al.,* 1976a,b) (Fig. 3). Over and above the covalent regulation of the glycogen-metabolizing enzymes, allosteric controls are superimposed on these enzymes, and the nature and extent of allosteric modulation is determined by the state of phosphorylation of the proteins. Phosphorylase *b* is activated by AMP even at high substrate concentrations, whereas phosphorylase *a* is activated only at very low concentrations of the substrate. Similarly, glycogen synthase *a* requires high levels of glucose-6-phosphate for activity, whereas glycogen synthase *b* is catalytically active even in the absence of the activator (Villar-Palasi and Larner, 1960). In addition, because these enzymes are themselves substrates for phosphorylation–dephosphorylation reactions, the allosteric effectors are also able to indirectly modulate the activities of the phosphorylating and dephosphorylating enzymes by inducing conformational changes in them. As an example, glucose-6-phosphate, an allosteric inhibitor of phosphorylase *b*, inhibits phosphorylase kinase by altering the interaction of the two enzymes (Graves *et al.,* 1974). In general, the allosteric controls of the glycogen-metabolizing enzymes are complementary to the covalent controls and seem to be designed to ensure that the biosynthetic and catabolic enzymes are not active at the same time in a cell.

5.2.1b Phosphorylation of Glycogen Synthase. As mentioned earlier, glycogen synthase, which catalyzes transfer of the glucose moiety of UDP-glucose to glycogen, exists in an inactive, phosphorylated *b* form and an active, dephosphorylated *a* form. The conversion of the *a* to the *b* form is catalyzed by a cyclic AMP-dependent protein kinase. This kinase

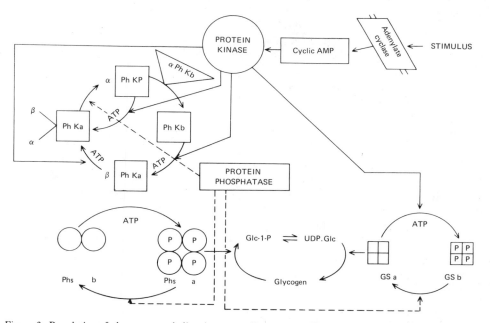

Figure 3. Regulation of glycogen metabolism in mammalian systems. The phosphorylations catalyzed by protein kinase are denoted by solid arrows. Dephosphorylations catalyzed by protein phosphatase are denoted by broken lines. Symbols and abbreviations: Ph Ka, phosphorylase kinase *a*; Ph Kb, phosphorylase kinase *b*; α Ph KP, α-phosphorylase kinase phosphatase; Phs b, dimeric phosphorylase *b*; Phs a, tetrameric phosphorylated (P) phosphorylase *a*; GS a, glycogen synthase *a*; and GS b, phosphorylated (P) glycogen synthase *b*.

also initiates the phosphorylation sequence leading to the formation of phosphorylase *a* from phosphorylase *b* (Söderling *et al.*, 1970; Cohen, 1978) (see Section 5.2.1c). Cyclic AMP itself is formed by hormonal activation of a membrane-bound adenylate cyclase (Fig. 3), and serves as an initiator for all the well-known effects of some hormones on glycogen metabolism (Walsh *et al.*, 1971). Glycogen synthase is an oligomer (Söderling *et al.*, 1970; Takeda and Larner, 1975), most probably a tetramer consisting of identical monomers (Cohen *et al.*, 1975; Cohen, 1976). Although estimates vary, the current consensus is that there are two to six phosphorylation sites per monomer (Takeda and Larner, 1975; Söderling, 1975; Cohen, 1976). The degree of dependence of glycogen synthase *b* activity on glucose-6-phosphate varies with the actual number of potential sites phosphorylated, the enzyme becoming almost totally dependent when all available sites are phosphorylated (Roach and Larner, 1976). What, then, determines the number of sites to be phosphorylated and the rate of phosphorylation? Some recent evidence, largely emanating from one laboratory (Cohen, 1976), suggests that glycogen synthase *a* is capable of being phosphorylated by a glycogen synthase kinase that is distinct in many respects from the cyclic AMP-dependent protein kinase. This newly discovered kinase apparently does not require cyclic AMP. It phosphorylates glycogen synthase *a in vitro* to a very limited extent (1 mole of phosphate incorporated per mole of enzyme) at a site distinct from the sites phosphorylated by the cyclic AMP-dependent protein kinase. Phosphorylation by glycogen synthase kinase does not convert glycogen synthase *a* to *b*—that is, the phosphorylated enzyme does not become glucose-6-phosphate-dependent—but the modified enzyme is capable of being fully phosphorylated at all potential sites at a greatly accelerated rate by the cyclic AMP-dependent protein kinase. Thus control of the *a* to *b* conversion, attended by activation of the enzyme, seems to be exerted by glycogen synthase kinase. It has been postulated (Cohen, 1976) that this "second site phosphorylation" (Cohen and Antoniw, 1973) may be an important control mechanism among covalent conversions. It may be pointed out, however, that control by glycogen synthase kinase has not yet been demonstrated *in vivo*; therefore, the physiological relevance of this concept will have to await such demonstration.

5.2.1c   Covalent Conversion of Phosphorylase.   Unlike glycogen synthase, which is phosphorylated directly by the cyclic AMP-dependent protein kinase, the phosphorylase *b* to *a* conversion proceeds through a cascading system in which the protein kinase first phosphorylates a specific phosphorylase kinase (Fig. 3) that, in turn, catalyzes the conversion of phosphorylase *b* into *a*. Phosphorylase kinase is capable of phosphorylating itself without cyclic AMP, but this autophosphorylation requires calcium ions and high concentrations of magesium-ATP. Whether this process participates at all in activating phosphorylase kinase physiologically is an open question. However, the important point to be made here is that the sequential activation (cascade phenomenon) of protein kinase, phosphorylase kinase, and phosphorylase results in amplification of the initial signal (cyclic AMP) generated by the action of certain hormones on adenylate cyclase (Fig. 3). In keeping with this strategy of control, the concentrations of the three sequentially acting proteins in skeletal muscle are approximately 0.03, 1, and 5–10% if the total soluble proteins, respectively (Krebs, 1972; Rubin and Rosen, 1975).

Phosphorylase kinase is composed of monomers of three different kinds, α, β, and γ, with the overall composition of α_4, β_4, γ_4 (Cohen *et al.*, 1975; Cohen, 1973). The kinase can be activated *in vitro* by incubation with trypsin (Krebs *et al.*, 1964). During proteolysis the α and β subunits are degraded, while the γ subunit and enzyme activity remain unaltered (Rubin and Rosen, 1975; Cohen, 1973). This observation suggests that γ is a catalytic subunit and the α and β are regulatory subunits. In keeping with this hypothesis, it has been found that both α and β subunits are capable of being phosphorylated by the cyclic AMP-dependent protein kinase, whereas the γ subunit is not susceptible to modification. The activation of phosphorylase kinase occurs by the phosphorylation of one site on the β subunit. When β is 50% phosphorylated the site on the α subunit begins to be phosphorylated at a rate equal to 25% of the initial rate of β subunit phosphorylation, but α subunit phosphorylation does not lead to an increase in the activity of phosphorylase kinase (Cohen et al., 1975; Cohen, 1973, 1976; Hayakawa et al., 1973). Phosphorylation of the α subunit seems to alter the conformation of phosphorylase kinase in such a way that the enzyme phosphorylated in both α and β subunits becomes dephosphorylated at a rate 50-fold greater than the rate of dephosphorylation of enzyme that is phosphorylated in the β subunit alone. In other words, α,β phosphorylated enzyme is a much better substrate for the β-phosphorylase kinase phosphatase than is kinase that is phosphorylated only in the β subunit (Cohen et al., 1975; Cohen, 1976). Thus it appears that phosphorylation of the α subunit controls dephosphorylation of the β subunit and the attendant inactivation of phosphorylase kinase. It has recently been suggested (Cohen et al., 1975) that dephosphorylation of the phosphorylated α and β subunits is catalyzed by two distinct protein phosphatases, an α-phosphorylase kinase phosphatase that dephosphorylates the α subunit, and a β-phosphorylase kinase phosphatase that dephosphorylates the β subunit (Fig. 3). The latter enzyme is most probably identical to the protein phosphatase that dephosphorylates glycogen synthase b (Killilea et al., 1976a). There are two major regulatory consequences of the dual mode of dephosphorylation: first, since dephosphorylated kinase is not readily attacked by the protein phosphatase (the enzyme that dephosphorylates the β subunit), the α-phosphorylase kinase phosphatase determines the duration and extent of hormonal activation of phosphorylase (b to a conversion); and second, hysteresis is established, which effectively prolongs the duration of hormonal stimulation of glycogenolysis. This follows from the facts that dephosphorylation of β-phosphorylated phosphorylase kinase does not occur readily unless the α subunit is phosphorylated and that the extent of this phosphorylation is controlled by the α subunit phosphatase (Cohen et al., 1975; Cohen, 1976). A lag must, therefore, intervene before the competition between dephosphorylation and phosphorylation can facilitate conversion of inactive phosphorylase b to active phosphorylase a. Hysteresis (Frieden, 1970) at the level of individual enzymes was discussed in Section 4.2.2b as one aspect of allosteric control; the above example shows how hysteresis can also be achieved by a combination of enzymes that have opposite effects on their target substrate.

The conversion of phosphorylase b to a by phosphorylase kinase occurs by phosphorylation of a single unique serine residue in each of the two monomers of phosphorylase b (Titani et al., 1975). In addition to activation of the enzyme, this conversion also leads to a tetramerization of the enzyme and to profound changes in the allosteric properties of the enzyme. Dephosphorylation of a to b is catalyzed by the same protein phosphatase (Fig. 3) that dephosphorylates glycogen synthase b and β-phosphorylated phosphorylase kinase (Cohen et al., 1975; Killilea et al., 1976a,b). Thus, the present evidence, which is by no means complete, indicates that the effect of a single cyclic AMP-dependent protein kinase (phosphorylation) is opposed by a single protein phosphatase (dephosphorylation) in the regulation of overall glycogen metabolism. The fact that the literature contains descriptions of several protein phosphatases of unique specificities probably stems from the fact that the enzyme as isolated can exist in multiple forms with various molecular weights. It has been postulated that these forms arise by interaction of a basic phosphatase subunit (molecular weight 35,000) with inhibitor proteins, followed by proteolysis to yield enzymes of various sizes (Killilea *et al.*, 1976*a*).

5.2.1d Regulation of the Pyruvate Dehydrogenase Complex. Linn *et al.* (1969) first reported that the pyruvate dehydrogenase aggregate from animal tissues is regulated by a cycle of phosphorylation and dephosphorylation catalyzed by a magnesium-ATP-dependent kinase and a magnesium-dependent phosphatase, respectively. Both of these enzymes are found closely associated with the multienzyme pyruvate dehydrogenase complex (Reed *et al.*, 1974). The catalytic complex itself consists of three enzymes, pyruvate dehydrogenase, dihydrolipoyl transacetylase, and dihydrolipoyl dehydrogenase, that act sequentially in the order given. The target of phosphorylation–dephosphorylation is the pyruvate dehydrogenase component of the aggregate. This enzyme component is itself a tetramer consisting of subunits of two types ($\alpha_2\beta_2$). The phosphorylation of only one α subunit in the tetramer results in inactivation of the enzyme. Inactivation correlates with the modification of one of two seryl residues in the subunit, although the second seryl residue also becomes phosphorylated after the first one has been modified (Reed *et al.*, 1974). The functional significance of the second phosphorylation site is not known, but by analogy with the multisite phosphorylation in the case of glycogen synthase and phosphorylase kinase discussed in Section 5.2.1b, it may have a regulatory function. To date, full details of the association of the pyruvate dehydrogenase kinase and the pyruvate dehydrogenase phosphate phosphatase with the multienzyme aggregate are not known, but the latter enzyme can be readily separated from the aggregate and requires calcium ions for binding (probably) to the transacetylase component (Reed *et al.*, 1974). The calcium-stimulated binding lowers the apparent K_m of the phosphatase for the phosphorylated pyruvate dehydrogenase about 20-fold. The interaction of transacetylase and kinase also increases the affinity of the latter enzyme for the pyruvate dehydrogenase component.

Since the activity of the pyruvate dehydrogenase aggregate is dependent on the degree of phosphorylation, any factor that is capable of modulating the activity of the kinase or the phosphatase is of potential regulatory significance. There are indications that the activity of the kinase may be regulated *in vivo* by pyruvate and by the ATP/ADP ratio and that the activity of the phosphatase may be regulated by free magnesium and calcium ions (Reed *et al.*, 1974). Recent evidence suggests that the kinase and phosphatase do not regulate pyruvate dehydrogenase activity in an on–off manner. Rather, they apparently maintain a steady-state level of dehydrogenase activity, and this level is modulated indirectly by the various metabolites that affect the kinase and the phosphatase.

5.2.2 Regulation by Nucleotidation–Denucleotidation

Covalent linkage of some nucleotides to acceptor groups on enzymes may cause modification of those enzymes, with concomitant changes in enzymic activity in a manner akin to phosphorylation. The three major types of modifications discovered in recent years are adenylylation–deadenylylation, in which AMP is the modifying nucleotide (Kingdon *et al.*, 1967; Ginsburg and Stadtman, 1975); uridylylation–deuridylylation, in which UMP is used in covalent linkage (Ginsburg and Stadtman, 1975); and ADP-ribosylation, in which the ADP-ribose moiety of NAD^+, in either a monomeric or polymeric form, is transferred to acceptor groups on some proteins (Honjo *et al.*, 1969; Honjo and Hayaishi, 1973; Hayaishi, 1976). Control by adenylylation and uridylylation has been studied extensively, but so far has been found to occur only in glutamine synthase of *E. coli*. The nucleotidation reactions constitute a cascading system much like that of the phosphorylations associated with the modification of phosphorylase b activity described in Section 5.2.1c (Ginsburg and Stadtman, 1975). How the cascading control of glutamine synthase increases the capacity for fine allosteric control will be discussed in the following sections.

    5.2.2a Cascade Control of Glutamine Synthase. Glutamine synthase of *E. coli* catalyzes the reaction

$$\text{L-Glutamate} + \text{ATP} + \text{NH}_4^+ \rightarrow \text{L-glutamine} + \text{ADP} + \text{P}_i \tag{11}$$

It plays a central role in nitrogen metabolism. The activity of this enzyme is controlled by cumulative feedback inhibition (Woolfolk and Stadtman, 1964). The native enzyme is constituted of 12 identical subunits that are arranged in two face-to-face hexagonal rings. A specific tyrosyl residue in each of the 12 subunits can be adenylylated by an adenylyltransferase to form a 5'-adenylyl-O-tyrosyl derivative. Adenylylated glutamine synthase differs from the unmodified form of the enzyme in two important respects: first, a changed metal ion specificity (Kingdon *et al.*, 1967); and second, an increase in the inhibition caused by various feedback inhibitors (Kingdon *et al.*, 1967; Shapiro *et al.*, 1967). Glutamine synthase requires magnesium ions for maximal catalytic activity; manganese has little effect. In contrast, the adenylylated enzyme is inactive with magnesium ions but active in the presence of manganous ions. The requirement for either of these two cations is a function of the pH of assay and the average extent of adenylylation of glutamine synthase. Increasing adenylylation produces a reciprocal inactivation of magnesium-dependent activity and activation of manganese-dependent activity (Kingdon *et al.*, 1967). The metabolic advantages of the changed metal ion specificity are not understood.

    The sensitivity of adenylylated glutamine synthase toward feedback inhibition is much greater than that of the unmodified enzyme. This is particularly evident with the feedback inhibitors, AMP, L-tryptophan, L-histidine, and CTP when the adenylylated enzyme is assayed in the presence of manganous ions (Kingdon *et al.*, 1967; Shapiro *et al.*, 1967).

    The adenylylation and deadenylylation of glutamine synthase are controlled by an unusual metabolite-regulated cascade system. Both the adenylylation and the deadenylylation are catalyzed by a single adenylyltransferase [Eqs. (5) and (6)] in the presence of either magnesium or manganous ions (Anderson *et al.*, 1970; Ginsburg and Stadtman, 1975). In the adenylylation reaction [Eq. (5)] the two terminal phosphates of ATP are released as PP_i and the 5'-adenylyl group is transferred to the enzyme. The reaction is reversible *in vitro* (Mantel and Holzer, 1970). Deadenylylation [Eq. (6)] occurs by a phosphorolytic cleavage of the adenylyl-O-tyrosyl bond and is apparently an irreversible process.

    The presence of only one enzyme catalyzing both the adenylylation and deadenylylation reactions poses the problem of "short-circuiting" or establishment of a futile cycle in which glutamine synthase uselessly oscillates between the modified and native form. Largely through the efforts of Stadtman and his co-workers (Ginsburg and Stadtman, 1975), the controls that prevent the aimless recycling of glutamine synthase have now been worked out. The controls consist of both an array of allosteric regulatory devices and a control mechanism of a fascinating new type, in which the direction and specificity of the adenylyltransferase is controlled by a small regulatory protein P_{II} (Shapiro, 1969). The allosteric controls of adenylyltransferase are based on the same strategy employed by cells to prevent occurrence of other futile cycles, namely, to have a metabolite or effector produce opposite effects on the opposing enzyme activities. In keeping with this strategy, glutamine stimulates, and α-ketoglutarate inhibits, adenylylation, whereas exactly the opposite happens with deadenylylation (glutamine inhibits and α-ketoglutarate stimulates). UTP and P_i inhibit the activity of the adenylyltransferase in the adenylylation reaction and ATP activates the deadenylylation reaction (Fig. 4).

The protein that regulates the specificity of adenylyltransferase, P_{II}, has a molecular weight of 46,000 and is apparently composed of four identical subunits (Ginsburg and Stadtman, 1975; Brown *et al.*, 1971). The regulatory protein is capable of existing in two forms, a native form, $P_{II}A$, and a uridylylated form, $P_{II}D$. In the latter form, four UMP residues per tetramer are attached covalently to specific tyrosyl residues. Uridylylation of $P_{II}A$ is catalyzed by a uridylyltransferase (distinct from adenylyltransferase) that requires UTP as substrate, is activated by α-ketoglutarate and ATP, and is inhibited by glutamine and P_i. It may be recalled that these very effectors cause similar allosteric modulation of the deadenylylation reaction [Eq. (6)]. This fact is of some relevance in the metabolite-controlled cascade regulation of glutamine synthase discussed below. Deuridylylation of $P_{II}D$ occurs by a hydrolytic cleavage of UMP by a manganese-dependent uridylyl-removing enzyme. To date, the uridylyltransferase and uridylyl-removing enzymes have not been separated from one another, and a possibility exists that both of these activities belong to one and the same enzyme. If that is the case, then more metabolite controls will perhaps be discovered that prevent futile cycling of $P_{II}A$ and $P_{II}D$.

The unmodified regulatory protein, $P_{II}A$, interacts with adenylyltransferase and stimulates the adenylylation reaction, while the uridylylated protein, $P_{II}D$, stimulates the deadenylylation reaction. The stimulation is due to protein–protein interactions (noncova-

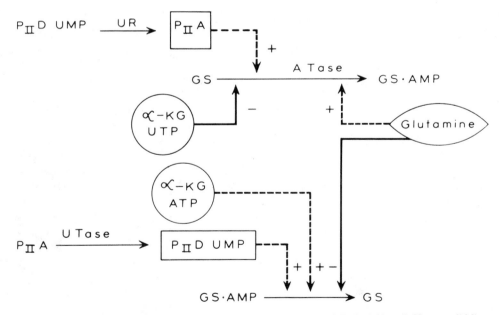

Figure 4. The allosteric and covalent regulation of glutamine synthase of *Escherichia coli*. Heavy, solid lines with (−) sign denote inhibition, and the broken lines with (+) sign denote activation. Symbols and abbreviations: GS, glutamine synthase; GS·AMP, adenylylated glutamine synthase; ATase, adenylyl transferase; $P_{II}A$, a small regulatory protein; $P_{II}D$ UMP, uridylylated regulatory protein; UR, uridyl-removing enzyme; UTase, uridylyl transferase; and α-KG, α-ketoglutarate. Details are given in the text.

lent). As indicated briefly elsewhere, the control of the direction of adenylyltransferase reaction by the regulatory protein, P_{II} is potentiated by the allosteric controls of enzymes that interconvert $P_{II}A$ and $P_{II}D$ as well as of the adenylyltransferase itself. This results in the establishment of a metabolite-controlled cascade system of regulation in which the capacity for allosteric control of the target enzyme, glutamine synthase, is greatly enhanced (Ginsburg and Stadtman, 1975). Inactivation of glutamine synthase begins with the deuridylylation of $P_{II}D$-UMP (Fig. 4). The product stimulates adenylylation of glutamine synthase by interacting with adenylyltransferase. Glutamine activates and α-ketoglutarate inhibits this reaction. Activation by glutamine ensures that no more of this metabolite is produced when it is present in excess. Conversely, when α-ketoglutarate is present in larger amounts, such as occurs during nitrogen starvation (when the need for glutamine exists), inhibition of the adenylyltransferase ensures that glutamine synthase will not be inactivated. A similar cascade system exists during activation of the partially active glutamine synthase. Here the sequence is initiated (Fig. 4) by the uridylylation of $P_{II}A$. The uridylylated product interacts with adenylyltransferase to produce an active, deadenylylated glutamine synthase. The enzymes of this cascade are individually susceptible to modulation by allosteric means. Thus, glutamine inhibits and α-ketoglutarate activates deadenylylation. In addition, uridylyltransferase is also stimulated by α-ketoglutarate. The net effect of these strategically directed allosteric controls is to produce a deadenylylated glutamine synthase (active) when need for production of glutamine is the greatest. It is clear from the above discussion and from Fig. 4 that the two oppositely directed cascade systems are controlled by the same (α-ketoglutarate and glutamine) or functionally equivalent (ATP and UTP) metabolites in exactly opposite ways. The result is that the response of the last enzyme in each of the cascades to the primary effectors is amplified in the direction demanded by cellular requirements.

5.2.3 ADP-Ribosylation

A unique mechanism of covalent modification of some enzymes and other proteins that has recently come to light is ADP- or poly(ADP)-ribosylation (Honjo and Hayaishi, 1973; Hayaishi, 1976; Hayaishi and Ueda, 1977). In ADP-ribosylation the ADP-ribose moiety of NAD^+ is transferred, by a specific ADP-ribosyltransferase, to a protein, with the release of nicotinamide and a proton. The reaction is reversible (Fig. 5). So far, a separate enzyme that produces NAD^+ from nicotinamide and protein-ADP-ribose has not been demonstrated; it is very likely that the same enzyme is responsible for ADP-ribosylation and de-ADP-ribosylation. If this is so, and if ADP-ribosylation has any regulatory significance, control systems are yet to be discovered that determine the directional properties of ADP-ribosyltransferase as we have seen for adenylylation–deadenylylation reactions in the last section. Till recently all ADP-ribosylations were thought to be catalyzed by enzyme systems extrinsic to the cells themselves. In other words, ADP-ribosylation did not appear to be an endogenous cellular control mechanism. However, an endogenous ADP-ribosyltransferase has recently been discovered (Moss and Vaughan, 1978) in avian erythrocytes which ADP-ribosylates adenylate cyclase. Discovery of ADP-ribosylation was made during elucidation of the mode of action of diptheria toxin. It was demonstrated by Collier and Pappenheimer (1964) that inhibition of mammalian protein synthesis by the toxin requires NAD^+. The target of the toxin was later found to be elongation factor 2, an enzyme involved in protein synthesis (Collier, 1967). Subsequently, Honjo et al. (1969) showed that the inactivation of elongation factor 2 was caused by the transfer of an ADP-ribose moiety to the protein.

An identical mechanism is now known to intervene in the toxicity of *Pseudomonas aeruginosa* toxin (Iglewski and Kabat, 1975), and it appears also that cholera toxin brings about its effects by ADP-ribosylation of some essential component of the adenylate cyclase system (Hayaishi and Ueda, 1977; Moss and Vaughan, 1977).

The examples cited above are in eukaryotic systems. However, ADP-ribosylation also is known to be brought about, again by extrinsic agents, in prokaryotes. Thus, the α-polypeptide of *E. coli* RNA polymerase is ADP-ribosylated, using NAD^+ as a donor, within 4 min after infection of the cell by the bacteriophage T4. The modification involves the guanidonitrogen of a specific arginine of the α subunit to which ADP-ribose seems to be linked through its terminal ribose (Goff, 1974). It is noteworthy, however, that the RNA polymerase activity is not significantly altered by this covalent modification. Recently, it has been demonstrated that infection of *E. coli* by phage N4 alters, probably by ADP-ribosylation, no less than 30 different host proteins (Pesce *et al.*, 1976). The enzyme ADP-ribosyltransferase is enclosed within the virion itself and is injected along with its DNA into the host cell.

Poly(ADP)-ribose, a linear homopolymer of repeating ADP-ribose units (Fig. 5), was discovered independently by several groups of workers (for review, see Hayaishi, 1976) in the nuclei of eukaryotes. It has not been found so far in plants or in prokaryotes. Poly(ADP)-ribose consists of 2 to 50 units of ADP-ribose, linked together by ribose-to-ribose (1'→2') bonds (Fig. 5). The polymer is covalently attached to histonelike nuclear proteins and may play a role in transcriptional control (Honjo and Hayaishi, 1973). Formation of poly(ADP)-ribose is catalyzed by a synthase that is associated with chromatin. In a highly purified, solubilized form this enzyme exhibits an absolute dependence for DNA or poly(dA-dT) for activity (Ueda *et al.*, 1975). The DNA-dependent activity is further stimulated by histones. The polymer is degraded *in vivo* by a poly(ADP)-ribose

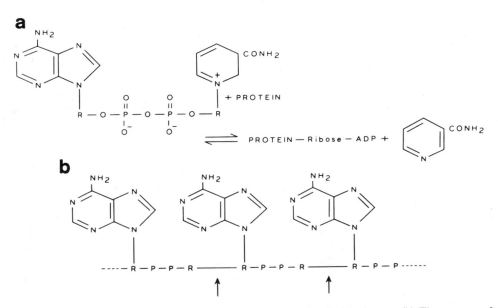

Figure 5. (a) The reaction catalyzed by ADP-ribose transferase described in the text. (b) The structure of poly(ADP)-ribose; R and P are ribose and phosphoric acid residues, respectively. Arrows denote the bond hydrolyzed by glycohydrolase.

glycohydrolase that specifically cleaves the 1-2-glycosidic bonds (Fig. 5). The enzymes that catalyze poly(ADP)-ribosylation of proteins are little understood. Indeed, their existence in almost all instances is assumed rather than proven. Poly(ADP)-ribosylation as an *enzymic* control mechanism is also known. In the calcium- and magnesium-dependent endonuclease of rat liver nuclei (Yoshihara *et al.*, 1975), ADP-ribosylation of the enzyme is thought to lead to an inhibition of activity; in its modified form the enzyme cannot generate primer sites on DNA.

6   Conclusions

From the foregoing account it is clear that a wide array of controls is utilized in biological systems for the modulation of enzyme activities. Perhaps the most interesting question here is why so many varieties of control systems exist. The two major categories of controls, allosteric controls and enzyme modifications, are utilized by cells in response to stimuli of two different kinds, metabolic and hormonal. The metabolic stimuli are provided by the myriad small precursors and end products of individual pathways. The control system utilized in response to stimuli of this kind is predominantly the allosteric one, and it persists as long as the stimulus continues to exist. In contrast, hormones, particularly those operating via cyclic AMP (Fig. 3), elicit controls that involve predominantly covalent changes, such as structural modifications of enzymes. The initial hormonal stimulus is magnified by the use of cascade systems and the enzyme alterations last long after the stimulus itself has disappeared. The peculiar nature of hormone action—that is, synthesis in an organ different from that of the target tissue—demands a relatively long-lasting modulation of enzyme activity, and this is effectively provided by covalent modifications far better than by allosteric changes. Despite this rationalization, there is still the question of why there exists such variety even among covalent modifications—phosphorylation in some cases, adenylylation and uridylylation in other cases, and ADP-ribosylation and poly(ADP)-ribosylation in still others. A reasonable explanation is that the use of covalent conversion as a mechanism of regulation was acquired late during evolution and that each kind of modification originated independently of the others. The nature of modification needed for a given enzyme was perhaps dictated by steric considerations and the need to bring about optimal configurational changes of the active site within the constraints imposed by an already evolved quaternary structure. Though this conclusion is conjectural, there seems to be general agreement among workers in this area that regulatory devices have evolved later than the enzymes themselves (Engel, 1973).

Apart from questions about evolution, it may also be pertinent to ask whether any new conceptual advances in the field of enzymic control mechanism are likely to be forthcoming, or whether the work that remains to be done in this area will instead prove to be just a "stamp-collecting" exercise. Some recent findings suggest that a few surprises may still await an observant researcher. These are, first, the mechanisms involved in the regulation of the specificity of enzyme active sites, and, second, the gradations that are beginning to emerge between the purely biochemical controls, on the one hand, and the purely genetic ones on the other. Several examples have been cited in which either the allosteric ligands (such as those of ribonucleoside diphosphate reductase) or regulatory proteins (such as the P_{II} regulatory protein in the glutamine synthase system of *E. coli*) are capable of changing the substrate specificity of the target enzymes. These findings are remarkable in view of the biochemical dogma of long standing that one enzyme catalyzes only one specific (or group-specific) reaction. The second conceptual advance referred to above is the involvement of some regulatory enzymes in the genetic control process. The allosteric enzyme is central to biochemical control just as the repressor is central to genetic regulation. The indication that, in a few pathways, the allosteric enzyme of the pathway may also act as a regulatory protein at the genetic level has produced an interesting link between the two parallel processes of control (Sanwal *et al.*, 1971; Goldberger, 1974; Magasanik *et al.*, 1974; Foor *et al.*, 1975; Savageau, 1979). Similarly, the discovery of antizyme regulation, which resembles end-product repression but is caused by protein–protein interaction (Heller *et al.*, 1976), and epienzymic controls (Wiame, 1971) have helped to bridge the gap and produce a pleasing continuum between the mechanisms of feedback repression on the one hand and feedback inhibition on the other. Elucidation of the molecular mechanisms involved in interactions of these types will be of great interest.

References

Anderson, W. B., Hennig, S. B., Ginsburg, A., and Stadtman, E. R., 1970, Association of ATP-glutamine synthetase adenylyltransferase activity with the P_I component of the glutamine synthetase deadenylylation system, *Proc. Natl. Acad. Sci. U.S.A.* **67**:1417.

Atkinson, D. E., and Walton, G. M., 1967, Adenosine triphosphate conservation in metabolic regulation, *J. Biol. Chem.* **242**:3239.

Atkinson, D. E., 1968, The energy charge of the adenylate pool as a regulatory parameter. Interaction with feedback modifiers, *Biochemistry* **7**:4030.

Atkinson, D. E., 1970, Enzymes as control elements in metabolic regulation, in: *The Enzymes* (P. Boyer, ed.), Vol. 1, pp. 461–489, Academic Press, New York.

Beavo, J. A., Bechtel, P. J., and Krebs, E. G., 1975, Mechanisms of control for cAMP-dependent protein kinase from skeletal muscle, *Adv. Cyclic Nucleotide Res.* **5**:241.

Bellamy, G., and Bornstein, P., 1971, Evidence for procollagen, a biosynthetic presursor of collagen, *Proc. Natl. Acad. Sci. U.S.A.* **68**:1138.

Blangy, D., Buc, H., and Monod, J., 1968, Kinetics of the allosteric interactions of phosphofructokinase from *Escherichia coli, J. Mol. Biol.* **31**:13.

Blobel, G., and Potter V. R., 1966, Relation of ribonuclease and ribonuclease inhibitor to the isolation of polysomes from rat liver, *Proc. Natl. Acad. Sci. U.S.A.* **55**:1283.

Bloch, K., and Vance, D., 1977, Control mechanisms in the synthesis of saturated fatty acids, *Annu. Rev. Biochem.* **46**:263.

Bonsignore, A., De Flora, A., Mangiarotti, M. A., Lorenzoni, I., and Alema, S., 1968, A new hepatic protein inactivating glucose 6-phosphate dehydrogenase, *Biochem. J.* **106**:147.

Brew, K., Vanaman, T. C., and Hill, R. L., 1968, The role of α-lactalbumin and the A-protein in lactose synthetase; A unique mechanism for the control of biological reactions, *Proc. Natl. Acad. Sci. U.S.A.* **59**:491.

Brostrom, M. A., Bromstrom, C. O., Breekennidge, B. M., and Wolff, D. J., 1976, Regulation of adenylate cyclase from glial tumor cells by calcium and a calcium binding protein, *J. Biol. Chem.* **251**:4744.

Brown, M. S., Segal, A., and Stadtman, E. R., 1971, Modulation of glutamine synthetase adenylylation and deadenylylation is mediated by metabolic transformation of the P_{II}-regulatory protein, *Proc. Natl. Acad. Sci. U.S.A.* **68**:2949.

Brown, N. C., and Reichard, P., 1969, Role of effector binding in allosteric control of ribonucleoside diphosphate reductase, *J. Mol. Biol.* **46**:39.

Brown, N. C., Larsson, A., and Reichard, P., 1967, On the subunit structure of ribonucleoside diphosphate reductase, *J. Biol. Chem.* **242**:4272.

Buchanan, B. B., and Schürmann, P., 1973, Ribulose 1,5-diphosphate carboxylase: A regulatory enzyme in the photosynthetic assimilation of carbon dioxide, *Curr. Top. Cell. Regul.* **7**:1–20.

Burgess, R. R., Travers, A. A., Dunn, J. J., and Bautz, E. K. F., 1969, Factor stimulating transcription by RNA polymerase, *Nature (London)* **221**:43.

Carlson, C. A., and Kim, K. H., 1974, Regulation of hepatic acetyl coenzyme A carboxylase by phosphorylation and dephosphorylation, *Arch. Biochem. Biophys.* **164**: 478.

Caskey, C. T., Ashton, D. M., and Wyngaarden, J. B., 1964, The enzymology of feedback inhibition of glutamine phosphoribosylpyrophosphate amidotransferase by purine ribonucleotide, *J. Biol. Chem.* **239**:2570.

Challand, G. S., and Rosemeyer, M. A., 1974, The correlation between the apparent molecular weight and the enzymic activity of lactose synthetase, *FEBS Lett.* **47**:94.

Chamberlin, M. J., 1974, Bacterial DNA-dependent RNA polymerase, in: *The Enzymes* (P. D. Boyer, ed), 3rd ed., Vol. 10, pp. 333–374, Academic Press, New York.

Changeux, J. P., 1961, The feedback control mechanism of biosynthetic L-threonine deaminase by L-isoleucine, *Cold Spring Harbor Symp. Quant. Biol.* **26**:313.

Chen, A. K., Ashok, B., Hopper, S., Abrams, R., and Franzen, J. S., 1974, Substrate and effector binding to ribonucleoside diphosphate reductase of *Lactobacillus leichmanni*, *Biochemistry* **13**:654.

Cheung, W. Y., 1969, Cyclic 3′,5′-nucleotide phosphodiesterase. Preparation of a partially inactive enzyme and its subsequent stimulation by snake venom, *Biochem. Biophys. Acta* **191**:303.

Cheung, W. Y., 1980, Calmodulin plays a pivotal role in cellular regulation, *Science* **207**:19.

Cohen, G. N., 1965, Regulation of enzyme activity in microorganisms, *Annu. Rev. Microbiol.* **19**:105.

Cohen, G. N., 1969, The aspartokinases and homoserine dehydrogenases of *Escherichia coli*, *Curr. Top. Cell. Regul.* **1**:183–231.

Cohen, P., 1973, The subunit structure of rabbit skeletal muscle phosphorylase kinase and the molecular basis of its activation reactions, *Eur. J. Biochem.* **34**:1.

Cohen, P., 1976, The regulation of protein function by multisite phosphorylation, *Trends Biochem. Sci.* **1**:38.

Cohen, P., 1978, The role of cyclic-AMP dependent protein kinase in the regulation of glycogen metabolism in mammalian skeletal muscle, in: *Current Topics in Cellular Regulation* (B. L. Horecker and E. R. Stadtman, eds.),Vol. 14, pp. 117–196, Academic Press, New York.

Cohen, P., and Antoniw, J. F., 1973, The control of phosphorylase phosphatase by "second site phosphorylation"; a new form of enzyme regulation, *FEBS Lett.* **34**:43.

Cohen, P., Antoniw, J. F., Nimmo, H. G., and Proud, C. G., 1975, Structure and regulation of enzymes for the degradation and resynthesis of glycogen, *Biochem. Soc. Trans.* **3**:849.

Cohen, P., Burchell, A., Foulkes, J. G., Cohen, P. T., Vanaman, T. C., and Nairn, A. C., 1978, Identification of the Ca^{2+}-dependent modulator protein as the fourth subunit of rabbit skeletal muscle phosphorylase kinase, *FEBS Lett.* **92**:287.

Collier, R. J., 1967, Effect of diphtheria toxin on protein synthesis: Inactivation of one of the transfer factor, *J. Mol. Biol.* **25**:83.

Collier, R. J., and Pappenheimer, A. M., Jr., 1964, Studies on the mode of action of Diphtheria toxin. II. Effect of toxin on the amino acid incorporation in cell-free systems, *J. Exp. Med.* **120**:1019.

Conway, A., and Koshland, D. E., Jr., 1968, Negative cooperativity in enzyme action. The binding of diphosphopyridine nucleotide to glyceraldehyde 3-phosphate dehydrogenase, *Biochemistry* **7**:4011.

Cori, G. T., Colowick, S. P., and Cori, C. F., 1938, The formation of glucose-1-phosphoric acid in extracts of mammalian tissues and of yeast, *J. Biol. Chem.* **123**:375.

Dahlquist, F. W., and Purich, D. L., 1975, Regulation of *Escherichia coli* glutamine synthetase. Evidence for the action of some feedback modifiers at the active site of the unadenylylated enzyme, *Biochemistry* **14**:1980.

Darnall, D. W., and Klotz, I. M., 1975, Subunit constitution of proteins: A table, *Arch. Biochem. Biophys.* **166**:651.

Datta, P., and Gest, H., 1964, Control of enzyme activity by concerted feedback inhibition, *Proc. Natl Acad. Sci. U.S.A.* **52**:1004.

Datta, P., Gest, H., and Segal, H. L., 1964, Effect of feedback modifiers on the state of aggregation of homoserine dehydrogenase of *Rhodospirillum rubrum*, *Proc. Natl. Acad. Sci. U.S.A.* **51**:125.

Davie, E. W., and Fujikawa, K., 1975, Basic mechanisms in blood coagulation, *Annu. Rev. Biochem.* **44**:799.

Davis, B. D., 1961, The teleonomic significance of biosynthetic control mechanisms, *Cold Spring Harbor Symp. Quant. Biol.* **26**:1.

Deutch, C. E., and Soffer, R. L., 1975, Regulation of proline catabolism by leucyl, phenylalanyl—tRNA—protein transferase, *Proc. Natl. Acad. Sci. U.S.A.* **72**:405.

Döbeln, U. von, and Reichard, P., 1976, Binding of substrates to *Escherichia coli* ribonucleotide reductase, *J. Biol. Chem.* **251**:3616.

Duffy, J. J., Petrusek, R. L., and Geiduschek, E. P., 1975, Conversion of *Bacillus subtilis* RNA polymerase activity in vitro by a protein induced by phage SPO1, *Proc. Natl. Acad. Sci. U.S.A.* **72**:2366.

Duncan, B. K., Diamond, G. R., and Bessman, M. J., 1972, Regulation of enzyme activity through subunit interaction, *J. Biol. Chem.* **247**:8136.

Ebner, K. E., and Magee, S. C., 1975, Lactose synthetase, in: *Subunit Enzymes* (K. E. Ebner, ed.), pp. 137–179, Marcel Dekker, New York.

Engel, P. C., 1973, Evolution of enzyme regulator sites: Evidence for partial gene duplication from amino-acid sequence of bovine glutamate dehydrogenase, *Nature (London)* **241**:118.

Engel, P. C., and Dalziel, K., 1969, Kinetic studies of glutamate dehydrogenase with glutamate and norvaline as substrate, *Biochem. J.* **115**:621.

Ferdinand, W., 1966, The interpretation of non-hyperbolic rate curves for two-substrate enzymes, *Biochem. J.* **98**:278.

Fitzerald, D. K., Broadbeck, U., Kiyosawa, I., Mawal, R., Colvin, B., and Ebner, K. E., 1970, α-Lactalbumin and the lactose synthetase reaction, *J. Biol. Chem.* **245**:2103.

Foor, F., Janssen, K. A., and Magasanik, B., 1975, Regulation of synthesis of glutamine synthetase by adenylylated glutamine synthetase, *Proc. Natl. Acad. Sci. U.S.A.* **72**:4844.

Fox, T. D., 1976, Identification of phage SPO1 proteins coded by regulatory genes 33 and 34, *Nature (London)* **262**:748.

Frieden, C., 1963, Glutamate dehydrogenase. V. The relation of enzyme structure to the catalytic function, *J. Biol. Chem.* **238**:3286.

Frieden, C., 1967, Treatment of enzyme kinetic data, *J. Biol. Chem.* **242**:4045.

Frieden, C., 1970, Kinetic aspects of regulation of metabolic processes. The hysteretic enzyme concept, *J. Biol. Chem.* **245**:5788.

Frieden, C., 1971, Protein–protein interaction and enzymatic activity, *Annu. Rev. Biochem.* **40**:653.

Gerhart, J. C., and Pardee, A. B., 1961, Separation of feedback inhibition from activity of aspartate transcarbamylase, *Fed. Proc.* **20**:224.

Gerhart, J. C., and Pardee, A. B., 1962, The enzymology of control by feedback inhibition, *J. Biol. Chem.* **237**:891.

Gerhart, J. C., and Schachman, H. K., 1965, Direct subunits for the regulation and catalytic activity of aspartate transcarbamylase, *Biochemistry* **4**:1054.

Gerhart, J. C., and Schachman, H. K., 1968, Allosteric interactions in aspartate transcarbamylase. II. Evidence for different conformational states of the protein in the presence and the absence of specific ligands, *Biochemistry* **7**:538.

Gerlt, J. A., Rabinowitz, K. W., Dunne, C. P., and Wood, W. A., 1973, The mechanism of action of 5'-adenylic acid-activated threonine dehydrase, *J. Biol. Chem.* **248**:8200.

Gill, G. N., and Garren, L. D., 1969, A cyclic 3',5'-adenosine monophosphate dependent protein kinase from the adrenal cortex: Comparison with a cyclic AMP binding protein, *Biochem. Biophys. Res. Commun.* **39**:335.

Ginsburg, A., 1969, Conformational changes in glutamine synthetase from *Escherichia coli*. II. Some characteristics of the equilibrium binding of feedback inhibitors to the enzyme, *Biochemistry* **8**:1726.

Ginsburg, A., and Stadtman, E. R., 1970, Multienzyme systems, *Annu. Rev. Biochem.* **39**:429.

Ginsburg, A., and Stadtman, E. R., 1975, Glutamine synthetase of *Escherichia coli*: Structure and regulation, in: *Subunit Enzymes* (K. E. Ebner, ed.), pp. 43–84, Marcel Dekker, New York.

Goff, C. G., 1974, Chemical structure of a modification of the *Escherichia coli* ribonucleic acid polymerase α-polypeptide induced by bacteriophage T_4 infection, *J. Biol. Chem.* **249**:6181.

Goldberger, R. F., 1974, Autogenous regulation of gene expression, *Science* **183**:810.

Goldie, A. H., and Sanwal, B. D., 1980, Allosteric control by calcium and mechanism of desensitization of phosphoenol pyruvate carboxykinase of *Escherichia coli*, *J. Biol. Chem.* **255**:1399.

Goldin, B. R., and Frieden, C., 1971, L-Glutamate dehydrogenase, *Curr. Top. Cell. Regul.* **4**:77–114.

Goulian, M., and Beck, W. S., 1966, Purification and properties of Cobamide-dependent ribonucleotide reductase from *Lactobacillus leichmannii*, *J. Biol. Chem.* **241**:4233.

Graves, D. J., Martensen, T. M., Tu, J.-I., and Tessmer, G. M., 1974, The use of alternative substrates in the study of phosphorylase phosphatase and phosphorylase kinase, in: *Metabolic Interconversion of*

Enzymes 1973 (E. H. Fischer, E. G. Krebs, H. Neurath, and E. R., Stadtman, eds.), pp. 53–61, Springer-Verlag, New York.

Haldenwang, W. G., and Losick, R., 1979, A modified RNA polymerase transcribes a cloned gene under sporulation control in *Bacillus subtilis, Nature (London)* **282**:256.

Hayaishi, O., 1976, Poly ADP-ribose and ADP-ribosylation of proteins, *Trends Biochem. Sci.* **1**:9.

Hayaishi, O., and Ueda, K., 1977, Poly (ADP-ribose) and ADP-ribosylation of proteins, *Annu. Rev. Biochem.* **46**:95.

Hayakawa, T., Perkins, J. P., and Krebs, E. G., 1973, Studies on the subunit structure of rabbit skeletal muscle phosphorylase kinase, *Biochemistry* **12**:574.

Heller, J. S., Fong, W. F., and Canellakis, E. S., 1976, Induction of a protein inhibitor to ornithine decarboxylase by the end product of its reaction, *Proc. Natl. Acad. Sci. U.S.A.* **73**:1858.

Hill, R. L., and Brew, K., 1975, Lactose synthetase, *Adv. Enzymol.* **43**:411.

Hirata, M., Tokushige, M., Inagaki, A., and Hayaishi, O., 1965, Nucleotide activation of threonine deaminase from *Escherichia coli, J. Biol. Chem.* **240**:1711.

Honjo, T., and Hayaishi, O., 1973, Enzymatic ADP-ribosylation of proteins and regulation of cellular activity, *Curr. Top. Cell. Regul.* **7**:87–127.

Honjo, T., Nishizuka, Y., and Hayaishi, O., 1969, Diphtheria toxin-dependent adenosine diphosphate ribosylation of aminoacyl transferase II and inhibition of protein synthesis, *J. Biol. Chem.* **243**:3553.

Hoppe, J., and Wagner, K. G., 1979, cAMP-dependent protein kinase I, a unique allosteric enzyme, *Trends Biochem. Sci.* **4**: 282.

Horecker, B. L., Melloni, E., and Pontremoli, S., 1975, Fructose 1,6-biphosphatase: Properties of the neutral enzyme and its modification by proteolytic enzymes, *Adv. Enzymol.* **42**:193.

Horvitz, H. R., 1973, Polypeptide bound to the host RNA polymerase is specified by T_4 control gene 33, *Nature (London), New Biol.* **244**:137.

Huestis, W. H., and Raftery, M. A., 1972*a*, A study of cooperative interactions in hemoglobin using fluorine nuclear magnetic resonance, *Biochemistry* **11**:1648.

Huestis, W. H., and Raftery, M. A., 1972*b*, Observation of cooperative ionizations in hemoglobin, *Proc. Natl. Acad. Sci. U.S.A.* **69**:1887.

Iglewski, B. H., and Kabat, D., 1975, NAD dependent inhibition of protein synthesis by *Pseudomonas aeruginosa* toxin, *Proc. Natl. Acad. Sci. U.S.A.* **72**:2284.

Issaly, I. M., and Issaly, A. S., 1974, Control of ornithine carbamoyltransferase activity by arginase in *Bacillus subtilis, Eur. J. Biochem.* **49**:485.

Kakiuchi, S., Yamazaki, R., Teshima, Y., and Uenishi, K., 1973, Regulation of nucleoside cyclic 3′ :5′-monophosphate phosphodiesterase activity from rat brain by a modulator and Ca^{2+}, *Proc. Natl. Acad. Sci. U.S.A.* **70**:3526.

Kemp, R. G., and Krebs, E. G., 1967, Binding of metabolites by phosphofructokinase, *Biochemistry* **6**:423.

Killilea, S. D., Brandt, H., and Lee, E. Y. C., 1976*a*, Modulation of protein function by phosphorylation: The role of protein phosphatase, *Trends Biochem. Res.* **1**:30.

Killilea, S. D., Brandt, H., Lee, E. Y. C., and Whelan, W. J., 1976*b*, Evidence for the coordinate control of activity of liver glycogen synthetase and phosphorylase by a single protein phosphatase, *J. Biol. Chem.* **251**:3263.

Kingdon, H. S., Shapiro, B. M., and Stadtman, E. R., 1967, Regulation of glutamine synthetase. VIII. ATP-glutamine synthetase adenyltransferase an enzyme that catalyzes alterations in the regulatory properties of glutamine synthetase, *Proc. Natl. Acad. Sci. U.S.A.* **58**:1703.

Kirschner, K., Eigen, M., Bittman, R., and Voight, B., 1966, The binding of nicotinamide adenine dinucleotide to yeast D-glyceraldehyde-3-phosphate dehydrogenase: Temperature–jump relaxation studies on the mechanism of an allosteric enzyme, *Proc. Natl. Acad. Sci. U.S.A.* **56**:1661.

Klee, W. A., and Klee, C. B., 1972, The interaction of α-lactalbumin and the A protein of lactose synthetase, *J. Biol. Chem.* **247**:2336.

Klemme, J.-H., 1976, Unidirectional inhibition of phosphoenolpyruvate carboxykinase from *Rhodospirillium rubrum* by ATP, *Arch. Microbiol.* **107**:189.

Klingsøyr, L., and Atkinson, D. E., 1970, Regulatory properties of phosphoribosyladenosine triphosphate synthetase. Synergism between adenosine monophosphate, phosphoribosyladenosine triphosphate, and histidine, *Biochemistry* **9**:2021.

Knudsen, J., Clark, S., and Dils, R., 1976, Purification and some properties of a medium-chain acylthioester hydrolase from lactating rabbit mammary gland which terminates chain elongation in fatty acid synthesis, *Biochem. J.* **160**:683.

Koshland, D. E., Jr., 1970, The molecular basis of enzyme regulation, in: *The Enzymes*, Vol. 1 (P. D. Boyer, ed.), pp. 341–396, Academic Press, New York.

Koshland, D. E., Nemethy, E., and Filmer, D., 1966, Comparison of experimental binding data and theoretical model in protein containing subunits, *Biochemistry* **5**:365.

Krebs, E. G., 1972, Protein kinases, *Curr. Top. Cell. Regul.* **5**:99–133.

Krebs, E. G., and Beavo, J. A., 1979, Phosphorylation–dephosphorylation of enzymes, *Annu. Rev. Biochem.* **48**:923.

Krebs, E. G., and Fischer, E. H., 1956, The phosphorylase b to a converting enzyme of rabbit skeletal muscle, *Biochem. Biophys. Acta* **20**:150.

Krebs, E. G., Love, D. S., Bratvold, G. E., Trayser, K. A., Meyer, W. L., and Fischer, E. H., 1964, Purification and properties of rabbit skeletal muscle phosphorylase b kinase, *Biochemistry* **3**:1022.

Krebs, H. A., 1969, The role of equilibria in the regulation of metabolism, *Curr. Top. Cell. Regul.* **1**:45–76.

Kun, E., Zimber, P. H., Chang, A. C. Y., Puschendorf, B., and Grunicke, H., 1975, Macromolecular enzymatic product of NAD$^+$ in liver mitochondria, *Proc. Natl. Acad. Sci. U.S.A.* **72**:1736.

Lane, M. D., Polakis, S. E., and Moss, J., 1975, Acetyl CoA carboxylase, in: *Subunit Enzymes* (K. E. Ebner, ed.), pp. 181–221, Marcel Dekker, New York.

LeJohn, H. B., 1968, Unidirectional inhibition of glutamate dehydrogenase by metabolites: A possible regulatory mechanism, *J. Biol. Chem.* **243**:5126.

Levitzki, A., 1975, Subunit interactions in proteins, in: *Subunit Enzymes* (K. E. Ebner, ed.), pp. 1–41, Marcel Dekker, New York.

Levitzki, A., and Koshland, D. E., Jr., 1972, Role of an allosteric effector. Guanosine triphosphate activation in cytosine triphosphate synthetase, *Biochemistry* **11**:241.

Levitzki, A., and Koshland, D. E., Jr., 1976, The role of negative cooperativity and half of the sites reactivity in enzyme regulation, *Curr. Top. Cell. Regul.* **10**:1–40.

Levitzki, A., and Schlessinger, J., 1974, Cooperativity in associating proteins. Monomer–dimer equilibrium coupled to ligand binding, *Biochemistry* **13**:5214.

Levy, H. R., Raineri, R. R., and Nevaldine, B. H., 1966, On the structure and catalytic function of mammary glucose 6-phosphate dehydrogenase, *J. Biol. Chem.* **241**:2181.

Libertini, L. J., and Smith, S., 1978, Purification and properties of a thioesterase from lactating mammary gland which modifies the product specificity of fatty acid synthetase, *J. Biol. Chem.* **253**:1393.

Linn, T. C., Pettit, F. H., Hucho, F., and Reed, L. J., 1969, α-Keto acid dehydrogenases complex. XI. Comparative studies of regulatory properties of the pyruvate dehydrogenase complexes from kidney, heart, and liver mitochondria, *Proc. Natl. Acad. Sci. U.S.A.* **64**:227.

McNamara, D. J., and Rodwell, V. W., 1975, Regulation of hepatic 3-hydroxy-3-methylglutaryl coenzyme A reductase, *Arch. Biochem. Biophys.* **168**:378.

Maeba, P., and Sanwal, B. D., 1966, The allosteric threonine deaminase of *Salmonella*. Kinetic model for the native enzyme, *Biochemistry* **5**:525.

Maeba, P., and Sanwal, B. D., 1969, Phosphoenolpyruvate carboxylase of *Salmonella*. Some chemical and allosteric properties, *J. Biol. Chem.* **244**:2549.

Magasanik, B., Prival, M. J., Brenchley, J. E., Tyler, B. M., DeLeo, A. B., Streicher, S. L., Bender, R. A., and Paris, C. G., 1974, Glutamine synthetase as a regulator of enzyme synthesis. *Curr. Top. Cell. Regul.* **8**:119–138.

Mansour, T. E., 1963, Studies on heart phosphofructokinase: Purification, inhibition and activation, *J. Biol. Chem.* **238**:2285.

Mantel, M., and Holzer, H., 1970, Reversibility of the ATP : glutamine synthetase adenyltransferase reaction, *Proc. Natl. Acad. Sci. U.S.A.* **65**:660.

Mark, D. E., and Richardson, C. C., 1976, *Escherichia coli* thioredoxin: A subunit of bacteriophage T$_7$ DNA polymerase, *Proc. Natl. Acad. Sci. U.S.A.* **73**:780.

Matthews, B. W., and Bernhard, S. A., 1973, Structure and symmetry of oligomeric enzymes, *Annu. Rev. Biophys. Bioeng.* **2**:257.

Meunier, J. C., Buc, J., Navarro, A., and Ricard, J., 1974, Regulatory behavior of monomeric enzymes. 2. A wheat-germ hexokinase as a mnemonical enzyme, *Eur. J. Biochem.* **49**:209.

Millar, R. E., and Stadtman, E. R., 1972, Glutamate synthase from *Esterichia coli*—An iron-sulfide flavoprotein, *J. Biol. Chem.* **247**:7407.

Miyamoto, E., Petzold, G. L., Harris, J. S., and Greengard, P., 1971, Dissociation and concomitant activation of adenosine 3',5'-monophosphate-dependent protein kinase by histone, *Biochem. Biophys. Res. Commun.* **44**:305.

Monod, J., and Jacob, F., 1961, General conclusions: Teleonomic mechanisms in cellular metabolism, growth, and differentiation, *Cold Spring Harbor Symp. Quant. Biol.* **26**:389.

Monod, J., Changeux, J. P., and Jacob, F., 1963, Allosteric proteins and cellular control systems, *J. Mol. Biol.* **6**:306.

Monod, J., Wyman, J., and Changeux, J. P., 1965, On the nature of allosteric transitions: A plausible model, *J. Mol. Biol.* **12**:88.

Moss, J., and Vaughan, M., 1977, Mechanism of action of choleragen. Evidence for ADP-ribosyl transferase activity with arginine as an acceptor, *Proc. Natl. Acad. Sci. U.S.A.* **74**:5440.

Moss, J., and Vaughan, M., 1978, Isolation of an avian erythrocyte protein possessing ADP-ribosyltransferase activity and capable of activating adenylate cyclase, *Proc. Natl. Acad. Sci. U.S.A.* **75**:3621.

Murphy, T. A., and Wyatt, G. R., 1965, The enzymes of glycogen and trehalose synthesis in silk moth fat body, *J. Biol. Chem.* **240**:1500.

Murray, A. W., Froscio, M., and Rogers, A., 1974, Dissociation of rabbit muscle cyclic AMP-dependent protein kinase into catalytic and regulatory subunits by *p*-chloromercuribenzoate and methylmercuric chloride, *FEBS Lett.* **48**:238.

Neet, K. E., and Ainslie, G. R., 1976, Cooperativity and slow transitions in the regulation of oligomeric and monomeric enzymes, *Trends Biochem. Sci.* **1**:145.

Neurath, H., Walsh, K. A., and Gertler, A., 1973, Inhibition of physiological function by limited proteolysis, in: *Metabolic Interconversion of Enzymes 1973* (E. H. Fischer, E. G. Krebs, H. Neurath, and E. R., Stadtman, eds.), pp. 301–312, Springer-Verlag, New York.

Niles, E. G., and Westhead, E. W., 1973, *In vitro* adenylylation of lysine-sensitive aspartylkinase from *E. coli*, *Biochemistry* **12**:1723.

Nordlie, R. C., 1976, Multifunctional hepatic glucose-6-phosphate and the "tuning" of blood glucose levels, *Trends Biochem. Sci.* **1**:199.

Numa, S., 1974, Regulation of fatty acid synthesis in higher animals, *Ergeb. Physiol.* **69**:53.

Panagou, D., Orr, M. D., Dunstone, J. R., and Blakeley, R. L., 1972, A monomeric, allosteric enzyme with a single polypeptide chain. Ribonucleotide reductase of *Lactobacillus leichmannii*, *Biochemistry* **11**:2378.

Pardee, A. B., 1971, Control of metabolic reactions by feedback inhibition, *Harvey Lect.* **65**:59.

Paulus, H., and Gray, E., 1964, Multivalent feedback inhibitions of aspartokinase in *Bacillus polymyxa*, *J. Biol. Chem.* **239**:Pc 4008.

Penninckx, M., Simon, J.-P., and Wiame, J.-M., 1974, Interaction between argniase and L-ornithine car-bamoyltransferase in *Saccharomyces cerevisiae*, *Eur. J. Biochem.* **49**:429.

Pesce, A., Casoli, C., and Schito, G. C., 1976, Rifampicin-resistant RNA polymerase and NAD transferase activities in coliphage N_4 virions, *Nature (London)* **262**:412.

Pike, M. C., Kredich, N. M., and Snyderman, R., 1978, Requirement of *S*-adenosyl-L-methionine-mediated methylation for human monocyte chemotaxis, *Proc. Natl. Acad. Sci. U.S.A.* **75**:3928.

Pogell, B. M., Tanaka, A., and Siddons, R. C., 1968, Natural activators for liver fructose 1,6-diphosphatase and the reversal of adenosine 5'-monophosphate inhibition by muscle phosphofructokinase, *J. Biol. Chem.* **243**:1356.

Pontremoli, S., Grazi, E., and Accorsi, A., 1968, Fructose diphosphatase from rabbit liver. X. Isolation and kinetic properties of the enzyme adenosine monophosphate complex, *Biochemistry* **7**:3628.

Purich, D. L., and Fromm, H. J., 1972, A possible role for kinetic reaction mechanism dependent substrate and product effect in enzyme regulation, *Curr. Top. Cell. Regul.* **6**:131–167.

Ratner, D., 1974, Bacteriophage T_4 transcriptional control gene 55 codes for a protein bound to *Escherichia coli* RNA polymerase, *J. Mol. Biol.* **89**:803.

Raybin, D., and Flavin, M., 1975, An enzyme tyrosylating α-tubulin and its role in microtubule assembly, *Biochem. Biophys. Res. Commun.* **65**:1088.

Reed, L. J., Pettit, F. H., Roche, T. E., Butterworth, P. J., Barrera, C. R., and Tsai, C. S., 1974, Structure, function and regulation of the mammalian pyruvate dephdrogenase complex, in: *Metabolic Interconversion of Enzymes 1973* (E. H. Fischer, E. G. Krebs, H. Neurath, and E. R. Stadtman, eds.), pp. 99–116, Springer-Verlag, New York.

Reeves, R. E., and Sols, A., 1973, Regulation of *Escherichia coli* phosphofructokinase *in situ*, *Biochem. Biophys. Res. Commun.* **50**:459.

Reitzer, L. J., and Neet, K. E., 1974, Regulatory kinetics of yeast hexokinase *in situ*, *Biochem. Biophys. Acta* **341**:201.

Ricard, J., Meunier, J.-C., and Buc, J., 1974, Regulatory behaviour of monomeric enzymes, *Eur. J. Biochem.* **49**:195.

Roach, P. J., and Larner, J., 1976, Regulation of glycogen synthase: A relation of enzymic properties with biological function, *Trends Biochem. Sci.* **1**:110.

Rosen, O. M., Erlichman, J., and Rubin, C. S., 1975, Molecular structure and characterization of bovine heart protein kinase, *Adv. Cyclic Nucleotide Res.* **5**:253.

Rubin, C. S., and Rosen, O. M., 1975, Protein phosphorylation, *Annu. Rev. Biochem.* **44**:831.

Rübsamen, H., Khandker, R., and Witzel, H., 1974, Sigmoid kinetics of the monomeric ribonuclease I due to ligand-induced shifts of conformational equilibria, *Hoppe Seyler's Z. Physiol. Chem.* **355**:687.

Sanwal, B. D., 1970, Allosteric controls of amphibolic pathways in bacteria, *Bacteriol. Rev.* **34**:20.

Sanwal, B. D., Meaba, P., and Cook, R. A., 1966, Interaction of macroions and dioxane with the allosteric phosphoenolpyruvate carboxylase, *J. Biol. Chem.* **241**:5177.

Sanwal, B. D., Kapoor, M., and Duckworth, H. W., 1971, The regulation of branched and converging pathways, *Curr. Top. Cell. Regul.* **3**:1–115.

Sarngadharan, M. G., Watanabe, A., and Pogell, B. M., 1969, Binding of adenosine 5'-monophosphate and substrate by rabbit liver fructose 1,6-diphosphatase, *Biochemistry* **8**:1411.

Savageau, M. A., 1979, Autogenous and classical regulation of gene expression, in: *Biological Regulation and Development* (R. F. Goldberger, ed.), Vol. I, pp. 57–108, Plenum Press, New York.

Sawula, R. V., and Suzuki, I., 1970, Effect of enzyme concentration on the kinetics of D-lactate dehydrogenase from *Aerobacter aerogenes, Biochem. Biophys. Res. Commun.* **40**:1096.

Serrano, R., Gancedo, J. M., and Gancedo, C., 1973, Assay of yeast enzymes *in situ*: A potential tool in regulation studies, *Eur. J. Biochem.* **34**:479.

Shapiro, B. M., 1969, The glutamine synthetase deadenylylating enzyme system from *Escherichia coli.* Resolution into two components, specific nucleotide stimulation, and cofactor requirements, *Biochemistry* **8**:659.

Shapiro, B. M., Kingdon, H. S., and Stadtman, E. R., 1967, Regulation of glutamine synthetase. VII. Adenyl glutamine synthetase: A new form of the enzyme with altered regulatory and kinetic properties, *Proc. Natl. Acad. Sci. U.S.A.* **58**:642.

Shen, L. C., and Atkinson, D. E., 1970, Regulation of pyruvate dehydrogenase from *Escherichia coli.* Interactions of adenylate energy charge and other regulatory parameters, *J. Biol. Chem.* **245**:5974.

Shill, J. P., and Neet, K. E., 1975, Allosteric properties and the slow transition of yeast hexokinase, *J. Biol. Chem.* **250**:2259.

Smando, R., Waygood, E. B., and Sanwal, B. D., 1974, Cooperative interaction in the binding of allosteric effectors to phosphoenolypyruvate carboxylase, *J. Biol. Chem.* **249**:182.

Smith, S., and Abraham, S., 1975, The composition and biosynthesis of milk fat, in: *Advances in Lipid Research* (R. Paoletti and D. Kritchevsky, eds.), Vol. 13, pp. 195–239, Academic Press, New York.

Söderling, T. R., 1975, Regulation of glycogen synthetase. Specificity and stoichiometry of phosphorylation of the skeletal muscle enzyme cyclic 3',5'-AMP-dependent protein kinase, *J. Biol. Chem.* **250**:5407.

Söderling, T. R., Hickinbottom, J. P., Reimann, E. M., Hunkler, F. L., Walsh, D. A., and Krebs, E. G., 1970, Inactivation of glycogen synthetase and activation of phosphorylase kinase by muscle adenosine 3',5'-monophosphate-dependent protein, *J. Biol. Chem.* **245**:6317.

Soffer, R. L., 1974, Aminoacyl tRNA transferase, *Adv. Enzymol.* **40**:91.

Sols, A., and Marco, R., 1970, Concentration of metabolites and binding sites. Implication of metabolic regulation, *Curr. Top. Cell. Regul.* **2**:227–273.

Springer, M. S., Goy, M. F., and Adler, J., 1979, Protein methylation in behavioural control mechanism and in signal transduction, *Nature (London)* **280**:279.

Srere, P. A., 1967, Enzyme concentrations in tissues, *Science* **158**:976.

Stadtman, E. R., 1970, Mechanism of enzyme regulation in metabolism, in: *The Enzymes* (P. Boyer, ed.), Vol. 1, pp. 398–444, Academic Press, New York.

Stadtman, E. R, and Ginsburg, A., 1974, The glutamine synthetase of *Escherichia coli*: Structure and control, in: *The Enzymes*, 3rd ed. (P. D. Boyer, ed.), pp. 755–807, Academic Press, New York.

Stebbing, N., 1974, Precursor pools and endogenous control of enzyme synthesis and activity in biosynthetic pathways, *Bacteriol. Rev.* **38**:1.

Stevens, F. C., Walsh, M., Ho, H. C., Teo, T. S., and Wang, J. H., 1976, Comparison of calcium binding proteins, *J. Biol. Chem.* **251**:4495.

Strong, C. R., Carey, E. M., and Dils, R., 1973, The synthesis of medium-chain fatty acids by lactating rabbit mammary gland studied *in vito*, *Biochem. J.*, **132**:121.

Sweeny, J. R., and Fisher, J. R., 1968, An alternative to allosterism and cooperativity in the interpretation of enzyme kinetic data, *Biochemistry* **7**:561.

Takeda, Y., and Larner, J., 1975, Structural studies on rabbit muscle glycogen synthase. II. Limited proteolysis, *J. Biol. Chem.* **250**:8951.

Taketa, K., and Pogell, B. M., 1965, Allosteric inhibition of rat liver fructose 1,6-diphosphate by adenosine-5'-monophosphate, *J. Biol. Chem.* **240**:651.

Talkington, C., and Pero, J., 1978, Promoter recognition by phage SPO1-modified RNA polymerase, *Proc. Natl. Acad. Sci. U.S.A.* **75**:1185.

Thelander, L., 1973, Physicochemical characterisation of ribonucleoside diphosphate reductase from *Escherichia coli*, *J. Biol. Chem.* **248**:4591.

Thelander, L., and Reichard, P., 1979, Reduction of ribonucleotides, *Annu. Rev. Biochem.* **48**:133.

Titani, K., Cohen, P., Walsh, K., and Neurath, H., 1975, Amino-terminal sequence of rabbit muscle phosphorylase, *FEBS Lett.* **55**:120.

Tomkins, G. M., Yielding, K. L., Curran, J.F., Summers, M. R., and Bitensky, M. W., 1965, The dependence of substrate specificity on the conformation of crystalline glutamate dehydrogenase, *J. Biol. Chem.* **240**:3793.

Trotta, P. P., Pinkus, L. M., Haschemeyer, R. H., and Meister, A., 1974, Reversible dissociation of the monomer of glutamine-dependent carbamyl phosphate synthetase into catalytically active heavy and light subunits, *J. Biol. Chem.* **249**:492.

Ueda, K., Okayama, H., Fukushima, M., and Hayaishi, O., 1975, Purification and analysis of the poly ADP-ribose synthetase system, *J. Biochem. (Tokyo)* **77**:1.

Umbarger, H. E., 1956, Evidence for a negative feed back mechanism on the biosynthesis of isoleucine, *Science* **123**:848.

Vagelos, P. R., Alberts, A. W., and Martin, D. B., 1963, Studies on the mechanism of activation of acetyl coenzyme A carboxylase by citrate, *J. Biol. Chem.* **238**:533.

Villar-Palasi, C., and Larner, J., 1960, Insulin-mediated effect on the activity of UDPG-glycogen transglucosylase of muscle, *Biochem. Biophys. Acta* **39**:171.

Walsh, D. A., Perkins, J. P., Brostrom, C. O., Ho, E. G., and Krebs, E. G., 1971, Catalysis of the phosphorylase kinase activation reaction, *J. Biol. Chem.* **246**:1968.

Wang, J. H., and Waisman, D. M., 1979, Calmodulin and its role in the second messenger system, *Curr. Top. Cell. Regul.* **15**: 47–107.

Watterson, D. M., Harrelson, W. G., Jr., Keller, P. M., Sharief, F., and Vanaman, T. C., 1976, Structural similarities between the Ca^{++}-dependent regulatory proteins of 3':5'-cyclic nucleotide phosphodiesterase and actomyosin ATPase, *J. Biol. Chem.* **251**:4501.

Waygood, E. B., Mort, J., and Sanwal, B. D., 1976, The control of pyruvate kinase of *E. coli*. III. Binding of substrate and allosteric effectors to the enzyme activated by fructose 1,6-diphosphate, *Biochemistry* **15**:277.

Weber, A., and Murray, J. M., 1973, Molecular control mechanisms in muscle contraction, *Phys. Rev.* **53**:612.

Weber, K., 1968, New structural model of *E. coli* aspartate transcarbamylase and the amino-acid sequence of the regulatory polypeptide chain, *Nature (London)* **218**:1116.

Weitzman, P. D. J., and Hewson, J. K., 1973, *In situ* regulation of yeast citrate synthase. Absence of ATP inhibition observed *in vitro*, *FEBS Lett.* **36**:479.

Wiame, J.-M., 1971, The regulation of arginine metabolism in *Saccharomyces cerevisiae*: Exclusion mechanism, *Curr. Top. Cell. Regul.* **4**:1–38.

Wiley, D. C., and Lipscomb, W. N., 1968, Crystallographic determination of symmetry of aspartate transcarbamylase, *Nature (London)* **218**:1119.

Woolfolk, C. A., and Stadtman, E. R., 1964, Cumulative feed back inhibition in the multiple end product regulation of glutamine synthetase activity in *Escherichia coli*, *Biochem. Biophys. Res. Commun.* **17**:313.

Wright, J. A., and Sanwal, B. D., 1969, Regulatory mechanisms involving nicotinamide adenine nucleotides as allosteric effectors. II. Control of phosphoenolpyruvate carboxykinase, *J. Biol. Chem.* **244**:1838.

Wright, J. A., and Sanwal, B. D., 1971, Regulatory mechanisms involving nicotinamide adenine nucleotides as allosteric effectors. IV. Physicochemical study and ligand binding to citrate synthase, *J. Biol. Chem.* **246**:1689.

Wyman, A., and Paulus, H., 1975, Purification and properties of homoserine transacetylase from *Bacillus polymyxa*, *J. Biol. Chem.* **250**:3897.

Wyman, A., Shelton, E., and Paulus, H., 1975, Regulation of homoserine transacetylase from *Bacillus polymyxa*, *J. Biol. Chem.* **250**:3897.

Yates, R. A., and Pardee, A. B., 1956, Control of pyrimidine biosynthesis of *E. coli* by a feed-back mechanism, *J. Biol. Chem.* **221**:757.

Yoshinhara, K., Tanigawa, Y., Burzio, L., and Koide, S. S., 1975, Evidence for adenosine diphosphate ribosylation of Ca^{2+}, Mg^{2+}-dependent endonuclease, *Proc. Natl. Acad. Sci. U.S.A.* **72**:289.

Zillig, W., Fujiki, H., Blum, W., Janekovie, D., Schweiger, M., Rahmsdorf, H. J., Ponta, H., and Kauffmann, M. H., 1975, *In vivo* and *in vitro* phosphorylation of DNA-dependent RNA polymerase of *Escherichia coli* by bacteriophage T_7 induced protein kinase, *Proc. Natl. Acad. Sci. U.S.A.* **72**:2506.

2

Protein Folding
Evolutionary, Structural, and Chemical Aspects

KENNETH A. THOMAS and ALAN N. SCHECHTER

1 Introduction

Proteins are used by organisms for structural, catalytic, and regulatory purposes. These functions require that the polypeptide chains fold into rather specific three-dimensional conformations, which are determined by both the sequential order of amino acids in the polypeptide chain and the chemical environment of the molecule. Under normal biological conditions proteins attain conformations compatible with functional efficiency. This results in the proper and relatively stable juxtaposition of certain amino acid residues that are necessary for the specific biological role of the protein.

The term "protein folding" has come to refer to both the detailed description of the spatial arrangement of the amino acid residues in the functional protein and the mechanism or time-dependent changes of these arrangements in space, by which the polypeptide chain achieves that conformation.

In recent years there has been increasing interest in both aspects of this subject. The demonstration by Anfinsen and his colleagues (see Anfinsen, 1973) that denatured proteins could spontaneously regain their native structure—the "thermodynamic hypothesis"—has made feasible studies of the prediction of protein structure from amino acid sequences and the synthesis of proteins by chemical means. The elucidation of the X-ray structure of myoglobin (Kendrew *et al.,* 1960) and of hemoglobin (Perutz *et al.,* 1968) initiated a large-scale study of protein structures at high resolution. At present over 100 protein structures have been solved by crystallographic methods. As a result, discussions of protein function are increasingly formulated in terms of detailed atomic geometry.

KENNETH A. THOMAS ● Merck, Sharp and Dohme Research Laboratories, Rahway, New Jersey 07065   ALAN N. SCHECHTER ● Laboratory of Chemical Biology, National Institute of Arthritis, Metabolism, and Digestive Diseases, National Institutes of Health, Bethesda, Maryland 20205

Most reviews of protein folding have concentrated on one aspect or the other of this subject. We believe that this separation tends to limit a comprehensive understanding of the interrelationships between the folding process and the final complex, ordered structure. In this chapter we will initially summarize current knowledge of covalent and noncovalent forces that stabilize the structures of globular proteins. Next we will review knowledge of protein structure at the secondary, tertiary, and quaternary levels. It is in the examination and comparison of proteins at these structural levels that major evolutionary insights have occurred during the last several years, which complement previous studies based on the comparisons of amino acid sequences—the primary structure. Finally, we will summarize solution studies and theoretical analyses that have attempted to penetrate the mechanism by which an unfolded polypeptide chain may achieve its functional structure in times consistent with biosynthetic rate processes.

1.1 Protein Structure and Evolution

Investigation of protein structure can yield information about the stability of these macromolecules and the evolutionary relationships among them. Historically, the study of biological evolution has focused on the taxonomy of current organisms and the fossils of extinct species. These studies generated a detailed model of the evolutionary development of multicellular organisms. In the last 20 years, however, the biological characterization of evolutionary relationships has extended to the molecular level. Comparison of nucleotide sequence similarities in DNA molecules and of amino acid sequences in proteins have made it possible to study evolution at the level of genes and their end products. The most recent extension of the characterization of evolutionary relationships at the molecular level involves examination of the three-dimensional folded structures of these protein sequences as determined by X-ray diffraction crystal structure analysis. The comparison of both very similar and quite different protein structures can be used to indicate taxonomic relationships among proteins.

By identifying which features of protein structures are best conserved and which are most variable, it is also possible to infer the relative importance of various chemical interactions within these macromolecules. Common features, especially those observed among evolutionarily distant proteins, are assumed to be the result of constraints imposed by the necessity of the polypeptide chain to fold properly and to attain a reasonable structural stability. For those proteins having common functions, additional functionally related structural similarities may also be maintained. From this comparative approach we shall consider a variety of recognized stabilizing interactions and the resulting ordered polypeptide conformations.

2 Stabilization Interactions

2.1 Covalent Bonds

2.1.1 Polypeptide Backbone

The most obvious type of stabilizing interaction in proteins is covalent bonding. The amino acids are polymerized by a dehydration reaction coupling the amino group of one amino acid to the carboxyl group of another with the elimination of a water molecule to form a substituted amide, or peptide, bond. Once polymerized, the individual amino acids are referred to as residues. Globular protein molecules generally have fewer than 1000 residues in any one polypeptide chain. Each amino acid residue in a protein molecule contains a central tetrahedral carbon atom, C^α, to which is bonded a nitrogen, N, a carbonyl group carbon, C', a hydrogen, H, and a functional side group, R, that differs for each type of amino acid residue.* This group of atoms can be arranged around the C^α atom in two ways. When viewed from the hydrogen to the C^α atom, the amino acid residue is in the L configuration if C', R, and N occur in clockwise order. Alternatively, they can be arranged in counterclockwise order in which case they are in the D configuration. The L and D configurations are nonsuperimposable mirror images, or enantiomorphs. The inability to superimpose L and D configurations results from the nonplanar arrangement of the four different atoms or groups about the C^α atom. The glycine residue C^α atom that has two hydrogen atoms bonded to it, the second one being the R "group," is the only nonenantiomorphic residue. Only the L configuration of the other 19 amino acids is utilized to synthesize proteins *in vivo*. It is possible that the observed C^α enantiomorphic uniformity results from the disruptive effect of amino acid residues of one configuration on secondary structures composed of residues of the opposite configuration. For example, it has been demonstrated with homopolymers that right-handed helices composed of L-amino acid residues are more stable than those containing a mixture of amino acid residues with both the D and L configurations (Blout *et al.*, 1957). Although inclusion of D-amino acids would allow polypeptide conformations not accessible to pure L sequences, the nonenantiomorphic glycine residue can accommodate structural requirements for backbone conformations accessible to the D isomers. The biological utilization of L-amino acids instead of the mirror image D configuration could be largely the result of chance. However, recent experiments have shown that polarized radiation, either radioactive β-decay (Bonner *et al.*, 1975) or ultraviolet solar radiation (Norden, 1977), causes D-amino acids to decompose at a slightly faster rate than L-amino acids. Thus low levels of naturally occurring polarized radiation acting over long periods of time may have increased the abundance of the L over the D stereoisomers and biased the chance that the L-amino acids would be used.

The peptide linkage has partial double-bond character and is usually quite planar. This planarity is a result of electron delocalization, the spreading of the peptide carbonyl group π-electron orbitals into the adjacent $C'-N$ bond. Therefore, the polypeptide backbone may be considered as a sequence of intersecting planar peptide units. Rotational freedom of the polypeptide backbone is localized in the two single bonds on either side of the C^α atoms. The torsion or rotational angles around the $C^\alpha-N$ bond and the $C^\alpha-C'$ bond are denoted ϕ and ψ, respectively. Both angles may vary from $-180°$ to $+180°$. The positive rotational direction is defined by a clockwise angle of the peptide unit as viewed from the C^α atom. The angle locations and the two conformational extremes are shown in Fig. 1. If all the peptide bonds are planar, then the sequence of ϕ and ψ angles characterize the conformation of a polypeptide backbone.

Since many rotational angles about protein covalent bonds are unfavorable because of both steric and electrostatic interactions, the single bonds may be considered to be centers of "restricted" rotation with high energies disfavoring many angles. These torsional angles

---

*The IUPAC-IUB convention with the atom position Greek letter (α, β, γ, etc.) as a superscript following the atom type letter (C, N, O, S, etc.) is followed in the text. Other conventions also appear in the literature. These alternative notations place the atom position Greek letter preceding, following, or as a following subscript to the atom type letter. For example, the β-carbon, C^β, is also written as βC, $C\beta$, or C_β.

KENNETH A.
THOMAS and ALAN N.
SCHECHTER

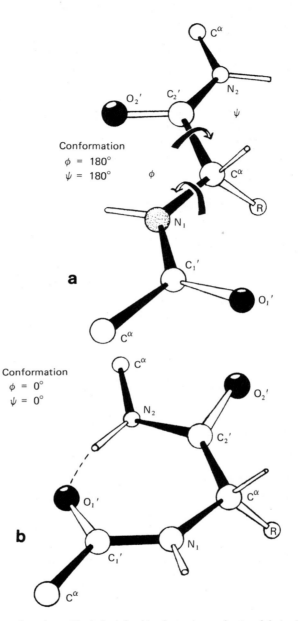

Figure 1. Conformations of a polypeptide chain defined by the torsion angles ϕ and ψ. As shown in (a) positive increases in the angles are accomplished by rotating the peptide unit clockwise (when viewed from the α-carbon) around the $C^\alpha-H$ and $C^\alpha-C'$ bonds while the α-carbon is held in the same position. (a) An extended chain with $\phi = 180°$, $\psi = 180°$, and (b) an eclipsed conformation with $\phi = 0°$, $\psi = 0°$ which involves an unfavorable contact (indicated by a dashed line) between the carbonyl oxygen, O', of one peptide and the amino hydrogen of the next. From Blundell and Johnson (1976) with permission.

have been considered in some detail for both the polypeptide backbone and side chains (Pullman and Pullman, 1974; Ramachandran and Sasisekharan, 1968). In addition, crystallographic evidence for polypeptide backbone carbonyl oxygen vibration has recently been presented (Epp *et al.,* 1975). Along with bond rotation within both the polypeptide backbone and the side chains, such vibration must contribute to the disorder or entropy of the dynamic folded state as compared to an "ideal" static structure. Small stable angular distortions are also observed. An extreme example is the tetrahedral distortion of the carbonyl carbon atom of the peptide bond of trypsin inhibitor that sits in the active site of trypsin in the inhibitor–protease complex (Deisenhofer and Steigemann, 1975). This distortion is likely to be similar to that which the enzyme stabilizes in the substrate molecule (Huber *et al.,* 1975; Huber *et al.,* 1974; Sweet *et al.,* 1974). The ideal α-carbon tetrahedral angle of 109.5° ($N–C^\alpha–C'$) is often slightly larger when examined in experimentally determined structures. For example, this angle has an average value of about 112° in carboxypeptidase A (Hartsuck and Lipscomb, 1971), α-chymotrypsin (Birktoft and Blow, 1972), elastase (Sawyer *et al.,* 1978), and pancreatic trypsin inhibitor (Deisenhofer and Steigemann, 1975).

Small distortions of peptide bonds from planarity are known to occur (Deisenhofer and Steigemann, 1975; Watenpaugh *et al.,* 1973; Winkler and Dunitz, 1971). Usually, however, these bonds have not been examined in detail so that currently there are insufficient data to form reliable general conclusions about the frequency, magnitude, and structural importance of these distortions. The cis–trans peptide bond isomer, however, has been the subject of more extensive examination. The two forms of the peptide bond are shown in Fig. 2. In small model compounds the trans form is more stable than is the cis conformation (Ramachandran and Sasisekharan, 1968). In dipeptides the steric crowding between the C^α and the preceding C^β group in the cis conformation favors the trans form. Furthermore, in a longer sequence of trans peptide units this energy difference may be greater as a result of steric repulsion of trans peptides on opposite sides of the cis isomer (Ramachandran and Mitra, 1976). Nevertheless, cis peptides have been identified in carboxypeptidase A between Ser 197 and Tyr 198 at a bend between two β chains (Quiocho and Lipscomb, 1971), and between Ala 207 and Asp 208 near the calcium binding site in concanavalin A (Reeke *et al.,* 1978).

A more common cis peptide isomer is found adjacent to proline residues. As seen in Fig. 2, the cyclic proline ring C^δ atom that is bonded to the peptide nitrogen causes steric crowding with the preceding C^β atom in the trans conformation. This effect is similar to that produced by the proline C^α atom in the cis conformation and thus decreases the energy difference between the two isomers (Ramachandran and Mitra, 1976). All reported cis proline residues occur in the third position of sharp turns that reverse the direction of the polypeptide chain (ribonuclease 93, 114; thermolysin 51; erythrocuorin 74; carbonic anhydrase 30, 202; immunoglobulin fragment REI* 8, 95; and subtilisin 168), since a trans proline in this position would crowd a preceding C^β group (Huber and Steigemann, 1974). Recently it was proposed that proline isomerism is important in the kinetics of protein folding. The activation energy of 20 kcal/mole for proline isomerism (from the trans conformation) is in accord with the measured activation energies for protein denaturation occurring at low temperatures. Inasmuch as the isomerism involves little or no enthalpy change and produces no direct optical spectral contribution, it is possible that isomerization would go undetected by the common thermodynamic tests for two-state (folded or completely unfolded) behavior (Brandts *et al.,* 1975).

---

*Immunoglobulin molecules are designated by the initials of the individual or myeloma cell line from which they were isolated.

KENNETH A.
THOMAS and ALAN N.
SCHECHTER

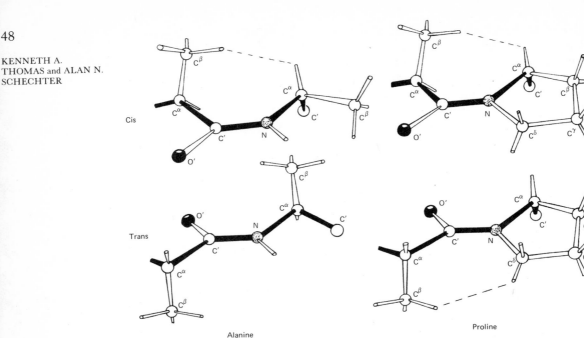

Figure 2. Cis and trans arrangements of planar peptides of alanine and proline. For alanine, an unfavorable contact (indicated by a dashed line) occurs between hydrogens bound to the β-carbon of one amino acid residue and the α-carbon of the next in the cis arrangement; this does not occur in the more stable trans peptide. In proline, however, unfavorable contacts occur in both cis and trans peptides. This decreases the difference in stability between the two proline isomers. Thus cis arrangements are rarely found in polypeptide chains, except in the case of peptides involving proline. From Blundell and Johnson (1976) with permission.

It is generally accepted that the information specifying the three-dimensional conformation of a globular protein is determined by the interactions among the amino acids in the linear sequence and the solvent. Prosthetic groups, although often noncovalently bound to the polypeptide chain, may in some cases be linked to amino acid side chains. These may also make significant contributions to the stability of some proteins. The integrity of all of the covalent bonding along the polypeptide backbone, however, is not always mandatory for proper formation of tertiary structure, as is demonstrated by both ribonuclease S (Richards and Wyckoff, 1971) and staphylococcal nuclease T (Anfinsen *et al.,* 1971*b*). These proteins, which are produced *in vitro* by single peptide bond cleavages by proteolytic enzymes, share four attributes: cleavage at loops between elements of secondary structure; retention of enzymic activity; reversible refolding of the two complementary polypeptide fragments to form the orginal cleaved structure; and decreased stability. The chain cleavages might destabilize the structures by increasing the entropy of the molecules in the unfolded state.

2.1.2 *Amino Acid Side Chains*

The atoms of each type of side chain are designated by sequential Greek letters. Since the side chains are bonded to the polypeptide backbone C^α atom, the lettering starts at β. Side-chain branching generates asymmetric C^β atoms in threonine and isoleucine. The conformation of each side chain is defined by a set of torsion angles, denoted by χ, with superscript numbers to indicate the specific bond. These numbers start at 1 for the bond between

C^α and C^β and increase going out the side chain. χ values range from $-180°$ to $+180°$. Positive angles correspond to clockwise angular rotations, with $\chi = 0$ being an eclipsed conformation. The complete nomenclature for protein conformation is detailed elsewhere (IUPAC–IUB, 1970). For steric reasons, the tetrahedral angles of side chains favor the staggered conformation, with χ values of $-60°$, $+60°$, and $\pm180°$. Torsion angles involving a trigonal carbon atom, such as the aromatic amino acid side-chain χ^2 angles, show a preference for $\pm90°$. In this conformation both aromatic C^δ atoms are as far away from the C^α atom as possible and do not eclipse the C^β hydrogens. In folded protein structures most, but not all, side chains are observed to have sterically favored torsion angles. Presumably the energies of these less advantageous conformations are outweighed by other more energetically favorable interactions.

The amino acid side chains of proteins may be divided into functional categories based on their physical and chemical characteristics. Side chains with either similar or different properties may be substituted at any sequential location as a result of point mutations. Single nucleotide changes at the third position of a triplet codon, however, frequently do not alter the amino acid specified, in which case they are indistinguishable at the amino acid sequence level. The genetic code also minimizes the chemical effect of nucleotide changes in the first codon position. An alteration at this position leads to the substitution of an amino acid of similar molecular weight, polarity, number of dissociating groups, pK, isoelectric point, and α-helix-forming potential. These six characteristics are only weakly correlated with each other (Alff-Steinberger, 1969). The informational redundancy in two of the three codon positions would be expected to minimize the effect of single nucleotide mutations on the protein structure and folding mechanism.

2.1.3 Disulfide Bonds

Two conformationally preferred orientations of disulfide bridges are recognized. As illustrated in Fig. 3, the bridge is right-handed if, looking down the $S^\gamma-S^\gamma$ bond, the far $S^\gamma-C^\beta$ bond is within $180°$ clockwise angle from the near $S^\gamma-C^\beta$ bond; it is left-handed

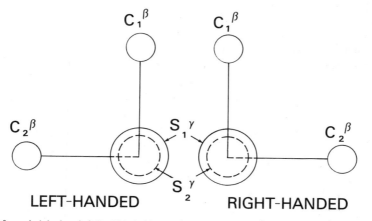

Figure 3. Left- and right-handed disulfide bridge conformations. The C^β and S^γ atoms of the two cysteines in the disulfide bridge are subscripted by 1 and 2, although, in general, these would not be expected to be adjacent amino acids in a protein sequence. The near S_1^γ atom is denoted by a solid circle and the far S_2^γ atom by a smaller dashed circle. The conformation is right-handed if the far $S_2^\gamma-C_2^\beta$ bond is within a $180°$ clockwise (or positive) angle of the near $S_1^\gamma-C_1^\beta$ bond. If this angle is within a $180°$ counterclockwise (or negative) angle, the disulfide bridge is left-handed. Note that the hand remains the same when viewed from either the S_1^γ to S_2^γ or S_2^γ to S_1^γ atoms.

if within a 180° counterclockwise angle. The two $S^{\gamma}-C^{\beta}$ bonds are generally rotated by about 90° from coplanar, with no strong bias toward one or the other rotational direction (Birktoft and Blow, 1972; Blundell *et al.*, 1972; Imoto *et al.*, 1972; Richards and Wyckoff, 1971). For any given disulfide bridge in a globular protein the hand is usually uniquely fixed. However, in the REI immunoglobulin fragment the internal disulfide bridge within the monomer between residues 23 and 88 occurs in approximately equal populations that differ by 162° (Epp *et al.*, 1975), thereby providing a clear example of a conformational isomer.

The formation of covalent disulfide bridges between cysteine residues contributes to the resistance of the protein structure to denaturation, perhaps largely by decreasing the entropy of the unfolded state (Anfinsen and Scheraga, 1975). The formation of disulfide bonds, however, is probably not obligatory for the folding of a protein to a specific conformation. This idea is compatible with the fact that evolutionarily homologous protein structures show variation in the occurrence of disulfide bridges. For the homologous serine proteases chymotrypsin, trypsin, elastase, and thrombin, there are a total of seven unique bridge locations, only three of which are common to all four proteins (Hartley and Shotton, 1971). Disulfide bonds specific to antibody Fab fragments of various classes could also be easily accommodated without major chain movement in the conformation determined for Fab New (Poljak *et al.*, 1973). Similarly, the single disulfide bridge in bullfrog cytochrome *c* could easily fit into the fold of the homologous protein from horse (Dickerson *et al.*, 1971).

Once formed, however, the disulfide bonds can protect a specific polypeptide conformation from significant structural alteration following proteolytic cleavage. For example, neither insulin nor α-chymotrypsin, after proteolytic activation from their zymogen precursors, is able to spontaneously reform the correct disulfide pairs in significant yield on reduction and reoxidation in solution (Givol *et al.*, 1965). In contrast, both the single polypeptide chains of proinsulin (Steiner and Clark, 1968) and chymotrypsinogen (Givol *et al.*, 1965) reform the native structure upon reduction and reoxidation of the disulfide bonds. Thus in the proteolytically activated forms of these molecules the disulfide bonds may be maintaining metastable polypeptide conformations.

2.2   Ionic Interactions

Ionic interactions between oppositely charged protein groups (lysine, arginine, and histidine residues, and the amino terminus on the one hand; and glutamic and aspartic residues and the carboxyl terminus on the other) are referred to as ion pairs or salt bridges. Many of these ion pairs are located on the highly solvent exposed protein exterior. In fact, some examples are known in which a water molecule is located between the two charged groups of the salt bridge (Fermi, 1975; Bode and Schwager, 1975). Ion pairs that are exposed to solvent may make a contribution to the stability of proteins. In the case of ferredoxins, for example, the increased thermal stability of the thermophile protein over the mesophile counterpart has been attributed to the addition of one surface ion pair in the former molecule (Perutz and Raidt, 1975). A number of salt bridges, some involving water linkages, are observed to be substantially or completely buried in the solvent inaccessible protein interior (Eklund *et al.*, 1976*b*; Delbaere *et al.*, 1975; Holbrook *et al.*, 1975; Colman *et al.*, 1972; Hartsuck and Lipscomb, 1971). It is also observed that solvent-inaccessible salt bridges occur between subunits of glyceraldehyde-3-phosphate dehydrogenase from a thermophilic bacterium but not from the homologous enzyme from lobster (Biesecker *et al.*, 1977). These salt bridges might be expected to make a rather substantial contribution to the stability of the protein in the folded state (Mavridis *et al.,* 1974) because the attractive force between the opposite charges would be greater in the protein interior where the dielectric constant is lower than in water, where it is quite high (Kauzmann, 1959). In the cases of α-chymotrypsin (Birktoft and Blow, 1972), trypsin (Bode and Schwager, 1975), and carp myogen (Kretsinger and Nockolds, 1973) the buried charge pairs are noted to interact with surrounding hydrogen-bonding groups. These interactions are proposed to decrease the effect of the locally strong electrostatic fields of the salt bridges (Kretsinger and Nockolds, 1973).

Single side-chain groups that are normally charged do occasionally become buried. In chymotrypsinogen, for example, the Asp 194 residue is buried and hydrogen-bonded to the side chain of His 40. This aspartic residue side chain forms an internal salt bridge with the newly formed amino terminal group of Ile 16 on proteolysis of this zymogen to α-chymotrypsin (Kraut, 1971*a*), as shown in Fig. 4. In both the serine proteases (Birktoft and Blow, 1972; Stroud *et al.,* 1972; Hartley and Shotton, 1971) and subtilisin (Kraut, 1971*b*) another buried aspartic residue, Asp 102 (in the α-chymotrypsin sequence numbering), binds to His 57 to form part of the catalytic center. Although His 57 was originally presumed to also hydrogen-bond to Ser 195 to form a "charge relay system" shown in Fig. 4, recent results (Matthews *et al.,* 1977) have indicated that the serine side-chain oxygen is probably too far from the imidazole ring to facilitate a strong interaction. Ironically, the His–Ser hydrogen bond may exist in the zymogen forms of these proteases. Chicken eggwhite lysozyme also contains a buried aspartic residue that hydrogen bonds to surrounding protein groups, which in turn hydrogen-bond to other groups. As in the case for the previously mentioned ion pairs, this network of surrounding hydrogen bonds may act to disperse some of the buried charge and thereby help to stabilize a potentially unfavorable

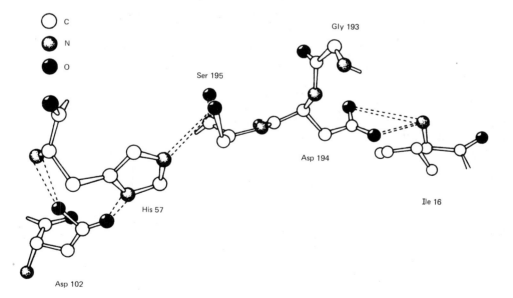

Figure 4. The active center of the enzyme α-chymotrypsin with the buried Asp 102 residue. Asp 102 forms a buried salt bridge to His 57. The adjacent buried salt bridge between Asp 194 and the terminal amino group of Ile 16, produced by proteolytic activation of the chymotrypsinogen precursor, is also shown. The probable hydrogen bonds are denoted by double dashed lines. The Ser 195 to His 57 hydrogen bond, although probably present in the zymogen form of the enzyme, is now thought to be very weak or absent in the active enzyme. The viewpoint is outside of the molecule looking toward the interior. From Blow (1971) with permission.

energetic state (Imoto *et al.*, 1972). It might be expected, however, that buried charges would destabilize a structure. In fact, the alkali denaturation of hemoglobin A has been attributed to the ionization of internal cysteine and tryosine residue side chains. Consistent with this interpretation is the increased resistance to denaturation by alkali of fetal hemoglobin, in which the equivalent positions contain threonine and tryptophan residues, respectively (Frier and Perutz, 1977).

Nonprotein cations and anions can also interact strongly with proteins. In addition to metallic cations (Liljas and Rossmann, 1974), anions such as Cl^- (Imoto *et al.*, 1972), NO_3^- (Moult *et al.*, 1976), SO_4^{2-} (Takano, 1977; Timkovich and Dickerson, 1976; Tulinsky *et al.*, 1973; Adams *et al.*, 1973), and PO_4^{2-} (Delbaere *et al.*, 1975) are observed by crystallographic analyses to interact with charged and hydrogen-bonding protein groups. The ions that bridge sequentially distant amino acid residue ligands can increase both thermal stability (Matthews *et al.*, 1974) and perhaps the cooperativity of the folding–unfolding transition (Ginsburg and Carroll, 1965).

2.3 Hydrogen Bonds

Numerous hydrogen bonds are present in all protein structures. Surface polar groups would be expected to interact with solvent to some extent since water is an excellent hydrogen bond donor and acceptor and is in very high concentration. This is particularly manifest in hen egg-white lysozyme, since the protein packing contacts are entirely different in the two crystal forms that have been analyzed. Every free polar group on the enzyme surface can be seen to interact with solvent in at least one of the two structures (Moult *et al.*, 1976). Surface protein–protein hydrogen bonds are generally thought to afford little or no stability, since they compete with protein–water hydrogen bonds (Némethy *et al.*, 1963). If this is true, then the occasionally observed surface protein–protein hydrogen bonds (Takano, 1977; Chothia, 1976; Huber *et al.*, 1971) may reflect either the effects of the crystallization medium or local steric restrictions of the donor and acceptor group orientations.

Some water molecules are observed to bridge sequentially distant protein groups (Sawyer *et al.*, 1978; Burnett *et al.*, 1974; Birktoft and Blow, 1972). These may be suspected of contributing to the cooperativity of the protein folding–unfolding transition in a manner analogous to that of the bridging ions. A second layer of solvent, containing water molecules that are not hydrogen-bonded directly to the protein molecule, is rarely observed. This probably reflects both the steric restrictions imposed on the solvent structure by adjacent protein molecules in the crystal lattice (Moews and Kretsinger, 1975) and the likelihood that water molecules in the second solvation layer, having no particular stable position, are fluid. In support of this observation, recent computer simulation of dynamic water structure surrounding a protein molecule has generated a model in which the water becomes more ordered at or very near the protein surface (Hagler and Moult, 1978). A rather extensive solvent structure is observed, however, in the active site of carbonic anhydrase C (Lindskog *et al.*, 1971). In general, nonpolar groups constitute about 50% of the solvent-accessible surface area of proteins (Chothia, 1976; Lee and Richards, 1971). No organized water structure is observed around these groups (Sawyer *et al.*, 1978; Birktoft and Blow, 1972).

Water molecules are sometimes found buried within a protein structure, forming as many as four hydrogen bonds with the protein (Epp *et al.*, 1975; Deisenhofer and Steigemann, 1975; Quiocho and Lipscomb, 1971). The water molecules often hydrogen-bond to polar groups of the protein that would otherwise not be able to form any satisfactory hydrogen bonds with other polar protein groups. Although their relative importance is unclear, it has been noted that many internal water molecules occupy equivalent positions in the evolutionarily homologous serine proteases trypsin, chymotrypsin, and elastase (Sawyer *et al.,* 1978; Bode and Schwager, 1975). Occasionally internal water molecules may also hydrogen-bond to each other, as in pancreatic trypsin inhibitor (Deisenhofer and Steigemann, 1975), trypsin (Bode and Schwager, 1975), elastase (Sawyer *et al.,* 1978), α-chymotrypsin (Birktoft and Blow, 1972), carboxypeptidase A (Hartsuck and Lipscomb, 1971), papain (Drenth *et al.,* 1971), and the immunoglobulin REI fragment (Epp *et al.,* 1975). In the case of the REI fragment, it appears that a cluster of three internal water molecules occurs in a hydrophobic protein pocket between the two subunits of the dimeric molecule without forming any hydrogen bonds to the surrounding protein.

Internal protein–protein hydrogen bonds probably make a significant contribution to the internal geometry of proteins. About 40% of the side chains of a protein are polar (Chothia, 1975) and about 50% of these are internal. Over 90% of the internal residues are hydrogen-bonded (Chothia, 1974) indicating that a buried dipole may destabilize the structure of a native protein molecule (Perutz and Lehmann, 1968). About 80% of the internal hydrogen bonds are within segments of secondary structure (Chothia, 1976). These hydrogen-bonding patterns have been noted in some cases to be evolutionarily conserved (Sawyer *et al.,* 1978; Bode and Schwager, 1975). Since the average dielectric constant of protein interiors is thought to be significantly less than that of water, the free energy of formation of internal hydrogen bonds is thought to be negative. Values of about -5 kcal/mole have been estimated (Némethy *et al.,* 1963).

2.4 Nonpolar Interactions

One of the principal driving forces of protein folding is thought to result from the entropically unfavorable interactions between nonpolar hydrocarbon amino acid side chains and water (Kauzmann, 1959). This leads to the clustering of many of the nonpolar side chains with formation of a tightly packed solvent-inaccessible hydrophobic protein interior or core. Nonpolar side chains that remain at the solvent surface are frequently observed to be oriented so that their contact with water is minimized. Side chains that have both polar and nonpolar groups are often located near the solvent-exposed surface of a protein with the nonpolar portion buried and the polar groups exposed (Birktoft and Blow, 1972; Blundell *et al.,* 1972). Mutations that decrease the polarity of the side-chain residues at the protein–solvent surface also decrease the exposure of the side chains to solvent (Banyard *et al.,* 1974; Dickerson *et al.,* 1971).

About one-third to one-half of the total surface area of the unfolded polypeptide chain becomes inaccessible to solvent by being buried in the protein interior upon folding. Over one-half of the surface area that becomes buried is that of polar atoms. Once hydrogen-bonded, the effective hydrophobicity of these polar oxygen and nitrogen atoms may increase and might be considered to be roughly equivalent to that of hydrogen methyl groups (Chothia, 1975), although this view has been questioned (Finney, 1978). It is observed, furthermore, that with increasing molecular weight the buried surface area increases, whereas the fraction of buried polar groups remains approximately constant. Therefore, for the sample of proteins studied, the percentage of buried nonpolar surface increases with increasing molecular weight (Chothia, 1976).

Examination of side-chain groups alone reveals that the polar side-chain atoms are

KENNETH A.
THOMAS and ALAN N.
SCHECHTER

about 3.5 times more accessible to solvent than are the nonpolar side-chain atoms in a folded protein. The net change in solvent-accessible surface area on protein folding, however, is less than this value, since in the extended polypeptide chain the polar side-chain atoms are twice as exposed as are the nonpolar side-chain atoms. Therefore, the net change of exposure on folding indicates that nonpolar side-chain atoms decrease the surface area of solvent exposure by almost twice as much (3.5/2) as do polar side-chain atoms (Lee and Richards, 1971).

Internal clusters of nonpolar side chains are commonly found sandwiched between elements of secondary structure. The extent to which solvent exposure is thus decreased depends on the type of secondary structure involved. On the average, a residue buries 77 Å^2 of surface area when transferred from an extended conformation to the middle of a helix. This transfer provides about -2 kcal/mole of hydrophobic stabilization energy. A residue in the interior of a β sheet, however, buries an average of about 50% more surface, 117 Å^2, which provides about -3 kcal/mole. Thus, residues located in the interior of β sheets appear to make a greater average contribution per residue to the total hydrophobic stabilization energy than do residues occurring in the middle of helices (Chothia, 1976).

The concept of burying nonpolar side chains to form a hydrophobic core does not directly explain the specific packing of the side chains. It is obvious, nevertheless, that the arrangement of the internal side chains is not only precise but also remarkably efficient. If the internal volume is compared to the sum of the volumes of the constituent side chains, the interior of the protein is found to be packed at about the same density as are solid crystalline amino acids, with internal holes being small and relatively rare (Richards, 1974; Shrake and Rupley, 1973; Drenth et al., 1971). Introduction of internal cavities may destabilize a core, perhaps by increasing the chance of burying a non-hydrogen-bonded polar water molecule (Perutz and Lehmann, 1968). Selection against such cavities may be accomplished in at least three ways. First, if hydrogen-bonding groups are available, the cavity may be filled by one or more hydrogen-bonded water molecules (Sawyer et al., 1978). Second, the groups surrounding the internal hole may reorient so as to "collapse" into the vacancy, with main chain movement, if necessary (Timkovich and Dickerson, 1976; Anderson, 1975; Epp et al., 1975). Finally, it appears that amino acid substitutions involving internal side chains may facilitate acceptance of additional mutations that lead to compensating volume changes among other internal residues in the immediate vicinity (Sawyer et al., 1978; Frier and Perutz, 1977; Bedarkar et al., 1977; Ekland et al., 1976a; Kozitsyn and Ptitsyn, 1974; Lee and Richards, 1971; Fitch and Markowitz, 1970; Browne et al., 1969; Wyckoff, 1968). The requirement for efficient packing of internal groups probably contributes to the increased evolutionary conservation of residues in the protein interior, as compared to those on the surface (Sawyer et al., 1978; Dickerson and Timkovich, 1975; Kabat et al., 1975; Banyard et al., 1974; Hendrickson and Love, 1971). Ultimately, evolutionary substitutions may lead to the reorientation of a cluster of internal side chains so that its orientation may vary considerably from that of a homologous structure. This is the case for one locationally equivalent cluster in elastase and α-chymotrypsin (Hartley and Shotton, 1971).

The relative contributions of various potential nonbonded "weak interaction" orienting forces in protein interiors are poorly understood. Although short-range and weak, van der Waals forces are generally considered to be a source of internal stabilization based on the large number of intramolecular contacts (Ramachandran and Sasisekharan, 1968). Initial calculations indicate that there is a possibility of significant contributions by weak electrostatic interactions (Johanin and Kellershohn, 1972). In addition, it has been suggested that donor–acceptor interactions, which are the result of the partial or complete transfer of electrons from an electron-donor to an electron-acceptor group, may exist in proteins. This type of interaction has been proposed to occur between the side chains of tryptophan and histidine (Vandlen and Tulinsky, 1973; Shintzky and Goldman, 1967), arginine and tyrosine (Sweet *et al.*, 1974), and either cysteine or methionine and π-bonded amino acid residues (Morgan *et al.*, 1976). Although the relative contributions of various weak interactions must be resolved before protein folding can be fully understood, lack of progress in this area is not entirely surprising, as the relative contribution of these forces is not always clear even for the physicochemical description of less complex molecular systems.

3 Structural Components

3.1 Secondary Structure

Secondary structure is characterized by periodic hydrogen-bonding patterns and repetitive ϕ and ψ torsion angle values of the polypeptide backbone. There are two categories of structures that meet these criteria: the helix and the β-pleated sheet.

3.1.1 Helix

The α helix, shown in Fig. 5, was postulated before the completion of the first protein crystal structure (Pauling and Corey, 1951*a*). A variety of additional helical conformations

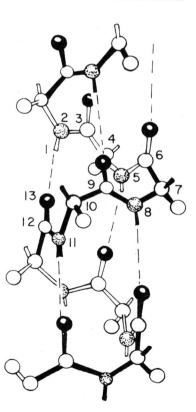

Figure 5. The right-handed α helix conformation. The bonds near the viewer are illustrated as broad solid lines and those farther away are drawn as double lines. Oxygens appear as solid black circles, nitrogens as dotted circles, and carbons as open circles. Hydrogen atoms are not shown, although the bonds to them are included. The hydrogen bonds between the peptide nitrogens and carbonyl oxygens are denoted by dashed lines. A single turn of α helix requires 3.6 residues and utilizes the 13 numbered atoms (including the hydrogen-bonding proton on the peptide nitrogen). Thus the α helix is also referred to as a 3.6_{13} helix. From Blundell and Johnson (1976) with permission.

KENNETH A.
THOMAS and ALAN N.
SCHECHTER

have since been identified that differ in pitch and in hydrogen-bonding arrangement (Dickerson and Geis, 1969; Némethy et al., 1967). In observed protein structures, however, these parameters and the corresponding hydrogen-bonding patterns frequently vary not only among different helices but also among different parts of the same helix (Remington et al., 1978; Birktoft and Blow, 1972; Hartsuck and Lipscomb, 1971; Hendrickson and Love, 1971; Lindskog et al., 1971). Apparently, either many of the altered structures are nearly isoenergetic (Robson and Pain, 1971) or the stable distortions permit energetically favorable interactions of the helix with other parts of the structure.

With one known exception, helices in globular proteins are right-handed. The only complete turn of left-handed helix noted to date is in thermolysin (Asp 226–Asn 227–Gly 228–Gly 229) (Matthews et al., 1974). The right-handed bias is thought to be the result of energetically unfavorable interactions between the main chain peptide oxygen and side-chain C^β atoms in the left-handed conformation (Kotelchuck and Scheraga, 1968; Leach et al., 1966b). An obvious exception to such a generalization is glycine, which does not have a C^β atom. Individual residues containing C^β atoms (Anfinsen and Scheraga, 1975; Matthews et al., 1974; Carter et al., 1974), especially asparagine (Chou and Fasman, 1974b; Crawford et al., 1973), are occasionally observed, however, in the left-handed helical conformation in proteins. For asparagine, this conformation may result in favorable electrostatic interaction between the polar side-chain and main-chain groups (Kotelchuck and Scheraga, 1968). In addition, polyproline is known to form a non-hydrogen-bonded structure with a left-handed helical pitch (Cowan and McGavin, 1955).

The stability of the normal right-handed helical conformation is also affected by amino acid sequence (Chou and Fasman, 1974b) and polypeptide length (Chou and Fasman, 1974a). The previously mentioned residues with left-handed conformations may contribute to the distortion, destabilization, or termination of the right-handed structures. The extensive conformational flexibility of glycine residues may also be considered to destabilize a helix for entropic reasons (Némethy et al., 1966). Proline residues, although occasionally located in helical stretches (Padlan and Love, 1974; Hendrickson and Love, 1971), are generally considered to be destabilizing (Perutz and Lehmann, 1968), except in amino terminal turns (Chou and Fasman, 1974b), since they do not have hydrogen atoms on the peptide nitrogen atoms and, therefore, are unable to contribute to the helical hydrogen-bonding patterns. In a systematic study of the frequency of occurrence of different amino acid residues at the ends of helices it was observed that the charged hydrogen-bond proton acceptors, aspartic and glutamic residues, commonly occur at the amino terminal ends of helices where the polypeptide backbone has free proton donor peptide $N-H$ groups. Conversely, the charged proton donor residues of lysine, arginine, and histidine are observed to occur more frequently at the carboxyl terminal ends of helices in the vicinity of free polypeptide backbone hydrogen-bond acceptor carbonyl groups (Robson and Suzuki, 1976). The atoms of side chains such as those of the uncharged asparagine, serine, and threonine residues, which can easily form hydrogen bonds with backbone atoms, are also observed to disrupt the hydrogen bonding of helices (Anfinsen and Scheraga, 1975; Huber et al., 1971; Watson, 1969). It is also possible that the restrictions on the side-chain orientations of the β-branched side chains of valine, isoleucine, and threonine residues, resulting from the presence of the helical backbone, may be a destabilizing factor for the helical conformation (Leach et al., 1966b). However, small rotations from the preferred staggered side-chain angles or of the helical backbone torsion angles may still accommodate a large number of acceptable conformations (Leach et al., 1966a). Finally, charge repulsion of adjacent helical side chains is thought to prevent the charged form of polylysine or polyglutamic acid from forming stable helices and may act analogously in globular proteins (Chou and Fasman,

1974*b*). Extensive analysis of the correlation between the amino acid residue type and secondary structural environment has been presented elsewhere (Robson and Suzuki, 1976; Anfinsen and Scheraga, 1975; Chou and Fasman, 1974*a,b*).

The tertiary environment of a helix may have an effect on its stability (Suzuki and Robson, 1976). In solution, amino acid polymers become more helical as the polarity of the solvent decreases, presumably because the polypeptide hydrogen bonds are progressively more stabilized (Hermans, 1966). This may explain, in part, the stability of helices that are located in the protein interior (Fletterick *et al.*, 1975; Matthews *et al.*, 1972; Drenth, *et al.*, 1971; Kraut, 1971*b*). Similar effects may contribute to the stabilization of the helical conformation of the S-peptide of ribonuclease S by removal of one side of the surface-bound helix from solvent upon formation of a specific complex with the remainder of the protein (Brown and Klee, 1969).

An environmental effect upon helix stability in crystal structures has also been observed. The clearest example is that of one hexameric crystalline form of insulin that has a segment of helix in the B chain of three subunits, as opposed to another crystalline form which contains no helix in the equivalent segments of the B chains in any of the six subunits (Bentley *et al.*, 1976). Also, the helical conformation of the polypeptide hormone glucagon is stabilized by crystal lattice formation (Sasaki *et al.*, 1975).

Helices are observed to be in close contact with each other in many protein structures. For nearly perpendicular helices the residues in the contact region are small. This has the effect of allowing the helices to come close together and thereby optimize the surface area that becomes buried between them. As the size of the residues in the contact surface increases, the distance between the helical axes increases. To compensate for this increased distance, the angle between the helical axes decreases. This is presumably favored by the stabilization achieved by maintaining a large buried surface area. For large side chains, the axes of the helices can become nearly parallel (Richmond and Richards, 1978).

The dispensability of certain helices in a protein fold of a particular type may be inferred from comparison of the tertiary structures of evolutionarily homologous proteins. Variability in helix length, pitch, hydrogen-bonding pattern, and even the insertion or deletion of entire helices are seen among virtually all families of proteins containing helices that have been examined by crystal structure analysis (Sawyer *et al.*, 1978; Davies *et al.*, 1975; Dickerson and Timkovich, 1975; Rossmann *et al.*, 1975; Padlan and Love, 1974). In addition, some structurally equivalent helices appear to have shifted position or even to have rotated since the time of divergence of homologous proteins from a common ancestor (Dickerson and Timkovich, 1975; Rossmann *et al.*, 1975; Huber *et al.*, 1971). Such shifts are structurally analogous to functional movement of helices observed, for example, in lactate dehydrogenase (White *et al.*, 1976) and hemoglobin (Deatherage *et al.*, 1976; Arnone and Perutz, 1974; Pulsinelli *et al.*, 1973), where ligand-induced structural changes include translations and rotations of helices. The fact that helices are observed to move as units, both in the extremely long time scale of evolutionary change and in the very short scale of structural movements associated with biological function, supports the conclusion from solution model studies with nonglobular polyamino acids (reviewed in Chou and Fasman, 1974*a*) that the helix in globular proteins is a cooperative structure.

3.1.2 *Beta Structure*

The second major category of secondary structure, the β-pleated sheet, differs qualitatively from helical structures because it involves hydrogen bonding between sequentially distant residues. β-strand arrangements of two principal types are observed: parallel and antiparallel, in which adjacent β chains run in the same or opposite directions, respectively (Pauling and Corey, 1951 b). In the antiparallel β sheet the solid angle formed between the N—H donor and carbonyl acceptor groups of the N—H$\cdots$O=C hydrogen bond is $180°$. In the parallel β-sheet hydrogen-bonding geometry, however, this angle is about $165°$. The parallel β sheets observed to date occur in protein interiors and have five or more strands. Antiparallel β structure, on the other hand, occurs both in the interior and in contact with solvent on the surface of the protein and is observed with as few as two strands (Levitt and Chothia, 1976). The ability of the antiparallel β structure to withstand solvent exposure may indicate that the straight hydrogen-bonding geometry makes it the more stable of the two structures. Although on simple statistical grounds one would expect mixed β sheets containing both parallel and antiparallel hydrogen-bonding patterns to predominate, this is not observed. Mixed sheets may be disfavored both because they express the bias of parallel sheets for internal environments and because a β strand that is involved in anti-parallel hydrogen bonding on one side might less easily accommodate the somewhat differ-ent parallel hydrogen-bond geometry on the other side (Chothia, 1973).

The β sheets are characterized by the order and type of connecting strands. The two basic types of connections are illustrated in Fig. 6. The first is the "same end" connection that bridges two β chains that run in opposite directions. The second is the "crossover" connection. This is located between two β chains that run in the same direction, requiring that the connecting strand cross over the β sheet. The connections and patterns of observed β sheets are shown in Fig. 7. Sequential pairs of β strands within a sheet can be categorized on the basis of the number of intervening strands in the sheet and their connection type. If β-strand proximity is denoted by 1 for spatially adjacent strands, then in general two β strands with n intervening strands can be labeled an $n \pm 1$ connection. The numbers are preceded by the $\pm$ sign, since the arbitrary choice of the direction from which to view the sheet determines whether the second strand of the pair will be to the right $(+)$ or left $(-)$ of the first strand. Connections of the crossover type are denoted by the suffix X. Thus, the examples in Fig. 6 are ± 1 for the same end and $\pm 1X$ for the crossover connection. The frequencies of connection types for observed β sheets are given in Table I.

The sequential arrangement of β strands in a sheet is interesting both from a struc-tural and an evolutionary point of view. From the tabulation of the observed strand con-nections it can be seen that not all arrangements occur with equal frequency. The connec-

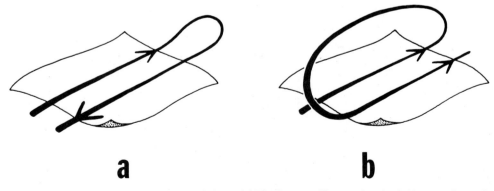

a   **b**

Figure 6. The two classes of connections in β sheets. (a) The "same-end" connection that bridges two β strands that run in opposite directions. (b) A "crossover" connection that occurs between β strands that run in the same direction. From Richardson (1976) with permission.

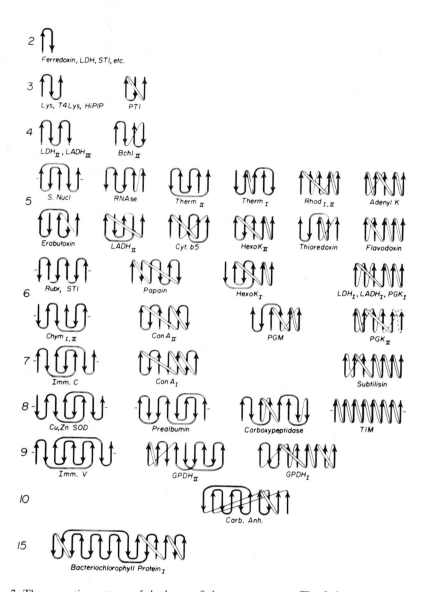

Figure 7. The connection patterns of the known β-sheet arrangements. The β sheets are grouped by their number of strands which is noted at the left. The vertical solid lines indicate the β strands with the arrows pointing from the amino to the carboxyl terminal end. Crossover connection chains in front of the β sheet are indicated by broad open lines, and those behind the β sheet are drawn as thin lines. Tentative connections are indicated by dashed lines. β sheets that close to form a β cylinder in three-dimensions are indicated by dashes at each end. Roman numeral subscripts to the protein name or initials denote different β sheets. The initials are LDH, lactate dehydrogenase; STI, soybean trypsin inhibitor; Lys, chicken egg-white lysozyme; T4 Lys, bacteriophage T4 lysozyme; HiPIP, high potential iron protein; PTI, pancreatic trypsin inhibitor; LADH, liver alcohol dehydrogenase; Bchl, bacteriochlorophyll protein; S. Nucl, staphylococcal nuclease; RNAse, pancreatic ribonuclease A; Therm, thermolysin; Rhod, rhodanese; Adenyl K, adenylate kinase; Cyt. b5, cytochrome b_5; HexoK, hexokinase; Rubr, rubredoxin; PGK, phosphoglycerate kinase; Chym, chymotrypsin; Con A, concan-avalin A; PGM, phosphoglycerate mutase; Imm. C, immunoglobulin constant domain; Cu,Zn SOD, Cu,Zn superoxide dismutase; TIM, triose phosphate isomerase; Imm. V, immunoglobulin variable domain; GPDH, glyceraldehyde-3-phosphate dehydrogenase; Carb. Anh., carbonic anhydrase. From Richardson (1977) with permission.

tion pattern distribution reveals a high frequency of close connections (Richardson *et al.*, 1976). Although many close connections would be expected in a random distribution of connection patterns, it has been determined that adjacent β-strand connections occur about twice as frequently as would be expected by chance (Sternberg and Thornton, 1977). This bias toward close connections might indicate a preference for interaction among sequentially adjacent segments of the polypeptide chain during the folding process. It has also been noted that the carboxyl terminal strands of the β sheets occur at the edge of the sheets with a greater frequency than that expected by chance. This may indicate that the carboxyl terminal strand is added to a rather well-formed β sheet near the end of the folding process (Sternberg and Thornton, 1977; Richardson, 1977). Another intriguing observation is that none of the observed β sheets is a knotted structure. This may reflect the selection against folding mechanisms that require the highly ordered pathway necessary to establish a knotted chain (that is, to thread the polypeptide through a previously formed circularly looped piece of chain) and perhaps implies that, in general, strand arrangements that are compatible with a larger number of folding pathways may have an evolutionarily greater selective advantage (Richardson, 1977). Extensively similar or identical strand arrangements involving large β sheets in different proteins are probably the result of divergent evolution from a common ancestor, as will be discussed in more detail in Section 4.1.

TABLE I. Connection-Type Frequencies in β Sheets[a]

Connection type	β-Sheet type[b]			Total
	Antiparallel	Mixed	Parallel	
± 1	63	39	—	102
± 1X	—[c]	12	32	44
± 2	—	9	—	9
± 2X	7	8	7	22
± 3	8	4	—	12
± 3X	—	8	5	13
± 4	—	2	—	2
± 4X	4	0	0	4
± 5	2	1	—	3
± 5X	—	0	0	0
± 6	—	0	—	0
± 6X	0	0	0	0
± 7	0	1	—	1
± 7X	—	1	0	1
± 8	—	0	—	0
± 8X	0	1	0	1
Sums:	84	86	44	214

[a]The frequency totals refer to the number of times a given connection type, specified in the left-hand column, occurs in Fig. 7. Identical β sheets that are repeated within one molecule, such as I and II in chymotrypsin and rhodanese, are each counted only once. The tentative connections in β sheet II of phosphoglycerate kinase are not counted.

[b]β Sheets are divided into those that have only antiparallel β strands, both antiparallel and parallel β strands (mixed) and only parallel β strands. The β sheets in staphylococcal nuclease and the immunoglobulin constant and variable domains are considered as pure antiparallel β sheets, although the two strands at the edges of the sheets are parallel to each other when the cylinders are formed.

[c]A dash indicates that this type of connection is not observed and would be very difficult or impossible to form. A 0 occurs when no examples of a presumably easily formed connection have been observed. In general, for pure antiparallel β sheets, same-end connections can occur for two β strands with an even number of intervening β strands and for crossover connections with an odd number of intervening β strands. Only crossover connections are allowed for pure parallel β sheets.

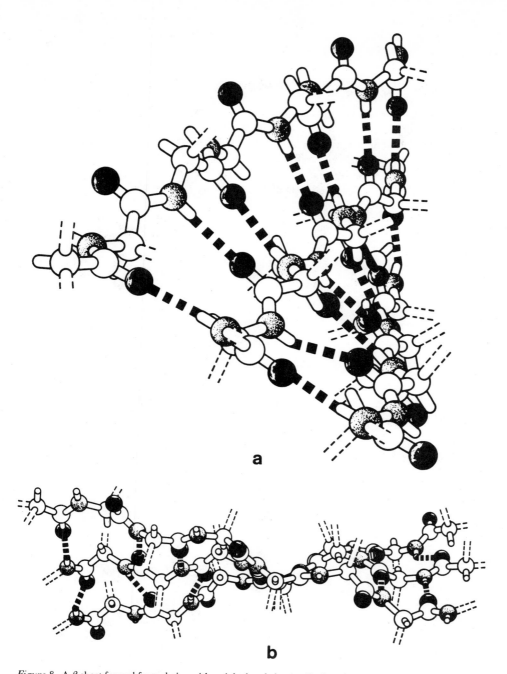

Figure 8. A β sheet formed from chains with a right-handed twist. Carbonyl oxygens are drawn as solid circles, peptide nitrogens as dotted circles, and backbone carbons as open circles. The dotted double lines indicate bonds to atoms not shown, either backbone or C^β side-chain atoms. The heavy dashed lines denote β-sheet hydrogen bonds. The first chain at the top in (a), or the front in (b), is antiparallel to the middle chain which is parallel to the last chain. View (a) along the direction of the last chain illustrates the right-handed twist. View (b) in the direction normal to (a) illustrates the left-handed twist of the β sheet. From Chothia (1973) with permission.

KENNETH A.
THOMAS and ALAN N.
SCHECHTER

The β sheets as a whole usually show a handed twist (right-handed when viewed along the polypeptide chain direction or left-handed about the axis normal to the polypeptide chain), as shown in Fig. 8. This twist moves the peptide carbonyl group away from the adjacent side-chain C^β atom. The effect is to generate backbone ϕ and ψ torsion angles for residues in the β sheet that are nearer to the energetic minimum of the polypeptide conformations so that the sheet may be expected to be more pliable and thus better able to locally deform to accommodate other portions of the tertiary structure (Chothia, 1973). The specific direction of the twist may be attributed to tetrahedral deformation of the bonds to the constituent peptide nitrogen atoms in combination with the introduction of slight nonplanarity of the peptide bond. These distortions would give the peptide a preferred directional twist which, when combined over sequential residues, results in the observed overall twist of the chain (Weatherford and Salemme, 1979).

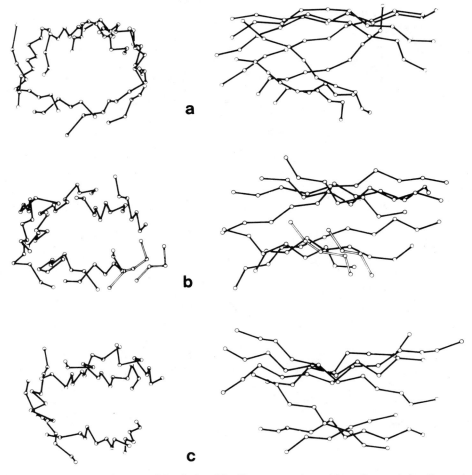

Figure 9. End-on and side views of β cylinders. The C^α atoms are shown by small open circles. Connections between different β strands are deleted. (a) The copper, zinc superoxide dismutase, (b) the heavy variable (V_H) region or domain from immunoglobulin McPC 603, and (c) a heavy constant (C_H1) region from the same immunoglobulin. For the antibody variable domain, β strands contributed by hypervariable chain segments are shown by open lines. From Richardson *et al.* (1976) with permission.

In addition to the typical sheet, β structure is observed to curl around and close upon itself to form antiparallel β cylinders in the serine proteases (Bode and Schwager, 1975; Delbaere *et al.*, 1975; Birktoft and Blow, 1972; Hartley and Shotton, 1971), thermolysin (Colman *et al.*, 1972), immunoglobulins (Richardson *et al.*, 1976), superoxide dismutase (Richardson *et al.*, 1975), and alcohol dehydrogenase (Eklund *et al.*, 1976a) and to form parallel β cylinders in triose phosphate isomerase (Banner *et al.*, 1975) and pyruvate kinase (Levine *et al.*, 1978). The closure may be considered to be the result of staggering the chains of a twisted sheet. Inasmuch as the β strands on the parallel β cylinder are connected by helices, the antiparallel β cylinders are the only observed case of a single unit of secondary structure being able to form the large hydrophobic core that is considered to be necessary for tertiary structural stability. Typical antiparallel β-cylinder polypeptide backbone conformations are shown in Fig. 9.

3.2 Nonsecondary Structure

We shall refer to the portions of the polypeptide chain that are involved in neither helix nor β sheet as nonsecondary structure. Conformations of this class lack the periodic hydrogen-bonding pattern and repetitive ϕ and ψ torsion angle values of the polypeptide backbone that characterize secondary structures. This structural category includes the short turns, longer loops, and extended chain conformations that often connect polypeptide chain segments involved in secondary structure.

3.2.1 Turns and Loops

Polypeptide chain segments frequently occur in conformations that result in directional reversal within three or four peptide bonds. These tight, hairpin or reverse turns can, in part, be classified on the basis of their hydrogen-bonding patterns and dihedral angles and have been discussed in detail elsewhere (Matthews, 1972; Némethy and Printz, 1972; Venkatachalam, 1968). Some types of hairpin turns require glycine residues for steric reasons. Glycine residues in these locations may be observed to be evolutionarily invariant. Many turns, however, do not fall into these categories, and some are observed to have no internal hydrogen bonds (Crawford *et al.*, 1973).

Tight turns are usually located on the exposed molecular surface (Kuntz, 1972) and may be regions of extensive solvation (Carter *et al.*, 1974), where the hydrophobicity of the sequence is at a local minimum (Rose, 1978). They are also sometimes partially disordered in the time-averaged electron density maps, indicating that they have no uniquely stable conformation and are probably quite flexible. Identical protein molecules in different crystalline geometries can also show different ordered turn conformations. This suggests that in some cases specific turn conformations can be easily altered by weak environmental perturbations (Moult *et al.*, 1976; Epp *et al.*, 1975). In addition, it has been noted by comparisons of homologous structures, that additions and deletions in the sequence are often located within or adjacent to turns (Dickerson *et al.*, 1976; Delbaere *et al.*, 1975). Turns are frequently observed to be flexible and are often subject to evolutionary alteration; therefore, they are probably permissive, rather than directive, participants in the folding mechanism. Although usually observed to be sequentially separated from each other by pieces of secondary structure, turn conformations also are recognized to exist in tandem linkage to generate longer stretches of "irregular" or "coiled" structure (Reeke *et al.*, 1975; Carter *et al.*, 1974).

KENNETH A.
THOMAS and ALAN N.
SCHECHTER

If the polypeptide alters direction over a longer stretch of chain than the tight turn, then it can be referred to as a loop. Long meandering stretches of polypeptide chain conformation are often found in proteins and can usually be considered to be composed of one or more loops. The ligand-binding or catalytic sites of proteins frequently utilize these structures. Ligand binding in ordered loop regions has in some cases been noted to alter the polypeptide conformation, suggesting that perhaps such regions are less rigid than are most of the helical and β structures (White *et al.*, 1976; Imoto *et al.*, 1972; Quiocho and Lipscomb, 1971). The flexibility or disorder of loop structures is also inferred from the weak appearance of these structures in electron density maps (Sweet *et al.*, 1974; Arnone *et al.*, 1971). Loop flexibility may play an important role in the control of protein function. In the case of the activation of trypsinogen to trypsin, three disordered loops in the zymogen become fixed to form the binding site pocket in the active enzyme (Fehlhammer *et al.*, 1977).

3.2.2 Extended Chains

Relatively straight segments of ordered polypeptide chain that are not involved in β structure are rarely noted in proteins. A recognized exception, however, exists in α-chymotrypsin. This protein contains an ordered extended chain conformation having roughly periodic ϕ and ψ angles but no internal hydrogen bonds. Although the lack of hydrogen bonds prevents this conformation from being categorized as a secondary structure in the classical sense, it has been termed the ϵ helix because the repetitive backbone conformation generates a helical pitch (Srinivasan *et al.*, 1976).

Some protein structures are observed to have disordered segments of chain at one or both ends of the polypeptide. These amino and carboxyl terminal segments of polypeptide chain probably extend out into the solvent and presumably allow rotation around single bonds within these segments. The disordered ends occur most frequently among terminal sequences that are devoid of larger nonpolar residues (Ploegman *et al.*, 1978; Fehlhammer *et al.*, 1977; Timkovich and Dickerson, 1976; Moews and Kretsinger, 1975; Blake *et al.*, 1974; Arnone *et al.*, 1971). Thus the lack of association between terminal segments and the external surface of the globular protein may be a direct result of the absence of stabilizing hydrophobic interactions.

4 Tertiary Structure

The tertiary structure of a native protein or subunit is commonly defined by the complete spatial arrangement of all the constituent atoms of the polypeptide chain. This three-dimensional atomic array, however, can be conceptually subdivided into smaller units, including supersecondary structure and structural domains.

4.1 Supersecondary Structure

Small polypeptide units containing similar sequential orders and three-dimensional arrangements of secondary structural elements are termed supersecondary structures (Rao and Rossmann, 1973). These can be observed to recur both within a protein molecule and among a variety of functionally different proteins. Supersecondary structures are also called "folding units" as they often involve sequentially adjacent strands of secondary structure that could be imagined to rapidly "diffuse" together to form early elements of tertiary structure in hypothetical kinetically controlled folding mechanisms (Levitt and Chothia, 1976). These conformational units are probably not very stable thermodynamically, as their small size precludes the formation of large hydrophobic cores.

An example of internal secondary structural repetition is found in carp myogen, a calcium-binding muscle protein. Each of the two calcium ions in the molecule is bound by a structurally equivalent arrangement formed by two roughly perpendicular helices connected by a loop of polypeptide chain. These small supersecondary structures are also related to each other by a pseudo twofold rotation axis.* The protein is completed by a third similar fold having the calcium ion replaced by an internal salt bridge. It is possible that the three structures, which contribute nonpolar side-chain residues to a common hydrophobic core, arose by means of gene triplication, although little or no sequence homology exists among them. These three structures and the entire molecule are shown in Fig. 10. The time since gene triplication is estimated to be $3–7 \times 10^8$ years (Kretsinger, 1972).

A supersecondary structure containing sequentially alternating parallel β strands and helices ($\beta\alpha\beta\alpha\beta$) occurs twice in the lactate, malate, alcohol, and glyceraldehyde-3-phosphate dehydrogenases. Two of these $\beta\alpha\beta\alpha\beta$ structures combine, using a pseudo twofold rotation axis, to form a structural lobe containing a six-stranded parallel β sheet with helices on both sides, as illustrated in Fig. 11. Although the β-strand spatial connections are well conserved among these structures, the helices are more vulnerable to distortions (Rossmann et al., 1974). The lobe binds the coenzyme dinucleotide NAD^+, with one $\beta\alpha\beta\alpha\beta$ unit binding the AMP portion at the carboxyl terminal end of the β strands and the other unit binding the NMN half of the coenzyme in a locationally equivalent position (Rossmann et al., 1975). The $\beta\alpha\beta\alpha\beta$ structure, therefore, is often referred to as a mononucleotide-binding unit. The same basic "coenzyme-binding" fold subsequently has been reported in phosphorylase (Fletterick et al., 1976) and probably has been identified in phosphoglycerate kinase (Blake and Evans, 1974). The crystals of phosphorylase bind adenosine and those of phosphoglycerate kinase bind ADP in sites that are locationally equivalent to the NAD-binding sites of the dehydrogenases. These repeated structures may have arisen by gene duplication of an ancestral mononucleotide binding unit (Rossmann et al., 1975).

The flavin-binding sites in flavodoxins, the aromatic side-chain binding site of the protease subtilisin (Rao and Rossmann, 1973) and both the NADP and FAD binding sites of glutathione reductase (Schulz et al., 1978) utilize equivalent locations in similar single $\beta\alpha\beta\alpha\beta$ structures. The possibility of distinguishing between convergent and divergent evolution among each other or with the two mononucleotide binding units of the dehydrogenase coenzyme binding folds seems more difficult. Furthermore, it would appear unlikely that either the triose phosphate isomerase (Banner et al., 1975) or the pyruvate kinase (Levine et al., 1978) structures, each of which contain eight pairs of alternating β and

---

*Rotation axes relate identical objects to each other by the rotation of one object through the angle around the axes that is required to superimpose it on another identical object. Thus, a twofold rotation axis relates two identical objects, such as protein subunits, by a $360°/2 = 180°$ rotation. In general, an n-fold rotation axis relates n objects by $n - 1$ successive $360°/n$ rotations. If the objects that are being related are similar, but not identical, then an axis of rotation that leads to maximum overlap or superposition of the objects is termed a pseudorotation axis. Thus a pseudo twofold rotation axis relates two similar, but not identical, objects by a $180°$ rotation.

helical strands $(\beta\alpha)_8$, bear any discernable evolutionary relationship to the mononucleotide binding folds contained within them, although they may be related to each other. Finally, if one allows additions, deletions, or rearrangements of strands in the β sheet, the number of possible evolutionary relationships with other proteins increases but, in the absence of independent evidence, at the expense of the confidence in any conclusions (Steitz *et al.*, 1976; Campbell *et al.*, 1974).

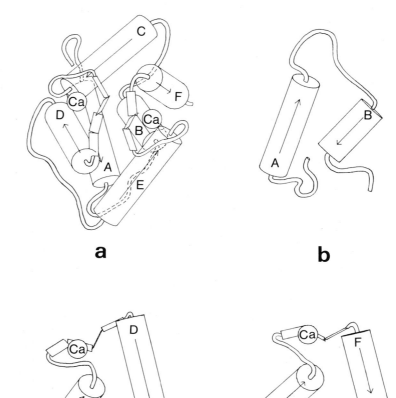

Figure 10. The carp muscle calcium-binding protein, or carp myogen, and its component substructures. Cylinders represent helices with internal arrows pointing toward the carboxyl terminal end, and linked rectangles indicate the planes of the peptide bonds in β strands. Helices are lettered sequentially from the amino to the carboxyl terminus. Connecting main chain segments that are not in any secondary structure are drawn as open double lines. Dashed lines denote features obstructed from direct view by other parts of the molecule. (a) Illustrates the entire 108 residue molecule looking down the pseudo twofold rotation axis that relates the C–D and E–F calcium-binding regions. The three component parts, rotated into similar orientations are as follows: (b) the A–B helix (residues 4–36), (c) the C–D helix (residues 40–69), and (d) the E–F (residues 79–108) helix-containing substructures.

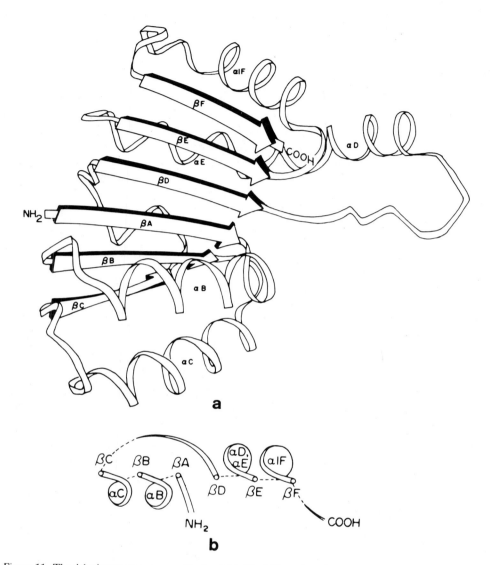

Figure 11. The dehydrogenase coenzyme-binding fold. The fold consists of two $\beta\alpha\beta\alpha\beta$ units. The first $\beta\alpha\beta\alpha\beta$ unit is composed of β strands A, B, and C and helices $\alpha\beta$ and αC. The second $\beta\alpha\beta\alpha\beta$ fold is made of β strands D, E, and F and helices αD-αE and α1F. The αD-αE helix can be considered to be either two adjacent helices or a single bent helix. αD is present in the lactate and malate dehydrogenases and absent in both the alcohol and glyceraldehyde-3-phosphate dehydrogenases. The two $\beta\alpha\beta\alpha\beta$ units are related by a pseudo twofold rotation axis between and approximately parallel to β strands A and D. Part (a) shows the structure with the β strands represented by arrows pointing from the amino to carboxyl terminus. Part (b) schematically illustrates the β sheet and helices as viewed perpendicular to (a) down the β-sheet strands drawn as straight rods and the helices represented by single loops. From Rossmann *et al.* (1975) with permission.

The $\beta\alpha\beta\alpha\beta$ unit can be considered as the overlap of two $\beta\alpha\beta$ units, a structure observed in many presumably unrelated proteins. The $\beta\alpha\beta$ or the more general β–coil–β fold approaches the lower limit of polypeptide chain length capable of folding back upon itself to create a small hydrophobic cluster. These three-stranded structures can occur in either a right- or left-handed looped conformation, as seen in Fig. 12. With one exception, in subtilisin, all such looped connections are right-handed. A number of explanations have been proposed to explain this supersecondary conformational bias of the $\beta\alpha\beta$ fold. One explanation is that the left-handed connection would require a longer crossover chain between the β strands of a twisted β sheet than would a right-handed connection, thus favoring right-handed crossovers for the shorter loops (Sternberg and Thornton, 1976, 1977). A second possible explanation is that the right-handed bias is a necessary result of the close packing of side chains of a helix and β strand (Nagano, 1977). In addition, it has been postulated that the right-handed sense of the helix may bias the direction of the turns between it and the two sequentially adjacent β strands during the folding process. Finally, it has been hypothesized that the preferred twist in an extended polypeptide chain might influence the folding direction of not only the $\beta\alpha\beta$ fold but also the longer loops in the β–coil–β structures (Richardson, 1976). As previously mentioned, this chain twist could be the result of a biased distortion of the peptide bonds from planarity (Weatherford and Salemme, 1979). Whatever the reason, the net result is that the similar right-handedness of the connections in the $\beta\alpha\beta\alpha\beta$ units cannot be used as evidence for common ancestry.

Another biased structure is a special strand arrangement observed in the four-stranded β-sheet segment shown in Fig. 13. This has been referred to as a "Greek Key" structure because of its similarity to that pattern found upon ancient Greek vases. This strand arrangement occurs more often than expected on the basis of the frequencies of occurrence of the component connections (Richardson, 1977). No stabilization advantage has yet been recognized for this structure.

Finally, although the possibly incomplete identification of the hydrogen-bonding pattern in ferredoxin precludes definitive assignment of the secondary structure, a repeated conformational unit is clearly observed. As seen in Fig. 14, the short chain has a pseudo

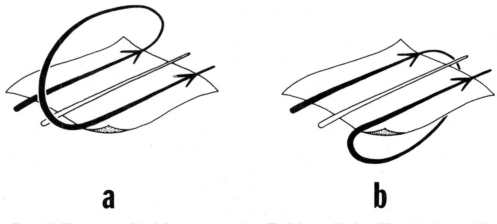

a   **b**

Figure 12. The two types of handed crossover connections. The light strand in the middle can be either parallel or antiparallel to the two adjacent side strands. Thus the connections are of the $\pm 2X$ type. (a) A right-handed connection and (b) a left-handed connection. Left-handed crossover connections are rarely observed in protein structures. From Richardson (1976) with permission.

twofold rotation axis relating two quite similar structures, each composed of three roughly parallel strands that interpenetrate to form a common hydrophobic core. A total of eight cysteine residues participate in the ligation of the two iron–sulfur clusters. Each unit contributes three cysteine residues to one cluster and one to the other. The two halves of the molecule are sequentially homologous to each other and, therefore, are considered to be the result of gene duplication (Adman *et al.*, 1973). The extensive interaction between the two halves of the molecule, however, would seem to preclude the possibility that each half is structurally independent of the other.

4.2 Structural Domains

The small supersecondary structures are found as part of larger globular units, referred to as domains, which contain well-formed hydrophobic cores. Proteins or subunits are formed from one or more of these structural domains.

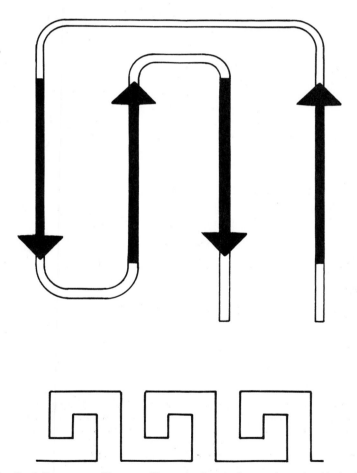

Figure 13. The Greek Key pattern. The upper illustration is a single repeating unit with the direction of the β strands indicated by the arrows. The lower diagram shows three consecutive repeating units of this motif.

KENNETH A.
THOMAS and ALAN N.
SCHECHTER

Small proteins usually appear to be made of single domains. For the porphyrin-binding proteins, such as those of the cytochrome or globin classes or the extreme example of the bacteriochlorophyll protein (Fenna and Matthews, 1975), the porphyrin group may contribute significantly to the stability of the core region. The diversity of observed polypeptide chain conformations indicates that many folds are compatible with the formation of a hydrophobic core. Thus, the similarity of tertiary polypeptide backbone structure, especially among proteins of similar function, suggests that the folds result from divergent evolution.

This is certainly the case for proteins showing strong overall sequence homologies such as the eukaryotic cytochrome *c* molecules. Nevertheless, when examined in a wide variety of species, it is evident that over one-half of the amino acid positions in this protein family

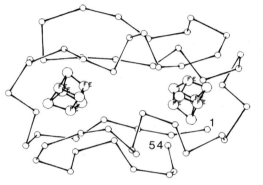

a

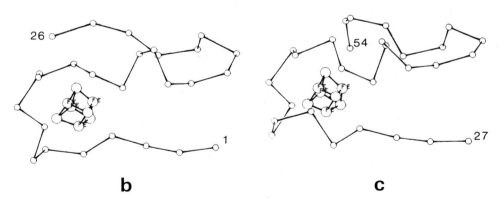

b   **c**

Figure 14. The two structurally similar halves of ferredoxin. The C^α positions along the polypeptide chain are indicated by open circles. The residue numbers at the ends of the chain segments are labeled. Each iron–sulfur cluster contains four labeled iron atoms and four inorganic sulfur atoms. (a) The entire 54 residue polypeptide chain with the two iron–sulfur clusters as viewed down the pseudo twofold rotation axis, (b) the first half of the polypeptide chain with one iron–sulfur cluster in the orientation of (a), and (c) the second half of the polypeptide chain with the other iron–sulfur cluster after rotation of 180° from the orientation in (a) so that it is in an orientation similar to (b).

are subject to point mutational changes, although in many cases the substituted amino acids may be constrained to be chemically similar (Dickerson and Timkovich, 1975). This is not to say that in any one sequence mutations can be tolerated at any of these locations. It is estimated that only 5 to 10 locations are variable in any one cytochrome c sequence. Once a substitution occurs at one of these positions, however, the set of locations elsewhere in the molecule that can tolerate a point mutation is thought to change. Thus some of the previously variable residue positions become unable to accept amino acid substitutions without serious adverse effects. Conversely, some of the previously invariable residue locations can now successfully accommodate different amino acids. Thus, by a series of point mutations much of the sequence can become subject to substitution at one time or another (Margoliash, 1972).

Perhaps even more interesting is the demonstration that the evolutionarily distant bacterial photosynthetic c_2 cytochromes have sequence and tertiary structural similarities to both eukaryotic c and bacterial c_{550} respiratory cytochromes. These molecules are illustrated in Fig. 15. This observation has been considered to be strong evidence in support of the hypothesis that the entire photosynthetic and respiratory electron transport chains are evolutionarily related. Furthermore, this fold appears to be as old as bacterial photosynthesis, or about 2.5×10^9 years, and could conceivably be as old as cellular life, that is approximately 3×10^9 years (Dickerson et al., 1976).

For more rapidly changing sequences, the similarities among three-dimensional folds may be more easily recognized than among the one-dimensional amino acid orders. For example, vertebrate and invertebrate globins have virtually the same fold (Love et al., 1972), as shown in Fig. 16. A preliminary low resolution crystal structure report has also documented the structural similarity of these animal globins with plant leghemoglobin (Vainshtein et al., 1975). Some of these similar folds have remarkably different sequences (Love et al., 1972). In fact, in all globins currently sequenced only three residues have remained invariant out of a total of about 140 to 150 residues (Vainshtein et al., 1975). Apparently, extensive sequence alterations by amino acid addition, deletion, and substitution are compatible with remarkably similar polypeptide backbone conformations, indicating that within the limits of recognizability of a tertiary structure there is extensive sequence degeneracy for a given fold.

Perhaps more intriguing, however, is the possibility of a distant evolutionary relationship between the globins and the cytochrome b_5 soluble fragment suggested by the partial structural similarity. In spite of the lack of convincing amino acid sequence homology, five of the eight globin helices occur in the same sequential order and similar relative orientation as five of the six cytochrome b_5 helices, as shown in Fig. 17. In the globin helical sequential lettering scheme (A to H) the b_5 equivalent helices are A, B, E, F, and G. Two of the three globin helices, C and D, that have no b_5 counterpart, comprise the region of greatest structural variability within the globin family, with the D helix even missing in both the mammalian α chain and annelid worm hemoglobins. The globin and b_5 carboxyl terminal helices are oriented differently enough to be considered structurally inequivalent. The orientation of maximum overlap of the five common helices orients the heme planes to within $10°$ of parallel to each other with the iron atoms displaced less than 1.4 Å from equivalent positions in spite of a $53°$ rotation of the heme in its plane. The most prominent difference involves the addition of a five-stranded β sheet in cytochrome b_5. It might seem somewhat unlikely that structural similarity in the common helical regions is the result of random resemblance. Alternatively, if this is caused by selective convergent evolution, it

KENNETH A.
THOMAS and ALAN N.
SCHECHTER

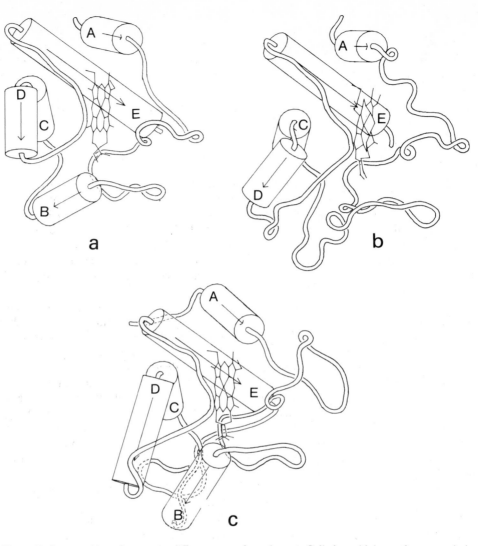

Figure 15. Structural homology among different types of cytochromes. Cylinders with internal arrows pointing toward the carboxyl terminal end represent helices. These are lettered sequentially from the amino to carboxyl terminus so that structurally equivalent helices in different cytochromes have the same letter. This results in helix B missing in (b). Dashed lines denote nonhelical features that are obstructed from direct view by the helices. All three molecules are in a similar orientation looking at the heme almost edge on. The molecules are (a) a respiratory cytochrome *c* from tuna, (b) a photosynthetic cytochrome c_2 from the bacterium *Rhodospirillum rubrum*, and (c) a respiratory cytochrome c_{550} from the bacterium *Paracoccus denitrificans*.

might imply the existence of some unrecognized set of restrictions on the ways in which a polypeptide chain may fold around a heme group. The remaining possibility is that these proteins are the result of divergent evolution. If electron carrier b_5 cytochromes and the oxygen-binding globins did diverge from a common ancestor, it is estimated that this occurred at least 1.5×10^9 years ago (Rossmann and Argos, 1975).

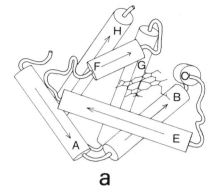
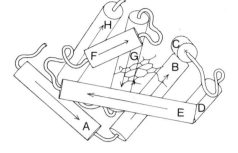
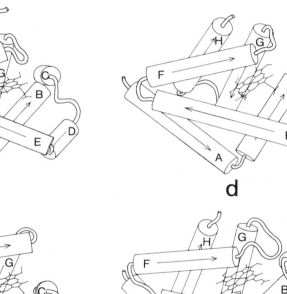

Figure 16. Structural homology among different types of globins. Cylinders with internal arrows pointing toward the carboxyl terminal end represent helices. These are lettered sequentially from the amino to carboxyl terminus so that structurally equivalent helices in different globins have the same letter. This results in helix D missing in (a) and (f). All 6 globin chains are in a similar orientation. The molecules are (a) the human deoxyhemoglobin α chain, (b) the human deoxyhemoglobin β chain, (c) sperm whale myoglobin, (d) lamprey hemoglobin, (e) insect *Chironomus* larva hemoglobin, and (f) annelid worm *Glycera dibranchiata* hemoglobin.

KENNETH A.
THOMAS and ALAN N.
SCHECHTER

Many larger monomeric proteins and protein subunits are composed of two or more structural domains. Globular domain substructures usually are made of contiguous pieces of polypeptide chain of less than about 200 residues in length. Whereas there is virtually no evidence that an isolated element of supersecondary structure has sufficient stability to remain primarily in the folded state, there is some indication that single domains of multiple domain proteins may have this ability.

One of the clearest examples of polypeptide chains composed of discrete domains is seen in the immunoglobulins. An IgG molecule, for instance, is composed of two identical light chains (L), each with a variable (V) and constant (C) domain (V_L, C_L), and two identical heavy chains (H) each with one variable and three constant domains (V_H1, C_H1, C_H2, C_H3). Stability does reside to some extent in the isolated domains. Separate V_L and C_L domains, or perhaps domain dimers, are stable in solution (Bjork et al., 1971). Denaturation studies on light chains support the model of independent denaturation of the V_L and C_L halves (Rowe and Tanford, 1973). The V_L domain dimer crystal structure clearly shows the characteristic V_L domain conformation that is observed in both the IgG and Fab fragment crystal structures (Epp et al., 1975). The C_H2 domains, which are not involved in dimer contacts in the Fc fragment, are also reported to be stable monomers in solution (Deisenhofer et al., 1976). The remarkably similar conformations of all six unique domains (an example of which is shown in Fig. 19) imply that the V_H1, C_H1, and C_H3 segments may also have independent structural stability. Presumably all six unique domains can also fold independently of each other. The structural similarity of the domains also supports the hypothesis that antibody structure is the result of multiple gene duplications and fusions (Hill et al., 1966; Singer and Doolittle, 1966).

The pancreatic serine proteases trypsin, chymotrypsin, and elastase and a microbial serine protease all have two internal domains, each composed of a β cylinder of quite sim-

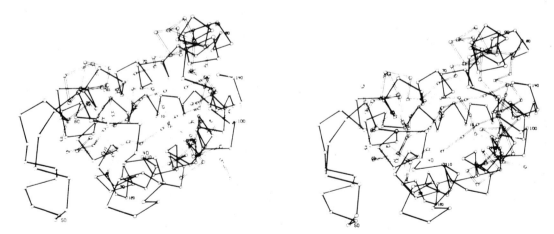

Figure 17. Structural similarities between the globins and cytochrome b_5. The horse hemoglobin β chain (dark lines) is superimposed on cytochrome b_5 (light lines) as shown in stereo. The small circles correspond to the C^α atoms. The numbering refers to the hemoglobin β-chain sequence. From Rossmann and Argos (1975) with permission.

ilar polypeptide backbone conformation. Although there is extensive sequence similarity among the pancreatic enzymes, there is little homology between the pancreatic and bacterial proteins (Delbaere *et al.,* 1975). Since these polypeptide chains, or their zymogen precursors, are continuous within each molecule, it seems reasonable that they too might be the products of gene duplication and fusion of a common ancestral domain. The lack of interdomain sequence homology would then be interpreted to be the result of a rate of amino acid sequential change that is rapid enough to obliterate the original sequence similarities (Birktoft and Blow, 1972). The acid proteases of the pepsin family also have two domains of similar conformation, each comprised of β structure. The two domains are related by a 175° rotation and a 1 Å translation. No homology has been demonstrated between the two domain sequences (Tang *et al.,* 1978).

Two extremely similar domain conformations are also observed in rhodanese. Each domain contains a five-stranded parallel β sheet flanked by two helices on one side and three helices on the other. The two similar domains are related to each other by a twofold rotation axis. The two domain sequences require a minimum of 1.27 nucleotide base changes per codon to be interconverted. Although two random sequences would require about 1.6 base changes per codon to become identical, it has been shown that this value is about 1.25 for two sequences having similar structural constraints (Dickerson, 1971). Therefore, although the possibility that both domains arose from a common ancestor by gene duplication and fusion is attractive, especially because they are in the same polypeptide chain, no evidence remains in the sequence which would substantiate this hypothesis (Ploegman *et al.,* 1978).

Finally, in spite of the virtual absence of secondary structure, conformational similarity has been seen among the four domains within each subunit of wheat germ agglutinin. Furthermore, there are four equivalently positioned disulfide bonds within each approximately 40 residue domain. This structural similarity, in addition to partial sequence homology involving the cysteine and glycine residues, strongly indicates that gene quadruplication has occurred (Wright, 1977).

Most multidomain proteins contain two or more conformationally dissimilar domains. Structural similarities between proteins of this type may involve all or only some of these domains. Chicken lysozyme, for example, is composed of two domains (Imoto *et al.,* 1972). Sequence homologies and model building have indicated that an extensive structural similarity between chicken lysozyme and mammalian α-lactalbumin very probably occurs and would include both domains (Browne *et al.,* 1969). Quantitative structural comparisons between chicken and phage T4 lysozyme have also revealed a structural similarity. T4 lysozyme contains the same two domains found in chicken lysozyme. The primarily helical domain in T4 lysozyme, however, is significantly larger than is its counterpart in the chicken molecule and might actually be considered to be the result of the addition of a third domain. The other common domain is about the same size in both enzymes. The active sites are located in equivalent locations between the two common domains. The structures are illustrated in Fig. 18. In spite of these structural similarities there is no detectable sequence homology (Rossmann and Argos, 1976). Furthermore, the uncertain evolutionary origin of viruses makes it difficult to estimate how long ago these proteins might have diverged from a putative common ancestor.

The dehydrogenases have a double domain structure composed of a catalytic domain involved in binding the substrate and a domain that binds the coenzyme NAD^+. As mentioned previously, the lactate, malate, alcohol, and glyceraldehyde-3-phosphate dehydro-

KENNETH A.
THOMAS and ALAN N.
SCHECHTER

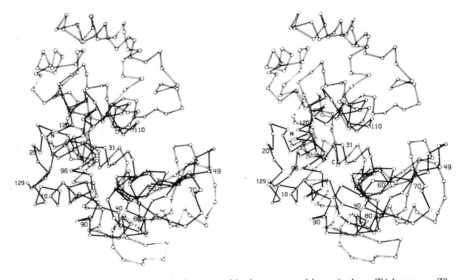

Figure 18. Structural similarities between hen egg-white lysozyme and bacteriophage T4 lysozyme. The hen lysozyme (dark lines) is superimposed on the T4 lysozyme (light lines) as shown in stereo. The small circles correspond to the Cα atoms. The numbering refers to the hen lysozyme sequence. From Rossmann and Argos (1976) with permission.

genases all have similar coenzyme-binding domains composed of two mononucleotide-binding supersecondary structures. The coenzyme-binding domain of glyceraldehyde-3-phosphate dehydrogenase has, in addition, three extra β chains, two at one edge and one at the other edge of the common six-stranded β sheet. Although the catalytic domains of the lactate and malate dehydrogenases are quite similar, they are distinctly different from the unique catalytic domains in either alcohol or glyceraldehyde-3-phosphate dehydrogenase (Rossmann *et al.*, 1975). It seems unlikely that the very different catalytic domain folds diverged from a common ancestor. Therefore, either the catalytic domain "grew out" of the coenzyme domain by a series of polypeptide insertions, or it involved the addition by gene fusion of a preformed catalytic domain that might have been used to bind the same or a similar substrate in a different protein. The latter mechanism may appear somewhat more appealing because it does not require the evolutionary creation of a new stable domain, and it would facilitate the rapid generation of functional diversity by combining preformed structural "modules" (Ohlsson *et al.*, 1974).

The binding of different substrates by the conformationally similar lactate and malate dehydrogenases and the ability of enzymes in both the serine protease and acid protease families to exhibit different substrate specificities shows that a given polypeptide conformation may bind a variety of substrate or ligand molecules. In addition, changes in the sequence of the antigen-binding hypervariable loops of the variable domains (V_L, V_H) in antibodies alter the binding specificity. A functionally unrelated protein, superoxide dismutase, has been observed to have the same basic polypeptide backbone fold as does the antibody domain, as shown in Fig. 19. The copper-containing active site of this enzyme utilizes two of the equivalent antibody hypervariable loops. Based on quantitative estimates of the similarity of these two folds it appears likely that they are the result of divergent evolution from a common ancestor. Since no convincing sequence homology exists, however, the alternative of convergent evolution cannot be ruled out. Still, this lack of sequence homology does not support the possibility of structural convergence, as the calculated rate of accepted amino acid substitutions within the immunoglobulins indicates that little or no homology would be expected between current antibody domain sequences and that of a common ancestral domain (Richardson *et al.*, 1976). It seems that the basic domain folds may change both sequence and function without marked alterations in the backbone conformation. This suggests that it is biologically more economical to modify the function of a preexisting fold than to evolve a new fold for each biochemical function.

Interdomain contacts appear to utilize the same general types of interactions, such as hydrophobic contacts, salt bridges, and buried hydrogen bonds (Chothia, 1976), that occur within a domain. At least for the serine proteases, many of these hydrogen bonds involve buried water molecules (Sawyer *et al.*, 1978; Bode and Schwager, 1975; Birktoft and Blow, 1972). Additionally, chain termini from one domain are sometimes observed to extend across the domain interface to form contacts with the adjacent domain, as is seen in papain (Drenth *et al.*, 1971), T4 lysozyme (Matthews and Remington, 1974), alcohol dehydrogenase (Eklund *et al.*, 1976a), and the serine proteases trypsin (Stroud *et al.*, 1972), chymotrypsin (Birktoft and Blow, 1972), and elastase (Sawyer *et al.*, 1978).

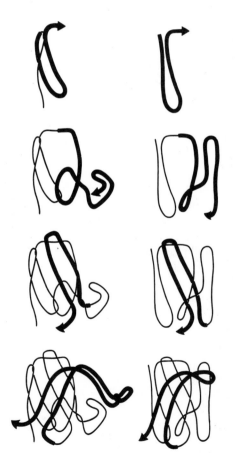

Figure 19. Structural similarities between Cu,Zn superoxide dismutase and an immunoglobulin variable domain. The comparison of the step-by-step amino- to carboxyl-terminal buildup of the backbone conformations for the dismutase (down the left side) and the immunoglobulin variable domain (down the right side) is shown. New backbone added at each step is illustrated by heavy arrows. From Richardson *et al.* (1976) with permission.

Interactions among domains can be directly correlated with biochemical functions. For example, in virtually all multidomain proteins examined, substrate or ligand binding sites span two or more domains. The movement of domains with respect to each other may also be involved in protein function. It has been observed that in proteolytic conversion of the zymogen chymotrypsinogen to α-chymotrypsin, the two β cylinders in each molecule undergo a shift of 1.5 Å relative to each other. The active site spans the interface between these two domains. The alteration of the hydrogen-bonding geometry between the serine and histidine residues in the active site that accompanies the domain movement may, in part, be responsible for the large increase in catalytic activity of α-chymotrypsin relative to chymotrypsinogen (Birktoft et al., 1976). In addition, it has been proposed that domain movement in antibodies may be caused by antigen binding. Such movement could facilitate transmission of information about the ligation state of the Fab binding sites to the complement-fixing Fc region (Huber et al., 1976).

5   Quaternary Structure

The spatial arrangements and contact interactions among subunits define the quaternary structure of oligomeric proteins. Subunit interactions are particularly relevant to the regulatory process, since allosteric interactions can occur through intersubunit contacts. Again, as is the case with intrapolypeptide interactions, intersubunit contacts involve hydrophobic clustering, ion pairs, and hydrogen bonds. In the case of subunit contacts, however, the effect of specific alterations in the contact surface can be more easily studied if the oligomeric subunit binding equilibrium is altered. For instance, the study of subunit association of various normal and mutant mammalian hemoglobins has demonstrated the destabilizing effect of the burial of not only unpaired charged groups (Perutz, 1974) but also single polar groups that lack a hydrogen-bonding partner (Perutz and Lehmann, 1968).

Stable subunit interactions are observed to change on the long evolutionary time scale. The quaternary structures of the dehydrogenases, shown in Fig. 20, demonstrate the variety of subunit arrangements observed in this protein family. As can be seen from the locations of the active sites, the mode of association of the dimeric malate and alcohol dehydrogenases is quite dissimilar. The tetrameric lactate and glyceraldehyde-3-phosphate dehydrogenases can be considered to be formed by the association of two dimers, each having the malate dehydrogenase subunit arrangement. They are associated in opposite arrangements, however, so that the two pairs of active sites point away from each other in lactate dehydrogenase and face each other in glyceraldehyde-3-phosphate dehydrogenase (Rossmann et al., 1975). A variety of oligomeric states also exists for hemoglobins from different species (Dayhoff et al., 1972). Furthermore, in the immunoglobulins, different sides of the variable and constant domains are used for the subunit contacts (Davies et al., 1975). The quaternary diversity of similar tertiary structures indicates that the rate of evolutionary change of subunit association can exceed that of major chain-folding pattern modifications within subunits and domains. The rates of accepted mutational changes in the subunit contact areas are nevertheless still lower than those for the solvent-exposed surfaces that are not directly involved in the functions of the proteins (Eklund et al., 1976b; Buehner et al., 1974). As demonstrated by guinea pig and coypou insulins, however, once a contact is dissociated as a result of mutation, this surface again tolerates amino acid changes at the more rapid rates characteristic of the nonfunctional solvent-exposed surface (Blundell *et al.*, 1972).

In some cases β-structural hydrogen-bonding patterns span the subunit interface. For example, in both insulin (Blundell *et al.*, 1972) and concanavalin A (Reeke *et al.*, 1975) an antiparallel β sheet crosses a subunit contact so that along the subunit interface a β strand from one subunit hydrogen bonds to a β strand from the adjacent subunit in an antiparallel manner. Similarly, in prealbumin the two subunits pair through external sur-

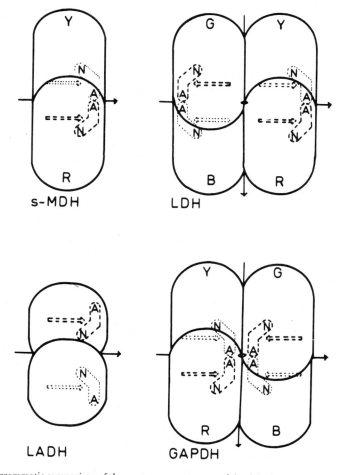

Figure 20. Diagrammatic comparison of the quaternary structures of the dehydrogenases. Dashed lines indicate near side, and dotted lines denote far side features on the subunits. The NAD binding sites contain the letters A for the AMP and N for the NMN parts of the coenzyme. Broad arrows within the subunit boundaries indicate equivalent subunit orientations. Thin arrows outside the subunits show the locations of the twofold rotation axes relating the subunits. The small double crescent at the centers of the lactate (LDH) and glyceraldehyde-3-phosphate (GAPDH) dehydrogenases denote twofold rotation axes that are perpendicular to the page. Thus the monomers (Y and R) or the malate dehydrogenase (s-MDH) dimer are related by a twofold rotation axis but in a different manner than those in the liver alcohol dehydrogenase (LADH) dimer. Two s-MDH-like dimer structures (Y–R and G–B) associate through opposite faces in LDH and GAPDH. Since the pairs of coenzyme binding sites are on opposite sides of the LDH tetramer, they all occur close together in GAPDH. From Rossmann *et al.* (1975) with permission.

KENNETH A.
THOMAS and ALAN N.
SCHECHTER

face β-structural hydrogen bonding to form two continuous β sheets in each of the two dimeric components of the tetramer (Blake *et al.*, 1974). The internal six-stranded parallel β sheet in each alcohol dehydrogenase subunit continues across the subunit contact by an antiparallel strand arrangement to form a 12-stranded β sheet (Eklund *et al.*, 1976a). If these subunits can fold independently, then this implies that multiple folding units could contribute to the formation of large continuous pieces of β structure within single subunits and monomeric proteins. Such component folding units within a single polypeptide chain might be difficult to recognize, since they would contribute to the formation of a common secondary structure.

As was seen with interdomain contacts, dimerization may be accompanied by chain crossover from one subunit to the other. For example, the crossover arm in one of the lactate dehydrogenase contacts is absent in malate dehydrogenase, and the corresponding subunit contact does not occur (Hill *et al.*, 1972). Presumably, either the arm is needed for stability or there is no selective advantage for its preservation once the subunit contact is broken for other reasons.

Catalytic sites may occur in or near subunit contact regions. The active sites of glutathione reductase (Schulz *et al.*, 1978), phosphorylase (Fletterick *et al.*, 1976), glucose-6-phosphate isomerase (Shaw and Muirhead, 1976), alcohol dehydrogenase (Eklund *et al.*, 1976a), and glyceraldehyde-3-phosphate dehydrogenase (Moras *et al.*, 1975) are all at or directly adjacent to the subunit contact region. These could presumably be effected by movement in the subunits or alteration of the subunit association. Similarly, both the immunoglobulin antigen-binding sites (Davies *et al.*, 1975) and the hexokinase ATP effector-binding site span the subunit contact regions. The hexokinase ADP/ATP effector binding is thought to control the hexokinase monomer–dimer equilibrium, which may be important in the regulation of catalytic activity (Anderson and Steitz, 1975).

The solution characterization of the dimerization of bovine pancreatic ribonuclease indicates that one histidine residue side chain from each subunit contributes to each active site. Thus, although the monomers are active, the catalytic sites in the dimeric enzyme contain functional side-chain residues from each subunit (Crestfield and Fruchter, 1967). This illustrates the feasibility of the evolutionary conversion of monomeric proteins to oligomers having active sites constructed from residues originating in different subunits.

The aggregation of subunits frequently utilizes point symmetry.* This often leads to small solvent accessible cavities at the center of the symmetrical array as, for example, in the dimeric malate dehydrogenase (Banaszak and Bradshaw, 1975) and the tetrameric concanavalin A (Reeke *et al.*, 1975), prealbumin, and vertebrate hemoglobins. These cavities are used as binding sites for thyroxine by prealbumin (Blake *et al.*, 1974) and organic phosphates by the vertebrate hemoglobins (Arnone and Perutz, 1974; Arnone, 1972). The latter case is exceptionally relevant to the regulatory process, since the binding of organic phosphates influences the effect of oxygen concentration on the cooperativity among the surrounding subunits. A significantly larger cavity is formed by the six pairs of regulatory and catalytic subunits in *Escherichia coli* aspartate transcarbamylase (Evans *et al.*, 1973), and extremely large cavities are created by the 24 subunits of apoferritin (Banyard *et al.*, 1978) and the 180 subunits of tomato bushy stunt virus (Winkler *et al.*, 1977).

In contradiction to initial speculations (Monod *et al.*, 1965), not all allosteric proteins utilize point symmetry, as is most dramatically illustrated by the inequivalent packing of

---

*Point symmetry occurs in a multimeric protein if all of the rotation axes relating the different subunits intersect at a common point, the center of gravity of the oligomer.

the two hexokinase subunits shown in Fig. 21. One subunit is related to the other, not by a 180° rotation, but rather by a 156° rotation coupled with a 13.8 Å translation (Steitz *et al.*, 1976). Although cooperative subunit interactions in allosteric proteins involve perturbation of the subunit contacts, transmission of the ligand effect to these regions is mediated by complicated tertiary structural changes, which are best documented in the mammalian hemoglobin system (Deatherage *et al.*, 1976; Anderson, 1975; Anderson, 1973; Perutz and TenEyck, 1972). Even for this case, however, the exact order of events in the mechanism of the mechanical transmission of the effects of oxygen binding is not entirely understood.

The structural distinction between nonidentical subunit and protein–protein interactions appears to be rather arbitrary. The crystal structures of the complexes of porcine trypsin with soybean trypsin inhibitor (Sweet *et al.*, 1974) and of bovine trypsin with bovine pancreatic trypsin inhibitor (Huber *et al.*, 1974) reveal that the contact surfaces utilized for these very tightly bound complexes are only a small part of the entire available surface of each molecule. Apart from the tetrahedral distortion of the carbonyl carbon atom of the inhibitor peptide bond that sits in the protease active site, the contact surfaces between the proteins look very much like those that occur between subunits in oligomeric proteins.

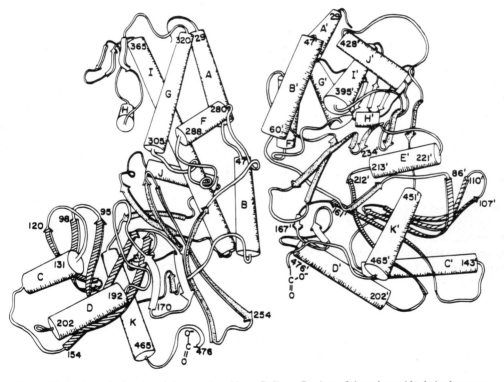

Figure 21. A schematic drawing of the yeast hexokinase B dimer. Portions of the polypeptide chain that are α-helical are represented by cylinders that are lettered sequentially from the amino to carboxyl terminus. β strands are indicated by arrows pointing toward the carboxyl end. The numbers represent an approximate residue number counting from the amino terminus. The interaction of the two subunits is asymmetric. They are related to each other by a 156° rotation about an axis running approximately vertically up the page between the two subunits and a 13.8 Å translation along this axis. The subunit on the right side is higher than the one on the left. From Steitz *et al.* (1976) with permission.

In this case, however, the importance of complementary geometry can be demonstrated. Subtilisin, a protease with an active site very similar to that of trypsin, is not inhibited by the trypsin inhibitors. This is presumably a result of steric interference between the subtilisin loop containing residues 97 to 102 and the inhibitor recognition site. This loop does not exist in trypsin (Robertus *et al.*, 1972). The loop probably prevents the efficient packing necessary to exclude water from these hydrophobic intermolecular surfaces (Janin and Chothia, 1976; Chothia and Janin, 1975; Sweet *et al.*, 1974).

The highest level of protein structural organization occurs in the interaction of multiple nonidentical polypeptide chains, such as in ribosomes. Because of their size and complexity, little direct evidence exists as to the details of such structures. There is no reason to doubt, however, that the interpolypeptide contacts in these particles utilize the same types of interactions as those observed in currently described protein interiors and oligomeric contacts.

6 Mechanism of Folding

The study of the mechanism of protein folding has several origins. Examination of the reversibility of the gross structural changes, called denaturation, induced by variations of temperature, pH, or solvent conditions led to the formulation of thermodynamic and kinetic studies on the mechanisms of these changes (Tanford, 1968, 1970; Lapanje, 1978). It was, however, during analyses of the disruption and subsequent reformation of disulfide bonds in ribonuclease (Sela *et al.*, 1957; Anfinsen and Haber, 1961) that these questions began to be posed in a biological context—that is, in relation to the mechanism by which a nascent polypeptide newly synthesized on a polyribosome folds to the biologically active species. These studies by Anfinsen and his colleagues (Epstein *et al.*, 1963) demonstrated that, for several globular proteins, it was possible to reduce the disulfide bonds and denature the polypeptide chain and then to change the solution conditions so as to reform the disulfide bridges and regain an enzymically active species. These results led to the hypothesis that "the particular configuration that a protein assumes under any specific set of conditions, is the one that is thermodynamically the most stable" (Epstein *et al.*, 1963). Although the time required for such renaturation was long as compared to that expected for biosynthetic processes, the discovery of an enzyme system which catalyzed the reoxidation process (Venetianer and Straub, 1963; Goldberger *et al.*, 1963; Freedman and Hawkins, 1978) has made it likely that refolding governed by the information in the amino acid sequence is the major determinant of this aspect of the translation of genetic information into biological function.

A chemical consequence of this hypothesis is that it should be possible to examine the mechanism of the biologically significant folding of proteins by suitable equilibrium and kinetic studies *in vitro*.* These experiments have been of three major types: (1) the use of spectroscopic techniques to characterize the initial and final states of structural transitions

---

*It should be noted that the hypothesis assumes that the native state of a protein is thermodynamically determined—that is, it is a global free-energy minimum and not a local minimum determined by a particular kinetic pathway. Although not proved, and very difficult to prove, this aspect of the hypothesis is still widely accepted. Nucleation-controlled pathways, however, may never reach a true energy minimum.

and to analyze the interconnecting pathways, (2) the use of chemical quenching methods to isolate and characterize intermediates in the refolding of disulfide-linked proteins, and (3) the study of complementing fragment systems. We will discuss each of these approaches.

6.1 Spectroscopic Intermediates

It has been known for some time that equilibrium measurements of denaturant-induced structural transitions for a number of systems are fit well by a two-state approximation (Tanford, 1968, 1970; Baldwin, 1975). In addition, such transitions tend to be highly cooperative in that most of the measured spectral or related chemical change occurs over a very narrow range of temperature, pH, or perturbant concentration. An example of such a sharp transition for staphylococcal nuclease is shown in Fig. 22. A result of this type implies that under equilibrium conditions the concentration of stable intermediates is very low. Thus when methods were developed for the measurement of these transitions in the millisecond and faster range (for a review see Schechter, 1970) it was surprising to find

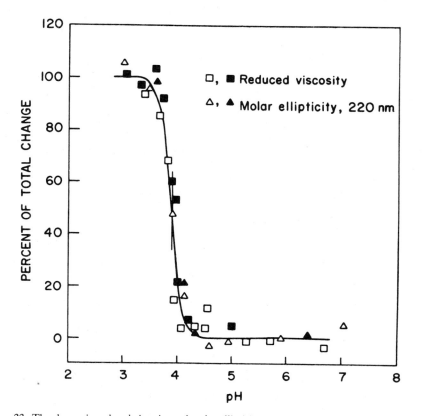

Figure 22. The change in reduced viscosity and molar ellipticity at 220 nm as a percentage of the total change of each measurement in the transition from the native to the acid-denatured form, plotted as a function of pH. Open symbols refer to the measurements made during the addition of acid; filled symbols, to measurements made during the addition of base. From Anfinsen et al. (1971b) with permission.

KENNETH A.
THOMAS and ALAN N.
SCHECHTER

that many folding and unfolding processes for small globular proteins exhibit complex kinetic behavior (Schechter *et al.*, 1970; Epstein *et al.*, 1971*a;* Tsong *et al.*, 1971; Ikai *et al.*, 1973).

Intense work has gone on in the last decade to characterize these kinetic intermediates (see Baldwin, 1975, 1978). Sophisticated mathematical analyses have been used to identify the number of kinetic intermediates and their positions on a reaction pathway. Unfortunately once the analyses go beyond one intermediate, the results frequently become degenerate, and it is very difficult to choose rigorously among large numbers of alternative pathways. Despite this it has been possible, by systematic variation of solution conditions and of spectral probes, to make tentative structural assignments for some of the intermediate steps. One such is the identification of cis–trans isomerization of proline residues as the cause of slow phase in the reaction pathway of a number of proteins (Brandts *et al.*, 1975, 1977). Thus the rate of the slow-folding form of denatured ribonuclease A is probably determined by the requirement that all of the proline residues attain the appropriate cis or trans form compatible with the properly folded conformation (Schmid and Baldwin, 1978).*

An alternative approach has been the use of nuclear magnetic resonance and related techniques to identify regions of proteins whose equilibrium properties vary across the structural transition region. Thus it was possible to show for the highly cooperative acid-induced transition of staphylococcal nuclease (see Fig. 22), that the region around one particular histidine residue shows changes at a pH value at which viscosity, circular dichroism, and fluorescence measurements show no evidence of conformational changes (Fig. 23) (Epstein *et al.*, 1971*b*). Similar findings have been reported for the alkaline transition of staphylococcal nuclease (Jardetsky *et al.*, 1971) and for the acid–heat denaturation of ribonuclease A (Westmoreland and Matthews, 1973).

Recently these approaches have been greatly extended into the use of the nuclear magnetic resonance methods to look for fluctuations, or the dynamics of residues within globular proteins under "physiological" conditions (Allerhand *et al.*, 1971; Campbell *et al.*, 1975; Wagner *et al.*, 1976). This approach is beginning to change fundamentally the way proteins in solution are viewed, since it is clear that there is normally a great range of modes, or fluctuations of structure, due to thermal energy. These concepts, stemming originally from hydrogen exchange studies, but now involving absorption, fluorescence, and X-ray diffraction spectroscopy as well as nuclear magnetic resonance spectroscopy (see review by Gurd and Rothgeb, 1979), are a bridge between the static pictures of crystallography and the kinetic studies of the large conformational transitions that we have been discussing.

It is worth noting, however, one other aspect of protein folding that becomes evident from these spectroscopic and chemical studies—namely, that folding is fast. The structural transitions that we have been discussing in many cases have a half-time much less than 1 sec. As has been pointed out (Levinthal, 1966, 1968), if a moderately long polypeptide chain were to try out all the conformations allowable, the time required for folding—using reasonable rate estimates for bond rotations and translational motions—would be much greater than the age of the universe. Irregardless of the accuracy of this estimate, such considerations have contributed to the development of the concepts of preferred kinetic pathways and, in particular, nucleation sites for folding.

---

*The agreement of the equilibrium and kinetic properties of the protein with model compounds is very strong. There is a suggestion, however, of kinetic coupling between early stages of the protein folding and the isomerization process.

The best chemical evidence for nucleation sites comes from detailed studies of intermediates in the reoxidation of disulfide-linked proteins. Wetlaufer and his colleagues initially observed preferential formation of certain disulfides in the reoxidation of lysozyme and concluded that a nucleation-limited search had occurred (for review see Wetlaufer and Ristow, 1973). More recently this group has produced a detailed folding pathway for the mechanism of glutathione regeneration of lysozyme (Anderson and Wetlaufer, 1976). Under somewhat different conditions, however, there is evidence that no one of the four disulfide bonds of native lysozyme is obligatory in the formation of the other three native disulfide bonds (Acharya and Taniuchi, 1977). This result, which suggests a more "random" pathway, is in accord with previous results for lysozyme (Bradshaw *et al.*, 1967) and ribonuclease (Hantgan *et al.*, 1974; Creighton, 1977*b*). In these studies it has been found that there is rapid formation of a large number of incorrectly paired disulfide bridges and that the slow attainment of biological activity is due to the repeated breaking and reforming of these bonds until the correctly paired disulfide-linked structure is formed.

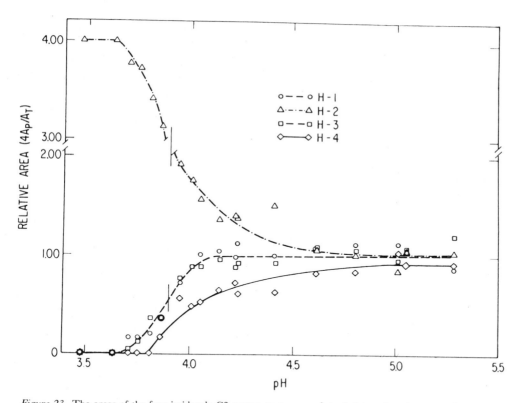

Figure 23. The areas of the four imidazole C2 proton resonances of staphylococcal nuclease as a function of pH. The changes of areas, which are measures of relative concentrations in this fast-exchange process, are very similar for three of the histidine residues (the resonances collapse on acidification to be identical to that of H-2). The histidine residue whose resonance is H-4 has a different pH dependence. From Epstein *et al.* (1971*b*) with permission.

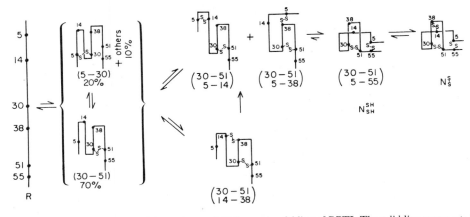

Figure 24. Schematic diagram of the pathway of folding and unfolding of BPTI. The solid line represents the polypeptide backbone, with the positions of the cysteine residues indicated. R is the fully reduced BPTI with no disulfide bonds. The configurations of species N_{SH}^{SH} and N_S^S approximate the conformation of the native inhibitor; those of the others are relatively arbitrary, except for the relative positions of the cysteine residues involved in disulfide bonds. The brackets around the single-disulfide intermediates indicate that they are in rapid equilibrium; only the two most predominant species are depicted. From Creighton (1977a) with permission.

The most completely and elegantly studied protein with respect to trapping and characterizing of disulfide intermediates is bovine pancreatic trypsin inhibitor (BPTI), a protein of 6500 molecular weight. Creighton has worked out in great detail the mechanism of folding of this protein (for review see Creighton, 1978). Figure 24 shows a summary of this pathway. The salient feature of this pathway is that a number of one-disulfide-linked intermediates, in rapid equilibrium with each other, are quickly formed. The one with the correct sulfhydryl pairing, residue 30 to residue 51, is predominant. These species rearrange to form two two-disulfide intermediates, each with an incorrect pairing, The two-disulfide species with correct pairing formed at this stage seems, surprisingly, to be an abortive form and not a true intermediate. The other two species rearrange to give the correct two-disulfide species and then the fully and correctly formed three-disulfide or native BPTI. These results, which are far from what would be expected from a "random" search, also indicate that the slowest step is very late in the pathway, after many compact but only partly folded species have formed. It is assumed that the polypeptide collapses to a globular arrangement, due to hydrophobic interactions, and this allows the formation of the particular disulfide bonds that characterize the pathway and their slow rearrangement as noncovalent and covalent bonds are corrected.

Since there are intrinsic difficulties in chemical quenching methods, there is reason to believe that the detailed results will be dependent on the experimental conditions and it would be premature to generalize too greatly from these comprehensive studies. It seems now, however, that preferential, but perhaps parallel and alternative pathways of folding will be established with respect to the formation of the disulfide linkages in the folding of proteins.

6.3   Protein Fragments

Chemical studies on proteins without disulfide bonds have also relied on the study of complementing fragments. This work is based on the original finding that subtilisin will cleave ribonuclease A into two fragments (the large S-protein and the small S-peptide), which will combine noncovalently to give a fully active enzyme, ribonuclease S (Richards and Vithayathil, 1959; Richards and Wyckoff, 1971). The chemical basis of this interaction has been studied extensively using S-peptide analogues and has been attributed to particular noncovalent interactions. The energetics of this interaction have been studied experimentally and theoretically (especially by measurements of solvent accessibility) (Richards, 1977) and the structure of the complex has been solved to high resolution (Richards *et al.,* 1971).

Analogous studies have been done with staphylococcal nuclease, a protein which can be cleaved into several fragment systems which combine to give enzymically active complexes (Fig. 25) (Taniuchi and Anfinsen, 1969, 1971). These systems are particularly interesting because the fragments may overlap with respect to primary structure, yet apparently the extra sequence does not interfere with folding and because the fragments may be used in several ways to form the ordered structure (Fig. 26) (Taniuchi and Anfinsen, 1971; Andria *et al.,* 1971; Taniuchi *et al.,* 1977). Taniuchi's group has studied the energetics of these fragment interactions in great detail and have, in addition, established a similar fragment system with cytochrome *c*, including a heme-containing fragment (Parr *et al.,* 1978).

An approach that differs from the use of complementing fragments has been the use of antibodies to study the conformation of these fragments. It was observed that each of these fragments alone in solution is essentially devoid of ordered structure, as tested by the usual spectroscopic and hydrodynamic methods; when added together, however, they complement to give ordered, active structures (Anfinsen *et al.,* 1971*a,b*). Evidence that these fragments have a small, but detectable, occupancy of folded forms was obtained from measuring the immunological cross-reactivity of the fragments with antibodies produced to the native protein (for reviews see Anfinsen and Scheraga, 1975; Schechter, 1976). The antibodies made to the entire protein were fractionated sequentially by immunoadsorption on columns of agarose to which peptides were attached so as to obtain antibody subpopulations specific for small regions of the surface of the protein. These subpopulations are nonprecipitating; branching lattices cannot form; and they allow formation of soluble complexes with the nuclease (Sachs *et al.,* 1972*a*). The equilibrium and kinetic parameters of the interactions of these monospecific antibodies with the native staphylococcal nuclease were measured (Sachs *et al.,* 1972*b*) and then the extent to which the fragments could compete

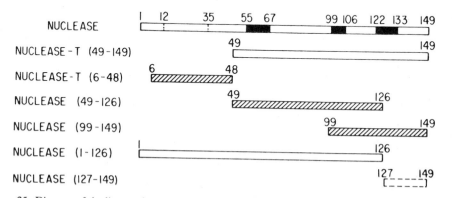

Figure 25. Diagram of the linear relationship of the amino acid sequence of nuclease and its fragments which yield productive complementation. The numbers indicate residues. The X-ray crystallographic study (Arnone *et al.,* 1971) has shown β structures between residues 12–35 and three α-helical parts (black bars). Nuclease (127–149) does not bind to nuclease (1–126). From Andria *et al.* (1971) with permission.

KENNETH A.
THOMAS and ALAN N.
SCHECHTER

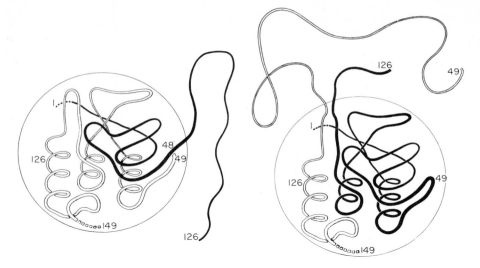

Figure 26. Schematic representation of the two types of complementation formed by noncovalent interactions of nuclease (1–126) (dark line) and nuclease (49–149) (open line). The ordered structures are surrounded by the circles. From Taniuchi and Anfinsen (1971) with permission.

with the native protein for binding to the antibody subpopulation was ascertained (Sachs *et al.*, 1972c). It was found that the fragments could compete, but would usually require several thousandfold molar excess over protein, to displace 50% of the protein. These molar ratios, or actually relative affinity constants, were used to define a K_{conf} which is a measure of the folding of any fragment in solution relative to the folding of that region of the protein in its native state (Sachs *et al.*, 1972c). This analysis is in many ways analogous to the Monod–Wyman–Changeux model for allosteric interactions (Monod *et al.*, 1965) in that it assumes the existence of an equilibrium between two conformational states, one of which has a higher affinity for the ligand. In particular, it assumes that each fragment exists in an equilibrium between nativelike conformation and unfolded ("random") conformations (Fig. 27), and that antibodies made to the native protein and then fractionated to be specific to a surface region recognize efficiently only the native conformation.* By these methods it was found, for example, that the fragment nuclease (99–149) had a K_{conf} of 1.9×10^{-4}, equivalent to about 0.02% of the molecules being folded at any one time (Sachs *et al.*, 1972c).

A very interesting staphylococcal nuclease fragment that was studied is nuclease (1–126). Although disordered in solution, this fragment has about 0.1% of the activity of the native protein; however, it was difficult to ascertain if this represented contamination with the intact protein. Antibodies to the native protein, purified to be specific to region (127–149) were found to inactivate nuclease but to have no effect on the activity of nuclease (1–126) (Sachs *et al.*, 1974). This suggests that tbe activity is intrinsic to the fragment and raises the possibility that the loss of relative activity is related to loss of stabilizing bonds

---

*Strictly speaking, nativeness refers to antigenic properties and not necessarily other structural features of the peptide. In view of the usual cooperativity of structural transitions (discussed in Section 6.1) it is likely that there is good concordance between different methods and that the assumption that this is a two-state system is a good one.

for the folded structure and not due to loss of active-site residues. Decreased activity would then be a result of decreased occupancy of a folded state by this protein fragment derivative. A similar model has been used to explain the very early appearance of enzymic activity in the refolding of reduced ribonuclease A (Garel, 1978).

A series of experiments of the reverse type were also done in which antibodies were made to a fragment of nuclease, nuclease (99–149) in particular, and then purified to monospecificity (Furie *et al.*, 1974). An equilibrium binding assay of these antibodies for the fragment was established. It was found that the native staphylococcal nuclease could compete for this binding but that it took a 2900-fold molar excess to displace one-half of the radiolabeled fragment from the antibody at 25°C, a K_{conf} of 2.9 $\times$ 10^3. At 4°C this value was 3.9 $\times$ 10^3 and at 39°C was 4.0 $\times$ 10^2; addition of the stabilizing ligands calcium ions and thymidine diphosphate raised K_{conf} by more than an order of magnitude (Furie *et al.*, 1973). These results are consistent with the idea that the native protein can unfold partially and, in this state, react effectively with the antibodies to the unfolded fragment. These ideas are thus very comparable to those of protein fluctuations or dynamics discussed above.

These immunological methods are subject to the validity of the above indicated assumptions but are internally consistent and are useful for detecting and quantitating conformational states which cannot be studied by the usual spectroscopic and chemical methods. These approaches have recently been applied to myoglobin (Hurrell *et al.*, 1977), albumin (Teale and Benjamin, 1977), and ribonuclease (Chavez and Scheraga, 1979). The ordered structures detectable in fragments immunologically have been suggested to correspond to nucleation sites in the complementation of fragments and in the folding of proteins (Anfinsen and Scheraga, 1975).

6.4   Theoretical Aspects

The concept of nucleation-controlled folding mechanisms has been alluded to several times in this chapter. This concept comes from polymer chemistry in which reaction rate processes may be analyzed in terms of separate rate-limiting nucleation and subsequent

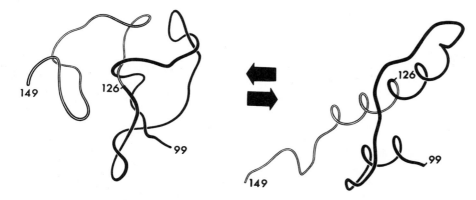

Figure 27. A diagram of the peptide, nuclease (99–149), shown in equilibrium between one unfolded conformation (on left) and a conformation similar to that in the native protein (on right). The dark line represents the region 99–126 to which a population of antibodies was purified so as to measure the K_{conf} of this region. From Sachs *et al.* (1972c).

growth phases (Flory, 1953). From the time it was realized that "random" folding would be too slow for a long polypeptide, nucleation mechanisms have been considered. Figure 28 shows an idea of this type in which it is suggested that local regions of a protein, perhaps corresponding to helices or sequentially adjacent β strands or β sheets, fold up rapidly and then diffuse together to form the beginning of a folded, globular protein (Anfinsen, 1973). Such a mechanism has been put on a more quantitative basis in a "diffusion-collision" analysis (Karplus and Weaver, 1976).

Another major approach to the folding problem is the use of conformational energy calculations to study the mechanism and predict the end state of the folding process. Despite the enormous difficulty of the computations involved and the approximations of the energy functions used, this approach has made significant progress recently, but we are still far from an *ab initio* prediction of protein structure (for review see Anfinsen and Scheraga, 1975; Némethy and Scheraga, 1977). Simplified approaches, using empirical predictions of protein conformation based on the relative appearances of particular residues in different secondary structures of known proteins, have also been responsible for some progress (for review see Chou and Fasman, 1978), but there is controversy as to how good they really are. It is unlikely that such methods will go beyond the secondary structure level to considerations of tertiary and higher structure or of interesting regions of proteins such as those around active sites.

Computer simulation of protein folding has been of considerable interest since the original studies (Levinthal, 1966, 1968) and its more recent revival (Levitt and Warshel, 1975). These approaches have significant problems (Hagler and Honig, 1978) as many simplifying assumptions are necessary, however, and a formidable amount remains to be done. One of the many functions of the computer work has been to make major improve-

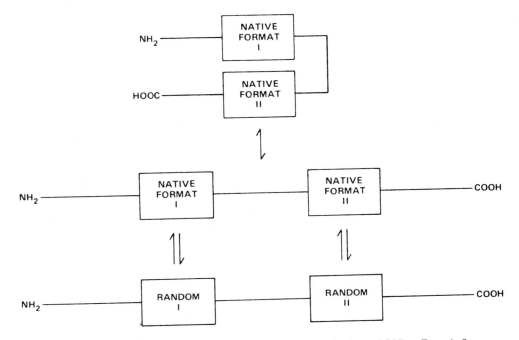

Figure 28. Schematic view of how protein chains might undergo nucleation and folding. From Anfinsen (1973) with permission.

ments in the representation of protein structures (Schulz and Schirmer, 1979). This has allowed perceptions of the topological aspects of protein structure discussed in the first sections of this article and has refined the nature of kinetic questions posed in the folding studies discussed in this section.

7 Conclusions

We have tried to recapitulate some of the recent advances in the understanding of stabilizing interactions, structural components, tertiary and quaternary structure, and the mechanism of folding of proteins. It is clear that knowledge of the atomic anatomy of proteins is in an exponentially growing phase and that studies of the mechanism of folding are also making major progress. Indeed, the rapidly developing field of protein dynamics, or the study of microscopic structural fluctuations, promises significant new understanding from the merger of what have previously been two separate approaches. Progress in understanding the basic physical chemistry of stabilizing forces and of interactions with solvent, especially water, has been less clear.

In any case, these studies of the structural and chemical bases of the folding of proteins into their biologically active structures—complemented by similar work on nucleic acids—opens a new level of description for probing the mechanisms of the flow of information in the cell that we call molecular biology. In addition, our understanding of the evolutionary relationships that exist among prokaryotic and eukaryotic organisms has been given new depth by the fact that we can now make comparisons among three-dimensional structures and perceive relationships not apparent in the linear sequence of the amino acids.

ACKNOWLEDGMENTS

K. A. T. thanks Terri Thomas for typing and proofreading early drafts and Ralph Bradshaw for providing support during the completion of this chapter.

References

Acharya, A. S., and Taniuchi, H., 1977, Formation of the four isomers of hen egg white lysozyme containing three native disulfide bonds and one open disulfide bond, *Proc. Natl. Acad. Sci. U.S.A.* **74**:2362.

Adams, M. J., Liljas, A., and Rossmann, M. G., 1973, Functional anion binding sites in dogfish M_4 lactate dehydrogenase, *J. Mol. Biol.* **76**:519.

Adman, E. T., Sieker, L. C., and Jensen, L. H., 1973, The structure of a bacterial ferredoxin, *J. Biol. Chem.* **248**:3987.

Alff-Steinberger, C., 1969, The genetic code and error transmission, *Proc. Natl. Acad. Sci. U.S.A.* **64**:584.

Allerhand, A., Doddrell, D., Glushko, V., Cochran, D., Wenkert, E., Lawson, P., and Gurd, F., 1971, Conformation and segmental motion of native and denatured ribonuclease A in solution: Application of natural abundance carbon-13 partially relaxed Fourier transform nuclear magnetic resonance, *J. Am. Chem. Soc.* **93**:544.

Anderson, L., 1973, Intermediate structure of normal human haemoglobin: Methaemoglobin in the deoxy quaternary conformation, *J. Mol. Biol.* **79**:495.

KENNETH A.
THOMAS and ALAN N.
SCHECHTER

Anderson, L., 1975, Structures of deoxy and carbonmonoxy haemoglobin Kansas in the deoxy quaternary structure, *J. Mol. Biol.* **94**:33.

Anderson, W. F., and Steitz, T. A., 1975, Structure of yeast hexokinase. IV. Low-resolution structure of enzyme–substrate complexes revealing negative co-operativity and allosteric interactions, *J. Mol. Biol.* **92**:279.

Anderson, W. L., and Wetlaufer, D. B., 1976, The folding of reduced lysozyme, *J. Biol. Chem.* **251**:3147.

Andria, G., Taniuchi, H., and Cone, J. L., 1971, The specific binding of three fragments of staphylococcal nuclease, *J. Biol. Chem.* **246**:7421.

Anfinsen, C. B., 1973, Principles that govern the folding of protein chains, *Science* **181**:223.

Anfinsen, C. B., and Haber, E., 1961, Studies on the reduction and reformation of protein disulfide bonds, *J. Biol. Chem.* **236**:1361.

Anfinsen, C. B., and Scheraga, H. A., 1975, Experimental and theoretical aspects of protein folding, *Adv. Protein Chem.* **29**:205–300.

Anfinsen, C. B., Cuatrecasas, P., and Taniuchi, H., 1971*a*, Staphylococcal nuclease, chemical properties and catalysis, in: *The Enzymes* (P. D. Boyer, ed.), 3rd ed., Vol. 4, pp. 177–204, Academic Press, New York.

Anfinsen, C. B., Schechter, A. N., and Tanuichi, H., 1971*b*, Some aspects of the structure of staphylococcal nuclease: Part II. Studies in solution, *Cold Spring Harbor Symp. Quant. Biol.* **36**:249.

Arnone, A., 1972, X-ray diffraction study of binding of 2,3-diphosphoglycerate to human deoxyhaemoglobin, *Nature (London)* **237**:146.

Arnone, A., and Perutz, M. F., 1974, Structure of inositol hexaphosphate–human deoxyhaemoglobin complex, *Nature (London)* **249**:34.

Arnone, A., Bier, C. J., Cotton, F. A., Day, V. W., Hazen, E. E., Richardson, D. C., Richardson, J. S., and Yonath, A., 1971, A high resolution structure of an inhibitor complex of the extracellular nuclease of *Staphylococcus aureus*, *J. Biol. Chem.* **246**:2302.

Baldwin, R. L., 1975, Intermediates in protein folding reactions and the mechanism of protein folding, *Annu. Rev. Biochem.* **44**:453.

Baldwin, R. L., 1978, The pathway of protein folding, *Trends Biochem. Sci.* **3**:66.

Banaszak, L. J., and Bradshaw, R. A., 1975, Malate dehydrogenases, in: *The Enzymes* (P. D. Boyer, ed.), 3rd ed., Vol. 11, pp. 369–396, Academic Press, New York.

Banner, D. W., Bloomer, A. C., Petsko, G. A., Phillips, D. C., Pogson, C. I., Wilson, I. A., Corran, P. H., Furth, A. J., Milman, J. O., Offord, R. E., Priddle, J. D., and Waley, S. G., 1975, Structure of chicken muscle triose phosphate isomerase determined crystallographically at 2.5 Å resolution using amino acid sequence data, *Nature (London)* **255**:609.

Banyard, S. H., Blake, C. C. F., and Swan, I. D. A., 1974, The high resolution X-ray study of human lysozyme: A preliminary analysis, in: *Lysozyme* (E. F. Osserman, R. E. Canfield, and S. Beychok, eds.), pp. 71–79, Academic Press, New York.

Banyard, S. H., Stammers, D. K., and Harrison, P. M., 1978, Electron density map of apoferritin at 2.8 Å resolution, *Nature (London)* **271**:282.

Bedarker, S., Turnell, W. G., Blundell, T. L., and Schwabe, C., 1977, Relaxin has conformational homology with insulin, *Nature (London)* **270**:449.

Bentley, G., Dodson, E., Dodson, G., Hodgkin, D., and Mercola, D., 1976, Structure of insulin in 4-zinc insulin, *Nature (London)* **261**:166.

Biesecker, G., Harris, J. I., Thierry, J. C., Walker, J. E., and Wonacott, A. J., 1977, Sequence and structure of D-glyceraldehyde 3-phosphate dehydrogenase from *Bacillus stearothermophilus*, *Nature (London)* **266**:328.

Birktoft, J. J., and Blow, D. M., 1972, Structure of crystalline α-chymotrypsin. V. The atomic structure of tosyl-α-chymotrypsin at 2 Å resolution, *J. Mol. Biol.* **68**:187.

Birktoft, J. J., Kraut, J., and Freer, S. T., 1976, A detailed structural comparison between the charge relay system in chymotrypsinogen and in α-chymotrypsin, *Biochemistry* **15**:4481.

Bjork, I., Karlsson, F. A., and Berggard, I., 1971, Independent folding of the variable and constant halves of a lamda immunoglobulin light chain, *Proc. Natl. Acad. Sci. U.S.A.* **68**:1707.

Blake, C. C. F., and Evans, P. R., 1974, Structure of horse muscle phosphoglycerate kinase, *J. Mol. Biol.* **84**:585.

Blake, C. C. F., Geisow, M. J., Swan, I. D. A., Rerat, C., and Rerat, B., 1974, Structure of human plasma prealbumin at 2.5 Å resolution, *J. Mol. Biol.* **88**:1.

Blout, E. R., Doty, P., and Yang, J. T., 1957, Polypeptides. XII. The optical rotation and configuration stability of α-helices, *J. Am. Chem. Soc.* **79**:749.

Blow, D. M., 1971, The structure of chymotrypsin, in: *The Enzymes* (P. D. Boyer, Ed.), 3rd ed., Vol. 3, pp. 185–212, Academic Press, New York.

Blundell, T. L., and Johnson, L. N., 1976, *Protein Crystallography*, pp. 18–58, Academic Press, New York.

Blundell, T., Dodson, G., Hodgkin, D., and Mercola, D., 1972, Insulin: The structure in the crystal and its reflection in chemistry and biology, *Adv. Protein Chem.* **26**:279–402.

Bode, W., and Schwager, P., 1975, The refined crystal structure of bovine-β-trypsin at 1.8 Å resolution. II. Crystallographic refinement, calcium binding site, benzamidine binding site and active site at pH 7.0, *J. Mol. Biol.* **98**:693.

Bonner, W. A., Van Dort, M. A., and Yearian, M. R., 1975, Asymmetric degradation of DL-leucine with longitudinally polarised electrons, *Nature (London)* **258**:419.

Bradshaw, R. A., Kanarek, L., and Hill, R. L., 1967, The preparation, properties, and reactivation of the mixed disulfide derivative of egg white lysozyme and L-cystine, *J. Biol. Chem.* **242**:3789.

Brandts, J. F., Halvorson, H. R., and Brennan, M., 1975, Consideration of the possibility that the slow step in protein denaturation reactions is due to cis–trans isomerism of proline residues, *Biochemistry* **14**:4953.

Brandts, J. F., Brennan, M., and Lin, L.-N., 1977, Unfolding and refolding occur much faster for a proline-free protein than for most proline-containing proteins, *Proc. Natl. Acad. Sci. U.S.A.* **74**:4178.

Brown, J. E., and Klee, W. A., 1969, Conformational studies of a series of overlapping peptides from ribonuclease and their relationship to the protein structure, *Biochemistry* **8**:2876.

Browne, W. J., North, A. C. T., Phillips, D. C., Brew, K., Vanaman, T. C., and Hill, R. L., 1969, A possible three-dimensional structure of bovine α-lactalbumin based on that of hen's egg-white lysozyme, *J. Mol. Biol.* **42**:65.

Buehner, M., Ford, G. C., Moras, D., Olsen, K. W., and Rossmann, M. G., 1974, Three-dimensional structure of D-glyceraldehyde-3-phosphate dehydrogenase, *J. Mol. Biol.* **90**:25.

Burnett, R. M., Darling, G. D., Kendall, D. S., LeQuesne, M. E., Mayhew, S. G., Smith, W. W., and Ludwig, M. L., 1974, The structure of the oxidized form of clostridial flavodoxin at 1.9 Å resolution, *J. Biol. Chem.* **249**:4383.

Campbell, I. D., Dobson, C. M., and Williams, R. J. P., 1975, Proton magnetic resonance studies of the tyrosine residues of hen lysozyme-assignment and detection of conformational mobility, *Proc. R. Soc. London, Ser. B* **189**:503.

Campbell, J. W., Watson, H. C., and Hodgson, G. I., 1974, Structure of yeast phosphoglycerate mutase, *Nature (London)* **250**:301.

Carter, C. W., Jr., Kraut, J., Freer, S. T., Xuong, N., Alden, R. A., and Bartsch, R. G., 1974, Two-angstrom crystal structure of oxidized chromatium high potential iron protein, *J. Biol. Chem.* **249**:4212.

Chavez, L. G., and Scheraga, H. A., 1979, Location of the antigenic determinants of bovine pancreatic ribonuclease, *Biochemistry* **18**:4386.

Chothia, C., 1973, Conformation of twisted β-pleated sheets in proteins, *J. Mol. Biol.* **75**:295.

Chothia, C., 1974, Hydrophobic bonding and accessible surface areas in proteins, *Nature (London)* **248**:338.

Chothia, C., 1975, Structural invariants in protein folding, *Nature (London)* **254**:304.

Chothia, C., 1976, The nature of the accessible and buried surfaces in proteins, *J. Mol. Biol.* **105**:1.

Chothia, C., and Janin, J., 1975, Principles of protein–protein recognition, *Nature (London)* **256**:705.

Chou, P. Y., and Fasman, G. D., 1974a, Conformational parameters for amino acids in helical, β-sheet, and random coil regions calculated from proteins, *Biochemistry* **13**:211.

Chou, P. Y., and Fasman, G. D., 1974b, Prediction of protein conformation, *Biochemistry* **13**:222.

Chou, P. Y., and Fasman, G. D., 1978, Empirical predictions of protein conformation, *Annu. Rev. Biochem.* **47**:251.

Colman, P. M., Jansonius, J. N., and Matthews, B. W., 1972, The structure of thermolysin: An electron density map at 2.3 Å resolution, *J. Mol. Biol.* **70**:701.

Cowan, P. M., and McGavin, S., 1955, Structure of poly-L-proline, *Nature (London)* **176**:501.

Crawford, J. L., Lipscomb, W. N., and Schellman, C. G., 1973, The reverse turn as a polypeptide conformation in globular proteins, *Proc. Natl. Acad. Sci. U.S.A.* **70**:538.

Creighton, T. E., 1977a, Conformational restrictions on the pathway of folding and unfolding of the pancreatic trypsin inhibitor, *J. Mol. Biol.* **113**:275.

Creighton, T. E., 1977b, Kinetics of refolding of reduced ribonuclease, *J. Mol. Biol.* **113**:329.

Creighton, T. E., 1978, Experimental studies of protein folding and unfolding, *Prog. Biophys. Mol. Biol.* **33**:231.

Crestfield, A. M., and Fruchter, R. G., 1967, The homologous and hybrid dimers of ribonuclease A and its carboxymethylhistidine derivatives, *J. Biol. Chem.* **242**:3279.

KENNETH A.
THOMAS and ALAN N.
SCHECHTER

Davies, D. R., Padlan, E. A., and Segel, D. M., 1975, Three-dimensional structure of immunoglobulins, *Annu. Rev. Biochem.* **44**:639.

Dayhoff, M. O., Hunt, L. T., McLaughlin, P. J., and Jones, D. D., 1972, Gene duplications in evolution: The globins, in: *Atlas of Protein Sequences and Structure* (M. O. Dayhoff, ed.), Vol. 5, pp. 17–30, The National Biomedical Research Foundation, Silver Spring, Maryland.

Deatherage, J. F., Loe, R. S., Anderson, C. M., and Moffat, K., 1976, Structure of cyanide methaemoglobin, *J. Mol. Biol.* **104**:687.

Deisenhofer, J., and Steigemann, W., 1975, Crystallographic refinement of the structure of bovine pancreatic trypsin inhibitor of 1.5 Å resolution, *Acta Cryst.* **B31**:238.

Deisenhofer, J., Colman, P. M., Huber, R., Haupt, H., and Schwick, G., 1976, Crystallographic structural studies of a human Fc-fragment. I. An electron-density map at 4 Å resolution and a partial model, *Hoppe-Seyler's Z. Physiol. Chem.* **357**:435.

Delbaere, L. T. J., Hutcheon, W. L. B., James, M. N. G., and Thiessen, W. E., 1975, Tertiary structure differences between microbial and pancreatic serine enzymes, *Nature (London)* **257**:758.

Dickerson, R. E., 1971, Sequence and structure homologies in bacterial and mammalian-type cytochromes, *J. Mol. Biol.* **57**:1.

Dickerson, R. E., and Geis, I., 1969, *The Structure and Action of Proteins,* pp. 24–34, W. A. Benjamin, Menlo Park, California.

Dickerson, R. E., and Timkovich, R., 1975, Cytochromes c, in: *The Enzymes* (P. D. Boyer, ed.), 3rd ed., Vol. 11, pp. 397–547, Academic Press, New York.

Dickerson, R. E., Takano, T., Eisenberg, D., Kallai, O. B., Samson, L., Cooper, A., and Margoliash, E., 1971, Ferricytochrome c, *J. Biol. Chem.* **246**:1511.

Dickerson, R. E., Timkovitch, R., and Almassy, R. J., 1976, The cytochrome fold and the evolution of bacterial energy metabolism, *J. Mol. Biol.* **100**:473.

Drenth, J., Jansonius, J. N., Koekoek, R., and Wolthers, B. G., 1971, The structure of papain, *Adv. Protein Chem.* **25**:79–115.

Eklund, H., Nordström, B., Zeppezauer, E., Söderlund, G., Ohlsson, I., Boiwe, T., Söderberg, B., Tapia, O., and Brändén, C., 1976a, Three-dimensional structure of horse liver alcohol dehydrogenase at 2.4 Å resolution, *J. Mol. Biol.* **102**:27.

Eklund, H. C., Brändén, C., and Jörnvall, H., 1976b, Structural comparisons of mammalian, yeast and bacillar alcohol dehydrogenases, *J. Mol. Biol.* **102**:61.

Epp, O., Lattman, E. E., Schiffer, M., Huber, R., and Palm, W., 1975, The molecular structure of a dimer composed of the variable portions of the Bence–Jones protein REI refined at 2.0 Å resolution, *Biochemistry* **14**:4943.

Epstein, C. J., Goldberger, R. F., and Anfinsen, C. B., 1963, Genetic control of tertiary protein structure: Studies with model systems, *Cold Spring Harbor Symp. Quant. Biol.* **28**:439.

Epstein, H. F., Schechter, A. N., Chen, R. F., and Anfinsen, C. B., 1971a, The folding of staphylococcal nuclease: Kinetic studies on two processes in acid renaturation, *J. Mol. Biol.* **60**:499.

Epstein, H. F., Schechter, A. N., and Cohen, J. S., 1971b, Folding of staphylococcal nuclease: Magnetic resonance and fluorescence studies of individual residues, *Proc. Natl. Acad. Sci. U.S.A.* **68**:2042.

Evans, D. R., Warren, S. G., Edwards, B. F. P., McMurray, C. H., Bethge, P. H., Wiley, D. C., and Lipscomb, W. N., 1973, Aqueous central cavity in aspartate transcarbamylase from *E. coli, Science* **179**:683.

Fehlhammer, H., Bode, W., and Huber, R., 1977, Crystal structure of bovine trypsinogen at 1.8 Å resolution. II. Crystallographic refinement, refined crystal structure and comparison with bovine trypsin, *J. Mol. Biol.* **111**:415.

Fenna, R. E., and Matthews, B. W., 1975, Chlorophyll arrangement in a bacteriochlorophyll protein from *Chlorobium limicola, Nature (London)* **258**:573.

Fermi, G., 1975, Three-dimensional Fourier synthesis of human deoxyhaemoglobin at 2.5 Å resolution: Refinement of the atomic model, *J. Mol. Biol.* **97**:237.

Finney, J. L., 1978, Volume occupation, environment and accessibility in proteins. Environment and molecular area of RNase-S, *J. Mol. Biol.* **119**:415.

Fitch, W. M., and Markowitz, E., 1970, An improved method for determining codon variability in a gene and its application to the rate of fixation of mutations in evolution, *Biochem. Genet.* **4**:579.

Fletterick, R. J., Bates, D. J., and Steitz, T. A., 1975, The structure of a yeast hexokinase monomer and its complexes with substrates at 2.7 Å resolution, *Proc. Natl. Acad. Sci. U.S.A.* **72**:38.

Fletterick, R. J., Sygusch, J., Semple, H., and Madsen, N. B., 1976, Structure of glycogen phosphorylase a at 3.0 Å resolution and its ligand binding sites at 6 Å, *J. Biol. Chem.* **251**:6142.

Flory, P. J., 1953, *Principles of Polymer Chemistry*, Cornell University Press, Ithaca, New York.

Freedman, R. B., and Hawkins, H. C., 1978, Enzyme-catalyzed disulfide interchange and protein biosynthesis, *Biochem. Soc. Trans.* **5**:348.

Frier, J. A., and Perutz, M. F., 1977, Structure of human foetal deoxyhaemoglobin, *J. Mol. Biol.* **112**:97.

Furie, B., Schechter, A. N., Sachs, D. H., and Anfinsen, C. B., 1973, An immunological approach to the conformational equilibrium of staphylococcal nuclease, *J. Mol. Biol.* **92**:497.

Furie, B., Sachs, D. H., Schechter, A. N., and Anfinsen, C. B., 1974, Antibodies to the unfolded form of a helix-rich region of staphylococcal nuclease, *Biochemistry* **13**:1561.

Garel, J.-R., 1978, Early steps in the refolding of reduced ribonuclease A, *J. Mol. Biol.* **118**:331.

Ginsburg, A., and Carroll, W. R., 1965, Some specific ion effects on the conformation and thermal stability of ribonuclease, *Biochemistry* **4**:2159.

Givol, D., DeLorenzo, F., Goldberger, R. F., and Anfinsen, C. B., 1965, Disulfide interchange and the three-dimensional structure of proteins, *Proc. Natl. Acad. Sci. U.S.A.* **53**:676.

Goldberger, R. F., Epstein, C. J., and Anfinsen, C. B., 1963, Acceleration of reactivation of reduced bovine pancreatic ribonuclease by a microsomal system from rat liver, *J. Biol. Chem.* **238**:628.

Gurd, F., and Rothgeb, T. M., 1979, Motions in proteins, *Adv. Protein Chem.* **33**:74.

Hagler, A. T., and Honig, B., 1978, On the formation of protein tertiary structure on a computer, *Proc. Natl. Acad. Sci. U.S.A.* **75**:554.

Hagler, A. T., and Moult, J., 1978, Computer simulation of the solvent structure around biological macromolecules, *Nature (London)* **272**:222.

Hantgan, R. R., Hammes, G. G., and Scheraga, H. A., 1974, Pathways of folding of reduced bovine pancreatic ribonuclease, *Biochemistry* **13**:3421.

Hartley, B. S., and Shotton, D. M., 1971, Pancreatic elastase, in: *The Enzymes* (P. D. Boyer, ed.), 3rd ed., Vol. 3, pp. 323–373, Academic Press, New York.

Hartsuck, J. A., and Lipscomb, W. N., 1971, Carboxypeptidase A, in: *The Enzymes* (P. D. Boyer, ed.), 3rd ed., Vol. 3, pp. 1–56, Academic Press, New York.

Hendrickson, W. A., and Love, W. E., 1971, Structure of lamprey haemoglobin, *Nature (London), New Biol.* **232**:197.

Hermans, J., 1966, The effects of protein denaturants on the stability of the α helix, *J. Am. Chem. Soc.* **88**:2418.

Hill, R. L., Delaney, R., Fellows, R. E., Jr., and Lebovitz, H. E., 1966, The evolutionary origins of the immunoglobulins, *Proc. Natl. Acad. Sci. U.S.A.* **56**:1763.

Hill, E., Tsernoglou, D., Webb, L., and Banaszak, L. J., 1972, Polypeptide conformation of cytoplasmic malate dehydrogenase from an electron density map at 3.0 Å resolution, *J. Mol. Biol.* **72**:577.

Holbrook, J. J., Liljas, A., Steindel, S. J., and Rossmann, M. G., 1975, Lactate dehydrogenase, in: *The Enzymes* (P. D. Boyer, ed.), 3rd ed., Vol 11, pp. 191–292, Academic Press, New York.

Huber, R., and Steigemann, W., 1974, Two *cis*-prolines in the Bence-Jones protein REI and the *cis*-pro-bend, *Fed. Eur. Biochem. Soc. Lett.* **48**:235.

Huber, R., Epp, O., Steigemann, W., and Formanek, H., 1971, The atomic structure of erythrocruorin in the light of the chemical sequence and its comparison with myoglobin, *Eur. J. Biochem.* **19**:42.

Huber, R., Kukla, D., Bode, W., Schwager, P., Bartels, K., Deisenhofer, J., and Steigemann, W., 1974, Structure of the complex formed by bovine trypsin and bovine pancreatic trypsin inhibitor. II. Crystallographic refinement at 1.9 Å resolution, *J. Mol. Biol.* **89**:73.

Huber, R., Bode, W., Kukla, D., Kohl, U., and Ryan, C. A., 1975, The structure of the complex formed by bovine trypsin and bovine pancreatic trypsin inhibitor. III. Structure of the anhydro–trypsin–inhibitor complex, *Biophys. Struct. Mech.* **1**:189.

Huber, R., Deisenhofer, J., Colman, P. M., Matsunhima, M., and Palm, W., 1976, Crystallographic structure studies of an IgG molecule and an Fc fragment, *Nature (London)* **264**:415.

Hurrell, J. G. R., Smith, J. H., and Leach, S. J., 1977, Immunological measurements of conformational motility in regions of the myoglobin molecules, *Biochemistry* **16**:175.

Ikai, H., Fish, W. W., and Tanford, C., 1973, Kinetics of unfolding and refolding of proteins: II. Results for cytochrome *c*, *J. Mol. Biol.* **73**:165.

Imoto, T., Johnson, L. N., North, A. C. T., Phillips, D. C., and Rupley, J. A., 1972, Vertebrate lysozymes, in: *The Enzymes* (P. D. Boyer, ed.), 3rd ed., Vol. 7, pp. 665–868, Academic Press, New York.

IUPAC–IUB Commission on Biochemical Nomenclature, 1970, Abbreviations and symbols for the description of the conformation of polypeptide chains, *J. Biol. Chem.* **245**:6489.

Janin, J., and Chothia, C., 1976, Stability and specificity of protein–protein interactions: The case of the trypsin–trypsin inhibitor complexes, *J. Mol. Biol.* **100**:197.

Jardetzky, O., Thielmann, H., Arata, Y., Markley, J. L., and Williams, M. N., 1971, Tentative sequential model for the unfolding and refolding of staphylococcal nuclease at high pH, *Cold Spring Harbor Symp. Quant. Biol.* **36**:257.

Johanin, G., and Kellershohn, N., 1972, An estimate of intraproteic electrostatic field values originated by the peptide groups in α-chymotrypsin, *Biochem. Biophys. Res. Commun.* **49**:321.

Kabat, E. A., Padlan, E. A., and Davies, D. R., 1975, Evolutionary and structural influences on light chain constant (C_L) region of human and mouse immunoglobulins, *Proc. Natl. Acad. Sci. U.S.A.* **72**:2785.

Karplus, M., and Weaver, D. C., 1976, Protein folding dynamics, *Nature (London)* **260**:404.

Kauzmann, W., 1959, Some factors in the interpretation of protein denaturation, *Adv. Protein Chem.* **14**:1–63.

Kendrew, J. C., Dickerson, R. E., Strandberg, B. E., Hart, R. G., Davies, D. R., Phillips, D. C., and Shore, U. C., 1960, Structure of myoglobin, *Nature (London)* **185**:422.

Kotelchuck, D., and Scheraga, H. A., 1968, The influence of short-range interactions on protein conformation. I. Side chain–backbone interactions within a single peptide unit, *Proc. Natl. Acad. Sci. U.S.A.* **61**:1163.

Kozitsyn, S. A., and Ptitsyn, O. B., 1974, The structure of hydrophobic cores of globins, *Mol. Biol.* **8**:427.

Kraut, J., 1971*a*, Chymotrypsinogen: X-ray structure, in: *The Enzymes* (P. D. Boyer, ed.), 3rd ed., Vol. 3, pp. 165–183, Academic Press, New York.

Kraut, J., 1971*b*, Subtilisin: X-ray structure, in: *The Enzymes* (P. D. Boyer, ed.), 3rd ed., Vol. 3, pp. 547–560, Academic Press, New York.

Kretsinger, R. H., 1972, Gene duplication in carp muscle calcium binding protein, *Nature (London), New Biol.* **240**:83.

Kretsinger, R. H., and Nockolds, C. E., 1973, Carp muscle calcium-binding protein, *J. Biol. Chem.* **248**:3313.

Kuntz, I. D., 1972, Protein folding, *J. Am. Chem. Soc.* **94**:4009.

Lapanje, S., 1978, *Physicochemical Aspects of Protein Denaturation*, Wiley, New York.

Leach, S. J., Némethy, G., and Scheraga, H. A., 1966*a*, Computation of the sterically allowed conformations of peptides, *Biopolymers* **4**:369.

Leach, S. J., Némethy, G., and Scheraga, H. A., 1966*b*, Intramolecular steric effects and hydrogen bonding in regular conformations of polyamino acids, *Biopolymers* **4**:887.

Lee, B., and Richards, F. M., 1971, The interpretation of protein structures: Estimation of static accessibility, *J. Mol. Biol.* **55**:379.

Levine, M., Muirhead, H., Stammers, D. K., and Stuart, D. I., 1978, Structure of pyruvate kinase and similarities with other enzymes: Possible implications for protein taxonomy and evolution, *Nature (London)* **271**:626.

Levinthal, C., 1966, Molecular model-building by computer, *Sci. Am.* **214**:42.

Levinthal, C., 1968, Are there pathways for protein folding? *J. Chim. Phys.* **65**:44.

Levitt, M., and Chothia, C., 1976, Structural patterns in globular proteins, *Nature (London)* **261**:552.

Levitt, M., and Warshel, A., 1975, Computer simulation of protein folding, *Nature (London)* **253**:694.

Liljas, A., and Rossmann, M. G., 1974, X-ray studies of protein interactions, *Annu. Rev. Biochem.* **43**:475–507.

Lindskog, S., Henderson, L. E., Kannon, K. K., Liljas, A., Nyman, P. O., and Strandberg, B., 1971, Carbonic anhydrase, in: *The Enzymes* (P. D. Boyer, ed.), 3rd. ed., Vol. 5, pp. 587–665, Academic Press, New York.

Love, W. E., Klock, P. A., Lattman, E. E., Padlan, E. A., Ward, K. B., Jr., and Hendrickson, W. A., 1972, The structures of lamprey and bloodworm haemoglobins in relation to their evolution and function, *Cold Spring Harbor Symp. Quant. Biol.* **36**:349.

Margoliash, E., 1972, The molecular variations of cytochrome c as a function of the evolution of species, *Harvey Lect.* **Series 66**:177.

Matthews, B. W., 1972, The γ turn. Evidence for a new folded conformation in proteins, *Macromolecules* **5**:818.

Matthews, B. W., and Remington, S. J., 1974, The three-dimensional structure of the lysozyme from bacteriophage T4, *Proc. Natl. Acad. Sci. U.S.A.* **71**:4178.

Matthews, B. W., Jansonius, J. N., Colman, P. M., Schoenborn, B. P., and Dupougue, D., 1972, Three-dimensional structure of thermolysin, *Nature (London), New Biol.* **238**:37.

Matthews, B. W., Weaver, L. H., and Kester, W. R., 1974, The conformation of thermolysin, *J. Biol. Chem.* **249**:8030.

Matthews, D. A., Alden, R. A., Birktoft, J. J., Freer, S. T., and Kraut, J., 1977, Re-examination of the charge relay system in subtilisin and comparison with other serine proteases, *J. Biol. Chem.* **252**:8875.

Mavridis, A., Tulinsky, A., and Liebman, M. N., 1974, Asymmetrical changes in the tertiary structure of α-chymotrypsin with change in pH, *Biochemistry* **13**:3661.

Moews, P. C., and Kretsinger, R. H., 1975, Refinement of the structure of carp muscle calcium-binding parvalbumin by model building and difference Fourier analysis, *J. Mol. Biol.* **91**:201.

Monod, J., Wyman, J., and Changeux. J. P., 1965, On the nature of allosteric transitions: A plausible model, *J. Mol. Biol.* **12**:88.

Moras, D., Olsen, K. W., Sabesan, M. N., Buehner, M., Ford, G. C., and Rossmann, M. G., 1975, Studies of asymmetry in the three-dimensional structure of lobster D-glyceraldehyde-3-phosphate dehydrogenase. *J. Biol. Chem.* **250**:9137.

Morgan, R. S., Gushard, R. H., Carpenter, K. L., and Chalfin, S., 1976, Chains of alternating S- and π-bonded atoms in proteins, in: *American Crystallographic Association Winter Meeting, Program and Abstracts*, Vol. 4, Ser. 2, p. 15.

Moult, J., Yonath, A., Traub, W., Smilansky, A., Podjarny, A., Rabinovich, D., and Saya, A., 1976, The structure of triclinic lysozyme at 2.5 Å resolution, *J. Mol. Biol.* **100**:179.

Nagano, K., 1977, Logical analysis of the mechanism of protein folding. IV. Super-secondary structures, *J. Mol. Biol.* **109**:235.

Némethy, G., and Printz, M. P., 1972, The γ turn, a possible folded conformation of the polypeptide chain, *Macromolecules* **5**:755.

Némethy, G., and Scheraga, H. A., 1977, Protein folding, *Quant. Rev. Biophys.* **10**:239.

Némethy, G., Steinberg, J. Z., and Scheraga, H. A., 1963, Influence of water structure and of hydrophobic interactions on the strength of side-chain hydrogen bonds in proteins, *Biopolymers* **1**:43.

Némethy, G., Leach, S. J., and Scheraga, H. A., 1966, The influence of amino acid side chains on the free energy of helix-coil transitions, *J. Phys. Chem.* **70**:998.

Némethy, G., Phillips, D. C., Leach, S. J., and Scheraga, H. A., 1967, A second right-handed helical structure with the parameters of the Pauling–Corey α-helix, *Nature (London)* **214**:363.

Norden, B., 1977, Was photoresolution of amino acids the origin of optical activity in life? *Nature (London)* **266**:567.

Ohlsson, I. B., Nordström, B., and Brändén, C., 1974, Structure and functional similarities within the coenzyme binding domains of dehydrogenases, *J. Mol. Biol.* **89**:339.

Padlan, E. A., and Love, W. E., 1974, Three-dimensional structure of hemoglobin from the polychaete annelid, *Glycera dibranchiata*, at 2.5 Å resolution, *J. Biol. Chem.* **249**:4067.

Parr, G. R., Hantgan, R. R., and Taniuchi, H., 1978, Formation of two alternative complementing structures from a cytochrome *c* fragment (residues 1 to 38) and the apoprotein, *J. Biol. Chem.* **253**:5381.

Pauling, L., and Corey, R. B., 1951*a*, Atomic coordinates and structure factors for two helical configurations of polypeptide chains, *Proc. Natl. Acad. Sci. U.S.A.* **37**:235.

Pauling, L., and Corey, R. B., 1951*b*, Configurations of polypeptide chains with favored orientations around single bonds: Two new pleated sheets, *Proc. Natl. Acad. Sci. U.S.A.* **37**:729.

Perutz, M. F., 1974, Mechanism of denaturation of haemoglobin by alkali, *Nature (London)* **247**:341.

Perutz, M. F., and Lehmann, H., 1968, Molecular pathology of human haemoglobin, *Nature (London)* **219**:902.

Perutz, M. F., and Raidt, H., 1975, Stereochemical basis of heat stability in bacterial ferredoxins and in haemoglobin A2, *Nature (London)* **255**:256.

Perutz, M. F., and TenEyck, L. F., 1972, Stereochemistry of cooperative effects in hemoglobin, *Cold Spring Harbor Symp. Quant. Biol.* **36**:295.

Perutz, M. F., Muirhead, H., Cox, J. M., and Goaman, L. C. G., 1968, Three-dimensional Fourier synthesis of horse oxyhaemoglobin at 2.8 Å resolution: The atomic model, *Nature (London)* **219**:131.

Ploegman, J. H., Drent, G., Kalk, K. H., Hol, W. G. J., Heinrikson, R. L., Keim, P., Weng, L., and Russell, J., 1978, The covalent and tertiary structure of bovine liver rhodanese, *Nature (London)* **273**:124.

Poljak, R. J., Amzel, L. M., Avey, H. P., Chen, B. L., Phizackerley, R. P., and Saul, F., 1973, Three-dimensional structure of the Fab' fragment of a human immunoglobulin at 2.8 Å resolution, *Proc. Natl. Acad. Sci. U.S.A.* **70**:3305.

Pullman, B., and Pullman, A., 1974, Molecular orbital calculations on the conformation of amino acid residues of proteins, *Adv. Protein Chem.* **28**:348–526.

Pulsinelli, P. D., Perutz, M. F., and Nagel, R. L., 1973, Structure of hemoglobin M Boston, a variant with five-coordinated ferric heme, *Proc. Natl. Acad. Sci. U.S.A.* **70**:3870.

Quiocho, F. A., and Lipscomb, W. N., 1971, Carboxypeptidase A: A protein and an enzyme, *Adv. Protein Chem.* **25**:1–78.

Ramachandran, G. N., and Mitra, A. K., 1976, An explanation for the rare occurrence of *cis* peptide units in proteins and polypeptides, *J. Mol. Biol.* **107**:85.

Ramachandran, G. N., and Sasisekharan, V., 1968, Conformation of polypeptides and proteins, *Adv. Protein Chem.* **23**:283–437.

Rao, S. T., and Rossmann, M. G., 1973, Comparison of supersecondary structures in proteins, *J. Mol. Biol.* **76**:241.

Reeke, G. N., Jr., Becker, J. W., and Edelman, G. M., 1975, The covalent and three-dimensional structure of concanavalin A. IV. Atomic coordinates, hydrogen bonding and quaternary structure, *J. Biol. Chem.* **250**:1525.

Reeke, G. N., Jr., Becker, J. W., and Edelman, G. M., 1978, Changes in the three-dimensional structure of concanavalin A upon demetalization, *Proc. Natl. Acad. Sci. U.S.A.* **75**:2286.

Remington, S. J., Anderson, W. F., Owen, J., TenEyck, L. F., Grainger, C. T., and Matthews, B. W., 1978, Structure of the lysozyme from bacteriophage T4: An electron density map at 2.4 Å resolution, *J. Mol. Biol.* **118**:81.

Richards, F. M., 1974, The interpretation of protein structures: Total volume, group volume distributions and packing density, *J. Mol. Biol.* **82**:1.

Richards, F. M., 1977, Areas, volumes, packing and protein structure, *Annu. Rev. Biophys. Bioeng.* **6**:151.

Richards, F. M., and Vithayathil, P. J., 1959, The preparation of subtilisin-modified ribonuclease and the separation of the peptide and protein components, *J. Biol. Chem.* **234**:1459.

Richards, F. M., and Wyckoff, H. W., 1971, Bovine pancreatic ribonuclease, in: *The Enzymes* (P. D. Boyer, ed.), 3rd ed., Vol. 4, pp. 647–806, Academic Press, New York.

Richards, F. M., Wyckoff, H. W., Carlson, W. D., Allewell, N. M., Lee, B., and Mitsui, Y., 1971, Protein structure, ribonuclease-S and nucleotide interactions, *Cold Spring Harbor Symp. Quant. Biol.* **36**:35.

Richardson, J. S., 1976, Handedness of crossover connections in β sheets, *Proc. Natl. Acad. Sci. U.S.A.* **73**:2619.

Richardson, J. S., 1977, β-Sheet topology and the relatedness of proteins, *Nature (London)* **268**:495.

Richardson, J. S., Thomas, K. A., Rubin, B. H., and Richardson, D. C., 1975, Crystal structure of bovine Cu, Zn superoxide dismutase at 3 Å resolution: Chain tracing and metal ligands, *Proc. Natl. Acad. Sci. U.S.A.* **72**:1349.

Richardson, J. S., Richardson, D. C., Thomas, K. A., Silverton, E. W., and Davies, D. R., 1976, Similarity of three-dimensional structure between the immunoglobulin domain and the copper, zinc superoxide dismutase subunit, *J. Mol. Biol.* **102**:221.

Richmond, T. J., and Richards, F. M., 1978, Packing of α-helices: Geometrical constraints and contact areas, *J. Mol. Biol.* **119**:537.

Robertus, J. D., Kraut, J., Alden, R. A., and Birktoft, J. J., 1972, Subtilisin: A stereochemical mechanism involving transition-state stabilization, *Biochemistry* **11**:4293.

Robson, B., and Pain, R. H., 1971, Analysis of the code relating sequence to conformation in proteins: Possible implications for the mechanism of formation of helical regions, *J. Mol. Biol.* **58**:237.

Robson, B., and Suzuki, E., 1976, Conformational properties of amino acid residues in globular proteins, *J. Mol. Biol.* **107**:327.

Rose, G. D., 1978, Prediction of chain turns in globular proteins on a hydrophobic basis, *Nature (London)* **272**:586.

Rossmann, M. G., and Argos, P., 1975, A comparison of the heme binding pocket in globins and cytochrome b_5, *J. Biol. Chem.* **250**:7525.

Rossmann, M. G., and Argos, P., 1976, Exploring structural homology of proteins, *J. Mol. Biol.* **105**:75.

Rossmann, M. G., Moras, D., and Olsen, K. W., 1974, Chemical and biological evolution of a nucleotide-binding protein, *Nature (London)* **250**:194.

Rossmann, M. G., Liljas, A., Branden, C., and Banaszak, L. J., 1975, Evolutionary and structural relationships among dehydrogenases, in: *The Enzymes* (P. D. Boyer, ed.), 3rd ed., Vol. 11, pp. 61–102, Academic Press, New York.

Rowe, E. S., and Tanford, C., 1973, Equilibrium and kinetics of a homogeneous human immunoglobulin light chain, *Biochemistry* **12**:4822.

Sachs, D. H., Schechter, A. N., Eastlake, A., and Anfinsen, C. B., 1972a, Antibodies to a distinct antigenic determinant of staphylococcal nuclease, *J. Immunol.* **109**:1300.

Sachs, D. H., Schechter, A. N., Eastlake, A., and Anfinsen, C. B., 1972b, Inactivation of staphylococcal nuclease by the binding of antibodies to a distinct antigenic determinant, *Biochemistry* **11**:541.

Sachs, D. H., Schechter, A. N., Eastlake, A., and Anfinsen, C. B., 1972c, An immunological approach to the conformational equilibria of polypeptides, *Proc. Natl. Acad. Sci. U.S.A.* **69**:3790.

Sachs, D. H., Schechter, A. N., Eastlake, A., and Anfinsen, C. B., 1974, An immunologic distinction between possible origins of enzymatic activity in a polypeptide fragment of staphylococcal nuclease, *Nature (London)* **251**:242.

Sasaki, K., Dockerill, S., Adamiak, D. A., Tickle, J. J., and Blundell, T., 1975, X-ray analysis of glucagon and its relationship to receptor binding, *Nature (London)* **257**:751.

Sawyer, L., Shotton, D. M., Campbell, J. W., Wendell, P. L., Muirhead, H., Watson, H. C., Diamond, R., and Cadner, R. C., 1978, The atomic structure of crystalline porcine pancreatic elastase at 2.5 Å resolution: Comparison with the structure of α-chymotrypsin, *J. Mol. Biol.* **118**:137.

Schechter, A. N., 1970, Measurement of fast biochemical reactions, *Science* **170**:273.

Schechter, A. N., 1976, The conformation of peptides and proteins in solution: Immunochemical studies, in: *Hormone and Antihormone Action on the Target Cell* (J. H. Clark, W. Klee, A. Levitski, and J. Wolff, eds.), pp. 29–38, Dahlem Konferenzen, Berlin.

Schechter, A. N., Chen, R. F., and Anfinsen, C. B., 1970, Kinetics of renaturation of staphylococcal nuclease, *Science* **167**:886.

Schmid, F. Y., and Baldwin, R. L., 1978, Acid catalysis of the formation of the slow-folding species of RNase A: Evidence that the reaction is proline isomerization, *Proc. Natl. Acad. Sci. U.S.A.* **75**:4764.

Schulz, G. E., and Schirmer, R. H., 1979, *Principles of Protein Structure,* pp. 131–148, Springer-Verlag, New York.

Schulz, G. E., Schirmer, R. H., Sachsenheimer, W., and Pai, E. E., 1978, The structure of the flavoenzyme glutathione reductase, *Nature (London)* **273**:120.

Sela, M., White, F. H., Jr., and Anfinsen, C. B., 1957, Reductive cleavage of disulfide bridges in ribonuclease, *Science* **125**:691.

Shaw, P. J., and Muirhead, H., 1976, The active site of glucose phosphate isomerase, *Fed. Eur. Biochem. Soc. Lett.* **65**:50.

Shintzky, M., and Goldman, R., 1967, Fluorometric detection of histidine–tryptophan complexes in peptides and proteins, *Eur. J. Biochem.* **3**:139.

Shrake, A., and Rupley, J. A., 1973, Environment and exposure to solvent of protein atoms. Lysozyme and insulin, *J. Mol. Biol.* **79**:351.

Singer, S. J., and Doolittle, R. L., 1966, Antibody active sites and immunoglobulin molecules, *Science* **153**:13.

Srinivasan, R., Balasubramanian, R., and Rajan, S. S., 1976, Extended helical conformation newly observed in protein folding, *Science* **194**:720.

Steiner, D. F., and Clark, J. L., 1968, The spontaneous reoxidation of reduced beef and rat proinsulins, *Proc. Natl. Acad. Sci. U.S.A.* **60**:622.

Steitz, T. A., Fletterick, R. J., Anderson, W. F., and Anderson, C. M., 1976, High resolution x-ray structure of yeast hexokinase, an allosteric protein exhibiting a non-symmetric arrangement of subunits, *J. Mol. Biol.* **104**:197.

Sternberg, M. J. E., and Thornton, J. M., 1976, On the conformation of proteins: The handedness of the β-strand-α-helix-β-strand unit, *J. Mol. Biol.* **105**:367.

Sternberg, M. J. E., and Thornton, J. M., 1977, On the conformation of proteins: An analysis of β-pleated sheets, *J. Mol. Biol.* **110**:285.

Stroud, R. M., Kay, L. M., and Dickerson, R. E., 1972, The crystal and molecular structure of DIP-inhibited bovine trypsin at 2.7 Å resolution, *Cold Spring Harbor Symp. Quant. Biol.* **36**:125–140.

Suzuki, E., and Robson, B., 1976, Relationship between helix-coil transition parameters for synthetic polypeptides and helix conformation parameters for globular proteins. A simple model, *J. Mol. Biol.* **107**:357.

Sweet, R. M., Wright, H. T., Janin, J., Chothia, C. H., and Blow, D. M., 1974, Crystal structure of the complex of porcine trypsin with soybean trypsin inhibitor (Kunitz) at 2.6 Å resolution, *Biochemistry* **13**:4212.

Takano, T., 1977, Structure of myoglobin refined at 2.0 Å resolution I. Crystallographic refinement of met-myoglobin from sperm whale, *J. Mol. Biol.* **110**:537.

Tanford, C., 1968, Protein denaturation, Parts A and B, *Adv. Protein Chem.* **23**:122.

Tanford, C., 1970, Protein denaturation, Part C, *Adv. Protein Chem.* **24**:2.

Tang, J., James, M. N. G., Hsu, I. N., Jenkins, J. A., and Blundell, T. L., 1978, Structural evidence for gene duplication in the evolution of the acid proteases, *Nature (London)* **271**:618.

Taniuchi, H., and Anfinsen, C. B., 1969, An experimental approach to the study of the folding of staphylococcal nuclease, *J. Biol. Chem.* **244**:3864.

Taniuchi, H., and Anfinsen, C. B., 1971, Simultaneous formation of two alternative enzymically active structures by complementation of two overlapping fragments of staphylococcal nuclease, *J. Biol. Chem.* **246**:2291.

Taniuchi, H., Parker, D. S., and Bohnert, J. L., 1977, Study of equilibration of the system involving two alternative, enzymically active complementing structures simultaneously formed from two overlapping fragments of staphylococcal nuclease, *J. Biol. Chem.* **252**:125.

Teale, J. M., and Benjamin, D. C., 1977, Antibody as immunological probe for studying refolding of bovine serum albumin, *J. Biol. Chem.* **252**:4521.

Timkovich, R., and Dickerson, R. E., 1976, The structure of *Paracoccus denitrificans* cytochrome c $_{550}$, *J. Biol. Chem.* **251**:4033.

Tsong, T. Y., Baldwin, R. L., Elson, E. L., 1971, The sequential unfolding of ribonuclease A: Detection of a fast initial phase in the kinetics of unfolding, *Proc. Natl. Acad. Sci. U.S.A.* **68**:2712.

Tulinsky, A., Vandlen, R. L., Morimoto, C. N., Mani, N. V., and Wright, L. H., 1973, Variability in the tertiary structure of α-chymotrypsin at 2.8 Å resolution, *Biochemistry* **12**:4185.

Vainshtein, B. K., Harutyunyan, E. H., Kuranova, I. P., Borisov, V. V., Sosfenov, N. I., Pavlovsky, A. G., Grebenko, A. I., and Konareva, N. V., 1975, Structure of leghaemoglobin from lupin root nodules at 5 Å resolution, *Nature (London)* **254**:163.

Vandlen, R. L., and Tulinsky, A., 1973, Changes in the tertiary structure of α-chymotrypsin with change in pH, *Biochemistry* **12**:4193.

Venetianer, P., and Straub, F. B., 1963, The enzymic reactivation of reduced ribonuclease, *Biochim. Biophys. Acta.* **67**:166.

Venkatachalam, C. M., 1968, Stereochemical criteria for polypeptides and proteins. V. Conformation of a system of three linked peptide units, *Biopolymers* **6**:1425.

Wagner, G., De Marco, A., and Wuthurch, K., 1976, Dynamics of the aromatic amino acid residues in the globular conformation of the basic pancreatic trypsin inhibitor (BPTI): 1. ^{3}H NMR studies, *Biophys. Struct. Mech.* **2**:139.

Watenpaugh, K. D., Sieker, L. C., Herriott, J. R., and Jensen, L. H., 1973, Refinement of the model of a protein: Rubredoxin at 1.5 Å resolution, *Acta Cryst.* **B29**:943.

Watson, H. C., 1969, The stereochemistry of the protein myoglobin, *Progr. Stereochem.* **4**:299.

Weatherford, D. W., and Salemme, F. R., 1979, Conformations of twisted parallel β-sheets and the origin of chirality in protein structures, *Proc. Natl. Acad. Sci. U.S.A.* **76**:19.

Westmoreland, D. G., and Matthews, C. R., 1973, Nuclear magnetic resonance study of the thermal denaturation of ribonuclease A: Implications for multistate behavior at low pH, *Proc. Natl. Acad. Sci. U.S.A.* **70**:914.

Wetlaufer, D. B., and Ristow, S., 1973, Acquisition of three-dimensional structure of proteins, *Annu. Rev. Biochem.* **42**:135.

White, J. L., Hackert, M. L., Buehner, M., Adams, M. J., Ford, G. C., Lentz, P. J., Jr., Smiley, J. E., Steindel, S. J., and Rossmann, M. G., 1976, A comparison of the structures of apo dogfish M_4 lactate dehydrogenase and its tertiary complexes, *J. Mol. Biol.* **102**:759.

Winkler, F. K., and Dunitz, J. D., 1971, The non-planar amide group, *J. Mol. Biol.* **59**:169.

Winkler, F. K., Schutt, C. E., Harrison, S. C., and Bricogne, G., 1977, Tomato bushy stunt virus at 5.5 Å resolution, *Nature (London)* **265**:509.

Wright, C. S., 1977, The crystal structure of wheat germ agglutinin at 2.2 Å resolution, *J. Mol. Biol.* **111**:439.

Wyckoff, H., 1968, (Comments on cow vs rat RNAse), *Brookhaven Symp. Biol.* **21**:252.

3

Regulation of Structural Protein Interactions as Revealed in Phage Morphogenesis

JONATHAN KING

1 Introduction

Advances in cellular and molecular biology continue to reveal new aspects of the structural organization within cells. A multiplicity of protein–protein interactions must be involved in the control of the assembly and function of the microtubules, microfilaments, nuclear pores, centrioles, and other components of cytoskeletal networks. These structures must assemble in precise locations, correctly linked to the overall cellular structure, and generally at a particular stage of the cell cycle. The characteristics of such structure-forming processes differ in many ways from the classical models for the control of intermediary metabolism, such as the induction and repression of enzyme synthesis.

Aspects of structure-forming processes which I include in the term "regulation" include: (1) initiation and termination of polymerization processes, (2) the location of initiation and termination structures, (3) determination of the dimensions of repeating (or nonrepeating) structures, (4) catalysis of protein polymerization and noncovalent bond formation, (5) the role of jigs or templates in the formation of complexes for which random diffusion processes are too slow, (6) feedback loops between protein assembly processes and protein synthesis, and (7) switching of subunits from inactive to active forms by means of aggregation or assembly reactions.

JONATHAN KING • Department of Biology, Massachusetts Institute of Technology, Cambridge, Massachusetts 02139

Despite the extensive genetic analysis of eukaryotes such as *Drosophila,* until recently very little was known of the genetic control of morphogenesis, with a few notable exceptions (Lewis, 1963; Garcia-Bellido, 1977). With the development of the genetics of smaller eukaryotes, such as *Chlamydomonas,* the nematode *Caenorhabditus elegans,* and the slime molds *Dictyostelium* and *Physarum,* the genetic control of morphogenesis, and therefore of assembly processes, is emerging as an area accessible to experimentation.

Nonetheless the most developed and precise models for both the genetic and physiological control of morphogenetic processes are based on work with prokaryotes, most notably in the area of bacterial virus assembly. Unfortunately, the very success of this work has led to its incorporation into textbooks in a simplified form, which brings out clearly the multistep character of the processes, but otherwise obscures all the regulatory processes underlying the precision of the overall reaction. For example, the central feature of the phage assembly pathways (see Fig. 4) is that no two proteins interact with each other unless another pair of proteins has previously interacted. Though implicit in the strict sequential nature of the pathway, it is quite easy to misread the picture as due to a *temporal* appearance of the proteins, rather than the very particular set of regulatory interactions that activate each protein as it is incorporated into the preceding substrate structure.

2   Genetic Analysis of Protein Interactions

Whatever may be the assembly mechanisms involved, the information for the specification of these processes must be encoded in genes. In a number of the examples described below, proteins contributing to certain assembly processes are not incorporated into the mature virus. These proteins were initially discovered through analysis of mutants defective in the genes coding for them. Similarly, a number of important proteins in the initiation of the assembly of virus heads and tails are present in very small numbers. These proteins were not identified through biochemical analysis of the structures, but again through studies on mutants defective in the genes coding for these proteins. Once identified, such proteins can often be purified and studied directly.

Unfortunately, the very success in genetically identifying almost all the proteins involved in bacteriophage assembly processes has rendered the subject somewhat arcane and inaccessible to investigators not raised on a diet of prokaryotic molecular genetics. This chapter attempts to make available to readers interested in the general problem of the control of cellular and organelle assembly some of the mechanisms operating at the molecular level that have been unearthed in the investigation of bacterial viruses. Rather than a detailed summary of the literature it is an overview colored by my experiences working with phages P22, T4, and lambda.

The control of assembly has been studied in detail for a number of double-stranded (ds) DNA phages, including T4, T3, T7, lambda, P2, and P4 of *Escherichia coli,* P22 of *Salmonella typhimurium,* and ϕ29 of *Bacillus subtilis.* These studies have been reviewed comprehensively and readers interested in detailed references and further information should consult the articles by Casjens and King (1975), Showe and Kellenberger (1976), Murialdo and Becker (1978), and Wood and King (1979). Also of interest are the studies on the single-stranded phages, particularly ϕX174 (Hayashi, 1978), and the filamentous phages f1 and fd, which acquire their protein coats in the course of passing through the *E. coli* membrane (Webster and Cashman, 1978). These studies on membrane-associated protein will not be covered here.

Before proceeding with specific examples drawn from the analysis of phage assembly, it will be useful to review some features of the techniques involved.

R. S. Edgar, R. H. Epstein, E. Kellenberger, and their colleagues (Epstein *et al.,* 1963) recognized that the identification of all of the genes of T4 essential for viral reproduction would be a powerful method for dissecting the character of this process. This proved to be particularly true for the assembly of the three-dimensional structure of the virus particles. To identify genes for essential functions these researchers developed the two well-known systems of conditional lethal mutations: temperature-sensitive mutations, which result in proteins that mature and function at low temperature, but are defective in maturation or function at high temperature; and chain termination or amber mutations, which result in synthesis of only a fragment of the mutant protein in the restrictive (suppressor⁻) host, but which support synthesis of a functional protein in the permissive (suppressing) host (Edgar, 1969). Extensive labors resulted in the development of a library of phage strains, each carrying a conditional lethal mutation in a different essential gene. With respect to morphogenetic processes, this allows one to remove from infected cells one protein at a time, by mutation, and examine the consequences. Do the remaining proteins assemble into structures? What happens to those proteins that do not assemble? Can one identify structural intermediates in the morphogenetic process? From the initial analyses which relied primarily on electron microscopy and serology it was clear that the gene products interacted through defined pathways, rather than just "coming together" (Edgar and Lielausis, 1968; King and Wood, 1969).

The succeeding discussions focus on infections of suppressor bacteria with phage carrying an amber mutation in a morphogenetic gene of interest. For all practical purposes the protein specified by the mutant gene is absent from the cell, and we concentrate on what happens to the proteins that are synthesized.

A critical tool in the establishment of morphogenetic pathways was the *in vitro* complementation assay (Edgar and Wood, 1966; Wood *et al.,* 1968), in which the structural intermediates and unassembled proteins accumulating in mutant-infected cells were assembled into viable virus *in vitro.* This established that many morphologically identifiable structures, and also unassembled proteins, represented precursors in the assembly process. In general, the only protein lacking in cells infected with mutants defective in morphogenesis was the protein product of the mutant gene. Any structure whose assembly is dependent on that protein will also be missing, but not the other component proteins that, as we will see, often accumulate as soluble precursors. Under conditions in which the assembly reactions proceed *in vitro,* addition of a sample containing the precursor form of the missing protein, or the intermediate structure incorporating it, leads to the production of viable phage. This therefore provides an activity assay for morphogenetic proteins of the virus, which often have no enzymic or other assayable activity.

The insoluble nature of many structural proteins initially limited our ability to catalogue them. Laemmli's development (1970) of high-resolution acrylamide gel electrophoresis in the presence of sodium dodecyl sulfate (SDS) made it possible to resolve even minor protein species. Proteins synthesized after phage infection could be identified by autoradiography of infected cells, regardless of whether or not they were assembled into phage particles. The chain termination mutants provided a unique means for correlating polypeptide chains with the genes that specify them. Because the chain termination mutations prevent the synthesis of complete polypeptide chains, the absence of a band from its normal position in the gel pattern of amber mutant-infected cells generally identifies that band as the normal product of the mutant gene. Once these gene–protein identifications have been made, one can isolate various structures, for example, the tail of T4, and by gel analysis of the polypeptide chains (Fig. 2), identify the gene products incorporated (or not incorporated) into the structure (King and Laemmli, 1973).

3 Self-Regulated Assembly: General Mechanism in the Control of Protein Polymerization

If the structural proteins of phage particles (or other organelles) simply aggregated spontaneously as they were released from ribosomes, a great variety of complex structures would certainly assemble, but it is unlikely that they would be very closely related to the structures needed for a functional virus. For example, the polymerization of the tail sheath subunits of T4 (Fig. 4) either unconnected to the baseplate, or without a tube within, would not be very useful. Some mechanism is required to prevent sheath subunits from interacting with themselves prior to their interactions with the baseplate. The interaction of such regulatory processes is what underlies the existence of precise pathways of protein interactions in morphogenesis.

Such regulation of the polymerization of protein subunits requires the existence of states of the protein that do not spontaneously polymerize. Fibrinogen and tubulin, in the presence of calcium, are examples of nonpolymerizing precursor forms of structural proteins in eukaryotes. Such precursor forms of structural proteins are quite general and are not limited to precursors of proteolysis or to dependence on small effectors. Many phage structural proteins do not undergo proteolytic cleavage, but nonetheless are synthesized in a nonaggregating form, and must be converted to another form, a comformationally different state, in order to aggregate. What controls this conversion? The general mechanisms of control that emerge from the examples described in the following sections include the following features: (1) many phage structural proteins are synthesized as "inactive" subunits that do not spontaneously interact with each other; (2) these subunits are activated by binding to an active site of an organized multiprotein intermediate in the assembly pathways; (3) the growing structure induces a conformational change in the subunit, switching it into a form that becomes part of a new growing site—that is, the growing structure has the capacity of catalyzing conformational changes in the newly bound subunit; and (4) the energy for these transitions is built into the proteins during the polymerization of amino acids, but is only released in the course of assembly of the proteins into organized structures (Caspar, 1976).

I will call such processes, in which polymerization is regulated through the interaction of free subunits with an organized structure, *self-regulated assembly*. A prerequisite for control of this kind is that more than one stable conformational state of the protein exist. It is worth noting that the ability of the growing sites to catalyze conformational changes is not associated with any single component of the structure but develops as the outcome of the interactions. A number of these ideas, particularly the notion of protein switching, were inaugurated by D. L. D. Caspar (1976) who used the term *self-controlled assembly* to describe these reactions.

3.1 Protein Switching in Flagellin Assembly

The best-developed example of the protein switching mechanism is to be found in the elegant work of Asakura and colleagues on the polymerization of *Salmonella* tubulin *in vitro* into bacterial flagella (Asakura, 1970). The disaggregation reaction leaves the tubulin monomers in the nonpolymerizing state; that is, the subunits show no tendency to react with each other when they are free in solution. When small flagella fragments (seeds) are added to the solution, polymerization proceeds rapidly by addition of subunits to the growing tips. The seeds are made up of the identical tubulin subunits and can be themselves dissociated into inactive subunits, which, in turn, repolymerize into flagella only in the presence of additional seeds. Circular dichroism and other techniques reveal that the subunits are in a different conformational state from that of the polymerized subunits in the seed. Growth is in fact a two-step process: first, the growing tip of the tube binds an inactive subunit; and second, the newly bound subunit is switched into the active conformation, generating a new growing site (Uratani *et al.,* 1972). The conformational transition takes about 0.3 sec at 37°C and is the rate-limiting step in polymerization.

The *in vivo* assembly process is substantially more sophisticated, as the subunits do not find the tip of the growing flagella by diffusion from the medium, but rather are transported from the bacterial cytoplasm through the lumen of the flagellin and thus reach the tip from within. Whether the free subunits derived by dissociation are identical with the form of the protein transported *in vivo* is not known. The *in vitro* dissociation process converts the subunits from the "sticky" state to the nonaggregating precursor state. If this were not the case, aggregation would proceed spontaneously in the absence of seed.

The flagellin example involves homopolymerization—the polymerization of one kind of subunit into a structure in which all the subunits make identical contacts. As a result of the switching mechanism, the overall process displays cooperativity—subunits add preferentially to preexisting structures. But in most organelles, interactions and binding occur with many different proteins. For example, the tubulins that form the microtubules of eukaryotic cilia and flagella interact with dynein and radial spokes, in addition to interacting with each other. What ensures that the bonds are formed in the right sequence? What selects from many possible interactions the appropriate one for a given stage in the process? The analysis of phage assembly brings out clearly that the same switching mechanisms operate, through interactions among different proteins. This leads to heterocooperativity (Wood, 1979), in which particular kinetic pathways are favored according to previous assembly interactions. This phenomenon operates in all aspects of phage assembly, but is most clearly demonstrated in the assembly of the tail of bacteriophage T4.

3.2 Protein Switching and Self-Regulated Assembly of the Tail Phage T4

3.2.1 Tail Structure and Function

The contractile tail of phage T4 is an organelle for the transport of a double-stranded DNA molecule across the cell envelope of *E. coli.* This requires recognition of, attachment to, and penetration of the outer membranes of the bacterial cell. The multistep character of this process is reflected in the complex morphology of the tail (Fig. 1).

Because the tail is a rather specialized organelle, and therefore highly differentiated morphologically, one can identify the functional roles of many of the components, such as the tail tube and tail sheath proteins and many of the baseplate proteins. The hexagonal baseplate, about 400 Å in diameter, constitutes the attachment organelle (Fig. 2). Projecting from this structure is a 1000-Å hollow tail tube, surrounded by the contractile sheath. A small connector terminates the tail and joins it to the head. This structure is assembled independently of the head, requiring that the DNA must be threaded down the tube at

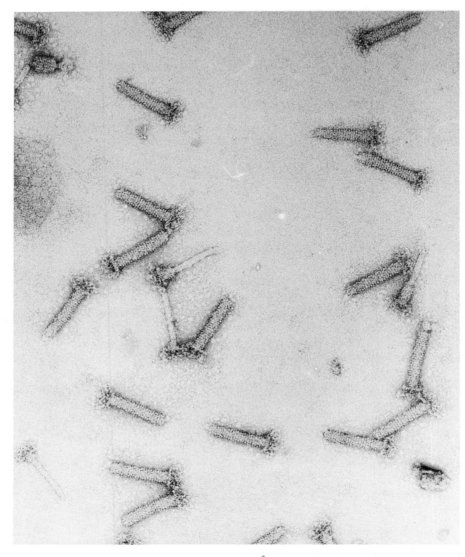

Figure 1. Tails of bacteriophage T4. These structures, 1000 Å in length, have been purified by sucrose gradient centrifugation from cells infected with chain termination mutants of the major head protein. The sheath is composed of 24 annuli of six subunits, each subunit being an 80,000-dalton monomer, the product of T4 gene 18 (King and Laemmli, 1973). Some of the tails have lost their sheath, exposing the tail tube, which is also composed of 24 annuli of six subunits each (Moody, 1971), each subunit being a 20,000-dalton monomer, the product of gene 19. The complex baseplate is shown more clearly in Figs. 2 and 3. During contraction of the sheath, the baseplate releases its bonds with the tail tube, and the sheath subunits and expanded baseplate slide up past the tube, which penetrates the cell envelope.

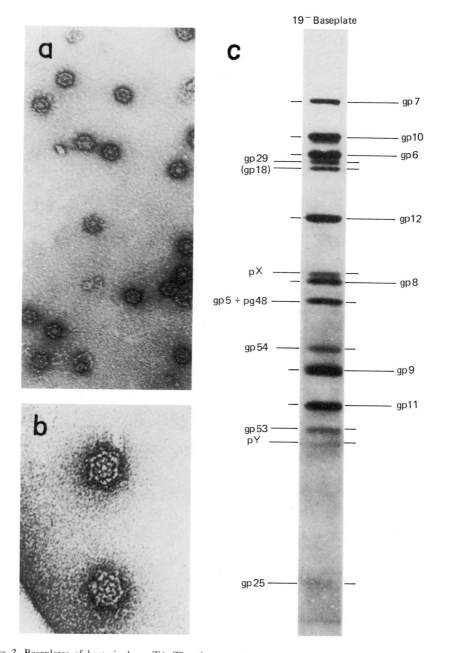

Figure 2. Baseplates of bacteriophage T4. The electron micrographs are of baseplates isolated from cells infected with chain termination mutants of gene 19, which specifies the tail tube subunits. The protein composition of these structures is shown in (c). Baseplates from cells labeled with [¹⁴C]amino acids were dissociated in SDS and electrophoresed through a 10% acrylamide gel. Almost all the protein bands have been correlated with the genes specifiying them, and are named accordingly; "gp" stands for gene product, followed by the gene number. The various proteins are probably present in multiples of six (Kikuchi and King, 1976). The darker bands are proteins of the outer rim and constitute the skeleton of the structure. (Figure from Kikuchi and King, 1975*a*, reprinted with permission.) The pathway for the assembly of these proteins into the baseplate is shown in Fig. 4.

some stage after the head–tail joining reaction. The 1400-Å long tail fibers are attached to the baseplate in the last step in phage assembly (see Fig. 10).

The cell-attachment and chromosome-transport functions of the phage tail are initiated by the prior binding of the long tail fibers of the phage to the cell surface. Once bound the fibers activate the baseplate, which expands from a hexagonal to a star-shaped configuration (Fig. 3), releasing the tip of the tail tube and triggering contraction of the sheath (Crowther *et al.*, 1978). The released tail tube is then forced through the outer layers of the cell envelope and a channel is open for DNA release.

The complete phage tail, sedimenting at about 130 S, can be isolated from cells infected with phage mutants blocked in head assembly. The tails are made up of at least 17 different species of polypeptide chains, all of which have been identified with the genes that encode them (Kikuchi and King, 1975a). Most of these proteins form the baseplate (Fig. 2). The hollow tail tube is composed of 144 copies of a 20,000-dalton subunit, the product of gene 19, whereas the contractile tail sheath, surrounding the tail tube, is composed of 144 copies of an 80,000-dalton subunit, the gene 18 protein. Two other proteins, present in fewer copies, form the small protuberance that terminates the tail and forms the site for head attachment (King and Mykolajewycz, 1973). In addition to the structural proteins the baseplate contains a number of molecules of a phage-specific folic acid conjugate (Nakamura and Kozloff, 1978) and two phage-coded enzymes, thymidylate synthetase and dihydrofolate reductase (Capco and Matthews, 1973). In the early stage of phage infection these enzymes are involved in DNA synthesis. Their later incorporation into the

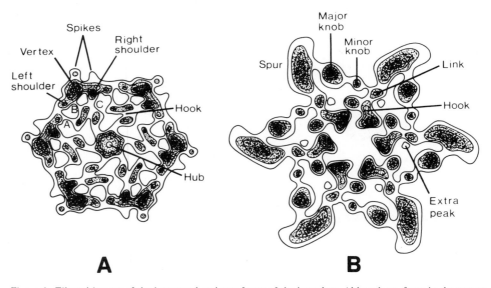

A **B**

Figure 3. Filtered images of the hexagonal and star forms of the baseplate. Although we focus in the text on protein interactions during assembly of organelles such as the baseplate, the later functional interactions are important to the comprehension of the assembly pathway. In the mature phage the long tail fibers, binding to their cell receptors, activate the baseplate, which attaches to other receptors and expands to a star shape, triggering sheath contraction and releasing the tail tube (Simon and Anderson, 1967; Arscott and Goldberg, 1976). This expansion process occurs spontaneously in free baseplates destabilized by the absence of certain proteins (Crowther *et al.,* 1978). These are drawings of computer-filtered images of electron micrographs of individual baseplates in the two states. Some of the features have been named to facilitate description of the morphology. The location of particular proteins in the structure are discussed in Crowther *et al.* (1978) and Dawes (1979).

baseplate presumably reflects some role in triggering the tail contraction process on attachment to the host cell surface.

3.2.2 Tail Tube Assembly

The pathway for the assembly of most of the tail tube components into a functional phage tail is shown in Fig. 4. Proteins are named for the genes that encode them and are shown above the arrows depicting the steps in which they participate. The left-hand part of the figure illustrates the formation of the hexagonal baseplate. The baseplate is assembled by the radial polymerization of six arm components around a central 22 S hub component. Both the arm and hub components are themselves the products of assembly pathways involving a number of proteins. I will describe some of the features of baseplate morphogenesis in the next section. The important point for the present is that the entire baseplate serves as the initiation complex for tail tube polymerization.

The regulation of the initiation of tail tube assembly is seen most clearly when the formation of the baseplate is blocked by mutation in any of the genes coding for major baseplate proteins. Under these conditions the tail tube subunits continue to be synthesized and accumulate within the cell. However, they do not aggregate with each other but instead accumulate as soluble 20,000-dalton monomers. When complete baseplates are added to a concentrated extract of cells containing the tube monomers, the monomers polymerize on the added baseplates (King, 1971; Kikuchi and King, 1975a). Thus the baseplate functions in the same manner as the seed in flagella assembly. After the first annulus is bound, the tube itself becomes the active site that sequesters soluble subunits. This binding must switch the tube subunits from nonpolymerizing to polymerizing conformation in order for the active site to propagate as with flagellin subunits.

Though the growing site is continually reproduced during the addition of successive annuli, this process clearly stops at the 24th annulus. I will discuss this more explicitly in Section 4.

Tail tube precursor subunits have been purified from phage-infected cells by Wagenknecht and Bloomfield (1978). These subunits show no tendency to aggregate with each other in vitro, paralleling the situation in vivo. Addition of purified baseplates results in rapid polymerization of the tube subunits onto the baseplate. This polymerization stops precisely at 1000 Å, establishing that regulation of the polymerization process is a property of the purified components. This reaction does not require energy, nor does it require any proteolytic activation.

Tail tube subunits have also been isolated by dissociation of mature tail tube (To et al., 1970; Poglazov and Nikolskaya, 1969). These subunits, when dialyzed back to physiological conditions, polymerize spontaneously in the absence of baseplates. These subunits therefore differ from the precursor subunits, which do not polymerize in the absence of baseplates. Presumably they represent the switched "sticky" state of the subunits. Such results point out that though the two forms of tube subunit differ only in conformation, the barrier to the transition must be quite high; the organized baseplate or growing tip is required to catalyze the transition.

Of course the energy barrier between interconversion of two conformations might be much lower for other structures. In the case of the tail tube subunit, only a one-way conversion from nonpolymerizing to polymerizing form is required. For subunits that participate in reversible polymerization, such as tubulin subunits, the interconversions probably involve smaller energy barriers.

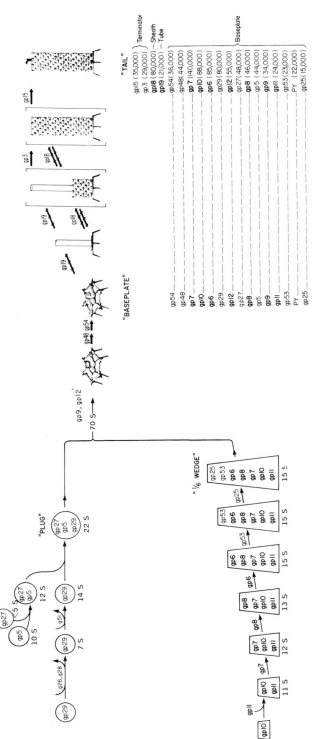

Figure 4. Pathway for T4 tail assembly. This diagram summarizes the known steps in the assembly of the phage tail from precursor proteins. The proteins are named for the genes that code for them ("gp" = gene product) and are shown above the steps in which they participate. The molecular weights of the polypeptide chains incorporated into the mature structure are given in the list beneath the completed tail. No protein cleavage has been detected in any of these proteins during assembly. Most of these reactions proceed *in vitro* in mixtures of extracts of mutant-infected cells (Kikuchi and King, 1975*a,c*). Polymerization of the tube subunits is irreversible in that intermediate length tubes are stable. Sheath polymerization is reversible, until the formation of the terminal connector that stabilizes the structure and forms the site for head attachment. Many of the intermediates in baseplate assembly have not been visualized in the electron microscope and are shown schematically and not to scale with the organized baseplate. The earliest arm precursor has been purified and looks like a vertex (Berget and King, 1978). The sedimentation coefficients of intermediates in baseplate assembly are shown beneath them. The "plug" structure at the center of the baseplate is referred to as a *hub* in the text, and the "wedge" is now called the *arm*. The phage-specific folate conjugate (Kozloff and Lute, 1973) is probably incorporated into the hub. Figure reprinted from Kikuchi and King (1975*c*) with permission.

3.2.3 Initiation Proteins

Sites on the baseplate bind six tail tube subunits and switch them into a form in which they then repeat the same process. It seems reasonable that these initiation sites on the baseplate resemble or reproduce conformational features of the tube subunits. Two protein species in the baseplate, specified by genes 48 and 54, are not needed for the assembly of baseplate, but are required to render it competent for sheath and tube polymerization. If these are not incorporated, the baseplates, though morphologically normal, do not bind tube subunits (King, 1971; Crowther *et al.*, 1978). Experiments with particles carrying thermolabile variants of the gene 48 protein suggests that this protein mediates the interaction between the baseplate and the first sheath annulus (Dawes, 1979). One can visualize one end of these molecules as resembling the sheath protein and the other end as being the binding site for incorporation into the baseplate. The gene 54 protein would then represent the binding protein for the tube subunit—again, one end mimicking the tube subunit-binding site, and the other end representing the baseplate-binding site.

3.2.4 Assembly of the Contractile Sheath

The assembly of the sheath displays features similar to those for tail tube assembly: subunit polymerization is initiated through binding of sheath subunits by a baseplate with the tail tube. The baseplate without the tail tube does not initiate polymerization of sheath subunits. Presumably sites on the baseplate together with sites on the first annulus of tube subunits bind the first six sheath subunits. This binding must switch the sheath subunits into the active conformation for binding six more subunits. As with the tube subunits, no proteolytic cleavage of sheath subunits occurs during polymerization, indicating that the activation involves conformational alteration (King and Laemmli, 1973). The sheath subunits polymerize to the length of the tube and no further. This indicates that the tube annuli participate in binding not only the first annulus of sheath subunits, but in the binding of subsequent annuli of sheath subunits (Tschopp *et al.*, 1979).

After the last annulus of sheath subunits has added to the tube, two other proteins, specified by genes 3 and 15, form a cap on the structure, terminating the tail and forming a site for head attachment. In the absence of either of these proteins the sheath is only reversibly polymerized; the diluted subunits come off sequentially from the head end down to the baseplate (King, 1968, 1971). Genes 3 and 15 proteins form a special set of bonds between the 24th sheath annulus and the 24th tube annulus. In a sense these assembly features are not surprising, given the functions of the sheath. When the sheath contracts subunits achieve a different conformation in which they are very strongly bonded to each other and in this process must slide past the rigid tail tube, forcing it through the cell envelope. Except for the top annulus, which must remain bound to the top of the tube if force is to be generated, the bonds between sheath and tube used during assembly are only transient, and the extended form of the sheath is only a metastable structure. Although the sheath appears to be most stable when it is contracted, this is not the conformation which the protein takes upon synthesis.

Tschopp *et al.* (1979) have shown that purified sheath subunits show similar behavior *in vitro*. Within a certain range of subunit concentration, sheath polymerization is dependent on the presence of tube baseplates. The polymerization *in vitro* was highly cooperative in that most of the structures formed were complete sheaths, with few intermediate structures. In the model proposed by Arisaka *et al.* (1979), the binding of the first annulus of sheath subunits by the tube baseplate is very slow with respect to propagation, and the 24th annulus is much more stable than previous annuli. Since two proteins, specified by genes 3 and 15, act at the tip of the tail, differentiating it from the rest of the tube, this is not an unreasonable suggestion. The mechanism allowing the termination proteins to distinguish between the 24th annulus and all previous annuli is discussed in Section 4.

The actual polymerization pathway remains obscure. For example, is the sixth subunit of an annulus required to be in place before addition of subunits to form the next annulus? It seems likely that some mechanism must be built in to ensure this requirement, otherwise gaps could be easily left in the structure. (Gaps have not been observed.) Formation of hexameric annuli first, followed by their polymerization would avoid this problem, but requires the annuli to loop over and down the tail tube. Inasmuch as the sheath subunits accumulate as monomers (Kikuchi and King, 1975a; Tschopp et al., 1979), this seems unlikely.

The switching of the sheath subunit from precursor to active form occasionally occurs in the absence of the core baseplate. If cells are incubated under conditions in which very high concentrations of subunits accumulate the subunits polymerize into an aberrant form, the polysheath, which is very long (often from one end of the cell to the other). This is also the case with purified subunits when the critical subunit concentration is exceeded (Tschopp et al., 1979; Arisaka et al., 1979).

3.2.5 Self-Regulation of Baseplate Assembly

Tube and sheath polymerization involve repeated binding of soluble subunits by their organized forms and switching them into the organized conformational state. This pathway effectively limits the binding sites to the growing structure, so that new tubes or sheaths are not continually initiating, unconnected to the baseplate. This switching of soluble subunits by an organized structure is not limited to polymerization of identical subunits but in fact characterizes almost all the protein–protein interactions in tail assembly and, as I will demonstrate below, in all of phage assembly. It is perhaps most evident in the assembly of the baseplate.

The arms of the baseplate are themselves the end products of a sequential pathway, as can be seen in Fig. 4. If the first protein in the arm pathway, specified by gene 10, is removed by mutation, all of the subsequent proteins are synthesized, but they remain as soluble subunits. If this protein is added back to a concentrated extract lacking it, assembly is initiated and all the subsequent assembly steps proceed *in vitro,* leading to the production of complete tails and thence viable phage (Kikuchi and King, 1975a). If a later protein in the pathway is removed by mutation—for example, gp6—the proteins to the left assemble into a complex, whereas the protein downstream of the block remains soluble and unassembled. Addition of the missing protein allows the assembly process to proceed to completion. Presumably the same binding and switching processes are regulating these interactions, with each intermediate binding and activating the next protein in the pathway and switching it into a form that contributes to the formation of the binding site for the next protein.

Though most of the baseplate proteins are synthesized in the inactive form and switched into the active form, this cannot be the case for the very first proteins in the pathway. In fact, in baseplate arm assembly, the first three proteins, gp10, gp7, and gp11, interact with each other directly, initiating the rest of the assembly reactions. The arms are built essentially from the outside in. The tail spikes projecting down from the baseplates are composed of the gene 11 product, most likely a dimer of 24,000-dalton chains. The gene 10 protein, a dimer of 88,000-dalton polypeptide chain, forms the vertex of the baseplate (Crowther *et al.*, 1978). The topological location of the 140,000-dalton gene 7 protein is less clear, but it probably extends in from the periphery.

The interaction of gp10 and gp11 is not one of the regulatory interactions. If gp11 is removed by mutation, gp10 stills binds gp7, and baseplate and phage assembly proceed normally, yielding particles lacking only the gp11 tail spikes (Edgar and Lielausis, 1968; Crowther *et al.*, 1978). Addition of a source of gp11 to the 11⁻ particles results in the formation of viable phage.

For the moment then we can think of the key interaction in the initiation of baseplate arm assembly as the binding of the vertex protein, gp10, with gp7. This reaction proceeds spontaneously. In contrast, though the next protein, gp8, binds efficiently to the complex, it does not complex to either protein in the absence of the other one (Kikuchi and King, 1975*a*). In this formulation the first two proteins, gp10 and gp7, activate each other through their initial interaction.

In the normal pathway the tail spike protein, gp11, complexes with the vertex protein in the first steps in assembly, even though that is not an obligatory interaction. This complex has been purified to homogeneity; it sediments at 9.3 S and is highly asymmetric as visualized in the electron microscope (Berget and King, 1978). There appears to be no tendency for the complex to dissociate.

The baseplate is assembled by a branched pathway, and the complete hub forms and accumulates in cells infected with mutants blocked in the assembly of the radial arms (Kikuchi and King, 1974*c*). The assembly of the 22 S hub itself is complex, involving proteins that are probably not incorporated into the structure. These proteins are specified by genes 26, 28, and 51; their function is necessary for the 75,000-dalton gene 29 protein to be competent for hub assembly. Kozloff and co-workers have assembled evidence that these proteins function enzymically in the synthesis of the folic acid that is incorporated into the baseplate. This suggests that the folate is a cofactor required for the maturation of the gene 29 protein. Though it may function in the triggering of hexagon-to-star transition during infections, it is in fact an obligate step in the assembly of the baseplate (Kikuchi and King, 1975*b,c*).

We have not visualized the hub directly in the electron microscope, but given its sedimentation coefficient of 22 S and the sixfold symmetry of the baseplate, we presume that is is hexameric. A model for the underlying organization would be that the competent mature form of gp29 aggregates into a hexamer, which then binds six of the complexes formed of the gene 5 and gene 27 proteins (Fig. 4). In the transition from hexagon to star (Fig. 3) the central part of the baseplate is opened out (Crowther *et al.*, 1978). It is tempting to think of this complex as a miniature diaphragm.

Though the hub may provide the "brains" of the baseplate, it also serves as an organizer for arm assembly. When the hub is prevented from assembling by mutation, the 15 S arm complexes accumulate. Since the arms must form edge-to-edge contacts with each other, the initial interaction of the arm with the hub probably activates the radial interactions in the normal assembly pathway. The activation energy for this transition is probably small, because the arms do aggregate with low efficiency in the absence of the hub. The aberrant hexagons they form have a hole in the center, but cannot be converted to viable baseplates by addition of the missing component. Apparently the hub must be incorporated during polymerization, and not after (Kikuchi and King, 1975*b,c*).

Aberrant hexagons lacking the hub also form in the absence of the last protein in the arm pathway, gp25. This protein probably forms the site for the binding of the arm to the hub. If the penultimate protein, gp53, is missing, no radial structures of any kind form, indicating that this protein either activates the side-to-side binding, or is the linking protein itself.

4   Determination of the Length of Tubular Structures

4.1   The Length of T4 Tail Tubes

A central question in morphogenesis is how dimensions are specified. For example, what determines the precisely specified distances between Z bands in skeletal muscle, or the length of actin and myosin filaments themselves? To my knowledge the only case in which we know the mechanism is in the length of rod-shaped viruses, notably tobacco mosaic virus (TMV), in which the length of the nucleic acid determines the length of the virus particle.

The tails of T4 particles are precisely 1000 Å long. No one has reported observing T4 particles with longer or shorter tails. I have never seen such a particle, and not for lack of searching. This length is a property of the tail tube polymerization, as noted above. If sheath assembly is blocked by mutation, the tail tubes still have a precise length distribution (King, 1968). Wagenknecht and Bloomfield (1977) have shown that in the *in vitro* assembly of the tail tube from purified baseplates and tail tube subunits, polymerization stops at 1000 Å. As noted above, the property of assembling into a proper length tail is limited to the precursor subunits; subunits obtained by dissociation of completed tubes polymerize in the absence of baseplates to a range of lengths (Poglazov and Nikolskaya, 1969). [Extra long tubes joined to baseplates have been observed, but under conditions in which the additional subunits are derived from breakdown of some tubes followed by reassembly (Tschopp and Smith, 1977).]

Edward Kellenberger proposed that the length could be determined by an "induced strain" mechanism, in which each subsequent annulus was incorporated in a conformation that slightly reduced the affinity of binding for the next annulus. This model predicts that the rate of polymerization should decrease sequentially from the initiation of the tube to its termination. Wagenknecht and Bloomfield (1977) tested this directly and found instead a uniform rate of polymerization.

My own preference is for a template- or length-determining molecule(s), akin to the RNA in TMV, but probably a protein. The symmetry of the baseplates suggests that three or six molecules are involved. Protein chains of 40,000 daltons stretch 1000 Å in their extended configuration. Any of the protein species of this size or larger in the baseplate would be candidates for such length-determining molecules. Such molecules have not been observed in the electron microscope, but they might not be detectable in negatively stained preparations. These chains can be imagined as facilitating tail tube polymerization, just as the tail tube facilitates sheath polymerization. Tube polymerization would stop at the chain terminus, and this would be the site recognized by the gene 3 protein.

Yoshiko Kikuchi and I tested a variety of such models by treating baseplates with degradative enzymes. Tube polymerization was insensitive to RNase and DNase. We screened a series of proteinases to see if any inactivated the free baseplate, but not the tube baseplate (in which the putative length-determining molecules would be protected). Because we suspected that the length determiner would have a repeating sequence we included collagenase in our experiments. Collagenases were found to inactivate free baseplates, but not baseplates with the tail tube. On mixing tube subunits with the treated baseplates we were unable to demonstrate a correlation between collagenase treatment of the baseplate and decreased tube length. Wagenknecht and Bloomfield (1978) treated baseplates with very high concentrations of RNase to test for a double-stranded RNA length determiner. They found some evidence for decrease in length, and this was apparently due to protease contamination in their enzyme preparation. Analysis of baseplate treated in this way revealed proteolysis of the gene 48 and 54 proteins, both of which are required for tail tube polymerization.

4.2 The Length of Lambda Tails

The tail of phage lambda is also made of repeating annuli, composed of the 31,000 dalton product of gene V. The tail tube is assembled on a small initiating structure, the tail tip, whose formation requires the products of seven additional lambda gene products (Katsura and Kuhl, 1975a,b). The tail of wild-type lambda is about 1500 Å long comprising 34–35 annuli. However, in the absence of the gene U product, extra-long tails of indefinite length accumulate. This initially indicated that gpU was determining the length of the tail. More careful studies (Katsura, 1976) revealed that though gpU normally terminates the tail at the correct length, it does not set the length. In the absence of gpU, gpV polymerizes to the normal length. However, as more tail subunits accumulate, a further aberrant polymerization of gpV past the normal termination point occurs. Thus gpU is somewhat similar in this respect to gp3 of T4: they both specifically recognize the terminal annulus of the wild-type tail. This leaves open the question of what determines which annulus becomes the last annulus.

4.2.1 Genetic Control of Length

To identify the gene or genes that specify the length of the tail, Phillip Youderian and I took advantage of the fact that phage φ80, a relative of lambda, has a longer tail. Youderian (1978) performed mixed infections of restrictive bacteria with φ80 and lambda phages with one parent carrying an amber mutation in one of the genes specifying the tail proteins. Such experiments were carried out for each of the genes involved in tail assembly. We then examined the length of the tails of the progeny phages in the electron microscope, to see whether they were φ80-length, lambda-length, or both. This allowed us to exclude genes U and V as candidates for the length-determining process. Youderian then isolated hybrids between lambda and φ80, and compared the length of their tail, with the genetic contribution of each parent. Analysis of the hybrids revealed that tail length was specified by a region of the genome including genes G, H, I, and L of phage lambda. These genes specify proteins needed for the formation of the tail tip, the initiation structure for gpV polymerization (Katsura, 1976). Gene H specifies one of the very large proteins of the lambda tail. GpH is cleaved during tail morphogenesis to 78,000 daltons. The equivalent cleaved protein is larger—95,000 daltons—in φ80. One simple interpretation of these results is that the size of the mature gene H protein specifies the length of the tail of lambdoid phages.

5 Scaffolding Proteins and Shell Assembly

5.1 Shell Assembly

The capsids of all the well-studied double-stranded DNA phages are assembled first as protein shells empty of DNA. Once completed, the shells then take an active part in the encapsulation of the DNA. These precursor shells are in essence DNA packaging organelles, and only later in the life cycle of the virus do they play the passive role of protecting the nucleic acid in the passage from one host to the next. (Casjens and King, 1975; Showe and Kellenberger, 1976; Murialdo and Becker, 1978).

The identification of precursor shells means that the mature capsid is not the direct product of subunit polymerization; the precursor shell is the structure that is assembled from free subunits. In all of the dsDNA phages substantial reorganization of the precursor shells occurs during the transition from the precursor to the mature capsid (Casjens and King, 1975; Showe and Kellenberger, 1976). These transitions, described in Sections 5.4 and 6.1, involve the regulation of interactions of subunits organized into an ordered structure.

Though the geometry of the packing of subunits in isometric shells has been well understood since the work of Caspar and Klug (1962), the *pathway* of the polymerization of the free subunits into closed shells has only recently begun to emerge. The construction of properly dimensioned closed shells, incorporating appropriate DNA packaging and injecting apparatus, requires a very precise sequence of interactions. These are controlled by the same switching interactions regulating bonding properties of subunits as those described for the tail proteins in Sections 3.1, 3.2.2, and 3.2.5 and for TMV (Caspar, 1976).

In all the double-stranded DNA phages that have been analyzed the formation of the shell requires an auxiliary protein in addition to the capsid protein (Casjens and King, 1975).

These auxiliary proteins share the following properties: (1) they are part of the structure of the precursor shell, (2) in their absence the coat proteins form aberrant aggregates, rather than the properly organized shell, and (3) they are absent from the mature phage. Such proteins have been called *assembly core* or *scaffolding proteins* (Showe and Black, 1973; King et al., 1973). In all cases the genes for these proteins are adjacent to the gene for the coat protein.

The scaffolding proteins must be removed from the precursor shell prior to the packaging of the viral DNA. These reactions are complex and are coupled to rearrangements of the coat protein lattice, described in Section 5.3. The extensive proteolytic cleavages of T4 capsid structural proteins are associated with these stages of shell maturation (Laemmli, 1970; Laemmli et al., 1976).

In *Salmonella* phage P22 the scaffolding protein is the product of gene 8, and the coat protein is the product of gene 5. In the normal course of assembly the two species copolymerize into a precursor shell containing about 420 copies of the 55,000-dalton coat protein, and about 200 copies of the 42,000-dalton scaffolding protein (Fig. 5). When examined in negatively stained preparations, such as that shown in Fig. 6, these structures look like a shell with a ball of protein within them (Botstein et al., 1973; King et al., 1973). In thin sections they appear to be double shells. Low-angle X-ray scattering studies have established in fact a thick shell of scaffolding protein within a thinner outer shell of coat protein (Earnshaw et al., 1976; Earnshaw and King, 1978). At a later stage of assembly, the scaffolding molecules exit from this structure and are reused in further cycles of assembly (King and Casjens, 1974).

The precursor shells of bacteriophage T4 are not isometric, and so cannot be studied by X-ray scattering, but they have been intensively studied by electron microscopy. Their structure is similar to that of P22 proheads, with the assembly core protein (the product of gene 22) forming an inner shell within the coat protein shell (Paulson and Laemmli, 1977). Analysis of bacteriophage lambda procapsids (Earnshaw *et al.*, 1979) have shown a similar double-shell structure for the precursor shells in phage lambda morphogenesis.

In the absence of the scaffolding proteins, the coat proteins do polymerize, but less efficiently, and the structures they form are incorrectly shaped. The scaffolding proteins in some sense regulate the polymerization of the coat proteins. The P22 scaffolding proteins do not aggregate with themselves in the absence of the coat protein *in vivo*. This suggests that the pathway for shell assembly is as follows: First, a mixed complex between coat and scaffolding protein forms. This mixed interaction activates both protein species, which now bind further subunits, activating them. Though we do not know whether the shell grows by incorporation of mixed complexes or of free subunits, the assembly process presumably involves the same self-regulation and switching phenomena as those described above for flagella assembly and tail assembly. The outcome of these processes is the limitation of

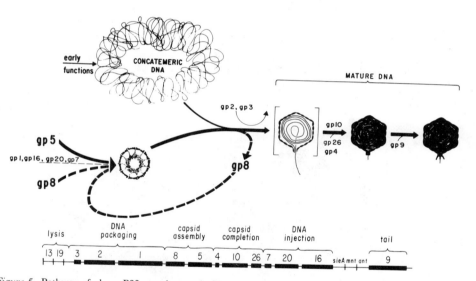

Figure 5. Pathway of phage P22 morphogenesis. The sequence of genetically identified steps in P22 assembly is shown with a map of the region of the phage chromosome coding for the morphogenetic proteins (King *et al.,* 1976; Susskind and Botstein, 1978). A precursor shell is assembled first and subsequently filled with DNA. The dashed line in the pathway represents the exit and recycling of the gene 8 scaffolding protein from the structure. Once filled with DNA, the capsid is stabilized by the addition of three additional protein species, which also form the site for the attachment of the tail spike protein. The minor proteins incorporated into the precursor shell probably form the vertex around which the shell is built, and through which the DNA will be first encapsulated, and later ejected (Murialdo and Becker, 1978; Hendrix, 1979*b*; Eiserling, 1979). The text discussion focuses on the assembly of the precursor shell and its transformations during the DNA packaging process. More detailed descriptions of the assembly of the capsids of phages lambda, T4, φ29, and T7 can be found in the reviews of Murialdo and Becker (1978) and Wood and King (1979).

active sites to growing structures, so that once initiated, free subunits are incorporated into that structure in preference to initiating new structures. The overall reaction is highly cooperative. Figure 8 provides a simplified model of the overall shell-forming reaction, which shows conformational changes in the later steps. It does not include conformational changes that we postulate underly the assembly reactions (Fuller and King, 1980*b*).

5.2   Initiation of Shell Assembly

Though all of the capsids of viruses have a high degree of symmetry, they all have a unique vertex, marked in the mature structure by the tail. This vertex plays a central role in the morphogenetic process. It is the site through which the DNA exits and through

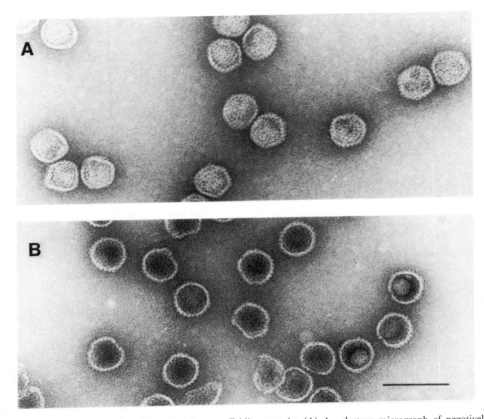

Figure 6. Precursor shells with and without scaffolding protein. (A) An electron micrograph of negatively stained P22 proheads, isolated from cells blocked in DNA packaging. These particles are composed of about 420 copies of the 55,000-dalton gene 5 coat protein, and about 200 copies of the 42,000-dalton gene 8 scaffolding protein. (B) Particles that have lost their scaffolding protein as a result of treatment with 0.8% SDS. These shells contain only coat protein. As described in the text, we believe that the scaffolding molecules depart through the coat protein lattice. The scaffolding protein prepared this way is active in prohead assembly *in vitro* (Fuller and King, 1980*a*). The structures of these particles as determined by low-angle X-ray diffraction are shown in Fig. 7. Figure reprinted from Casjens and King (1974) with permission.

which the DNA is threaded into the precursor shell (Murialdo and Becker, 1978). Furthermore, this unique vertex is the structure that initiates the assembly of the capsid (Hsiao and Black, 1978a,b). Though the capsid proteins of lambda, P22, and T4 can form closed shells in the absence of the initiation complex, these shells are incapable of packaging DNA. Thus the polymerization of capsid proteins into closed shells must initiate around the appropriate initiation vertex, for viable phage to form. Similarly, the microtubules of the mitotic apparatus must be anchored to centromeres or centrioles to be useful in mitosis.

The structure of the initiation complex is well illustrated with bacteriophage T4. *In vivo* the initiation of capsid assembly occurs on the cell membrane (Simon, 1972). The initiation complex, presumably a fivefold vertex, is built of the 64,000-dalton gene 20 product, together with a number of other minor protein species that are needed later in the DNA packaging process. Normally the scaffolding and major coat protein assemble on this initiation complex to form a prolate shell. If the gene 20 protein is removed by mutation, the coat and scaffolding protein accumulate, and eventually an aberrant initiation occurs followed by polymerization into cylindrical polyheads, which run the length of the cell. The gene 20 protein is presumably required to select the correct lattice for the initiation of coat and scaffolding polymerization (Laemmli *et al.,* 1970; Paulson and Laemmli, 1977).

The formation of the initiation complex requires an additional factor at elevated growth temperatures. The product of gene 40, not essential for growth at 30°C, becomes essential for growth at 42°C. At the high temperature cells lacking gp40 have the same defective phenotype as cells lacking gp20—aberrant polymerization of the coat and scaffolding proteins into polyheads. Gp40 appears to bind to the *E. coli* membrane and help organize the gp20 complex at high temperature (Hsiao and Black, 1978a,b). A host protein is also required for the formation of this complex, the product of *E. coli* gene *groE*. This 75,000 dalton protein forms a 14-subunit double donut complex with sevenfold symmetry (Hendrix, 1979b; Hohn *et al.,* 1979). Its role in initiation is unclear, but it is needed for both T4 and lambda head morphogenesis.

The assembly of T4 proheads proceeds *in vitro* with partially purified proteins (Van Driel and Couture, 1078a). Leaving out the gene 20 protein results in *in vitro* assembly of double-walled long polyheads. Addition of gp20 results in the assembly of prolate shells. We can envisage gp20 as forming a ring that correctly organizes the first annulus of the scaffolding protein. This structure then binds the coat protein. The course of the propagation of the polymerization reaction remains unclear. Nonetheless the reactions must involve the binding of free subunits to a growing structure, with concomitant switching of the subunits themselves into the active state (Fuller and King, 1980b).

There are many T4 capsid proteins other than those discussed here. Some are needed for the DNA injection process, others probably play auxiliary roles in the assembly process, and others are needed for the binding of the head to the tail. A more complete description can be found in Eiserling (1979) or in the reviews referred to in Section 1.

5.3 The Control of Shell Dimensions

5.3.1 Isometric Shells

In the isometric phages, such as lambda, P22, and T3, the proper radius for the shell is determined by the interaction between the coat and scaffolding protein. The coat subunits alone can polymerize into isometric shells, though with reduced accuracy. In the absence of the scaffolding protein, the coat protein of P22 forms shells smaller than normal, shells of the normal dimensions, and spiral or nested shell structures in which one side of the growing shell overlaps and continues past the other side (Earnshaw and King, 1978). For icosahedral shells the radius is determined by the location of fivefold coordinated subunits among the sixfold coordinated subunits. The P22 spiral structures represent cases in which the fivefold vertices are irregularly located among the sixfold coordinated subunits. It is easy to visualize a second layer of subunits more precisely specifying the radius of curvature of the shell than a single layer, but the actual topology of the interactions is not known.

The control of shell radius is most sharply posed in the construction of the shells of phage P2 and its satellite phage P4. The mature isometric shell of phage P2 has a diameter of 595 Å. However, if P2 cells are coinfected with the satellite phage P4, the P2 capsids proteins form a considerably smaller shell of diameter 455 Å (Diana et al., 1978). The actual mechanism of this control is unclear, though Geisselsoder et al. (1978) have shown that it operates at the transcription level (see Section 7). One possibility is that P4 changes the ratio of the synthesis of the P2 coat and (as yet unidentified) scaffolding protein.

5.3.2 Nonisometric Shells

Whereas a single radius fully specifies the shape of isometric shells, the nonisometric shells, such as those of T4 and ϕ29, require that two dimensions be precisely specified: a cylindrical radius and a length. For T4 it is clear that the proper radius is specified by the interaction of the scaffolding core and the major capsid protein (Laemmli et al., 1970). These two proteins are not sufficient to specify the length. Both in vivo and in vitro the presence of initiation protein gp20 is necessary to specify the second dimension, the length of the prolate shell (Van Driel and Couture, 1978b). The vertices of the mature phage are in fact formed by a fourth protein, the product of gene 24 (Muller-Salamin et al., 1978). In the absence of this protein, shells of correct dimension still form, but they lack protein at the vertices, as well as aberrant polyheads. Gp24 appears to be an auxiliary factor that increases the accuracy of a process controlled by the interaction of the initiation, coat, and scaffolding proteins.

Paulson and Laemmli (1977) proposed a helical vernier model for the determination of the length. They showed that the scaffolding core within T4 precursor shells was composed of five or six helices of subunits, with a pitch different from the arrangement of coat subunits in the outer shell. Assuming that the first set of vertices from the initiation complex formed correctly, the second set would be determined by the coming into phase of the lattices of the inner scaffolding and outer shell. The initiating vertex composed of gp20 would in this case control the initial relationship of the two sets of helices.

These models view the shell as composed of two hemispherical caps with a cylindrical shell connecting them. The organization of both the scaffolding and coat subunits in the cap regions is uncertain. That such models of the phage are relevant comes from the existence of "giant" phage, particles that are many times longer than they are wide. (These particles can be induced by use of amino acid analogues, or by infection with certain mutants in the genes for the coat protein or scaffolding protein.) The length of these particles varies independently of the width. They are discussed in detail by Cummings and Bolin (1976).

5.4   Dissociation of the Scaffolding from the Precursor Shell

121

PROTEIN
INTERACTIONS IN
PHAGE
MORPHOGENESIS

5.4.1   Recycling Scaffolding Protein of P22

The scaffolding or assembly core proteins have to be removed from the precursor shell prior to the packaging of DNA. In phage P22, the scaffolding molecules exit intact from the shell, and then take part in further rounds of prohead assembly (King and Casjens, 1974). Experiments with antibodies to scaffolding protein suggest that the gp8 molecules (42,000-dalton monomers) exit through the coat protein lattice, rather than through the hole through which the DNA presumably enters (Shea, 1977). The scaffolding monomer is a long thin molecule with an axial ratio of about 9 (Fuller and King, unpublished), making this notion palatable.

Within the prohead the scaffolding protein appears to be stably bonded to the coat protein and to itself.

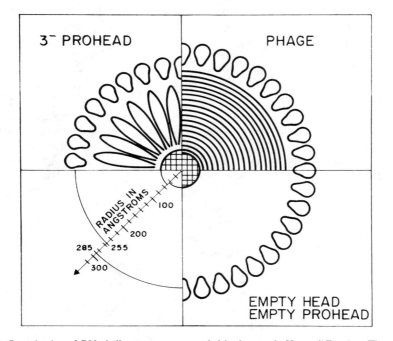

Figure 7. Organization of P22 shell structures as revealed by low-angle X-ray diffraction. The upper right panel shows the organization of the mature phage. The shape of the coat subunit is arbitrary. DNA molecules are locally parallel and concentrically coiled, but with only 5 or 6 concentric shells (Earnshaw and Harrison, 1977). The precursor proheads of small radius contain only protein and show an inner shell, illustrated as composed of elongated scaffolding protein molecules. This is in keeping with measurements of the axial ratio of the scaffolding protein (Fuller and King, unpublished). When the scaffolding protein exits, either *in vivo* during DNA packaging, or *in vitro* by treatment described in Fig. 6, the shell enlarges to the mature radius. [From Earnshaw *et al.* (1976), reprinted with permission.] Casjens (1979) has proposed that the coat protein probably has at least two distinct domains, joined by a hinge, accounting for the expansion. The protein must be synthesized with the hinge "closed," as coat protein assembles into shells of prohead radius with or without the scaffolding protein (Earnshaw and King, 1978).

The mature shell that has lost its scaffolding protein and gone through the DNA packaging process has a radius about 10% larger than the precursor shell radius (Earnshaw *et al.*, 1976). This enlargement of the precursor shell before or during the DNA packaging process is a feature found in all the dsDNA phages (Murialdo and Becker, 1978). Presumably alteration in the coat protein lattice is coupled to the release of the scaffolding protein. These events in P22 are coupled to the DNA packaging process. DNA packaging requires an interaction between a precursor shell containing the gene 1 protein (presumably organized into a fivefold vertex) with a complex of two other proteins, gp2 and gp3, which probably interact with the packaging site on the replicating DNA (Jackson *et al.*, 1978; Poteete *et al.*, 1979). One feature of this interaction would then be the triggering of the rearrangement of coat and scaffolding subunits, and the propagation of this change throughout the entire shell.

The exit of the P22 scaffolding protein can be triggered *in vitro* by treatment with low concentrations of denaturing agents, such as SDS or guanidine hydrochloride (Earnshaw *et al.*, 1976). The resulting empty shells, an electron photomicrograph of which is shown in Fig. 6, have the expanded lattice characteristic of the mature shell. This lattice transformation appears to be built into coat protein because shells made only of coat protein, isolated from cells infected with chain termination mutants of the scaffolding protein, also undergo expansion upon similar treatment (Earnshaw and King, 1978).

Both X-ray diffraction analysis and electron-microscopic studies suggest that the surface arrangements of the coat protein subunits in the precursor shell is geometrically equivalent to the arrangement in the mature shell (Earnshaw and King, 1978; Casjens, 1979). As pointed out by Casjens (1979), a simple explanation for the shell expansion, with preservation of geometry, would be an increase in the angle between two ends of the coat protein subunits. This implies the existence of a hinge between two domains of the subunit. Such hinges have been described for the coat protein of tomato bushy stunt virus, whose structure has been solved by X-ray diffraction to high resolution (Winkler *et al.*, 1977). It would be of considerable interest to know whether such molecular hinges are also involved in the switching interactions controlling the polymerization process.

6 Catalytic Proteins in Morphogenesis

Some of the proteins discussed in this chapter, such as the P22 scaffolding protein, can be thought of as morphogenetic enzymes. In the case of the scaffolding protein, some 200 molecules are needed for the formation of one product, the mature virus shell (Fig. 8). Each molecule takes part in 6–7 rounds of shell assembly in the normal infectious process (King and Casjens, 1974). The stability of the "enzyme–substrate" complex renders this way of thinking of limited use. A number of other proteins fulfill much more sharply enzymic functions in assembly.

6.1 Proteolytic Cleavage in Morphogenesis

Although most enzymology still focuses on the activity of proteins free in solution, work on mitochondrial membranes, muscle, cilia, flagella, and even DNA synthesis demonstrates the importance of catalytic processes that occur as part of organized structures. A particularly clear example of the control of catalytic activity according to organizational state is the control of the cleavage of the T4 coat proteins during morphogenesis. As noted above, the scaffolding of the T4 procapsid is removed by proteolysis within the procapsid. The protein responsible for these maturational cleavages is the product of gene 21 of T4. Showe *et al.* (1976*a,b*) isolated this protein and found it to be an inactive zymogen. It is incorporated into the procapsid during assembly. After the completion of the prohead some as yet undefined change in the prohead results in its activation. It is cleaved down to a chain of smaller size, which is active and apparently cleaves both the coat protein and the scaffolding protein. Associated with these major cleavages is an expansion and transformation of the shell lattice into a precursor of the DNA-packaging reaction. The gene 21 product—"T4 prehead proteinase"—is further autocatalytically cleaved to small fragments itself, and is not found in the mature phage (Showe *et al.*, 1976*b*).

Prolate-shaped proheads are formed in the absence of prehead proteinase, but these particles are irreversibly dead and cannot be converted to mature heads by addition of the missing protein. Presumably gp21 must be assembled within the precursor shell and cannot be added after the shell is topologically completed. Given the highly organized character of the shell it seems unlikely that the proteinase molecules randomly diffuse from one substrate protein to another. It would seem more likely that the molecules are precisely located within the shell, and that each one acts only on its neighboring substrate molecules, or else the proteins migrate along a precise two-dimensional pathway following the disintegrating lattice of the inner core (Paulson and Laemmli, 1977).

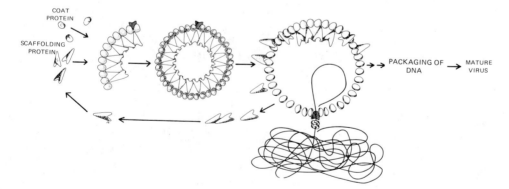

P22 CAPSID MORPHOGENESIS

Figure 8. Polymerization of P22 shell proteins. This cartoon illustrates the overall character of the assembly and disassembly of the double-shell precursor capsid, but does not show the changes in protein conformation that we believe accompany the interaction of coat and scaffolding protein, and the incorporation of the active complex into the growing shell. Although this reaction proceeds *in vitro* with purified subunits (Fuller and King, 1980*a,b*), we do not know if it is free subunits or coat-scaffolding complexes that add to the growing shell. Van Driel and Couture (1978*a,b*) have assembled prolate T4 precursor shells *in vitro,* and they found evidence that the T4 scaffolding protein can self-assemble, suggesting the formation of an inner shell first, followed by the outer shell. With the P22 scaffolding protein, we find no evidence for self-assembly in the absence of coat protein.

6.2   Catalysis of the Formation of Noncovalent Bonds

The clearest case of the catalysis of the formation of noncovalent bonds between structural proteins is the role of the gene 63 product of T4 in the tail fiber attachment reaction. This protein catalyzes the joining of the long slender tail fibers to the baseplate of the phage particle. Neither the tail fibers nor the particles can be activated by preincubation with the attachment enzyme. None of the proteins involved in the reaction undergoes alteration in molecular weight, and there are no known cofactors for the reactions. In the absence of gp63 the fibers do join to the particles, but at only 1% of the rate found in the presence of gp63 (Wood and Henninger, 1969; Wood and King, 1979).

Wood has proposed that the protein either forms a hydrophobic cavity or pocket in which the fiber–baseplate bond forms, or else forms a transient complex with one of the components that induces or stabilizes a conformational intermediate in the reaction. It is worth noting that the fiber–baseplate bond is quite novel, in that it is a molecular swivel (Wood, 1979).

An additional finding, not easily assimilated into the above scheme, is that the gene 63 protein is responsible for the RNA ligase activity found in T4-infected cells (Snopek *et al.,* 1977). The RNA ligase and tail fiber-attaching activities of the protein have different cofactor requirements, and respond differently to inhibitors, suggesting that they are independent activities of a bifunctional enzyme (Wood, 1979).

6.3   Assembly Jigs and Templates

The scaffolding and assembly core proteins that function in head assembly can be thought of as transient jigs for assembly, and if long molecules do determine the length of phage tails they would be acting as templates. The gene 40 protein, which is required for the initiation of T4 shell assembly at high temperature is not incorporated into the shell and can also be thought of as providing a jig for the initiation of the shell on the host cell membrane (Hsiao and Black, 1978b). The need for such devices is more sharply defined in the joining or locating of large components.

As structures get larger, and their diffusion coefficients decrease, the probability of two very small sites finding each other decreases sharply. The long slender fibers of T4 are joined through one tip to the vertices of the phage baseplate. An odd feature noted early in the study of T4 morphogenesis was the inability of completed fibers to attach to completed tails that had not joined to a head. It has long been known that the fiber–baseplate joint was a kind of hinge, and the fibers could swivel from a fully extended position to a position in which they were folded up alongside the head. This suggested that the completed particle might serve as a jig for the initial lining up of the tail fiber, followed by the action of the tail fiber-attaching enzyme, gp63 (Terzaghi *et al.,* 1979). The true character of this process subsequently emerged when it was discovered that there was a previously unidentified set of fibers *(whiskers)*, emerging from the neck of the phage. When mutants defective in this protein were isolated, the whiskerless particles were found to be a poor substrate for tail fiber attachment. These whiskers appear to provide an extended site for initial weak binding of the elbow of the tail fibers. Once bound in two dimensions, formation of the fiber–baseplate bond is presumably highly efficient, although still, of course, requiring the action of gp63 (Conley and Wood, 1975; Coombs and Eiserling, 1977). This process is illustrated in Fig. 9.

6.4   Proteins Needed for Protein Folding

Though it has been shown that in many cases protein folding is a spontaneous event, there are several cases involving structural proteins in which this is not the case. Three proteins of T4, fibrous in nature, require the function of an additional protein to achieve their mature conformations. The proteins are the distal and proximal halves of the tail fibers, the gene 34 and gene 37 proteins, and the short fibers of the baseplate, the gene 12 protein. The tail fiber proteins have been particularly well characterized (Wood and King, 1979). The individual polypeptide chains are very large, 120,000 and 150,000, respectively, and the mature fibers are probably dimers of the chains, organized as parallel chains, for example, as in the rod portion of myosin, but they are not alpha helical.

Gene 57 of T4 specifies a small protein, which has not been found in phage particles. However, if this protein is removed by mutation, the tail fiber polypeptides are unable to achieve their mature conformation. The polypeptides are synthesized, but behave as if they are denatured chains (King and Laemmli, 1971). They do not form morphologically recognizable fibers, nor do they display the antigenicity normally associated with the half-fibers. Similar results are found with respect to the short baseplate fibrils, the gene 12 protein.

The function of gp57 in aiding three structural proteins into their mature conformations is unclear. Formally speaking it behaves as a protein that catalyzes protein folding or subunit assembly (Fig. 10).

Another protein which probably functions in a manner similar to that of gp57 is gp38, a 28,000-dalton protein absent from both mature tail fibers and mature phage but required

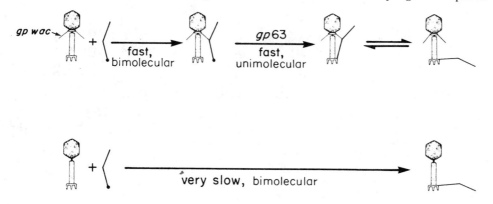

Figure 9.   The jig function of the T4 whisker proteins. The fibers emanating from the neck region of T4 are composed of the product of gene *wac,* gp*wac.* In the absence of this protein phage particles can still be formed, but the rate of attachment of tail fibers is severely decreased (Wood, 1979; Terzaghi *et al.,* 1979). In the mature phage the interaction between these whiskers and the elbow of the tail fiber controls whether the fibers are up or down, and therefore whether the particle is infectious or not. Thus they act as a device that permits the phage to alter its infectivity, depending on the properties of the media (Conley and Wood, 1975).

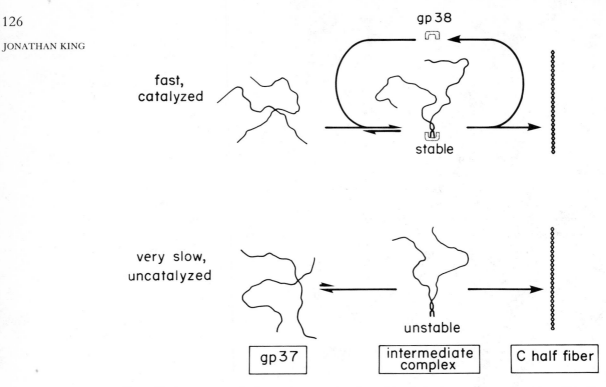

Figure 10. Accessory proteins in protein maturation. A possible role for the nonstructural protein gp38 in T4 tail fiber assembly is to catalyze the dimerization of the two 120,000-dalton chains of gp37 into the distal half-tail fiber (Bishop and Wood, 1976; Wood, 1979).

for fiber assembly. In cells lacking gp38, the two major proteins of distal and proximal half-fibers behave as though they were accumulating as unfolded polypeptide chains. Bishop and Wood (1976) described "bypass" mutants of T4 in which the fibers could be assembled in the absence of gp38. These mutants mapped in the distal end of gene 37. Figure 10 shows a hypothetical mechanism by which gp38 promotes the formation of the gp37 half-fiber by stabilizing or lining up the C-terminal ends of the chains. Perhaps the protein functions in a manner similar to the registration peptides of procollagen (Fessler and Fessler, 1978).

6.5 *DNA-Packaging Enzymes*

A central problem in virus assembly is the mechanism of the packaging and condensation of phage DNA into the precursor shell. Work on bacteriophage lambda, P22, and T4 has clearly shown the role of specific proteins in the packaging process, some of which probably function as do polymerases in being able to move or ratchet DNA, while others may cut the headful of DNA from the replicating concatamer. The catalytic activities of such proteins must be regulated by interactions with the precursor shells. Discussions of these proteins can be found in reviews of Murialdo and Becker (1978) or Earnshaw and Casjens (1980). Studies of the organization of DNA within phage heads show clearly that it is highly organized into concentric shells or coils and that these coils have a unique axis (Earnshaw and Harrison, 1977; Earnshaw *et al.*, 1978; Kosturko *et al.*, 1979). Although it is not difficult to imagine how a polymeraselike complex could sit at the DNA packaging vertex of a shell and rachet in DNA, it is less obvious how the local high order within the shell is achieved, considering how highly constrained the packed DNA is. Hendrix (1979*b*) has proposed that the neck or collar structures, through which the head–tail connection is made, can rotate with respect to the head. Perhaps the DNA is twisted as it is pumped in, resulting in tight internal coils. But proteins within the particle would still be necessary to orient the coils with respect to the shell.

T4 and P22 package DNA by a "headful" cutting mechanism in which the volume of the head determines the length of DNA packaged. The protein(s) which makes the double-stranded scission separating the newly packaged DNA from the replicating concatemer, must not act until the head is filled. The mechanism by which such enzymic activity is regulated by supramolecular organization is at present a mystery.

7 *Regulation of Protein Ratios*

Structural proteins of viruses and other organelles are required in precise proportions with respect to each other. In phage-infected cells structural proteins are generally synthesized in the same ratios, or very similar ratios, as those in which they appear in the virus particle or intermediate structure (King and Laemmli, 1973; Ray and Pearson, 1974). These ratios are determined not at the transcriptional level, but probably by differences in the efficiencies with which the various mRNAs are translated. For example, Ray and Pearson (1975) showed that the late messengers for lambda structural proteins are all synthesized in the same molar ratio, even though the proteins themselves are synthesized in very different numbers. Their results suggested that ribosome initiation or related events determine the rates of polypeptide chain synthesis.

Given that the ratios are determined translationally, we can still ask, are the various proteins synthesized constitutively or is there some feedback regulation, for example, between the assembly pathway itself and the synthesis of the structural proteins. Given the multistep character of the phage assembly pathways, one might imagine feedback loops between the levels of intermediate structures and the synthesis of proteins downstream. From a strictly formal point of view, structural intermediates might serve as inducers of the synthesis of subsequent proteins in the pathway. In the course of identifying the proteins in T4 tail assembly, Laemmli and I looked for such feedback loops (King and Laemmli, 1973). In fact the tail proteins were synthesized at the same rates regardless of what stage in tail assembly had been blocked by mutation. This has generally been the case in the other phage systems examined.

In most of the phage morphogenesis pathways, the major structural proteins are incorporated into the particles and are thus used up in the reactions. The scaffolding protein of P22 is an exception, in that it recycles. In fact, investigation of the control of the synthesis of the P22 coat and scaffolding protein reveals the existence of a feedback loop that couples the rate of synthesis of scaffolding protein to the overall assembly pathway (King *et al.*,

1978). In a formal sense the feedback loop behaves as if unassembled scaffolding subunits autogenously repress their own further synthesis. Thus when scaffolding protein is recycling, subunits will be available both for assembly and for regulation, and new synthesis will be at a low rate. When DNA packaging is blocked by mutation, the scaffolding protein accumulates within the organized prohead shell where it presumably is without repressor activity. Under these conditions the rate of new synthesis increases substantially (King *et al.*, 1978) permitting continued synthesis of prohead shells, in the absence of gp8 recycling. Gene dosage experiments indicate that the feedback loop maintains the concentration of available subunits in the proper ratio for correct prohead assembly.

An alternative hypothesis would involve interaction between the nascent coat protein and the newly synthesized or recycling scaffolding protein. If interaction with the nascent coat protein was required for maintenance of the rate of scaffolding synthesis, then recycling molecules could compete with nascent scaffolding chains for the coat protein. Similarly, mutations that prevented formation of a functional scaffolding protein, such as chain termination mutations, would prevent interaction with the coat protein, resulting in a basal level of scaffolding protein synthesis. In fact, amber fragments of the scaffolding protein are synthesized at a basal rate (King *et al.*, 1978). Nevertheless, the amber fragments may have repressor activity, even though they are inactive in the assembly process.

ACKNOWLEDGMENTS

The development of our understanding of the genetic control of and physiological control of phage morphogenesis has come through the contributions, ideas, and interactions of a large group of molecular biologists, few of whom are adequately recognized in this review. Discussions with my colleagues W. B. Wood, Uli Laemmli, Edward Kellenberger, Helios Murialdo, D. L. D. Caspar, Michael Showe, Richard Goldstein, and Roger Hendrix, were particularly helpful in preparing this chapter. Thanks are due also to my co-workers Sherwood Casjens, William Earnshaw, Ruth Shea, Jonathan Jarvik, Samuel Kayman, Phillip Youdarian, David Goldenberg, Donna Smith, Margaret Fuller, and Peter Berget.

The preparation of this review was supported by Grant No. 17980 from the National Institute of General Medical Sciences and PCM7813209 from the National Science Foundation.

References

Arisaka, F., Tschopp, J., Van Driel, R., and Engel, J., 1979, Reassembly of the bacteriophage T4 tail from the core-baseplate and the monomeric sheath protein p18: A cooperative association process, *J. Mol. Biol.* **132**:369.

Arscott, P. G., and Goldberg, E. B., 1976, Cooperative action of the T4 tail fibers and baseplate in triggering conformational change and in determining host range, *Virology* **69**:15.

Asakura, S., 1970, Polymerization of flagellin and polymorphism of flagella, *Adv. Biophys. (Tokyo)* **1**:99.

Berget, P. B., and King, J., 1978, Isolation and characterization of precursors in T4 baseplate assembly: The complex of gene 10 and gene 11 products, *J. Mol. Biol.* **124**:469.

Bishop, R. J., and Wood, W. B., 1976, Genetic analysis of T4 tail fiber assembly. I. A gene 37 mutation that allows bypass of gene 38 function, *Virology* **72**:244.

Botstein, D., Waddell, C. H., and King, J., 1973, Mechanism of head assembly and DNA encapsulation in *Salmonella* phage P22, I. Genes, proteins, structures, and DNA maturation, *J. Mol. Biol.* **80**:669.

Capco, G. R., and Mathews, C. K., 1973, Bacteriophage-coded thymidylate synthetase. Evidence that the T4 enzyme is a capsid protein, *Arch. Biochem. Biophys.* **158**:736.

Casjens, S., 1979, Molecular organization of the bacteriophage P22 coat protein shell, *J. Mol. Biol.* **131**:1.

Casjens, S., and King, J., 1974, P22 morphogenesis: Catalytic scaffolding protein in capsid assembly, *J. Supramol. Struct.* **2**:202.

Casjens, S., and King, J., 1975, Virus assembly, *Annu. Rev. Biochem.* **44**:555.

Caspar, D. L. D., 1976, Switching in the self-control of self-assembly, in: *Proceedings of the Third John Innes Symposium* (R. Markham and R. W. Horne, eds.), pp. 85–99, North-Holland, Amsterdam, New York, and London.

Caspar, D. L. D., and Klug, A., 1962, Physical principles in the construction of regular viruses, *Cold Spring Harbor Symp. Quant. Biol.* **27**:1.

Conley, M. P., and Wood, W. B., 1975, Bacteriophage T4 whiskers: A rudimentary environmental sensing device, *Proc. Natl. Acad. U.S.A.* **72**:3701.

Coombs, D. H. and Eiserling, F. A., 1977, Studies on the structure protein composition and assembly of the neck of bacteriophage T4, *J. Mol. Biol.* **116**:375.

Crowther, A. R., Lenk, E. V., Kikuchi, Y., and King, J., 1978, Molecular reorganization in the hexagon to star transition of the baseplate of bacteriophage T4, *J. Mol. Biol.* **166**:489.

Cummings, D. J., and Bolin, R. W., 1976, Head length control in T4 bacteriophage morphogenesis: Effect of canavanine on assembly, *Bacteriol. Rev.* **40**:314.

Dawes, J., 1979, Functions of baseplate components in bacteriophage T4 infection. III. The functional organization of the baseplate, *Virology* **93**:1.

Diana, C., Deho, G., Geisselsoder, J., and Goldstein, R., 1978, Viral interference at the level of capsid size determination by satellite phage P4, *J. Mol. Biol.* **126**:433.

Earnshaw, W., and Harrison, S. C., 1977, DNA arrangement in isometric phage heads, *Nature (London)* **268**:598.

Earnshaw, W., and King, J., 1978, Structure of phage P22 coat protein aggregates formed in the absence of the scaffolding protein, *J. Mol. Biol.* **126**:721.

Earnshaw, W., Casjens, S., and Harrison, S. C., 1976, Assembly of the head of bacteriophage P22: X-ray diffraction from heads, proheads, and related structures, *J. Mol. Biol.* **104**:387.

Earnshaw, W., King, J., Harrison, S. C., and Eiserling, F. A., 1978, The structural organization of DNA packaged within the heads of T4 wild type, isometric and giant bacteriophages, *Cell* **14**:559.

Earnshaw, W., Hendrix, R., and King, J., 1979, Structural studies of bacteriophage lambda heads and proheads by small angle X-ray diffraction, *J. Mol. Biol.* **137**:575.

Edgar, R. S., 1969, The genome of bacteriophage T4, *Harvey Lect.* **63**:263.

Edgar, R. S., and Lielausis, I., 1968, Some steps in the assembly of bacteriophage T4, *J. Mol. Biol.* **32**:263.

Edgar, R. S., and Wood, W. B., 1966, Morphogenesis of bacteriophage T4 in extracts of mutant-infected cells, *Proc. Natl. Acad. Sci. U.S.A.* **55**:498.

Eiserling, F. A., 1979, Bacteriophage structure, in: *Comprehensive Virology,* Vol. 13 (H. Fraenkel-Conrat and R. R. Wagner, eds.), pp. 543–580, Plenum Press, New York.

Epstein, R. H., Bolle, A., Steinberg, C. M., Kellenberger, E., Boy de la Tour, E., Chevalley, R., Edgar, R. S., Sussman, M., Denhardt, G. H., and Lielausis, A., 1963, Physiological studies of conditional lethal mutants of bacteriophage T4, *Cold Spring Harbor Symp. Quant. Biol.* **28**:375.

Fessler, J. H., and Fessler, L. I., 1978, Biosynthesis of procollagen, *Annu. Rev. Biochem.* **47**:129.

Fuller, M., and King, J., 1980a, Purification of the coat and scaffolding proteins of bacteriophage P22 procapsids, *Virology* (in press).

Fuller, M., and King, J., 1980b, Regulation of coat-protein polymerization by the scaffolding protein of bacteriophage P22, *Biophys. J.* (in press).

Garcia-Bellido, A., 1977, Homeotic and atavic mutations in insects, *Am. Zool.* **17**:613.

Geisselsoder, J., Chidambaram, M., and Goldstein, R., 1978, Transcriptional control of capsid size in the P2:P4 bacteriophage system, *J. Mol. Biol.* **126**:447.

Hayashi, M., 1978, Assembly of ϕX174, in: The Single-Stranded DNA Phages (D. T. Denhardt, D. Dressler, and D. S. Ray, eds.), Cold Spring Harbor Laboratories, N.Y.

Hendrix, R. W., 1979a, Purification and properties of gro E, a host protein involver in bacteriophage assembly, *J. Mol. Biol.* **129**:375.

Hendrix, R. W., 1979b, Symmetry mismatch and DNA packaging in large bacteriophages, *Proc. Natl. Acad. Sci. U.S.A.* **75**:4779.

Hohn, A., Hohn, B., Engel, A., Wurtz, M., and Smith, P., 1979, Isolation and characterization of the host protein gro E involved in bacteriophage lambda assembly, *J. Mol. Biol.* **129**:359.

Hsiao, C. L., and Black, L. W., 1978a, Head morphogenesis of bacteriophage T4 I. Isolation and characterization of gene 40 mutants, *Virology* **91**:2.

Hsiao, C. L., and Black, L. W., 1978b, Head morphogenesis of bacteriophage T4. II. The role of gene 40 in initiating prehead assembly, *Virology* **91**:15.

Jackson, E. N., Jackson, D. A., and Deans, R. J., 1978, *Eco* R I analysis of bacteriophage P22 DNA packaging, *J. Mol. Biol.* **118**:365.

Katsura I., 1976, Morphogenesis of bacteriophage lambda tail polymorphism in the assembly of the major tail protein, *J. Mol. Biol.* **107**:307.

Katsura I., and Kuhl, P. W., 1975a, Morphogenesis of the tail of bacteriophage lambda. II. *In vitro* formation and properties of phage particles with extra long tails, *Virology* **63**:238.

Katsura, I., and Kuhl, P. W., 1975b, Morphogenesis of the tail of bacteriophage. III. Morphogenetic pathway, *J. Mol. Biol.* **91**:257.

Kikuchi, Y., and King, J., 1975a, Genetic control of bacteriophage T4 base plate morphogenesis. I. Sequential assembly of the major precursor *in vivo* and *in vitro*, *J. Mol. Biol.* **99**:645.

Kikuchi, Y., and King, J., 1975b, Genetic control of T4 baseplate morphogenesis. II. Mutants unable to form the central part of the baseplate, *J. Mol. Biol.* **99**:673.

Kikuchi, Y., and King, J., 1975c, Genetic control of bacteriophage T4 base plate morphogenesis. III. Formation of the central plug and overall assembly pathway, *J. Mol. Biol.* **99**:695.

Kikuchi, Y., and King, J., 1976, Assembly of the contractile tail of bacteriophage T4, in: *Cell Motility* (R. Goldman, T. Pollard, and J. Rosenbaum, eds.), Book A, pp. 71–91, Cold Spring Harbor Laboratories, New York.

King, J., 1968, Assembly of the tail of bacteriophage T4, *J. Mol. Biol.* **32**:231.

King, J., 1971, Bacteriophage T4 tail assembly: Four steps in core formation, *J. Mol. Biol.* **58**:693.

King, J., and Casjens, S., 1974, Catalytic head assembling protein in virus morphogenesis, *Nature (London)* **251**:112.

King, J., and Laemmli, U. K., 1971, Polypeptides of the tail fibers of bacteriophage T4, *J. Mol. Biol.* **62**:465.

King, J., and Laemmli, U. K., 1973, Bacteriophage T4 tail assembly: Structural proteins and their genetic identification, *J. Mol. Biol.* **75**:315.

King, J., and Mykolajewycz, N., 1973, Bacteriophage T4 tail assembly: Proteins of the sheath, core, and baseplate, *J. Mol. Biol.* **75**:339.

King, J., and Wood, W. B., 1969, Assembly of bacteriophage T4 tail fibers: The sequence of gene product interaction, *J. Mol. Biol.* **39**:583.

King, J., Lenk, E., and Botstein, D., 1973, Mechanism of head assembly and DNA encapsulation in *Salmonella* phage P22. II. Morphogenetic pathway, *J. Mol. Biol.* **80**:697.

King, J., Botstein, D., Casjens, S., Earnshaw, W., Harrison, S., and Lenk, E., 1976, Structure and assembly of the capsid of bacteriophage P22, *Phil. Trans. R. Soc. London, Ser. B* **296**:37.

King, J., Hall, C., and Casjens, S., 1978, Control of the synthesis of phage P22 scaffolding protein is coupled to capsid assembly, *Cell* **15**:551.

Kosturko, L. D., Hogan, M., and Dattagupta, N., 1979, Structure of DNA within three isometric bacteriophages, *Cell* **16**:515.

Kozloff, L. M., and Lute, M., 1973, Bacteriophage tail components. IV. Pteroyl polyglutamate synthesis in T4D-infected *Escherichia coli* B. *J. Virol.* **11**:630.

Kozloff, L. M., and Lute, M., and Crosby, L. K., 1970, Bacteriophage tail components. III. Use of synthetic pteroyl hexaglutamate for T4D tail plate assembly. *J. Virol.* **6**:754.

Laemmli, U. K., 1970, Cleavage of structural proteins during the assembly of the head of bacteriophage T4, *Nature (London)* **227**:680.

Laemmli, U. K., Molbert, E., Showe, M., and Kellenberger, E., 1970, Form-determining function of the genes required for the assembly of the head of bacteriophage T4, *J. Mol. Biol.* **49**:99.

Laemmli, U. K., Amos, L. A., and Klug, A., 1976, Correlation between structural transformation and cleavage of the major head protein of T4 bacteriophage, *Cell* **7**:191.

Lewis, E. B., 1963, Genes and developmental pathway, *Am. Zool.* **3**:33.

Moody, M. F., 1971, Application of optical diffraction to helical structures in the bacteriophage tail, in: *New Developments in Electron Microscopy* (H. E. Huxley and A. Klug, eds.), Royal Society, London.

Murialdo, H., and Becker, A., 1978, Head morphogenesis of complex double-stranded deoxyribonucleic acid bacteriophages, *Microbiol. Rev.* **42**:529.

Muller-Salamin, L., Onorato, L., and Showe, M. K., 1977, Localization of minor protein components of the head of bacteriophage T4, *J. Virol.* **24**:121.

Nakamura, K., and Kozloff, L. M., 1978, Folate polyglutamates in T4D bacteriophage and T4D-infected *Escherichia coli, Biochem. Biophys. Acta* **540**:313.

Paulson, J. R., and Laemmli, U. K., 1977, Morphogenetic core of the bacteriophage T4 head. Structure of the core in polyheads, *J. Mol. Biol.* **111**:459.

Poglazov, B. F., and Nikolskaya, T. J., 1969, Self-assembly of the proteins of bacteriophage T2 tail cores, *J. Mol. Biol.* **43**:231.

Poteete, A. R., Jarvik, V., and Botstein, D., 1979, Encapsulation of phage P22 DNA *in vitro, Virology* **95**:550.

Ray, P. N., and Pearson, M. L., 1974, Evidence for post-transcriptional control of the morphogenetic genes of bacteriophage lambda, *J. Mol. Biol.* **85**:163.

Ray, P. N., and Pearson, M. L., 1975, Functional inactivation of bacteriophage lambda morphogenetic gene mRNA, *Nature (London)* **253**:647.

Shea, R., 1977, Genetic identification of phage P22 antigens and their structural location, Ph.D. thesis, Massachusetts Institute of Technology, Cambridge, Mass.

Showe, M. K. and Black, L. W., 1973, Assembly core of bacteriophage T4: An intermediate in head formation, *Nature (London), New Biol.* **242**:70.

Showe, M., and Kellenberger, E., 1976, Control mechanisms in viral assembly, in: *Control Processes in Virus Multiplication* (D. C. Burke and W. C. Russel, eds.), pp. 407–438, Cambridge University Press, Cambridge.

Showe, M. K., Isobe, E., and Onorato, L., 1976a, Bacteriophage T4 prohead proteinase. I. Purification and properties of a bacteriophage enzyme which cleaves the capsid precursor protein, *J. Mol. Biol.* **107**:35.

Showe, M. K., Isobe, E., and Onorato, L., 1976b, Bacteriophage T4 prehead proteinase. II. Its cleavage from the product of gene 21 and regulation in phage infected cells, *J. Mol. Biol.* **107**:55.

Simon, L. D., 1972, Infection of *Escherichia coli* by T2 and T4 bacteriophages as seen in the electron microscope: T4 head morphogenesis, *Proc. Natl. Acad. Sci. U.S.A.* **69**:907.

Simon, L. D., and Anderson, T. F., 1967, The infection of *E. Coli* by T2 and T4 bacteriophages as seen in the electron microscope. I. Attachment and penetration, *Virology* **32**:279.

Snopek, R. J., Wood, W. B., Conley, M. P., Chen, P., and Cozzarelli, N. R., 1977, *Proc. Natl. Acad. Sci. U.S.A.* **74**:3355.

Susskind, M. M., and Botstein, D., 1978, Molecular genetics of bacteriophage P22, *Microbiol. Rev.* **42**:385.

Terzaghi, B. E., Terzaghi, E., and Coombs, D. H. 1979, The role of the collar/whisker complex in bacteriophage T4D tail fiber attachment, *J. Mol. Biol.* **127**:1.

To, C. M., Kellenberger, E., and Eisenstark, A., 1970, Diassembly of T-even bacteriophage into structural parts and subunits, *J. Mol. Biol.* **46**:493.

Tschopp, J., and Smith, P. R., 1977, Extra long bacteriophage T4 tails produced under *in vitro* conditions, *J. Mol. Biol.* **114**:281.

Tschopp, J., Arisaka, F., Van Driel, R., and Engel, J., 1979, Purification, characterization and reassembly of the bacteriophage T4D tail sheath protein P18, *J. Mol. Biol.* **128**:247.

Uratani, Y., Asakura, S., and Imahori, K., 1972, A circular dichroism study of *Salmonella* flagellin: Evidence of conformational change on polymerization, *J. Mol. Biol.* **67**:85.

Van Driel, R., and Couture, E., 1978a, Assembly of bacteriophage T4 head related structures. II. *In vitro* assembly of prehead-like structures, *J. Mol. Biol.* **123**:115.

Van Driel, R., and Couture, E., 1978b, Assembly of the scaffolding core of bacteriophage T4 preheads, *J. Mol. Biol.* **123**:713.

Wagenknecht, T., and Bloomfield, V. A., 1977, *In vitro* polymerization of bacteriophage T4D tail core subunits, *J. Mol. Biol.* **116**:347.

Wagenknecht, T., and Bloomfield, V. A., 1978, Bacteriophage T4 tail length is controlled by its baseplate, *Biophys. Res. Biophys. Res. Commun.* **82**:1049.

Webster, R. E., and Cashman, J. S., 1978, Morphogenesis of the single stranded filamentous phage, in: *The Single Stranded DNA Phages* (D. T. Denhardt, D. Dressler, and D. S. Ray, eds.), Cold Spring Harbor Laboratories, New York.

Winkler, F. K., Schutt, C. E., Harrison, S. L., and Brilogne, G., 1977, Tomato bushy stunt virus at 5.5 Å resolution, *Nature (London)* **265**:509.

Wood, W. B., 1979, Bacteriophage T4 assembly and the morphogenesis of subcellular structure, *Harvey Lect.* **73**:203.

Wood, W. B., and Henninger, M., 1969, Attachment of tail fibers in bacteriophage T4 assembly: Some properties of the reaction *in vitro* and its genetic control, *J. Mol. Biol.* **39**:603.

Wood, W. B., and King, J., 1969, Assembly of bacteriophage T4 tail fibers: The sequence of gene product interaction, *J. Mol. Biol.* **39**:583–602.

Wood, W., and King, J., 1979, Genetic control of complex bacteriophage assembly, in: *Comprehensive Virology,* Vol. 13 (H. Frankel-Conrat and R. R. Wagner, eds.), pp. 581–633, Plenum Press, New York.

Wood, W. B., Edgar, R. S., King, J., Henninger, M., and Lielausis, I., 1968, Bacteriophage assembly, *Fed. Proc.* **27**:1160.

Youderian, P., 1978, Genetic analysis of the length of the tails of lambdoid bacteriophages, Ph.D. thesis, Massachusetts Institute of Technology, Cambridge, Mass.

Mechanisms Regulating Pattern Formation in the Amphibian Egg and Early Embryo

J. C. GERHART

1 Introduction

Interest in pattern regulation arose with the first experiments of embryology almost a century ago. The hypotheses of His and Weismann (reviewed by Wilson, 1896) predicted that the fertilized egg behaves as a mosaic of autonomously differentiating parts, due to either the invisible promorphology of its cytoplasm (His) or the preexisting architecture of its chromosomes (Weismann). Roux's first experiments in 1888 seemed to support these hypotheses: he pricked a frog egg *(Rana esculenta)* at the two-cell stage with a hot needle so as to inactivate one cell, hence half the egg, and found the other cell continuing to divide and in many cases developing into a half-embryo, usually a right or left half. Thus, the active cell failed to adjust its development to compensate for its defective neighbor, at least under these experimental conditions. The frog egg appeared to meet the definition of a mosaic egg, that is, one showing no pattern regulation.

On the other hand, Driesch (1892) obtained the opposite result when he separated the cells of a 2- or 4-cell sea urchin egg, finding the individual cells to develop in some cases to complete embryos of one-half or one-fourth size. The sea urchin egg showed pattern regulation, a new phenomenon and one invalidating the His and Weismann hypotheses of total mosaicism. Shortly thereafter, Hertwig, Enders, and Herlitzka separated cells of the 2-cell amphibian egg and observed complete pattern regulation by the individual parts.

J. C. GERHART ● Department of Molecular Biology, University of California, Berkeley, California 94720

Further, Morgan (1895) repeating the Roux experiment, inverted the frog egg after the hot needle puncture, thereby creating conditions for the living half to develop to a complete (but half-size) embryo. Clearly, the extent of pattern regulation depends profoundly on the conditions of treatment of the egg. In time the categorization of eggs as mosaic or regulative lost value because all eggs behaved to some extent as both, depending on the stage of development and the conditions of treatment. Operationally, the amphibian egg and embryo at the early stages are relatively easy to induce to adjust pattern in response to surgical removal or rearrangement of parts, more so than the eggs of ascidians, nemertine worms, molluscs, and some insects, but perhaps less so than the mammalian egg. Therefore the amphibian egg is referred to as "regulative," even though no deep distinction is implied.

At the present time it is not possible to give an incisive analysis of pattern regulation as represented by regulative eggs and embryos, or even to identify exactly the level of the problem. The subjects of metabolic and genetic regulation are in a far more orderly state of understanding because the basic mechanisms of metabolism and gene expression are at least well outlined. In metabolic regulation one would no longer think of suggesting that enzymic catalysis regulates metabolism, even though this is accurate in the larger sense of an existential statement about metabolism. Instead, one would discuss allosteric effectors, covalent modification of enzymes, protein turnover, and other mechanisms by which the rate of catalysis and number of enzyme molecules are affected, while taking for granted the existence of catalysis. But in pattern regulation the discussion of experimental results frequently concerns mechanisms that "regulate" patterns only in the sense of allowing their existence in the first place—that is, in the sense that catalysis "regulates" metabolism. Less frequently, we meet experimental results about mechanisms by which pattern formation is self-adjusting in the way metabolism is self-adjusting, (through feedback inhibition, for example). In the long run, however, it may turn out that the mechanisms regulating pattern formation are conceptually distinguishable from the mechanisms of pattern formation themselves, but are experimentally less separable than are catalysis and allosteric control, given the exceeding intricacy of spatial patterns and the likelihood of a close interplay between patterning and its control.

It still challenges embryologists to comprehend the experimental effects on regulative eggs and embryos as a way to illuminate and evaluate mechanisms of pattern formation and their regulation. During the past century there has accumulated a large body of observations from such experiments as well as a body of interpretations from the foremost experimentalists such as Spemann, Holtfreter, and Nieuwkoop. These comprise the material of the present review, whereas less emphasis is given to general theoretical considerations of pattern formation (see Child, 1915; Dalcq, 1949; Turing, 1952; Wolpert, 1969, 1971; Goodwin and Cohen, 1969; Meinhardt, 1978; and Kauffman et al., 1978).

Pattern regulation resembles the better characterized regulations of metabolism and gene expression in that it, too, contributes to the homeostasis of the organism, in this case to the homeostasis of pattern, that is, the spatial array of cell types at the multicellular stages of the organism and the spatial array of intracellular materials at the unicellular stages. In the case of pattern regulation by the egg or early embryo, as opposed to pattern regulation by the regenerating adult, the pattern toward which homeostasis is directed is not explicit, since early development comprises a ceaseless progression through a variety of forms, each with a relatively indistinct pattern compared with that of the differentiated adult or larva. As a common embryological convenience, the egg and embryo are described in terms of a "fate map" of "prospective" adult or larval parts. That is, the egg is known for what it will become rather than for what it is. If the egg and embryo are too literally regarded as a latent larva or adult, it becomes a small step to view pattern regulation at the embryonic stages as homeostatically directed toward the adult pattern, as if the embryo has foreknowledge of its fate or seeks the ideal of adulthood.

As the alternative used in this review, early development is regarded as a series of defined morphological and physiological stages, each with its own pattern, albeit difficult to discern, and each pattern serving as the spatial condition for the transformation of that stage to the next by a limited set of morphogenic mechanisms. These stages and morphogenic mechanisms have become increasingly well described in recent years with the advances in microscopy and biochemistry. Morphogenic mechanisms include asymmetric cell division, ion transport, tight junction formation, cell motility and specific cell adhesion, to mention a few well-understood examples, and include inductions, self-organization, and dominance to mention less well-understood ones. Each mechanism presumably has built-in self-adjusting properties, that is, a capacity for regulation applied at the time that mechanism is in use in the embryo. Accordingly, homeostasis of pattern would concern only the present state, not the future. There would be no long-range homing of the embryo toward the adult except insofar as this is an apparent consequence of continuous self-adjustment by morphogenic mechanisms acting to restore aspects of the immediate pattern. Pattern regulation by the egg and embryo can be seen as the composite effect of the regulatory properties of stage-specific morphogenic mechanisms transforming stage-specific patterns, ever correcting the developing organism to the path of its life cycle.

Innumerable experimental treatments have been applied to eggs and embryos in an effort to affect development. These fall into a few classes: first, there are chemical, physical, or surgical treatments that change the existing pattern of the egg and embryo, for example, centrifugation of the egg or removal of parts of the embryo; second, there are those treatments that specifically alter a morphogenic mechanism, for example, exposing the egg to isotonic salt solutions to inhibit osmotic swelling of the blastocoel; and third, there are those treatments that affect both pattern and morphogenic mechanism, exemplified by the excision of the prospective chordamesoderm of the early gastrula. The last class demonstrates that morphogenic mechanisms are themselves part of a pattern. They not only transform pattern, but also express the existing pattern insofar as they differ from one region to the next of the egg or embryo.

Thanks to the recent studies of Nieuwkoop and his colleagues (1969a,b, 1973, 1977), there has been an important modification of the long-accepted picture of pattern formation in the early amphibian embryo, as originating from the experiments of Spemann and Mangold (1924). These classic experiments had demonstrated the ability of the dorsal lip region (the prospective chordamesoderm and somite region) of the early gastrula to "organize" a quite complete embryo from surrounding host tissue when transplanted to a new site at the ventral margin between animal and vegetal hemispheres of a recipient early gastrula. This organizing capacity did not belong to other regions of the early gastrula, entitling the dorsal lip to be called the "organizer" of the embryo. The organizer was said to work through inductions of neighboring tissue, particularly the induction of the precursors of the central nervous system from overlying ectoderm, a process referred to as the "primary induction." The origin of the organizer region was unknown, although numerous researchers identified the gray crescent region of the uncleaved fertilized egg as the most likely precursor, because of its topographic coincidence with the later organizer region. Dalcq and Pasteels (1937) carried this idea furthest in their hypothesis of a "cortical field" centered at the gray crescent in the egg surface and subjacent cytoplasm and organizing the dorsal lip region of the early gastrula. Nieuwkoop's recent results indicate that well before gastrulation it is not the gray crescent, but a region of the vegetal half of the cleaving egg that harbors the potential for establishing the dorsal lip region in the animal half. The gastrula organizer is organized at the blastula stage by a region outside itself, and not by the gray crescent. The gastrula organizer is itself a product of an induction from the vegetal half, long before the organizer exerts its primary induction.

What, then, is the role of the gray crescent? In this review Nieuwkoop's suggestions (1969a,b) are adopted and extended: the crescent is assumed to have no role, but only to mark the position at which an asymmetric cortical contraction had caused the displacement of vegetal cytoplasmic materials to one side shortly after fertilization, producing a vegetal heterogeneity or cytoplasmic localizations on which the dorsal–ventral axis of the embryo will be based much later via several intermediate steps. During the mid and late blastula stages, the pattern of vegetal localizations is used to pattern the animal hemisphere by inductions of dorsal, lateral, and ventral mesoderm. The vegetalizing inductions are assumed to differ quantitatively from one region of the vegetal half to another (Weyer *et al.*, 1978). On the basis of these results and interpretations from Nieuwkoop and his colleagues, many of the old experiments on pattern regulation at the gray crescent and blastula stages in amphibia are reexamined in this review.

The oocyte and egg have recently received less theoretical attention as far as their patterning than have the later embryonic multicellular stages such as the gastrula, neurula, and tailbud stages, or has the regenerating adult. These theories tend to emphasize intercellular mechanisms for pattern formation for which the unicellular oocyte and egg may not qualify. Likewise, the full-grown oocyte, the egg, and the early blastula stages are not easily included in theories of gene expression in early development, as in amphibia these stages accomplish major transformations of pattern and pattern regulation without genetic intervention, using only cytoplasmic agents and circuits established in oogenesis. On the other hand, the early stages offer the opportunity to examine separately these cytoplasmic mechanisms of pattern formation, on the chance that at early developmental stages these are circumscribed enough to study and interpret along the lines of current cell biology and biochemistry. It has been possible to connect pattern formation to cell biology in the instances of maturation, fertilization, and blastula formation in amphibia. This review includes information on the cell biology of the early stages wherever related to questions of patterning.

In summary, this review will cover information on pattern regulation in the oocyte, egg, and early embryo of amphibians, a class of organisms on which there has been extensive experimentation over the past century. Each stage will be represented as having a distinctive spatial pattern and a set of morphogenic mechanisms orginating within the existing pattern and transforming it into the pattern of the subsequent stage. Pattern regulation is reviewed as the sum of the self-adjusting properties of the individual morphogenic mechanisms.

2   Outline of Amphibian Development

2.1   Definitive Axes of the Tadpole

Embryonic development in amphibia leads to the tadpole as the first free-living developmental stage. In *Xenopus laevis*, for example, the tadpole hatches 3 days after fertilization at 20°C, but only begins feeding at 7 days when the gut is completed. The elongated

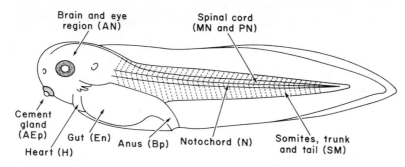

Figure 1. Xenopus laevis, stage 41, hatched larva, length approximately 7 mm, approximately 0.5×10^6 cells. Diagrammatic external view, with some internal organs visible owing to the transparency of the surface. A few organs are identified by name and by embryonic origin in parentheses referring to regions of the gastrula fate map shown in Fig. 4. Abbreviations are explained in Table I.

tadpole has approximate bilateral symmetry, so its anatomical pattern can be referred to two "definitive" body axes. At the histological level, the patterns comprise the spatial array of definitive cell types in their specific tissue and organ associations. The commonly used body axes are first, the head–tail axis, and second, the dorsal–ventral axis (back to belly axis), as shown in Fig. 1. The dorsal side of the tadpole is noteworthy for its axially organized musculature, nerve cord, and skeleton, the mark of a true vertebrate. It is usually ignored that the right and left halves of the body are not quite perfect mirror images due to asymmetry of the circulatory system and the coiling of the gut. In addition, it is sometimes useful to speak of the inside–outside axis comprising the endodermal (innermost), mesodermal (intermediate), and ectodermal (outermost) layers of tissue. Endodermal structures include elements of the digestive and respiratory systems; mesodermal structures, the elements of the musculature (for example, the somites), the skeletal system (for example, the notochord of the tadpole), and the circulatory system; and ectodermal structures, the skin, the sensory organs, and nervous system. These examples are mentioned because they will be traced to their origins at earlier developmental stages. Details of *X. laevis* tadpole anatomy can be found in Nieuwkoop and Faber (1975), Weisz (1945), and Deuchar (1975).

2.2 The Late Neurula–Early Tailbud Stages (Stages 18–26, X. laevis, 24-Hours Postfertilization, 10^5 Cells)

Here the embryo (Fig. 2) has achieved the pattern of the tadpole insofar as the dorsal–ventral and anterior–posterior axes are definitively positioned; however, the definitive cell types are not yet present. The molecules and structures that provide the various cell types with their specific characteristics emerge after the tailbud stage; for example, actomyosin appears (Nass, 1962) in striated muscle cells, larval hemoglobin appears in red blood cells, pigment in retinal cells and chromatophores, neural cells acquire axonal extensions; and cells in general undertake synthesis of materials such as mitochondrial enzymes (Boell *et al.*, 1947; R. A. Wallace, 1961; Williams, 1965), mitochondria (Chase and Dawid, 1972), ribosomes (Brown and Litna, 1964a,b), and total nonyolk protein (Nass, 1962) in the postneurula stages. It is clear that morphogenesis is completed well in advance of the terminal

TABLE I. Abbreviations Used in Figures 1 through 7

A	Anterior end
AAF	Anterior archenteron floor
AAR	Anterior archenteron roof
AC	Activation contraction
AEp	Anterior epidermis
AN	Anterior neural tube
AP	Animal pole
AW	Activation wave
BC	Bottle cells
Blc	Blastocoel
Bp	Blastopore
D	Dorsal side
DE	Deep ectoderm
DYFC	Dorsal yolk-free cytoplasm
En	Endoderm
GC	Germ cells (indicated by asterisks)
GrC	Gray crescent
IM	Invaginated mesoderm
H	Heart mesoderm
HM	Head mesoderm
LM	Lateral mesoderm
MAF	Midarchenteron floor
MAR	Midarchenteron roof
MEp	Midepidermis
MN	Midneural tube
MIZ	Mesoderm invagination zone
N	Notochord
P	Posterior end
PAF	Posterior archenteron floor
PAR	Posterior archenteron roof
PEp	Posterior epidermis
PFW	Postfertilization wave
PhEn	Pharyngeal endoderm
PIM	Preinvagination mesoderm
PN	Posterior neural tube
SbEn	Subblastoporal endoderm
SCW	Surface contraction wave
SE	Superficial ectoderm
SEP	Sperm entry point
SpEn	Superblastoporal endoderm
SM	Somite mesoderm
V	Ventral side
VLM	Ventral lateral mesoderm
VM	Ventral mesoderm
VP	Vegetal pole
X	Last point of endoderm to be invaginated in gastrulation

differentiation of cell types. The tail is not yet present and will later grow out from a small bud of actively dividing cells at the posterior end of the embryo, as an almost separate line of morphogenesis from that already completed in the head and trunk. Furthermore, the tailbud cells are among the first to increase their mass in step with their number, instead of just subdividing the parental cell mass as in cleavage.

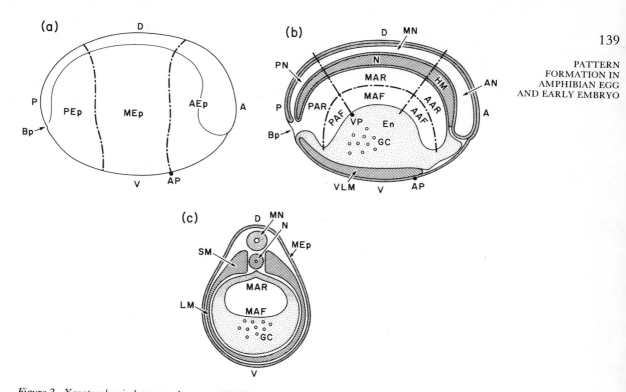

Figure 2. Xenopus laevis, late neurula, stages 18–20, length 2 mm, approximately 5 × 10⁴ cells. (a) Fate map of external surface. Note absence of prospective neurula tissue from the surface. Note location on the surface region over the heart of the original animal pole of the egg. Along the dorsal surface is indicated the bulge produced by the enlarging brain and spinal chord rudiments in the neurula interior. The tail is not yet present; it will form from a bud of actively dividing cells located just above the blastopore. The figure is redrawn from the illustrations of Keller (1975) with the omission of data points. (b) Internal structures displayed in a dorsal median section. Note that the lining of the neural tube and archenteron derives from the original external surface of the early gastrula. Note absence of blastocoel, last present in the region of the animal pole. Redrawn as a composite from the figures of Keller (1976), Nieuwkoop and Florschutz (1950), and Whitington and Dixon (1975), with the omission of data points and cell outlines. (c) Internal structures displayed in a transverse section at the level of the midarchenteron. Layers of internal tissue are shown separated from one another, whereas they are normally apposed. Redrawn as a composite from the figures of Nieuwkoop and Florschutz (1950) and Whitington and Dixon (1975), with the omission of cell outlines.

2.3 The Mid- and Late Gastrula Stages (Stages 11–12½, 12- to 15-Hours Postfertilization in X. laevis)

As shown in Fig. 3, the axes are already in place, but still missing are the neural tube, the material precursor of brain, eyes, spinal chord, and other elements of the central nervous system. The major externally visible event in the transformation of the gastrula to the neurula is neural tube formation from the dorsal ectoderm directly covering the head mesoderm and notochord mesoderm. Internally, the mesodern also guides the development of the archenteron, the central tubular precursor of the gut. Thus, the head mesoderm and chordal mesoderm have a main role in organizing the anterior–posterior and dorsal–ventral axes of the embryo at these stages.

(a)   **(b)**

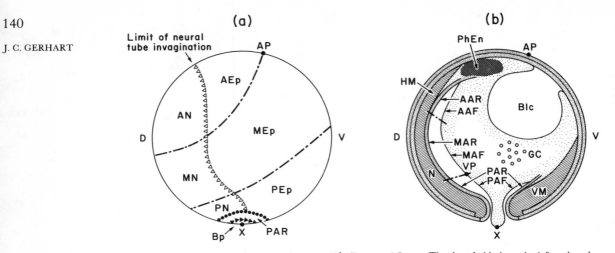

Figure 3. Xenopus laevis, late gastrula, stage 12½, diameter 1.5 mm. The dorsal side is to the left rather than the top due to the internal location of the liquid-filled blastocoel and the yolky endoderm, both affecting the center of gravity of the embryo. (a) Fate map of the external surface. Note the presence of prospective neural tissue, not yet removed to the interior by neurulation. Note the absence of archenteron tissue, already removed to the interior by gastrulation. The original pigmented hemisphere of the egg now covers the entire surface except for a small region at the blastopore. Redrawn from the figure of Keller (1975) with the omission of data points. (b) Internal structures displayed in a dorsal median section. Note the upward extension of the notochord, head mesoderm, and archenteron. Note the position of the original vegetal pole of the egg now on the floor of the archenteron. Redrawn from the figures of Keller (1975, 1976), Nieuwkoop and Florschutz (1950), and Whitington and Dixon (1975), with the omission of data points and cell outlines.

2.4   Axes of the Late Blastula and Early Gastrula (Stages 9 and 10, X. laevis, 15,000 Cells, 9- to 10-Hours Postfertilization)

The axes of the tadpole are unrecognizable at these stages (Fig. 4), which have their own set of axes derived from the egg. The "fate map" of the gastrula displays the relation of the egg axes to the future axes of the tadpole.

The major axis of the late blastula is the animal–vegetal axis, running vertically, around which the blastula has approximate radial symmetry. The upper hemisphere (the animal hemisphere) contains small cells, with a small amount (16% of their volume) of relatively small ellipsoidal yolk platelets (about 4 μm in length), whereas the lower hemisphere (the vegetal hemisphere) contains fewer and larger cells, with a large amount of yolk (55% by volume) made up of large platelets (about 12 μm in length). The animal hemisphere contains abundant dark pigment granules in its most superficial cells; the vegetal cells contain many fewer pigment granules. The animal hemisphere also contains a large fluid-filled cavity, the blastocoel. On one side the late blastula, the surface becomes indented by the first trace of the blastopore. This trace defines the early gastrula and is the indication of the start of gastrulation, the process of massive cell migration transforming the late blastula to the late gastrula, and replacing the egg axes with the embryonic axes. The course of cell migration was first revealed by the vital staining experiments of Vogt (1925, 1929) on *Triturus* (a urodele), as follows: a small dye mark is made on the early gastrula at a precise position noted with respect to the animal–vegetal axis and the blas-

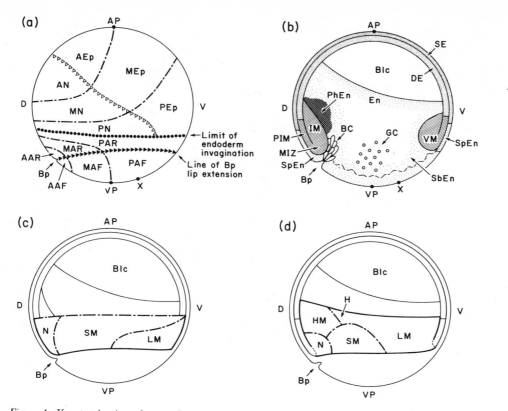

Figure 4. Xenopus laevis, early gastrula, stage 10, diameter 1.5 mm, approximately 20,000 cells. The blasto-pore lip has just appeared. (a) Diagrammatic external view shown as a fate map. Cell boundaries are not indicated. The surface is subdivided into regions identified by their destination in the embryo after gastrulation and neurulation are completed. See Table I for listing of abbreviations. Path of extension of the blastopore lip indicated by (▲); the limit of invagination of surface cells into the interior in gastrulation indicated by (●); and the limit of invagination of surface cells into the interior in neurulation indicated by (△). The original pigmented surface of the egg reaches down to the invagination limit. Redrawn from the figure of Keller (1975), with the omission of the data points. (b) Diagrammatic median section through the blastopore, to show the fate map for internal regions. Note completely internal position of prospective mesoderm, and involution of more than half of the prospective mesoderm to give a two-layered band prior to the appearance of the blastopore. See Table I for explanation of abbreviations. Redrawn from the figures of Keller (1976) and Whitington and Dixon (1975), with the omission of the data points and cell outlines. (c) Fate map of outer layer of mesoderm (migrating downward, not yet involuted), lying directly underneath gastrula surface. Note absence of prospective head mesoderm from this layer. Redrawn from the figures of Keller (1976), with the omission of the data points. (d) Fate map of inner layer of mesoderm (already involuted, migrating upward). Note presence of prospective head mesoderm but little prospective notochord. Redrawn from Keller (1976), with the omission of the data points.

topore. During gastrulation and finally in the tadpole, the new position of the mark is noted. Cells of the early gastrula can be related to cells of the tadpole and all intermediate stages. When the future positions are projected back onto the early gastrula a "fate map" is produced, in which the developmental future of each region is displayed. Keller (1975, 1976) has recently obtained very accurate fate maps of *X. laevis.* Some of his findings are highlighted here, since they will be discussed later.

2.4.1 Internalization of the Surface

As shown in Fig. 4, the surface of the early gastrula comprises future ectoderm and endoderm, but not mesoderm. In gastrulation, the entire endoderm is engulfed. The limit of invagination of surface cells reaches the edge of the pigmented animal hemisphere. The blastopore extends progressively around the embryo at a constant latitude, starting from one side. As endoderm cells invaginate, they move more deeply inward to form a fluid-filled cavity, the archenteron connecting to the blastopore. The original vegetal pole of the egg comes to reside on the posterior floor of the archenteron. Later, in neurulation, the neural ectoderm will be engulfed, leaving a layer of cells descended from only one-fourth of the early gastrula surface as the epidermis covering the tadpole. This is an indication of the cell-surface expansions and patterns of invagination accompanying these stages of development.

2.4.2 Rearrangement of Axes

The blastopore begins on the future dorsal side. As the dorsal endoderm originally above the blastopore turns inward through the blastopore, it expands enormously to form the archenteron roof, extending forward internally to a position almost beneath the original animal pole of the egg, which in the tadpole will lie external to the heart. The animal-vegetal axis of the gastrula does not directly correspond to the anterior–posterior axis of the tadpole. The vegetal pole comes to rest on the floor of the posterior third of the archenteron.

2.4.3 The Mesoderm

Prospective mesoderm is located internally in a ring around the edge of the blastocoel of the early gastrula (Fig. 4b). By the time the blastopore appears, half the mesoderm has already moved over itself and upward on the blastocoel wall, an internal "cryptic" gastrulation. These findings for X. laevis differ from the classic descriptions for Triturus (Vogt, 1925, 1929) and for other urodeles, and even from Rana species and other anurans, where at least some of the prospective mesoderm derives from surface blastomeres. Lovtrup (1975) questions whether the early fate maps of these other species are accurate, since the early researchers may not have controlled carefully the depth of their staining and therefore inadvertently mistook internal cells for surface cells. Keller (1976) discusses this point also, but considers that there may be real variation among the amphibia in the amount of surface-derived mesoderm. Urodele and anuran early gastrulae differ slightly in their architecture and extent of invagination. X. laevis may be an exteme case in having no surface mesoderm at all.

2.4.4 Disappearance of the Blastocoel

The blastocoel is slowly displaced toward the ventral side by incoming cells, for which it provides space (Figs. 3b and 4b). As the blastocoel collapses, the fluid is thought to escape to the archenteron through intercellular spaces and then to the outside through the blastopore (Tuft, 1961a).

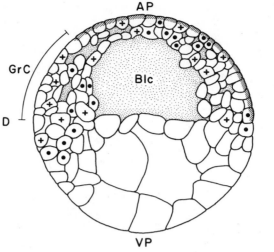

Figure 5. Xenopus laevis midblastula, stage 8, diameter 1.4 mm, approximately 1024 cells. Semidiagrammatic view of Feulgen-stained median section (7 μm) through the gray crescent region, to show animal–vegetal differences in cell size, the position of the blastocoel in the animal hemisphere, and distribution of cells in interphase (●) or in mitosis (+). Many cells do not contain a nucleus in the plane of the section. Redrawn from Bonanno (1977).

2.5   Midblastula (Stage 8, 5-Hours Postfertilization, 10^3 Cells)

A cross section of a midblastula of *X. laevis* is shown in Fig. 5. The animal–vegetal gradation of cell size is clear. The blastocoel is already present. The animal hemisphere is marked by dark surface pigment; its limit is approximately midway between the animal and vegetal poles, higher than in the early gastrula, where cell expansion of the animal hemisphere surface had brought the pigment border downward, an event called epiboly. The blastopore lip is not yet present; the only sign of the future dorsal side is the slightly lighter pigmentation on one side of the animal hemisphere, the "gray crescent." Vital staining shows the animal–vegetal margin on the gray crescent side (about the level of the blastocoel floor) to comprise the cells later originating the blastopore.

2.6   Fertilized Egg (Stage 1, 1 Cell)

The animal–vegetal axis of the egg is essentially the same as that of the late blastula as far as pigment and yolk platelet distributions. This axis is formed in early oocytes growing in the ovary, long before fertilization, as shown in Fig. 6. The gray crescent of the egg is absent at fertilization, making its appearance before the first mitotic cell division of the egg ("cleavage"). The blastocoel is absent, forming between the new cell membranes produced by the rapid cell divisions (20–30 min each) prior to the midblastula stage.

The developmental fate of regions of the egg can be predicted even at this early stage. First, with respect to the animal–vegetal axis, the animal hemisphere is destined for ectodermal structures of the tadpole, the yolky vegetal half is destined for endoderm in the tadpole, and the equatorial zone at the meeting of the hemispheres is destined for mesodermal structures. Second, the surface of the egg gives rise to the embryonic external and internal surfaces such as the epidermis, archenteron boundary, and internal surface of the neural tube. In some anurans, such as *X. laevis,* the egg surface probably gives rise only

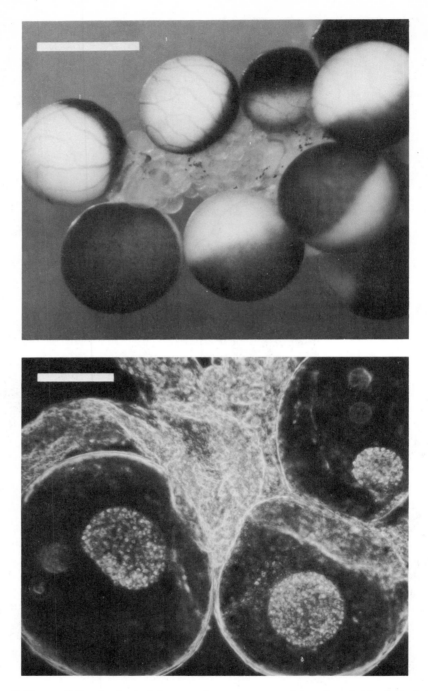

Figure 6. Oocytes of *Xenopus laevis*. A fragment of ovary was removed through an incision in the body wall of an adult female. The top panel shows oocytes of different sizes. Size correlates with the stage in oogenesis. Note the full-grown stage 6 oocytes (diameter 1.2–1.4 mm) in the postvitellogenic period with a reduced vascularization, ready for maturation and ovulation. Note the stage 4 and 5 oocytes (0.8–1.2 mm) in active vitellogenesis with an extensive capillary network in the follicle, and smaller oocytes, including transparent previtellogenic examples (less than 0.3 mm) clustered in the center of the ovary fragment. Note the animal–vegetal

to embryonic surfaces, whereas in other amphibia the egg surface is thought to produce a few deep tissues as well, namely, part of the notochord and head mesoderm—the "organizers" of the late gastrula and neurula. And third, the side of the egg bearing the gray crescent is destined to form dorsal structures such as notochord, somites, and neural tube. In some species such as *X. laevis* and *R. pipiens,* the dorsal–ventral fates can be predicted in the egg even before gray crescent formation, as soon as the sperm enters at random but uniquely through the animal hemisphere, leaving a pigmented surface spot indicative of the future ventral side of the embryo. The gray crescent subsequently forms on that side of the animal hemisphere opposite the sperm entrance point.

In summary, this outline has traced structures of the tadpole back to their origins in regions of the egg. While fate maps of the egg, blastula, or gastrula accurately reveal the spatial relations and rearrangements of these stages, they provide no information about the commitment of cells to their developmental fate. Whereas the map only affirms that the developmental use of various regions is largely the same from one undisturbed embryo to the next, the experiments of surgical transplantation, rearrangement, and explantation have addressed the question of commitment and have permitted an evaluation of which parts of the embryo determine the future of other parts. Also, these experiments have revealed the extent to which a rearranged or diminished embryo can alter its fate map and achieve regulation toward the normal larval pattern.

3 From the Full-Grown Oocyte to the Unfertilized Egg

We now proceed to the detailed architecture of the early developmental stages, the morphogenic mechanisms which transform one stage to the next, and the regulatory processes that endow these transformations with homeostatic properties, insofar as these processes are known. In the next 5 sections, development will be traced in the forward direction, particularly for the case of *X. laevis,* from the full-grown oocyte to the unfertilized egg, then to the fertilized egg, the two-cell stage, the 2000-cell midblastula, the 20,000-cell early gastrula, and the neurula. Finally in the last section we will examine the early oocyte and germ cell, the ultimate precursor of the embryo, to review the origins of egg architecture and of the morphogenic mechanisms of the later stages.

3.1 Morphology of the Full-Grown Oocyte

The full-grown oocyte is the direct precursor of the egg. It is approximately the same size as the egg—that is, 1.2–1.4 mm for *X. laevis,* 1.6–1.8 mm for *R. pipiens,* and 1.8–2.2 mm for the axolotl, to mention frequently used species. An external view of *X. laevis* oocytes is shown in Fig. 6 where the dark animal hemisphere and light vegetal hemisphere are apparent on the larger members of the population. Internally there are four unmistakable zones of material in a full-grown oocyte.

---

axis of the medium and large oocytes, revealed by the pigmented animal hemisphere. The bar length indicates 1.0 mm. The lower panel shows stage 2 previtellogenic oocytes at higher magnification. The largest internal body in the cytoplasm is the germinal vesicle containing hundreds of nucleoli. The mitochondrial cloud is the smaller opaque sphere in the cytoplasm. Two of the oocytes show a ring-shaped structure, perhaps the residual ring canal of the oocyte's connection to a sister cell at the last oogonial division (Kinderman and King, 1973). The bar length indicates 100 μm. Dark-field illumination was used.

J. C. GERHART

3.1.1 The Plasma Membrane and Submembrane "Cortex"

In the full-grown oocyte, the surface is greatly enlarged by microvilli interdigitating with the macrovilli of the surrounding, flattened follicle cells of the ovary. Occasional desmosomes connect the microvilli to macrovilli. Between the microvilli are deposits of glycoprotein material of the extracellular vitelline membrane (Wischnitzer, 1966; Steinert *et al.*, 1977) later to cover the ovulated egg. In *Pleurodeles, Triturus,* and *X. laevis,* the microvilli resemble those of intestinal brush border epithelial cells in containing bundles of parallel core microfilaments issuing from a "continuous web" of microfilaments in the cell cortex (Franke *et al.*, 1976). The oocyte cortex differs from that of other cells in containing dense nonordered aggregates and paracrystals of microfilaments, which Franke *et al.* (1976) consider to be storage forms of actin reserved for use in the egg. The microfilaments have been identified as actin-containing by immunofluorescence microscopy of cortical fragments torn manually from the oocyte surface and washed to remove cytoplasmic contaminants.

The plasma membrane engages in a high level of pinocytosis at earlier stages of oogenesis when yolk deposition occurs, a process discussed in Section 8. But the full-grown oocyte has entered a postvitellogenic phase (diameter 1.2 mm in *X. laevis* oocytes), where the level of pinocytosis is greatly reduced (Holland and Dumont, 1975; R. A. Wallace *et al.*, 1970; R. A. Wallace and Bergink, 1974). Some coated vesicles and pinocytotic pits may still be present as a sign of residual surface activity.

As a specialized zone, the cortex is usually defined as the submembrane cytoplasm, which excludes yolk platelets. Its thickness of 1–10 μm is particularly difficult to estimate in the oocyte due to the contours of the microvilli. In addition to the network of actin-containing microfilaments, the cortex harbors most of the pigment granules (melanosomes) and the cortical granules of the oocyte, as well as glycogen granules, ribosomes, mitochondria, and vesicles, which abound as well in the deeper cytoplasm. The pigment granules and cortical granules are membrane-bounded organelles derived from the Golgi complexes of the young oocyte (Balinsky and Devis, 1963; Wischnitzer, 1966). The former are unusually dense due to the deposits of melanin (1.88 g/cm^3), providing a basis for purification (Eppig, 1970). Their size is 0.5–2 μm, depending on the species. Cortical granules seem to occur in anuran but not in urodele oocytes (Wischnitzer, 1966). In *X. laevis,* there are two kinds of cortical granules: (1) *a* granules predominating in the animal hemisphere cortex and identified by their 1-μm diameter and compact, deep-staining appearance in electron micrographs; and (2) *v* granules of the vegetal hemisphere, 2.5 μm in diameter, irregularly shaped, with flocculent contents. Cortical granules of both kinds are thought to contain approximately 20% by weight neutral and sulfated mucopolysaccharides and 80% proteins, the entire contents released by exocytosis at fertilization (Kemp and Istock, 1967; Campanella and Andreuccetti, 1977; Wolf, 1974*b;* Grey *et al.*, 1974). The basis for the selective association of the granules with the cortex is unknown, but presumably it involves the same mechanisms that bring exocytotic vesicles to the surface in other cells due to special materials of the vesicle membrane. Likewise, the basis for the localization of melanosomes and *a* cortical granules to one hemisphere and *v* granules to the other is unknown.

3.1.2 The Germinal Vesicle

This massive nucleus (300–400 μm diameter) comprises 3–4% of the oocyte volume (Gurdon, 1976; Wischnitzer, 1976). The nuclear membrane is perforated by closely packed pores (Scheer, 1973) and underlain by a nuclear "cortex" (Schatten and Thoman, 1978), a meshwork perhaps of actin, which abounds in the germinal vesicle (Clark and Merriam,

1977). Near the center are the partially condensed chromosomes in diakinesis, the last stage of meiotic prophase. Gruzova and Parfenov (1977) describe an intranuclear membrane system surrounding the partially condensed chromosomes. Between the chromosomal center and the nuclear cortex reside the numerous nucleoli (more than 1000 per germinal vesicle in *X. laevis*; Perkowska *et al.*, 1968), each of large size (3–10 μm diameter). The nucleoli no longer associate with the nuclear membrane as they did earlier in the rapid growth phase of the oocyte. The remaining volume of the nucleus comprises a nuclear sap, perhaps rich in sulfhydryl proteins, since Brachet (1968) reported intense nuclear staining with nitroprusside and Hill *et al.* (1974) stress the need for high concentraitons of mercaptoethanol to maintain the solubility of germinal vesicle proteins. The latter authors identify more than 25 kinds of proteins by polyacrylamide gel electrophoresis, some localized in nucleoli and some in the chromosomes. Actin comprises 6% of the protein (Clark and Merriam, 1977) and RNA polymerase localizes totally in the vesicle (Wasserman *et al.*, 1972). Histones are present in the vesicle well in excess of the available DNA (Bonner, 1975a; Adamson and Woodland, 1974).

In some amphibian species, the single large germinal vesicle is absent, replaced by several hundred separate small nuclei (Del Pino and Humphries, 1978).

3.1.3   The Animal Hemisphere Cytoplasm

The germinal vesicle is located in the animal hemisphere, the half of the oocyte identified externally by its pigmented surface, as shown in Fig. 6. The cytoplasm of this region contains "small" ellipsoidal yolk platelets, 2–4 μm in the long axis in *X. laevis* (Nakatsuji, 1976; Robertson, 1978), occupying 10–20% of the cytoplasmic volume. Each platelet has a crystalline core, a thin amorphous superficial layer, and a unit membrane (Karasaki, 1963). The crystalline core is a lipoprotein complex of lipovitellin and phosvitin, a future source of essential amino acids, lipids, and phosphate for embryonic biosynthesis. The animal hemisphere cytoplasm also contains large amounts of glycogen granules (Ubbels, 1977; Perry, 1966) and fat droplets (lipochondria), ribosomes, mitochondria, numerous unidentified vesicles, and annulate lamellae (Balinsky and Devis, 1963). In general, the organelles resemble those of other cells, except for the presence of yolk platelets.

3.1.4   The Vegetal Hemisphere Cytoplasm

The vegetal hemisphere is characterized by a nonpigmented or less-pigmented surface, and the presence of "large" yolk platelets (10–14 μm in the long axis in *X. laevis*) (Nakatsuji, 1976; Robertson, 1978). These platelets may or may not be of the same composition as the small platelets, and this point of comparison has not been made in detail (Panijel, 1950). The platelets occupy more than half the cytoplasmic volume (54%: Robertson, 1977), approaching close-packing. Between the platelets are organelles of approximately the same appearance as those in the animal hemisphere. The animal and vegetal hemispheres define the main axis of the oocyte and future egg. Wittek (1952) has reported that oocytes of some species also show a bilateral symmetry, owing to the displacement of vegetal yolk platelets to one side relative to the germinal vesicle. Pasteels (1937) considers this oocyte axis to presage directly the dorsal–ventral axis of the embryo in such species as *R. esculenta* and *Discoglossus pictus*.

In addition to these four major regions there are other areas less well identified but perhaps no less important. First, there is a transition zone of intermediate-size platelets between the animal and vegetal hemispheres, marked in *R. fusca* and the axolotl by brown and gray yolk, respectively (Pasteels, 1948; Banki, 1929). Wittek (1952) in particular has distinguished the platelets of this zone as unique on the basis of their origin from a perinuclear source rather than from the oocyte periphery, which is the origin of the other platelets. These interesting proposals of Wittek remain to be reexamined by modern methods. Czolowska (1969) notes an unusual yolk-free body in the intermediate zone at the base of the germinal vesicle in full-grown oocytes of *X. laevis*. Second, T. M. Harris (1964) subdivides the vegetal hemisphere of *Ambystoma maculatum* into two zones on the basis of natural pigmentation, the more vegetal and darker region occupied by the largest platelets, the ones destined to demarcate the level of the blastopore much later as gastrulation. (In Section 4.5 the heterogeneity of the vegetal hemisphere contents will be used in proposals about dorsalization of the egg.) Third, near the vegetal pole reside small yolk-free islands of mitochondria and fibrous material, likely the precursors of the germinal plasm, the cytoplasmic markers included in germ cells of the next generation (Williams and Smith, 1971; Mahowald and Hennen, 1971). Directly at the vegetal pole is a thin layer of cytoplasm with small platelets, more rich in organelles and ribosomes than the deeper vegetal yolk (Czolowska, 1969; Daniel and Yarwood, 1939). Finally, in the animal hemisphere near the germinal vesicle is a region more basophilic and yolk-free, perhaps through accumulation of exported nuclear products (Czolowska, 1969; Brachet *et al.*, 1970).

In summary, major cytoplasmic localizations are present in the full-grown oocyte; many of these persist in the fertilized egg and even the early gastrula as components of the animal–vegetal axis. In the case of familiar cellular organelles we can surmise their function from studies of the genetics and metabolism of other cells; in the case of unique organelles, such as the yolk platelets, we must learn their function from studies of the egg and embryo. Information on the developmental role of these components and their localization will be discussed later for postfertilization stages because there are few experiments directed toward the rearrangement or removal of materials of the oocyte.

3.2 Morphology of the Follicle

The oocyte is contained in a three-layered envelope, the follicle, which attaches to the continuous tissue sheet of the ovary through a stalk or pedicle (Rugh, 1951; Wischnitzer, 1966). The inner layer of the follicle comprises a monolayer of small loosely meshed follicle cells adherent to one another and to the oocyte surface (Dumont, 1972). Each follicle cell extends long macrovilli contacting the microvilli of the oocyte through desmosomes and tight junction (Wallace and Bergink, 1974). Van Gansen and Weber (1972) report a slightly different appearance and arrangement of the follicle cells over the animal and vegetal hemispheres of the oocyte, including the presence of doughnutlike protrusions at the vegetal pole. A thick deposit of extracellular glycoproteins occupies spaces between the oocyte and follicle cells, this being the future vitelline membrane of the unfertilized egg, and after modification, the fertilization membrane of the fertilized egg and embryo. Eventually, the tadpole escapes from the fertilization membrane after weakening it by the action of a proteolytic "hatching enzyme" (Carroll and Hedrick, 1974; Katagiri, 1975).

The middle layer of the follicle comprises fibroblastic cells and capillaries, the latter entering and leaving via the stalk and arranged in a network over the oocyte surface, as is visible in Fig. 6. Through the capillaries arrive massive amount of vitellogenin, the lipoprotein procursor of the yolk proteins, massive amounts of metabolic precursors for glycogen and fat-droplet synthesis, as well as the usual metabolites for macromolecule biosynthesis and energy production, attesting to the fact that the oocyte develops in the ovary as a nutritionally dependent cell of the mother. Also, pituitary hormones arrive via the capillaries, estrogens leave, and waste products are removed. The outer layer of the follicle is the theca interna, composed of smooth muscle and connective tissue, and is continuous with the rest of the ovary (Wischnitzer, 1966).

3.3 Meiotic Maturation and Ovulation

The full-grown oocyte is competent to undergo conversion to an unfertilized egg in response to steroid hormones such as progesterone synthesized and released by the follicle cells (Fortune *et al.,* 1975) in response to circulating pituitary hormones of the luteinizing hormone (LH) class (see Schuetz, 1974; Smith, 1975; for general reviews). The conversion involves four major changes before normal fertilization and development can occur: nuclear changes, surface changes, cytoplasmic changes, and ovulation.

3.3.1 Nuclear Changes

The nuclear events of meiosis begin with the migration of the germinal vesicle to the animal pole, where the nuclear membrane disintegrates first from its more vegetal end (Brachet *et al.,* 1970), releasing voluminous sap into the animal hemisphere cytoplasm, thereby creating a new cytoplasmic localization of major importance. Simultaneously the chromosomes condense and the cell proceeds through the entire meiotic first cycle, entering the second meiotic cycle, finally to arrest in metaphase of the second meiotic cycle. These nuclear events involve many of the same cytological changes as mitosis, that is, nuclear membrane disintegration, chromosome condensation, spindle formation, and cytokinesis, a comparison suggesting that meiosis preserves aspects of the more general phenomenon of cell division (Masui and Clarke, 1979) while adding modifications for the reductive distribution of chromosomes. These modifications occur also in spermatogenesis and concern centromere doubling and separation.

3.3.2 Surface Changes

The surface events such as the disengagement of oocyte microvilli from the follicle cells, the preparation of the cell surface for fertilization and activation, and the conversion of the surface from its normal state of ion permeability to the unusual state of impermeability, which allows the amphibian egg to survive in the hypotonic environment of pond water. In preparation for fertilization and activation, cortical granules move closer to the surface, a layer of membrane-bound vesicles collects close to the plasma membrane (Balinsky and Devis, 1963), the plasma membrane becomes capable of fusion with the sperm plasma membrane, and the surface gains the ability to propagate from the point of sperm entry a "fast arrest" to polyspermy and a signal for cortical granule expulsion.

3.3.3 Cytoplasmic Changes

There are metabolic changes consistent with the future of the egg as a nutritionally independent organism nourished by its own reserves after months of dependence on the metabolism of the mother. General reactions of biosynthesis accelerate as exemplified in *X. laevis* by a twofold increase in polysome content and a 16-fold increase in histone synthesis (Adamson and Woodland, 1977) despite a decrease in the variety of proteins synthesized

(Ballantine *et al.*, 1979). Legname and Buhler (1978) report progesterone-induced changes of carbohydrate metabolism, with an increase in the pentose phosphate steps and a decrease in the Embden–Meyerhof route. Also, the cytoskeletal structure of the cytoplasm changes as reflected by a decrease in cytoplasmic viscosity (Merriam, 1971), although a still larger reduction will occur after fertilization.

3.3.4 Ovulation

Ovulation involves release of the maturing oocyte through the stalk of the follicle and into the body cavity where it is carried anteriorly by cilia of the body wall to the oviduct. In the oviduct, the egg surface is modified further to become sensitive to sperm (Brun, 1974, 1975) and is covered with three layers of jelly, at least one of which participates in sperm activation and a slow block to polyspermy (Wyrick *et al.*, 1974).

3.4 Mechanisms of Maturation

Thanks to the work of Masui (1967), Schuetz (1967), and Subtelny *et al.* (1968), full-grown oocytes of a variety of amphibia can be matured *in vitro,* either exposed in the follicle to pituitary hormones, or removed from the follicle and exposed to progesterone. This has permitted detailed studies of the steps of maturation, to some extent at the biochemical level. In *X. laevis,* oocytes smaller than 0.9 mm in diameter are unable to mature in response to steroids (Hanocq-Quertier *et al.*, 1976), whereas larger ones reach second meiotic metaphase, are activated by needle pricking and support development to postneurula stages if injected with a diploid nucleus (Smith and Ecker, 1969; Drury and Schorderet-Slatkine, 1975). However, they are not fertilized by sperm due to the lack of jelly and interaction with the oviduct. If allowed to pass through the oviduct after insertion into the body cavity, they are fertilizable (Smith *et al.*, 1968; Brun, 1975). There are probably treatments that would bypass the need for the oviduct, since Elinson (1975) and Katagiri (1975) have been able to fertilize body cavity eggs (not yet reaching the oviduct) at low frequency after treatment of the egg surface with proteolytic enzymes and treatment of the sperm with jelly extracts and ficoll, and Subtelny and Bradt (1961) accomplished the fertilization of body cavity eggs to which jelly capsules were added manually.

In *X. laevis in vitro* maturation requires 3–6 hr from the time of exposure to progesterone until the germinal vesicle breaks down, whereas in *R. pipiens* a period of about 16 hr is required. Approximately the same time is required for *in vitro* maturation. During this period, maturation cannot be blocked by inhibitors of RNA synthesis such as actinomycin, and in fact even removal of the germinal vesicle does not retard many of the nonnuclear changes, such as retraction of the microvilli and preparation of the surface for activation by needle puncture. Transcription appears dispensable by the time the oocyte is full grown. Protein synthesis does seem necessary at least for the first 60% of the time interval, based on the inhibitory effects of cycloheximide and puromycin. After the 60% period has passed, the *X. laevis* oocyte cannot be blocked by cycloheximide or puromycin as far as germinal vesicle breakdown is concerned, although later events on the path to second meiotic metaphase remain sensitive to these inhibitors (Wasserman and Masui, 1976*a*).

Masui and Markert (1971) and L. D. Smith and Ecker (1971) found that shortly before germinal vesicle breakdown, a factor appears in the oocyte cytoplasm that can induce maturation in a recipient oocyte to which it is transferred by microinjection, even though the recipient had never been exposed to progesterone. The interval until maturation is shortened to approximately half that required for the progesterone effect. This maturation-promoting factor (abbreviated "MPF") induces maturation to the stage of germinal vesicle breakdown even in cycloheximide-treated recipients (Wasserman and Masui, 1975*a*), a finding which explains why steriod-treated oocytes become insensitive to cycloheximide at a certain point: they have accumulated enough MPF to induce their own maturation. Also, MPF appears on schedule in enucleated oocytes, indicating its strict cytoplasmic origin. When recipient oocytes receive a small dose of MPF in an injection of cytoplasm from a matured oocyte (5% of the oocyte volume) and mature, they likewise become effective donors of MPF for a subsequent transfer to a new recipient; this serial transfer has been carried through 10 cycles, a large enough number to dilute out residual material from the original progesterone-treated donor (Masui and Markert, 1971; Reynhout and L. D. Smith, 1974). MPF appears to promote its own formation by an autoactivation or a multistep positive feedback system, a property which would suggest in the normal course of events after progesterone treatment that a threshold level of MPF is finally produced that allows autoactivation and a rapid conversion of the oocyte from a cytoplasmic condition of very little MPF to one of very high MPF, an all-or-none response. Wasserman and Masui (1975*a*) gained evidence of this autoactivation even in cycloheximide-inhibited oocytes, although Drury and Schorderet-Slatkine (1975) report contrary results under somewhat different conditions. If the former data are correct, then the autoactivation of MPF would seem to resemble the autoactivation of self-phosphorylating kinases or of proteases that activate their own zymogens by proteolysis. The extraction of the MPF system of *R. pipiens* has yielded a protease-sensitive material sedimenting at several positions in a sucrose gradient (with marker proteins at 80, 160, and 240 $\times$ 10^3 daltons) (Wasserman and Masui, 1976). Partial purification of the MPF system of *X. laevis* by 20- to 30-fold has yielded two components needed for full activity: a large ($>$10 S) heat-labile one and a small heat-stable one replaceable by ATP (Wu and Gerhart, 1980). The ATP stimulation is consistent with the idea that MPF involves an autophosphorylating kinase (see Maller and Krebs, 1977; Maller *et al.*, 1977).

A recent experiment of Wasserman and Smith (1978) strongly supports the possibility that MPF is a normal agent not only of meiosis, but also of mitosis. The authors removed cytoplasm from fertilized *R. pipiens* eggs at various times during the first and second mitotic division cycles and injected the cytoplasm into full-grown oocytes. This procedure provided an assay for MPF in eggs. The eggs contained MPF before fertilization but none within 30 min after fertilization, indicating its inactivation after fertilization as the egg completes meiosis. Then MPF appeared again shortly before the first division, after which it disappeared, a pattern repeated at the second division as well. MPF levels changed cyclically with mitosis. More recently, Sunkara *et al.* (1979) discovered the appearance of MPF-like agents in cultured HeLa cells shortly before mitotic prophase. Thus, it seems plausible that the meiotic and mitotic MPF systems are the same.

While the steps after MPF activation may be common to all dividing cells, the steps preceding it must be unique to the oocyte, a cell that completed its last DNA synthesis several months earlier and then arrested at G2 or early prophase for its prolonged period of enlargement and differentiation. Only a few cell types are known to use mechanisms of arrest at the G2 phase (Prescott, 1976), and only the ooctye is known to reverse the arrest through a steroid-dependent system. The oocyte probably has receptors for steroids, although their specificity is so broad as to include adrenal corticosteroids (deoxycorticosterone acting at least as strongly as progesterone), and their detection *in vitro* has been impeded by a high background of steroid binding to other cell materials (Coffman *et al.*,

1979) and by the ability of the oocyte to metabolize steroids (Reynhout and Smith, 1973). Cloud and Schuetz (1977) showed by localized application of progesterone that the oocyte is most responsive on its animal hemisphere surface. It is unknown whether this hemisphere abounds in receptors or in subsequent biochemical relays, either possibility perhaps related to the fact that the animal hemisphere contains approximately twice as much cytosol due to its smaller amount of yolk.

As one of the earliest responses to progesterone, the influx and efflux rates of Ca^{2+}, K^+, and Na^+ in the oocyte increase substantially within 3 min, in part because intracellular levels of these ions increase (O'Connor et al., 1977) and in part because of changes in ion permeability of the plasma membrane. Several authors have suggested that Ca^{2+} is a key signal in the chain of commands from steroid reception to MPF to explain the fact that maturation can be induced by a wide variety of artificial agents such as La^{3+}, Ca^{2+}, or Mg^{2+} added with divalent-ion ionophores, Zn^{2+}, and mercurials, which may all facilitate Ca^{2+} entry or mimic Ca^{2+} (Wasserman and Masui, 1975b; Schorderet-Slatkine et al., 1976; Brachet et al., 1975).

As another early response, the level of cAMP in the oocyte appears to drop within 30 min after exposure to progesterone (Morrill et al., 1977; Speaker and Butcher, 1977), a finding suggesting that high cAMP levels are antagonistic to maturation. Exogenous dibutyryl cAMP is, in fact, known to block maturation in mouse oocytes (Cho et al., 1974) and to block mitosis in cultured cells, although not at G2 (Prescott, 1976). Evidence of an antagonism between high cAMP levels and high Ca^{2+} levels has been obtained by Bravo et al. (1978), who found it is possible to reverse the cAMP and theophylline inhibition of oocyte maturation in X. laevis by increasing the level of extracellular calcium ions.

Maller and Krebs (1977, 1980) have attempted to incorporate these findings into a model in which a progesterone-dependent calcium ion flux leads to a decrease in cAMP, a plausible suggestion in light of recent information from studies of muscle and brain on a heat-stable calcium-binding regulator protein affecting adenylate cyclase and cAMP phosphodiesterase. The low cAMP level is proposed to release the prophase arrest of the oocyte by inactivating the cAMP-dependent protein kinase, the immediate agent enforcing the arrest. For this model Maller and Krebs (1977) have obtained indirect proof, finding that maturation is induced in oocytes by injections of two known specific protein inhibitors of cAMP-dependent protein kinase and is inhibited by an injection of the kinase itself when progesterone is present. More direct evidence of the Ca^{2+}- and cAMP-sensitive factors of oocytes and their eventual coupling to MPF formation has not yet been obtained. At one stage of the coupling, protein synthesis must be required if the cycloheximide sensitivity is to be explained. Bravo et al. (1978) report that injected cAMP-dependent protein kinase inhibits protein synthesis by phosphorylating an initiation factor. If applicable to the normal oocyte, the results would suggest that the high cAMP levels of the oocyte lead, through the kinase, to a partial suppression of protein synthesis [the oocyte is unusual in having only 3% of its ribosomes in polysomes (Woodland, 1974)] and an inability to produce some protein needed for activation of a low but threshold level of MPF. Thereafter, autoactivation would take over and bring MPF to a high titer capable of triggering the numerous enzymes responsible for the nuclear, cytoplasmic, and surface changes of meiotic maturation.

The fact that oocytes injected with MPF in the presence of cycloheximide are capable of undergoing germinal vesicle breakdown and surface changes demonstrates that MPF does not need newly synthesized proteins to activate the enzymes of these steps and that the enzymes themselves must have been present in a completely synthesized but inactive form. This is the physiological situation in which enzyme activation by a covalent modification of the protein would be expected. Such a scheme may be reflected in the finding of Maller *et al.* (1977) of a 2.5-fold increase in the level of protein phosphorylation in the maturing oocyte shortly before germinal vesicle breakdown. This increase in phosphoproteins occurs in MPF-injected oocytes in the presence of cycloheximide, indicating a close association of phosphorylation with maturation.

3.5   Pattern Regulation

These studies of maturation demonstate the highly differentiated state of the oocyte as a large single cell prepared to receive external signals and to respond with preestablished specificity. It is noteworthy that the full-grown oocyte does not require gene expression for maturation to an unfertilized egg. This is the first step in a prolonged independence from gene expression, reaching to the midblastula stage. The events of maturation, fertilization (activation), gray crescent formation, cleavage, cytoplasmic ingression, blastocoel formation and perhaps epiboly occur under the auspices of cytoplasmic control circuits laid down in oogenesis and systematically played out later. Even though the establishment of these circuits probably required gene expression during oogenesis, their operation spatially and temporally during development does not.

In contrast to the numerous experiments on the regulation of pattern formation after fertilization, the periods of oogenesis and maturation are almost unexplored. Experiments in these periods are now feasible with oocytes *in vitro,* for it would be possible in principle to force the rearrangement of the cytoplasmic pattern of the oocyte, then to mature the oocyte *in vitro* and inseminate it. The insemination procedure remains somewhat difficult, requiring addition of jelly coats or proteolytic modification of the egg surface, or nuclear transplantation.

At present the experiments on these stages concern not pattern, but the essentiality of the germinal vesicle and the commitment of the oocyte to stage-specific responses that cannot be taken out of order. On the first point, Smith and Ecker (1969) removed the germinal vesicle from the full-grown oocyte of *R. pipiens,* matured the enucleated oocytes in progesterone *in vitro,* and then introduced a single diploid nucleus by transplantation. The "eggs" activated but did not cleave, whereas nucleated, matured, control oocytes cleaved to form postneurula embryos. Smith and Ecker (1969) could restore the capacity for cleavage in 12–16% of the matured enucleated oocytes by injecting a small volume of germinal vesicle sap from nonmatured oocytes. The germinal vesicle provides materials for cleavage, although it is unknown at what step cleavage fails without these materials.

On the second point, Elinson (1977) has obtained precociously released eggs of *R. pipiens,* which were coated with jelly but still in the first meiotic cycle. These "eggs" did not activate and became multiply inseminated. The several sperm nuclei condensed to metaphase figures, some of which imbedded in the cortex. Later the eggs completed maturation and could be activated by pricking, in which case several polar bodies were extruded. Clearly, the normal responses of activation are scheduled by the egg and early-arriving sperm are entrained in the maturation events of the egg. Subtelny and Bradt (1961) had shown previously that mature body cavity eggs, taken from the frog before the addition of jelly coats in the oviduct, supported normal development of tadpoles when transplanted with a diploid nucleus. Smith and Ecker (1969) obtained equivalent development with oocytes matured *in vitro,* as mentioned above. These results demonstrate the unimportance of the jelly coats for development, despite their essentiality for normal sperm entry.

Because the oocyte itself possesses the complete machinery for responding to external agents such as steroid hormones via ion fluxes and cyclic nucleotides, and most materials of the oocyte are passed to the egg and embryo, it is worth anticipating that the responses of embryonic cells to external inductors at the midblastula and gastrula stages are based on the cytoplasmic circuitry received directly from the oocyte. The artificial neuralizing and vegetalizing inductors of gastrula ectoderm cells are often the same as those effective in the artificial induction of maturation of the oocyte (see Section 7.3.1 for further discussion).

4 From the Unfertilized Egg to First Cleavage

4.1 Morphology of the Unfertilized Egg

The egg preserves the basic morphology of the full-grown oocyte, though there are minor modifications. The most apparent modification is the new region of animal hemisphere cytoplasm mixed with the sap of the disintegrated germinal vesicle. This central area is recognizable by its low yolk content, high glycogen and ribosome content (D. pictus, Klag and Ubbels, 1975; X. laevis, Ubbels, 1977; A. mexicanum, Ubbels and Hengst, 1978) and high sulfhydryl content (Brachet, 1968) the latter presumably arising from the sulfhydryl-rich proteins identified by Hill et al. (1977) in the germinal vesicle. The nucleoli disintegrate in the course of maturation, although the rDNA may persist (Brachet et al., 1970). At the animal pole resides the second meiotic metaphase spindle imbedded in the cortex and carrying the egg chromosomes. The animal pole is in fact defined by the position of the spindle. The unusual barrel shape of the spindle and absence of spindle pole asters probably reflect the absence of centrioles (Wilson, 1925; Fankhauser, 1932; Fankhauser and Moore, 1941a; Brachet et al., 1970). The spindle also is reported to contain sufficient carbohydrate material to stain with PAS (Brachet et al., 1970), perhaps a result of its arrested state. Externally the animal pole is identifiable as a light spot [about 200 μm diameter in X. laevis (Rzehak, 1972)] owing to the local dispersal of pigment granules from the area of the spindle. The first polar body also resides in the light area after its budding from the oocyte surface in the extremely asymmetric cell division of the first meiotic cycle.

In the yolky cytoplasm near the vegetal pole the finely dispersed small islands of germinal plasm underwent limited coalescence during maturation. These yolk-free islands contain mitochondria, ribosomes, and the ring-shaped germinal granules (200-nm diameter in X. laevis) later associated with the blastula cells destined to become primordial germ cells, the precursors of future eggs and sperm (Whitington and Dixon, 1975). The coalescence occurs even in enucleated maturing oocytes (Williams and Smith, 1971). In the egg of Ambystoma maculatum with naturally pigmented large yolk granules in the lower half of the vegetal hemisphere, maturation brings a visible rearrangement of the pigmented yolk, concentrating it nearer the vegetal pole by unknown means (Harris, 1962). Moen and Namenwirth (1977) have achieved mass sectioning of oriented R. pipiens eggs and demonstrate a variety of differences in amounts of specific proteins along the animal–vegetal axis.

Finally, the egg surface is relatively smooth after the microvilli retract from the follicle cells during maturation. The cortical granules almost contact the plasma membrane. Campanella and Andreuccetti (1977) make the interesting observation of the association in X. laevis eggs of the cortical granules with an extensive system of membranous cisternae and endoplasmic reticulum, which, according to these authors, may conduct the ionic stimulus for cortical granule discharge at egg activiation, by a mechanism similar to the mediation of muscle contraction by the sarcoplasmic reticulum. Grey *et al.* (1974) had also commented on the close association of vesicles and flattened sacs with dehiscing cortical granules. Dick and Fry (1975) have suggested a comparable anastomosing membrane system in oocytes on the basis of electron microscopy and postulated ionic compartments consistent with data from efflux and influx studies with radioactive sodium ion. Balinsky and Devis (1963) and Williams and Smith (1971) note at maturation the conversion of masses of annulate lamel-lae of the oocyte cytoplasm to small vesicles that move to the egg surface. Perhaps these rearrangments are involved in the preparation of the egg surface for propagation of the activation stimulus from the point of sperm entry to the entire surface, a circumferential distance of 2 mm to the opposite side in the case of a *X. laevis* egg. The egg membrane itself is unusual in its distribution of intramembranous particles (IMP) associated with the outer rather than inner face of the lipid bilayer (Bluemink and Tertoolen, 1978), an arrangement also known for the membranes of bladder cells, which, like the egg, are highly impermeable.

Nieuwkoop (1956) distinguishes at least six cytoplasmic regions in light micrographs of *X. laevis* eggs: the plasma membrane with an associated thin cortex of yolk-free cytoplasm containing cortical and pigment granules, a subcortical plasm of small platelets and dispersed pigment granules in the animal hemisphere, a deeper plasm of medium-size platelets still in the animal hemisphere, a central plasm of smaller less-frequent platelets at the equatorial level where the germinal vesicle had been located in the ooctye, a transitional zone between animal and vegetal hemispheres, and a large cup-shaped mass of large platelets in the vegetal hemisphere. These regions derive from those of the full-grown-oocyte, with modifications introduced by germinal vesicle movement and breakdown, as well as surface changes. Harris (1962) identifies five major cytoplasmic areas of the *A. maculatum* egg, where distinctions are aided by natural pigmentation differences of the yolk in different regions. The structural and functional individuality of some of these zones is clear; namely, of the yolk-free cortical layer, the animal hemisphere cytoplasm as a whole, and the vegetal hemisphere cytoplasm as a whole. The intermediate cytoplasmic zone may be only an intergrade of the animal and vegetal areas or may be functionally distinct. In *R. fusca,* this zone is more darkly pigmented, and in *X. laevis* it contains yolk platelets of intermediate size (3 μm; Ubbels, unpublished). For speculative purposes later, the functional uniqueness of this intermediate zone will be assumed.

The macroviscosity of the unfertilized egg is sufficiently great that these regions preserve their position despite the random orientation of the egg with respect to gravity. Rotation of the unfertilized egg is prevented by the adherence of the plasma membrane to the extracellular vitelline membrane, which in turn adheres to the jelly coats stuck to fixed objects or other eggs. Unfertilized eggs can remain for several days randomly oriented in the ovisac without detriment to later development (*R. pipiens,* Witschi, 1934). Exposure of the unfertilized egg to colchicine allows gravitational rearrangement of the internal contents, suggesting that microtubules preserve egg structure (Manes *et al.,* 1978). The effect of rearrangement on development has not been tested.

4.2 Sperm Entry

Newport in 1854 and Roux in 1888 used localized insemination to show that some frog species develop the bilateral axis of the embryo in the plane defined by the sperm entry point and the animal and vegetal poles, with the anterior end of the embryo directed toward the side of sperm entry. Roux (1888) observed that the gray crescent formed on the side opposite sperm entry and was a reliable indication of the future dorsal side. Subsequent authors confirmed these results in general outline with other species (Tung, 1933; Ancel and Vintemberger, 1948; see reviews by Pasteels, 1964; and Brachet, 1977). These classical results demonstrated the ability of the egg to establish its fate map with the dorsal side at any position 360° around the equator, and to select one position in some way depending on the point of sperm entry. In such eggs a prelocalization before fertilization of materials for dorsalization either does not exist or is easily overcome. It must be emphasized that other frog species such as *R. esculenta* and *Discoglossus pictus* do *not* orient the bilateral axis according to the point of sperm entry, nor do most urodele eggs, which are usually polyspermic (*T. palmatus* averaging three entries per egg, with up to 10 still allowing normal development; Fankhauser, 1932; Fankhauser and Moore, 1941*a*). These eggs appear to have an asymmetry before fertilization, or to establish asymmetry later by a mechanism unrelated to sperm entry. Even in the classical cases, the point of sperm entry only approximately coincides with the future bilateral axis; 5–10% of the eggs may be 90° off and 30% may be 30–60° off (Tung, 1933). The same approximate relation has been found for *X. laevis* (Kirschner *et al.*, 1979). The relation of sperm entry to bilaterality is good but far from perfect. It is possible that all these eggs eventually generate the same cytoplasmic localization needed for axis determination, but exploit different spatial signals and make commitments at different times, a hypothesis that will be discussed later.

Elinson (1975) has characterized animal–vegetal differences of the egg in susceptibility of its surface to sperm penetration. When the egg is uniformly exposed to a sperm suspension, the point of entry can be located usually by the formation of a small light spot (*R. pipiens*) in the dark background, or a dark spot of concentrated pigment (*X. laevis*, Palaček *et al.*, 1978) in the slightly less pigmented background, or by small pits in the surface (urodeles, Fankhauser, 1932; Fankhauser and Moore, 1941*a*). Elinson (1975) established that in *R. pipiens* the point of sperm entry is almost always in the animal hemisphere and that the success of entry is highest near the animal pole when comparing sperm entries per unit surface area. By localized application, sperm can be made to enter as much as 30° below the equator, but rarely or never at the vegetal pole. Sperm entering below the equator are reported to activate the egg, but lead only to abortive cleavage, probably because the sperm centriole never reaches the animal hemisphere where the egg nucleus awaits. Thus the egg expresses an animal–vegetal surface gradation in the frequency of sperm entry. The step responsible for the animal–vegetal difference is not known: eggs can be artificially activated by needle puncture anywhere on the surface, including the vegetal pole, so the activation system can receive the stimulus and propagate it from anywhere on the surface. Perhaps the vegetal surface deficiency occurs at steps of sperm adsorption or of egg fusion with the sperm membrane. As mentioned earlier (Section 3.4) the vegetal surface of the oocyte is also much less responsive to progesterone than is the animal surface in the induction of maturation.

4.3   Activation

An unknown early step of sperm and egg interaction triggers the egg to set in motion a wide variety of processes prepared in oogenesis and poised in maturation. Many of these processes can be triggered as well by needle puncture, electric shock, osmotic shock, or by divalent-ionophore exposure; namely, the fast block to polyspermy, the discharge of cortical granules, the separation of the vitelline membrane from the plasma membrane, the swelling of the perivitelline space between the two membrances, the rotation of the egg to a position with vegetal end downward, the cortical contraction, the reinitiation of meiosis with the expulsion of the second polar body and reconstitution of the haploid egg pronucleus, the initiation of DNA synthesis in the egg pronucleus, and the appearance of the gray crescent. These artificial means of activation do not allow the egg to form a normal mitotic spindle and undergo normal cleavage; they initiate only a shallow and regressing division furrow at the animal pole. Nonetheless, some form of nuclear division continues without cleavage (Ramirez and Huff, 1967).

Guyer showed in 1907 that frog eggs would cleave if activated by puncture with a bloody needle rather than a clean one, and this result led Shaver (1953) and Fraser (1971) to test a wide variety of tissues and species for the presence of a "cleavage initiation factor" (CIF). Any nucleated animal cell seemed an adequate source and Fraser and Ramirez and Huff have suggested that the factor is the centriole. Fraser obtained a partially purified fraction from chick brain capable of allowing not only cleavage but parthenogenesis in *R. pipiens* to a haploid tadpole. Heidemann and Kirschner (1975) purified centrioles in the form of ciliary basal bodies from *Chlamydomonas* and showed by injection that these allow deep furrows to form in *X. laevis* eggs, although parthenogenesis did not occur. High doses of basal bodies caused the cytoplasm of the egg to fill with spherical asters and to initiate supernumerary furrows. Maller *et al.* (1976) used sea urchin sperm as a source of centrioles and obtained an active fraction of demembranated, disrupted sperm heads carrying the centriole and the fossa region of the nuclear membrane. This fraction allowed parthenogenesis in *X. laevis,* in some cases to tadpoles. It is unclear why these sperm fractions allow parthenogenesis when basal bodies do not. It is possible that the basal body centriole, being plasma-membrane associated, lacks some material for associating with the egg nuclear membrane, such as the fossa region, or that the centriole of *Chlamydomonas* cannot undergo duplication cycles in *X. laevis* cytoplasm, whereas the sea urchin centriole can. In any event, these studies suggest that the amphibian egg is unable to mobilize for spindle formation its own centriole, which had been present in the dividing germ cell ancestors, and that the classical notion (Wilson, 1896) of the sperm as the strict donor of the functional centriole in development maybe close to the truth in amphibians. On the other hand, Ramirez and Huff (1967) have examined activated noncleaving eggs of *R. pipiens* after 3–4 days and report that asters do eventually form. They suggest that the amphibian egg is just very slow to mobilize its own aster formation. In activated *R. pipiens* eggs, a small aster is reported to form even shortly after activation (Manes and Barbieri, 1977). It is unknown whether these asters contain centrioles, and why they fail to function in spindle formation. Ramirez and Huff (1967) showed that the cytoplasm of aged activated eggs is rich in "CIF," and by implication is rich in centrioles.

Haploid parthenogenesis ("gynogenesis") has been obtained in several amphibia in which eggs were activated with CIF preparations, and it is clear that the sperm nucleus contributes nothing irreplaceable for development as far as its chromosome complement is concerned. Similar conclusions are affirmed by the successful haploid development of eggs fertilized by sperm inactivated by X irradiation or chemicals (such as aziridinyl urea, Jones *et al.,* 1975).

We now return to the question of what the sperm or its substitute does to activate the egg. This aspect of amphibian cell physiology is less well studied than for the cases of the sea urchin or medaka egg, and the suggestions for amphibians draw heavily on the findings for these other organisms. Maeno (1959) found that *Bufo* eggs undergo a rapid transient positive-going change of membrane potential to $+20$ mV at fertilization or artificial activation. This spike appeared due to a sudden increase in the Cl^- permeability of the membrane. Morrill (1966; Morrill *et al.,* 1975) observed a similar spike in *R. pipiens* eggs. Over a period of hours, the potential slowly drops to levels of -70 mV, a typical level for many cell types. More recent studies have shown that the unfertilized egg starts from an unusually positive potential, -10 to -20 mV, apparently due to its high membrance impermeability to K^+. L. A. Jaffe (1976) considers that in the sea urchin egg the transient positive potential is the egg's mechanism for a fast block to polyspermy, and supports her idea by the result of obtaining polyspermy under conditions in which the potential is kept "artificially" negative by a voltage clamp during fertilization and by the result of preventing fertilization under conditions where the egg is given an "artificial" positive potential by external ion conditions. Grey and Schertel (1978) consider the positive-going chloride potential as the source of the fast block to polyspermy in *X. laevis,* based on the increased incidence of polyspermy in eggs fertilized in the presence of a high concentration of chloride ion. Whatever the mechanism, the frog egg does achieve high-frequency monospermy under conditions of dilute ions, despite high sperm levels such as those used in artificial fertilization (Wolf and Hedrick, 1971), whereas in urodeles polyspermy is the rule; it is unknown if urodeles achieve a transient positive potential at fertilization.

A series of slower activation reactions are also initiated from the point of sperm penetration, and from analogy to other systems several authors have suggested that these are Ca^{2+}-triggered reactions or H^+-inhibited reactions (see Epel, 1978, for the sea urchin data). Exocytosis of cortical granules spreads as a wavefront, probably identified in cinematography as the "activation wave" in *X. laevis* at approximately 7 min after mixing egg and sperm (Hara *et al.,* 1979; Fig. 7), when the sperm has finally penetrated the jelly layers and vitelline membrane. Then 2–3 min elapse between the initiation and the completion of this wave on the opposite side of the egg. The exocytosis reaction involves the fusion of the cortical granule membranes with the egg plasma membrane, increasing the plasma membrane surface area perhaps by 50% [estimated from the electron micrographs of Balinsky (1966) and Grey *et al.* (1974)]. It is unknown to what extent the cortical granule membrane modifies the ion permeability of the egg membrane.

Hollinger and Schuetz (1976) have shown that Ca^{2+} injected locally even in oocytes can lead to cortical granule discharge in the region of injection; Wolf (1974a–c) showed that Ca^{2+} chelators are necessary for the successful isolation of *X. laevis* cortical granules, which can then be disrupted by Ca^{2+} addition. Thus it seems likely that calcium ion is the immediate and local signal for cortical granule discharge. In *X. laevis,* urethan (Wolf, 1974c) and ionophores of divalent ions (Schroeder and Strickland, 1974; Wolf *et al.,* 1976) cause activation in the absence of external calcium ion, suggesting that the egg has internal stores. These reservoirs are not known, although, as mentioned previously, Campanella and Andreuccetti (1977) have detected an extensive membranous reticulum connecting cortical granules and suggest its roles in triggering cortical granule breakdown as well as propagating the activation stimulus, roles analogous to that of the Ca^{2+}-releasing and sequestering sarcoplasmic reticulum of striated muscle. As yet, there has been no report of a direct detection of calcium ion fluxes at fertilization of an amphibian egg as visualized by luminescence of injected aequorin, as has been accomplished with the equally large medaka egg (Ridgeway *et al.,* 1977) where a band of intracellular free calcium ion appears to advance from the point of sperm entry across the egg within a period of several minutes. The geom-

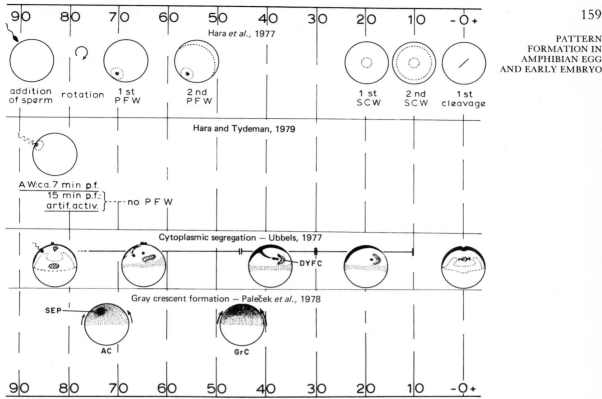

Xenopus laevis. Cortical contraction waves and cytoplasmic displacement.

Figure 7. "Normal table" of external and internal rearrangements of the egg of *X. laevis* between fertilization and first cleavage, summarized by G. A. Ubbels (1977) from her own studies, Palacek *et al.* (1978), Hara *et al.* (1977) and Hara and Tydeman (1979). Sperm and eggs are mixed at zero time, and several minutes are required for sperm to penetrate the egg jelly layers and the vitelline membrane; the actual fertilization probably occurs at 7 min when the activation wave (AW, row 2) begins. See text for explanation. Abbreviations are listed in Table I. Note the activation contraction of the animal hemisphere surface at 12 min after mixing, the two postfertilization waves at 20 and 35 min after mixing (perhaps related to growth of the sperm aster), the asymmetric cortical contraction of gray crescent formation starting at 45 min, the two surface contraction waves at 70 and 80 min (an autonomous cyclical behavior of the egg cytoplasm and cortex), and the appearance of the furrow at 90 min, at a temperature of approximately 18°C.

etry and kinetics of the luminescent band in the medaka egg indicate an autocatalytic release of calcium ion followed by a rapid removal into compartments inaccessible to aequorin. Comparable results have been obtained for the sea urchin egg (Steinhardt *et al.,* 1977). By analogy to these results, the amphibian egg may also propagate the activation stimulus by way of a Ca^{2+}-excitable calcium ion reservoir, perhaps the membranous reticulum connecting cortical granules, which would only need a single local exposure to external calcium ion at the point of sperm entry to initiate propagation and release. Wolf (1974*c*) and others have observed that activation by needle puncture in contrast to ionophore or urethane activation does require the presence of calcium in the external medium, a finding consistent with the notion that a small local calcium ion influx is necessary to trigger the primed relays. It should also be mentioned that urodele eggs lack cortical granules (Wischnitzer, 1966); their dependence on external Ca^{2+} for activation is not known.

The reinitiation of meiosis and the entry of the egg into mitotic interphase are also thought to be triggered by calcium ion. In this area, work on amphibians is far ahead of that on other species, thanks to the investigations by Y. Masui and his colleagues. The unfertilized egg is arrested in the second meiotic metaphase, where it may remain for several days until activation. Masui and Markert (1971) found that the cytoplasm of unfertilized eggs, as well as of oocytes matured *in vitro* contain a powerful metaphase-arrest agent, which can be assayed by withdrawing cytoplasm in a micropipette and injecting it into a fertilized egg of *R. pipiens* at the 2-cell stage. The injected blastomere arrests at mitotic metaphase, whereas the companion continues cleaving even to the late blastula stage. Meyerhof and Masui (1977) have partially purified this "cytostatic factor" (CSF), as they call it, and report it is a protease-sensitive material, inactivated *in vitro* by 1 mM Ca^{2+}. In their view CSF appears during maturation shortly after MPF formation and arrests the egg in the second meiotic metaphase. It is unclear why it does not affect the first meiotic metaphase, although the metaphase–anaphase transition for the two meiotic cycles is certainly different in terms of centromere duplication, a plausible target for CSF action. Then, according to Meyerhof and Masui, at the time of egg activation a wave of Ca^{2+} inactivates CSF and allows meiosis to continue. Apparently the subsequent mitotic cycles do not generate such large waves of Ca^{2+}, as the metaphase arrest of the blastomeres injected with CSF is not spontaneously reversed. In fact Ridgeway *et al.* (1977) detect only small calcium ion fluxes at cleavage in the medaka egg.

In addition to CSF destruction, the establishment of mitotic interphase would require destruction of MPF as well. As described earlier, Wasserman and Smith (1978) found MPF to disappear within 15 min of fertilization, not to reappear until shortly before first cleavage. Wasserman and Masui (1976) report that MPF is inactivated in *R. pipiens* extracts by Ca^{2+} addition, and in fact EGTA must be present to obtain active extracts. On this basis, Masui and Clarke (1979) have proposed that MPF is inactivated by Ca^{2+} at fertilization, clearing the way for interphase.

4.4 Events before Gray Crescent Formation

Pasteels (1964) has viewed the period prior to gray crescent formation as a "preparatory" period for that important event. As shown in Table II and Fig. 7 this period lasts about half the interval from fertilization to first cleavage and includes the completion of meiosis, surface changes at the sperm entry point, and rearrangements of the internal contents of the egg.

Shortly after exocytosis of the cortical granules, a prominent cortical contraction occurs, pulling the pigment of the animal hemisphere to within 60° of the pole, a 50% reduction in surface area occupied by pigment (Elinson, 1975; Palaček *et al.,* 1978; Hara *et al.,* 1979). This contraction has been observed in eggs of *X. laevis,* several *Rana* species, and *Bufo regularis,* and is called the "activation contraction." Within 10 min, relaxation sets in and the pigment returns to its earlier equatorial border, or nearly so.

Schroeder and Strickland (1974) treated unfertilized *X. laevis* eggs with the divalent cation ionophone A23187 and observed an immediate extreme activation contraction of such strength that the egg contents burst through the surface of the vegetal pole, and the

TABLE II. *Normal Table of Events from Fertilization to First Cleavage*[a]

161

PATTERN
FORMATION IN
AMPHIBIAN EGG
AND EARLY EMBRYO

Event	X. laevis	R. pipiens, R. fusca, and R. nigromaculata	T. palamatus and T. viridescens
1. Egg–sperm contact	0	0	0
2. Activation wave (cortical granule breakdown)	0.08–0.12		
3. Activation contraction	0.12–0.20	0.10–0.15	
4. Rotation (perivitelline space fills with liquid)	0.12–0.15		
5. 2nd polar body extruded	0.26	0.12	0.16
6. Cortical rigidity increases		0.18–0.50	
7. Sperm aster enlarges		0.18–0.66	0.25–0.65
8. 1st postfertilization wave	0.20–0.40		
9. DNA synthesis	0.32–0.58		
10. Gray crescent appears	0.50	0.50	Variable?
11. Pronuclei in contact	0.52	0.58	0.58
12. Yolk-free cytoplasm to dorsal side	0.55		
13. Pigment granules leave cortex, then return	0.55–0.68		
14. Sperm aster disintegrates		0.66	0.63
15. Bipolar spindle appears		0.72	0.65
16. Accessory sperm nuclei degenerate	N/A	N/A	0.68
17. Mitotic prophase	0.77	0.72	0.68
18. Pigment granules aggregate	0.78		
19. 1st surface contraction wave	0.80–0.90		
20. Mitotic metaphase	0.85	0.83	0.79
21. 2nd surface contraction wave	0.88–1.0		
22. Mitotic telophase, appearance of nuclear vesicles	0.92	0.92	0.92
23. DNA synthesis starts on 2nd cycle	0.96		
24. Cleavage furrow appears	1.0	1.0	1.0
Approximate interval (min) at 20°C	90	180	270

[a]Entries for each species indicate time on a relative scale from fertilization (0.0) to first cleavage (1.0). Figures are accurate to one decimal place only. N/A indicates event does not occur in this species. Absence of an entry indicates no datum available. Entries for *X. laevis* were obtained from Ubbels (1977), Hara *et al.* (1977, 1980), Palaček *et al.* (1978), and Graham (1966); for *R. pipiens* from Subtelney and Bradt (1963), Elinson and Manes (1978), Elinson (1976), and Malacinski *et al.* (1975); for *R. nigromaculata* from Kubota (1966, 1967, 1969); for *R. fusca* from Ancel and Vintemberger (1948) and Ancel and Calame (1959); for *T. palmatus* from Fankhauser (1934, 1937); and for *T. viridescens* from Fankhauser and Moore (1941).

pigmented cortex of the animal hemisphere could be retrieved as a small knot. This result indicates that a violent contraction can be provoked at the same time as cortical granule extrusion, well before the usual time of the activation contraction, but it is unclear whether ionophore-induced contraction mimics the activation contraction or is a global wound-healing response. Nonetheless, the authors suggest that the contraction is actin-driven and dependent on internal calcium ion reservoirs, as it occurs equally well in ionophore-treated eggs in calcium-free media.

The apparent orientation of the contraction toward the animal pole does not imply actual contraction of individual surface elements toward the pole, because a hemispherical surface contracting at all points would give the same geometry. The vegetal hemisphere cortex does not appear as capable of contraction, another example of an animal–vegetal difference in surface properties in addition to the differences in progesterone sensitivity and sperm penetration.

A number of functions have been suggested for this prominent contraction. Lovtrup (1965) noted its occurrence in the period of 10–20 min postfertilization when the egg is completing meiosis, and suggested a role for the contraction in cytokinesis of the second polar body. Elinson (1975) has suggested that the function of the activation contraction is to transport the sperm nucleus and centriole toward the animal pole where the egg pronucleus is reconstituted upon the completion of meiosis, thereby assisting pronuclear fusion. Although the sperm nucleus does seem to be transported upward at first, the pronuclei eventually meet deep in the egg near the vegetal yolk, according to Subtelny and Bradt (1963). Palacek *et al.* (1978) suggest that the activation contraction assists in the separation of the egg surface from the extracellular vitelline membrane, allowing rotation of the egg to a position where the vegetal hemisphere is downward. To this list of possible functions might be added another one, namely that the contraction transports not only the sperm pronucleus but also raises equatorial yolk platelets and other cytoplasmic materials as shown in Fig. 7, row 4, and in the figures of Klag and Ubbels (1975) for *Discoglossus* and of Ubbels (1977) for *X. laevis.* As yet there has been no functional test of the importance of the contraction, such as to paralyze the contraction to test the effect of its absence on subsequent development. However, Young *et al.* (1970) found that low-speed centrifugation of eggs of *X. laevis* in an upright position at this early time leads to the frequent production of twin embryos. The effect of the centrifugation has not been explored but might be expected to counteract the upward movement of cytoplasmic materials, a point discussed later.

As the relaxation proceeds in eggs of *X. laevis,* the perivitelline space between the vitelline membrane (the "fertilization envelope") and the egg surface fills with liquid and the egg becomes free to rotate vegetal-end down. Also, the sperm entry point becomes visible as a black spot, a locus of pigment concentration (Palaĉek *et al.,* 1978). Lovtrup (1965) viewed sperm entry as a localized wounding, and the evidence is good for this analogy because needle puncture can trigger local cortical contraction in the presence of calcium ions (Bluemink, 1972). Even without breakage of the surface, Gingell (1970) has obtained localized contractions around small spots of externally applied polylysine or detergents; since the contractions require external calcium ion, Gingell proposes the surface is locally permeabilized to calcium ion by the polynion setting off the contractile system. In *R. pipiens* the sperm entry point remains discernible for several hours by scanning electron microscopy as a bald patch in the midst of the short microvilli covering the egg surface (Elinson and Manes, 1978).

At about 20-min postfertilization in *X. laevis,* a "postfertilization wave" becomes clear by cinematography (Hara *et al.,* 1977) as indicated in Fig. 7, row 1. The wave advances slowly from the sperm entry point, reaching the opposite side in about 15–20 min, just about the time of gray crescent formation. A second wave is seen in many eggs. The physical basis for the visibility of the wave is not known, but presumably involves a change in surface or subsurface fine structure of the egg. It correlates in time with three interesting events. First, Elinson and Manes (1978) detect in *R. pipiens* a wave of shortening of the egg surface microvilli starting from the sperm entry point (a "bald" spot) and spreading during the same time period as the postfertilization wave of Hara *et al.* (1977) when *X. laevis* and *R. pipiens* are normalized to the same time scale. Second, Kubota (1967) has attempted to quantify a time-dependent increase of cortical rigidity in eggs of *R. nigromaculata* by centrifuging eggs at low speed just sufficient to drive pigment granules from their cortical position into the interior of the cytoplasm. He found that before polar body extrusion the pigment granules are held firmly in the cortex, whereas just afterward they are easily displaced, indicating a reduction in cortical structure, perhaps accompanying the end of the activation contraction. Then, the granules become again firmly held but only in a limited area enlarging progressively from the point of sperm entry and reaching the opposite side just before gray crescent formation. Third, as meiosis is completed the sperm centriole organizes a steadily enlarging aster, which fills the entire cytoplasm of the animal hemisphere just before gray crescent formation, as shown, for example, by Subtelny and Bradt (1963) for *R. pipiens*. Kubota (1967) correlated the increase in cortical rigidity with the growth of the aster and from the good correlation suggests that the aster affects the cortex. Elinson and Manes (1978) likewise consider the growing aster to affect the structure of microvilli through its effects on the surface. Conceivably, a change in surface refractility due to changes of the microvilli or the cortex might be visible as the postfertilization wave. Recently Kirschner and Hara (unpublished) showed that fertilized eggs of *X. laevis* do not emit the postfertilization wave if treated with vinblastine or colchicine, two antimicrotubule drugs, in direct support of the above postulates. Lovtrup (1965) and Elinson and Manes (1978) proposed that the sperm entry point, as a surface singularity, and the enlarging aster, as an internal singularity, provide the early asymmetries on which the orientation of the gray crescent is based. These proposals are discussed in Sections 4.2 and 4.5.

At the same time as the aster enlarges, it moves progressively toward the center of the egg, with the centriole and sperm nucleus close to the level of the vegetal yolk mass (Winterbert, 1933; Subtelny and Bradt, 1963). The egg nucleus descends from the animal pole, meeting the sperm nucleus near the yolk at a time approximately halfway through the period from fertilization to first cleavage (Table II). In artificially activated eggs, in which there is no sperm aster, the egg nucleus descends only one-third of the distance accompanied by a small asterlike yolk-free area (Fankhauser, 1937; Manes and Barbieri, 1977; Ramirez and Huff, 1967). As mentioned before, it is not known if this egg aster contains a centriole, and at what step the aster fails in spindle formation. It seems likely that in fertilized eggs the sperm astral rays guide the two pronuclei to their meeting (Wilson, 1928). Simultaneously, cytoplasmic materials rearrange perhaps under the influence of the aster; in *Discoglossus pictus*, the central cone of cytoplasm rich in germinal vesicle sap shifts in a complicated pattern, which finally delivers the bulk of the material to the prospective dorsal side of the egg (Klag and Ubbels, 1975). In *X. laevis,* there is a small yolk-free region centered above the vegetal yolk mass (called the "central cytoplasm") and this regions shifts dorsally during aster migration (Ubbels, 1977) to become the "dorsal yolk-free cytoplasm," rich in mitochondria and membrane-bounded vesicles (Herkovits and Ubbels, 1979). Brachet (1950) had earlier demonstrated that sulfhydryl-rich proteins (nitroprusside staining) and nucleic acids (basophilia) of the germinal vesicle are transported preferentially to the gray crescent side of the egg, in line with the more recent results. Ubbels (1977) has observed that the central cytoplasm does not move from its central position in artificially activated eggs or in fertilized eggs injected with colchicine or treated with vinblastine (Kirschner and Ubbels, unpublished). These movements may involve microtubules organized by the sperm aster. In addition to these movements of materials of the animal hemisphere, vegetal yolk is also displaced, although the displacements in this early period are not well catalogued [see figures in Ubbels (1977); Klag and Ubbels (1975); Ancel and Vintemberger, (1948)]. It is assumed later in this review for purposes of explication that these vegetal movements are the critical ones for gray crescent orientation.

In summary, the egg at activation spends 20–30 min releasing the metaphase arrest, completing meiosis, and entering interphase. Arrival at interphase is marked by the activation contraction. Thereafter the sperm centriole organizes a large moving aster around which various internal materials rearrange. At the same time, and perhaps by independent means, the cortex of the egg changes, perhaps arriving at a state of contractility similar to that it possessed just before the activation contraction. Perry (1975), Bluemink (1972), and others have visualized actinlike microfilaments in the cortical region, presumably those visualized at the oocyte surface by Franke *et al.* (1976). Perhaps these undergo a buildup and breakdown at the activation contraction and require a period of reorganization leading to a new poised network ready to contract again at the moment of gray crescent formation. The *schedule* of this buildup may not depend on the aster, although the *orientation* of the contraction may, as discussed in the next section.

4.5 Gray Crescent Formation

Formation of the gray crescent is an old and fascinating subject of amphibian embryology, starting with Roux's observation (1888) of lighter pigmentation of the fertilized egg *(R. esculenta)* on that side of the animal hemisphere opposite the side of sperm entry and culminating in the "cortical field" hypothesis of Dalcq and Pasteels (1937) as supported by the cortical transplantation experiments of Curtis (1960, 1962a). The gray crescent, or its equivalent (the clear crescent of the axolotl), marks the egg's change from radial symmetry to bilateral symmetry in many anurans and urodeles. As pointed out earlier, the sperm entry point is a rather reliable but not a perfect indicator of the future ventral side of the embryo in some anuran specispecies, and in other anuran and most urodele species it has no clear relation to the future dorsal–ventral axis. Now, in those species with a clearly identifiable gray crescent, the crescent and the future dorsal side correlate well, even in those species in which the site of sperm entry had been a useless indicator. The good correlation concerns eggs under normal conditions. Schechtman (1936) has reported that in *T. torusus* the crescent is only a moderately reliable indicator, with the dorsal side being frequently displaced by 30°. Nonetheless, in many normal cases the subequatorial zone beneath the gray crescent becomes the future site of the blastopore's first appearance many hours and many cell divisions after crescent formation. The surface area of the crescent itself corresponds on the fate map to the future chordamesoderm, head mesoderm, and somites, those regions which in the early gastrula show strongest "organizer" properties in inducing a neural tube and new body axis on transplantation to a ventral site of a recipient early gastrula. These fate map relationships do not hold for *X. laevis* in which prospective mesoderm is wholly internal and not included by the gray crescent cortex.

Recent thorough reviews of the gray crescent have been presented by Dollander (1962), Lovtrup (1965), Clavert (1962), Pasteels (1964), and Brachet (1977). The following discussion divides the results into three areas, namely, (1) the mechanism by which the crescent actually forms, (2) the factors that determine the orientation of the crescent, and (3) the role of the crescent in dorsalization of the embryo.

4.5.1 The Mechanism of Crescent Formation

There are two recent views about the mechanics of the appearance of a less-pigmented area (the crescent) on one side of the egg in the upper hemisphere. The older view is that of Ancel and Vintemberger (1948, 1949), in which the entire cortex of the egg is thought to shift by approximately 30° relative to the internal contents, particularly the vegetal yolk mass. By shifting, one side of the pigmented hemisphere would be raised above the equator and the other side would be lowered from its previous position. The raised side would now allow a visualization of deep cytoplasm previously masked by the pigmented surface. The grayness of the cytoplasm would give the crescent its coloration. The surface layers of the egg would presumably not change composition but just position. The active process of movement is not specified, although the authors consider an asymmetric weighting of the cortex around the sperm entrance point as a possibility. They called this movement the "rotation of symmetrization"—that is, the rotation established bilateral symmetry in the egg.

The more recent view is that of Lovtrup (1965), who suggested a cortical contraction on the sperm entrance side of the animal hemisphere as the basis for removal of pigment granules from the other. Again, the removal of pigmentation would allow a glimpse of the deeper cytoplasm, the coloration of which would affect our description of the crescent. The Lovtrup view is easier to comprehend—a localized contraction toward the ventral side needs no further assumptions as to the motive force, whereas the cortical shifting of Ancel and Vintemberger still leaves open the question of what moves the entire surface relative to the contents. Even before crescent formation, the cortex has already displayed its ability to contract at the time of polar body cytokinesis with the strong activation contraction, as well as with the localized pigment movement toward the sperm entry point or a wound. We do not have a description of exactly what moves the pigment granules during crescent formation. Presumably the granules are associated with a contractile protein matrix that moves. Pigment migration in retinal cells is known to involve filamentous systems and microtubules (Murphy and Tilney, 1974; Schliwa and Bereiter-Hahn, 1975). But we do not know which membrane and endoplasmic proteins are carried along as well. We also do not know the limits of cortical displacement in the vegetal hemisphere where there are insufficient pigment granules to mark the movements. Banki (1929) marked the equator of an axolotl egg with vital dye spots prior to crescent formation and observed the stretching of the spots on the crescent side. The vital dyes stain internal materials such as yolk platelets and pigment granules even though applied externally, and the stretching of the spots records the displacement of internal materials but not membrane materials. If the cortical contraction of crescent formation resembles the contractions of wound healing, the membrane itself would be thrown into folds and the cortex would thicken on the ventral side (Gingell, 1970; Bluemink, 1972). Brachet (1977) has shown scanning electron micrographs of the gray crescent and opposite surfaces of the axolotl egg and interprets these micrographs in terms of a thin and thick cortex, respectively. Dollander has investigated the penetration of the surface by the vital stain Nile blue, and finds rapid deep penetration in the vegetal hemisphere and poor penetration on the animal hemisphere, with the gray crescent area allowing a more vegetal type of penetration (Dollander, 1962; Dollander and Melnotte, 1952). These data are consistent with the idea of a "thin cortex" in the crescent region. Modern techniques with specific nonpenetrating surface reagents would allow a more complete description of the molecular reorganization of the exterior egg surface during crescent formation. In summary, crescent formation probably reflects an oriented contraction of cortical filamentous proteins, carrying along pigment granules and associated materials of the deeper cytoplasm. On the whole, though, the contraction has been incompletely described.

4.5.2 Orientation and Timing of Crescent Formation

There are two recent views as to which asymmetrically arranged structures of the egg precede the crescent and determine the orientation of the contraction of the surface. Lovtrup (1965) suggested that the sperm entry point is the asymmetric and ventral focus of the contraction, a view considered more recently by Elinson (1975), Elinson and Manes (1978), and Palaĉek et al. (1978). The argument in favor of this view is that a localized contraction occurs at the entry point well before the large-scale contraction involved in crescent formation. The other candidate for the preceding asymmetric center is the sperm centriole with its aster, moving from the entry point on a course toward the center of the egg. The postfertilization waves, microvillus shortening, and change of surface rigidity seem to follow this organelle, which itself has a close relation to the sperm entry point. Manes and Barbieri (1976, 1977) have injected unfertilized *Bufo* eggs with suspensions of disrupted sperm and shown that the gray crescent forms opposite the injection point. They propose that the active agent is the centriole, earlier regarded by Ramirez and Huff (1967) and Fraser (1971) to be the parthenogenic factor needed for cleavage initiation. In the Manes–Barbieri experiments the sperm homogenate was always delivered close to the point of injection through the egg surface, so one cannot separate the role of the penetration point from that of the aster. Manes and Barbieri (1977) repeated the classic 1911 study of A. Brachet to show that *Bufo* eggs injected with buffer form the crescent *at any position* around the equator, with no relation to the injection point, thus supporting the conclusion that the centriole is involved in orienting the crescent but *not* in forming it.

Kubota (1967) studied artificially activated eggs of *R. nigromaculata,* measuring their cortical rigidity by his centrifugation assay as described earlier. These eggs were identical to fertilized eggs as far as their high cortical rigidity before polar body elimination and low rigidity afterward. The artificially activated eggs were slower to develop their new cortical rigidity after polar body elimination and they showed no pattern of having a locally enlarging area of rigidity. Just before the normal time of crescent formation, they suddenly became rigid over the entire cortex of the animal hemisphere, and resembled the fertilized egg. Their crescents formed on time. Thus, the increasing rigidity of the cortex does not require the aster for its formation, but is somehow affected kinetically and spatially by the aster, as if the aster "seeds" the crystallization of the fibrous polymers of the cortex. Interestingly, colchicine prevents the appearance of a gray crescent (Manes *et al.,* 1978) as if microtubules of the cortex rather than of the aster are needed for the cortical contraction. It is unknown what step in the buildup or discharge of the cortex is blocked.

For several anuran species, the crescent has been reported to appear at a time approximately halfway through the period from fertilization to first cleavage (Ancel and Vintemberger, 1948; Pasteels, 1964). At this time, pigment granules move from the crescent region, and dye marks placed at the equator on the prospective dorsal side are transported upward. In *R. fusca,* crescent formation starts at 70 min and continues until 130 min, with cleavage at 150 min (Ancel and Calame, 1959). In *X. laevis* crescent formation may begin earlier as judged from pigment movements (Palaček *et al.,* 1978), but in this species it would be important to know which pigment displacements are really associated with crescent formation and which displacements represent continued accumulation of pigment around the sperm entrance point. At approximately the same time as crescent formation, the pronuclei meet and the large sperm aster breaks down and is replaced by a small bipolar spindle, presumably attended by centriole duplication and separation (Wintrebret, 1933; Fankhauser, 1932, 1934a; Fankhauser and Moore, 1941a,b; Subtelny and Bradt, 1963). This breakdown might itself seem a plausible trigger for the cortical contraction, but we must further inquire what it is that signals aster breakdown at this time and whether this signal could trigger the cortical contraction as well, as part of a general interphase–prophase transition. Evidence for a timing mechanism in the egg will be considered in a later section on furrow initiation.

It is clear from several studies that artificially activated eggs are perfectly able to form a crescent despite the absence of the sperm centriole and aster or of the sperm entrance point. Other indications of the dispensability of these asymmetric centers are that (1) electrically activated eggs have no specific puncture point, and these also form a crescent (Ancel and Vintemberger, 1948); and (2) urodele eggs are polyspermic and normally form a crescent with an orientation unrelated to either the multiple sperm entry points or to the aster (Fankhauser, 1930; Banki, 1927, 1929; B. C. Smith, 1922). Therefore, if we are to have a general explanation for the identity and role of the asymmetric signal used to orient gray crescent formation in amphibian eggs, we should look for some agency other than the centriole and sperm entry point, for these seem to be, at most, sources of weak bias, overwhelmed in a variety of conditions, natural and artificial. There are several lines of experimentation which suggest that immediate control of crescent position is exerted by the vegetal yolk mass, and these results will be discussed in Sections 4.6 and 4.8. If this suggestion is correct, then the influence of the sperm in orienting the crescent under normal conditions in certain anuran eggs would be achieved indirectly through its effects on the geometry of the vegetal yolk mass.

4.5.2a   Bilateral Symmetry in Unfertilized Eggs.   In axolotl eggs, Banki (1929) recognized a slight tilting of the pigmented cap and maturation spot of the animal hemisphere to one side even in freshly fertilized eggs just after rotation and long before gray crescent formation. The higher side of the pigment border invariably became the gray crescent and dorsal side, irrespective of the position of sperm entry. In sections, the vegetal yolk was already slightly displaced away from the high side. He marked such eggs with Nile blue spots on the equator and observed the spots to stretch vertically on the gray crescent side during crescent formation, but not on the ventral side. The dye had penetrated the surface and stained yolk granules, which were pulled up the side of the egg in a thin layer. Pasteels (1937) made similar observations on certain batches of unfertilized eggs of *D. pictus* and *R. esculenta,* in which the pigment was raised on one side relative to the animal–vegetal axis—that is, the perpendicular line through the animal pole and center of the spherical egg, thereby giving the egg bilateral symmetry even before fertilization. The dorsal side of the embryo usually developed from the side of the egg with the raised pigment. Wittek (1952) makes the interesting observation that even oocytes of *R. esculenta* show bilateral symmetry, detectable internally by the displacement of the vegetal yolk mass to one side.

In species of *Triturus* the egg possesses bilateral symmetry at least soon after fertilization if not before, but the material basis of the axis is not known. As a means of detection, Fankhauser (1930) and Streett (1940) used fine nylon thread to constrict fertilized uncleaved eggs into two equal-size fragments at very early times, as little as 10 min after fertilization. For a fragment to cleave normally it had to receive both the egg nucleus and a sperm centriole (always with the nucleus), so less than half the fragments met this requirement. Of the ones that did cleave well, approximately 20% failed to form a good blastopore and arrested as so-called "ventral pieces" lacking dorsal structures such as a notochord, somites, and a central nervous system. Fankhauser (1930) reasoned that insofar as the constriction was done at random, one-fifth of the fragments had by chance not received a prearranged essential factor for dorsalization, even though they received the sperm entry point and centriole. He proposed the factor must have been distributed within four-fifths of a half circle around the equator (or another latitude), soon after fertilization, if not before. These results would fit well with Banki's and Pasteel's findings, described above, of an initial vegetal yolk asymmetry causing a slight tipping of the egg axis. The eggs of some *Triturus* species are known to form a crescent, visible with difficulty but well correlated with the dorsal side (Dollander, 1950). Fankhauser (1930) and Streett (1940) did not correlate the developmental behavior of their pieces or the crescent position with aspects of egg promorphology.

4.5.2b   The Ancel–Vintemberger Experiments.   Ancel and Vintemberger (1948) achieved a penetrating analysis of the effects of gravity in determining the position of the gray crescent in artificially activated eggs of *R. fusca*. This is an important case because the sperm centriole and sperm entry point are absent. When they rotated an activated egg with its fertilization membrane so that its vegetal pole was 135° elevated from its downward position, the egg then rotated back to the upright position over a period of minutes, the speed of return depending on the degree of hydration of the perivitelline space. After this treatment, the gray crescent reliably formed in the animal hemisphere above the equator on the side that had been highest during the rotation. Sections of eggs fixed after rotation showed that vegetal yolk had moved slightly to one side while the egg was gravitationally off balance, pulling down from the equator at the uppermost side and accumulating on the lower side. These rearrangements by gravity occur after fertilization but not before, as fertilization seems to trigger a large decrease in cytoplasmic macrovisocisty (Merriam, 1971). In the Ancel–Vintemberger experiment, the vegetal yolk did not pull cleanly away from the subequatorial cortex but left a thin adsorbed layer. This was the zone above which the crescent formed. The authors showed that a 60° rotation was inadequate, a 90° rotation marginally effective, and a 135° rotation fully effective in orienting the crescent. In repeated rotations in opposite directions, the last rotation determined the position. Of course, once the crescent had formed, it was not possible to test subsequent effects of gravity on dorsal–ventral polarity because artificially activated eggs fail to cleave or gastrulate. These experiments show clearly that the distribution of cytoplasmic materials can affect the position of the crescent, and at a force of one gravity the vegetal yolk mass is the material most clearly redistributed. In a parallel set of experiments, Ancel and Vintemberger (1948) examined fertilized eggs as well. The sperm exerted a strong orienting influence on the gray crescent, such that a single 135° rotation was not sufficient to overcome the effect. Four successive rotations, each of 90°, did orient the gray crescent in all cases, dominating the effect of the sperm. The position of the equator on the gray crescent side reliably became the locus of the future blastopore. Once the crescent had formed, the four successive 90° rotations were ineffective in relocating the future blastopore, but the authors did not study more severe treatments for accomplishing rearrangements of vegetal yolk, as can, in fact, be achieved in the Born and Schultze procedures (see Section 5.4). It may be more accurate to conclude that the egg becomes more refractory to oblique orientation when the "rotation of symmetrization"—that is, gray crescent formation—occurs, but that the egg can still be affected by stronger treatments. Ancel and Vintemberger also tested urodele species *(Triturus alpestris)* and several other anurans and got the same results as with *R. fusca*. Recently these experiments were repeated with *X. laevis* (Kirschner *et al.*, 1980); eggs kept on their sides for 20 min during the period before crescent formation subsequently form the crescent on the side which had been uppermost and form the blastopore there in all cases, even when the sperm entry point is on the same side. Gravity readily determines the dorsal–ventral axis in *X. laevis*. These experiments differ from the Ancel–Vintemberger procedure only insofar as the egg was in a fixed position on its side, due to dehydration of the perivitelline space by Ficoll, and could not rotate back until the exposure period was intentionally terminated. It is noteworthy that gravity seems to be the major factor determining axis polarity in fish and bird eggs (Clavert, 1962; Kochav and Eyal-Giladi, 1971).

As an explanation of these effects in the precrescent period, it seems possible that both the sperm and gravity affect the same target in the egg, inducing in it some asymmetry, which then determines the direction of cortical contraction and therefore crescent orientation. In the case of gravity, the vegetal yolk mass shifts slightly to one side when the egg is off balance; perhaps the sperm accomplishes the same shift in the course of the migration of its aster across the equatorial surface of the vegetal yolk (Subtelny and Bradt, 1963), since asters appear to expel yolk granules from their vicinity (see figures of Heidemann and Kirschner, 1975). A shift of intermediate platelets away from the sperm aster has been observed in *X. laevis* during the precrescent period (Ubbels, 1977). However, we are far from knowing how an asymmetric disposition of platelets could orient the cortical contraction involved in crescent formation, perhaps by inhibiting the contraction of the region (prospective dorsal) with more platelets. Apparently the cortex itself does not cumulatively develop an irreversible commitment to one direction of contraction during the preparatory period, at least in activated eggs; Ancel and Vintemberger (1948) could use their rotation procedure a second time to establish new orientations right up to the moment of crescent formation. The timing of the contraction seems to be a cortical property; the results of Kubota (1967) on cortical rigidity in artificially activated eggs suggest that the cortex builds up its structure abruptly just before crescent formation (starting at 45 min in *X. laevis,*) whereas fertilized eggs increase their cortical rigidity gradually after polar body elimination. It is not known if fertilized eggs become less sensitive to second rotations in the precrescent period.

There is at least one other internal rearrangement occurring in the preparatory period before crescent formation and that is the movement of the central yolk-free cytoplasm (perhaps germinal vesicle contents) toward the prospective dorsal side in *X. laevis* and *D. pictus*. In the former species, this region associates with the egg and sperm pronuclei, all slightly to the prospective dorsal side of the animal hemisphere by the time of crescent formation (Ubbels, 1977). Since these movements do not occur in artificially activated egg (Ubbels, 1977) which nonetheless form a crescent, it may be concluded that the specialized plasm does not control the orientation of the asymmetric contraction and, vice versa, that the contraction does not transport the specialized plasm to the dorsal side. The association of this plasm with the pronuclei suggests that it also follows the sperm aster migration. At present, there has been no direct test of the involvement of this "dorsal yolk-free cytoplasm" in dorsalization of the embryo.

4.5.2c Internal Movements during Crescent Formation. When the contraction or "rotation of symmetrization" does occur, its direct consequence is to pull a layer of deep vegetal yolk upward on the gray crescent side, to form what Pasteels (1948, 1964) calls the "vitelline wall." This wall contains densely packed large platelets of the size found earlier in the lower third of the egg near the vegetal pole. The platelets are closely associated with the cortex in the region of the vitelline wall. As shown by Klag and Ubbels (1975) for the eggs of *D. pictus,* the deep vegetal yolk of the vegetal hemisphere is moved upward on the prospective dorsal side of the egg, through a middle layer of intermediate-size yolk platelets, and is brought into contact with animal hemisphere cytoplasm—that is, cytoplasm with sparse small platelets and abundant sap from the germinal vesicle of the oocyte. This new contact of the most animal and most vegetal materials is presumably caused by the same motive force which forms the gray crescent. This analysis by Klag and Ubbels (1975) is currently the most complete description we have of the internal movements accompanying gray crescent formation. In a sense, the original animal–vegetal axis of the egg becomes bent to one side, a point returned to later. A relevant issue now is the relationship of the small yolk asymmetry *before* gray crescent formation—that is, the cue perhaps determining the direction of the contraction—to the large yolk asymmetry completed by the end of crescent formation. These asymmetries appear to have the same orientation (Banki, 1929; Pasteels, 1937) but differ in magnitude. If this comparison is accurate, gray crescent formation would have the role of an *amplifying mechanism* capable of converting a small difference in yolk distribution (due to gravity, oogenesis, or aster migration) into a large difference, the vitelline wall. Later it will be suggested that the region of the platelets of the vitelline wall is directly involved in dorsalization as the main component of Nieuwkoop's vegetal dorsalizing center. If the important rearrangement for dorsalization is really vitelline wall formation, the gray crescent itself would have no function, but only mark the position and direction of the asymmetrical cortical contraction. The contraction would be the important event, as a morphogenic mechanism, and the vitelline wall would be its product. The crescent would only be an epiphenomenon. This set of proposals is not at all consistent with the cortical field hypothesis of Dalcq and Pasteels (1937), in which the cortical region of the crescent itself is considered to play a continued role in dorsalization due to unique potentialities it acquired during crescent formation. Their hypothesis will be discussed in Section 4.6.

It should be mentioned that Ancel and Vintemberger (1949) did not place great emphasis on the crescent but considered that the "rotation of symmetrization"—called here the asymmetrical cortical contraction—accomplished three important rearrangements: (1) the shift of a special "premesoblast" cytoplasmic region from an internal equatorial position to one higher on the dorsal side and lower on the ventral side; (2) an adsorption of vegetal yolk platelets onto the cortex on the dorsal vegetal and equatorial surface, but not on the ventral side, and (3) a shift of the fused pronuclei toward the dorsal side. They did not ascribe special properties to the gray crescent cortex.

4.5.2d Other Effects on Axis Orientation. Three other treatments of eggs in the precrescent period affect crescent position and dorsalization, namely, low-speed centrifugation, compression, and UV irradiation. Although these effects are incompletely studied at present, they can be reviewed in light of the above proposals.

Young *et al.* (1970) observed that low-speed centrifugation (40 × g for 15 min) of *X. laevis* eggs in the normal upright position in the period before crescent formation led to the production of many embryos with double axes—that is, twins. The frequency was over 30%. After crescent formation centrifugation produced only normal single-axis embryos. The eggs were not examined for crescents (double crescents or no crescent?) or for the distribution of internal materials; the force was perhaps great enough to affect the position of the vegetal yolk mass or to make the yolk more resistant to displacement by the sperm aster. The twinning is an interesting example of the egg's failure to make singular its founding of a vegetal dorsalization center. It would be informative to know if the twin axes were always separated by 120° or more (regarding the 30% frequency of twins) on the chance that the egg is able to unify closer vegetal centers.

In the case of compression, Ancel and Calame (1959) showed that artificially activated eggs of *R. fusca* compressed on two sides by parallel glass slides form the gray crescent in 100% of the instances within 45° of the longer axis of the egg, that is, on one of the sides not touching the glass. They did not study the effectiveness of lateral compression in orienting the crescent of fertilized eggs. In preliminary studies of compressed eggs of *X. laevis,* the sperm still controlled the dorsal–ventral axis (J. C. Gerhart, M. W. Kirschner, and G. A. Ubbels, unpublished), indicating that compression is a relatively weak reorienting influence. Vertical compression has not been investigated, but Ancel and Calame (1959) caution that early authors such as Banki (1927) and Tung (1933) had used such eggs and might have inadvertently introduced orienting forces other than the sperm, invalidating their conclusions about the absence of a strong sperm effect. There has been no study of the effect of compression on the position of internal contents of the egg or on the ability of a region of cortex to contract when its surface tightly contacts the immobile glass plate. These conditions of compression may immobilize parts of the surface and cortex, preventing vegetal displacements by preventing the cortical contraction in that region. Local applications of polylysine of A23187 might be expected to cause local contraction and affect the orientation of the dorsal–ventral axis.

In the case of UV irradiation we encounter a different result, for as Grant and

Wacaster (1972) showed, irradiation of the vegetal hemisphere of *R. pipiens* eggs shortly after fertilization causes the development of embryos defective in gastrulation and neurulation. Dorsal axial structures such as notochord, somites, and central nervous system are reduced, and in the extreme, missing. Development resembles that of "ventral pieces." Malacinski *et al.* (1977) have documented carefully in *X. laevis* that increasing UV dose correlates well with increased retardation in the appearance of the blastopore, with decreased invagination of the archenteron endoderm, and with decreased neural induction by the "organizer" region of mesoderm. In short, UV has the effect of inhibiting dorsalization, a quite different effect from the reorientation of the dorsal–ventral axis by gravity or compression. More recently, Scharf and Gerhart (1980) have shown that the UV-induced delay of gastrulation still occurs at times when the dorsalization is no longer inhibited; thus these effects are separable.

Malacinski *et al.* (1975) have demonstrated that the UV-sensitive period of *R. pipiens* eggs falls within 90 min after fertilization—that is, until gray crescent formation. This would seem to suggest that UV irradiation impairs the egg's ability to form a crescent or a vitelline wall, and Elinson and Manes (1979) report that UV-irradiated *R. pipiens* eggs form crescents poorly. In *X. laevis* eggs, the effect on the crescent is less clear (Scharf and Gerhart, 1980). The exact region of sensitivity occupies the equatorial and vegetal surface, not the animal hemisphere, as learned from localized irradiation and from the calculation that UV cannot penetrate below the outer layers due to absorbance by the cytoplasm. Grant and Wacaster (1972) found that the pronuclei of the exposed egg are undamaged because their nuclear descendants at the blastula stage support good nuclear transplantation. They note increased leakage and penetration of radioactive sodium ion after irradiation, and Grant and Youngdahl (1974) report severe inhibition of cleavage furrow formation in the vegetal region. Beal and Dixon (1975) and Zust and Dixon (1975) confirm the inhibition of cleavage, although it is not clear whether the time period of sensitivity for this effect coincides with the sensitive period for inhibition of dorsalization. Eventually furrows do form in the normal number and arrangement (Malacinski *et al.*, 1977). There are several steps in furrow formation, such as activation of a cortical inducer, competence of the surface to respond to that inducer, and actual formation of the contractile ring of microfilaments (see Sections 4.7 and 5.2), but the UV-sensitive step is not known. In general, UV irradiation seems to damage the egg surface. Perhaps the irradiated surface is unable to contract in an integrated and concerted way to displace vegetal materials optimally, or perhaps the irradiated cortex does not adhere to the underlying vegetal materials to drag them along as it contracts.

If the net effect of the cortical contraction is to rearrange internal materials and thereby initiate dorsalization, it should be possible to rescue UV-irradiated eggs from their adorsal fate by rotating them according to the procedure of Ancel, Vintemberger, and Born in order to create artificially a vegetal yolk asymmetry suitable for dorsalization. Recently this expectation has been realized (Scharf and Gerhart, 1980). If UV-irradiated eggs of *X. laevis* are held obliquely at 90° for 40 min immediately after irradiation, they develop completely normally, whereas without the rotation treatment they give "ventral pieces" lacking all dorsal structures. The rotation treatment seems to do more than cue the egg as to the direction to apply its axis-forming mechanisms; the treatment seems to replace these mechanisms in the uncleaved egg. This result provides strong indirect evidence that gravity-driven displacements of the vegetal yolk mass may mimic the normal ones formed in unirradiated eggs by the cortical contraction. It was also found that the rescue procedure can work effectively even after first cleavage, that is, long after gray crescent formation. This is further evidence that gravity is bypassing the egg's normal axis-determining mechanism.

As far as the molecular targets of UV in the egg surface, Brachet (1977) suggests protein sulfhydryl and disulfide groups, rather than nucleic acids. Absorption by groups of these kinds is well documented in model compounds. The egg jelly itself is thought to contain numerous disulfide linkages on the basis of its solubilization by cysteine, mercaptoethanol, or DTT (Gusseck and Hedrick, 1971) and UV irradiation is used as a dejellying agent (Gurdon, 1960). Other cell types have been studied with regard to their membrane sulfhydryl groups and disulfide linkages (Rothstein, 1970). Reagents affecting such structures might mimic UV irradiation inhibiting dorsalization; numerous authors have observed the gastrulation-blocking action of mercaptoethanol in amphibia (Seilern-Aspang, 1959; Brachet, 1962) although the time course of exposure often extended into the post-gray crescent period so that porcesses other than crescent formation may also have suffered. Preliminary experiments reveal that brief exposure to the strong disulfide-breaking reagent DTT at high pH before crescent formation produces cleaving eggs which arrest or retard at gastrulation with a syndrome resembling the UV-irradiation syndrome (J. C. Gerhart, unpublished). And L. J. Barth (1956) has reported changing patterns of regional sensitivity of the *R. pipiens* egg to reagents specifically derivatizing sulfhydryl groups (pMB, iodoacetamide) and has noted higher resistance to toxic effects in the gray crescent region. The sensitivity of crescent formation was not studied.

4.6   The Role of the Gray Crescent in Dorsalization of the Embryo

Long ago, Dalcq and Pasteels (1937; Pasteels, 1964; Brachet, 1977) proposed that the two morphological axes of the amphibian egg (the animal–vegetal and dorsal–ventral axes) are, in fact, the sources of two material gradients, which together provide the quantitative information used by embryonic cells for selecting a path of eventual differentiation. The animal–vegetal axis of the egg established the "vitelline gradient" of a morphogen "V" from the yolk or the yolk-associated materials of the egg. The highest intensity of V would occur in the vegetal hemisphere. The dorsal–ventral axis was regarded as a cortical axis establishing a gradient of the morphogen "C" derived from materials of the plasma membrane and cortex. The highest level of "C" would occur in the gray crescent cortex near the equator. In their terminology, the egg surface carried a "cortical field." The morphogens C and V were said to react together to form the inductor "organisine," which, at its highest level, induced formation of the dorsal lip of the blastopore. This level would occur in the territory of high V and C, namely, at the equatorial surface position where vegetal yolk and the crescent meet. Intermediate levels of C · V (concentrations multiplied together) would promote mesoderm formation and would predominate on the dorsal side where the "C" concentration is higher. Lowest levels of C · V would promote ectoderm formation, as at the animal pole. Simultaneously, cells would interpret the ratio of C/V to subdivide the mesodermal and ectodermal zones into dorsal–ventral regions, for example, into chordamesoderm, somites, and lateral mesoderm. With the two quantities C · V and C/V, Dalcq and Pasteels (1937) could approximate the fate map of the urodele early gastrula.

Taking the two axes and gradients one at a time, we find strong experimental support for a role of the animal–vegetal axis in the topography of the fate map. The 180° inversion experiments of Motomura (1935) and Pasteels (1941) are very compelling and important. For example, Pasteels inverted eggs of *R. fusca* prior to gray crescent formation, held them in position by dehydrating the perivitelline space, and centrifuged them (460 × g, 1–3 min) to accomplish the total displacement of vegetal yolk from one hemisphere into the other hemisphere. Reciprocally, the less-yolky cytoplasm of the original animal hemisphere was driven upward into the old vegetal half. The eggs remained in the inverted position after removal and hydration. Although the description of these eggs is not extensive, they seem to have remained with their unpigmented surface up and pigmented surface down, just the opposite of normal eggs. Cleavages seem to have begun at the top of the unpigmented hemisphere. A similar inversion with respect to cleavage patterns can also be achieved with inverted *X. laevis* eggs (Gerhart and Bluemink, unpublished). In Pasteel's results, 38 out of 40 inverted eggs developed normally, with the yolk mass still forming endoderm even though it now occupied the hemisphere of pigmented cortex. The entire animal–vegetal axis of the egg had been inverted successfully! These important results demonstrate that *before* crescent formation, the egg surface has little or no patterning of a fixed sort, although it is probably capable of reestablishing animal–vegetal differences in the cortex (for example, cortical thickness) within a short time, as discussed later. Of course, one could argue that normal and inverted eggs share the same equatorial cortex, which might be the important zone of patterning. But Pasteels (1941) also turned eggs 90° and centrifuged them to bring the yolk mass to a position at right angles to the original animal–vegetal axis; these eggs also developed normally, indicating that any region of the surface can be used for any region of the fate map if the rearrangement is done at this early time.

Earlier evidence for the substantial role of the vegetal yolk mass in fixing the animal–vegetal orientation of the fate map had come from the results of Schultze (1894), Penners and Schleip (1928a,b) and Penners (1929, 1936) on eggs inverted (but not centrifuged) at the two-cell stage; the vegetal yolk fell to new positions in the egg, sometimes making contact with regions of the pigmented cortex. These regions of contact often became the foci for secondary blastopores and new dorsal axes, leading Penners and Schleip to emphasize the importance of yolk for the positioning of the blastopore. The main conclusions regarding the yolk mass are as follows: (1) wherever it resides, that region becomes endoderm; (2) wherever the yolk mass meets the less yolky animal hemisphere cytoplasm, that zone becomes mesoderm; and (3) wherever the animal hemisphere cytoplasm is out of contact with the yolk mass, that region becomes ectoderm. These venerable conclusions are fully explained by Nieuwkoop's recent discovery (1969a) of the induction of mesoderm in the animal half of the egg by the vegetal half—that is, by the yolk mass, as discussed later. The conclusions are also embodied formally in the Dalcq and Pasteels (1937) proposal of a vitelline gradient leading at highest intensity to endoderm, middle intensity to mesoderm, and lowest intensity to ectoderm. There is good evidence for the action of the yolk mass in orienting the fate map in the animal–vegetal direction, and for the hypothesis in formal terms of a vitelline gradient, although several authors have pointed out that yolk distribution better approximates a discontinuity than a gradient (Holtfreter and Hamburger, 1955).

The novel contribution of Dalcq and Pasteels (1937) concerned the other axis and gradient, namely, the cortical field centered at the gray crescent and expressed by a cortical morphogen "C" of decreasing intensity at increasing distances from the center. The field, once established, was supposed to be long lasting. The exact time of its effect was not specified. It was to be a fixed pattern of the cortex, imperturbable by gravity and yolk movements. In fact, insensitivity to gravitational effects was the test of a cortical localization. There are four main results cited by Pasteels (1964) in favor of the cortical field idea: (1) rotations of *R. temporaria* eggs *after* crescent formation, to achieve movement of the yolk mass at right angles to the crescent position, followed by identification many hours later of the position of the blastopore relative to the original position of the crescent and yolk mass; (2) rotations of axolotl eggs *after* gray crescent formation, to achieve yolk mass displacements in the same plane as the crescent, an experiment closely related to the first, and with a similar analysis; (3) inversions of eggs at the two-cell stage according to the procedures of Schultze (1894) and Penners and Schleip (1928a,b) with subsequent identification of blastopore position, an experiment to be discussed later after the events of cleavage have been reviewed; and (4) the cortical transplantation experiments of Curtis (1960, 1962a, 1963), the most direct and readily understood test. These results will now be reviewed critically in an attempt to dismiss the cortical field idea and replace it with an idea closer to the current view of Nieuwkoop (1972, 1977), namely, that the vegetal yolk mass is the carrier of the dorsal–ventral axis as well as the animal–vegetal axis at these early stages.

4.6.1   Rotations of R. temporaria Eggs

Pasteels (1948, 1964) used the procedures of Ancel and Vintemberger (1948) as a first step in the experiment to establish in fertilized eggs a predictable position of the gray crescent on the equator and then *after* crescent formation rotated the eggs to a new orientation so an equatorial point 90° to the right or left of the crescent would be uppermost. In this position the eggs were held between glass plates and the vegetal yolk slid downward leaving a trail of large vegetal yolk platelets behind on the cortex, in a configuration closely resembling the vitelline wall of the original gray crescent. This new wall was slightly translucent, allowing the grayness of the deeper, slightly pigmented and less yolky cytoplasm to show through in a crescent-shaped area. This new crescent is called "Born's crescent," recalling the original observations by Born (1885) that the yolk mass of inverted eggs streams to a new position and the bilateral axis of the embryo falls in the plane of streaming. In Pasteels' experiment, 170 eggs were rotated after gray crescent formation to form a Born's crescent to the right or left of the gray crescent. Of these, 49% initiated the blastopore many hours later at the position of the gray crescent, 22% at the position of Born's crescent, and 29% at a position between the two crescents. Pasteels observed that on further development bilateral organs tended to be enlarged on the gray crescent side in those embryos with axes originating from the Born's crescent or intermediate positions. He concluded that the gray crescent cortical field, though affected by the yolk rearrangements, remains "stronger" than the new organization of the Born's crescent (Pasteels, 1964, p. 376). It seems at least as striking that rotation of the egg after gray crescent formation can still shift the prospective dorsal–ventral axis; 51% of the eggs were shifted. By the Dalcq–Pasteels double gradient idea, one would expect no shifting of blastopore position after this particular rotation since "V" as well as "C" would be unchanged at the original crescent position. Perhaps additional gravitational force (centrifugation or greater angle of inclination) could shift all of them. Recently, Kirschner et al. (1980) have found that X. laevis eggs turned 90° off-axis shortly before first cleavage and well after crescent formation will in many cases shift the position of the dorsal–ventral axis. Pasteels notes that Born's crescent has not only the appearance of the gray crescent, including the vitelline wall, but also has its impact on dorsalization. And yet it can be formed *after* the gray crescent, when the cortical field has supposedly become fixed. To preserve the cortical field as essential to dorsalization, one must conclude that the field is not uniquely and irreversibly established at the time of crescent formation, but can be modified later by gravity. But then one is invoking the properties of the "vitelline gradient" of the yolk mass and losing the distinguishing property of the cortical field, namely, resistance to the effects of gravity.

The alternative is to focus on the vitelline wall rather than on the crescent. As Pasteels demonstrated, the wall is a layer of large platelets drawn up on the cortex and exposed to the cytoplasm of the animal hemisphere. The wall is normally formed by the cortical contraction of crescent formation (or better of "wall formation"), which occurs only once. The wall can, perhaps, be erased and reformed any number of times after the contraction if gravity is employed. Perhaps wall position could be affected well into the cleavage stages until new plasma membrane introduces a barrier to rearrangement, for example, at the eight-cell stage when the first horizontal cleavage separates animal and vegetal materials. Also, it may be necessary to increase gravitational force at the cleavage stages due to increased viscosity of the cytoplasm and adherence of cell materials to the cleavage membranes. If dorsalization at these early stages really depends on the exposure of large vegetal platelets to the animal hemisphere cytoplasm—that is, a cytoplasmic localization rather than a cortical field—it should be possible to alter the fate map by gravity well into the cleavage stages, or to transfer platelets by microinjection to initiate secondary axes.

4.6.2   Rotation of Axolotl Eggs after Gray Crescent Formation

This is basically the same experiment as the preceding one, but the eggs were turned in the same plane as that of the gray crescent rather than at right angles to it. The crescent was marked and the eggs were embedded in 10% gelatin, of which the entire solidified bed was set at the chosen angle. Pasteels (1946, 1964) chose positions so the yolk mass would slide across the crescent cortex. At an angle of 45°, the blastopore lip appeared many hours later on the side of the yolk mass toward the marked crescent (nine cases), no change from the unrotated control. At 90° (one case), a double axis embryo (twins) resulted with one blastopore lip on the leading edge of the yolk and one on the trailing edge. In this case, Pasteels concludes that the two edges of the yolk mass are equally distant from the center of the crescent cortex, each experiencing a sufficiently high C · V level to induce a blastopore. At 135°, the embryo formed later a single blastopore on the trailing side of the yolk mass—that is, the edge now nearest the crescent (7 cases). And at 180°, the same result was obtained (three cases). Pasteels concludes that only the trailing edge is close enough to the crescent to receive a high C · V dose. In the discussion of these results, Pasteels (1946) does not refer to Born's crescent, even though the eggs are submitted to the same treatment as in the right-angle series described before. In terms of vegetal yolk displacements, it seems plausible that the 45° rotation did not obliterate the old vitelline wall, that the 90° rotation formed a new wall but did not obliterate the old one, and that the 135° and 180° rotations both obliterated the old wall and formed a new one. The critical experiment on Pasteels' part would have involved rotating the eggs 90°, 135°, and 180° in the *opposite* direction; he would presumably predict in the 180° case that the blastopore would appear on the yolk edge *toward* the crescent, whereas by the vitelline wall idea the blastopore would presumably appear on the edge *away* from the crescent. Pasteels did only a 45° rotation away from the crescent, with results similar to control eggs. Thus, the experiment is incomplete.

4.6.3   The Cortical Transplantations of Curtis (1960, 1962a)

In his most crucial experiment, Curtis (1960, 1962a) dissected, from an uncleaved fertilized egg of *X. laevis,* a small piece of gray crescent cortex and membrane (150 μm on an edge by 10 μm deep) that was free from subcortical yolk, and transplanted it into the surface of the prospective ventral side of a recipient egg just starting cleavage. The recipient egg developed a double axis (twins, 11/11 cases). The piece of gray crescent cortex apparently directed the formation of a new axis, which formed in addition to the recipient's own axis originating from its own crescent. The same result was obtained when gray crescent cortex was taken from an 8-cell stage donor. Unfortunately, there is no true negative control for this important experiment: in all transplantations into the ventral side of an uncleaved recipient, a double axis embryo resulted. Curtis did not report a transplant of *ventral* cortex into the ventral side to demonstrate the absence of a second axis. Thus, there was no test for cortex specificity. As a control, he slit the recipient egg in the ventral side and let it heal and develop, resulting in a single axis embryo, but the manipulations of this control differ from those of the experiment in that a cortex implantation was not made. The importance of the negative transplantation control is this: the eggs of other amphibian species are well known to undergo double-axis formation when the egg is inverted shortly after first cleavage, for example, the famous Schultze inversion experiment (1894, 1900) as studied by Penners and Schleip (1928a,b). Penners (1929) found that uncleaved eggs of *R. fusca* would develop double axes in response to inversion before first cleavage but after crescent formation, and the frequency could reach 20–50% of the inverted eggs. Furthermore, in the study by Pasteels on axolotl eggs rotated after crescent formation (discussed in Section 4.6.2), the single case turned *ventral side up* (90°) gave a double-axis embryo. Likewise, Kubota (1967) obtained 20–40% twin embryos after turning eggs 90° to a position with ventral side up. Also, aged eggs of *R. pipiens* are reported to increase greatly their frequency of spontaneous twinning (Witschi, 1934, 1952; Zimmerman and Rugh, 1941). The egg of *X. laevis* was not checked to see if it twins when perturbed by flattening on an agar surface ventral side up without its vitelline membrane for support, followed by surgery at the equator where the yolk mass might be distorted or divided. Recently it has been found that such treatment of *X. laevis* eggs causes twinning in 10–70% of the eggs, even without transplantation of the cortex (Kirschner *et al.* 1980). At present, rather than conclude that the cortical dorsalizing center is transplantable, it might be safer to conclude that Curtis discovered conditions for twinning *X. laevis* eggs at high frequency. The latter is itself an interesting conclusion since there have been few experimental opportunities to learn about the mechanisms by which the large amphibian egg normally suppresses secondary axes and unifies its early cytoplasmic movements to achieve a single axis. However, this is an invitation to experimentation, not a conclusion about the dorsalizing action of the gray crescent.

As a further criticism of the original experiments, Curtis did not mark the gray crescent to verify in his single-axis cases that the axis derived from the gray crescent and not from a "Born's crescent." And finally, the crescent in *X. laevis* is not easy to recognize, and Curtis did not assess in marked controls of unrotated eggs his accuracy in identifying the crescent as the location of the prospective blastopore lip or as the source of his cortical graft. In a related experiment, Curtis, (1962a, 1963) removed the crescent and found that the eggs failed to develop beyond the blastula stage. However, he did not do a control of removing ventral cortex to show such eggs would develop normally.

These important experiments have not been repeated in their original outline, although Tompkins and Rodman (1971) describe a related study in which gray crescent cortex was injected into the blastocoel of late blatulae and the induction of secondary axes was subsequently scored. It should be recalled that as gastrulation proceeds, the blastocoel is squeezed to a ventral equatorial position, bringing the cortex piece directly beneath ventral ectoderm, which, if induced, forms a secondary axis of mesoderm and neural tube. They report successful induction of strong secondary axes in 50–70% of the blastulae receiving gray crescent cortex. Their control blastulae, which received animal pole cortex, produced 10–20% cases of small secondary axes. The more appropriate control would have been injection of ventral or vegetal cortex into the blastocoel, on the chance that adherent yolk granules might have an inductive effect, as has been observed by Faulhaber and Lyra (1974), Wall and Faulhaber (1976), and Asashima (1975). Recently, Malacinski *et al.* (1980) have repeated the conditions of the experiment, but failed to observe secondary axes in any of a large number of trials. Thus, it has not been possible to extend the original observation.

In summary, the evidence for the gray crescent as a dorsalizing determinant in the amphibian embryo remains inconclusive and can perhaps be explained at least as well on the basis of a cytoplasmic, and not cortical, localization—such as the exposure of deep vegetal yolk material to animal hemisphere cytoplasm. For either explanation, critical experiments are needed. Curtis himself (1962a) concluded that the gray crescent is dispensable by the 8-cell stage because surgical removal of it at this stage did not interfere with later axiation. He suggested that either the surrounding cortex was capable of regulative reconstruction of a cortical field by that stage, or the cortex had registered its effect on some cytoplasmic material beneath the cortex. Here we are suggesting that the gray crescent itself is not important at any time, but that the cortical contraction of crescent formation is important as a force to shift the deep vegetal yolk mass to one side to form the vitelline wall and the vegetal dorsalizing center. The dorsal–ventral axis can, perhaps, be perturbed after the time of crescent formation, provided the force is sufficient to overcome the cytoplasmic asymmetry already established by the cortical contraction and to replace it with a new asymmetry.

4.7 *Preparation for First Cleavage*

The cortex is intimately involved not only in the contractions of activation and crescent formation but also in the cleavage process. The furrow of cleavage is itself a contraction of the surface, promoted by a thick contractile ring of actin filaments built up locally in the direction of the furrow (Selman and Perry, 1970; Perry, 1975; Bluemink and deLaat, 1977). Once the furrow has indented the egg by approximately one-third its diameter, a second step begins in which new plasma membrane is deposited from vesicles, a very different means of cytokinesis. These are both late steps in cleavage, following a series of events that determine the time, origin, and direction of cleavage, involving noncortical agents.

It is believed that normal cleavage requires the presence of a centriole from the sperm as a natural "cleavage-initiating factor," as artificially activated amphibian eggs develop only a short, shallow abortive furrow at the animal pole. In contrast, eggs injected with fractionated brain homogenates (Shaver, 1953; L. R. Fraser, 1971), lysed sperm heads (Maller *et al.*, 1976), basal bodies (Heidemann and Kirschner, 1975), or a transplant diploid nucleus with its centriole (Subtelny and Bradt, 1963) develop normal deep furrows. Artificially activated eggs contain a small asterlike yolk-free region near the egg nucleus but fail to produce a full-size aster or a bipolar spindle (Subtelny and Bradt, 1963; Ramirez and Huff, 1967; Manes and Barbieri, 1977). This is also true in the egg fragments obtained by Fankhauser (1937), using a nylon noose to constrict the fertilized *Triton* egg into two pieces, one of which lacked the sperm aster and nucleus; in these "gynomerogons," asters were absent, spindles were anastral (like meiotic spindles), and furrows were weak. As discussed in Section 4.3, the centriole is thought to allow formation of an astral spindle, which at metaphase determines the position of the furrow. Although the case for the centriole is good, there may be other noncortical agents necessary as well; Briggs and King (1953) made the interesting observation that manually enucleated eggs of *R. pipiens*, when injected with homogenates of "cleavage-initiating-factor" fail to form a deep furrow, in contrast to the furrow of nucleated injected controls. Thus, some element of the nucleus may be necessary for complete cleavage. Briggs *et al.* (1951) obtained good cleavage with enucleated eggs fertilized with heavily X-irradiated sperm, where the cleavage nuclei were almost devoid of chromosomal material as detected by the Feulgen stain; thus the nuclear component is certainly not the chromosomal arms, leaving as possibilities the centromere region or a specialized region of the nuclear membrane such as the fossa, either of which may function as a nuclear membrane organizing center (B. G. Smith, 1929).

The period of direct preparation for cleavage may begin about the time the pronuclei meet and the cortical contraction of crescent formation begins (Table II, Fig. 7). At this time the large sperm aster dies back and is replaced by a small bipolar spindle, indicating centriole splitting (Subtelny and Bradt, 1963). In *Triton* the asters of the supernumerary sperm also die back at this time even in the absence of an egg nucleus for pronuclear pairing (Fankhauser and Moore, 1941*a*.) Thus, the timing of astral disintegration and centriolar splitting is probably not controlled by the aster or the nucleus, but by an unknown cytoplasmic or cortical timing mechanism. It is interesting that at the same time, an MPF-like agent appears in the cytoplasm of the *R. pipiens* egg (Wasserman and L. D. Smith, 1978), presumably initiating the transition from interphase to prophase in the egg, just as it initiated germinal vesicle breakdown, meiotic chromosome condensation, and cytokinesis of the polar body in the ococyte. Possibly MPF appearance marks the moment when the sperm aster dies back in the fertilized egg.

Shortly thereafter, a "surface contraction wave" originates from the area of the animal pole and progresses downward, reaching the vegetal pole within 15–20 min, as visualized by cinematography (Hara *et al.*, 1977). A second wave can sometimes be detected 10 min behind the first. In side view eggs removed from the vitelline membrane can be seen to round themselves up from a flattened shape and to relax again later after cleavage (Harvey and Fankhauser, 1933; Selman and Waddington, 1955; Sawai, 1979; Hara *et al.*, 1980). About the time the first wave reaches the vegetal pole and the egg is most rounded, the furrow appears at the animal pole. Hara *et al.* (1980) have shown these interesting waves to occur in artificially activated eggs of *X. laevis* and in fertilized eggs blocked from cleavage by colchicine or vinblastine. These results indicate that the waves originate and propagate without the sperm centriole or microtubules. Hara *et al.* (1980) and Sawai (1979) also found that the waves occur in activated egg fragments lacking the egg pronucleus; thus nuclei are also not involved. Remarkably, the waves continue to appear cyclically, at least 6 or 7 times in these fragments, on the same schedule as in normal fertilized eggs where a cleavage regularly follows in sequence each round of waves. These experiments give us a new view of the cycling properties of the amphibian egg, and bring it close to the earlier work on sea urchin eggs with cyclical changes in sulfhydryl–disulfide ratios, in thymidine kinase activation, and in calcium-dependent ATPase activity, independent of the presence of nuclei or centrioles or furrows (Harris, 1978). It is also known that the "maturation-promoting factor" appears cyclically in amphibian eggs lacking nuclei or cleavage furrows (Wasserman and Smith, 1978); as discussed previously, this factor may be the cytoplasmic activator of the enzymes of mitosis or meiosis. But the cyclic initiation of MPF activity and of the surface contraction wave presumably has its origin in the cyclic behavior of yet another oscillator, perhaps the subcortical membranous network considered by Campanella and Andreuccetti (1977) to propagate the calcium-dependent activation wave of cortical granule discharge at the time of fertilization. Harris (1978) has suggested that calcium-pumping vesicles underlie the cell cycle. On the other hand, Kauffman and Wille (1975) have argued that an explanation of the cyclic behavior of cells should not be sought in the cyclic behavior of one component, but in the coupled oscillation of at least two steady-state processes. In any event, the involvement of a membranous network is unknown; in the amphibian egg the subsurface network undergoes reorganization after fertilization as many vesicles collect in the subcortical region (Balinsky, 1966; Selman and Perry, 1970; Ubbels and Hengst, 1978). The calcium dependence of the surface contraction waves is not known.

The surface contraction wave may be identical with the wave of surface stiffness detected in *T. pyrrhogaster* eggs by Sawai and Yoneda (1974) by means of a sucking pipette surface elastimeter. The wave of stiffness started at the animal pole ahead of the furrow and moved downward as a latitudinal band around the entire egg circumference. As mentioned previously in this section, several authors have observed that the amphibian egg without its vitelline membrane rounds up shortly before cleavage, as if the surface becomes more rigid. It is well known that cultured cells round up prior to division as well.

For the actual initiation of the furrow, evidence from several systems gives the metaphase spindle the role of triggering the furrow as well as determining its position and direction. First, in amphibian eggs the direction of the furrow may be determined by lateral compression of the egg until metaphase has passed (Zotin, 1964); the furrow orients in the shorter axis of the egg (Sachs' Law), meaning that the spindle had oriented in the longer axis. While in anurans the furrow normally aligns in the plane of the sperm entry point and animal pole (the old meiotic spindle site), this is a very weak bias, the basis of which is unknown. In urodeles the direction of the furrow is thought to be unrelated to the sperm entry point in polyspermic eggs (although see Ancel and Vintemberger, 1948, for *T. alpestris*). Second, Kubota (1969) has removed the spindle from the fertilized egg of *R. nigromaculata* by micropipette and finds that after metaphase the furrow appears nonetheless, whereas before metaphase the same treatment prevents furrow appearance. This is similar to the experiment and result of Rappaport on sea urchin eggs. And third, Kubota (1969) prevented contact of the spindle with the cortex by injecting a drop of oil between them, and found that after metaphase the furrow formed despite the injection, whereas before metaphase it did not. Thus, this step in cleavage seems to be spindle-dependent and therefore centriole-dependent. Masui and Clarke (1979) have summarized the evidence suggesting that the metaphase–anaphase transition of mitosis comprises a release from a metaphase arrest similar to the release from meiotic metaphase arrest accomplished by activation of the unfertilized egg when a calcium flux inactivates the cytostatic factor preventing anaphase. This release may be cortically controlled in both meiosis and mitosis.

The final step in furrow formation concerns furrow extension, which follows downward the wave of stiffness (Sawai and Yoneda, 1974). Kubota (1969) has reported that if the egg surface ahead of the furrow is rubbed to displace subcortical cytoplasm laterally, then the furrow forms with a lateral displacement, as if a committed prefurrow cytoplasm lies ahead of the actual furrow. Sawai (1972, 1974) has reported elegant transplantations of subcortical cytoplasm from the advancing tip of the furrow to a new subcortical location of the same latitude, where that cytoplasm will induce a new furrow, which then proceeds to extend downward. When the same subcortical cytoplasm is transplanted to a new location of a different latitude (by 30°), a furrow is not induced. Sawai concludes the surface must be receptive and at the same time must experience an induction from subcortical cytoplasm of the approaching furrow in order for extension to occur.

Furrow formation continues long after interphase has begun again. In *X. laevis,* the second cycle of DNA synthesis begins even before the first furrow is visible (Graham, 1966), and the second furrow begins before the first one is finished.

4.8 Pattern Regulation

In the fertilized uncleaved egg, pattern formation takes place through rearrangements of the cytoplasmic and cortical materials of this large single cell, along both its animal–vegetal and prospective dorsal–ventral axes. Pattern regulation would concern the egg's ability to restore its cytoplasmic and cortical arrangements after experimental disturbance; of course it is assumed that this ability is also exercised against the less obvious disturbances of normal development. It is surprising that the large egg succeeds in establishing in its whole volume a single animal–vegetal axis and a single prospective dorsal–ventral axis under normal conditions and even more surprising after rotation or inversion, especially after UV irradiation and rotation. The ability of the egg to unify its contents deserves attention here. This ability is finally exceeded in certain conditions where twin axes arise.

The vegetal yolk mass seems to determine the vegetal end of the animal–vegetal axis of the egg and the position of endoderm in the fate map. These conclusions are demonstrated by the inversion experiments of Motomura (1935) and Pasteels (1941), as described previously, where the yolk mass was centrifuged into the opposite hemisphere, exchanging positions with the nonyolky cytoplasm. Since the fate map was entirely reversed by the reversal of the internal contents, we learn first that the endoderm is determined by the position of the yolk mass; second, that the ectoderm is determined by the position of the nonyolky cytoplasm, and third, that the cortex does not have an irreversible animal–vegetal pattern of importance to the fate map at this early stage, a point made by Fankhauser (1948). As a more direct demonstration of the determination of ectoderm by the nonyolky cytoplasm, Streett (1940) obtained by constriction a nucleated fragment comprising the animal hemisphere of a freshly fertilized egg of *T. pyrrhogaster* and observed the fragment to cleave well, forming an enlarged sphere with a blastocoel but failing to gastrulate or develop tissues other than an epidermal sheet, even though it contained part of the gray crescent cortex. Numerous experiments described later affirm that the animal hemisphere develops to ciliated epidermis, unless it interacts with the vegetal half of the egg. Nieuwkoop (1973, 1977) has summarized these findings in a model of the amphibian egg as two unique interacting hemispheres, discontinuous at their equatorial interface rather than comprising a continuous animal–vegetal gradient. This model is adopted in the present review.

The coherence of the yolk mass may explain the ability of the egg to establish a single animal–vegetal axis instead of several in its large volume. The yolk mass comprises large and intermediate sized platelets, each a membrane-bounded organelle; coherence of the mass presemably results from the affinity of the platelets for each other or from their containment in a common cytoskeletal matrix. As a demonstration of this coherence, it is very difficult to draw vegetal yolky cytoplasm from a fertilized egg into a micropipette, whereas the animal hemisphere cytoplasm can be easily pipetted. In contrast, before fertilization the egg cytoplasm is more jelled and the contents do not rearrange despite a gravitationally unstable position. A further pattern regulating mechanism of the egg is its ability to rotate freely in the perivitelline space after fertilization when the egg becomes sensitive to rearrangements of the yolk mass. Without this freedom of rotation to a gravitationally stable position, the egg would probably derive twin or multiple axes at considerable frequency as in the inversion experiments of Schultze, of Penners and Schleip, and of Pasteels. Apparently the yolk mass becomes sensitive to splitting in the period approaching first cleavage, perhaps as the furrow begins to bisect it.

Regarding the prospective dorsal–ventral axis at the 1-cell stage, we have discussed the role of the cortical contraction of gray crescent formation as the means of forcing a lateral rearrangement of the vegetal yolk mass, bringing large yolk platelets upward from a deep vegetal position through the layer of intermediate platelets, and into the region of the gray crescent to form the vitelline wall. By this movement the largest platelets are exposed to the less yolky cytoplasm of the animal hemisphere. Pasteels (1948, 1964) had noted this movement clearly. Later in the midblastula the region of the vitelline wall approximates topographically the "vegetal dorsalizing center" of Nieuwkoop (1969a,b), that center which "organizes the organizer." It is the region of strongest inductive action of the vegetal hemisphere when it acts on the animal hemisphere, evoking dorsal mesoderm and pharyngeal endoderm from cells of the animal hemisphere. For simplicity, it is suggested that the largest and most closely packed platelets of the lower vegetal part of the egg, or materials associated with these platelets, have the latent ability to establish inductive activity in the blastomeres receiving them. The cortical contraction would normally shift the largest platelets to the prospective dorsal side and bring them into contact with the animal hemisphere, the future target of their inductive action. Perhaps in a period shortly after vitelline wall formation, the animal hemisphere cytoplasm modifies the large platelets in some way, and "activates" them for future induction, making them a finished "cytoplasmic localization" in the dorsal vegetal region. This model illustrates a possible multistep process in forming such a cytoplasmic localization and the use of reciprocal interactions between different parts of the egg—that is, the aster affecting the yolk, which affects the cortex, which affects the yolk, which is also modified by animal hemisphere cytoplasm, the future inductive target of the yolky region. On the ventral side this contact and activation would be reduced by the intervening zone of intermediate platelets. Although this proposal is almost completely speculative, it is a way to visualize early steps of dorsalization consistent with the findings of Nieuwkoop (1973, 1977) for the midblastula stage and with a minimal role for the gray crescent.

According to this view we can discuss pattern regulation again in terms of the coherence of the yolk mass within itself and also its adherence to the vegetal cortex, a point made by Pasteels (1964). This adherence would assure the lateral shift of the platelets with the cortical contraction. Also, we can discuss the behavior of the animal hemisphere cortex as a unit; its cortical contraction is unified in the sense of achieving the formation of a single gray crescent and a single vitelline wall, rather than several. This unification presumably resides in the ordered buildup of contractile fibrils in the cortex and perhaps in the uniform arrangement of the subcortical vesicular reticulum, which might trigger contraction. But why the cortex contracts more strongly on the prospective ventral side is unknown. The rotation experiments of Ancel and Vintemberger (1948) on activated eggs suggest that the position of the yolk mass is a more direct spatial cue to the direction of contraction than is the position of the sperm aster or sperm entry point that are absent in such eggs. In any event, the large egg does manage to unify its cortical contraction, just as it manages to unify its yolk mass.

Even if the cortex has no role in patterning the fate map at this early stage, it enters actively into numerous cellular processes such as the two cortical contractions of activation and crescent formation, the three waves of activation, postfertilization, and surface contraction, and polar body extrusion and cleavage furrow formation. Actin filaments participate in several of these processes. A few of these processes are probably patterned in oogenesis, namely, the position of polar body extrusion and the activation contraction, based on the position of the meiotic spindle left from maturation. For most of the processes the egg cortex probably exists temporarily in a labile and alterable state where pattern is imposed from the underlying cytoplasm. This is indicated by the inversion experiments of Motomura (1935) and Pasteels (1941), exemplified by recent results on *X. laevis* eggs (J. C. Gerhart and J. G. Bluemink, unpublished); inverted and lightly centrifuged eggs start their first cleavage from the pole of the unpigmented hemisphere, previously the vegetal pole but now the "animal pole" as defined by its position in the nonyolky hemisphere after centrifugation has interchanged the contents of the hemispheres. The furrow carries a thick contractile ring and vesicles of the type normally seen in the animal hemisphere of control eggs. Thus, the cytoplasm determines the position and appearance of the furrow. Presumably the surface contraction waves also now begin at the unpigmented pole after centrifugation, whereas in controls they originate at the opposite pole. Of course the cortex reaches a stage in which it is "mosaic" and insensitive to rearrangements, as shown by Kubota (1969). The long period of lability of the amphibian egg cortex contrasts with the findings on invertebrate eggs. The more relevant comparison may come from the mammalian egg where there seems to be no inherent cytoplasmic or cortical asymmetry brought from oogenesis, not even as asymmetry from yolk platelets, which they lack.

5   From First Cleavage to the Midblastula

The midblastula stage, when the embryo contains approximately 1000 cells, is marked by the first gene expression as detected by RNA synthesis. Until that time the developmental events are independent of new gene expression. The architecture of the midblastula and, in many respects, of the late blastula can be generated by cellular processes prepared in oogenesis. These stages are reached by eggs of lethal hybrids from intergenus or interspecies matings (Subtelny, 1974), by eggs exposed to inhibitors of RNA synthesis such as actinomycin (Brachet *et al.*, 1964; Wallace and Elsdale, 1963) and α-amanitin (Brachet *et al.*, 1972), and even by eggs lacking most if not all chromosomal material (Briggs *et al.*, 1951). In the last experiment *R. pipiens* eggs were fertilized with heavily X-irradiated sperm and then pricked at the animal pole to remove the egg meiotic spindle carrying the egg chromosomes. The eggs cleaved well and formed a blastocoel, achieving a blastula appearance, but failed to gastrulate. Cytological analysis showed many cells to lack nuclei with material stained by the Feulgen reaction for DNA. Fankhauser (1934b) had likewise obtained occasional partial blastulae lacking nuclei from fragments of *Triturus* eggs constricted shortly after fertilization.

5.1   Mitotic Rate and the Animal–Vegetal Gradation of Cell Size

The pattern of cleavage of axolotl eggs has been carefully analyzed cinematographically by Hara (1977) and in a series of timed photographs by Signoret and Lefresne (1971, 1973). A study of the latter type has been carried out with *X. laevis* eggs by Satoh (1977) and K. Hara (unpublished). The interval between early cleavages shortens successively, reaching a minimum by third cleavage (85 min in axolotl, 35 min in *X. laevis* at 21°C). This minimum is sustained until the 10th or 11th cleavage. Rough estimates of cell numbers from fixed and sectioned eggs of *X. laevis* are shown in Fig. 8. Graham (1966) established by [³H]thymidine injection and autoradiography of eggs of *X. laevis* that the inter-

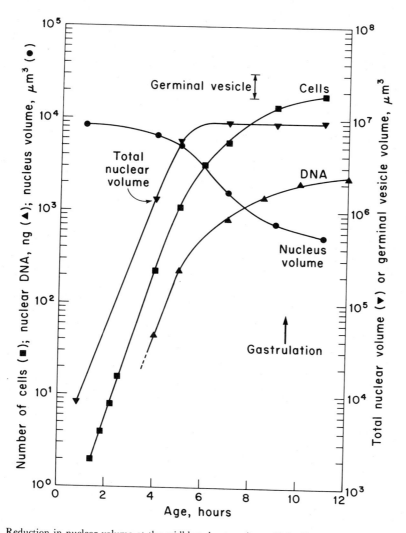

Figure 8. Reduction in nuclear volume at the midblastula stage (stage 8) in *Xenopus laevis.* The age of the embryo refers to hours postfertilization, at approximately 22°. Stage 8, when cleavage slows and desynchronizes, occurs at approximately 5 hr. Data are compiled from several sources: nuclear DNA values are those of Dawid (1965); measurements prior to 4 hr are omitted because of the large background value of mitochondrial DNA, approximately 3.5 ng per embryo. Note reduction in rate of DNA synthesis at stage 8. Cell counts are those of Bonanno (1977) calculated from counts of nuclei in Feulgen-stained 7-μm sections of embryos from different times in development. Nuclear counts in a section are converted to nuclei per embryo based on the volume of the embryo, with a correction for the number of successive sections in which a nucleus will appear due to its large diameter. Note that cell counts and DNA values agree closely, based on the value of 6.3 pg of DNA (prereplication value) per diploid (2C) nucleus (Dawid, 1965), divided by (ln 2) to correct for the exponential age distribution of cells between the 2C and 4C DNA values. Average nuclear volume was calculated from the average nuclear diameter measured in stained 7-μm sections (Bonanno, 1977). This is designated "nucleus volume" in the figure. The value for nuclear volume at 1 hr was taken from Graham (1966), and for the germinal vesicle volume from Gurdon *et al.* (1976). Total nuclear volume was calculated as the product of individual nucleus volume and the number of nuclei per embryo at a given stage. Note that the total nuclear volume reaches a maximum value at stage 8, and does not increase from 5 to 10 hr of development. As the number of nuclei increases, the volume of each is proportionately smaller. Note that total nuclear volume of the midblastula stage approaches within two- or threefold the volume of the germinal vesicle of the full-grown oocyte.

mitotic period is spent almost entirely in S, with M occupying approximately 2 min, and no detectible G_1 or G_2 period. An S period of 28 min is, of course, remarkably short for a eukaryotic cell, although surpassed by the 8-min S period of *Drosophila* early embryos, where it has been found that the great speed results not from faster moving replication forks but from two factors: (1) the replication of DNA from many forks, proportionately more than in later stages with longer S periods, and (2) the synchronous initiation of replication, which becomes increasingly asynchronous at later stages with longer S periods (Kriegstein and Hogness, 1974; Blumenthal *et al.*, 1973; McKnight and Miller, 1977). In amphibia the analyses are much less advanced, although Hyode and Flickenger (1973) have observed a high frequency of DNA replication forks even in gastrula cells of *R. pipiens*.

Gurdon (1967, 1974) demonstrated that full-grown *X. laevis* oocytes do not support the synthesis of DNA of injected nuclei from liver, brain, or blastulae, but that synthetic capacity arises during maturation. Matured enucleated oocytes do not support DNA synthesis, indicating a contribution from the germinal vesicle. Benbow *et al.* (1975, 1977) developed an *in vitro* assay for DNA synthesis in liver nuclei exposed to egg and oocyte extracts and concluded that the germinal vesicle releases an initiator of DNA synthesis whereas DNA polymerases of three types abound in the cytoplasm of all stages from the unfertilized egg to the gastrula (Benbow and Ford, 1975). One polymerase (type β) increases during maturation. At the present time, this initiator of DNA synthesis is a foremost candidate for the factor allowing the rapid replication of the cleavage stages prior to midblastula.

By the midblastula stage the animal hemisphere contains 4- to 8-fold smaller cells than the vegetal hemisphere, and in 1888 Balfour explained this gradation in terms of *rates* of cell division in the two hemispheres. In Balfour's rule, the rate of division was said to depend inversely on the amount of yolk contained in a blastomere. Recently, Hara (1977) scrutinized this assumption by following division times in the two hemispheres of axoltl eggs and found that by the 10th cleavage, the vegetal blastomeres lagged less than one division behind the animal blastomeres, and that embryos dissociated just before the 11th cleavage contained exactly 1024 cells, as expected for 10 cleavages in all cell lineages. Satoh (1977) and Hara (1977) have noted that the early cleavages are not exactly synchronous after the third cycle, but are metachronous; that is, systematically staggered with one region always dividing first. Satoh (1977) considered the prospective dorsal side to have a slight lead, whereas Hara (1977) found the animal pole blastomeres slightly ahead. In an important experiment, Hara (1977) produced separated blastomeres by removing the vitelline membrane and incubating the fertilized axoltl egg in 67 mM phosphate buffer, in which medium the cells fail to cohere. The dissociated embryo showed metachronous cleavage in a pattern just like that of the intact embryo. The synchrony does not depend on cell contacts or junctions; it results simply from the fact that the egg started as a single cell. As discussed earlier, Hara *et al.* (1980) have found evidence for a cortical or cytoplasmic oscillator affecting the whole egg and set in motion at fertilization. Presumably all blastomeres receive material of this oscillator and therefore follow its timing.

Aside from small animal–vegetal differences in the time of cleavage initiation, vegetal blastomeres do not divide more slowly than animal hemisphere blastomeres; therefore, the clear difference in cell size results from asymmetric division planes. Balfour's rule must be recast in the form that a blastomere cleaves away from the side of more yolky cytoplasm. This is easily seen even at the third cleavage, the first latitudinal cleavage, which is well above the egg equator, cutting off four small upper blastomeres from four large lower ones.

In addition, in most eggs of *X. laevis* there is a slight gradation of cell size from the dorsal to ventral side; whereas the first cleavage bisects the gray crescent vertically and divides the egg into equal blastomeres, the second cleavage occurs at right angles to the first and splits the egg vertically into two slightly smaller gray crescent blastomeres and two larger ventral ones (Morgan and Boring, 1903; Nieuwkoop and Faber, 1975; Landstrom and Lovtrup, 1977). Lovtrup *et al.* (1978) have attached great significance to these dorsal–ventral differences in cell size in their hypothesis of dorsalization in which the smallest blastomeres are considered as the first to begin preparations for gastrulation. Hara's films indicate that the dorsal–ventral differences are not due to differences in division frequency, but just furrow position. This position at early stages can be readily changed by rotation or by compression between parallel surfaces, perhaps allowing a future critical test of the proposals of Lovtrup *et al.* (1978).

5.2 Completion of the Cleavage Furrow and the Beginning of Blastocoel Formation

The contractile ring of actin filaments in the egg cortex accomplishes an indentation of the surface to a depth of about one-third the egg diameter at the animal pole and only one-sixth or less at the vegetal pole. The contractile ring appears less well-developed in the yolky hemisphere (Bluemink and deLaat, 1973; Selman and Perry, 1970; Sanders and DiCaprio, 1977; Kalt, 1971b). There appear to be one and perhaps two subsequent important steps in cytokinesis in these large cells; these involve extension of the furrow by membrane addition from vesicles and Golgi bodies rather than complete constriction by the contractile ring. On completion of cytokinesis there may be three distinct regions of the plasma membrane between blastomeres: the indented region of the original egg membrane, carrying short microvilli; the animal hemisphere new membrane identifiable as smooth, and thickly covered with extracellular osmiphilic deposits; and the vegetal hemisphere new membrane, identifiable as smooth and less covered with deposits (Singal and Sanders, 1974b; Kalt, 1971b).

The contractile ring circumvents the egg surface within 30 min after the beginning of the first cleavage, but the complete separation of the two blastomeres requires 1 hr, in the case of *X. laevis*. The new membrane added from vesicles and Golgi bodies is clearly different from the original egg surface of the animal hemisphere in terms of its lack of pigmentation, as seen when the blastomeres are allowed to separate in cleaving eggs removed from the vitelline membrane. At the boundary between these two kinds of membrane is a row of long microvilli probably involved in the initial steps of blastomere association, leading to tight junction formation (Singal and Sanders, 1974a,b). Thanks to the position of these junctions, the new membrane is not exposed to the external medium. According to Singal and Sanders (1974 a,b), new membrane of the animal hemisphere is deposited from large vesicles containing fibrous material, whereas new membrane of the vegetal part is from small vesicles free of fibrous material and closely associated with Golgi bodies. These two kinds of new membrane meet at the level of the vegetal yolk mass where traces of the spindle midbody remain. At their border microvilli appear (Singal and Sanders, 1974b). This is the region that will form the floor of the blastocoel. There may be tight junctions between vegetal blastomeres at the border of the new vegetal membrane and old vegetal egg surface (Sanders and Zalik, 1972).

The main functional difference of the new membrane between blastomeres as compared with the membrane of the original egg surface concerns its ion permeability. This difference has great importance for blastocoel formation. The egg surface is remarkably impermeable to ions and solutes, consistent with the ability of the egg to develop in pond water or even distilled water (Slack *et al.*, 1973; Morrill *et al.*, 1975). Numerous researchers have reported the lack of penetration of radioactive metabolites (for example, Bachvarova and Davidson, 1966; Johnson, 1977*b,c*) and of inhibitors (such as cytochalasin B, Bluemink and deLaat, 1973; Selman *et al.*, 1976), although Loeffler and Johnston (1964) claim that salt solutions strongly promote metabolite entry. Earlier, the full-grown oocyte had been normally permeable to metabolites and ions (O'Connor *et al.*, 1977), but in the course of meiotic maturation became impermeable at the time of germinal vesicle breakdown when the microvilli of the oocyte withdrew their interdigitation with macrovilli of the follicle cells. Simultaneously the uptake of yolk precursors stopped (Schuetz *et al.*, 1977). Enucleated oocytes also achieve impermeabilization, indicating that this is a cytoplasmic response. Morrill *et al.* (1975) have suggested that the basis of the sudden drop in permeability involves a loss in K^+-conductive channels, and report decreased levels of $Na^+–K^+$-dependent ATPase during maturation. The mechanism of this interesting inactivation is unknown.

Then, shortly after fertilization the surface permeability to water increases briefly (Lovtrup, 1962), perhaps in conjuction with the substantial increase of surface membrane due to the fusion of the cortical granule membranes with the surface. Still the egg remains insensitive to inhibitors such as cytochalasin B, which starts to inhibit only when the new cleavage membrane appears (Bluemink and deLaat, 1973). Several groups have investigated the ion permeabilities of this new cleavage membrane (Morrill *et al.*, 1975; Slack and Warner, 1973; DiCaprio *et al.*, 1974; deLaat and Barts, 1976; deLaat *et al.*, 1976; Bluemink and deLaat, 1977). It appears to display normal K^+ permeability, indicating that $Na^+–K^+$ ATPases are active. New membrane transports K^+ inward and Na^+ outward, whereas the original egg surface membrane does neither. deLaat *et al.* (1976) have been able to obtain values for the specific membrane resistance for new membrane (0.4 $k\Omega cm^2$) and for old membrane (74 $k\Omega cm^2$), a difference of over 100-fold. Numerous researchers report entry of radioactive metabolites through this new membrane (for example, Bachvarova and Davidson, 1966).

The formation of the blastocoel has been explained recently as the result of the following three physiological operations: (1) the ion-transporting properties of the newly formed plasma membranes in the egg interior, (2) the ion impermeability of the outer egg-surface plasma membrane, and (3) the close association of blastomeres through tight junctions after the completion of each cleavage. On the last point, Sanders and Zalik (1972) and Kalt (1971*b*) observed in electron micrographs the close apposition of neighboring blastomeres at the borderline where the indented external surface meets the new internal membrane. These tight junctions form within a few minutes after cleavage. They are readily disrupted by media lacking Ca^{2+} (Holtfreter, 1943*a,b*; Johnson, 1970). Normally, their association is so effective that the internal fluid of the blastocoel makes no direct contact with the external medium, as evidenced by the absence of ion passage or entry of radioactive metabolites or inhibitors (Slack and Warner, 1973). The complete sealing of the surface from interior may not be accomplished for four or five cleavages; before that time, some of the rounded blastomeres still show visible liquid channels between each other, giving the egg the appearance of a grape cluster, hence the name of "morula" stage.

The actual inflation of the blastocoel with liquid is thought to proceed as follows (Morrill *et al.*, 1975; Slack and Warner, 1973). As new membrane is added on the inner faces of blastomeres, the pumping of sodium ion begins as a property of the new membrane.

Sodium ion is pumped from the cell interior into the intercellular space, whereas potassium ion lost passively into the intercellular space is pumped back into the cells. Sodium ion comes from reservoirs within the cell, a major source being the membrane-bounded yolk platelets, which previously in oogenesis had formed by endocytosis, taking in a large sampling of extracellular ions such as sodium and calcium along with vitellogenin, the highly anionic macromolecular precursor of yolk proteins (Redshaw, 1972; Wallace and Bergink, 1974). As sodium ion is pumped from the cytosol into the intercellular space of the blastula, it is replaced in the cytosol by sodium ion from the yolk reservoirs.

In more quantitative terms Slack et al. (1973) measured a total of 60–70 nmoles of sodium ion per unfertilized egg of X. laevis and the same amount of potassium ion. The amphibian egg is unusual in containing so much sodium. The intracellular free sodium ion registers as approximately 14 mM on ion-selective electrodes (see also estimates by Century et al., 1970). Most of the sodium appears bound or otherwise sequestered, consistent with suggestions of a platelet reservoir. During the cleavage period the total amount of sodium in the egg drops dramatically to 20–25 nmoles by the late blastula period, one-third the original amount. Most of this drop matches the accumulation of sodium in the blastocoel, the fluid of which contains 104 mM sodium ion and 1.1 mM potassium ion in a volume of 350 nl, 30% of the egg volume of the late blastula stage. Losses of ions to the external medium are small. Ions seem to be released from the reservoirs as needed, by an unknown mechanism. Morrill et al. (1977) have recently reported that the full-grown oocyte of R. pipiens contains over 90% of its sodium ion in a free state measurable in the cytoplasm by ion-selective electrodes, in contrast to the lower levels in R. pipiens eggs (Morrill et al., 1975). The source of this difference in measurements of free sodium ion is not known, but could reflect active transport of sodium into the platelets by pumps of the yolk platelet membrane at maturation. Morrill et al. (1975) discuss the changes in levels of free sodium ion at postfertilization stages in R. pipiens. Slack et al. (1973) and Slack and Warner (1973) discuss fluctuations of the free sodium ion level in blastomeres during the cleavage cycle and suggest a change in the rate of release of ions from their reservoirs. It is possible, that the platelet membrane is a site at which the ionic composition of the cytosol is controlled.

The actual accumulation of water in the blastocoel is likely to be an osmotic effect; the tightly joined surface blastomeres form a "semipermeable membrane," like a frog skin, allowing passage of water but not ions under conditions in which the ion concentration of the external pond water is far less than the ion concentration of the intercellular space. Therefore, water moves inward, inflating the space to form the blastocoel. Consistent with this view, blastocoel formation is suppressed by 110 mM NaCl in the external medium of the egg (Holtfreter, 1943b; Morrill et al., 1975), or by 220 mM sucrose (Tuft, 1962; Morrill et al., 1975), indicating a simple osmotic force. Also, blastocoel formation is inhibited by ouabain, a well-known inhibitor of the Na^+–K^+ ATPase pumping ions in other cell types (Slack and Warner, 1973; Morrill et al., 1975). Morrill et al. (1975) have correlated increases of this enzyme activity with the appearance of new membrane during cleavage and with formation of the blastocoel. Blastocoel enlargement seems well-explained by the model of water flow by osmosis following vectorial ion transport, an application of the model of Diamond for water movement in the gall bladder (Slack and Warner, 1973). The earlier model of Zotin (1965), based on water binding by glycogen in the blastocoel, seems less suited to the data as a whole, and the model of Tuft (1962) for energy-dependent water pumping is subsumed by the present notion of passive water flow following active sodium ion pumping.

The source and identity of the counterion of the 104 mM sodium ion of the blastocoel are not known. Based on Tuft's (1962) report that the osmolarity of the blastocoel fluid as approximately 220 mOsm, it is likely that the counterion is monovalent. Slack *et al.* (1973) and Slack and Warner (1973) have suggested bicarbonate as the candidate, deriving from the carbonic acid–carbon dioxide equilibrium of the external medium. Morrill *et al.* (1975) suggest that a blastocoel potential develops across the egg surface as the result of the selective uptake of bicarbonate from the external medium, with the exclusion of hydrogen ion. These proposals could be tested, but have not been. Also, it is not known what cation replaces sodium in the cell interior when sodium ion is pumped into the blastocoel. Potassium ion is a possibility when the egg is in potassium-containing medium, and Slack *et al.* (1973) measure an increase of 15–30% in intracellular total potassium from fertilization to gastrulation in Holtfreter's salt solution. The blastocoel also forms in eggs developing in distilled water. Perhaps hydrogen ion from carbonic acid serves as the replacement for sodium ion in the cell. If true, the intracellular pH should change in the pregastrula period, perhaps with interesting consequences for the cells pumping the most sodium ion.

The pH of the blastocoel fluid is high, 8.5 to 8.8 (Holtfreter, 1943*b*; Stableford, 1949, 1967) and the total calcium concentration is in the range of 0.5 mM (Stableford, 1967). The level of free calcium ion is not known. Holtfreter (1943*b*) has suggested that the blastocoel creates conditions favoring disaggregation of cells, namely, high pH and low calcium ion concentration, thereby allowing at gastrulation a degree of independent migration of cells reaching the blastocoel (see Section 7.3).

In addition to the change in ionic content, the blastocoel also accumulates polysaccharides. These polysaccharides consist in part of glycogen, according to Kalt (1971*b*) and Zotin (1965). The former author has observed PAS-staining vesicles in cells bordering the blastocoel and has suggested that they function in emptying glycogen from the cell interior. Numerous authors have noted the presence of large quantities of glycogen in the intercellular spaces (Ubbels and Hengst, 1978; Tarin, 1973). It is not known whether this glycogen is utilized as a source of glucose for the surrounding cells. There may also be acid mucopolysaccharides in the blastocoel, since Stableford (1967) found strong metachromasia by the dried contents. Protein is apparently not detectable (Stableford, 1967).

The next question concerns the position of the blastocoel in the animal hemisphere. The roof and walls of the blastocoel are derived from the blastomeres of the animal hemisphere, whereas the floor derives from large yolky vegetal blastomeres. The location is presumably a consequence of the low yolk content of the animal hemisphere and high content of the vegetal hemisphere in several ways: first, there is more intercellular membrane in the animal hemisphere as there are more cells there due to asymmetric cleavages away from the yolky regions; second, the intercellular membrane may be different, covered with extracellular deposits (Singal and Sanders, 1974*b*; Ubbels and Hengst, 1978) perhaps reducing cell–cell adhesion, whereas vegetal blastomeres may adhere more; and third, the small, less yolky cells of the animal hemisphere may be more malleable internally and more expandable at the egg surface, allowing the 30% increase in egg volume attending blastocoel formation by the gastrula stage. Thus, blastocoel location may follow automatically from the position of the vegetal yolk. On inversion of the egg contents by centrifugation, the blastocoel forms in the nonyolk hemisphere, irrespective of the pigmented surface (Pasteels, 1941).

If the transport of sodium ion to the blastocoel depends only on the presence of the new ion-pumping membrane, then the blastomeres of the animal hemisphere would be the main suppliers of sodium because they possess approximately two-thirds of the new membrane of the late blastula (see Fig. 5) owing to their small cell size. Many embryologists have observed the ability of isolated blastomeres of the animal hemisphere to form a blastocoel; for example, Streett (1940) constricted off the nucleated animal half of the egg of *T. pyrrhogaster* shortly after fertilization, and the fragment formed a large fluid-filled cellularized sphere, a "hyperblastula." Animal hemisphere blastomeres can pump ions from their reservoirs, but, the animal hemisphere contains only one-third the yolk platelet material of the vegetal blastomeres (16 vs. 54% cell volume occupied by platelets, respectively; Robertson, 1978) and perhaps only one-third of the sodium reserves if these are primarily located in the platelets. Now, if the animal blastomeres provide two-thirds of the sodium of the blastocoel (20–25 nmoles in *X. laevis*) from one-third of the egg reserves (total of 60–70 nmoles; Slack *et al.*, 1973), they may become relatively depleted for sodium while the vegetal hemisphere remains sodium-rich, producing an animal–vegetal gradient in intracellular sodium. Furthermore, if hydrogen ion replaces sodium during transport, the animal hemisphere would become proton-rich with a lower intracellular pH than the vegetal hemisphere, and a pH gradient would develop. So far there has been no study of regional ion abundance during early amphibian development.

Hamilton and Tuft (1972) have surveyed a variety of treatments affecting blastocoel formation, many causing persistence of the blastocoel into postgastrula stages when the blastocoel normally collapses and releases its fluid to the archenteron through spaces between vegetal cells separating the two cavities. In the tadpole from a treated egg, the usual syndrome is edema in the heart region, the last site of the blastocoel, leading to problems in heart development and often death. The treatments include exposure to mercaptoethanol or cyanide (also breaks disulfide linkages), or haploidy, or mutation in the *f* gene in axolotls (Humphrey, 1960; Malacinski and Brothers, 1974). High doses of UV irradiation on the uncleaved egg can also produce the same syndrome in tadpoles (Malacinski *et al.*, 1975, p. 106). Hamilton and Tuft (1972) have studied the effect of haploidy in *X. laevis*; they prepared closed multicellular vesicles from the blastocoel roof of haploid or diploid midblastulae (Stage 8) and found that the haploid vesicles inflate twice as fast as do the diploid. The vesicles eventually overinflate and collapse, 6 or 12 hr later, respectively. Inasmuch as the haploid vesicles contained twice as many cells, the authors suggest that the more inward movement of water is related to the greater amount of membrane. According to the osmotic model, this could mean that the inner membranes of the haploid vesicles have greater area and therefore pump sodium into the vesicle center more rapidly; or it could mean the haploid vesicles have more length of water-permeable junctions between tightly adherent surface cells, and so water enters more rapidly; or both factors may apply.

Other treatments have not been studied as thoroughly. Mercaptoethanol does not affect cell number but could increase the water permeability of the egg surface or of the junctional contacts. On the other hand, there may be a second class of effects. Tuft (1961*a,b*) found that mercaptoethanol-treated eggs of *X. laevis* inflate the blastocoel at normal rates through the blastula period but fail to collapse it by the early neurula period as do normal embryos. He suggested that blastocoel size is normally controlled by two opposing rates, one of entry of water through the animal hemisphere cells and the other of exit through the vegetal cells. The first process has been subsumed in the sodium transport-osmotic theory but the second has not. Tuft viewed the exit of water also as an energy-dependent transport, for which there is still no direct evidence. Water exit may be simple intermittent escape between cleaving blastomeres, occurring more in the vegetal hemisphere where cells may form tight junctions less rapidly. Cell adherence could govern the rate of water loss and provide a "safety valve" against overinflation of the blastocoel by osmotic flow. Perhaps, as Tuft suggests, some agents selectively affect water exit; perhaps agents reducing vegetal cleavage, such as UV irradiation, have this effect. In any event, it may be important to preserve Tuft's idea of water exit as a control of blastocoel size, even though the exact mechanism may differ from his original proposal.

In summary, the blastocoel appears to form as an automatic consequence of straightforward physiological processes: ion pumping by the internal cleavage membranes; ion impermeability by the original egg membrane; tight junction formation between neighboring surface cells preventing ion loss to the outside medium; and osmotically driven water movement through the semipermeable external surface. As more membrane forms by cleavage, the total rate of ion pumping accelerates. The blastocoel fluid comprises approximately 30% of the egg volume by the late blastula stage. The position of the blastocoel in the animal hemisphere is probably due to the presence of smaller cells in the animal half, as a result of asymmetric cleavages biased by vegetal yolk. The size of the blastocoel is perhaps controlled by rates of water loss as well as entry. Though blastocoel formation seems a simple example of pattern formation by cytoplasmic reactions established in oogenesis and used without gene expression, it is an important aspect of pattern formation in the pregastrula stages. The size, shape, and position of the blastocoel have consequences for the inductive patterning of the animal hemisphere by the vegetal hemisphere at the mid- and late blastula stages, and for the rearrangement of surface and interior cells at gastrulation (see Sections 6.2 and 7.3).

5.3   Vegetal Cortical and Cytoplasmic Ingression

This interesting rearrangement of vegetal cytoplasm was first recognized by Schechtman (1934), who marked the vegetal pole of an uncleaved fertilized egg of *T. torosus* with Nile blue and found that some of the dye appeared near the floor of the blastocoel by the midblastula stage. He called this pregastrular movement "unipolar ingression." Nicholas (1945, 1948) repeated the study with *A. punctatum* and stressed the importance of these movements. Ballard (1955) substantially advanced the subject by his study of ingression in a variety of urodele and anuran eggs. As his first point, he found that vegetal cells do not migrate inward but that cortical and subcortical cytoplasm within the early blastomeres moves upward along the cleavage furrows. We have a "cortical ingression" as he called it. As his second point, he found that equatorial and animal blastomeres also transport peripheral cortex and cytoplasm inward along their furrows, so the phenomenon is not unipolar, but rather "multipolar cortical ingression." The intracellular movements were visible because Nile blue, in contrast to other vital dyes, binds preferentially to cortical materials (Ballard, 1955). The greatest distance of transport is achieved in the vegetal hemisphere where the longest time elapses before a paratangential cleavage cuts off inner from outer cells and blocks further inward transport. As Ballard showed, this finally occurs in the vegetal hemisphere at the 64- to 128-cell stage in *R. sylvatica* when the dye from the vegetal pole has made its way 80% toward the blastocoel floor. Then a paratangential cleavage cuts off cells at this level, to form the cell layer of the blastocoel floor. One can say that the rearrangements of vegetal cytoplasm continue until this stage. These movements are regarded in this review as a further step, like the cortical contraction of gray crescent formation, involved in rearrangements of the vegetal hemisphere at these early stages, leading it to a final pattern of cytoplasmic localizations, a pattern to be impressed on the animal hemisphere by induction at the midblastula stage.

While Ballard (1955) could follow only cortical material with his dye, subsequent authors have followed parallel simultaneous movements of subcortical cytoplasm containing natural markers. Harris (1964) chose *A. maculatum*, a urodele the eggs of which have naturally pigmented zones of vegetal yolk. The most darkly pigmented zone of yolk is most vegetal. This material is transported upward along the first two furrows, eventually to produce a pigmented cruciform pattern just beneath the blastocoel floor by the midblastula stage. Apparently the amount of dark yolk transported in this period was sufficient for easy visualization, but was not otherwise quantified. As this material moved upward centrally, it was replaced by downward-moving pigmented yolk and nonpigmented equatorial yolk from the egg periphery. The border of the pigmented and nonpigmented yolk eventually comprised the level of blastopore formation—that is, the real division between animal and vegetal hemispheres. After gastrulation the pigmented yolk became the region of the posterior gut, a prime source of nutrients for the rest of the embryo prior to feeding. At the early stages Harris did not investigate dorsal–ventral differences in the amount of pigmented yolk transported upward by multipolar ingression.

Researchers interested in the movement of the germinal plasm granules have been able to map vegetal ingression because these granules are also transported. From the time of the discoveries of Bournoure in 1934, these granules in anurans (not in urodeles) have been regarded as indispensable to the lineage of blastomeres destined to form germ cells much later when the gonads are established (Williams and Smith, 1971; Mahowald and Hennen, 1971; Blackler, 1970). As mentioned earlier, full-grown oocytes carry dispersed islands of granulelike material in the yolky cytoplasm near the vegetal pole; these islands enlarge and perhaps coalesce during maturation. After fertilization, the islands definitely coalesce and move inward toward the first and second furrows, and then upward along the vertical membrane. Some of these granules reach the level of blastomeres beneath the blastocoel floor (Whitington and Dixon, 1975). Through continued coalescence, the number of islands of granules reduces to a small number, 10 to 20. At cleavage they stay with one spindle pole, partitioning to only one of the two blastomeres. By the late blastula stage, only 10 to 20 blastomeres possess granules and qualify for the germ cell lineage. These blastomeres are called "presumptive primordial germ cells"; their subsequent development will be described in Section 8. Again there is no clear evidence as to their dorsal–ventral distribution at early stages.

In summary, the importance of this upward transport of large vegetal yolk platelets, germinal plasm, organelles, and associated material is not understood, but deserves further attention in light of Nieuwkoop's evidence (1969a,b, 1973, 1977) for the vegetal yolk mass as the source of dorsalizing inductions in the mid- and late blastula embryo. It is possible that the ingression serves to carry vegetal yolk of a certain type from a deep vegetal position toward the equator, increasing its exposure to the animal hemisphere, the eventual target of induction by the vegetal hemisphere. However, at present there is really no critical test of the importance of these movements for normal pattern formation and gastrulation.

Among potentially relevant studies, Streett (1940) removed vegetal yolk at the 1-cell stage of *T. pyrrhogaster* by constricting off the vegetal pole region with a hair loop shortly after fertilization. The egg cleaved and developed to an abnormal gastrula with a very large yolk plug. The relationship of this syndrome to the absence of the most vegetal yolk is unknown. Also, UV-irradiated eggs (exposure of the vegetal half) fail to achieve substantial upward transport of the germinal plasm because the vertical furrows do not form until much later (Beal and Dixon, 1975). At moderate UV doses, these embryos gastrulate adequately but suffer later defects in the germ cell lineage, apparently because the presumptive primordial germ cells are located in unusual positions and arrive late in the gonad

(Zust and Dixon, 1975), in some way resulting in sterility. At higher doses, dorsalization and gastrulation itself fails (Malacinski *et al.,* 1977). The relationship between the abnormal yolk movements and failure to dorsalize or gastrulate is unknown. As mentioned earlier, though, irradiated eggs can be fully rescued if placed in an oblique position before first cleavage, to obtain gravity-driven rearrangements of internal materials (Scharf and Gerhart, 1980). It is not known if such eggs are also rescued in their cytoplasmic ingression.

5.4   The Schultze Inversion Experiment

Schultze observed in 1894 that frog eggs *(R. fusca)* held between horizontal glass plates and inverted at the two-cell stage occasionally give rise to double-axis embryos, even to twins, which never appear in control eggs. Extra axes appeared in as many as 25–50% of the cases. Both compression and inversion are required. Penners and Schleip (1928a,b) repeated the experiment in more detail, noting that until the 8-cell stage the inversion treatment could cause twinning. They studied the movement of the vegetal yolk during the inversion. In some cases a portion of the vegetal yolk fell completely through the animal hemisphere cytoplasm to make contact with the pigmented animal hemisphere cortex; in such eggs the pigment dispersed at the contact regions, leaving a light patch on the border of which a secondary blastopore would form. Penners and Schleip concluded that blastopores, whether primary or secondary and whether in the animal or vegetal hemisphere, can develop only at the edge of vegetal yolk. This accords well with what we know now from Nieuwkoop's results (1969a,b) of the yolk mass inducing pharyngeal endoderm and dorsal mesoderm in the adjacent animal blastomeres; namely, that these dorsal members will only appear near yolk.

Pasteels (1937) was the first to try to relate the dorsal–ventral axes of inversion twins or triplets to a dorsal–ventral axis present in the egg before inversion. He identified the gray crescent position in *R. temporaria* eggs held on top and bottom between glass plates, inverted the plates, and later noted the positions of the blastopore lips, especially in eggs with multiple lips. He confirmed the Penners and Schleip (1929a,b) observation that lips appear only at the edge of yolky patches on the egg surface, and added that the lips appear first on the side of the patch toward the old position of the gray crescent. Slightly later the lip would extend around the patch, or another lip would begin on the opposite side. But, according to Pasteels, the lip first appeared on the crescent side. Pasteels followed some of the inverted eggs to show that the first lips were dorsal lips. This interesting observation suggested to Dalcq and Pasteels (1937) that the crescent continued to exert a dorsalizing influence during the inversion period after first cleavage; that the dorsalizing influence must be based in a structure not reorganized by gravity (namely, the cortex); that the crescent must be able to act at a considerable distance, even a quadrant or more away from the original center; and that the cortical dorsalizing influence must interact with the vegetal yolk (which does reorganize during inversion) to generate the blastopore lip. The inversion results provided the earliest experimental evidence for their proposal of a cortical field centered in the gray crescent and acting through a dorsalizing gradient. In the following decade Pasteels did his series of rotations of eggs to oblique positions after crescent formation but before first cleavage (Pasteels, 1946, 1948), to collect further experimental support for the cortical field.

Other interpretations of the Schultze (1894) inversion experiment have been made without invoking a cortical field. Penners and Schleip (1928a,b) themselves did not emphasize the dorsal–ventral aspect of the multiple lips, but rather the fact that they occurred next to vegetal yolk. Spemann (1938) suggested that subcortical cytoplasm of the gray crescent region became dislodged during inversion and translocated to another region of the egg, there to organize a new dorsal center in the cytoplasm. In a similar vein, Nieuwkoop (1969a,b; 1973) has suggested that the Schultze inversion result could be explained by the forced translocation of dorsal vegetal yolk (near the gray crescent) to a new location at the two-cell stage, with this special yolk taking its dorsalizing activity with it. Nieuwkoop (1969b, 1973) explains Pasteels' (1937) correlation of blastopore orientation toward the gray crescent position as follows: even when the specialized dorsal yolk falls to a new location on inversion of the egg, the new location will still be relatively close to the gray crescent, the region from which it came, compared to the location of the remainder of the yolk. That is to say, inversion is not an effective means to achieve a critical rearrangement of yolk *vis à vis* the gray crescent cortex. What was dorsal yolk before inversion will tend to be dorsal yolk after inversion, just moved to the other hemisphere.

These explanations, including the hypothesis of the cortical field, preserve the idea that a dorsalizing cortex, subcrescent cytoplasm, or vegetal cytoplasm, is formed *uniquely* at the time of crescent appearance and that dorsalizing regions cannot be created later, but only divided or dispersed to new locations. There is another class of explanations of the Schultze result—namely, that new dorsal centers are created by inversion during the course of yolk movements. Penners and Schleip (1928a,b) seemed to think along these lines, for among their inversion embryos they noted some in which the secondary blastopore seemed to form not in the animal hemisphere but on the prospective ventral side, opposite the gray crescent. When Spemenn [summarized in Spemann (1938)] suggested that these results could be explained by the translocation of subcrescent cytoplasm to the ventral side, Penners (1936) devised an interesting refinement of the Schultze experiment in order to test Spemann's proposal. Penners chose only those *R. fusca* eggs which, by chance, cleaved frontally to the gray crescent, waited until these had almost completed first cleavage to have a natural barrier between the gray crescent cytoplasm and the ventral side, and then did the inversion. Still he obtained seven embryos forming dorsal blastopores on the ventral side, some as twins and some not. Therefore, he could conclude that it was unlikely that the gray crescent subcortex reached the ventral side through the cleavage membrane. The frequency was 7 out of 300 eggs, the same frequency as in eggs with median cleavage in these experiments. This is one of the few experiments addressing the question of the formation of new dorsal centers in the post-gray-crescent period. A more explicit test might be made by separating off the ventral blastomere by constriction at the two-cell stage in an egg that had been compressed laterally to obtain a high frequency of frontal cleavage (Penners just took the rare spontaneous frontal cleavages) and then to rotate or invert such a ventral blastomere to cause it to dorsalize.

Previously we discussed the dorsalizing action of the new vitelline wall formed with Born's crescent after rotation of the egg to an oblique position in the period between gray crescent formation and first cleavage. We emphasized movements of the vegetal yolk which might expose deep vegetal yolk to the activating animal cytoplasm through the equatorial layer of intermediate-sized platelets. If these rotations are effective until first cleavage, it seems plausible that they would be effective after first cleavage as well, just as long as deep vegetal yolk can be exposed to the less yolky cytoplasm of the animal hemisphere. If this were to occur on the prospective ventral side in the course of inversion, then a dorsal center might start there. If deep yolk were exposed in several places, then multiple blastopores would form. Also, the inversion might have the opposite effect and cover an exposed region with intermediate platelets, eliminating a dorsal center.

As a further speculation, increasing gravitational force may be required at later and later stages in order to achieve a major redistribution of yolk. The cleavage furrows and new plasma membrane may adhere to cytoplasmic materials as indicated by multipolar ingression, thus preventing their free movement. The egg cytoplasm may change in cell-cycle-specific ways, perhaps becoming more rigid at the time of cleavage, thereby resisting gravity-driven movements. Also, cleavages produce natural barriers to movement, and the cleavages occur at accelerating frequency. Then the third cleavage, which is latitudinal, cuts off the animal hemisphere from the vegetal. Penners and Schleip (1928b) found that inversion no longer caused twinning if started after the 8- to 16-cell stage. Experimentally induced rearrangements may need to be completed more rapidly the longer one waits. Perhaps these reasons explain why prolonged inversion is needed in the Schultze experiment. Also, Scharf and Gerhart (1980) were unable to rescue UV-irradiated eggs by 90° oblique orientation after second cleavage. Presumably, the third cleavage puts an end to major rearrangements thereafter. But whatever the stage up to third cleavage, the basic and effective experimental operation has been one of gravitational displacement of yolk, whether done just after fertilization with a transcient small rotation (Ancel–Vintemberger, 1948), shortly before first cleavage with a prolonged oblique rotation (Born, 1885), or after cleavage with prolonged inversion (Schultze, 1894).

Curtis (1962) reported that the 8-cell stage of *X. laevis* no longer produces twins after receiving a graft of gray crescent cortex in its ventral equatorial zone. Perhaps the latitudinal third cleavage just precludes major yolk rearrangment in the course of orienting the egg ventral side up, and submitting it to surgery.

5.5   Development of Isolated Blastomeres

After Driesch reported in 1891 the complete development of individual blastomeres separated from the sea urchin egg at the 2-cell stage, several embryologists undertook blastomere isolations with amphibian eggs to resolve the discrepancy of Driesch's finding with the earlier result of Roux (1888). In Roux's procedure, the inactivated blastomere had remained in place and did not disintegrate. Enders in 1895 and Herzlitzka in 1897 succeeded in separating the first two blastomeres of the newt egg *(T. taeniatus)* by gradually tightening a hair loop around the cleavage furrow. In some cases both blastomeres developed into normally proportioned embryos of half-size, in contrast to the result of Roux. When the blastomeres remained together, there was little or no pattern regulation, whereas when separated, regulation was complete.

In the period 1901–1903 Spemann extended these studies with larger numbers of *Triton* eggs and recognized that some eggs at the 2-cell stage contained a blastomere which cleaved normally when separated but failed to neurulate. It produced a ciliated sphere with a limited variety of ventral cell types such as blood cells, occasional pronephros, and mesenchyme, but no axial organs such as notochord, somites, or central nervous system. These embryos were called "belly pieces" or "ventral halves." In retrospect, Spemann (1938) concluded that the first cleavage furrow must have been perfectly frontal in some cases dividing the egg into cytoplasmically different portions. He had not been able to correlate the furrow position with the gray crescent position, given the difficulty of detecting it in this species. Ruud (1925) was even able to separate from one another the four blastomeres produced after the second cleavage and showed that usually two of them develop to quarter-size gastrulae and partial neurulae, whereas the other two devleop to ventral pieces. In some eggs she obtained three ventral pieces and only one dorsal embryo, suggesting that a critical cytoplasmic asymmetry of the egg covered perhaps less than one-fourth of the egg equatorial perimeter. These results agree in general with the findings of Fankhauser (1930) and Streett (1940) on the production of ventral pieces in *Triturus* eggs constricted into halves at the 1-cell stage, although the frequency of ventral pieces was lower (one out of five) at the earlier stage.

The classical results of Spemann (1928) and Ruud (1925) have been disputed by Dollander (1950), who repeated their experiments with *T. taeniatus* and with axolotl eggs, this time identifying the gray crescent at the time of constriction with a hair loop. Dollander reports 85% accuracy in predicting the dorsal side from the gray crescent positions. He chose eggs in which the first cleavage was frontal, dividing the egg into a crescent-containing half and a noncrescent half. These were presumably dorsal and ventral, respectively. To his surprise, many of the supposed ventral halves (no crescent) did form blastopores and did gastrulate completely, giving half-sized well-proportioned embryos. In addition, Dollander did obtain some ventral pieces, but these sometimes derived from the gray crescent half as well as from the noncrescent half. Thus, Dollander's results do not match the simple prediction of crescent position and dorsality in these species under these conditions.

The discrepancy of these and earlier results has not been resolved. Holtfreter and Hamburger (1955) comment that the crescent is not easily identified in this species, and therefore the halves may have been misidentified. They also suggest that the future dorsal side cannot be reliably predicted from the crescent position; for example, in eggs of *T. torosus,* the crescent diverges 25° to 90° from the bilateral axis of the gastrula in 27% of the cases (Schechtman, 1936). On the other hand, Dollander (1950) suggests that the constriction brings together the two extreme tips of the gray crescent on the ventral half, fusing them into an effective dorsal center. In support of this view he notes that the ventral pieces initiate gastrulation at the position of the constriction. But the question remains why Spemann and Ruud had obtained ventral pieces more frequently. Perhaps the discrepancy in results reflects the differences in manipulations with respect to the speed of constriction and the thickness of the hair noose. Ruud used a completely different and more gentle method, a glass rod weighing down on the furrow to separate blastomeres automatically as they cleaved. A still more gentle method, not yet tried, would involve blastomere dissociation at the 2-cell stage by exposure of the egg to phosphate buffer (Hara, 1977) after removal from the vitelline membrane. Perhaps Dollander simply perturbed his eggs more than did Spemann and Ruud. If one considers that the egg can still be dorsalized after the first cleavage, rather than just at the time of gray crescent formation, then it would be necessary to look more closely at constriction as a means to cause vegetal yolk displacements. Perhaps different manipulations can initiate, destroy, or preserve asymmetries in the vegetal regions of the halves. Whereas Dollander stressed the union of lateral extremities of the gray crescent by constriction, it is equally relevant to stress the bringing together of animal and vegetal regions that were previously separated, creating a closer contact for the deep vegetal yolk to exert its induction at a later time on the animal hemisphere. Without constriction this yolk would remain buried under intermediate platelets, or at best would ascend by vegetal ingression only to meet the blastocoel floor, far removed from receptive cells of the animal hemisphere. In this view of "dorsalization by constriction" the blastopore would be expected to appear at the constriction site, as in Dollander's interpretation. If gravity can dorsalize the ventral blastomere of an inverted 2-cell egg, as shown by Penners (1936) then constriction might also do so. Furthermore, if constriction is effective, one must also consider the surgical manipulations of Curtis (1962*b*) as potential means of dorsalizing the ventral portion of the cleaving recipient egg.

In a different series of experiments, Ruud (1925) succeeded in separating the animal and vegetal quartets of blastomeres at the 8-cell stage of *T. taeniatus;* the upper ones developed only to a ciliated sphere (a hyperblastula) lacking all axial structures, even though they carried a portion of the area corresponding to the gray crescent, which is poorly formed or absent in this species. Similar results were obtained by Vintemburger (1934) with the animal quartet of cells of the 8-cell stage in *R. fusca.* More recently, Grunz (1973) and Grunz *et al.* (1976) have isolated equivalent animal fragments and cultivated them in media favorable for advanced differentiation; the fragments developed to ciliated epidermis of normal appearance, but nothing else. These results support strongly Nieuwkoop's interpretation (1969*a,b*) of the animal half of the egg as limited to this fate unless induced by the vegetal half.

Ruud (1925) and Vintemberger (1934) also isolated the vegetal quartet of blastomeres at the 8-cell stage, and found them capable of initiating a blastopore and gastrulation, although the degree of invagination was limited, perhaps due to the small blastocoel and the lack of expanding animal hemisphere surface. These results already hinted at the importance of the vegetal half for dorsalization, at least as involved in blastopore localization, in contrast to the anaxial development of the isolates from the quartet of animal blastomeres. Similar results were obtained by Vintemberger (1936) who killed the two gray crescent-containing blastomeres in the animal hemisphere of the *R. fusca* egg at the 8-cell stage, and nonetheless obtained essentially normal development. He thought the egg had regulated its pattern to compensate for the loss of the crescent, but from Nieuwkoop's results (1969*a,b*) it is more likely that the crescent region had no dorsalizing properties at this stage anyway.

Stableford (1948), using a suction pipette, removed all the pigmented blastomeres of the animal half of an *R. pipiens* embryo at the early blastula stage (about 500 cells) and found that the vegetal half could still form a blastopore, although gastrulation did not proceed well. These early studies indicated, on the one hand, the ability of the vegetal blastomeres to initiate events of dorsalization by themselves, and on the other hand, the inability of the animal hemisphere blastomeres to do so even when containing part of the gray crescent area.

Although blastomeres of the animal hemisphere and gray crescent region lack the ability to develop mesodermal structures when isolated from egg at the 8-cell stage, they gain this ability by the 64- to 128-cell stage, as shown in *X. laevis* and *T. pyrrhogaster* by Nakamura and his colleagues (Nakamura and Takasaki, 1970, 1971*a,b*; Nakamura *et al.,* 1970). At the 32-cell stage the "ideal" egg of *X. laevis* is arranged as four tiers of eight cells each, designated tiers A, B, C, D starting from the animal pole. These cells were individually marked with vital dyes *in situ* to obtain a fate map (Nakamura and Kishiyama, 1971) revealing that the blastopore originates at the equatorial level within tier C where pigmented and nonpigmented hemispheres meet. Nakamura *et al.* (1970) removed the dorsal blastomeres of tiers B and C (two from each, four total), cultivated them in Holtfreter's medium for 2 days and found no mesodermal structures appearing. When blastomeres of this same area (eight total) were removed at the 64-cell stage, about half the explants developed mesoderm such as notochord, somites, blood cells, and mesenchyme. At the 128-cell stage explanted blastomeres (16 total) of the same area developed mesoderm in 75% of the cases. Cells of the gray crescent area gain the capacity for autonomous differentiation of mesoderm at the 64- to 128-cell stage. Even though the production of mesoderm does not occur at these stages, the *capacity* for production does. The same results were obtained with *T. pyrrhogaster,* at a slightly later stage of 128–256 cells (Nakamura and Takasaki, 1970). It is not known what changes in the early cleavage stages that establishes this autonomy. One possibility is that vegetal cytoplasmic ingression transports new materials into the region of the future equatorial blastomeres until the 64- to 128-cell stage (Ballard, 1955). A change of another kind, reported by Ubbels (1978), is an intensification of the staining of the vegetal yolk platelets with Alcian blue (pH 2.5) at the 32- to 64-cell stage. The origin or consequence of this interesting change is unknown.

As one final point about the establishment of a dorsal center at these early stages, Spemann (1938) and Fankhauser (1930, 1948) partially constricted *Triturus* eggs prior to the first cleavage so that the nuclei would be confined to one half even though the halves remain in cytoplasmic contact through a thin channel. Then as the blastomeres became smaller after four or five cleavages, a nucleus slipped through the cytoplasmic connection and promoted cleavage on the other side. The two sides developed into advanced embryonic stages as twins, with one embryo many hours behind the other. This result suggests the anucleate half preserves perfectly well its capacity to cleave, dorsalize, and develop despite the prolonged delay until cleavage began. It was not reported whether these halves gastrulated at the same or different times.

5.6   Pattern Regulation

During the period from first cleavage to the midblastula, there are three major events of pattern formation:

(1) There are the continuing changes in the arrangements of components of the vegetal cytoplasm, involving the mechanisms of cortical and cytoplasmic ingression along the new cleavage plasma membrane (not the old egg membrane), and perhaps also continued lateral shifting of contents in eggs where the cortical contraction of gray crescent formation is late or prolonged. For example, in *X. laevis* the crescentlike changes in pigmentation dorsally in the animal hemisphere seem to continue until the 4-cell stage, although photographic records of these pigment changes usually end at the time of first cleavage (Palaček *et al.*, 1978). The final pattern of cytoplasmic localizations in the vegetal hemisphere may be completed at the 64- to 128-cell stage when paratangential cleavage in the vegetal blastomeres cuts off further long-range translocations of materials. Presumably, the "pattern" of the vegetal hemisphere concerns mostly the pattern of the more equatorial level of that hemisphere—namely, the layer in contact with the animal hemisphere, the target of its inductive activity at the midblastula stage. The exact contribution of ingression to the pattern of this layer is unknown; it seems to bring deep vegetal materials such as large platelets and germinal plasm toward an equatorial level, perhaps into closer contact with the animal hemisphere. It is also unknown how this upward movement modifies, amplifies, or unifies the lateral displacements of the yolk mass achieved earlier in vitelline wall formation; in other words, how ingression contributes to the dorsal–ventral cytoplasmic differences of the egg at these early stages. As a general perspective, the stages prior to the midblastula should perhaps be considered as devoted to the patterning of the vegetal half through rearrangement of its cytoplasmic contents. In contrast, the stages after midblastula would be devoted to the patterning of the animal half by induction, based on the pattern of cytoplasmic localization of the vegetal half.

(2) There is the development of the animal–vegetal difference in cell size, directly produced by asymmetric cleavages of blastomeres away from the region of higher yolk content (Hara, 1977), rather than by differences in rates of cleavage.

(3) There is the formation of the blastocoel, a multicellular enterprise involving sodium pumping from intracellular reservoirs into the intercellular space across the new plasma membrane of the cleavage furrows, under conditions in which the egg surface is sealed by intercellular tight junctions to create a water-permeable, ion-impermeable sphere. The blastocoel inflates osmotically in the animal hemisphere where the less yolky cytoplasm has allowed the cleavage of smaller cells, more new ion-pumping membrane, and more intercellular space. As Nieuwkoop (1969a,b) has emphasized, the blastocoel is a very important element of the pattern of the midblastula, for it controls the extent of contact between the animal and vegetal halves; those blastomeres in the roof area are removed from the inductive action of the vegetal half and are not driven into archenteron–roof endoderm or mesoderm. Only blastomeres of the blastocoel walls experience induction. This cavity is more than just a space into which cells move at gastrulation, although this function is also important.

The animal–vegetal differences in cell size and the location of the blastocoel result so directly from the relative positions of the yolky and nonyolky cytoplasm, it is easy to see why these aspects of pattern reverse their orientation 180° when the egg is inverted and centrifuged lightly before the time of gray crescent formation, as demonstrated by Pasteels (1941). They are expressions of the cytoplasm rather than the cortex, even though the cortex is used extensively in their formation.

Pattern regulation at the period from first cleavage to the midblastula stage concerns the mechanisms by which the above examples of pattern formation are self-restoring in the face of normal and abnormal perturbations. Particularly, pattern regulation would concern the mechanisms by which the large amphibian egg unifies pattern formation to achieve a single vegetal dorsalizing center and a single blastocoel, thereby preventing twinning. Pattern regulation of blastocoel formation is illustrated by a comparison of the development of a lateral blastomere of the 2-cell stage either when remaining in the intact egg next to its inactivated neighbor (the Roux experiment) or when separated from its neighbor (the Enders–Herlitzka–Spemann–McClendon experiment). In the former, the blastomere cleaves, forms a half-blastocoel against the nondividing neighbor with which it presumably continues to form tight junctions, and then gastrulates via a dorsal lip appearing next to the inactive half—that is, on the median axis of the whole egg. A lateral half-gastrula results. The lateral blastomere develops without pattern regulation.

In the separated lateral blastomeres of the two-cell stage, pattern regulation begins immediately: the hemispherical half-egg rounds into a sphere and forms a normal internal blastocoel. Yet this blastomere brought with it on one side the ion-pumping new plasma membrane of the cleavage furrow, which would normally face the blastocoel. There is no information about how this membrane is modified to make it impermeable, as is the remainder of the egg surface; perhaps it is removed in a contractile wound-healing response as the old egg surface expands to replace it, or perhaps it is converted to "egg-surface" membrane by the same reactions that made the oocyte membrane impermeable at maturation.

Holtfreter (1943b) favored the idea that the blastomere surface could be modified by the cell in response to exposure to the harsh conditions of pond water, by deposition of a surface coat. For him, the inside–outside axis of the egg develops simply as the egg's regulative response, on the one hand, to the external pond water medium by making its exposed cell surfaces impermeable, adherant, and expansive, while on the other hand, to the internal medium of the blastocoel to make the surface permeable, nonadherent, and nonspreading. Although the functions he ascribed to the extracellular surface coat are probably due to the plasma membrane itself, the basic idea of a regulative surface remains valid and largely untested. Yet it may be an early aspect of pattern regulation by separated blastomeres. Whatever the mechanism for modifying this permeable membrane, the next steps of development show that all surface cells of the cleaving separated blastomere are able to make tight junctions and seal off the intercellular space for ion accumulation and osmosis of water, because a blastocoel formed. Additionally, the spherical blastomere must undergo rearrangements of its vegetal yolk mass. Furthermore, the normal shape of the blastocoel allows animal hemisphere cells to contact vegetal cells around the entire periphery of the equator, whereas in the Roux experiment (1888) the geometry of animal–vegetal contact was quite different. The comparison illustrates the role of surface pattern regulation and the importance of total geometry of the egg for blastocoel formation and hence animal-vegetal contact.

The next experiments concern pattern regulation in the formation of the vegetal dorsalizing center at these early stages. As shown by Schultze (1894), Penners and Schleip (1928*a,b*), and Pasteels (1937), inversion of eggs at the 2-cell stage causes the development of embryos with twin or multiple axes. The first cleavage furrow splits the yolk mass at this stage and perhaps allows it on inversion to redistribute itself as parts rather than as a unit. Nieuwkoop (1969*a,b*, 1973) has obtained evidence for the presence of a special vegetal dorsalizing center in the yolk mass and has suggested some material of this center falls into the opposite hemisphere on inversion of the egg, causing the founding of a secondary center. This may be one mechanism of twinning.

On the other hand, Penners (1936) was able to obtain secondary dorsal centers even on the ventral side in frontally cleaved eggs, a result perhaps indicating that inversion also causes the formation of new dorsalizing centers, not just the fragmentation of the old one. Once more we come to the question of what is the vegetal dorsalizing center; we are suggesting it is a region of contact of deep vegetal yolk (containing the large platelets) with the animal hemisphere cytoplasm, a contact optimizing the inductions beginning at the mid-blastula stage. It is suggested that the contact between the deep vegetal yolk and animal cytoplasm is initially prevented in unfertilized eggs of many species by the intervening equatorial layer of middle-size platelets. In normal vitelline-wall formation, the deep yolk is pulled through this layer and exposed to the animal hemisphere cytoplasm (that cytoplasm containing abundant germinal vesicle material and small, sparse platelets) by the cortical contraction at the time of gray crescent appearance. This configuration is apparent in *D. pictus* eggs when the crescent forms (Klag and Ubbels, 1975). But gravity can plausibly create regions of contact, as in the inversion experiments. Gravity might continue to have effects on inverted or oblique eggs until the horizontal cleavage of the 8-cell stage blocks further movement.

Likewise, constriction of the egg with a hair noose in the frontal plane might create a dorsalizing center in the ventral fragment, thereby explaining Dollander's (1950) high frequency of normal development by ventral blastomeres. Constriction might have effects at far later stages than gravity, as it provides a means to appose animal and deep vegetal blastomeres forcibly, even across the space of the blastocoel. The results of Huang and Dalcq (presented by Brice, 1958) will be discussed in this light (in Section 7.4.1). The main point here is to raise the possibility that dorsalizing centers can be created long after gray crescent formation and by means not involving the cortex.

What then is the egg's capacity for pattern regulation in the formation of the vegetal dorsalizing center? Perhaps cytoplasmic ingression is involved in unifying the center or in suppressing secondary centers, but too little is known of this process to discuss the possibility. There are few tests of the egg's ability to unify artificially arranged multiple centers, although a test could be made based on the fusion method of Mangold and Seidel (1927). These workers took two *Triturus* eggs at the 2-cell stage, stretched each to a dumbbell shape, and placed them across each other at right angles to make a double-sized 4-cell stage. These fusion products developed to embryos with one, two, three, or more dorsal axes. The particular species did not permit accurate identification of the gray crescent, so the eggs were combined randomly, and the results were interpreted retrospectively to surmise where the dorsal centers must have been at the time of fusion. This experiment could be repeated with a more favorable species and with eggs marked with a vital dye. Marking would reveal if stretching and fusion cause changes in the site of the blastopores relative to the centers of the blastomeres. Perhaps dorsal centers are united if within a certain distance. Pasteels (1946, 1948) made the interesting comment that amphibian species differ in the ease with which twinning can be induced experimentally, but the quantification or analysis of this difference has not been made.

6   From the Midblastula to the Early Gastrula

The midblastula stage, when approximately 1000 cells are present, is distinguished by the reduced rate of cleavage and the first synthesis of RNA. The architecture of the blastula is not quite complete as far as expansion of the blastocoel and as far as the animal–vegetal difference in cell size are concerned, but the processes for completion are those initiated at first cleavage and discussed previously. This architecture provides the anatomical context for gastrulation. In the midblastula period blastomeres autonomously acquire new properties of adhesiveness and motility, the animal hemisphere begins to expand downward in epiboly, and in some species internal cell migrations begin. Most importantly, during this period the vegetal yolky blastomeres exert an inductive effect on their animal hemisphere neighbors to become pharyngeal and archenteron endoderm as well as dorsal, lateral, and ventral mesoderm. This is the first step in which the vegetal hemisphere passes its dorsalizing role to a cell population of the animal hemisphere, which will subsequently "organize" the gastrula and neurula stages.

6.1   Reduced Mitotic Rates and the Onset of Asynchrony

With regard to cell-cycle changes, Hara (1977) and Signoret and Lefresne (1971, 1973) have recorded the intercleavage periods in the axolotl egg, reported by the latter investigators as increasing from 90 min at the 10th cleavage to an average of 127 min at the 11th and to 262 min by the 14th cleavage, shortly before gastrulation. In *X. laevis* the mitotic cycle increases also about fourfold before gastrulation (Satoh, 1977). As a result, the embryo achieves only about 15,000 cells instead of the 130,000 expected if cleavage had continued at the rapid rate of the early blastula stage. The mitotic index has long been known to drop sharply in the late blastula period, from an average of 50% mitotic cells in the midblastula (see Fig. 7) to approximately 5% mitotic cells (Bragg, 1938) in the early gastrula.

Signoret (1977) has recently analyzed the temporal pattern of the initiation of cleavage in surface blastomeres of the axolotl egg at the 11th cleavage, and reports an abrupt onset in asynchrony of division at this cycle. Whereas the time of initiation of cleavage varied less than 8 min out of 90 min at the 10th cleavage, the individual blastomeres varied from 85 to 215 min in their intermitotic period at the 11th cleavage. There would be a degree of error in the measurements because paratangential cleavages (ones parallel to the surface) could not be scored in cinematographic records of the egg surface, but this error should have entered as well at earlier cleavages. Chulitskaia (1970) has reported a desynchronization of mitoses in *R. temporaria* at the 10th to 12th cycle, and Hara (unpublished) has observed in time-lapse films desynchronization in cleavage initiation in *X. laevis* at the 10th to 12th cycle.

Signoret (1977) analyzed his data for asynchrony in the axolotl egg by a stochastic model of the type recently developed by Martin and Smith (1973) for the asynchrony of divisions in cultured cells; namely, with the assumption that a cell has two phases in its cell cycle, one of fixed duration and one of variable duration, the latter phase responsible for rapid desynchronization. For the fixed period, cells are assumed to differ only slightly and randomly according to a normal distribution with a small standard deviation; Signoret (1977) analyzed the average cleavage interval of 127 min at the 11th cycle in the axolotl egg and assigned a fixed period of 97.5 min with a standard deviation of 9 min, whereas for the variable period he assigned a half-time for escape of 16 min. That is to say, for one-half the axolotl cells the duration of the variable period was 0–16 min, for one-fourth the duration of the variable period was 16–32 min, for one-eighth it was 32–48 min, and so on in an exponential distribution of the population for the duration of this period. This is the distribution that skews the population strongly and accounts for the skewness in Signoret's observed distribution of cleavage intervals.

Martin and Smith (1973) have equated the fixed period of the cell cycle with the phases of S, G_2, mitosis, and part of G_1, whereas the variable period would be contained entirely in the G_1 phase. If applicable to blastula cells, the Martin and Smith proposals would suggest that the blastomeres suddenly gain a variable period at the 11th cleavage, that this new period is responsible for their rapid desynchronization. A G_1 phase is indeed first detectible at the midblastula stage, as discussed later in this section. Shields (1978) has shown that the variable period has remarkable desynchronizing effects in cultured cells: daughter cells are no more synchronized with respect to their next division time than are nondaughter cells of the same age measured from the previous division. One would not expect even "pockets" of synchrony to persist in a midblastula after the appearance of the variable period. In fact, Bonnano (1977) has analyzed the spatial distribution of metaphase, anaphase, and telophase cells in sections of *X. laevis* eggs at the 11th and 12th cleavages (see, for example, Fig. 7) and finds no statistically significant clustering of mitotic cells.

Signoret (1977) has analyzed the 12th through 14th cleavages in axolotl eggs and suggests that the variable period retains its half-time of 16 min, whereas the fixed period lengthens greatly from 97 to 240 min, though still with a small standard deviation of 9 min. Graham and Morgan (1966) have shown for *X. laevis* that at the midblastula stage the lengthening of the cell cycle is caused not only by longer S and M periods, but by the appearance of a G_1 and G_2 period, both previously indetectible. For *R. pipiens*, Flickenger *et al.* (1967) identified a prominent G_1 period and a short G_2 period in addition to a lengthening S period at the late blastula stage. Desynchronization may reflect the appearance of a G_1 phase. The longer S period has not yet been related to a decreased number of growing forks of DNA synthesis or to an increased asynchrony of initiation of DNA synthesis at the different forks, as has been shown for *Drosophila* embryos (McKnight and Miller, 1977; Kriegstein *et al.*, 1978). By the late blastula stage, amphibian cells have clearly established the periods of the cell cycle familiar in larval and adult cells at later times.

Several authors have considered the cause and effect of the transition from synchronous or metachronous cleavage to asynchrony. G. Hertwig (quoted by Schonmann, 1938) was perhaps the first to note and speculate about the transition, and Detlaff (1964) drew recent attention to the phenomenon. She suggested that its importance lies in its connection to the embryo's initiation of gene expression needed in preparation for gastrulation (see Sections 6.4 and 7.1). This is generally accepted as the effect of desynchronization. The cause of the desynchronization and lengthening of the cell cycle has been suggested to be the exhaustion of one or more cytoplasmic materials needed for rapid cleavages (Detlaff, 1964; Lovtrup et al., 1978; Signoret and Lefresne, 1973). The "exhaustion hypothesis" is supported cytologically by the reduction of nuclear diameter and volume at the midblastula stage, as first noted by Schonmann (1938) in *Triturus palmatus* and illustrated in Fig. 8 for *X. laevis*. At stage 8 (midblastula) there begins a rapid 15-fold drop in average nuclear volume; the rate of reduction slows by gastrulation, but by that stage the mitotic rate has also decreased substantially. The decrease of nuclear volume apparently continues even to the hatching stage (Imoh and Sameshima, 1976). If one calculates the total volume of nuclear material from the data of Fig. 8 by multiplying the number of nuclei by the average individual nuclear volume, one obtains the interesting result that *total* nuclear material increases until the midblastula stage and then suddenly reaches a limit, increasing no further until well after gastrulation. In the mid- and late blastula periods, nuclear number increases directly at the expense of nuclear volume. The limit value for nuclear volume, as indicated in Fig. 8, is close to but less than the volume of the oocyte germinal vesicle, by a factor of two- to fourfold. From these cytological data, it is plausible that germinal vesicle sap provides the nuclear sap of the cleavage stages, and runs out when 1000 to 2000 nuclei have been formed. This idea has been proposed by several authors (see Chulitskaya, 1967). In contrast, it is unlikely that all materials for the cleavage nuclei are stockpiled in the germinal vesicle since Adamson and Woodland (1974) have demonstrated *de novo* histone synthesis after fertilization; the newly synthesized histones are calculated to contribute half the total nuclear histones of the gastrula, the remainder already accumulated in the oocyte. Also, Arms et al. (1968) found that radioactive proteins newly synthesized by the egg enter cleavage nuclei or injected liver nuclei.

Furthermore, there is direct evidence for *X. laevis* that germinal vesicle proteins enter cleavage nuclei and are necessary for nuclear activity (Gurdon, 1977) as follows: (1) somatic nuclei injected into oocytes swell rapidly and extensively if the germinal vesicle is broken (Gurdon, 1976), less rapidly if the germinal vesicle is intact, and only slightly if the oocyte has been enucleated prior to injection of the somatic nuclei; (2) the level of RNA synthesis in somatic cell nuclei injected into oocytes correlates closely with their degree of swelling (Gurdon, 1976; Gurdon et al., 1976); (3) the level of DNA synthesis in sperm nuclei injected into fertilized eggs correlates well with their extent of swelling (Graham, 1966); (4) eggs prepared from enucleated oocytes by maturation, do not support nuclear swelling, DNA synthesis (Skoblina, 1976), or cleavage (L. D. Smith and Ecker, 1970); (5) newly synthesized (radioactive) proteins of the oocyte enter the germinal vesicle and enter injected somatic nuclei (Merriam, 1969; Arms, 1968); (6) artificially labeled [^{125}I]- or [^{35}S]proteins isolated from the germinal vesicle and injected into oocytes selectively enter the germinal vesicle, whereas labeled cytoplasmic proteins do not (Bonner, 1975a,b; Feldherr, 1975; Feldherr and Pomerantz, 1978; De Robertis et al., 1979); and (7) Wassarman et al. (1972) find that RNA polymerase activity is localized in the germinal vesicle, from which it is released at maturation; the level of this enzyme activity does not increase until after gastrulation.

It is far from certain what materials are the critical ones exhausted or limiting at the midblastula stage. This problem extends also to somatic cells, where G_1 is viewed as a result of the deficiency of a material needed by the cell to enter S (Prescott, 1976) and synthesized in G_1. In strong support of this view, certain cell lines are found to lack a G_1 period (entering S immediately after M), as do early blastula cells, and to act dominantly in fusion hybrids with G_1 cells (Liskay and Prescott, 1978). These lines yield variants possessing a G_1 period, and pairwise fusion tests for complementation show that the loss of any one of at least five functions leads to the presence of a G_1 period. In cleaving eggs the candidate for the critical exhausted material leading to G_1 is the initiation factor of DNA synthesis studied by Benbow and his associates (Benbow *et al.*, 1977). This factor is thought to appear during oocyte maturation when the cytoplasm supports DNA synthesis from native templates for the first time since the oogonium. The two DNA polymerases are already present in the immature oocyte, although one may be inhibited by a specific agent. After maturation the activity of these polymerases remains constant until neurulation when a third form appears. The initiator can be assayed under defined conditions *in vitro;* it is absent in oocytes and larvae. Its role in the midblastula transition remains to be evaluated.

Lovtrup *et al.* (1978) and Landstrom *et al.* (1975) favor the idea of deoxynucleotides limiting the rate of DNA synthesis at the midblastula stage, as might happen if their reserve from oogenesis is depleted and *de novo* synthesis cannot match the demand. This explanation, especially dTTP limitation, had been tested some years earlier for the G_1–S transition in cultured cells, but the results were not encouraging (Prescott, 1976). In the case of *X. laevis* oocytes, Woodland and Gurdon tried unsuccessfully to stimulate DNA synthesis by injections of deoxynucleotides (Gurdon, 1974). Landstrom *et al.* (1975) have attempted to test their proposal by injecting deoxynucleotides into *X. laevis* fertilized eggs and then examining late blastula and early gastrula stages for the frequency of mitotic cells. By their prediction, deoxynucleotides would keep the embryo in its state of rapid cleavage, and therefore a higher mitotic index would be found. The mitotic index may or may not also reflect synchrony, depending on the statistical sampling. In fact, they did observe a higher mitotic index in embryos injected with the four deoxynucleotides, each in the amount of 16 times above the oocyte reserves, which comprise 50 pmoles per egg: in the experimental series, the index was 19–25%; and in controls of the same age, 7–9%. Unfortunately, the result is not definitive because "sick" cells may arrest or delay in mitosis and artificially increase the mitotic index. Embryonic cells of chick and guinea pig are known to block at metaphase after treatments such as cold or heat shock (Konishi and Kosin, 1974; Edwards *et al.*, 1974). Mitotic index is not an infallible criterion of mitotic rate. In the experiments of Landstrom *et al.* (1975), if the cleavages had actually continued at the normal rapid rate of the early cleavages of controls, one would expect many more cells in the injected embryos (for example, 16 times more), but these cell counts were only marginally higher (18%) in the injected embryos *vis à vis* the controls at the late blastula stage. The only definitive criterion would be a record of the cleavage patterns by cinematography, with both intact and dissociated embryos, as done by Hara (1977) and Signoret (1977). The experements of Chulitskaia (1970) are related and have the same drawback; working with sturgeon eggs, Chulitskaia prepared an egg cytoplasmic extract concentrated severalfold by drying, and injected it into fertilized eggs; at later stages, eggs were fixed, sectioned, and scored for mitotic figures. The mitotic index was higher in the experimental series, consistent with the conclusion that synchrony persists longer than in controls, but no verification of the good health of the cells was presented. This experiment would be well worth repeating in light of the plausible role of the germinal vesicle as the source of materials exhausted at the midblastula stage. It is now feasible to inject oocytes with the contents of one or more vesicles, to mature the oocytes, and fertilize them for development, and by cinematography observe whether the period of rapid sychronous cleavage is prolonged for one or more cycles; or the specific DNA-synthesis initiation factor of Benbow *et al.* (1977) could be injected.

Whatever the specific mechanism, the exhaustion hypothesis implies that the initiation of the midblastula stage depends on the egg's reaching a critical nucleocytoplasmic ratio due to the increase in nuclei in a cytoplasm of constant volume. The most careful test of this assumption has been made by Signoret and Lefresne (1973) with timed photographic records of cleavage in haploid and diploid axolotl eggs of equal size. The haploid eggs managed one more rapid cleavage cycle than did the diploid eggs before the onset of the slower rates of the posttransition period. The exact size of haploid nuclei at these stages is not known, but is assumed to comprise half the nuclear volume or half the nuclear membrane surface (Wilson, 1925) of the diploid nucleus. Also, Signoret and Lefresne compared large and small axolotl eggs, both diploid, and found the smaller eggs to switch from rapid to slow cleavages at an earlier cycle than the large ones, again indicating the importance of nucleocytoplasmic ratios consistent with the exhaustion hypothesis. There are many changes at the midblastula stage in addition to changing mitotic rates, and some of these, such as the start of the downward expansion of the animal hemisphere blastomeres and compensatory contraction of the vegetal surface (Fig. 9), may be independent of the nucleocytoplasmic ratio, and initiated by an autonomous cytoplasmic "clock" (see Section 6.4).

In summary, the midblastula transition is accompanied by a lengthened cell cycle, the appearance of G_1 and G_2 phases, and the desynchronization of the cleavages. These changes may result from the exhaustion of nuclear sap materials provided to the egg by the oocyte.

6.2   The Induction of Mesoderm and Archenteron-Roof Endoderm

Nieuwkoop reported in 1969 a series of experiments that have substantially deepened our understanding of pattern formation and its regulation in the amphibian embryo. These are the experiments used for the interpretative framework of this review. In the first series, he cut an axolotl midblastula embryo (about 1000 cells) around the equator, separating the animal and vegetal hemispheres, and then removed the regions bordering the equator, to eliminate cells contained in the mesoderm area of the fate map. In that way he obtained a cap of prospective ectoderm and a base of prospective endoderm. When these were cultured separately in a balanced salts solution, the cap eventually differentiated into hollow spheres of ciliated flattened cells, the atypical epidermis produced by animal hemisphere blastomeres under the direction of their own cytoplasmic contents without inductive interactions with other parts of the embryo or an "inducing" medium. The vegetal hemisphere base in Nieuwkoop's study also failed to differentiate by itself; it remained a mound of large yolky cells similar to the nutritive yolk cells occupying the lumen of the gut in a normally developing embryo. Neither piece produced mesodermal structures. These were the controls for the recombinates he made by covering the endodermal base with the ectodermal cap. These recombinates formed abundant mesoderm, neural tubes, and gut structures; in fact, some of them developed to rather well-proportioned tadpoles (Nieuwkoop, 1969a). It was clear that mesoderm originated by an interaction of the two hemispheres. Similar results were obtained with *X. laevis* (Sudarwati and Nieuwkoop, 1971). When cap and base from embryos of different urodele species were used, it was possible to recognize that the mesoderm arose exclusively from the animal hemisphere cells in the recombinate (Nieuwkoop, 1969b). Mesoderm originated from prospective ectodermal cells by an inductive interaction with endoderm. In further documentation of this point, Nieuwkoop and Ubbels (1972) marked animal hemisphere cells of the axolotl with [3H]thymidine and showed by auto-

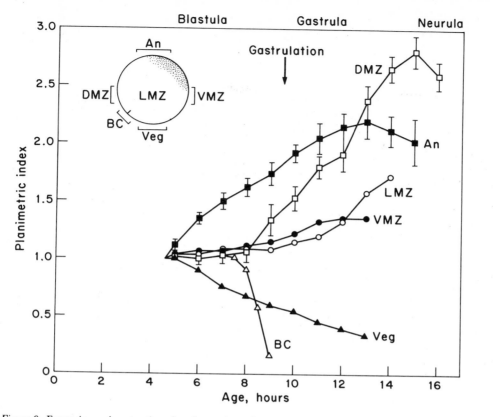

Figure 9. Expansion and contraction of surface regions of the embryo blastula and gastrula of *X. laevis.* The graph is redrawn, with permission of the author, from Keller (1978) who prepared and analyzed time-lapse films of the egg surface at the animal pole (AN), the vegetal pole (VEG), the dorsal, lateral, and ventral marginal zones (DMZ, LMZ, and VMZ, respectively) and the bottle cells (BC) of the dorsal blastopore lip. Surface area of a region is reported at the "planimetric index," PI, on the *y* axis above, which is defined as the planar area enclosed by a set of cell boundaries chosen at 4.5 hr after fertilization and followed until 14–16 hr. At 4.5 hr the area is designated 1.0 unit of area, for normalization, and later values for area are given relative to this. The planar area is a projected area from the films; it does not take into account the surface contours of the ever increasing cells within the chosen boundaries. Note the early and compensatory changes of area of the animal and vegetal regions (epiboly), the later strong expansion of the dorsal marginal zone just before gastrulation starts, and the rapid contraction of the outfacing surface of the bottle cells as the blastopore lip forms and the bottle cells extend inwardly. Age is given as the hours after fertilization at 22°C. Stage 8 occurs at approximately 5 hr.

radiography that in recombinates the mesoderm of this species also derived from animal blastomeres. Close analysis of the marked recombinates showed that suprablastoporal endoderm (for example, archenteron roof and head endoderm) also derived from animal hemisphere cells. It is not known exactly where is the limit between the induced and the inducing cells in the recombinates; possibly even the smaller cells of the upper half of the blastopore lip derive from cells of the animal hemisphere by induction, although this has not been examined. Because the induction directs ectoderm to endoderm, it is called a "vegetalizing induction." Mesoderm is considered a partial vegetalization, and persisting ectoderm would be uninduced.

Nieuwkoop (1969*a,b,* 1973, 1977) proposed that the intact amphibian egg be considered as two discontinuous and unique hemispheres, animal and vegetal, rather than an animal–vegetal gradient of materials. Alone, each hemisphere would have a very limited capacity for development; together, they form mesoderm with one hemisphere as the inductor, the other as the inductee. The animal hemisphere is the source of responsive cells. The mesoderm produced by this induction then affects the course of ectodermal and endodermal development by further inductions of its own at the gastrula and neurula stages.

From the results of recombinates of cap and base of different developmental stages from the midblastula onward, Boterenbrood and Nieuwkoop (1973) concluded that the vegetal cells are maximally inductive at the midblastula stage, whereas by the early gastrula stage their effect is weaker. Asashima (1975) made similar observations with vegetal pole explants of different stages in *T. alpestris*. The animal hemisphere cells, on the other hand, remained equally responsive from midblastula through early gastrula, and decreased their competence to produce mesoderm only in the mid- to late gastrula stages.

Previously, it was mentioned that the equatorial regions of a normal midblastula already have the capacity to produce mesoderm autonomously when cultured as explants in a balanced salt solution (Nakamura *et al.*, 1970) and that autonomy is gained at the 64- to 128-cell stage. Autonomy for long-term development of mesoderm in explants does not, of course, indicate that the inductions are already occurring by the 64- to 128-cell stage; it indicates only that equatorial cells then have all the components they need for subsequent inductions and autonomous expression. Now, in Nieuwkoop's experiments (1969a,b), these cells were removed, and yet mesoderm could still be formed in the recombinants, indicating that the necessary contacts could be established again.

Nieuwkoop's second line of inquiry (1969b) concerned determination of the dorsal–ventral axis at the midblastula stage. This also concerns mesoderm induction, as dorsal mesoderm is the "organizer" of the embryonic axis of the gastrula and neurula. Nieuwkoop (1969b) prepared recombinates in which the cap was rotated 180° to bring its prospective dorsal side diametrically opposite the dorsal side of the vegetal piece. Surprisingly, these recombinates developed axes based on the dorsality of the *vegetal* piece, despite the fact that the animal hemisphere cap still contained a portion of the gray crescent region. It would have been informative to test in this same way the dorsal–ventral bias of the two hemispheres of earlier stages, but unfortunately the earlier stages did not endure the surgical removal of equatorial material. Nieuwkoop concluded that *at least* by the midblastula stage it is the vegetal half that determines the dorsal–ventral axis. We suggested earlier that it is this hemisphere that harbors the dorsal–ventral differences right from the time the vitelline wall appears. If the dorsalizing center is located in the vegetal hemisphere, we can understand why removal of the animal dorsal blastomeres of the gray crescent region at the 8-cell stage (Votquenne, 1933) or midblastula stage (Curtis, 1962b) did not disrupt development. Asashima (1975) has tested mesoderm induction by explanted vegetal material from the pole region of *T. alpestris* fertilized eggs and from later stages, and concluded that the midblastula explant is stronger than those of earlier stages, although the earlier explants decomposed and probably did not register their full effectiveness.

In order to survey the pattern of inductive abilities of the vegetal half of the axolotl midblastula stage, Boterenbrood and Nieuwkoop (1973) divided the half into four pieces, dorsal, ventral, and two lateral, and covered each with prospective ectoderm from the blastocoel roof. These recombinates developed as if the yolk mass were a mosaic of inductors arranged in a "prepattern" of the future mesoderm: the ventral piece induced ventral mesodermal elements such as blood cells, pronephros, and germ cells (a mesodermal derivative in urodeles); the lateral pieces induced pronephros, germ cells, and muscle cells with a stronger induction on the left side (a finding related to the minor bilateral asymmetry of the tadpole); and the dorsal piece induced dorsal mesoderm such as notochord, head mesoderm, and somites. This remarkable finding indicates that the future pattern of the mesoderm of the animal half, including the organizer region itself, is generated from inductive differences already patterned in the vegetal half by the midblastula stage. That region of the vegetal half responsible for the induction of the dorsal mesoderm, the classical organizer, is called the "vegetal dorsalizing region." Its exact location has not been defined except that it partially remains even when the outer third of the dorsal side of the yolk mass has been removed, and therefore must extend rather far into the egg. Even the vegetal blastomeres of the blastocoel floor have a good ability to induce lateral mesoderm, suggesting that the entire vegetal half has at least some inductive strength.

An important question immediately arises: What is the material basis for the regional differences in inductive capacity of the vegetal half? It is not possible to infer from the experiments of Nieuwkoop and his colleagues whether the vegetal pattern of inductions reflects regional quantitative or qualitative differences (or both) in the actual inductors. Arguments for a quantitative difference will be presented later, in connection with the model of Weyer *et al.* (1978), in which the distribution of mesoderm in the fate map of the gastrula is accounted for by assumptions about the location and geometry of the source of inductor, about the rate of diffusion of a single kind of inductor, and about the rate of breakdown of the inductor by the animal hemisphere cells. Since inductive activity by the vegetal blastomeres seems responsible not only for the animal–vegetal orientation of the mesoderm between the endoderm and ectoderm on the fate map of the late blastula (Fig. 4), but also for the dorsal–ventral axis with more and perhaps different mesoderm on the dorsal side, it seems possible that the dorsal–ventral axis may be a modification of the original animal–vegetal axis of the egg. Perhaps the dorsal–ventral axis represents a bending of the animal-vegetal axis to the dorsal side, rather than the establishment of a wholly new axis, such as a cortical axis. As discussed previously, this modification of the original axis might involve a slight displacement of the vegetal platelets to the dorsal side in the elevated region of the vitelline wall, and an "activation" of these platelets by cytoplasmic agents of the animal hemisphere. And finally, one more factor would determine the ultimate effectiveness of vegetal blastomeres in the induction process at the mid- and late blastula stages, namely, their proximity to the target cells in the animal hemisphere.

Despite the evidence for induction at the midblastula stage, there has been only a limited search for the inductor. Faulhaber and her colleagues have attempted to obtain inductive preparations from extracts of late blastulae of *X. laevis* (Faulhaber, 1972; Faulhaber and Lyra, 1974; Wall and Faulhaber, 1976). Fractions containing the membranes of yolk platelets were weakly active as vegetalizing agents, promoting the appearance of mesenchyme tails in *T. alpestris* ectoderm; however, membrane-containing fractions from microsomes were also weakly inductive, and the best fractions were less active than the purified vegetalizing inductor from chick embryo extracts. The specificity and strength of the inductions are not especially convincing in light of the long history of heterologous and artificial inductors. On the other hand, it is difficult to know ahead of time what quantitative criteria are appropriate to evaluate results on induction: see the discussion of neural induction in Section 7.3.

There is no reason *a priori* to expect yolk or other vegetal materials to act directly as inductors on animal hemisphere cells. These materials may act *within* vegetal cells, affecting their gene expression; it may be the gene products that exert the inductive effects. There has been no study to learn the level of viability needed by vegetal cells to exert their induction; perhaps vegetal halves treated at the midblastula stage with inhibitors of DNA, RNA, or protein synthesis could still induce mesoderm in recombinates with untreated animal hemisphere cells.

J. C. GERHART

Bachvarova and Davidson (1966) split early embryos of *X. laevis* into prospective dorsal and ventral halves, which they exposed to [^{3}H]uridine in a balanced salts solution. After 1 hr of incubation, the halves were fixed, sectioned, and autoradiographed to visualize the level of incorporation of radioactivity into the nuclei. Control digestions of the sections with RNAse were used to evaluate the radioactivity in RNA, a necessary treatment, for [^{3}H]uridine can also enter DNA, which is synthesized very rapidly at these stages. Before stage 8, incorporation into RNA was barely detectible above background, whereas by stage $8\frac{1}{2}$ incorporation was readily detected, an increase of approximately 20-fold. The midblastula period is the time of greatly increased RNA synthesis. The increase was regionally dependent in an interesting way: nuclei from the blastocoel roof showed little labeling, whereas nuclei from the vegetal hemisphere were well labeled (0.8–1.8 vs. 9.3–17.3 grains/ nucleus, respectively). Within the vegetal hemisphere, the dorsal nuclei were higher than the ventral (17.3 vs. 9.3 grains/nucleus). The highest incorporation occurred in nuclei of the dorsal subequatorial zone, that is, dorsal to the blastocoel floor (17.3 grains/nucleus). This pattern of animal–vegetal and dorsal–ventral differences became more accentuated at the late blastula stage. As it turns out, the region of most intense labeling is exactly the region of the vegetal dorsalizing center of Nieuwkoop, as located in the explant studies of Boterenbrood and Nieuwkoop (1973), and of the induced dorsal mesoderm. Yet at the time of the analysis by Bachvarova and Davidson (1966), the fate map of *X. laevis* was not well known (Nieuwkoop and Florschutz, 1950; Keller, 1975, 1976) nor was the developmental significance of these regions for pattern formation appreciated. Bachvarova and Davidson (1966) even recorded that in the dorsal equatorial region the highest labeling is internal, whereas the surface cells were poorly labeled; now we know that the surface cells are not mesoderm at all but future archenteron roof or neural ependyma.

These data are important in providing the earliest regionally different biosynthetic pattern in the amphibian embryo. As soon as gene expression starts, it is regionally different. The pattern is somewhat surprising. First, there is a strong animal–vegetal difference of incorporation; the vegetal nuclei are much more active (10-fold or more) in nuclear RNA synthesis. Second, there is a detectable (almost 2-fold) dorsal–ventral difference in the nuclear incorporation of the vegetal half, but not the animal half. Presumably this pattern of quantitative differences in incorporation reflects a preexisting pattern of cytoplasmic differences in the midblastula. It is proposed here that the large animal–vegetal difference in incorporation is related to the large difference in yolk platelet composition of the cells (both number and size of platelets). Secondarily, it is proposed that the vegetal hemisphere itself differs regionally in a dorsal–ventral sense in its platelet composition, perhaps not so much in abundance of platelets but in regional "activation" of platelets. There is no information about how platelets differ or how they could affect the level of RNA synthesis, although a few possibilities will be discussed later. It is not known whether animal, vegetal, dorsal, and ventral cells synthesize different RNA species in the midblastula, or just different quantities of the same RNA. Also, it is unknown whether gene expression is needed for the vegetal cells to become inductively active, or for the animal hemisphere cells to respond, or both. Nakamura and Takasaki (1971) have treated the midblastula and later stages of *T. pyrrhogaster* with actinomycin D and concluded that a short period of RNA synthesis at the midblastula stage is sufficient to allow the embryo to initiate gastrulation, assuming the inhibitor penetrates the cells rapidly. An experiment of this kind has not been done with *X. laevis* yet.

At the midblastula stage nuclear RNA synthesis increases abruptly, probably representing the first gene expression of the embryo. Similar conclusions were reached by Carroll (1974) on the basis of [³H]uridine incorporation into RNA in axolotl midblastulae. The RNA synthesized at the mid- and late blastula stages has been only partially characterized. Davidson *et al.* (1968*b*) concluded from hybridization experiments of newly synthesized radioactive RNA from *X. laevis* late blastulae competing with nonradioactive RNA of other stages, that most of the newly made sequences differ not only from those supplied by the oocyte to the early embryo but also from those made in the early gastrula. The RNA appears to be stage-specific, made only at the mid- and late blastula stages. This result contrasts with recent conclusions about sea urchins where newly synthesized RNA of gastrulae has a close sequence similarity to oocyte RNA, at least in terms of the "complex class" of polysomal mRNA species (Galau *et al.*, 1977). At the time of the original study on amphibia (Davidson *et al.*, 1968*b*), methods were adequate to analyze only total RNA, as opposed to polysomal RNA, and those RNA sequences complementary to moderately repeated DNA, rather than to the complex, more unique sequences of DNA. Thus the early work does not allow conclusions about the RNA species reaching the cell cytoplasm for translation into proteins, namely, those species most likely to affect cell function.

As far as the identity of the RNA species made at the late blastula–early gastrula stages, Woodland and Gurdon (1969) found one-third to one-half as much radioactivity in tRNA as in heterogeneous RNA in both animal and vegetal halves labeled for 2 hr with [³H]uridine or [³H]guanosine, whereas rRNA was synthesized at a much lower rate and only in the animal halves at these early stages. Prospective mesoderm was probably included with both halves. These results were confirmed by Knowland (1970) who used high-resolution gel electrophoresis to resolve DNA from the RNA species extracted from *X. laevis* whole embryos previously injected with [³H]uridine and [³H]guanosine, and found tRNA and heterogeneous RNA appearing at stages 8–9 but no rRNA until stage 12, well into gastrulation. Several authors agree the vegetal production of heterogeneous RNA and tRNA exceeds that of the animal regions; there has been no comparison of the RNA sequence complexity from the dorsal and ventral sides, much less from the vegetal dorsalizing region, although these would now be feasible with the procedures of Moen and Namenwirth (1977) for large-scale sectioning of oriented embryos.

In addition to synthesis, degradation of RNA species carried over from the oocyte may also contribute to the uniqueness of the midblastula RNA population. Crippa *et al.* (1967) reported a 20–30% reduction in sequence variety among oocyte RNA species at the mid- to late blastula stages. The significance of this degradation is not known—that is, whether for processing as part of activation of mRNA precursors or true elimination of already functioning sequences.

As for the cytoplasmic events which cause the initiation of large-scale RNA synthesis at the midblastula period, several authors have noted the coincidence of the decreasing rate of cleavage and DNA synthesis with the increasing rate of RNA synthesis. Dettlaff (1964) was one of the first to draw attention to the midblastula period as the beginning of preparations for gastrulation, and Bachvarova *et al.* (1966) stressed gene expression as the new event. Flickenger *et al.* (1970) and Gurdon and Woodland (1969) suggest a direct antagonism between DNA and RNA synthesis as the mechanism allowing gene expression as replication slows down. The early cleavage stages might suffer this inhibition more strongly than do nonembryonic cells in their S period, because the cleavage cells have many more growing forks simultaneously in operation—approximately 20 times as many if one compares the lengths of the S periods. This straightforward relationship is easily added to the exhaustion hypothesis: a cytoplasmic material needed for rapid DNA synthesis would become limiting, and the cell would acquire the "waiting period" of G_1 and G_2. These periods would allow initiation of RNA synthesis, as would the lengthening S period. Flickenger *et al.* (1967, 1970) favor the G_2 phase for the initiation of new gene expression. This had been the phase (or a variation thereof) of arrest for the oocyte while massive gene expression occurred from the lampbrush chromosomes. Also, the G_2-like phase of the oocyte allows changes in patterns of transcription as found by Gurdon *et al.* (1976) for liver cell nuclei injected into oocytes. Thus, a plausible hypothesis can be made for the initiation of RNA synthesis based essentially on the acquisition of a critical nucleocytoplasmic ratio.

On the other hand, it seems unlikely that the initiation of specific transcription at the midblastula stage can be due simply to an exhaustion of a material needed for DNA synthesis, as indicated by the recent results of Brothers (1977) with the *o* mutant of the axolotl. These experiments suggest a positive control factor as well, as follows. In this maternal-lethal mutant, o/o homozygous females (offspring of heterozygous parents) lay eggs that arrest at or before gastrulation, no matter what the genetic constitution of the sperm, whereas eggs from o/+ heterozygous females gastrulate normally. Eggs from o/o females (called *o* eggs) do gastrulate if injected with a small amount of germinal vesicle sap from a +/+ or o/+ oocyte. This injection result provides an assay for the essential *o* substance, which has been partially purified and characterized as a protein (Briggs and Cassens, 1966; Briggs and Justis, 1968). The *o* substance seems to be required in some way as a positive effector for the initiation of RNA synthesis (Carroll, 1974), without which gastrulation fails.

In her recent studies with nuclear transplantation, Brothers (1977) established that wild-type nuclei taken from early blastulae and put into enucleated *o* eggs do not overcome the defect, a result consistent with the performance of *o* eggs fertilized with wild-type sperm. Surprisingly, though, wild-type nuclei from normal late blastulae *do* rescue the *o* condition; a completely unexpected result. These nuclei had themselves passed the midblastula transition. After transplantation the egg developed without need for the cytoplasmic *o* substance, apparently because of a specific heritable state of the nucleus. RNA synthesis was successfully initiated by the nucleus. As a control for the possible coinjection of wild-type cytoplasm, Brothers did transplantations of a late blastula wild-type nucleus and its nuclear descendants through several enucleated *o* eggs in series, each time removing a nucleus before the midblastula stage for transplantation into a fresh *o* egg. After 30 cycles of cleavage, a nucleus was put into an *o* egg, which was then allowed to develop beyond the midblastula stage; it gastrulated normally, indicating that the nucleus had retained completely its independence of the *o* substance through enough serial transfers to dilute greatly the original wild-type egg materials. Also, the independence of the nucleus from the *o* requirement persisted, even though gene expression had been very low or absent in the early blastula cytoplasms in which it replicated, when the *o* gene would have had to express itself if it were to preserve its independence by replenishing the proteinaceous *o* substance.

These results suggest but do not yet prove that the wild-type nucleus undergoes a permanent alteration of its DNA at the midblastula stage in the presence of *o* substance; a base sequence change alone would be heritable in the absence of RNA synthesis. Such alterations in known instances include insertions, inversions, sequence amplifications, restriction–modification patterns (see Holliday and Pugh, 1975), translocations, and eliminations. The gene rearrangements in the immunoglobulin system force us to consider related mechanisms for the *o* effect. The possibility of DNA elimination has serious application to the axolotl case in light of the report by Signoret and Lefresne (1969) that in this species the long arm of the number 7 chromosome displays a cold-induced shortening in metaphase figures from cells taken *after* the midblastula stage. The change affects all nuclei by the midgastrula stage. Signoret and Lefresne suggest that the shortening proceeds by the actual loss of a small end knob previously attached by a constriction to the terminus of the long arm—that is, by chromosome elimination of a small piece of the number 7 chromosome. If this is a true elimination, rather than a step of heterochromatin formation, it would fit the transplantation results of Brothers (1977). Carroll (1974) could not detect differences in the constricted chromosomes of *o* mutants and wild-type midblastulae, although shortening was not measured. Also, there is no information relating the normal chromosome shortening to RNA synthesis or to the action of *o* substance. Whatever the alteration of the DNA, we are faced with the "dilemma of Weismann," for the future germ-line nuclei must either escape the permanent heritable modification or reverse the heritable change, otherwise the *o* substance would never be necessary and the *o* mutant would never have been found.

As a final point from her experiments, Brothers examined the fate of wild-type nuclei from normal early blastula cells, when they are transplanted into an *o* egg and left there through the midblastula period. As mentioned earlier, RNA is not synthesized and gastrulation fails. Also, it is interesting that in the mutant the mitotic frequency drops almost to zero at the mid- to late blastula stage, whereas the wild-type egg maintains a low but significant level (for example, 5–7%) at these stages (Carroll, 1974). Brothers found that wild-type nuclei passing the midblastula transtion in *o* cytoplasm become irreversibly damaged in some way, for when transferred back to wild-type enucleated eggs, they fail to support gastrulation. It would be interesting to know if this damage is a consequence of the failure to synthesize RNA, a distinction that could be made with transplantations of nuclei in normal eggs allowed to reach the late blastula stage in the presence of α-amanitin. Other cases of chromosome damage starting at the late blastula stage are known: Schönmann (1938) observed severe aneuploidy and chromosome breakage in certain urodele hybrids only after the midblastula stage, coincident with the sharp reduction in nuclear volume.

At present the action of the *o* substance is far from known. In its presence nuclei do not suddenly stop dividing in the late blastula period as happens in the mutant (Carroll, 1974). Perhaps the substance is needed to establish one of the new periods of the cell cycle, a period allowing the initiation of RNA synthesis. It is not known in which phase of the cell cycle the cells of the *o* mutant arrest, except that the arrest does not occur in mitosis. While the results of Brothers demonstrate the requirement by the nucleus for a cytoplasmic protein to undergo a stable modification allowing RNA synthesis, a normal cell cycle, and gastrulation, the interdependence of these midblastula changes has evaded understanding. The solution to these phenomena would likely benefit molecular cell biology at large, not just embryology.

Thanks to the work of Woodland and his colleagues, we have a sketch of the changes in protein synthesis at the early stages of development of at least one amphibian, *X. laevis*. It is, of course, necessary to recognize changes at this level of synthesis if we are to comprehend changes of cell function involving gene expression and translational control. Ballantine *et al.* (1979) have examined ^{35}S-labeled newly synthesized proteins by 2-D gel electrophoresis, which allows the discrimination of about 500 protein species, the most abundant proteins of the cell; presumably, there are several times as many less abundant proteins that are not detectable. According to these authors, the oocyte produces a great variety of proteins, almost exceeding the resolution of the technique. Then, at maturation the variety of synthesized proteins drops noticeably, even though the rate of protein synthesis increases about two-fold in step with the doubling of polysomes (Woodland, 1973). Among the proteins no longer made are β- and γ-actin (nonmuscle actins) and H1 histone, even though other histones are now made 10–20 times faster than before maturation (Adamson and Woodland, 1974, 1977; Woodland and Adamson, 1977). The variety of newly made proteins reaches a minimum in the unfertilized egg and remains at this level until the midblastula stage when, surprisingly, the variety increases and approaches that of the oocyte. The H1 histone and β- and γ-actin are made again in detectible quantities. This increase in the variety of newly synthesized proteins is the result not only of the synthesis of new mRNA, but also the mobilization of stored oocyte mRNA, as shown by Woodland (personal communication) for the case of H1 histone from the analysis of androgenetic haploids from enucleated *X. laevis* eggs fertilized with sperm from *X. mulleri*, a frog species in which the H1 histone is electrophoretically distinguishable from that of *X. laevis*. At the midblastula stage of this haploid egg, *both* the *laevis* and *mulleri* H1 histones appear, demonstrating synthesis of new mRNA and mobilization of preexisting mRNA. The synthesis of *laevis* H1 histone ceased at the gastrula stage, as if the stored mRNA were exhausted. It is not presently known what mechanism activates or reactivates stored oocyte mRNA, although in related work Ruderman *et al.* (1979) found that the cessation in H1 histone synthesis at maturation is accompanied by a decrease in the amount of polyadenylated H1 histone mRNA, a degradative reaction; perhaps renewed polyadenylation mobilizes stored mRNA. While the midblastula stage restores the oocyte pattern of protein synthesis, it is only the gastrula stage that synthesizes wholly new protein species not seen in the oocyte, fertilized egg, or blastula (Ballantine *et al.*, 1979). The midblastula transition does not seem to depend on the appearance of totally unique protein species in large amounts, although one might have expected new proteins from the evidence for unique RNA species at this stage. However, there may be unique small differences from stage to stage even before gastrulation, as noted by Bravo and Knowland (1979) after careful scrutiny of 2-D gels; they observed a few proteins appearing just at the 2- and 4-cell stages and at the midblastula stage. The synthesis of proteins from newly synthesized mRNA would fit easily into the postulated sequence of events of the midblastula transition discussed previously; namely, the rapid increase in nuclei at cleavage exhausts a material supplied to the egg by the oocyte, the cell cycle lengthens with the appearance of G_1 and G_2 phases, RNA synthesis begins, and now new translation products follow. It is possible that the mobilization of oocyte mRNA at the midblastula stage is also a consequence of the egg's critical nucleocytoplasmic ratio, but in the absence of data we cannot eliminate the alternative possibility that the mobilization is autonomously timed by the cytoplasm alone and would occur punctually in unfertilized activated eggs or in fertilized eggs arrested with colchicine or vinblastine. Also it would be informative to know if the mobilization requires mRNA synthesis, as tested in α-amanitin-treated eggs unable to synthesize mRNA.

As for regional differences in protein synthesis, Ballantine *et al.* (1979) have compared animal and vegetal halves at the midblastula and later stages but have found no clear difference in synthesized proteins until the gastrula and neurula stages; for example, the animal half began the synthesis of α-actin (striated muscle actin) in the gastrula, and the vegetal half began the synthesis of a unique protein of unknown function at the neurula stage. However, there was no detectible qualitative difference in protein synthesis in the two halves at the early stages when RNA synthesis is more intense in the vegetal half and when the induction of mesoderm is in progress, an interesting result. Dorsal–ventral differences were not surveyed, much less the synthetic pattern of the vegetal dorsalizing center.

6.4 Preparations for Gastrulation

Be definition, gastrulation begins with the apperance of the blastopore on the future dorsal side of the embryo, followed immediately by massive cell movements from the embryonic surface to the interior. But well in advance of blastopore formation, preparations for gastrulation proceed, and small movements of cells are detectable. These preparations and movements of the mid- and late blastula periods include: (1) epiboly—that is, the expansion of cells of the animal hemisphere and the compensatory contraction of the vegetal surface; (2) the induction of mesoderm and head endoderm in animal hemisphere blastomeres bordering on the vegetal yolk mass, as already discussed; (3) changes in the properties of cell adhesiveness and motility; (4) the migration of mesoderm upward on the inner wall of the blastocoel, especially in *X. laevis* where a substantial "cryptic gastrulation" occurs; and (5) the formation of bottle cells at the blastopore and the first receptivity of the embryo to bottle cell invagination.

While embryos with defective gene expression are known to arrest before gastrulation, there has been no thorough analysis of the dependence of individual pregastrulation changes on gene expression, largely because the changes have been identified only in recent years, especially at the cellular and molecular level. Therefore, our understanding of the role of RNA syntehsis in gastrulation remains incomplete. An exception is the study of Johnson (1970) on cell motility and adhesiveness in various lethal hybrids of *Rana* species (discussed below). In hybrids it is uncertain exactly what level of RNA synthesis occurs; it might be more definite to analyze α-amantin-inhibited embryos because these lack RNA synthesis and yet cleave to apparently normal late blastula stages (Brachet *et al.,* 1973; Crippa *et al.,* 1973; K. Hara and M. Kirschner, unpublished).

6.4.1   Epiboly

Epiboly and a reciprocal vegetal contraction begin at the time of the midblastula transition in *X. laevis,* as shown by the cinematographic measurements of Keller (1978) reproduced in Fig. 9. The animal hemisphere, at least in the polar region, expands by approximately two-thirds its area prior to gastrulation, whereas the vegetal pole surface contracts by about two-thirds its area. The equatorial (marginal) zone maintains its original size, but is necessarily transported in the vegetal direction. The expansion and contraction are symmetrical about the animal–vegetal axis, without evidence of a dorsal–ventral bias at the pregastrula stages. Keller has followed individual cells of the surface and concludes that no inner cell adds to the animal surface during its expansion, and no surface cell leaves the vegetal area as it contracts. The movements are accomplished strictly by changes of cell shape. Keller also finds that the ratio of radial (perpendicular to the surface) to paratangential (parallel to the surface) divisions increases in areas of stretching with the outcome that the surface area occupied by single stretched cell (at the animal pole) is no greater than that of a single nonstretched cell (at the equator). This result could simply reflect the cell's continued observance of the "Sachs–Hertwig rule"—namely, the spindle typically orients to the longest axis of the cell, and division tends to cut this axis transversely (Wilson, 1925). As a corollary of this relationship of surface area and division planes, regions of epibolic expansion would be expected to release fewer cells into internal cell layers by division, and hence to have fewer cell layers, compared to nonexpanding regions.

Small fragments of blastocoel roof are capable of expansion (Holtfreter, 1943*a,b*), whereas vegetal fragments seem incapable of contraction (Stableford, 1948); therefore the normal contraction in the vegetal surface of intact blastulae may be a compensation for the expansion. The animal hemisphere surface cells are considered all to expand, and each to expand in all directions (Holtfreter, 1943). Net displacement toward the vegetal pole would follow as the geometric consequence of an expanding hemisphere. Epibolic movements are reported to occur in intergeneric and interspecies amphibian hybrids, which nonetheless arrest at gastrulation (Johnson, 1970; Subtelny, 1974). This observation suggests that epiboly is independent of new gene expression after the midblastula transition.

Holtfreter (1943a) arrived at the same conclusion when he observed an epibolylike expansion of the pigmented surface of the animal hemisphere in unfertilized *R. pipiens* eggs remaining in a salt solution (110 mM NaCl) for 1 or 2 days. The vegetal pole simultaneously contracted until the entire surface was pigmented. At the vegetal pole the surface buckled inward to give a blastoporelike invagination. These same effects have been achieved by L. D. Smith and Ecker (1970) and by Baltus *et al.* (1973) with *R. pipiens* eggs matured from oocytes *in vitro* and by Malacinski *et al.* (1978) with unfertilized eggs of *R. nigromaculata* left in a full-strength salt solution (about 240 mOsm). These subcellular arrangements have been termed "pseudogastrulation" by Holtfreter (1943a), to indicate their similarity not only to epiboly but also to invagination. Even enucleated oocytes of *R. pipiens,* when matured *in vitro,* show pseudogastrulation (L. D. Smith and Ecker, 1970). Cytologically, it involves a segregation of yolk-free cytoplasmic islands and vesicles to the cortex to form a cortical layer, which then advances to the vegetal pole, forming a "yolk plug" filled not with yolk but with basophilic yolk-free cytoplasm (Baltus *et al.,* 1973; discussed by Brachet, 1977). Thus, the egg cortex seems capable of an epibolylike expansion with no contribution from cellularization or gene expression at the midblastula stage. At the biochemical level, the cortical expansion is perplexing because actin-dependent movements are usually viewed as contractions rather than expansions. Perhaps epiboly is the relaxation of a much earlier contraction, one made in early oogenesis when the oocyte transformed from a uniformly pigmented sphere to one that is only half covered by pigment. Or perhaps the vegetal hemisphere is the center of a major active contraction.

It is not known what triggers the start of epiboly at the midblastula stage. Holtfreter (1943b) suggested that isotonic salt solutions used for pseudogastrulation may "soften" the surface of the egg. Malacinski *et al.* (1978) demonstrated that 0.12 M NaCl alone is sufficient to stimulate pseudogastrulation, whereas 0.06 M NaCl is not. Holtfreter favored the idea of a cortical timing mechanism because pseudogastrulation starts in uncleaved eggs at approximately the same time as does epiboly in cleaving eggs. In the cleaving egg, the surface layer of cells does undergo a modification of cell contacts in the late blastula period when tight junctions may be supplemented by full-fledged desmosomes (Sanders and Zalik, 1972). Brachet (1962) observed that epiboly was blocked in embryos treated continuously with 10 mM mercaptoethanol or 0.1 mM lipoic acid, both reagents capable of breaking disulfide linkages. But beyond these chance observations, the membrane and cortical changes accompanying epiboly are uncharacterized, as are their stimuli. They have not been related to the cyclic surface contraction waves known to occur in vinblastin-arrested eggs of *X. laevis* in step with the cleavages of normal eggs (M. W. Kirschner and K. Hara, unpublished). These waves imply an autonomous cytoplasmic or cortical timing mechanism, used perhaps not only at cell division, but also for longer intervals, such as some midblastula changes and for blastopore formation.

6.4.2 *The Acquisition of Cell Motility*

Hara (1977, unpublished) has prepared a series of remarkable time-lapse films of fertilized axolotl eggs removed from the vitelline membrane at the 1-cell stage and incubated in 67 mM phosphate buffer, in which the blastomeres separate as they cleave. Divisions continue at the normal rapid rate with high synchrony until the 10th or 11th cycle, and then display the effects of the midblastula transition in dividing more slowly. Surprisingly, at the time of the transition, some blastomeres begin unmistakable cell-surface activity, extending and retracting blebs, ruffles, and pseudopodia in random directions. Although the cells do not achieve real migration under the conditions of incubation, it seems likely that the membrane activity is at least a prerequisite for true locomotion. Membrane activity is one of the cellular acquisitions of the mid- and late blastula periods. The films show clearly that surface activity is an autonomous expression of individual blastomeres, not requiring interactions with neighbors for its establishment. Thus surface activity seems to arise in parallel with the new G_1 and G_2 phases of the cell cycle and with RNA synthesis. All blastomeres gain surface activity except for the largest, yolky, vegetal members, in which it may be difficult to observe whether such activity occurs. No test with α-amanitin-treated eggs has been made to determine whether RNA synthesis is actually needed; perhaps a G_1 or G_2 phase is sufficient to allow surface activity: in cultured cell lines distinctive surface activity is known to correlate with each phase of the cell cycle (Porter et al., 1973). These data fit well with the earlier observations of Johnson (1976b) to be discussed later.

The locomotor behaviors of individual cells from late blastulae and early gastrulae were first observed and categorized by Holtfreter (1943a,b, 1944, 1946, 1947a). His method was to dissociate embryos (R. pipiens, T. viridescens) by brief exposure to pH 10.5 solutions, to transfer the cells to a neutral solution, and to provide a glass substrate for cell attachment and movement. Locomotor behavior of four types occurred: (1) large, rounded yolk-free lobopodia extending from globular, unattached cells with a yolky center; the lobopodium rotated around the periphery at a speed of approximately 1 rpm, a behavior referred to as the "circus" or "limicola" movement; (2) a slow creeping or flowing directional movement by elongated cells weakly attached to the glass; the cells experienced rhythmic peristaltic waves of constriction passing from the yolk-free end of the cell to its yolky end, in the direction of movement; (3) single or multiple extensions and retractions of fingerlike processes, from weakly attached cells; and (4) highly flattened, large, fan-shaped extensions from strongly attached cells; the fans possessed radial ribs. He stressed that the surface activity concerns only those parts of the cell containing the new cleavage membrane, but not the rigid impermeable egg surface membrane. Holtfreter (1947a) also noticed that attached cells could phagocytize carbon and carmine particles, whereas the unattached ones could not. These four behaviors were shown by cells from all regions of the late blastula or early gastrula, except for the largest yolky blastomeres, which seemed inactive. There was no clear correlation of migratory behavior with developmental fate at this stage in these amphibia. Satoh et al. (1976) have refined and enlarged the categorization with blastomeres dissociated from X. laevis early gastrulae by EDTA treatment, recognizing in addition to Holtfreter's types certain unattached cells with bubbling blebs, and attached cells with branched, fanlike processes. Furthermore, these authors claim to find some correlation between the quantitative level of surface activity of isolated cells and the position of their embryonic origin; particularly, prospective mesoderm cells just before gastrulation show the highest frequency (20–30% of the cells) of circus movement, ahead of prospective ectodermal (10% frequency) and endodermal cells. All in all, though, the number of cells to display directional movement is low, perhaps because few cells attach well to glass, a prerequisite for directed movement. It would be interesting to provide these dissociated blastomeres with a more "natural" substratum such as a monolayer of X-irradiated "feeder" cells, as was recently done by Wylie and Roos (1976) and Heasman et al. (1977) for X. laevis presumptive primordial germ cells. When plated on glass or collagen surfaces, these cells extend lobopodia without clear directional movement, whereas on monolayers of X. laevis embryonic cells or of adult kidney or mesentery the germ cells attach firmly and move by elongation and contraction, with extensions of filopodia and retraction of a long "tail," in a manner similar to their behavior *in vivo*. Nonetheless, on glass surfaces at least some prospective mesoderm cells of early gastrulae of *R. pipiens* can move in a "fibroblast" manner with lamellipodia, microspikes, and ruffles (Johnson, 1976a). When crossing paths, these cells display contact inhibition of locomotion.

Johnson (1976b) has made a quantitative study of two important aspects of cell motility in *X. laevis* embryos: its dependence on the developmental stage and its dependence on the position of origin of the cells in the embryo. On the first point, Johnson found that only 5–8% of the cells showed circus movements and blebbing when isolated at the mid-blastula stage, but the frequency increased sharply in the late blastula period, reaching 42% by gastrulation and 93% by the late gastrula stage. The increase in frequency of motile cells correlated well with the decrease in mitotic index. Johnson noted that the isolated cells halt their circus movements at the time of division and the daughter cells begin the movements again, thus supporting the possibility that movement is a cell property of certain phases of the cell cycle which are absent in the early blastula.

On the second point, Johnson found that the highest frequency (80%) of motile cells in the late blastula occurred in isolates from the prospective mesoderm and head endoderm area—that is, the dorsal lip—the lowest frequency (20%) from the prospective endoderm of the blastocoel floor, and intermediate frequency (50–60%) from the prospective ectoderm of the blastocoel roof. It would be interesting to have similar data from a slightly earlier stage in view of the fact that *X. laevis* engages in mesoderm migration prior to the appearance of the blastopore. But the data already show that the prospective mesoderm comprises an actively motile population of cells by the late blastula stage.

The mechanisms by which cells modulate their surface membrane activity are not well understood but presumably involve interactions of membrane proteins with the contractile and cytoskeletal elements of the cell. These have not been characterized by biochemical or immunofluorescent methods in amphibian cells gaining motility at the midblastula transition. Johnson and Smith (1976) have examined the "capping" reaction of *R. pipiens* cells labeled with fluorescent concanavalin A and report that early gastrula cells develop single or multiple discrete "caps," whereas blastula cells do not. The latter merely retained the fluorescent protein uniformly over the surface. This difference indicates an increased lateral mobility of receptors for concanavalin A in the membrane or the apperance of a new class of mobile receptors, in cells of the late blastula stage.

As for the locomotor activity of blastomeres *in vivo* during the mid- and late blastula stages, there have been several recent analyses by scanning electron microscopy (Schonwolf and Keller, 1978; Monroy *et al.*, 1976) and by light microscopy (Nakatsuji, 1974, 1975, 1976). These demonstrate pseudopodia and filipodia on cells bordering the blastocoel, especially in the area of prospective mesoderm, and have led several authors to suggest the importance of active migration of mesodermal cells across the blastocoel walls and roof as a major motive force in gastrulation. This will be discussed further in the next section. Here it is relevant to note that in many amphibian species signs of cell movement can be found *in vivo* well in advance of blastopore lip formation. The foremost case is perhaps *X. laevis* where the prospective mesoderm turns over itself at its lower boundary and begins an upward migration unusually early, with at least half the mesoderm having reversed direction by the time of blastopore appearance (Nieuwkoop and Florschutz, 1950; Keller, 1976; see Fig. 4). Also, in *X. laevis* the dorsal equatorial cells on the egg surface begin a strong expansion 45 min to 1 hr before gastrulation, as recognized in time-lapse films by Keller (1978; see Fig. 9). This expansion is an interesting analogue of the epibolic expansion of the animal hemisphere region, but delayed several hours and centered at the dorsal equator.

The coordination between the external and internal movements of the late blastula is unknown. *X. laevis* is unusual in having its entire mesoderm located internally where cells were never strongly attached by tight junctions and desmosomes. In other species, the pro- spective mesoderm derives at least in part from surface blastomeres and is confined to that cell sheet until after the sheet enters the blastopore, at which time mesodermal cells break loose and migrate in a more fibroblastic manner, as does the *X. laevis* mesoderm.

6.4.3 Changes of Cell Adhesiveness

Changes of cell adhesiveness are first detectable in the late blastula period, although the main effects occur subsequently in gastrulation. Johnson (1970) established a standard set of assays for cell adhesion and then compared the performance of cells from different normal embryonic stages of *R. pipiens* and from embryos of different interspecies matings chosen for their increasing severity of developmental arrest. The assays included: (1) the rate of release of cells from embryos exposed to Ca^{2+}- and Mg^{2+}-free medium containing EDTA, (2) the fraction of dissociated cells attaching to a glass surface in 30 min under standardized conditions of divalent cations, (3) the rate of aggregation when the dissociated cells are returned to a medium containing Ca^{2+} and Mg^{2+} and shaken gently, (4) the ability of a piece of test ectoderm to spread over a standard piece of endoderm from a normal late blastula, and (5) the appearance of cell–cell contacts in intact embryos observed by trans- mission electron microscopy. Johnson found that after the midblastula stage cells of *X. laevis* become progressively better in their attachment to glass until shortly before gastru- lation; 90% of the cells attach in 30 min, compared to less than 10% at the midblastula stage. These figures correspond well to his more recent quantification of motility in *X. laevis* (1976). In *R. pipiens*, fewer cells attach to glass by the time of gastrulation (20%) and the 90% level was reached only well into gastrulation. This difference is perhaps again an expression of the relatively advanced state of development of the internal mesoderm of the *X. laevis* embryo by the time its blastopore appears. Certain of the *Rana* hybrid embryos failed to gain adhesiveness for glass, perhaps indicating a requirment for gene expression for the acquisition of cell adhesiveness. Parallel results were obtained for normal embryos in the other assays; dissociation of embryos in medium containing EDTA becomes progressively slower as gastrulation approaches, and reaggregation of cells becomes more rapid and extensive. These changes again fail to occur in lethal-hybrid embryos, even though cleavage progresses normally.

Johnson (1970, 1976, 1977) examined prospective ectoderm, mesoderm, and endo- derm to look for differences in adhesive properties; there were twofold differences in attach- ment to glass and in aggregation, the endoderm being least active in both respects. It would be interesting to determine whether blastomeres cleaving in dissociated embryos in 67 mM phosphate buffer change adhesive properties just as they gain motility and new phases of the cell cycle (Hara, 1977); that is, whether the changes of the surface adhesiveness also occur autonomously in cells, without need for cell–cell interactions in the embryo. If adhe- siveness is developed autonomously, the only role of cell–cell interactions in the embryo might be an inhibitory one, to delay or prevent autonomous expression, thereby creating a position-dependent pattern.

For a discussion of the question of specific cell–cell affinity as an aspect of gastrulation see Sections 7.2 and 7.3. The present concern will be the appearance of extracellular materials, which might be laid down by cells to modify their own surfaces for adhesion or to serve as substrata on which specific directed cell movement might occur, as suggested by Johnson (1977a,b,d) and by other authors. Tarin (1971a,b, 1973) distinguished intercellular materials of two main kinds by electron microscopy: 30 nm granules and fine fibrils. The former were present by the late blastula stage and the latter built up in the early neurula stage in *X. laevis*. The granules resemble glycogen particles in size; the active extrusion of glycogen into the intercellular space has been proposed (Zotin, 1964; Van Gansen and Schram, 1969; Kalt, 1971b; Ubbels and Hengst, 1978) in connection with blastocoel formation, as discussed previously. Tarin (1973) considers the granules to be ribosomes on the basis of their staining properties and removal by ribonuclease. The role of glycogen or ribosomes in the intercellular space remains unknown. Also, Tarin (1973) identifies the fibrils as hyaluronic acid or a related glycosaminoglycan (GAG) on the basis of staining and sensitivity to hyaluronidase.

A clear indication of the change in surface materials in the late blastula to neurula period comes from the microelectrophoresis studies of cell surface charge by H. Schaeffer *et al.* (1973) and B. Schaeffer *et al.* (1973), using blastomeres dissociated by EDTA treatment from various regions of early embryos of *R. pipiens*. In the late blastula they found two electrophoretic classes of cells: fast-moving, negatively charged, large cells from the vegetal hemisphere and slower-moving (by half), negatively charged, small cells from the animal hemisphere. Prospective mesoderm and ectoderm were indistinguishable at this stage. In the early gastrula, prospective mesoderm became more negatively charged as reflected by increased mobility, and matched the vegetal cells (prospective endoderm) in charge density on the surface. Prospective ectoderm was unchanged, as was endoderm. During gastrulation the mesoderm increased its surface charge substantially. Related, although less complete measurements, were made by MacMurdo-Harris and Zalik (1971) on *X. laevis* blastomeres.

Kosher and Searles (1973) suggested that these negatively charged surface materials include sulfated mucopolysaccharides such as glycosaminoglycans (GAG), because gastrulae of *R. pipiens* incorporate substantial amounts of [^{35}S]sulfate, an exclusive precursor of sulfated cell materials. As shown by light microscopic autoradiography, the incorporated radioactivity became associated with the cell surface, particularly of mesoderm engaged in migrating along the blastocoel roof after invagination through the blastopore. Curiously, the radioactivity also became associated with the yolk platelet periphery, a result confirmed by Robertson (1979) for *X. laevis* with electron microscopic autoradiography. These organelles carry a large amount of membrane derived from the oocyte surface in oogenesis. Overall, the results of Kosher and Searles are consistent with the formation of negatively charged GAG on the surface of mesoderm cells at gastrulation.

Recently, Johnson (1977a–d, 1978) made a thorough study of the intercellular material of *R. pipiens* early gastrulae. First, he documented by microscopy the increase in amount of intercellular material between invaginated cells of the dorsal blastopore lip area, as revealed by its metachromatic staining with Toluidine blue, a characteristic of polyanionic macromolecules such as GAG. These results agree well with those of Tarin (1973) for *X. laevis*. Also, as Johnson showed, the material stains with lanthanum as seen by electron microscopy, again a characteristic of negatively charged polymers (see also Moran and Mouradian, 1975). Cells surrounded by intercellular deposits contain abundant Golgi bodies and vesicles seemingly engaged in discharging materials into the intercellular space. To follow the movements of these materials in the cell, Johnson (1977c,d, 1978) exposed early gastrulae to [^{3}H]galactose and [^{3}H]fucose, two sugars frequently present in mucopolysaccharides. The radioactivity first entered polymers contained in the endoplasmic reticulum near the cell nucleus, then passed to Golgi bodies and secretory vesicles, and finally appeared outside the cell. Late blastulae were only one-fourth as active in incorporating radioactivity from these sugars, although the results did not allow a distinction as to whether only one-fourth as many cells incorporated with full activity or each of the same number of cells incorporated only one-fourth as much. The chemical structure of the mucopolysaccharides has not been worked out fully; according to Johnson (1977c) they are GAG of the mucin or hyaluronic acid type, differing in a secondary way from standard GAG preparations from adults on the basis of their sensitivity to degradative enzymes.

Johnson (1977d, 1978) has also shown that lethal hybrid embryos fail to produce extracellular materials at the late blastula and early gastrula stages, in some cases because the galactose- and fucose-containing polymers do not form in the cells and in other cases because the secretory vesicles fail to deliver their contents to the outside. These observations provide indirect evidence of the importance of RNA synthesis for this new cell function, although it is far from clear at what step gene expression is required. Hybrid embryos produce an unknown but probably variable amount of RNA; therefore it might be preferable to test these productions of new cell materials in the presence of α-amanatin to inhibit RNA synthesis to a reliably low level.

Another demonstration of surface mucopolysaccharides at the late blastula stage has been provided by Fraser and Zalik (1977) using sugar-specific lectins to agglutinate embryonic cells of X. laevis. Strong agglutination was caused by WGA (specific for N-acetylglucosamine groups) and by RCA (specific for galactose groups). After treatment of the cells with neuraminadase to remove sialic acid groups from the surface, RCA became a still more potent agglutinin. These procedures might provide an easy means for selecting those disaggregated cells with new surface materials; the authors did not compare the lectin responses of cells at various stages or positions in the embryo, so the usefulness of the methods for identifying early patterns of cell differences in the intact embryo remains to be established.

Tarin (1973) attempted to evaluate the importance of intercellular materials for neurulation (but not gastrulation) by injecting late gastrulae (stage $11\frac{1}{2}$ to 12) of X. laevis with specific degradative enzymes. Collagenase and testicular hyaluronidase had no effect on neural tube induction and formation; the former enzyme had a slight effect on the thickness of intercellular fine fibrils, and the latter abolished fibrils altogether, and rapidly. Ribonuclease did not have a reliable effect except to cause disaggregation of many of the embryos, a curious observation. Amylase had no effect on development or on intracellular fibrils or granules. Tarin concluded that the neurula stage of development does not depend on intercellular materials and suggested that their accumulation (at least the accumulation of fibrils) is related to early differentiation of notochord cells, which eventually will fill the notochord capsule with collagen and hyaluronatelike materials. Johnson (1977c,d) has reported that R. pipiens "hyaluronate" is not sensitive to the usual enzymes and recommends caution on conclusions about the absence of a role of the intercellular material in gastrulation and neurulation. The question of whether these materials just reflect development or in some way also affect its direction remains open.

In addition to mucopolysaccharides, embryonic cells of X. laevis begin to produce collagen at the midgastrula stage, based on the appearance of macromolecules containing hydroxyproline (Green et al., 1968). It would be informative to locate the productive cells by autoradiography, to identify them as epithelial, presomite, or chordamesoderm cells. The chordamesoderm cells may demonstrate by the midgastrula stage the attributes of mesenchymal, fibroblastic cells of later stages, namely, collagen and GAG production, migration by lamellipodia, and a flattened, extended morphology, and may qualify as the first cells of the embryo to display a differentiated phenotype characteristic of later stages. The notochord formed by the late gastrula stage, before the somites or neural tube (Mookerjee *et al.,* 1953). As yet there has been no biochemical comparison of dorsal and ventral mesoderm at the late blastula and early gastrula stages to see whether there are quantitative or qualitative differences in cell shape, motility, or synthesis of extracellular materials. We have no evaluation of dorsal–ventral polarity of the mesoderm in biochemical or cytological terms at the level of transcription, translation, or enzyme function.

6.4.4 *Readiness for Blastopore Formation*

When bottle cells first appear at the blastopore on the prospective dorsal side, the embryo as a whole has entered a permissive or receptive state allowing gastrulation to occur. This is concluded from experiments of Cooke (1972a) on *X. laevis* and earlier ones of Spemann (1938) on *Triturus,* in which a dorsal blastopore lip from an early gastrula is transplanted to the ventral equator of a late blastula, still 2 hr away from its expected time of gastrulation. The transplant heals in place but does not initiate premature gastrulation; 2 hr later gastrulation begins on the ventral (transplant) side at the same time as the resident lip begins gastrulation on the dorsal side. The cells surrounding the transplant in some way become receptive to the bottle cells, be they host or graft in origin. The initiation of this panembryonic receptive state is not known; it is plausibly related to changes in the adhesive properties of cells throughout the embryo at the late blastula stage, as measured by Johnson (1970), or to the cortical "clock" suggested by Holtfreter to explain the fact that pseudogastrulation in uncleaving unfertilized eggs occurs at the same time as gastrulation in cleaving, fertilized eggs.

6.5 *Regional Quantitative Differences as the Initial Basis for Patterning*

The Dalcq and Pasteels hypothesis (1937) for the patterning of the amphibian fate map at the early gastrula stage invoked regional quantitative differences of two morphogens, "V" related to the yolk distribution in the egg, and "C" related to the distribution of unknown materials in the cortex, the latter with the center of the gray crescent as the point of highest amount. The exact spatial distribution of the morphogens was, of course, unknown, although the positions of highest and lowest morphogen concentrations could be assigned plausibly. The intermediate values were largely chosen to fit the fate map, which they wished to explain. They proposed that embryonic cells could evaluate the local quantity of V and C in terms of $C \times V$ and C/V, and could use this quantitative information to select a path of differentiation. To explain how cells in one area give a uniform response, whereas cells on the other side of a sharp boundary give a different uniform response, Dalcq (1949) introduced the notion of "thresholds," critical concentrations of morphogen required by the cells to select a particular path of differentiation. While the morphogen varied continuously across the embryo, the response of the cells was spatially discontinuous. With certain assumptions about the shape of the cortical and vitelline gradients, and the threshold values of $C \times V$ and C/V used by cells, Dalcq and Pasteels (1937) could approximate the fate map of the urodele gastrula.

The core of their hypothesis is the cell's conversion of quantitative information to a qualitative response, an idea first formulated by C. M. Child (1915) as a solution to the historic paradox of the genesis of patterns—namely, did all patterns have to be preceded by equally complex patterns in an infinite regression? Child proposed the logical extreme that one quantitative gradient would be sufficient for all future patterning. This is a major theoretical contribution on his part. His choice for the material basis of the gradient was subtle and complex—namely, it was a metabolic or physiological gradient involving the cell's permeability, membrane potential, and overall metabolic intensity. During Child's time, there was no way for him to suggest the underlying control circuits stabilizing cells at different levels of overall metabolic activity, nor has that explanation been achieved in the meantime. He hypothesized that cells within a group (or even the cytoplasm of a large single cell) invariably establish a gradient of metabolic levels among themselves in response to any of a variety of external stimuli such as O_2, CO_2, pH, light, or metabolites. This is Child's second important theoretical contribution—namely, that groups of cells are capable of self-organization. According to Child's mechanism, the cells closest to the stimulus respond most strongly and adopt a level of most intense metabolism. These cells transmit (not transport) a stimulus to their neighbors, causing them to raise their metabolic intensity also, but only to a level slightly below that of the first cell. The neighbors would in turn stimulate *their* neighbors to a metabolic level slightly less than theirs, and so on. The stimulus would pass in a wave, like a nerve impulse, but attenuating with distance from the origin. Cells of higher metabolic level would "dominate" their neighbors of lower level, preventing their increase, and thereby stabilize the monotonic order. For Child, these metabolic differences seemed not only to underlie and precede the differentiated pattern but also to comprise in large part the final differentiated character of cells. He also saw genes as responding to the levels, furnishing the cells with different products for their cytoplasm. It is amazing that so many aspects of Child's original hypotheses remain with us today in modified, extended, and more explicit form in the contemporary hypotheses of Wolpert (1971) and Goodwin and Cohen (1969).

Dalcq and Pasteels (1937) did not concern themselves with mechanistic detail in their hypothesis, although Dalcq (1949) later applied the physiological competition hypothesis of Spiegelman (1948) to amphibian development in suggesting that "each presumptive region of the mesoderm strives to get a certain amount of the metabolites" utilizable for morphogenesis, thereby creating sharp boundaries between competing areas. But the main value of the Dalcq–Pasteels hypothesis is its identification of the two egg axes with two materially different gradients, one localized in the cortex.

6.5.1 Experiments with Unilateral Anaerobiosis

Among recent experimentalists, S. Lovtrup and his colleagues (see Lovtrup, 1958; Lovtrup *et al.*, 1978) have given the greatest attention to the question of regional quantitative differences of metabolism that might generate dorsal–ventral differences in the amphibian embryo. Lovtrup's first proposal (1958) was this: the egg has a relatively thin and permeable surface in the area of the gray crescent, across which oxygen readily enters. Cells in this area are allowed a greater oxidative metabolism than their diametric counterparts, and therefore are the first ones to complete preparations for gastrulation, such as blastopore formation and invagination. The lateral and ventral neighbors complete the preparations more slowly and gastrulate later, causing the tips of the blastopore to extend around the embryo. About subsequent steps Lovtrup is less explicit: presumably the dorsal members in some way preempt later dorsal development so that the ventral members never become dorsal no matter how long they persevere at their slower rate (dorsal dominance), or, alternatively, the dorsal members accede to dorsal development by their continued higher metabolism, while ventral cells never qualify. This early model is an admirable attempt on Lovtrup's part to make concrete the role of the gray crescent (oxygen permeability) and to use differences in developmental speed instead of quantitative material gradients as the basis for patterning. Thus, cytoplasmic localizations are minimally invoked. Ideas of this class have relevance particularly to mammalian embryos where cytoplasmic localizations may not exist initially (Tarkowski and Wroblewska, 1967).

Lovtrup reported a dramatic experimental result supporting his general idea (Lovtrup and Pigeon, 1958; Landstrom and Lovtrup, 1975): namely, that the dorsal–ventral axis of the embryo can be reversed by unilateral restriction of the oxygen supply to the gray crescent side of the embryo. The recent experiments with eggs of *X. laevis* will be discussed, since these are more detailed than the early tests with axolotl eggs, which gave the same general result. The egg was inserted into a tight-fitting capillary closed at one end, and was oriented with its prospective dorsal side to the closed end. The prospective dorsal side was identified as comprising at the 4-cell stage the pair of smaller and less-pigmented blastomeres. The liquid reservoir at the closed end was equal to or slightly less than the egg volume itself, whereas the open end contained a large volume of medium in contact with the air. Eggs were introduced at various times after the 4-cell stage and removed at stage 11 when the blastopore was circular, having formed completely. Landstrom and Lovtrup (1975) reported a high frequency of reversal of the axis: 37 out of 40 embryos, the other 3 developing abnormally. The latest stage allowing reversal is the mid-late blastula. After that, enclosure in the tube causes abnormal development and no reversal. The reversal occurs as long as the embryo serves out its period of mesoderm induction in the tube. The authors observed that the reversing embryo initiated the blastopore on the open side and at a time 2 hr delayed from the normal aerated controls. The blastopore lip extended laterally on both sides toward the closed end of the tube. Embryos remaining in the tube through neurulation formed the neural folds on the open side. This is an important observation as evidence that the lip on the open side is really a dorsal lip and not just a ventral lip in its normal location. Since the authors did not mark their embryos with vital dyes, this is the only evidence for true reversal of the axis. The authors did not present cytology of enclosed embryos, so it is not possible to tell the extent to which the anaerobic half was retarded, if at all. But the overall effect fits perfectly the expectation of Lovtrup's model.

As a second experiment, Landstrom and Lovtrup (1975) reported that a ventral half-embryo prepared by puncturing and removing the two dorsal blastomeres at the 4-cell stage, can be induced to form a blastopore and a dorsal–ventral axis if inserted in a closed tube, whereas such half-embryos fail to gastrulate when left unconfined. In this case, the enclosure in a tube does not just *reverse* a preexisting axis but seemingly creates an axis *de novo,* a possibility to be discussed later. If this interpretation is correct for the ventral halves, then the enclosed whole embryo may reverse its axis in two steps: one, suppression of the old dorsal side: and two, the induction of a new dorsal side. These dramatic results demonstrate the ability of *X. laevis* eggs to achieve a new dorsal–ventral polarity at the late blastula stage. Even though the Lovtrup hypothesis of dorsalization based on regional differences in oxidative metabolism may miss the mark, the results will nonetheless require inclusion in a comprehensive explanation of amphibian pattern formation. Alternate explanations will be discussed later. It is interesting to note that Vogt (discussed in Spemann, 1938) attempted similar experiments enclosing urodele eggs in pits in a wax-bottom dish;

even though he observed severely retarded development on the enclosed side, he did not achieve reversal of the dorsal–ventral axis.

In testing the idea of dorsal–ventral metabolic differences, Landstrom and Lovtrup (1976) measured oxygen consumption of cells of the dorsal or ventral side of late blastulae enclosed in tubes. Their methodology is perhaps the most refined yet applied to amphibian embryos, involving a magnetically balanced micro-Cartesian diver. They reported a 20–40% greater oxygen consumption rate by dorsal sides (three embryos studied). Earlier they had measured dorsal and ventral half-embryos (not enclosed in tubes), and found that the dorsal halves were approximately 50% more active, compared to the ventral halves, in terms of O_2 uptake per unit time and unit mass of embryo (Landstrom and Lovtrup, 1974). The validity of normalizing to unit mass has been debated for many years; the ventral "half" seems to be slightly larger and to contain slightly more total yolk and less metabolically active cytoplasm than the dorsal half so that normalization to "active cytoplasm" rather than to total mass abolishes the small dorsal–ventral differences (Boell, 1947). Thus, the nonyolky cytoplasm may be no more active on the dorsal side. Furthermore, it is difficult to conclude whether a demonstrated higher O_2 consumption is a cause or effect of dorsalization.

In a related study, Thoman and Gerhart (1979) attempted to detect dorsal–ventral differences in intermediary metabolism by injecting radioactive intermediates into individual blastomeres at the 4-cell stage of *X. laevis* and analyzing the rate and directions of conversion to other compounds in the period until midblastula. No significant dorsal–ventral difference was found for enzymatic steps of glycolysis, the citric acid cycle, the pentose phosphate shunt, the entry of phosphate into nucleotides and creatine phosphate, or the energy charge. The technique is sufficiently sensitive to detect 10–20% differences and automatically gives a result related to "active cytoplasm" rather than total mass. Overall, the egg during cleavage possessed a complete metabolic network, but one operating at a low level as if unchallenged by macromolecule biosynthesis. Thus, oxygen consumption data and *in vivo* assays of enzymic activities and turnover rates of metabolic pools give at best marginal support to the concept of a higher dorsal metabolism at these early stages. Landstrom *et al.* (1976) suggested that unilateral anaerobiosis of eggs in tubes allows the accumulation of lactate, an inhibitor of gastrulation, giving the aerated prospective ventral side a chance to gastrulate first and thereby become dorsal. In this variation of the model, it is not clear whether the enclosed half of the embryo is expected to divide as rapidly as when aerated but just to confront too much lactate to gastrulate, or whether division should be slower as well. The authors note that lactate could be the normal inhibitor of gastrulation in fully aerated embryos, with the less aerobic ventral side unable to keep the lactate level as low as the dorsal side. Instead of testing lactate directly on embryos either unilaterally in tubes or injected in the blastocoel, the authors prepared disaggregated blastomeres from late blastulae of *X. laevis* and exposed these cells to lactate in a culture medium. The frequency of appearance of elongated cells resembling bottle cells was reduced significantly by lactate. The identify of these cells as bottle cells is not certain, nor is the specificity of the inhibition. Lovtrup *et al.* (1978) comment that lactate suppresses other differentiations, such as that of cells of the blastocoel roof to ciliated epidermis and now regard lactate as an effector of the animal–vegetal axis, although it can have a role in reversing dorsal–ventral polarity. Lactate accumulation is an expected consequence of unilateral anaerobiosis, but we do not know if this is the immediate and sufficient agent for axis reversal. Another candidate is CO_2, which would accumulate in the closed end before oxygen runs out. This metabolic product passes cell membranes well and is capable of breaking electrical coupling between blastula cells of *X. laevis* (Turin and Warner, 1977) and of stimulating the formation of secondary blastopores in *R. pipiens* (Witschi, 1934).

Whatever the mechanism of the reversal may be, it is conceivable that the response has importance in a natural condition—namely, when eggs develop in large clusters. Eggs at the center of clusters readily become anaerobic and deteriorate; perhaps the eggs on the outer layers of the cluster respond to the unilateral anaerobiosis and exposure to toxic products from the center by gastrulating from the opposite side.

Lovtrup *et al.* (1978) beleive that lactate is involved in determining the animal–vegetal axis and consider deoxynucleotides to be the effectors of the dorsal–ventral axis. Their proposal harks back to an old observation that the dorsal blastomeres tend to be smaller than ventral ones (Morgan and Boring, 1903; Rugh, 1951; Nieuwkoop and Faber, 1967). They suggest that the smaller blastomeres have a higher nucleocytoplasmic ratio, assuming the nuclei are the same size dorsally and ventrally and the volume of "active" cytoplasm is different, neither point of which has been demonstrated experimentally. However, in their model, the nucleus means "DNA," which would be the same throughout the embryo. Essentially they suggest that a nucleus of a smaller dorsal blastomere will sooner exhaust its reserves of deoxynucleotides needed for rapid synchronous cleavage, with the consequence of starting earlier on the preparations for gastrulation. The dorsal cells are those gastrulating first because the blastomeres are smaller and reach the end of the rapid cleavage period first. But *why* are the blastomeres smaller on one side? The authors suggest that the astral microtubules of the spindle attach better to the less pigmented cortex of the gray crescent region than to the opposite pigmented cortex and shift the spindle toward the gray crescent side. Their evidence for deoxynucleotides as the materials limiting the rapid synchronous cleavages comprises the finding of an increased mitotic index in eggs injected with high levels of deoxynucleotides, as discussed before. Though pertinent to questions of the cause of the midblastula transition, the results do not in any way reveal whether dorsal blastomeres do in fact reach the midblastula transition first.

Previously Lovtrup (1965) related the smaller size of the dorsal blastomeres to their postulated higher metabolic activity, including rate of division. The recent careful photographic records and cinematography of Signoret and Lefresne (1971, 1973) and Hara (1977) with axolotl and *X. laevis* eggs indicate that the dorsal blastomeres do not divide more rapidly; rather, their divisions are asymmetric toward the dorsal side. Also, these photographic records give no evidence that the midblastula transition is reached earlier by the dorsal blastomeres, although small temporal differences would be hard to detect. Bonanno (1977) measured in *X. laevis* the dorsal–ventral difference in cell size: in midblastulae from eggs of several different frogs, dorsal surface cells occupied 20–30% less area than did their ventral counterparts, whereas the dorsal wall of the blastocoel had the same number of cells as the ventral wall, on the average. Thus, it is possible that the smaller dorsal surface blastomeres reflect slightly more frequent radial (perpendicular to surface) cleavage on the dorsal side. Keller (1978) did not mention dorsal–ventral differences in cell size in his cinematographic analysis of surface cells in *X. laevis* midblastulae, although he did not particularly investigate this point. But even granting that differences in cell size exist, we need an experiment to test whether this difference is important. The division planes of amphibian eggs can be easily rearranged by pressure (Zotin, 1964) and rotation; it should be possible to displace the second cleavage well away from the gray crescent side, giving two small ventral blastomeres, to see whether dorsal–ventral polarity is affected.

On a more general level, the evolving proposals of Lovtrup and his colleagues have retained the two original ideas that an early cortical asymmetry is the source of future dorsal–ventral differences in the egg, and that faster or earlier cells in the preparations for gastrulation accede to the dorsal state. The question of cortical asymmetry in gray crescent formation has been discussed in earlier sections; it is appropriate now to discuss the evidence concerning the question of whether prospective dorsal cells are in fact faster or earlier in their progress toward gastrulation, drawing on experiments other than the unilateral restriction of oxygen. There have been experiments of two types: those in which the egg is exposed to a temperature gradient in order to alter the relative rate of development of different regions of the egg, and those in which the nucleocytoplasmic ratio is varied to affect the onset of the midblastula transition and the invagination of the blastopore, the latter involving the appearance of bottle cells.

6.5.2   Experiments with Temperature Gradients

The classical experiments with temperature gradients have been reviewed by Huxley and DeBeer (1934) and by Spemann (1938), with the conclusion (particularly by Spemann) that the evidence for the importance of differences in developmental rate for dorsal–ventral polarity is equivocal or negative. Gilchrist (1927, 1933) was among the first to apply a lateral gradient, in his case a 6.5° gradient, across the eggs and embryos of *T. torosus* at stages from first cleavage to neurulation. His interest concerned mainly effects on neurulation, but he did describe, in passing, effects on earlier stages. He observed no axis reversal even though the rates of development on the dorsal and ventral sides of the eggs were very different. There was no cytological characterization of the embryos at early stages to estimate the difference in number of cell cycles on the opposite sides. He marked eggs with Nile blue and reported that a ventrally warmed embryo (Gilchrist, 1933) initiated a lip on the ventral side at the same time as on the dorsal side, and concluded that the lip was probably a precocious ventral lip rather than a new dorsal lip, on the basis of the shape of the neural plate formed at neurulation. These important experiments would bear repetition, with emphasis on the early stages.

More recently, Glade *et al.* (1967) did succeed in biasing the dorsal–ventral axis of eggs of *R. pipiens* exposed to a 2° horizontal gradient. Eggs randomly placed against the vertical heat source produced dorsal structures within the warmest quadrant of the egg in 60% of the cases, rather than in the expected 25% for random orientation. Neighboring quadrants had fewer than 25% dorsal cases, while the coolest quadrant gave about 25%. Axes were not really reversed, but were shifted by approximately 60° toward the warmth. Glade *et al.* identified two temperature-critical periods for axis shifting, one comprising the first 2 hr after fertilization (when gray crescent formation is afoot) and a second comprising the 4 hr before gastrulation (when mesoderm induction occurs). This latter period is the same stage when Landstrom and Lovtrup (1975) obtained axis reversal in *X. laevis* by unilateral oxygen restriction. For Glade *et al.* temperature control in the intermediate 16 hr period did not affect axis orientation. Eggs exposed to the gradient in only one of the two critical periods *failed* to change the direction of the axis, explaining why Gilchrist (1933) had missed reorientation in his studies of embryos after the first cleavage. The requirement for two periods is interesting and unexplained, except in representing the two most important periods of patterning. The effect of the gradient on the development of the gray crescent was not examined; the cortex on the cold side might build up contractility more slowly than on the warm side. Perhaps the cortex no longer contracts as a unit and now forms two crescents or at least a broad one. Nor were the internal movements of materials examined cytologically—for example, vitelline wall formation. Overall, however, the results of Glade *et al.* (1967) and Gilchrist (1933), showing that exposure of the egg to a temperature gradient in the later period is not a sufficient condition to reverse the dorsal–ventral axis, provide an indication that Lovtrup's first proposal (1958) about differential rates of progress toward gastrulation is not a complete explanation for dorsalization. From the temperature gradient results, there is no supportive evidence that dorsal blastomeres must develop ahead of ventral blastomeres in order to remain dorsal.

6.5.3 Experiments on Altering the Nucleocytoplasmic Ratio

In the second group of experiments several authors have tested the relation of cell size and nucleocytoplasmic ratio to the time of onset of the midblastula transition and the onset of gastrulation as defined by bottle-cell appearance. Haploid embryos and tadpoles are known to contain smaller cells and smaller nuclei than diploids, although both develop from eggs of the same size. Briggs (1949) examined the development of haploid and diploid eggs from a female of *R. pipiens* producing two sizes of eggs differing 1.8-fold in volume. Surprisingly, gastrulation was initiated at 27 hr after fertilization in all four cases (two sizes of eggs, each haploid or diploid), indicating no identifiable importance of the individual cell volume or number of cells to the exact time of gastrulation. Indirectly, this result suggests that the time of exhaustion of the factors for rapid synchronous cleavage—and therefore the time of onset of the midblastula transition—is not sufficient in itself for scheduling the time of gastrulation.

More recently, Signoret and Lefresne (1973) observed haploid and diploid axolotl eggs. Small eggs of equal volume were chosen for complete photographic records. The diploid eggs began to lengthen the cell cycle at the 10th to 11th cleavage, as discussed previously. The haploid eggs began to lengthen the cell cycle at the 11th to 12th cleavage—that is, approximately one cleavage later than the diploids. Nonetheless, gastrulation occurred at the same time. The authors concluded that the scheduling of gastrulation is a property of the cytoplasm, not the nucleus, whereas the midblastula transition may be a function of the nucleocytoplasmic ratio, as predicted from the exhaustion hypothesis. If RNA synthesis (as an effect of the transition) can start a cleavage later and still gastrulation starts at the normal time, it seems that the preparations for gastrulation can be completed with less than the normal time interval of the late blastula period. Of course, it should be recalled that the additional single rapid cleavage of the haploids consumes just 85 min more of a 13–14 hr mid- to late blastula interval in the axolotl; perhaps the time difference in the haploids and diploids is too small to discover the period required *minimally* to be ready for gastrulation. These preparations may be quite separate from the signal to gastrulate, which itself may relate to the cytoplasmic "clock" of Holtfreter (1943) originally invoked for pseudogastrulation, and perhaps also important for epiboly, for the panembryonic receptivity to grafts of the dorsal lip, and for the surface contraction waves. A clearer separation of the two processes needed for gastrulation (inductive patterning of mesoderm in the animal half, and the signal to invaginate) might be seen with eggs of higher ploidy; Fankhauser (1948) studied developing *Triturus* embryos up to the hexaploid level which are viable with cells roughly six times larger than the haploid. There is no published record of the gastrulation times of eggs of this ploidy series. Nor from Spemann's experiment (1938) of delayed nucleation of one lateral half of a constricted *Triturus* egg, is there a record of the gastrulation times of the twin halves which were many hours out of step in cleavage stages.

6.5.4 Models Based on Quantitative Differences

Since the appealing idea of dorsal–ventral differences in developmental rate has so little experimental support, we must turn to a class of ideas based instead on material differences along the dorsal–ventral axis—namely, on patterns of cytoplasmic localizations with or without cortical differences. The Dalcq–Pasteels hypothesis (1937) of a double gradient was already discussed and criticized in terms of its use of a cortical field centered in the gray crescent. In contrast, the other gradient emanating from the vegetal yolk mass is affirmed in a formal sense by many experiments, including those of Nieuwkoop. In addition, Nieuwkoop (1969a,b) found that the vegetal yolk mass harbors the dorsal–ventral axis at the midblastula stage, a role previously ascribed by Dalcq and Pasteels (1937) to the cortex. After the discovery by Nieuwkoop (1969a,b) of the importance of the vegetal blastomeres in mesoderm induction and in dorsalization at the midblastula stage, Weyer *et al.* (1978) proposed a model for patterning of the embryo as follows: the patterning of the embryo concerns special vegetal cells on the prospective dorsal side, comprising a "dorsalizing" center. The origin of this center is left unexplained but in some way results from the gray crescent at a very early stage (Nieuwkoop, 1969a,b). Weyer *et al.* (1978) consider two models. In the first, a morphogen diffuses out from the center and induces all vegetal cells to become secondary inductive sources equal in strength to the center. At the same time, morphogen diffuses to the animal half and induces mesoderm. Under the authors' conditions of diffusion rate and breakdown rate of the morphogen and the geometry of the center, no stable distribution of morphogen was obtained and the model was considered inapplicable. In the second model, the center was considered the sole source of the morphogen, with the rest of the vegetal half failing to produce the morphogen. With a certain geometry of the center, a stable diffusion pattern of morphogen resulted and the pattern resembled the arrangement of mesoderm on the urodele fate map. It is not surprising that the fate map can be fit when there is so much leeway in choosing the geometry of the source. If anything, the real source is more complex in geometry than Weyer *et al.* (1978) assume, since the experiments of Boterenbrood and Nieuwkoop (1973) show that the lateral and ventral regions of the vegetal half have inducing activity, although weaker than or different from that of the vegetal dorsal region. But significantly, Weyer *et al.* (1978) did obtain a fit equal to that of Dalcq and Pasteels (1937), while using only one gradient rather than two and omitting any cortical contribution. Assuming that the morphogen distribution must be set up within 13–14 hr (the period of the mid- and late blastula stages in the axolotl), Weyer *et al.* (1978) calculate that the morphogen must be smaller than a macromolecule. Mesoderm induction, dorsal and ventral, can be explained by an inductor of one kind diffusing from one source in the vegetal dorsal region of the embryo. In this way the vegetal half could pattern the animal half at the midblastula period.

There is a straightforward way to unify the findings of Nieuwkoop (1969a,b) on vegetal induction of mesoderm with the findings and ideas of Lovtrup on dorsal–ventral rate differences. The vegetal half may be viewed as a cytoplasmic mosaic at the midblastula stage in terms of its cytoplasmic localization of yolk platelets of various kinds and numbers (or whatever associated material makes them uniquely vegetal). There remain two major steps in the translation of this mosaicism to the completed patterning of the animal half and these steps may depend on metabolic rates sensitive to temperature or to energy availability through oxygen-requiring reactions. In the first step of inductive patterning, the vegetal nuclei are presumably stimulated by their surrounding vegetal cytoplasm to synthesize a high level of RNA. As pointed out earlier, the vegetal nuclei differ about twofold dorsally and ventrally in their incorporation of [^{3}H]uridine as soon as RNA synthesis begins at the midblastula stage, consistent with the idea of dorsal–ventral cytoplasmic differences preceding gene expression in the vegetal half. In contrast, the animal half blastomeres synthesize much less RNA at this stage (Bachvarova and Davidson, 1966) and contain much less yolk. If RNA synthesis is important for the production of an inductor from the vegetal half, then temperature gradients and unilateral oxygen availability should affect not only the absolute rate of inductor production but also the relative rates of production from the dorsal and ventral regions of the vegetal half. Presumably, unilateral treatments of the egg could act strongly enough to reverse the quantitative pattern of RNA synthesis and this could be a necessary first condition for axis reversal. A necessary second condition might concern reversing the pattern of response of animal hemisphere blastomeres to the inductor as discussed in the next paragraph. We have no estimate of the temperature dependence of vegetal RNA synthesis at the midblastula stage; therefore we cannot predict what temperature differential would suffice to cause the ventral side to synthesize twice as much RNA as the dorsal side—that is, a reversed pattern of synthesis. If the RNA polymerase reaction has an ordinary dependence of a twofold decrease for each 10° drop, then the 6.5° gradient of Gilchrist would be close to sufficient, whereas the 2° gradient of Glade *et al.* would be perhaps too slight. In the case of unilateral anaerobiosis, we again have no estimate of the sensitivity of RNA synthesis, although autoradiography of [^{3}H]uridine-labeled eggs enclosed in a tube by the Lovtrup procedure would give this information. It is known that gastrulation is very resistant to oxygen deprivation (Brachet, 1950; Harrison, 1965).

As the second step in the inductive patterning of the animal half by the vegetal half, we must consider the temperature- and oxygen-dependent reactions of the animal half as it receives inductor, quantifies the level of reception, and responds with altered cellular activities. We have no inkling of the way blastomeres quantify or respond to the inductor, except that the response may involve RNA synthesis since the autoradiograms of Bachvarova and Davidson (1966) show high [^{3}H]uridine incorporation in the region of prospective mesoderm. Whatever the reactions may be, it is likely that they are completed within the time period between the midblastula transition and the start of gastrulation (perhaps as signaled by the cytoplasmic and cortical "clock" proposed by Holtfreter). It is plausible that in selecting a fate, the animal blastomeres "add up" all the inductors received in this critical period: those receiving the least inductor choosing ectoderm, those receiving a low amount choosing ventral and lateral mesoderm, those receiving more inductor choosing dorsal mesoderm, and those receiving the most inductor choosing archenteron roof and pharyngeal endoderm. "Summation" during a critical period was found by Grigliatti and Suzuki (1973) for *Drosophila* imaginal disc determination. If this is true for the amphibian embryo, then unilateral temperature or oxygen gradients should be applied to the animal half in a direction to shorten the duration of the critical period on the dorsal side and to lengthen it on the ventral side. This direction for reversing the pattern of reception and response is *opposite* that needed for the vegetal half to reverse its pattern of production of inductor. This curious prediction might explain why unilateral temperature gradients in the late blastula period have been so ineffective in reversing polarity: if the dorsal side is cooled on both its animal and vegetal portions, the rate of inductor output by the vegetal region is reduced but at the same time the period of receptivity of the animal region is lengthened, leading to "temperature compensation." The optimal condition for reversal might be to upset on each side the *relative* activities of the animal and vegetal region: on the old dorsal side one should reduce the activity of the vegetal region while increasing the activity of the animal region, and on the old ventral side increase the activity of the vegetal region while reducing the activity of the animal region. Perhaps the only researcher to approach this experimental condition has been Penners (1936) who, in a preliminary report, achieved good reversal of the dorsal–ventral axis by cooling the late blastula of *R. fusca* overall *while heating it locally* with a small heating element placed at the ventral equator where the vegetal yolk meets the animal blastomeres and where ventral mesoderm induction normally occurs.

Thus, Lovtrup may have been on the correct track in thinking about relative developmental rates on the dorsal and ventral sides, but it is the data of Nieuwkoop which suggest that the rates contain inductive and receptive components and which reveal the pattern of these rates in the embryo. Further exploration of axis reversals at the midblastula stage might document the validity of ideas about a purely quantitative pattern of induction—that is, ventral mesoderm is just *less* induced than dorsal mesoderm but not *differently* induced. In a later Section the evidence will be discussed that the ventral mesoderm remains open to promotion to dorsal or lateral mesoderm well into the gastrula period, as indicated by assimilative induction by grafts of the dorsal lip or by artificial inductors.

In summary, the experimental treatments of eggs affecting dorsal–ventral polarity can be divided into two classes, at least in theory: those affecting the pattern of cytoplasmic localizations of the vegetal half, and those affecting the projection of this pattern onto the animal half by induction. In the former class are treatments effective before the midblastula transition—that is, prior to the onset of gene expression. The redistributions of internal contents by gravitation, as in the inversion and rotation experiments of Ancel and Vintemberger, Born, Schultze, and Pasteels fall into this class, as would disruptions of the gray crescent cortical contraction by a temperature gradient (the early critical period of Glade *et al.,* 1967), or by compression. In addition, some treatments in the mid- and late blastula period may belong to this class; for example, vegetal blastomeres are forcibly brought into new contact with animal blastomeres by constricting the embryo with a hair noose. In the second class of effects belong those directed toward the production of inductor by the vegetal blastomeres and the reception of inductor and response to it by the animal blastomeres, as already discussed in this section. Temperature gradients and unilateral anaerobiosis are members of this class; localized injections of specific inhibitors could probably be used in the same way. Thus, one can classify all the results on altered dorsal–ventral polarity in amphibian embryos, and approach their physiological bases.

6.6   Pattern Regulation

The main event of pattern formation at the mid and late blastula stages is clearly the inductive patterning of the animal half of the egg by the vegetal half, as already discussed. In addition, there are unified regional movements properly associated with gastrulation itself even though by definition gastrulation begins with the appearance of the blastopore. Some of the movements are consequences of induction—for example, in *X. laevis* the migration of internal prospective mesoderm over itself and upward along the wall of the blastocoel. Other movements are probably independent of induction and perhaps even of gene expression—for example, the epibolic expansion of the animal hemisphere. Aside from movements, the panembryonic permissiveness of the embryo surface to bottle cell invagination and also some steps of invagination itself, may develop independent of inductions. Thus, gastrulation would represent the intersection of two lines of preparations, one involving gene expression and one not. These have not been well separated experimentally, and their means of coordination have not been identified.

Pattern regulation at the mid- and late blastula stages concerns on the one hand the mechanisms by which the vegetal half generates a unified inductive pattern differing in strength or kind of inductor along the dorsal–ventral axis, and on the other hand the mechanisms by which the animal half unifies its response in producing mesoderm differing in amount or kind along the dorsal–ventral axis. The question is whether the vegetal half, with its mosaic of cytoplasmic localizations, has the burden of generating a quantitatively perfect pattern of inductor to which the animal half will passively submit in being patterned, or whether the animal half already has the means for "self-organization" in keeping with its size and geometry, and needs only a rough regional threshold level of inductor to get started rather than a detailed accurate pattern from the vegetal half. Little is known about self-organization at these stages, whereas ample evidence for it is available by the gastrula stage. The diffusion model of Weyer et al. (1978) assigns the burden for accurate patterning completely to the vegetal half; the animal half does participate in generating the final and stable distribution of morphogen (inductor) since its cells degrade morphogen at a uniform rate, but the animal half does not develop its own pattern in any way. Thus, at the mid- and late blastula stage we are already facing the problems which plagued students of primary induction for many years: whether the inductor is just a trigger for the receptive tissue to initiate use of its own powers of self-organization, perhaps employing its own morphogens, or whether the inductor provides a full pattern of quantitative detail. Wolpert (1969) and Cooke (1975) have recommended a distinction between inductors and morphogens, the latter being the agents of positional information and self-organization within a field of cells and the former being topographic cues to the field of cells to use its own positional information. This distinction is difficult in practice, as will be discussed later in connection with primary induction, where artificial inductors seem to act mostly as triggers, without providing detailed quantitative information.

Pattern regulation at the mid- and late blastula stages has not been directly separated out for study—for example, by healing together vertical halves of midblastulae cut so as to bring two vegetal dorsalizing centers within known angles of each other, and then examining the unified or disparate blastopores that appear several hours later. One of the problems in studying pattern formation at the mid- and late blastula stage is to recognize the pattern of inductive and induced cells; the latter only reveal themselves at gastrulation and neurulation, and the former do so only in induction tests when explanted and used in recombinates. At present, there is no simple cytological identification, although assays could be devised based on RNA synthesis, motility, or production of extracellular materials. Despite the absence of direct studies, there are many experimental indications of pattern regulation at work at the mid- and late blastula stages to be discussed below. Perhaps the ultimate demonstration of regulation is provided by Hara and Nieuwkoop (Nieuwkoop, 1973), who reported that axolotl midblastulae dissociated in calcium-free medium and then reaggregated, form cell clumps capable, at low frequency, of developing a coherent pattern, an embryoid with a main axis of notochord and neural tube. Interestingly, the embryoids rarely contain more than one axis, despite the random mixing of vegetal blastomeres presumably carrying portions of the vegetal dorsalizing center. It is not known whether unification of an axis occurs before, during, or after induction. Probably it occurs at all stages, but the better known and powerful pattern regulating mechanisms of the gastrula and neurula stages tend to overshadow those at work earlier.

One early demonstration of pattern regulation was Spemann's separation of a late blastula or even early gastrula into lateral halves by means of a hair noose, with each half developing to a bilaterally symmetric half-size larva (Spemann, 1902, 1938). Thus, even at these late stages the regulation of lateral pattern was complete. The act of constriction forced the blastocoel roof and floor together along the original bilateral plane of symmetry, giving each half-embryo a new but approximately normal architecture for a late blastula. As will be discussed in the next section lateral pattern regulation at this stage can be understood in terms of the individual events of gastrulation.

At this point, we will devote attention to the frontal constriction of the late blastula or early gastrula. As shown by Spemann (1902, 1922) and then by Ruud and Spemann (1922) and Schmidt (1932), both urodeles and anurans at these stages display the following phenomenon: the dorsal half is regulated normally, developing to a half-size whole larva. In contrast, the ventral half usually produces a "belly piece"—that is, a ciliated ball of cells in some cases able to gastrulate and in some cases not, but eventually containing mesenchyme, blood cells, pronephric tubules, and ciliated epidermis. From the existence of the belly piece we learn first that the ventral half of the egg at the late blastula stage is not in all species autonomously able to form the lateral and ventral portions of the blastopore and thereby to gastrulate; and second, that differentiated cell types of a limited variety can arise without gastrulation, but axial organization is absent (there is no dorsal–ventral or anterior–posterior axis, nor is there a notochord, central nervous system, or gut). From the full pattern regulation of the dorsal half, we learn that there are mechanisms of pattern regulation for the half-egg to apportion its material into half as much mesoderm despite the presence of a full vegetal dorsalizing center. That is, the mechanisms for size invariance of the final pattern have not been precluded at this stage.

This classic result has been challenged by Huang and Dalcq, as summarized by Brice (1958), regarding the difference of the dorsal and ventral halves in their ability to gastrulate. According to Brice (1958), Huang identified the dorsal and ventral sides of late blastulae of *T. alpestris* (the same species used by Spemann) and marked them with Nile blue and neutral red. Then he slowly constricted the embryos with a hair noose during a 1-hr period to obtain dorsal and ventral halves. The constriction collapsed the blastocoel roof onto the floor over a large area and pushed yolky vegetal cells into the blastocoel, establishing many new areas of contact between the hemispheres. A new equatorial zone was created where the hair noose constricted together the blastocoel roof and floor, a site marked by a sizable wound. Without doubt the previous arrangement of interactions between the hemispheres was severely disturbed by the constriction. Huang observed that gastrulation began 6–8 hr early in all constricted embryos and *began at the locus of the constriction and wound.* That is to say, the prospective dorsal pieces did not even gastrulate from their gray crescent side but from the opposite side, at the wound (five neurulae from eight dorsal halves). Surprisingly, the ventral pieces in many cases also initiated gastrulation and also from the position of the constriction (six neurulae from eight halves). In most cases, the neurulae developing from ventral halves were well axiated and at least as complete as their dorsal counterparts. The ventral half had regulated fully. As Brice (1958) and Dollander (1950) pointed out, Spemann (1938) had not been able to identify dorsal and ventral halves and had to postulate them retrospectively from his results. It could be argued that for Huang and Dalcq, gray crescent identification was unreliable in this species of newt and therefore supposed ventral halves might actually receive a substantial portion of dorsal material. However, Dalcq and Dollander (Brice, 1958) claimed 85% accuracy in recognizing the crescent (clear or yellowish in *T. alpestris*) and in predicting the blastopore location in intact eggs. Accepting their accuracy in distinguishing dorsal and ventral halves, Brice (1958) suggested that the ventral half contains the lateral tips of the gray crescent, which might constitute a new cortical dorsalizing center when brought together by the force of constriction. She also drew attention to the wound at the constriction site and to the forced apposition of animal and vegetal blastomeres at that site as additional possible sources of dorsalizing influences, since the dorsal halves themselves gastrulate from the constriction site despite their abundance of gray crescent cortex. This is a compelling point. Since the publication of these experiments, Nieuwkoop's evidence (1969a,b) for the vegetal half as the source of the prospective dorsal–ventral axis at the midblastula stage makes it reasonable to omit altogether Brice's explanation about fused lateral tips of the gray crescent and to emphasize instead her secondary proposals about new animal–vegetal contacts and about wounding as the causes for gastrulation of the dorsal and ventral halves from the constriction site.

As regards animal–vegetal contacts, we previously discussed the original vegetal dorsalizing center of the egg as a region of contact between deep vegetal yolk and the nonyolky cytoplasm of the animal hemisphere, originally separated by the equatorial layer of intermediate-size platelets but apposed by the cortical contraction of gray crescent formation at a time before cellularization of the egg by cleavage. In this context, constriction may just be an artificial means of bringing cytoplasmic regions together forcibly, *after* cellularization. By quite a different procedure, Pasteels may have demonstrated the same point (1941; see discussion by Nieuwkoop, 1969a): he centrifuged midblastulae so that the blastocoel roof was driven into close contact with the blastocoel floor; the embryos developed excessive mesoderm and secondary axes. In Brice's description (1958), the constriction seemed to produce even greater displacements, forcing deep vegetal blastomeres upward into the blastocoel and into contact with animal hemisphere cells of the blastocoel roof. Thus, there should be abundant opportunity for new contacts and new inductions of mesoderm.

As regards wounding as a stimulus for dorsalization, it is well known in the case of primary induction at the later stages that ions, pH, and osmotic shock can by themselves evoke from prospective ectoderm not only the development of neural and pigment cells, but even notochord, muscle and other mesodermal derivatives (Barth and Barth, 1974). Presumably, internal cells in the region of the constriction wound in the half-embryos of Huang and Dalcq (Brice, 1958) are suddenly exposed to pond water conditions rather than blastocoel conditions, and perhaps vegetal cells are "activated" as inductive sources or animal hemisphere cells are sensitized in their response to endogenous inductors. Nieuwkoop (1970) has studied the effect of lithium ion on mesoderm formation in animal–vegetal recombinates from late blastulae, and found that the animal blastomeres are sensitized by lithium to inductors from the vegetal blastomeres as revealed by the larger amounts of dorsal mesoderm than in controls without lithium ion. Sensitization could have two components: the direct inducing action of the ion on the cells but at a subthreshold level, and the slowing down of the animal blastomeres so that they sum up vegetal inductor over a longer period. The later effect coincides with the observation of Osborn and Stanistreet (1977) that lithium ion slows cell divisions to half in intact embryos. If the region of the constriction wound is a region of sensitized cells of the animal hemisphere, then even a moderate output of inductor by the newly apposed vegetal blastomeres might suffice to evoke dorsal mesoderm. Furthermore, wounding and abnormal ion fluxes may activate vegetal cells to release inductor at "dorsal" levels. Unfortunately there is too little information to compare this artificial activation of vegetal cells with the normal one proposed to occur at the 1-cell stage in the region of the vitelline wall.

After these considerations, it is worth asking once again why Landstrom and Lovtrup (1975) were able to reverse the dorsal–ventral axis of eggs inserted in a closed tube at the midblastula stage and were able to induce an axis in a ventral half-embryo by such enclosure. As a speculative answer, it is possible that several effects combined to create a new dorsalizing center in the egg on the aerobic (open) side. First, the egg may have experienced some rearrangements of its animal–vegetal contacts from being squeezed to a shape twice as long as wide in the tube (no cytology was presented on the extent of internal damage). And second, the egg may have been slightly tipped off the horizontal so that its animal blastomeres on the open end were more against the glass wall whereas the vegetal blastomeres were more against the glass wall on the closed end. The importance of this arrangement is as follows: the most anaerobic and most slowly dividing cells of the egg are those pressed against the glass wall of the tube, for they are larger in size than their counterparts on either the open or closed end (Bonanno, 1977). If these conditions apply to the Landstrom–Lovtrup results, then the ventral side of the egg would present a favorable situation for dorsalization: the vegetal blastomeres would produce inductor at the usual ventral rate but the animal blastomeres would be slow due to their contact with the wall of the tube, and would therefore accrue inductor over a longer period, perhaps reaching the "dorsal level" in amount. That is to say, the tube may have created conditions changing the relative rates of development of animal and vegetal blastomeres, not dorsal and ventral sides, in the direction of favoring dorsalization on the open side and disfavoring it on the closed side; this is different from saying that the animal and vegetal blastomeres are equally retarded on the closed side and equally fast on the open side, the essence of the Landstrom–Lovtrup explanation of unilateral anaerobiosis.

Taken as a whole, the results of Huang and Dalcq and of Landstrom and Lovtrup demonstrate that dorsalizing centers can be established in the egg long after gray crescent formation, perhaps until the late blastula or early gastrula stage. Huang and Dalcq (Brice, 1958) in fact obtained an indication of the cutoff time for dorsalization when they repeated their frontal constriction experiments on early gastrulae of *T. alpestris*. In contrast to the pattern regulation of blastulae, the ventral halves of gastrulae usually developed just to belly pieces (9 cases of 11), rather than to half-sized embryos. This is, in fact, the finding of Spemann, made years earlier (1902). Perhaps the discrepancy in results with late blastulae had been a problem of the stage of the egg used for constriction, or Huang and Dalcq may have constricted more vigorously. Also, for Huang and Dalcq the dorsal halves of early gastrulae often initiated a new blastopore at the constriction site, setting up a competition with the original blastopore already started, and giving complex gastrulae. Nonetheless, in four of eight cases the dorsal halves produced normal half-sized embryos, although two of them derived from the constriction blastopore. As the authors pointed out, Spemann could not have known the complexity of his results since he did not mark his eggs with vital stains. The stage comparison of Huang and Dalcq shows that there is a definite decline in the pattern-regulating abilities of the halves, especially the ventral halves, by the early gastrula stage. This coincides with the decline in inductive ability of the vegetal half, as found by Boterenbrood and Nieuwkoop (1973) from recombinates of midblastula animal blastomeres with vegetal blastomeres of different stages. In conclusion, new vegetal dorsalizing centers probably cannot be established after the late blastula stage because the vegetal blastomeres decline in inductive ability.

Whether the new dorsalizing center of the egg was created by constriction or by anaerobiosis, it seems to function in a sufficiently accurate way to allow full and proportionate patterning of the animal half in whole or half-sized eggs. This is surprising if the vegetal half is to have the burden of producing a quantitatively precise pattern for the animal half to receive passively. It seems unlikely that a vegetal dorsalizing center produced by as drastic a means as constriction could generate so normal a pattern of inductor. Rather, pattern regulation in these cases may really illustrate the abilities of the animal blastomeres to organize themselves, once stimulated to do so by the vegetal blastomeres. The extent and timing of this self-organization are unknown. Self-organization is exercised by the dorsal mesoderm by the early gastrula stage; perhaps self-organization starts even in the mid- and late blastula.

J. C. GERHART

The important periods of gastrulation and neurulation require approximately 16 hr in *X. laevis,* twice the interval from fertilization to gastrulation. Whereas the early gastrula retains the basic architecture of the oocyte or unfertilized egg (but containing a blastocoel), the neurula has the basic architecture of the tadpole as far as the arrangement of its body axis. Major morphogenic movements are accomplished in these periods; they are readily observed in living specimens, explaining the long history of observation and analysis of their morphological details and regulative properties.

7.1   *Blastopore Formation*

The blastopore first appears on the prospective dorsal side of the embryo. In species showing a gray crescent, the blastopore appears on the gray crescent side in normal unperturbed eggs. Vital staining of early stages, such as the 32-cell stage in *X. laevis* (Nakamura and Kishiyama, 1971), shows that the blastopore derives from cells containing the surface area of the original equator of the egg, approximately at the original interface of the animal and vegetal hemispheres. By the time of gastrulation, the equatorial region has been carried approximately one-third to halfway toward the vegetal pole by the epibolic expansion of the animal hemisphere (Keller, 1975). Penners and Schleip (1928*a,b*) found that secondary blastopores in gastrulae from eggs inverted at the 1- or 2-cell stage always arise on the egg surface at the interface of yolky and nonyolky cells. This interface is readily identified when vegetal yolk falls onto the pigmented cortex of the animal hemisphere because pigment disperses from the contact region, leaving a light-colored patch. The blastopore arises at the edge of such a patch. This localization of the blastopore can perhaps be explained by the results of Nieuwkoop and his colleagues (1973, 1977) showing the ability of prospective endoderm (yolky cells) to induce head endoderm, archenteron roof endoderm, and dorsal mesoderm in cells of the animal hemisphere. The upper lip of dorsal blastopore may itself be an induction product of the animal–vegetal interaction; this has not yet been examined in gastrulating recombinates of vegetal pieces and animal caps taken from the midblastula stage, where one partner had been labeled with [³H]thymidine (compare related experiments of Nieuwkoop and Ubbels, 1972). At present, supportive evidence comes from light microscopy of normal embryos, showing that the upper dorsal lip comprises small cells with intermediate and small-sized yolk platelets, whereas the lower lip comprises large cells with large closely packed platelets (Perry and Waddington, 1966; Baker, 1965; Cooke, 1972*a*). This configuration is consistent with the possibility that the upper cells of the blastopore are cells from the animal hemisphere induced by the vegetal hemisphere.

Even before the blastopore forms, there are special surface changes at the prospective site. As observed by Keller (1978) in time-lapse films, the epibolic expansion of the animal hemisphere in the mid- and late blastula stages is compensated by a contraction of the vegetal surface, without a decrease in cell number, as shown in Fig. 9. Then, about 1 hr before gastrulation, cells of the dorsal equatorial surface begin an expansion of their own, and the cells of the future blastopore site begin a strong contraction, still without a decrease in cell number. The contraction is unmistakeable since it leads to the formation of a short pigment line composed of the clustered tips of 10- to 20-fold contracted cells, just at the margin of large vegetal blastomeres and smaller animal hemisphere blastomeres. For some reason, the pigment granules (which were none too plentiful in this region) remain at the outer surface, an interesting demonstration of the integrity of the original egg cortex, and become more concentrated as the cells contract on the egg surface. Internally the same cells elongate greatly in a bottle- or flask-shaped extension from the embryo surface, from which the name "bottle cells" or "flask cells" derives. The unusual morphology of these cells has been examined in electron micrographs (Perry and Waddington, 1966; Baker, 1965). They have long microtubules in their necks, numerous vacuoles in the perinuclear region, and long cytoplasmic processes extending to the front. Lovtrup *et al.* (1978) have regarded bottle cells as the first differentiated cells of the embryo, although their differentiation has not been characterized in terms of new proteins or RNA species. The vegetal (lower) lip of the blastopore contains much less extended bottle cells. Thus, the upper cells seem more active in elongation and invagination.

As gastrulation proceeds, these bottle cells continue to originate at the lateral extremities of the extending blastopore, until eventually the blastopore extends entirely to the ventral side where the tips meet. In *X. laevis,* it takes about 2 hr for the blastopore to complete itself. The process of blastopore initiation on the dorsal side, and the process of blastopore extension around the border of the yolk mass are presumably closely related. As for initiation, Stableford (1948) showed that vegetal halves of *R. pipiens* midblastulae, from which animal blastomeres had been removed by suction, could form a blastopore lip but could not extend it. The cytological nature of the blastopore was not examined in sections, to see whether the upper and lower lips were normal in comprising small and large blastomeres, respectively. Recently, in a more complete study, Doucet-de-Bruine (1973) prepared vegetal halves of various pregastrulation stages of axolotl embryos, from which prospective mesoderm was carefully removed. By the midblastula stages (stage $7\frac{2}{3}$), the isolated vegetal half could later form a narrow blastopore, without extension, in 30% of the cases. Shortly thereafter (stage 8), 100% of the vegetal isolates formed a slightly extending blastopore. Only much later, by stage 9, would an isolate form an entire blastopore reaching to the ventral side. Then, in a further analysis, Doucet-de-Bruine (1973) separated the dorsal and ventral portions of the vegetal half, to find when the ventral portion became autonomous of the dorsal portion as far as its ability to form a blastopore. Only at stage $10\frac{1}{2}$, when the blastopore lip had just about reached the ventral side, was the ventral vegetal isolate autonomously able to form a blastopore, in which case it performed a limited gastrulation. These results suggest several steps in blastopore formation: (1) at the midblastula stage the dorsal vegetal blastomeres gain the capacity to form a short blastopore at the onset of gastrulation much later; (2) at the late blastula stage the lateral and ventral vegetal blastomeres gain the capacity to extend the blastopore at the onset of gastrulation, provided they retain contact with dorsal vegetal blastomeres, and this capacity is gained only if the vegetal half can interact with the animal half during the late blastula stage; and (3) during gastrulation the lateral and ventral vegetal blastomeres actually form a blastopore only if the already-present blastopore has approached within 45°. Steps 2 and 3 can be interpreted as the lateral and ventral regions first gaining the *competence* to be induced to form a blastopore and then much later receiving the actual *induction* from the existing blastopore. Nieuwkoop (1969a) had suggested that this competence in step 2 is gained from an interaction of the vegetal half with the mesoderm just itself formed by induction from the vegetal half—that is, a reciprocal effect of mesoderm on endoderm. Doucet-de-Bruine (1973) tried adding back fragments of animal hemisphere to the vegetal explants to find this reciprocity and observed some, but not compelling, enhancement in their capacity for lip extension. At present, it is not known what interactions promote the capacity for lip extension. Doucet-de-Bruine has proposed, instead of a reciprocal interaction, that the competence for blastopore formation is induced in the vegetal cells of the egg surface by the deeper inductive vegetal blastomeres, the same ones that induce mesoderm. In this way, mesoderm formation and movement are coordinated with bottle cell formation and invagination. However, if the inductions originate wholly within the vegetal half, the question remains why the vegetal half requires such prolonged contact with the animal half to gain competence.

Step 3 can be explained as an induction of bottle cells by neighboring bottle cells or their immediate precursors. An inductive step is well demonstrated in early gastrula embryos receiving a graft of the blastopore region into the ventral side, as in the classic transplantation experiment of Spemann and Mangold (1924). The graft and host blastopores located opposite each other begin to extend laterally with the graft blastopore inducing bottle cells in the neighboring ventral area several hours ahead of the time when these cells would have been affected by the normal host blastopore advancing from the dorsal side. Thus, the lateral and ventral margins contain cells receptive to bottle cell induction long before the induction actually occurs.

If these inferences about induction are correct, it is difficult to understand the report of Gilchrist (1933) that a ventral lip can appear precociously in embryos (two cases only) that have been heated on the ventral side to accelerate it relative to the dorsal side, since the dorsal lip of the blastopore was more than 90° separated from the precocious ventral side at the time the ventral lip appeared. It would be important to repeat these experiments in greater detail with careful marking and analysis of subsequent development to secure this point about induction. Also, it is important to analyze gastrulation in eggs heavily irradiated with UV shortly after fertilization, for these gastrulae seem to form no dorsal mesoderm (Malacinski *et al.*, 1977; Scharf and Gerhart, 1980). These eggs, paradoxically, gastrulate well, after a delayed initiation of the blastopore which appears abruptly at many sites on its circumference.

7.2   Holtfreter's Integrated View of Gastrulation

In two papers on the mechanics of gastrulation in 1943 and 1944, Holtfreter offered a remarkable synthesis of many disparate observations, many contributed by himself in the prior decade. To begin with, he suggested that the outer surface of the gastrula is covered by a "coat" which binds the individual blastomeres into an integrated cell sheet. The surface coat of the sea urchin egg had just been identified (see Citkowitz, 1971; Kane, 1973) as an exuded, extracellular glue, and Holtfreter considered the observation applicable to amphibian embryos, in which he had documented the tight adherence between blastomeres in isolated fragments of the blastula or gastrula surface. Cells in contact with the surface coat would for the most part be destined to line external or internal surfaces after gastrulation and neurulation—for example, the archenteron lining, the lumen of the neural tube, and the ventral epidermis which gradually spreads to cover the entire surface of the embryo. The expansion of the ventral egg surface at the neurula stage is the last of the series of major surface expansions which had started with epiboly at the mid- and late blastula stages, followed by the dorsal and later the lateral expansions at the gastrula stage. Holtfreter identified the two unique properties of surface blastomeres as their organization into a sheet due to their mutual adhesion through a self-produced surface coat, and their great capacity for spreading. According to Holtfreter, these properties were acquired by any blastomere having its membrane exposed to the external medium of low ionic strength and osmolarity, relatively low pH, and high calcium ion concentration. This interesting proposal for pattern regulation of the inside-outside axis of the egg has not been tested explicitly by modern means, following Holtfreter's report (1943a,b) of such responses by internal cells exposed to the medium by explantation. His overall emphasis on integration of the surface layer still has value even though the present microscopic evidence favors tight junctions and perhaps desmosomes (Sanders and Zalik, 1972) rather than an extracellular coat as the agents of cohesion.

Second, Holtfreter proposed that internal blastomeres are loosely attached to each other due to their exposure to the blastocoel fluid, a medium promoting partial disaggregation by its high pH, high ionic strength, and low calcium ion concentration. These cells are very different from the surface cells in being nonexpansive, and easily invaded by other cells. In Holtfreter's view, the bottle cells occupied an intermediate position between surface and internal cells by elongating and invading internal cells while retaining sufficient connection with the surface cells to pull the sheet inward to form the archenteron cavity.

Third, Holtfreter attributed the movements of gastrulation to two forces, the epibolic expansion of the surface cells and the tugging action of the bottle cells as they sink into the deep endoderm. The prospective mesoderm was not given a role; Holtfreter's views have been challenged on this point. In the 1940s mesoderm was thought to derive from the surface of the gastrula, from the gray crescent region, from which it would invaginate after the bottle cells. As is now known, mesoderm in *X. laevis* derives entirely from an internal position from which movements start well in advance of the appearance of bottle cells (Nieuwkoop and Florschutz, 1950; Keller, 1975, 1976). Lovtrup (1975) considers this arrangement to pertain to other anurans as well and at least in part to urodeles. Recently, several authors have ascribed a central role to the mesoderm in gastrulation movements, in light of the ability of prospective mesoderm cells to migrate actively by lamellipodia (Nakatsuji, 1974, 1975, 1976; Johnson, 1976a,b; Keller and Schonwolf, 1978), as observed by scanning electron microscopy of embryos and by cinematography of disaggregated cells. Recently Kubota and Durston (1978) achieved an important step in opening a gastrula of the axolotl and filming the movement of mesoderm cells along the roof of the blastocoel, their natural substrate. Keller and Schonwolf (1978) have suggested that not only does the mesoderm move itself, but also carries along the archenteron roof cells. Thus, the dependence has been reversed; mesoderm is now considered to be the mover rather than the moved.

And fourth, according to Holtfreter the final positions of cells after gastrulation depend to a large extent on the selective affinities of the individual cells for each other. Holtfreter had reached this conclusion from his observations of cell movements in pairwise combinations of tissues explanted from gastrulae, one of his most important experimental contributions. These experiments will be discussed later. Townes and Holtfreter (1955) concluded from a systematic comparison of combinations that mesoderm in the late gastrula is *prevented* from invading the endoderm because it has still stronger affinity for the overlying ectoderm. Thus, its adhesive qualities cause mesoderm to take up an intermediate position between ectoderm and endoderm.

In summary, Holtfreter viewed the early gastrula as comprising cells of different properties of adhesion and migration, occupying defined initial positions in the total population. Given the properties and positions of the cells, gastrulation moved forward inevitably, coordinated by the expanding sheet of surface cells. On the one hand, this sheet lined the new internal cavity of the archenteron and restrained the bottle cells from entering the endoderm completely and, on the other hand, this sheet covered the gastrula surface externally and restrained the mesoderm beneath it from entering the endoderm completely, thus controlling the final position of cells after gastrulation.

7.3 Pattern Formation by Induction and Self-Organization

Overwhelming evidence for the importance of induction in early amphibian development came in 1924, when Spemann snd Mangold discovered the ability of a transplanted dorsal lip of the blastopore of an early gastrula of *Triturus* to organize new dorsal–ventral and anterior–posterior axes from the ventral tissues of the equatorial site to which it had been grafted in a host embryo also of the early gastrula stage. Using combinations of grafts and hosts of different urodele species, they demonstrated that the prospective ventral cells of the host become true dorsal cells, such as neural tube, notochord, and somites, in accordance with the dorsal cells of the graft. Clearly, the ventral cells were still open to shifts of fate, at least in a dorsal direction, even at the early gastrula stage. These results indicate that the fate map of the entire animal half of the embryo is normally organized by the dorsal lip region that resides on one side of the animal half. This region was termed the "organizer region" since its position controlled the fate map of the animal half. More recent experiments have shown that the region also organizes the anterior–posterior axis of the underlying endoderm, inducing it to form gut structures (Okada, 1957; Tahara and Nakamura, 1961). Thus, the vegetal half eventually becomes subject to the action of the organizer region of the animal half. The locus of the organizer in the animal half of urodeles corresponds to the dorsal mesoderm areas of the normal early gastrula fate map, namely the areas of prospective head mesoderm, notochord, and somites. In some amphibians, these areas derive from none other than the gray crescent surface region of the fertilized egg, a topographical coincidence explaining why early embryologists such as Dalcq and Pasteels and Spemann, thought that the organizer had to gain its powers directly from the crescent cortex or subcortical cytoplasm. Only recently did Nieuwkoop (1969a,b) discover that the organizer was itself organized by a vegetal center outside the gray crescent area.

Since the inductive effect of the organizer region is so early and so important, it has been called the "primary induction." Inductions as such were well known to Spemann from his earlier work with the formation of the lens in late neurula epidermis making contact with the optic cup portion of the brain (reviewed in Spemann, 1938). This was classified as one of many "secondary inductions" by which embryonic regions became ever more locally organized. Of course, Nieuwkoop's results (1969a,b) demonstrated an important induction that occurs even before primary induction so the terms "primary" and "secondary" lose their literal accuracy. Taken together, the results of Spemann and Nieuwkoop indicate a spreading of inductive ability from the vegetal dorsalizing center to the dorsal mesoderm and finally to components of the neural ectoderm, an example of a chain reaction in which the induced tissue become inductive. As discussed below, the mechanism of induction is still not known in terms of the identity of the natural inductors, the means of transfer from cell to cell, and the level of information provided to the recipient cell. There is a coherent body of evidence supporting the idea that there are inductors of at least two kinds working in primary induction; one, a neuralizing agent and the other, a caudalizing or mesodermalizing agent (Saxen and Toivonen, 1961, 1962; Tiedemann, 1976; Toivonen, 1978). Also, there is compelling evidence that these inductors act as triggers to stimulate the competent and receptive cells to use their inherent mechanisms of self-organization—that is, the inducers do not provide detailed and spatially precise information for patterning, but only activate the processes which do generate this information.

7.3.1 Inductions

The clearest external manifestation of primary induction is the formation of the neural plate bounded by the neural folds. During neurulation the folds move together and fuse on the neurula surface while the plate rounds into a tube and sinks into the interior to form the brain and central nervous system. The most posterior regions of the plate also enter into tail formation, including the nervous system, somites and notochord of the tail, coordinated somehow with the vigorous outgrowth of the tail bud. Tail formation differs from head and trunk formation in that pattern formation occurs in a small group of actively dividing cells, as in limb outgrowth, whereas the head and trunk are patterned from preexisting cells which rarely engage in cell division (Bijtel, 1931; Hamilton, 1969; Cooke, 1975b, 1979a,b), and where cell division may not be important for patterning (Cooke, 1973a). Experiments with localized vital stains reveal the fate map of the neural plate—that is, the identification of those regions destined for the different parts of the brain and nerve chord. And surgical and histological studies indicate the correspondence of each fated region with an underlying region of mesoderm. Within a few years of the Spemann–Mangold experiment (1924), the question arose whether each region of mesoderm provides a specific inductor for its overlying target or whether a quantitative gradient of one or two inductors might pattern the entire neural plate. A related question concerned what level of detail is required of the inductor as far as informing the overlying tissue as to its fate. Those embryologists favoring many regionalized inductors also tended to favor the transmission of a high level of information, in the extreme even viruslike particles which might infect the target cells and change their genetic makeup. And on the other end, the proponents of single and double gradients tended to favor a low level of information from the inductor, with the recipient cells providing from within themselves the information to interpret different quantitative levels.

At present, Toivonen (1978) distinguishes three steps in primary induction. First, there is the formation of dorsal mesoderm and pharyngeal endoderm from cell groups of the animal hemisphere, induced in the mid- and late blastula stages by the blastomeres of the vegetal dorsalizing center. This is the induction discovered by Nieuwkoop (1969a,b), as discussed previously in this review. The inductor is regarded as a "vegetalizing" inductor since it directs potential ectoderm tissue into mesoderm and endoderm. Second, there is a neuralizing step during gastrulation as the dorsal prechordal mesoderm and pharyngeal endoderm migrate anteriorly under the roof of the blastocoel and induce the overlying ectoderm to become neural plate tissue. This step is regarded as the effect of a "neuralizing" inductor, different in action and substance from the vegetalizing inductor of the first step. And finally, the third step involves a "caudalization," or "regionalization," or "transformation" of neural plate cells to trunk and tail nervous tissue and to tail mesoderm. Without this induction, neuralized ectoderm would develop only to anterior neural structures such as the brain, but not to spinal cord or tail structures. The caudalizing inductor is thought to originate from the posterior mesoderm of the archenteron roof, that is, the last mesoderm to migrate upward during gastrulation, and to constitute a substance different from the vegetalizing or neuralizing inductors, or perhaps to be the same as the vegetalizing inductor since some mesoderm is produced. If the latter possibility is the case, the same inductor would have different effects in steps one and three presumably because the responding cells have a different competence, on the one hand, in the mid- and late blastula to become dorsal mesoderm and pharyngeal endoderm and, on the other hand, in the late gastrula to become trunk and tail neural tissue and mesoderm.

In the experimental studies of induction, more attention has been given to the chemical nature of the inductors than to the response of the competent cells. Once Bautzmann *et al.* (1932) had shown that partial secondary axes can be induced in the ventral equatorial region of an early gastrula even by implantations of *nonliving* dorsal lip material (fixed by boiling, acid, or ethanol treatment), the search began in many laboratories for the inductive agents. This chapter in the history of embryology is well known (see Saxen and Toivonen, 1962; and Nakamura *et al.*, 1978, for historical reviews). Methylene blue, fatty acids, steroids, iodine, and many other chemicals off the shelf turned out to induce neural tissue, but to induce notochord, somites, and tail less well. The neural tissue often constituted brain regions and nose, ear, or eye vesicles. Holtfreter (1947) introduced the unifying concept of "sublethal cytolysis" to explain the effectiveness of so many treatments as neuralizing inductors—namely, any agent causing ectoderm cells to suffer a shock or slight membrane damage as reflected by depolarization of the membrane potential and by ion leakage inward and outward, could set off the cells' autoneuralizing system. For him, the specificity of the response originated from the cell's competence—that is, its physiological commitment to a limited set of developmental options, from which it would choose one based on external stimuli. The inductor acted as just a stimulant. This view is very close to modern concepts of hormone action: for example, mast cells of the adult are preloaded with histamine-containing vacuoles, epinephrine binds to surface receptors and causes an internal flux of calcium ion which is thought to trigger vacuole discharge, the cell's specific response to hormone. The response by the mast cell can be evoked as well by agents causing internal increases in free calcium ion (for example, ionophores) or by chemicals mimicking calcium ion, such as lanthanum ion (Kanno *et al.*, 1973). These are agents of "sublethal cytolysis." Likewise, in the well-studied contractile response of muscle, the actomyosin machinery is poised as part of the cell's physiological competence, and responsive to transcient increases in internal calcium ion levels. Normally the calcium flux originates from the sarcoplasmic reticulum through the trigger of acetylcholine bound at the motor end-plate, but it can occur artificially due to exposure to lanthanum ion or ionophores, setting off contractility (MacLennan and Holland, 1975). And within the amphibian egg's own history, the oocyte is competent to mature with respect to meiosis and impermeabilization of the plasma membrane; the internal triggers are supposedly a calcium flux and a drop in cyclic nucleotides whereas the normal external effector is progesterone, releasing the internal circuitry (Masui *et al.*, 1977; O'Connor *et al.*, 1977). But lanthanum ion, mercurials, acetylcholine, hypertonic salt, or ionophores can bypass the progesterone receptors and elicit maturation (Tchou-Su, 1950; Kusano *et al.*, 1977; Schorderet–Slatkine *et al.*, 1976, 1977). Also, amphibian egg activation at fertilization is simulated by the sperm substitutes of urethan (in the presence of calcium ion) or ionophores, both membrane permeabilizing agents (Wolf, 1974*a*). Thus, Holtfreter's original idea for permeabilization of the cell membrane as a way to set off the specific response for which the cell is primed, has applicability not just to neural induction but to other developmental stages and to hormone-responsive cells in general. The neuralization response just seems more complicated than that of a muscle cell or mast cell since it involves new gene expression and gross changes of morphology, but the cytoplasmic circuitry used to elicit the cell's prepared response may be identical in all cases. As Holtfreter (1951) very explicitly pointed out, induction is different from self-organization, the pattern-forming potential of a cell group, as discussed later. For Holtfreter, detailed regional patterning is not the responsibility of the inductor but of the responding cells.

Perhaps the most thoroughly defined and analyzed conditions for artificial neural induction are those of Barth and Barth (1974) with ions. From their 20-year exploration of neural induction *in vitro* they arrive at the following protocol: presumptive ectoderm from the blastocoel roof of the early gastrula of *R. pipiens* is disaggregated into small clumps of approximately 100 cells each. The clumps are submitted to three periods of treatment: first, a shock period of 1 to 2 hr in a semitoxic ("sublethal") medium such as 80 mM $CaCl_2$, 40 mM $MgSO_4$, 40 mM LiCl, or 7 mM $ZnSO_4$, the medium brought up to normal osmotic strength (about 200 mOsm) with Barths' balanced salt solution containing calcium ion; second, a recovery period of 4 to 5 hr in a medium containing high sodium ion such as their own medium with 88 mM NaCl; and third, a culture period of 7 to 10 days in a balanced salt solution allowing the cell clumps to differentiate to definitive cell types. Under these conditions, the ectodermal clumps develop prominent nerve cells with axons in 100% of the cases, whereas clumps exposed continuously to balanced salt solution differentiate only ciliated epidermis and mucus-forming cells, the latter of the sort normally located in the sucker or cement gland of the tadpole. Intermediate levels of induction were obtained by reducing the time or intensity of the shock period or of the recovery period. In such cultures, melanocytes and astrocytes arise, in addition to a few nerve cells and a few epidermal cells. Although the Barths emphasized neural induction, their data indicate the production of even mesoderm, such as notochord and muscle in cultures submitted to the most intense shock treatments. Under a well-defined set of ionic conditions they were able to evoke a wide spectrum of differentiative responses from the prospective ectoderm. Unfortunately, there has not been a study of the effect of these treatments on cell permeability except for the observation by the Barths of increased entry of $^{22}Na^+$. It would be informative to relate the path of differentiation to the degree of permeability increase or intracellular calcium ion, as a further refinement of Holtfreter's original proposal and as a way to rank responses on a quantitative scale, which is still lacking.

The Barths considered their protocol explicable in terms of a transient calcium flux into the cells in the shock period, a fixation of the calcium signal—that is, a selection of a developmental option—during the recovery period during which high intracellular sodium ion is required, and finally a supportive culture condition for expression of the developmental choice. They go further in suggesting that regional differences in intracellular ionic conditions are important for inductive responses in the gastrula *in vivo*. For example, they cite the possibility that different blastomeres possess different intracellular sodium ion levels due to their yolk platelet content and their participation in blastocoel formation in which sodium ion is pumped from the cells. They also suggest that calcium ion levels may differ from cell to cell, putting some cells closer to the threshold of induction. These are interesting suggestions relating to the old questions of a cell's competence and threshold for induction, likely in the future to receive a chemical answer in which ions may figure. The Barths finally suggest that the extracellular ionic environment of cells also differs at various regions of the gastrula, so that ions themselves might be the real inductors of the neural plate or at least co-inductors; however, there is at present no evidence for this proposal.

Although the Barths emphasized ions, particularly calcium and sodium ion, parallel studies could be done with pH, since protons are currently being considered for intracellular regulatory circuits (Epel, 1978), and Holtfreter (1948) and Yamada (1950) showed the strong neuralizing effect of acids and bases, such as ammonia. The latter has been used by Picard (1975) for specific induction of mucous-gland cells. And cAMP could also be used in a parallel series since it has been assumed to play a role in the regulatory circuits of many cells and is known to neuralize ectoderm (Wahn *et al.,* 1976). Or NAD could be used (Rosenberg and Caplan, 1974). The problem is not a shortage of candidates for the role of intracellular effector of neuralization or a shortage of experimental effects, but a means to evaluate experimentally their relevance to neuralization. This problem attends the studies of other responses by other cell types where the effects of calcium ion, sodium ion, pH, NAD, and cyclic nucleotides interact in unknown ways.

There has been no equally detailed study of mesoderm induction under well-defined and systematically varied conditions comparable to that which the Barths did for neural induction. They observed mesoderm induction in passing, but did not emphasize it. Masui (1961) showed the effectiveness of lithium ion at high pH for mesoderm induction from prospective ectoderm and Yamada (1950) obtained very impressive inductions of dorsal mesoderm by exposure of prospective ventral mesoderm to ammonia. It remains to be distinguished whether mesoderm inductions belong on the same or a different quantitative scale as neural inductions; is the mesodermal response a "hyperneural" response or is it a response reached through a completely different set of intracellular circuits? The latter possibility is favored by experiments with purified proteinaceous inductors, described below, but there has been no comparative survey of the two responses with easily controlled conditions of pH and ions.

Whereas neural inductions have been more effectively studied with artificial conditions for "sublethal cytolysis," the mesodermal or "vegetalizing" inductions have been best obtained in experimental systems with purified mesodermal inductors from biological sources. The most powerful of these on amphibia is the fish swim bladder extract of Kawakami (1977), the purified fraction from chick embryos of Tiedemann (1976) and his colleagues, and the guinea pig bone marrow preparations of Saxen and Toivonen (1961). For example, the chick embryo agent appears to be a protein of approximately 3×10^4 daltons, active in purified preparations at the level of 0.1 ng per *Triturus* embryo (Geithe *et al.,* 1975). *In vitro,* cell clumps of *Triturus* ectoderm exposed to this material differentiate notochord, muscle, pronephros, intestine, and even presumptive germ cells (Kocher-Becker and Tiedemann, 1971). This result provides important evidence that mesoderm and endoderm induction respond on the same quantitative scale. Since the endodermal cells are favored by high concentrations of inductor, the material can properly be called a "vegetalizing inductor." The quantitative relatedness of mesoderm and endoderm is also a part of Nieuwkoop's view (1977) of the patterning of the animal half of the midblastula by the vegetal blastomeres: those animal blastomeres closest to the vegetal inductive source are driven to endodermal fates whereas those at a greater distance (and lower concentration) reach only mesodermal fates. The *in vitro* studies with various concentrations of vegetalizing inductor support this view (Tiedemann, 1976). There is no concentration of vegetalizing agent that causes neural structures to appear, demonstrating that inductions of the two kinds are nonoverlapping. Similar experiments with crude inductor preparations implanted into early gastrulae had led Chuang (1938), Yamada (1950) and Saxen and Toivonen (1961, 1962) to consider that there had to be inductors of at least two kinds at work in normal primary induction. So far, the amphibian blastula or gastrula has not provided a rich source of vegetalizing inductor (Tiedemann, 1976); active preparations can be obtained by the same procedures as those used for the chick embryo extract agent but the amount of starting material is relatively small and the yields poor. However, the preparations may indicate the presence of a similar material in amphibia. Since vegetalizing inductors can be purified from a variety of nonembryonic sources (for example, bone marrow), the question of specificity always remains even when an embryonic source is used.

The ability of artificial inductors to stimulate the formation of well-organized axes in the whole embryo is impressive. And in fact this is one of the pieces of evidence that led Holtfreter (1951) to emphasize the capacity of the induced tissue to organize itself when inductively stimulated since the inductor itself could not have provided a detailed quantitative pattern. Perhaps the most complete axes have been obtained by Saxen and Toivonen (1961, 1962) using localized inductors, as follows. First, they prepared a good neuralizing inductor (liver extract) in a solid noninductive carrier and implanted the small mass in the blastocoel of an early *Triturus* gastrula where it is passively pushed toward the ventral equator as the blastocoel collapses. The implant comes to rest beneath the ventral ectoderm in which it induces extensive brain and head structures, but not spinal chord and tail. The partial axis later protrudes from the belly of the tadpole. Second, they prepared their best vegetalizing inductor (from bone marrow) in carrier material and repeated the implantation, inducing tail structures which finally protrude from the belly of that tadpole. This preparation never induces head structures. Also, the preparation fails to induce spinal chord, a structure lacking from the neuralizing inductions as well. Third, Saxen and Toivonen implanted the two inductors side by side in the same gastrula, each contained in its solid carrier, and arranged them so that the neuralizing agent would be closer to the animal pole. With this arrangement of two topographically separated inductors, they obtained very complete secondary body axes with head, trunk, and tail. Some trunk structures, such as nerve chord, had not appeared with either inductor separately, but did appear with the pair. These authors proposed a double gradient model in which the precise selection of a commitment was made by a cell on the basis of two quantitative values, one from each inductor. Although this model for primary induction is comprehensive and useful, it appears that the process may be more complex, with neural plate induction as a first step and subdivision of the neural plate into brain areas and nerve chord as a second step, as discussed below. It is unclear in the model to what level of resolution the paired inductors must provide quantitative information. Perhaps the double gradient is roughly accurate in setting off the subdivisions of the neural plate, and then at the next level of refinement of pattern, the mechanisms of self-organization take over. This speculation is an attempt to escape the dilemma of requiring two artificial inductors in solid carriers implanted in the blastocoel to generate a quantitative and spatially exact pattern down to the level of an individual cell. It seems inevitable that at some level of resolution, other pattern-forming processes must take over from the inductors.

In recent years, the natural process of neural induction has been analyzed by Toivonen *et al.* (1976), using nucleopore filters inserted between explanted layers of mesoderm and overlying ectoderm of the *Triturus* gastrula during primary induction. These filters have pores with a highly uniform diameter and with straight smooth walls, in contrast to the tortuous path of the holes in Millipore filters which yielded ambiguous and no longer valid results. Even at the smallest pore size of the nucleopore filters (0.11 μm), good neural induction was achieved, suggesting that the natural neuralizing agent of the *Triturus* gastrula is a relatively small, diffusible macromolecule or small molecule. These pores are so small that no cell extension entered them, as assessed by electron microscopy of vertically sectioned preparations (Toivonen *et al.*, 1975). For the induction of forebrain and hindbrain in 100% of the cases, contact between mesoderm and ectoderm for 10 hr was required (Toivonen and Wartiovaara, 1976). This is our most complete description of neural induction *in situ,* as opposed to neural induction in artificial situations. As the authors further demonstrate, there are subsequent interactions of the mesoderm and neural ectoderm, by which the spinal chord and posterior regions of the neural plate are diverted from the path toward brain differentiation; these are achieved by an inductive system *not* capable of passing the small-pore nucleopore filters. Thus, the authors distinguish a second and quite different inductive effect, which would correspond more to a hypothetical "caudalizing" induction, in which neuralized ectoderm is induced to posterior or tail structures.

In considering now the responding cells of the ectoderm rather than the inductors, it is of interest to examine the time course of responsiveness of the cells to the two inductors. Leikola (1965) tested *Triturus* ectoderm exposed to artificial inductors in the blastocoel or in "sandwiches" or inductor and ectoderm *in vitro* and concluded that the competence for the mesodermal and endodermal transformations is lost by the midgastrula stage, whereas the competence for neural induction continues until the late gastrula stage. This time course is worth remembering in connection with the results on UV-irradiated eggs which gastrulate long after the normal time; if the ectoderm runs autonomously on schedule, its responses to the normal vegetalizing inductors of the late-arriving dorsal mesoderm might be diminished, perhaps accounting in part for the shortened axes.

A second interesting point about the responding ectoderm was raised by Ave *et al.* (1968) who claimed that these cells are a naturally heterogeneous population, comprising at least three subpopulations separable by electrophoresis—possibly the neural, epidermal, and mesodermal precursors intermixed throughout the ectoderm. From the time course of the changing sizes of these populations in the ectoderm, they suggested that primary induction involves the selective regional elimination of two of the three cell populations through cell death brought on by the inducer. This proposal, if true, would alter fundamentally our view of induction as an intracellular selection of a developmental path from a set of options comprising each cell's competence, a competence that would be the same for all ectodermal cells. Ave *et al.* (1968) were proposing that there are subpopulations of cells with different competences. Nieuwkoop (1973) has reviewed these proposals and cites the results of Dasgupta and Kung-Ho (1971), which indicate that the subpopulations were themselves induction products formed from the initially homogeneous ectoderm as it yielded mesoderm during gastrulation and neural plate cells later. Heterogeneity is thus more likely to be a result rather than cause of differentiation. In addition, cell death of the magnitude required to make regions homogeneous has never been detected in the ectoderm. Finally, the proposals of Ave *et al.* (1968) would not readily apply to *X. laevis,* since its mesoderm is topographically separate from the ectoderm.

As a third point, one concerning the early responses of ectodermal cells after exposure to a vegetalizing agent, Noda and Kawakami (1977) examined the cell cycle dynamics of *Triturus* gastrula explants cultured with the bone marrow vegetalizing extract. The explants begin with a mitotic index of 5% which was maintained for 24 hr in control cultures without inductor, but dropped to 0% with inductor after 4 hr. About 30% of the cells appeared to die in the treated explants. At 12 hr, the mitotic index jumped to 12%, and eventually leveled to 5% by 24 hr, equivalent to the controls. It was not established whether the early drop in mitotic index reflects only a selective lethal effect on a class of premitotic cells or a general temporary delay of the cell cycle of all surviving cells. Nor was it made clear whether subsequent increase of mitotic index represents a shorter cell cycle or a slightly synchronized population. The authors mentioned that the 30% cell death may be related to the proposals of Ave *et al.* (1968) for selective cell death in subpopulations of cells in normal ectoderm. In contrast, the death of cells in the treated explant could represent just the level of "lethal cytolysis" accompanying the "sublethal cytolysis" of the rest of the population. While the authors consider the changes in cell cycle to be important for the ectoderm's switching to a vegetalized path of development, we do not know what these changes really are, much less how they would affect or reflect vegetalization. Nonetheless, the study is one of few directed toward the cell biology of the responding tissue rather than toward the inductor. In a cytophotometric study of DNA levels in various tissues of *T. vulgaris* gastrulae, Lohmann (1974) has observed that the mesoderm has an increased G_1 and S period compared to prospective ectoderm. Cooke (1979*a,b*) has made similiar observations in *X. laevis* and considers that prospective notochord and somite cells arrest in G_2 prior to somitogenesis in the early neurula. Thus, an altered cell cycle is an early characteristic of mesoderm.

In summary, it seems likely that primary induction involves the regionalizing action of inductors of two kinds on the ectodermal sheet overlying the mesoderm at the gastrula stage. The neuralizing agent from the mesoderm is probably a macromolecule or small molecule, since it passes the smallest nucleopore filters. Its action can be simulated by artificial agents causing "sublethal cytolysis"—that is, a partial permeabilization of the cell membrane, leading to changes of internal pH and ion composition. The specificity of the response to neuralizing agents is determined by the ectoderm itself, probably according to a quantitative scale of thresholds of inductor required by the cell to enter on a particular line of "neural" development. These neural lines include brain, pigment cells, and epidermis but not mesoderm or spinal chord.

The caudalizing agent of primary induction does not seem to pass nucleopore filters and its mode of action is unknown. It may or may not be similar to the vegetalizing agent active in the induction of dorsal mesoderm in the mid- and late blastula. Very effective proteinaceous vegetalizing agents have been purified from embryonic and nonembryonic sources. These are usually tested for their inductive effect on early gastrula ectoderm, and are defined as vegetalizing inductors when mesoderm and endoderm are formed. When implanted in the blastocoel of early gastrulae, they induce tail structures. They are usually not tested on neuralized ectoderm to see if trunk and tail structures are formed, as would be required in the definition of a caudalizing inductor. It seems clear that the vegetalizing inductors are different from the neuralizing inductors. The cells responding to the vegetalizing inductor may have a quantitative scale of thresholds, the highest level equated with endodermal lines of development and lower levels with mesodermal lines. Neural lines of development are probably not included on this scale. That is, the neural and mesodermal scales are mutually exclusive. The intracellular basis for the two scales is unknown. Some lines of development may require inputs from both scales, especially posterior trunk structures, such as spinal chord. The neural plate may be roughly mapped out according to quantitative levels of the two inductors experienced in various regions of the ectoderm, but it seems unlikely that the inductors provide precise details of positional information beyond the level of segmental compartments (see Garcia-Bellido, 1977). The action of the inductors may be to induce the ectoderm cells to choose a general line of development and to use their powers of "self-organization" locally to achieve spatially precise patterns.

7.3.2 Self-Organization

It was emphasized long ago by C. M. Child (1915) that self-organization is a fundamental property of cell groups, invariably triggered by asymmetric external stimuli and expressed in the form of quantitatively graded metabolic differences within the group. According to Child, the cell group uses an impulse-transmitting mechanism to relay information about the location of the stimulus and a dominance mechanism to establish the stable metabolic differences grading off from a high point nearest the stimulus. When the organized cell group is disrupted, it can, in theory, reestablish its pattern of quantitative differences as a result of the tendency of cells to rise to a higher metabolic level in the absence of already higher and therefore dominant members. Thus, pattern regulation was seen by Child as a fundamental property of cell groups, as a homeostatic aspect of their mechanisms of self-organization. In the intervening years, Weiss, Huxley and DeBeer, and Waddington clarified the notion of a "field" of cells as a group capable of self-organization and pattern regulation (see historical review by Nieuwkoop, 1967a,b). More recently, Wolpert (1969, 1971) proposed in explicit terms a mechanism for the self-organization of a field of cells by "positional information"—namely, first a field of cells is defined by its ability to surround itself by a boundary and to develop a quantitative gradient of "morphogen" within its boundaries such that each cell experiences a local level of morphogen related to its position in the field; and second, all cells of the field start with the same competence (a set of developmental options accessible to them), and each cell accepts one option on the basis of its morphogen level and according to its inherent "code table" relating levels to options. Wolpert delineated the roles of position and genome in the patterning of cell groups. By postulating self-regulating properties for the steps of boundary and gradient formation, Wolpert provided the field with the capacity for pattern regulation. In general, Wolpert does not devote attention to the role of inductors in evoking self-organization in fields of cells or their role in defining a gross orientation for the field of cells vis à vis other fields, especially as might be required in complex three-dimensional arrangements such as the dorsal–ventral or anterior–posterior axes of the neurula. Other models for the patterning of cell groups have been proposed by Cooke and Zeeman (1976), Summerbell et al. (1973), and Goodwin and Cohen (1969), to mention a few; many of these are discussed by Cooke (1975c). Our experimental information lags far behind the models at this time.

Among the early impressive examples of self-organization in amphibian development, Holtfreter (1951) cites: (1) his own finding that the dorsal lip of the blastopore of an early gastrula can be disaggregated and reaggregated before implantation in the blastocoel of a young gastrula of Triturus, and the induced secondary axis will nonetheless contain perfectly organized head and trunk–tail structures; (2) his own finding that the prospective notochord or somite area of a young gastrula when explanted in a culture medium will differentiate into a miniature axial system with bilateral symmetry consisting of a median notochord and spinal nerve chord with muscle cells arranged on each side, and an enlarged brain vesicle at one end covered with epidermis; and (3) the fact that artificial neuralizing inductors in culture with ectoderm fragments or implanted in the blastocoel lead to well-organized brain areas with nasal placodes and eyes. To him, these results clearly indicate that the pattern-organizing ability lay not in the inductor but in the responding tissue. Holtfreter and Hamburger (1955) stress this same point. The inductor was only the "external asymmetric stimulus" in Child's terms.

More recently, the examples have been supplemented by Townes and Holtfreter (1955), who found that the explanted somite region of the young gastrula can organize an axial system even after disaggregation and reaggregation of the cells, and by Nieuwkoop (1963) and Boterenbrood (1962), who disaggregated and reaggregated neurectoderm of the early gastrula of R. pipiens or the axolotl, "neuralized" it by exposure to high pH, and found the cell clumps to organize themselves into brain regions with nasal placodes or eyes. As an analytical step, Nieuwkoop (1963) noted the basic simplicity of the self-organized structures as hollow spheres and tubes, with contact regions against cell sheets or other spheres or tubes. This raises the question of how much topological variety really exists at the level of self-organization, once the focus is on a very localized region of the embryo, and how many different patterns of cell behavior are needed to generate these structures. Nieuwkoop began the documentation of the intermediate steps of pattern formation by the reaggregated cells as they rearranged themselves into structures, and proposed time-lapse cinematography for future recording of these events. In general, though, the analysis of morphogenic mechanisms under experimental conditions for amphibian materials is well behind similar studies of morphogenesis *in vivo,* for example of the neural tube (Burnside and Jacobsen, 1968; Karfunkel, 1971; Schroeder, 1970), or of the salivary gland or pancreas in mammalian systems. This is an interesting situation since for years amphibian embryologists have had the "perfect system" comprising defined fragments of embryonic tissue in culture, a defined purified inductor or other external stimulus, and near-normal differentiation of complex structures; but the problem has so far defied analysis. Researchers of teratocarcinoma differentiation are approaching the same barrier. Holtfreter (1951) concluded that the burden of elucidating the problems of self-organization "rests more on the shoulders of the analytically minded morphologist than on the biochemist."

There is one area of analytical research on self-organization in which advances are considerable—namely, cell sorting and selective adhesion. Holtfreter (1939, 1943, 1944) laid the groundwork by examining the interactions of explants from different regions of the gastrula or neurula. He discovered evidence of selective "tissue affinities" among the following pairs: (1) from the neurula, prospective epidermis always spread over explants of neural tube, chordamesoderm, somite mesoderm, or deep endoderm to form a closed surface sheet; (2) archenteron-lining endoderm would do the same; (3) prospective epidermis and archenteron-lining endoderm would compete for surface area on other tissues but not cover each other; (4) deep endoderm engulfed mesoderm; and (5) bottle cells from the gastrula actively invaded deep endoderm, dragging the surface cell sheet into an invagination. In later work, Townes and Holtfreter (1955) traced tissue affinity to the selective affinity of individual cells by repeating their entire series of pairwise combinations with disaggregated explants mixed together randomly and allowed to reaggregate. By using explants from different amphibian species they could identify the source of each cell. They discovered the ability of the cells to sort out from random mixtures to give well-structured aggregates capable of advanced differentiations. The conclusions matched those for the tissue pairs described above. In addition, they tried threeway random mixes and found, for example, that mesoderm sorts out to a position between endoderm and epidermis, exactly its position in the embryo. Mesoderm seems to remain in the intermediate position in the gastrula because of the affinity of its cells for *both* endoderm and ectoderm, since with endoderm alone, the mesoderm ends up inside. Overall, the final arrangements of randomly aggregated cells after sorting out, strikingly resemble the gross architecture of the embryo with respect to the inside–outside axis. Thus this major aspect of pattern regulation in the gastrula and neurula stages seems governed simply and automatically by the selective affinities of the individual cells. Other aspects of pattern regulation that might be attributed to selective adhesion by the individual cells are discussed below.

Townes and Holtfreter (1955) considered two factors at work in the aggregates: one, an active migration of some of the cells to the periphery of the clump along an unknown gradient, perhaps of oxygen, or CO_2, or lactate; and two, the specific cell–cell adhesion which keeps homogeneous cell groups together. They also envisioned that heterologous cell associations exist and are weaker than homologous associations, thus providing a theoretical step toward an understanding of the hierarchy of cell types in aggregates. But Steinberg (1963, 1964, 1970) realized that the assumption of an unknown gradient is not essential to explaining the final equilibrium cell arrangements of aggregates after cell sorting. The equilibrium arrangements can be explained simply by *relative strengths* of associations of like and unlike cells.

As Steinberg stressed in 1970, his differential adhesion hypothesis contains no assumption about the molecular basis of cell–cell association; it predicts the equilibrium position of cells in the aggregate only from their hierarchy of relative affinities, with no reference to their material basis. One attractive possibility for a hierarchy of cell affinities is contained in Steinberg's original "site frequency model"—namely, that cell types differ in the number but not kind of surface sites for cell–cell association. Thus, if deep endoderm carries the most sites on its surface, mesoderm an intermediate number, and ectoderm the least, the predicted sorting out patterns correspond exactly to those found. Of course, such a gradient of sites on the cell surface parallels the animal–vegetal axis of the early gastrula in which the cell types arise. There is, however, no experimental evidence for the special hypothesis of the site frequency model, despite the abundance of evidence for the more formal and general differential adhesion hypothesis.

In conclusion, the mechanism of selective adhesion provides strong pattern regulation along the inside–outside axis of the gastrula and neurula, one of the few cases in which we can trace pattern regulation to a cellular, and perhaps even molecular, level.

7.3.3 Dynamic Determination

The amphibian embryo clearly engages in massive rearrangements of cells at the gastrula and neurula stages, long before definitive cell types are identifiable on a cytological or histological basis. That is, cell arrangements precede cell types in development. The early embryologists (Spemann, 1938; Holtfreter, 1951) distinguished "dynamic determination" from "material determination," the former concerning the commitment of a cell to a particular shape, motility, and adhesiveness at gastrulation and neurulation, and the latter concerning the cells final commitment to the characteristics of a terminally differentiated cell type well after the tail-bud stage. Were the gastrula and neurula cells to be considered transient but differentiated cell types and, if so, what would be their relation to definitive cell types? Although not explicitly codifying their thoughts into a theory, the embryological morphologists wondered whether a cell's behavior during morphogenesis did in some way enlist it for a later category of differentiation, and whether the context of cell shape somehow determines the content of gene expression.

The distinction of dynamic and material determination remains to be defined in modern terms. As discussed above, most cells of the late blastula acquire locomotory activity autonomously as the result of just passing the midblastula transition; perhaps the organization of the surface blastomeres into a cell sheet in the intact blastula actually inhibits these cells from the full expression of their locomotory potential, whereas the more disaggregated inner cells are permitted this expression. If any cells of the late blastula and early gastrula are to have additional "dynamic determination," they would be the prechordal and chordal mesoderm and the bottle cells, the mesoderm acquiring lamellipodia and fibroblastic movement and the bottle cells acquiring extended necks. Townes and Holtfreter (1955) have suggested that the primary response of embryonic cells to inductors involves transformations of cell shape and motility, long before distinctive patterns of gene expression occur—that is, inductors would first effect changes of dynamic determination. This idea could be more thoroughly analyzed nowadays, with the immunofluorescent methods for visualizing cytoskeletal and contractile fibers in cells, to follow the conversion of flattened prospective ectoderm into elongated fibroblastic mesoderm cells in the presence of a vegetalizing inductor. There is by now a body of observations on changes of cell morphology in cultured cell lines exposed to various chemicals such as butyric acid or cAMP (see, for example, Leder and Leder, 1975), both of which are also neuralizing agents, and the information on the changing architecture of these cells might provide background for the examples of inductions.

The other side of the distinction concerns the supposed tardiness of material determination *vis à vis* dynamic determination; the question arises whether the classical methods for identifying cell types cytologically and histologically give an accurate estimate of the time at which "material determination" sets in. Modern methods of gel electrophoresis and of fluorescent antibody detection of specific cell proteins characteristic of cell types would allow a better estimate of the time course, better even than the methods of bulk isolation or enzymatic assay of proteins in whole embryos as previously used. Ballantine *et al.* (1979) were able to detect newly synthesized muscle actin by 2-D gel electrophoresis and autoradiography at stages well before that at which histologists score somites or striated muscle. By analogy to the extensive work on chick myoblasts and chondroblasts, there are presumably also in amphibian embryos populations of fully committed precursor cells for certain kinds of differentiated cells and yet these precursors would show none of the patterns of differentiated cell proteins of the definitive cell type. The relation of such precursor cells as these to the "dynamically determined" cells of the blastula and gastrula remains to be identified.

As a concrete proposal for a dependence of material determination on dynamic determination, Wahn *et al.* (1976) have suggested that cells must migrate away from inhibitory neighbors before they can express their differentiation. The authors obtained preliminary supportive data from explants of neuralized *Pleurodeles* ectoderm in which cell migration was favored or hindered by the size of the explant; the smaller explants gave a higher frequency of nerve cell differentiation and of migration. Also Ballantine *et al.* (1979) observed a precocious expression of muscle actin in mesodermal cells of the animal half of the *X. laevis* egg separated from the vegetal half at the late blastula stage. Possibly mesoderm migration also accomplishes a separation from inhibitory vegetal cells, an effect simulated by surgery.

The author most intent on perceiving the relatedness and interconversions of cell types in the early amphibian embryo has been Lovtrup (1973; Lovtrup *et al.*, 1978). He extended the ideas of Willmer (1970) for classifying cells on the basis of their mode of locomotion, shape, and adhesiveness, and for using the evolution of cell types as a basis for predicting the ontogency of cell types. According to Lovtrup, the blastula contains essentially amoeboidal cells in the vegetal hemisphere and ciliated cells in the animal hemisphere, with solitary fibroblastic mesoderm cells arising through a transition of the ciliated cells. These morphologic types would generate other types to give a periodic table of related varieties. These speculations on relatedness of embryonic amphibian cells cannot be tested at present the way the speculations of Kauffman (1973) on the relatedness of *Drosophila* imaginal disc cells can be tested by transdetermination frequencies.

7.4 Mechanisms of Regulation

In the transformation of the late blastula to the late neurula, the tissues of the embryo reach the gross arrangement of the tadpole itself (Fig. 2). The dorsal–ventral and anterior–posterior axes are approximately in place. Regions of the egg originally far apart—even diametrically opposite—are brought together. The morphogenic mechanisms of the transformation include: (1) the timed series of expansions of the tightly connected sheet of surface blastomeres, starting with epibolic expansion from the animal pole, then a strong dorsal expansion, then more lateral expansions, and finally in neurulation the expansion of the ventral surface as the neural tube closes (see Fig. 9 for some of these); (2) the invagination of the bottle cells at the leading tip of the archenteron; and (3) the active migration of dorsal mesoderm along the blastocoel wall and roof, perhaps carrying the archenteron roof along with it. The movement of the dorsal mesoderm has particular importance because the dorsal mesoderm serves as the organizer of the gastrula and neurula stages by way of its inductive activity. During neurulation the main rearrangement of cells concerns neural plate and neural tube formation, initiated by inductions from the underlying dorsal mesoderm.

In this dynamic situation it is easy to attribute "strategic value" to the use of inductors in the patterning of the embryo. Many organs of the tadpole contain tissues brought together from distant parts of the egg. If these tissues are patterned as a mosaic while still separated, the morphogenic movements would have to occur with complete accuracy to insure the superposition of the parts of each organ, whereas with the use of inductions *after* morphogenesis the morphogenic movements need accomplish only the approximate apposition of inductive and responsive tissues. In the next step the inductor of one tissue will serve as a topographic cue to define the position of development of the other tissue of the pair. The amphibian gastrula has in a sense taken this strategy to the limit since it has really only one inductive tissue, the dorsal mesoderm, around which the rest of the embryo is patterned. It is a classical finding (Spemann, 1938) that the responsive tissue is receptive to inductor over a much larger area than ever actually used for the induction. Thus it is really the position of only one member, namely, the inductive partner, which defines the position of the multitissued organ. Spemann (1938) has commented on the mechanism of "double assurance" whereby the responsive tissue also has a pattern ahead of time insofar as being more receptive to inductor in certain regions, but still the margin of allowed error in alignment is considerable and well within the accuracy of the movements. Thus, inductors seem to be an important element of three-dimensional pattern formation, to define position and orientation for the more local mechanisms of self-organization which may be only two-dimensional patterning mechanisms. Cooke (1975c) has, in fact, confined the discussion of positional information to cases of patterning in a cell sheet, a two-dimensional arrangement.

In the case of the neural plate and neural tube, we have a very large sheet of cells which seems in the first step to be divided into smaller and different regions on the basis of local quantitative levels of two inductors, one a low-molecular-weight neuralizing agent (passing through the smallest nucleopore filters) from the underlying prechordal and chordal mesoderm, and the other a large vegetalizing agent (not passing filters) possibly from the posterior mesoderm or endoderm. Thus, the inductors may provide positional information according to their quantitative levels, but at a gross level of patterning. Their action would be to signal regions of the neural plate to embark on different kinds of local self-organization by way of more precise patterning mechanisms perhaps based on the kinds of positional information discussed by Wolpert (1971) or Cooke (1975c). If there are, in fact, two levels of patterning, one gross and one refined, in the development of the axially arranged organs of the tadpole, the overlap of the two levels remains to be seen. Even after the regionalization of the neural plate and later the neural tube or of the gut endoderm, there are numerous secondary inductions still operating in the postneurula stages. For example, Spemann (1938) discovered the induction of lens in the epidermis by the underlying optic cup extending from the brain. This induction, like the earlier ones, guarantees the superposition of two parts of an organ derived from spatially separate tissues, in this case the superposition of the image forming and image receiving parts of the eye.

In reviewing the mechanisms of pattern regulation at the gastrula and neurula stages, we can begin with the induction of the neural plate, as just discussed. In a sense, the inductive specialization of the dorsal mesoderm allows a postponement of the patterning of other parts of the embryo until after the morphogenic movements are completed. In fact, the ectoderm and ventral mesoderm remain impressively open to change until well into the gastrula period. Thus rearrangement or deletion of many parts of the gastrula has relatively little effect on the final pattern, unless the dorsal mesoderm is hit; and it too has regulative mechanisms. If the dorsal mesoderm has the burden of patterning the gastrula at least at the gross level, then we must ask about the ability of this one cell group to maintain its particular shape, movement, and integrity. For if the dorsal mesoderm were to split into several independent populations, the neurula would have multiple axes. Earlier, we discussed the movement of the mesoderm between the endoderm and ectoderm as a reflection of the selective affinity of the individual mesoderm cells. If the ectoderm is absent, as in an exogastrula, the mesoderm sinks into the endoderm; its normal position in the gastrula requires both ectoderm and endoderm as substrata. These are actually the two targets of inductive patterning by the dorsal mesoderm. Thus, pattern regulation on the inside–outside axis can be explained by selective adhesion of the dorsal mesoderm cells.

But in addition, there are pronounced anterior–posterior differences in the dorsal mesoderm, the prechordal cells moving as a broadly fanned population and the chordal cells as a narrow band. Clearly, there are cell contact properties allowing the population to move as an integrated whole with regional differences. These properties are not known, but recently Kubota and Durston (1978) made an important advance in recording cinematographically the movement of dorsal mesoderm across the blastocoel roof in an opened gastrula. They report a loose arrangement of moving cells where the leaders go forward only if contacted behind by followers. In addition, Johnson (1976) has noted contact inhibition of movement among mesoderm cells. Thus, there could be a rich variety of inhibitions and stimulations to movement between mesoderm cells themselves and with their ectodermal and endodermal substrata.

In addition to pattern regulation in the movement, shape, and integrity of the dorsal mesoderm cell population, there may be mechanisms regulating the pattern of output of neuralizing and caudalizing inductors. And even if the pattern of inductor output is irregular, there are the self-organizing properties of the ectoderm and endoderm available to promote normal patterning. As discussed earlier, these impressive pattern-regulating mechanisms allow normal head, trunk and tail structures to form in a secondary axis induced by disaggregated and reaggregated cells of the dorsal blastopore lip or by artificial neuralizing and vegetalizing inductors (Holtfreter, 1951) where the quantitative pattern of inductors is probably abnormal.

With this general outline, it is possible to assess some of the many experiments of pattern regulation at the gastrula and neurula stages. These stages are the ones best studied for pattern regulation because of the relative distinctness of their own morphological patterns and because of the attention given to the organizer region.

7.4.1 Lateral and Frontal Constriction Experiments

Spemann (1918, 1922) and Ruud and Spemann (1924) with *Triturus* and Schmidt (1933) with several anuran species, divided early gastrulae into halves with a baby hair or nylon noose. When divided laterally, so that each half received half the blastopore, regulation was usually complete and half-sized symmetrical embryos were obtained. Thus, the left half formed a complete right side and vice versa for the right half. Spemann (1938) noted a slight asymmetry in that the replaced side was somewhat smaller and interestingly the right half often underwent an inversion whereby the heart and convolutions of the gut became biased to the right rather than to the left side of the half-sized embryo.

In their repetition of these classic experiments, Huang and Dalcq (Brice, 1958) report that vital staining of the embryos shows that the course of gastulation in the halves can be more complicated than imagined by Spemann, at least under their conditions: during the constriction the blastocoel is largely displaced and collapsed, with the vegetal yolk mass pushed upwards to the blastocoel roof; the blastopore is pulled over to the constriction wound; and invagination can proceed either toward the animal pole as normal (55% of the cases) or toward the vegetal pole (45% of the cases), perhaps depending upon where the remnant of the blastocoel remained after constriction. In the latter case, the embryo inverts its bilateral symmetry. Spemann (1938) had already noted the great regulative capacity of the gastrula for pattern formation in the bilateral axis; Huang and Dalcq just took the conclusion further by showing that the bilateral axis can be inverted as well. In the cases of inversion, the position of the blastopore and the blastocoel must have been severely disturbed, so that prospective mesoderm could migrate toward the vegetal pole. Brice (1958) mentions a case in which nonpigmented surface blastomeres were incorporated into the neural plate. These results may just exemplify the singular organizing role of the dorsal mesoderm and the great adaptability of the remaining surface blastomeres to neural inductions, even those blastomeres which would otherwise have entered the archenteron lining. And the dorsal mesoderm seems to keep its integrity during movement in the abnormal direction and the surface blastomeres to exercise self-organization when induced.

The frontal constrictions have been discussed earlier. Spemann (1938) had obtained belly pieces from the ventral halves, and Huang and Dalcq (Brice, 1958) concurred on this, for early gastrulae. The absence of regulation in the ventral piece concerns its lack of dorsal axial structures—that is, of notochord, central nervous system, and gut. The ventral pieces are still unable to form the ventral and lateral portions of the blastopore autonomously, an ability gained slightly later in the midgastrula stage in the intact embryo. However, the ventral pieces are able to differentiate certain full-fledged cell types of a ventral sort, such as blood cells, pronephric tubules, some mesenchyme, and ciliated epidermis (Brauns, 1940; Spemann, 1902; Frankhauser, 1930, 1948). Thanks to the nutrient reserves of the individual cells, the ventral piece survives for several weeks.

7.4.2 Transplantation and Explantation Experiments

Spemann and Mangold (1924) discovered in the *Triturus* early gastrula the ability of a small piece of tissue from the dorsal region above the blastopore to promote the formation of a complete new and secondary axis when transplanted to the ventral side of another early gastrula. The new axis contained the cells of the transplant and in addition many cells of the host. These could be distinguished by interspecies transplants. The host cells enter notochord to some extent, somites to a large extent, and neural tube almost exclusively (Holtfreter, 1951). Thus, the transplanted piece organized a new axis from host tissue, which was drawn completely away from its fate as ventral mesoderm and epidermis. The host tissue is said to undergo "assimilative induction" by the graft, to enter a new local fate map, and to form dorsal cell types and structures rather than ventral ones. These experiments demonstrate: first, that in addition to its action on ectoderm and endoderm, the dorsal mesoderm can induce ventral mesoderm to its dorsal level, a pattern-regulating mechanism by which dorsal mesoderm can complete itself from its surroundings; and second, that the ventral mesoderm preserves the ability to achieve dorsal development and has not been irreversibly shunted into a ventral fate, even though by the gastrula stage it is autonomously able to differentiate ventral mesoderm given sufficient time in culture as an explant (Holtfreter and Hamburger, 1955).

Yamada (1950) discovered that prospective ventral mesoderm can easily be promoted to dorsal mesoderm by brief exposure of a gastrula explant to ammonia (pH 11–12, 4 min), in which case abundant well-organized somites and notochord are differentiated. The results illustrate the self-organizing powers of the mesoderm in the absence of a coherent quantitative pattern of inductor. Dorsalization by ammonia further indicates that prospective ventral mesoderm may be in some respect "subdorsal" mesoderm. Perhaps at this stage the dorsal–ventral difference is still quantitative, with the ventral mesoderm just less induced and less inductive than its dorsal counterpart. As discussed above, we have no criterion for comparing cells quantitatively. It is not even certain that dorsal and ventral mesoderm are two different cell populations (each one homogeneous). Possibly, each is a mixture of mesoderm and ectoderm cells, with the dorsal region having a higher frequency of mesoderm cells.

On the question of the irreversible commitment of dorsal mesoderm cells at the gastrula stage, Holtfreter (1951) noted the development of explants of the prospective somite region to miniature axial systems with not only notochord and parallel muscle cells but also neural tube and epidermis, both of these being ectodermal derivatives. In order to account for the origin of ectoderm from mesoderm, we can propose either that mesoderm regulates back to ectoderm in the absence of the ectodermal neighbors it would have in the embryo, or that the prospective somite mesoderm is heterogeneous at the gastrula stage, containing some cells with ectodermal competence and others with mesodermal competence. In the latter alternative, it is plausible that the ectodermal members are normally induced to mesoderm later as gastrulation proceeds, since dorsal mesoderm can even induce ventral cells to join its ranks in transplants. On the other hand, these alternatives can both be subsumed by the hypothesis that determination at the gastrula stage is not carried by individual cells but by groups of cells and that the continuing selection of developmental paths by a cell is guided by interactions of that cell with its dissimilar neighbors, by a complex intercellular circuitry of inhibitions and activations. Such ideas have been formulated by French *et al.* (1977) to explain the different patterns of bristles regenerated or duplicated by a piece of imaginal disc tissue when combined with various other disc regions. In this material, patterning by a region depends strongly on the identity of neighboring cells. Presumably, the explanted prospective somite region of the *Triturus* gastrula shows some of the same effects. Gurdon and Woodland (1970) suggested that in dividing differentiated cells the genome must be reinstructed each cell generation as to its pattern of gene expression, based on regulatory signals from the cytoplasm and from neighboring cells. Perhaps dorsal mesoderm cells of the gastrula require feedback from each other and from ectoderm and endoderm to remain active as the organizer.

In addition to the ability of the dorsal mesoderm to regulate back to ectoderm in explants, it may also regulate to endoderm, a possibility raised by Nieuwkoop (1970) in studies of vegetalization by lithium ion. In the presence of the vegetal blastomeres, the animal half produces endoderm in the presence of lithium ion, but in the absence of the vegetal half it does not. Thus, Nieuwkoop suggested that the vegetal half enhanced the formation of mesoderm which lithium ion could induce to endoderm, a two-step process.

In the course of Spemann's systematic transplantations of small regions of the entire animal hemisphere, he established that the only highly determined area is the prospective dorsal mesoderm comprising the prospective head mesoderm, notochord, and somites. In these experiments (1938) determination is defined as development according to the original fate map despite transplantation to a new position. Prospective lateral and ventral mesoderm, epidermis, neural tube, and gut regions are still open to regulation according to their new position after transplantation. This pattern of regulation is readily explained if these regions in the early gastrula are still naive or indifferent as to their fates despite their ability to give in some cases autonomous fated differentiations when cultured as an explant. That is, they contain no pattern to regulate. Thus the early gastrula may have only one major organizing center, the dorsal mesoderm, after the waning of the vegetal dorsalizing center.

In an interesting analysis of the ability of the gastrula to integrate its pattern when two blastopores are present, Cooke (1972a–c) placed a transplanted dorsal lip at various angles from the host lip and studied the movement of mesoderm and the arrangement of the subsequently induced neural plate. Although the dorsal blastopore lip induced a complete but smaller embryonic axis when implanted 180° away from the resident dorsal lip, the independence of the lips diminished as they were put closer together. Cooke (1971a) found that lips separated by 60° or less in *X. laevis* became unified at some step of gastrulation to give a single body axis with a head slightly broader than normal, the only effect on final pattern. Separation of the lips by more than 150° gave two nonoverlapping axes. Between 60° and 150°, the axis was split to increasing extents at the anterior end while still joined posteriorly. Cooke suggested that the regulation takes place at the step of mesoderm migration, due to the inherent tendency of mesoderm cells to aggregate or separate from one another. As mentioned above, Johnson (1976a) has noted contact inhibition of movement by mesoderm cells on glass surfaces and Kubota and Durston (1978) have recorded contact stimulation of movement *in vivo*. In addition to these possible cellular mechanisms of integration, there is the unifying action inherent in the dynamics of gastrulation itself, in that as the blastopore becomes more circular, the diameter of that circle becomes smaller until the blastopore closes to a point at the end of gastrulation. At the same time the cells entering the blastopore are derived from ever more equatorial origins on the egg surface and must converge ever more strongly as they approach the blastopore. This convergence automatically brings together the separate mesodermal territories from which the two organizers recruit their posterior mesoderm, thereby explaining the fusion of axes posteriorly. Cooke (1972b) tracked the movements of vitally stained cells of the transplanted lip as they entered the migrating mesodermal sheet and noted an aggregate of leading cells and a trail of followers. Analyses with marked cells would be well worth extending since there is so little information about the interactions of migrating mesoderm cells at the gastrula stage, especially when there are two organizers.

In embryos with two blastopores there is indirect evidence for important interactions between mesodermal cells during migration, as follows. When a secondary body axis develops under the direction of a transplanted organizer region, it is much smaller—one-fourth to one-half the size—than the primary axis organized by the host. Cooke (1973b) has found this inferiority obtains whether the secondary organizer is grafted into the embryo just at the start of gastrulation or several hours in advance of gastrulation, even at lowered temperatures chosen to allow more time for hypothetical morphogens to diffuse to the host cells neighboring the graft. Thus, as a first point, the length of time of contact of the graft and host before gastrulation is not important for the final size of the secondary axis, as if organizer influences are not effective at this period. This conclusion perhaps reflects Nieuwkoop's position (1973) that the vegetal dorsalizing center is the "organizing" region of the pregastrula stages whereas the dorsal mesoderm organizer gains its neuralizing and other inductive capacities only during gastrulation. It is possible that the dorsal mesoderm also begins its organization of the surrounding mesoderm only during migration. The smaller size of the secondary axis may simply reflect the fact that the graft contained less dorsal mesoderm in the first place than did the resident dorsal side, especially in *X. laevis* in which the dorsal mesoderm has begun migration before the appearance of the blastopore lip. In comparing Cooke's diagrams (1972*a–c*) of his explants with the fate map of Keller (1975, 1976), it is clear that only a fraction of the dorsal mesoderm was included in the grafts.

The only way that Cooke (1973*b*, 1975) found to increase the size of the secondary axis was to remove the primary blastopore and adjacent mesoderm by surgery at the start of gastrulation. Despite the operation, gastrulation continued from the primary as well as secondary side, indicating the autonomy of the invagination movements of the primary side at the time of gastrulation. At neurulation, however, the primary side was delayed several hours relative to the secondary side and finally, by the tailbud stage, the two axes had the same size. It seems likely that the primary side had less dorsal mesoderm due to surgical removal and was not able to recruit for itself as much neighboring mesoderm as it normally would, whereas the secondary dorsal mesoderm was now relatively effective in the competition for neighboring mesoderm. That is, the two dorsal organizers have been brought to the same effective size by surgery. The exact stage of the competition for lateral mesoderm is not known but would be identifiable from the paths of stained cells. There is a strong convergence of mesoderm not only during the closing of the blastopore but also at the end of gastrulation as lateral and ventral internal mesoderm moves toward the dorsal median line; this could be the time of the competition between organizers (Cooke, 1979,*a,b*). Incidentally, competition may be the developmental event closest to Dalcq's expectation of "dominance" expressed through competition for a limited resource, in this case the lateral mesoderm (Dalcq, 1949).

We know that the direction of migration of the mesoderm is not specified in detail by a polarized structure of the blastocoel roof, the cellular substrate of migration. This was shown by Spemann, Holtfreter (1951), and others, and most recently by Cooke (1973*b*) in surgical experiments with *X. laevis* in which a rectangle of roof was removed between the animal pole and equator and rotated 180° before replacement. Even though the roof cells were now "pointing" the opposite direction, the migration of mesoderm occurred normally. As a different approach, Kubota and Durston (1978) opened the blastocoel of an early gastrula and placed on the inner surface of the roof a small explant of dorsal mesoderm. The mesoderm cells migrated out slightly and stopped; the cell group achieved no directional movement. Thus, directed movement of the mesoderm seems to be a property of the population in its spatial context in the early gastrula. As mentioned above, Kubota and Durston (1978) have made a first step in identifying the contact behavior of individual mesoderm cells which would generate the population behavior shown in mesodermal migration. So far, three contact "rules" are apparent: first, isolated mesoderm cells do not move in a net direction on their substrate of blastocoel roof, but probably just engage in limicola surface activity; second, mesoderm cells contacted on all sides by neighbors do not move; and third, mesoderm cells contacted on just one side by neighbors move away. These three rules generate a coherently migrating population which would advance across the blastocoel roof where there are as yet no mesoderm cells. There are undoubtedly other rules as well. For example, the dorsal mesoderm must be given a leading role whereby lateral mesoderm converges toward it at one stage. Perhaps just as the bottle cells of the endoderm induce endodermal neighbors of the surface to invaginate, the dorsal mesoderm can induce neighboring mesoderm to move in a certain direction. Perhaps these stimuli spread by contact rather than by diffusion. There may be interactions between the endoderm cells of the invaginating archenteron and the migrating mesoderm, as part of the coordination of the external event of blastopore extension with the internal event of migration. As these rules are identified, it should be possible to return to embryos with two organizers and interpret at a cellular level the competition for lateral mesoderm.

In summary, the migration of mesoderm at the gastrula stage includes many possibilities for mechanisms of pattern regulation, insuring the integrity and structure of this special population of cells prior to the critical stage of neural plate induction when the embryo becomes strongly committed to a definitive body axis.

7.4.3 Ablation and Inhibition Experiments

Removal or "ablation" of the blastopore and dorsal organizer has been performed on a variety of amphibian embryos by Goerttler, Spemann, Tondury, Shen, Dalcq (reviewed by Spemann, 1938; Holtfreter, 1951), and most recently by Cooke (1973*b*, 1975) on *X. laevis,* as described above. Gastrulation continues on schedule despite the absence of the lip and most dorsal parts of the organizer, demonstrating the autonomy of the paradorsal regions of the gastrula for epiboly, invagination, dorsal expansion, and mesoderm migration. According to Cooke, the only noticible effect of the operation is the delay by several hours of neurulation, but thereafter axiation is normal and the final tadpole is remarkably normal in its pattern. In some way the late gastrula manages to regulate its dorsal mesoderm to give a normal inductive pattern to the neural plate for neural plate formation. The origin of the cells for the new dorsal mesoderm is unknown, although by the fate map the most likely candidate is the nearby somite mesoderm, itself quite dorsal. There is no information as to the regulative mechanism by which new dorsal mesoderm appears when old dorsal mesoderm is removed. Formally, the phenomenon illustrates the classical concept of "dominance" where one region suppresses its neighbors from achieving its developmental state, but when that region is removed, the neighbors automatically accede to the dominant state and inhibit their neighbors. While it is plausible that the somite mesoderm might achieve the state of chordamesoderm, there must be a limit to the change a cell can make when dorsal parts are removed since ventral halves of gastrulae by and large do not manage to regulate toward dorsal mesoderm formation nor to gastrulate. This is the result at the early gastrula stage whereas ventral halves of late blastulae do seem to regulate, as discovered by Huang and Dalcq (as presented by Brice, 1958).

In a more drastic ablation series, Paterson (1957) reported the effects of removing the animal one-third of the early gastrula of *R. pipiens* or *Pseudacris nigrita.* Gastrulation continued with almost no delay and normal tadpoles resulted. The healing of the blastocoel roof was not described but presumably occurred rapidly, restoring a surface for mesoderm migration. On the other hand, ablation of the vegetal one-third of the gastrula, just above the level of the blastopore lip, stopped gastrulation entirely and the embryoid product eventually resembled a ventral piece with no differentiation of the ectoderm or endoderm. But when Paterson removed the lower third *and* the upper third, gastrulation did occur after a long delay (16–18 hr) and finally a well-organized half-sized tadpole emerged in 40 out of 82 trials. There was no analysis of the steps of repatterning and no vital staining to identify the original dorsal side.

And finally, Cooke (1975*a*) initiated a line of experiments well worth extending— namely, the use of inhibitors to test the importance of cell division for morphogenesis in gastrulation and neurulation, and for regulation of pattern. He exposed very late blastulae to colcemid or mitomycin C, injected into the blastocoel as well as present in the medium, and observed that morphogenesis continued with remarkably little disruption to give tailbud embryos with rather normal morphology except for the poorly defined somites. There was no companion test for molecular differentiation of tissue-specific proteins such as muscle actomyosin to see if postmorphogenic development can occur. As pointed out earlier, morphogenesis normally precedes definitive molecular differentiation by many hours. As documentation of the effectiveness of the inhibitors, Cooke counted cells from disaggregated embryos and reported a reduction of about fivefold, almost to the number present at the time the inhibitor was added. But it is doubtful that the cells were inhibited fully; first, colcemid metaphase figures did not accumulate to high frequencies; second, cell migration continued although motility is not supposed to be a property of blocked mitotic cells (Johnson, 1976b); and third, morphogenesis would seemingly occur without microtubules, a possible though very unexpected result. More likely, cell division was reduced in rate while the cell cycle continued, perhaps leading to endomitosis and polyploidy. The mitotic rate is rather low at these stages anyway (Bragg, 1938; Lohmann, 1974; Cooke, 1979a), and Fankhauser (1948) showed that *Triturus* eggs up to the hexaploid level can gastrulate and neurulate despite their sixfold oversized cells. Thus, Cooke's findings correspond to those of Fankhauser, but with different methods. Both indicate cell size and division frequency are not crucial to morphogenesis. The second experiment of Cooke (1975a, 1979a,b) is more surprising, for he found that a secondary axis was organized around a grafted secondary blastopore in an inhibited embryo—that is, cell division was not crucial to pattern regulation at least at these stages of morphological patterning. Also, he examined transplanted embryos *not* exposed to inhibitors and found no stimulation of cell division in the region of the implanted dorsal lip. These conclusions are important to models of pattern formation based on cell divisions. The experiments would bear repeating under better conditions of inhibition and with other inhibitors, in order to test the importance of various phases of the cell cycle as opposed to just the one step of cytokinesis. Also, it would be informative to test inhibitors of mRNA synthesis at these stages since the cellular preparations made by the late blastula stage may be sufficient to carry morphogenesis through to the tailbud stage. Nakamura *et al.* (1973) did, in fact, expose *X. laevis* and *T. pyrrhogaster* to actinomycin in the very late blastula stage and did obtain normal late neurulae, but it is not certain how much time elapsed before the inhibitor penetrated the cells.

7.4.4 The Roux Experiment

Roux discovered in 1888 an important experimental condition under which the amphibian embryo fails to initiate pattern regulation. When one blastomere of a 2-cell egg of *R. esculenta* is inactivated by a puncture with a hot needle, the other blastomere continues to cleave normally, to gastrulate, and even to neurulate, forming a near-perfect lateral *half* tailbud embryo. Thus, the live blastomere develops according to the same fate it would have had in an unpunctured egg. Yet Roux' successors discovered that a blastomere separated from its partner at the two-cell state develops into a whole embryo of half-size; it does regulate fully. How can this difference in the regulative behavior of a blastomere be understood?

In Roux' drawings of the various stages of his half-embryos, it is clear there was no regulation to the transient pattern of any of the intermediate stages: the living blastomere remained a perfect vertical hemisphere in firm contact with the inactivated neighbor; during cleavages a perfect half-blastocoel was inflated, indicating the presence of tight junctions with the neighbor; at gastrulation the dorsal blastopore initiated at a position next to the boundary with the inactivated neighbor, indicating the first cleavage had divided the egg along the median plane; the invaginating dorsal mesoderm moved upward along the blastocoel roof, led by cells remaining near the old median plane; and after gastrulation one-sided mesoderm induced a half-neural plate. Essentially all steps occurred as if the other half were alive. As a minor point, there was in fact a small amount of regulation in the chordamesoderm and neural tube at later stages, as the notochord began to complete itself into a rod and traces of somite mesoderm appeared on the more median side. By and large, however, the geometry of the living side was not altered by the inactivation of the other lateral half of the egg. And apparently the living cells continued to develop in the geometrical context of a lateral half egg.

In the separated blastomeres, the very act of separation causes immediate pattern regulation—namely, restoring a spherical egg-like shape to the hemispherical blastomere. All subsequent events occur in this new geometric context; a half-sized but symmetric blastocoel inflates; mesoderm moves along the new median plane; and the neural plate is induced symmetrically from ectoderm on both sides of the axial mesoderm. If the restoration of a spherical shape triggers pattern regulation in the separated blastomeres, we can ask which aspect of pattern formation is particularly sensitive to shape. Three phases of pattern formation have been discussed: the formation of the vegetal dorsalizing center by rearrangements of cytoplasmic localizations in the vegetal half during the period before the midblastula stage; the induction of dorsal, lateral, and ventral mesoderm in the animal half by the vegetal half during the late blastula stage; and the induction of the neural plate by the dorsal mesoderm, the organizer region. Thus, it would seem reasonable to associate the shape-responsive patterning mechanism with the vegetal dorsalizing center and the organizer region. But the organizer is not in itself a source of bilateral symmetry for the embryo since Mayer (1935) has shown that transplantation of the left half of a dorsal lip into the ventral side of an early gastrula causes the development of a normal bilateral secondary axis, not a left half-axis. Thus, the organizer does not have an inherent shape or sidedness but "regulates" in relation to something else. As a parallel to Mayer's experiment (1935), some of Roux' half-eggs must have comprised a lateral half containing a left half-organizer, but under these conditions a left half-axis was formed.

Thus, the source of bilateral symmetry for the egg, as opposed to the source of a dorsalizing activity for the egg, probably resides in the geometrical background or "context" in which the organizer works. For the most part, this background concerns the walls and roof of the blastocoel, and therefore concerns blastocoel formation. There are three ways in which the blastocoel serves as context for the organizer: first, the blastocoel walls provide the cells of the animal hemisphere on which the vegetal half acts in mesoderm induction. The blastocoel walls are the region of animal–vegetal contact. The shape of the blastocoel affects the arrangement of the mesoderm. In Roux' half-eggs the living half formed the organizer near the old median axis, but possessed mesoderm only to one side of the organizer, for on the other side the blastocoel had separated animal and vegetal cells, preventing inductive contact. As the dorsal mesoderm migrates inward, it is followed by mesoderm on only one side. In contrast, the separated egg-half becomes symmetrical and forms mesoderm on both sides of the organizer. Second, the blastocoel roof provides the cellular substrate on which the mesoderm migrates. In Roux' cases this substrate was available on just one side of the organizer, again as a result of the shape of the blastocoel. And third, the blastocoel roof provides the material for the neural plate, and this material is available to the axial mesoderm on only one side in the Roux experiment because of the shape of the blastocoel.

The fact that only half of a neural plate forms indicates that self-organization by the neural plate concerns lateral half formations. Nieuwkoop (1963) commented on the formation of half-structures of the brain in his experiments with ectoderm neuralized *in vitro* during disaggregation and reaggregation; for example, a single eye or nasal placode accompanied the appropriate brain vesicle. Thus, the blastocoel provides the context within which the organizer acts. As for the blastocoel, the shape of the egg provides the context within which the mechanisms of blastocoel formation operate. Thus, we are back to the separation of the blastomeres at the two-cell stage as a trigger to pattern regulation by altering the geometry of the egg.

Another demonstration of the importance of egg geometry for pattern regulation is offered by T. H. Morgan's (1895) variation on the Roux experiment (1888), in which Morgan inverted the 2-cell egg as done by Schultze (1894), after killing one blastomere with a hot needle. In about 6% of the cases Morgan obtained rather complete symmetrical axes. No cytological analysis was reported, but we can surmise that the falling yolk in the inverted egg led to not only a new location for the vegetal dorsalizing center but also a new location for the blastocoel. There is a good chance that the blastopore lip no longer appeared on the old median axis of the egg but now at a position where the organizer could be followed by mesoderm from both sides and where the blastocoel roof was available on both sides for subsequent neural plate induction. The organizer was again brought to a region where it would act in a bilateral context, but by different means from the procedure of blastomere separation.

Thus, we see from the comparison of the Roux experiment and separation experiment that whereas small groups of cells and perhaps individual cells can carry a specific commitment to a level of the dorsal–ventral axis, they have little control over bilateral symmetry. The latter property derives from the shape of the egg as a whole as expressed in the shape of the blastocoel. This generalization concerns the gross patterning of the embryo for there is, in fact, a slight lateral asymmetry reflected in the coiling of the gut and the position of the heart. According to Boterenbrood and Nieuwkoop (1973) this asymmetry originates in a lateral difference in the inductive strength of the vegetal yolk mass, a difference about which we know nothing.

8   Oogenesis

The preceding five sections provide abundant evidence that the patterning of the early embryonic stages in amphibia is based largely on the distribution of the vegetal yolk mass and the less yolky animal hemisphere cytoplasm in the egg. This section concerns the origin of the animal–vegetal axis of the egg in oogenesis, and of the dorsal–ventral axis in those species in which the eggs are bilaterally symmetric before fertilization. Blackler (1970) has divided oogenesis into three major phases: (1) determination of the few (10–30) presumptive primordial germ cells, occurring in the embryonic stages; (2) proliferation of these cells to establish several thousand oogonia during the larval stages; and (3) the differentiation of oogonia into full-grown oocytes, occurring in the adult stages. The differentiation of a single oogonium needs several months during which time the cell volume increases greatly (for example, 20,000-fold in *X. laevis*) and the animal–vegetal axis becomes identifiable in terms of the arrangement of yolk platelets, pigment granules, and the germinal vesicle. As described at the beginning, oogenesis is terminated by the steroid-induced meiotic maturation of the full-grown oocyte into an ovulated egg which becomes covered with jelly layers in the oviduct and released into the water for fertilization.

8.1 Determination of the Germ Cells

As proposed by Bounoure in 1934 and more recently by Blackler (1970), the germ cells in anurans, but not in urodeles, derive exclusively from a small number of blastomeres receiving by chance the "germinal plasm" originating near the vegetal pole of the unfertilized egg. At the light microscopic level, this plasm has been identified as comprising yolk-free islands staining intensely with azan under conditions which probably detect a high RNA content, or with pyronine, a strain more specific for RNA (Czolowska, 1969). By electron microscopy, the plasm is identified as containing osmiphilic ring-shaped aggregates called "germinal granules," associated with numerous mitochondria, ribosomes, and glycogen granules (Williams and Smith, 1971; Mahowald and Hennen, 1971). An exact morphological definition of the germinal plasm is difficult to give because the rings change morphology at various stages and change association with cellular organelles. Also, the islands change in size and number through progressive coalescence during development. Shortly after fertilization in *R. pipiens,* each ring-shaped germinal granule is 0.2–0.5 μm in diameter. Tens of rings are found together in one yolk-free island, each a few micrometers in diameter, and a few hundred islands reside near the vegetal pole of the egg. During early cleavages the islands move upward along the vertical cleavage furrows as part of a general cytoplasmic ingression; they coalesce to form fewer, ever-larger islands finally reaching diameters of 40–50 μm each. Due to the coalescence of islands and their distribution to just one daughter cell at division, only 10–20 blastomeres of the *X. laevis* late blastula of 20,000 cells contain germinal plasm (Whitington and Dixon, 1975). These blastomeres are located almost as far inward as the floor of the blastocoel. During gastrulation and neurulation, these blastomeres remain in the posterior ventral endoderm and divide slowly at the rate of the surrounding endoderm cells (Ijiri and Egami, 1975).

The germinal granules of anurans (*R. pipiens* and *X. laevis*) strikingly resemble the polar granules of insect eggs; polar granules are the germ cell determinants located at the posterior end of these eggs (Mahowald, 1974). Such a similarity in remotely related phyla implies a very basic role of these granules in germ cell determination. However, even within the amphibia, the urodeles appear to derive their germ cells by an entirely different mechanism. Urodele eggs lack germinal granules at the vegetal pole, although Williams and Smith (1973) report granulelike aggregates in the equatorial cytoplasm of the axolotl, a urodele. In experiments with axolotl blastulae, Boterenbrood and Nieuwkoop (1973) and later Sutasurya and Nieuwkoop (1974) and Maufroid and Capuron (1977) demonstrated that presumptive germ cells arise from animal hemisphere blastomeres brought into contact with vegetal ventral blastomeres in recombinates, and therefore the authors concluded that germ cells arise by induction in the urodeles. All regions of the animal hemisphere have the capacity to form germ cells, but only the ventral vegetal blastomeres have the inductive potential. Notochord induction by the dorsal vegetal blastomeres appeared to antagonize germ cell formation. Furthermore, Kocher-Becher and Tiedemann (1971) claim to have induced *Triturus* germ cells in animal hemisphere blastomeres exposed to their purified vegetalizing agent from chick embryo extract. In all these cases, the identification of presumptive germ cells rests on their large size, their continued high yolk content, their large-lobed nucleus, and their general purple–brown coloration after a standardized staining procedure. Since urodele presumptive germ cells lack germinal granules, this identifying feature is not available (Ikenishi and Nieuwkoop, 1978). It is not known if these induced germ cells are functional. A definitive proof of their identity would involve transplantation into a living embryo at the germ cell migration stage or before, but this has not been done. Nieuwkoop and Sutasurya (1976) suggest that germ cell determination differs so greatly from anurans to urodeles that the amphibians must have had a polyphyletic origin. On the other hand, it is possible that germinal granules do not really function as germ cell determinants even in anurans and are themselves an effect of some more basic determinative mechanism.

There have been three functional tests of the requirement for germinal granules in germ cell determination in anurans. First, Buehr and Blackler (1970) punctured the vegetal pole of 4-cell eggs of *X. laevis* to release a small drop of vegetal cytoplasm carrying the granules. The wound healed and the eggs developed into tadpoles with few or no germ cells in their gonads. At adulthood the frogs were reportedly sterile. The same experiment had been reported earlier by Nieuwkoop and Suminski (1959), but with negative results.

Second, Smith (1966) repeated with *R. pipiens* Bounoure's experiment of producing sterile adults by irradiating the vegetal pole of eggs with UV light at the 2- and 4-cell stages and, in addition, showed that irradiated eggs could be rescued in 43% of the cases by an injection of vegetal cytoplasm from an unirradiated egg at the 4- to 8-cell stage. The UV sensitivity was high at the 2- and 4-cell stages when the granules pass close to the vegetal pole near the new furrows. In several of these experiments sterility was scored not in the adult but by the absence of germ cells in the germ ridge of newly hatched tadpoles. More recently, Ijiri (1977) has shown that even unfertilized eggs are fully sensitive to apparent UV sterilization whereas by the 32-cell stage the eggs are UV-resistant. The mechanism of the UV sterilization has been discussed by several authors; most recently Züst and Dixon (1975, 1977) have shown in *X. laevis* eggs that after UV irradiation the yolk-free islands coalesce more slowly, and do not move to a more internal position, probably because the irradiation impairs furrow formation and thereby blocks cytoplasmic ingression. They found that eventually there is a formation of blastomeres containing granules, but the location of these cells in the vegetal hemisphere is abnormal. At stage 38, shortly before hatching, when presumptive germ cells normally begin to migrate from the yolky endoderm to the gonad, these germ cells in UV-irradiated eggs are slow to migrate and arrived much later. These tadpoles do in fact produce fertile frogs, indicating the functionality of the germ cells, a result directly contradicting the earlier conclusions (Smith, 1966). It is unclear whether UV irradiation affects the granules directly or just damages the vegetal surface, impairing its ability to form a furrow which would establish cytoplasmic ingression of the granules, transporting them inward and bringing them into contact for coalescence. The prolonged dispersion of the granules may mean that no blastomere receives the critical number necessary to undertake germ cell activities at the normal time and intensity, such as migration to the gonad. Regardless of the transient inactivation mechanism, as Blackler (1970) points out, no blastomere succeeds in taking over the germ cell lineage in place of those containing the germinal granules, and therefore the granules must be essential determinants. However, since the UV-irradiated embryos ultimately do succeed in forming germ lines in many cases, this line of evidence cannot be used to support the conclusion that the germinal granules are essential cytoplasmic determinants.

The third test of the essentiality of the germinal granules is indirect, namely that of transplantation of presumptive primordial germ cells at the late neurula stage, as introduced by Blackler (1970). This approach shows that blastomeres containing granules give rise to germ cells. At the late neurula stage in *X. laevis,* there are 10–20 germ cells located deep in the ventral posterior endoderm, not scheduled to migrate to the gonad until stage 38, shortly before hatching, when the gonad rudiment has formed as a small sac in the coelmomic mesoderm. In the neurula the distance between the germ cells and the future gonad is great enough to allow the germ cells to be surgically removed with their neighboring tissue without removing the prospective gonadal site. Blackler used donor neurulae of *X. laevis,* carrying the *O-nu* allele (the anucleolate mutant) in the heterozygous condition so the graft cells could later be identified by their possession of one nucleolus per cell rather than the normal two nucleoli. The recipients were of normal genotype. After the operation, the graft healed into place, the germ cells migrated, and the hatching tadpole was found to contain *1-nu* germ cells in gonads composed of *2-nu* mesodermal tissue. This result further confirms the derivation of germ cells from the blastomeres with germinal granules, since these were transferred in the graft. However, the result does not prove the essentiality of the granules.

Once it was established that oocyte nucleoli are produced in large numbers in the oocyte by an unusual gene amplification mechanism (Brown and Dawid, 1968), H. Wallace *et al.* (1971) proposed that the germinal granules are none other than nucleolar cores, episomelike bodies preserved at meiotic maturation when the nucleoli are otherwise destroyed, and these bodies are segregated to the vegetal pole area of the egg where they partition to a few blastomeres which are then able to produce again a large number of nucleoli by gene amplification in the new line of oocytes. This novel idea fit the light-microscope cytological data of Czolowska (1969) for *X. laevis* in which she could not find islands of granules before maturation and where the disintegrating germinal vesicle released basophilic material from its vegetal end. However, Williams and Smith (1971) discovered granules in the vegetal area in electron micrographs of nonmatured oocytes, and found that the islands appear in maturing enucleated oocytes even though the nucleoli had been removed. Also, Blackler and Brown (1973) made a more comprehensive test by analyzing the ribosomal RNA made in oocytes of an interesting *X. laevis* frog derived from a *X. laevis borealis* egg into which an *X. laevis laevis* nucleus was transplanted. Thus, the egg carried the chromosomes of one species but the germinal granules of the other. The ribosomal RNA sequences of the two subspecies differ slightly, allowing identification of their origin from the granules (or from other cytoplasmic nucleotide sequences) or from the nucleus. The results showed clearly that the origin of the ribosomal RNA was the nuclear sequences, thus invalidating the proposal of H. Wallace *et al.* (1971); the germinal plasm is not used to provide base sequence information for the amplified nucleolar DNA in oogenesis. The role of the plasm remains unknown.

Perhaps the first distinctive behavior of the presumptive primordial germ cells is their migration to the gonads from the ventral endoderm. At stage 38 in *X. laevis,* the cells begin to move dorsally around the archenteron lumen and by stage 41 they reach the dorsal mesentery tissue suspending the gut within the coelomic cavity. The presumptive germ cells migrate up the mesentery into the paired gonads, then called the germinal ridges, mere parallel thickenings of mesoderm along the roof of the coelom (Whitington and Dixon, 1975). Wylie and Heasman (1976), and Wylie and Roos (1976) have observed the stages of migration, identifying the germ cells by their large size (20–30 μm diameter) compared with their mesodermal neighbors, their multilobed nucleus, their multiple (four to eight) nucleoli, and their large yolk platelets. The germinal plasm changes position within the cell, moving from the cytoplasm to a perinuclear position (Ikenishi and Kotani, 1975), releasing granular components, leaving an irregular stringlike residue at the nuclear outer surface. In light microscopy, the characteristic azan-staining property disappears simultaneously, perhaps indicating the loss of RNA. Thus, the initiation of migration seems to be associated with an internal change in the structure and location of the germinal granules. Kerr and Dixon (1974) have suggested that the initiation of migration may be one specific function of the germinal plasm. The means by which the migration is guided is not well established; chemotaxis toward the gonad or contact guidance by the cellular substratum of the mesentery are possibilities mentioned by Wylie and Heasman (1976). Either of these would indicate an early sensitivity of the germ cells to their surroundings—for example, via chemoreceptors in the case of chemotaxis—and hence the ability of the presumptive germ cells to receive extracellular signals. The migration is complete by approximately stage 41, the stage of hatching.

8.2   Proliferation of the Germ Cells and the Choice of Oogenesis

Once they have arrived in the gonad, the 20–30 primordial germ cells in $X.$ $laevis$ begin to multiply with an ever shorter generation time, until their number reaches approximately 1000 by the 40th day after fertilization (Ijiri and Egami, 1975). The tadpole begins feeding on the seventh day and metamorphoses approximately 4 months later. Eventually, the number of germ cells reaches tens of thousands in $X.$ $laevis$. Proliferation of male germ cells is somewhat slower than is proliferation of female germ cells, with rate differences apparent even at the time 100 germ cells are present. At these early stages, the gonad itself is "indifferent" with no indication of its fate as a testis or ovary. Likewise, the germ cells show no sexual dimorphism as yet and are referred to as "gonocytes." In anurans, the gonocytes are able to enter either oogenesis or spermatogenesis regardless of their own sex chromosome constitution, but depending on the maleness or femaleness of the mesodermal tissue of the gonad and perhaps of other somatic tissues. This point was demonstrated by Blackler (1970), using germ cell transplants at the neurula stage in $X.$ $laevis$. He exchanged germ cell grafts at random between embryos without regard to sex chromosome constitution, and when the tadpoles metamorphosed into frogs, he mated each male and female to a normal (nontransplant) frog of known sex chromosome constitution, and scored the sex ratio of the offspring. In $X.$ $laevis$, normal males have a ZZ sex chromosome constitution in all their cells and females are ZW, consistent with the 50 : 50 ratio of male and female offspring in normal matings. Half of Blackler's transplant females produced eggs which on fertilization by a normal male gave only male offspring, indicating that the eggs carried only the Z chromosome and no W chromosome. This is the expected result for a transplant frog in which "male" germ cells (ZZ sex chromosomes) had been introduced into a female soma, where they developed not into sperm, but into eggs carrying only the Z sex chromosome. These eggs were nonetheless normal in appearance and capacity for development. Thus, expression of the W chromosome in the germ cells is unnecessary for successful oogenesis. From the other half of the cases of transplant females, a normal sex ratio was obtained in offspring from matings with normal males, indicating that these were "normal" females having received at random grafts of ZW germs into their ZW soma.

These results are in full accord with the discovery of Chang and Witschi (1956) and others that entire populations of young $X.$ $laevis$ tadpoles exposed to the female sex steroid estradiol at 1 μg/ml in their pond water during the larval period, metamorphose into fertile female frogs, irrespective of their sex chromosome constitution. Thus, the germ cells appear to choose their course of oogenesis based wholly on extracellular cues, not on their chromosomal "mosaicism." This is surprising since oogenesis and spermatogenesis yield alternate cell types, the egg or sperm, which would seem to differ as much morphologically and biochemically as any two cell types in the developmental spectrum. Again, the presumptive germ cells demonstrate their definite responsiveness to their surroundings and ability to receive extracellular signals. The results of Chang and Witschi (1952) show in addition that the mesoderm of the indifferent gonad develops as an ovary based on hormonal cues.

Among Blackler's transplant male frogs, half of the cases gave normal sex ratios in the offspring from matings with normal females, but the other half fathered populations which comprised 75% females (Blackler, 1970). This abnormal ratio was explained by assuming that those males carried Z and W sperm, due to transplantation of ZW "female" germ cells into a ZZ soma, as would be expected in 50% of the transplant males. The population of offspring of these males mated to normal females would be 25% ZZ normal males, 50% ZW normal females, and 25% unusual WW offspring (which turned out to develop as females). Thus, the transplant males demonstrate that the ZZ constitution is not required in germ cells for spermatogenesis. It would be interesting to see if WW germ cells could also develop as sperm in a ZZ somatic background, and thereby to test whether even a single Z is needed in the germ cells for spermatogenesis. The existence of the WW females indicates that not even a single Z is needed by the germ cells for oogenesis. Apparently, neither Z nor W is uniquely needed in the germ line for oogenesis, indicating the unimportance of the sex chromosomes in the sex cells. Unfortunately, direct cytological demonstration of the Z and W sex chromosomes has not been possible in anurans; their similarity to other chromosomes is too great to allow an identification of chromosomal sexual dimorphism (Mikamo and Witschi, 1966). Sex conversion of tadpoles to males by exposure to androgenic steroids has been less successful than the conversion to females by estrogens. Tadpoles exposed to androgen acquire secondary male sex characteristics as adults, but do not form normal testes (Chang and Witschi, 1956). Apparently other factors are important in the establishment of the testis, as discussed by Witschi (1951).

8.3 Nuclear Differentiation during Oogenesis

The gonad begins differentiation into an ovary as a part of sexual differentiation at the stage when forelimbs erupt in the advanced tadpole (Niuewkoop and Faber, 1967 for *X. laevis*), an indicator of imminent metamorphosis. Not all the germ cells undertake differentiation simultaneously. While a large reserve or stem cell population of "primary oogonia" is maintained by mitosis, some primary oogonia divide to produce secondary oogonia, defined as the first cells irreversibly committed to the differentiative steps of oogenesis. The dynamics of the reserve cell population is not known—that is, it has not been determined whether intracellular control mechanisms insure that every division of the primary oogonium produces a secondary oogonium plus another primary one, as in *Drosophila* (Kinderman and King, 1973) or whether an intercellular feedback mechanism controls the population as a whole to provide a steady supply of secondary oogonia, randomly committing toward differentiation both daughter cells of some primary oogonia and neither of others. In some species the ovary contains simultaneously all stages of oogonial and oocyte development and the admission of oocytes to the developmental path is controlled at a very early step (Kalt, 1976; Wallace and Bergink, 1974).

The secondary oogonium undergoes four divisions to produce a nest of 16 daughters, at least some of which are connected by cytoplasmic bridges left from incomplete cytokinesis (Coggins, 1973). This pattern is strikingly like that of insect oogonia. Presumably, these divisions are scheduled by an internal control circuit. Its evolutionary conservatism is remarkable. In *Drosophila* the cytoplasmic bridge between each pair of the 16 oocytes of a nest becomes a specialized thickened structure called the "ring canal" (Kinderman and King, 1973). Such a ring canal has not been identified in amphibia, although the circular object in the transparent oocytes of Figure 6 certainly looks like a ring canal. In spermatogenesis in *X. laevis,* the secondary spermatogonium goes through five to eight sequential divisions to produce a nest of 32–200 spermatocytes also connected by cytoplasmic bridges (Kalt, 1973, 1976). In all these cases, it is not known whether the sequential divisions function only for the amplification of cell numbers, or whether they are needed as part of the differentiation process—for example to dilute or replace components of the stem cell. Kerr and Dixon (1974) have suggested that the germinal plasm functions in allowing the germ cells to establish this unusual division pattern.

At the end of the mitotic series, the oocytes proceed through a final S period of DNA synthesis lasting six to seven days in *X. laevis,* each arriving at the 4C level of DNA—that is, the replicated diploid complement. Then they enter four specialized stages of nuclear and chromosomal differentiation having to do with genetic recombination, gene expression, and gene amplification. These stages are remarkably similar in oocytes and spermatocytes, whose paths have not yet diverged. Recent studies of these stages in *X. laevis* oogenesis and spermatogenesis are available (Coggins, 1973; Kalt, 1976). First, with the nuclear membrane intact, the chromosomes become visibly condensed in leptotene (lasting 4–5 days, oocyte diameter of 100 μm); second, the chromosomes further thicken in zygotene (lasting 5–7 days), pairing as homologues engaged in synaptonemal complexes (Coggins, 1973); third, the chromosomes still further thicken and shorten in pachytene (lasting 18 to 20 days, oocyte diameter of 200 μm); and fourth, the chromosomes finally loosen, become thinner and partially separate in diplotene [lasting several months, during which time the homologous chromatids remain joined at several chiasmata (Muller, 1974)] and enter the lampbrush configuration. The recombination events are certainly prominent in germ cells; Kerr and Dixon (1974) have suggested that a third function of the germ plasm is to give the germ cells the ability to engage in meiotic recombination. However, while most pronounced in germ cells, recombination may not be unique to them since some organisms (for instance, *Drosophila*) display mitotic pairing and crossing over in their somatic cells as well. In general, the overlap of meiotic and mitotic mechanisms is not yet understood. At the zygotene stage, the chromosomes can be seen to associate with one side of the nuclear membrane toward the centriole pair, in an arrangement called the "bouquet stage" (Wilson, 1925; Kalt, 1976; Al-Mukhtar and Webb, 1971). This asymmetry may give an early indication of a basic cellular axis on which the animal–vegetal axis of the egg can be elaborated, as discussed below.

In addition to the chromosomal differentiation involved in genetic recombination, there is a specialized amplification of DNA sequences used for ribosomal RNA synthesis (Brown and Dawid, 1968). This process has been most thoroughly worked out for *X. laevis* but occurs in most if not all amphibia. Each haploid chromosomal gene set of *X. laevis* contains about 450 copies of the sequences for 18 S and 23 S ribosomal RNA or 1800 copies in the 4C state. Even primary oogonia and spermatogonia seem to have an unusually large number of nucleoli, 4 to 16, as opposed to the normal 2, perhaps indicating some amplification of ribosomal DNA sequences at that early stage. However, by the pachytene stage, production of these sequences accelerates via a rolling circle replication of extrachromosomal sequences, until completion in the diplotene stage when approximately 2×10^6 sequence copies are present, a 1000-fold increase in DNA available for ribosomal RNA transcription. These sequences eventually partition into 1000–1500 nucleoli per germinal vesicle (Perkowska et al., 1968). The total nucleolar DNA amounts to 30 pg per germinal vesicle, compared to the 12 pg of DNA associated with the 4C level of chromosomal DNA. At these early stages, the nucleolar DNA tends to cluster at one end of the germinal vesicle, opposite the chromosomal attachment area, forming a cap recognizable in light microscopy by basophilic staining (Van Gansen and Schram, 1972; Ficq and Brachet, 1971). This is another indication of localization of materials within the nucleus. The zygotene and pachytene stages also support the large-scale synthesis of an oocyte-specific 5 S RNA, not by a gene amplification mechanism, but by transcription from highly repeated DNA sequences in the chromosomes (10,000 copies per haploid genome). The sequences differ from those of the 5 S RNA transcribed in somatic cells (Brown and Brown, 1976), although a small amount of this RNA is made in the oocyte as well.

At the pachytene stage, the paths of spermatogenesis and oogenesis diverge. The meiotic events of leptotene and zygotene were similar, but then pachytene for the oocyte is 50% longer (18 days), presumably because of the amplification of rDNA in oocytes but not spermatocytes. After a diplotene period of one day, the spermatocyte enters true meiotic prophase; the chromosomes condense again (diakinesis), the nuclear membrane disintegrates, and the spermatocyte completes the first and second meiotic cycle to produce four haploid spermatids. The spermatids then differentiate in the 1C state. The oocytes, on the other hand, remain in the diplotene stage for at least one more month and undergo cytoplasmic differentiation in the 4C condition, only to enter meiosis afterward.

In diplotene oocytes, as the chromosomes extend and separate, a further differentiation of the chromosomes sets in; namely, the lampbrush configuration in which numerous, long, lateral loops of DNA protrude from the axis of each chromosome (for reviews see Wischnitzer, 1976; Davidson, 1976). In *Triturus cristatus*, there are about 5000 loops per haploid set, some loops extending 200 μm to the side (Callan, 1963). In *X. laevis*, the maximum extension of the loops is reached when the oocyte has a diameter of about 500 μm and pigment deposition just begins, after which time the loops retract. By the time the oocyte is full grown (1200–1300 μm), the chromosomes are quite condensed and occupy a small membranous enclosure in the center of the germinal vesicle (Gruzova and Parfenov, 1978). It is not appropriate to present here the large literature on the structure and function of the lampbrush chromosomes. Suffice it to say, the loops appear to be lengths of DNA on which active transcription of RNA is occurring. Earlier, it was thought from the time course of incorporation of radioactivity into RNA that the DNA moves through a single loop so that the sequences transcribed in the loop change with time (Callan, 1963; Wischnitzer, 1976). More recent data suggest, however, that the DNA probably does not change in a specified loop, but the active RNA polymerases progress unidirectionally along their template, trailing out ever longer RNA transcripts (Davidson, 1976).

It has long been assumed that the unusual lampbrush configuration of the chromosomes serves a special function in transcription, but this function remains unknown. The configuration perhaps allows a high rate of transcription and in fact this rate is estimated to be 100 to 1000 times higher than found in postgastrula embryonic somatic cells (Davidson, 1976). On the other hand, it is possible the configuration allows a particular class of DNA sequences to be transcribed. This speculation has received considerable experimental attention. As summarized by Davidson (1976), the lampbrush loops contain about 10% of the chromosomal DNA and the transcripts from the loops meet the criteria of heterogeneous nuclear RNA (hnRNA) on the basis of their large size (some over 9000 nucleotides in length; Anderson and Smith, 1978), their DNA-like base composition, their nuclear location, and their short half-life of 36 to 90 min in *X. laevis* stage 3 oocytes (Anderson and Smith, 1978). The sequence complexity of the transcripts approximates 10% of the chromosomal DNA complexity and represents both the unique and repetitive sequences of the DNA (Chamberlin *et al.*, 1975). Unusually large amounts of protein associate with the transcripts in the nucleus (Somerville and Hill, 1973).

In stage 3 oocytes of *X. laevis,* when the loops are maximally extended, 86 to 88% of this hnRNA is degraded, perhaps in the removal of interspersed sequences as the pre-messenger RNA is processed into messenger RNA ready for transport to the cytoplasm. Thus, only 12 to 14% of the RNA reaches the cytoplasm where it is moderately stable and probably contributes to the pool of maternal messenger RNA used in oogenesis and the stages of early development, at least to the midblastula stage. The sequence complexity of this cytoplasmic RNA represents about 1% of the DNA sequence complexity and represents mostly but not entirely unique sequences with an average of 2×10^6 copies of each sequence (Anderson and Smith, 1978). Much of this cytoplasmic RNA is probably messenger RNA since it carries poly(A) sequences, either long ones (40–60 nucleotides) or short ones (15–30). However, 10% or less of the poly(A)$^+$RNA is associated with polysomes and therefore engaged in protein synthesis. The remainder is presumably stored for later use. Thus it is plausible that the extended lampbrush configuration of the chromosomes allows the rapid and general transcription of DNA sequences for messenger RNA. Perlman and Rosbash (1977) suggest that this transcription is so general as to be indiscriminate because they were able to detect 2×10^5 copies of RNA for *adult* hemoglobin in the cytoplasm of *X. laevis* oocytes, a protein considered to have no function in the oocyte. Similarly, Denis *et al.* (1977) have found adult somatic 5 S RNA sequences in the oocyte cytoplasm. If these sequences have, in fact, no function in the oocyte, how can we rationalize indiscriminate transcription? Perhaps it is more economical for the oocyte to waste a small amount of synthesis on inappropriate and harmless messenger RNA species than to synthesize the necessary regulatory RNA and protein molecules to prevent this small waste. The level of hemoglobin RNA is low compared with that in an erythrocyte and low even compared to the level of the average cytoplasmic RNA in the oocyte.

However, Davidson (1976) points out that the lampbrush configuration may not just allow rapid general messenger RNA production because the oocyte accumulates poly (A)$^+$ RNA (presumptive messenger RNA) to a high level even before the lampbrush loops are fully extended and the level stays approximately constant thereafter due to continued synthesis and breakdown (LaMarca *et al.*, 1973; Dolecki and Smith, 1979). Even at stage 6, when the loops are considerably collapsed, the rate of poly(A)$^+$RNA synthesis and hnRNA synthesis is still high. Davidson (1976) proposes that the loops are special in allowing the synthesis and cytoplasmic accumulation of repetitive RNA sequences, not unique ones. These sequences comprise only 2% of the cytoplasmic RNA complexity (Hough and Davidson, 1972) and accumulate during the lampbrush stage. They seem to be rather unusual in cells other than oocytes and eggs, at least as cytoplasmic components. They could derive either from DNA sequences of repeated structural genes or from the interspersed repetitive DNA sequences. This latter possibility is preferred by Davidson (1976, p. 382) who proposes that the repetitive RNA species serve a regulatory role, being sequestered in specific regions of the egg cytoplasm where they later control differential gene expression in the nuclei which in the embryo come by chance to reside there. The mechanism of localization is not specified, but presumably would be one responsive to the effects of gravity, temperature gradients, unilateral anaerobiosis, constriction of the egg, transplantations, and the other treatments evoking pattern regulation, as discussed earlier. This proposal would link cytoplasmic localization to the lampbrush configuration. On the other hand, the repetitive sequences might be involved with the storage and mobilization of messenger RNA (E. H. Davidson, personal communication), that is, in a temporal rather than spatial control.

In summary, the early differentiation of the oocyte primarily concerns changes of chromosome structure and activity involved in genetic recombination, gene amplification of sequences for ribosomal RNA, high level transcription of 5 S sequences, and finally elaboration of the lampbrush configuration perhaps for high levels of transcription of a wide variety of sequences.

8.4 Cytoplasmic Differentiation during Oogenesis

Whereas the oogonium contains no identifiable yolk platelets, the full-grown oocyte contains massive amounts, comprising at least two-thirds of the total protein of the cell. Thus the process of vitellogenesis, or yolk deposition, is the major material differentiation of oogenesis: from this perspective, there are two periods designated in oogenesis, namely, the previtellogenic period and the vitellogenic period. In X. laevis, the former occupies a period of several weeks (30–40 days; Coggins and Gall, 1972) and comprises the stages of nuclear differentiation discussed in the previous section as well as important cytoplasmic changes to be presented here. At the end of this period, the oocyte is approximately 300 μm in diameter and already displays marked signs of polarity. Then begins the uptake of yolk precursors that were synthesized in the liver and transported by the bloodstream. During yolk deposition the true animal–vegetal axis of the oocyte originates, an axis based morphologically on platelet distribution. Other materials are deposited as well, to the extent that only one-eighth of the egg volume in R. pipiens is aqueous cytoplasm accessible to radioactive metabolites, the rest being yolk, lipids, glycogen, and vesicles (Melton and Smorul, 1973).

8.4.1 The Previtellogenic Period

In the previtellogenic period the oocyte of X. laevis enlarges from 50 μm to approximately 300 μm in diameter with a proportional enlargement of the germinal vesicle, which occupies approximately one-eighth of the cell volume (Peterson, 1971; Billet and Adam, 1976). The oocyte is transparent, allowing good visualization by dark field illumination, in which it is easily seen that a single large opaque body is free in the cytoplasm or adheres to one side of the germinal vesicle, giving the young oocyte a clearly polarized appearance (see Fig. 6). This is the mitochondrial cloud composed of small vesicles and numerous close-packed mitochondria in a phase of rapid proliferation. As Billet and Adam (1976) have found by light and electron microscopy, the mitochondria and vesicles collect at a depression in the nuclear membrane. In R. pipiens oocytes, this area contains a well-defined centriole pair (Al-Muktar and Webb, 1971) similar to the better studied case in chicken oocytes (Greenfield, 1966). The presence of a centriole and aster would explain the movement of organelles to this area of the nuclear surface. And the fact of one centriole pair in the cell would explain the presence of just one mitochondrial cloud, as is found in most oocytes (some have two clouds which later fuse). Thus, the inherent centriolar–nuclear axis of the cell might become elaborated as the mitochondrial cloud. Since the axolotl oocyte is said to lack a cloud, it may be an effect rather than a cause of polarity.

There is a rapid increase in mitochondrial DNA in this period, as shown by Webb and Smith (1977). Billet and Adam (1976) estimated that there are 10^5 closely packed mitochondria, assuming an average length of 2 μm which is typical of later mitochondria. They reported the presence of an unusual mitochondrial morphology at this stage with individual lengths of over 6 μm in the cloud. They suggested that the numerous small membraneous vesicles of 0.1 μm diameter derive from an endoplasmic reticulum system. Earlier, this aggregate had been incorrectly called the "yolk nucleus," anticipating its role in yolk formation; since the role of mitochondria in vitellogenesis remains uncertain, the name "mitochondrial cloud" is more appropriate. It is not currently justified to conclude that the mitochondrial cloud directly leads to the vegetal yolky hemisphere of the egg. In fact, at the end of the previtellogenic period the mitochondria disperse to the oocyte periphery, there to comprise a uniform layer beneath the plasma membrane (Wischnitzer, 1966; Balinsky and Devis, 1963; Al-Muktar and Webb, 1971). Mitochondrial proliferation stops by the time the oocyte is about 500 μm in diameter, as shown by the constancy of mitochondrial DNA (Webb and Smith, 1977), and mitochondrial DNA synthesis does not start again until stage 30, the tailbud stage of embryogenesis (Chase and Dawid, 1972). The factors controlling stage-specific mitochondrial proliferation are unknown.

During most of the previtellogenic period, the oocytes of *X. laevis* cluster in a nest of 16 daughter cells descended from a single secondary oogonium, and retain their cytoplasmic connections to one another. In this syncytium, the oocytes remain synchronous in their developmental progress. As early as the leptotene stage, follicle cells of the ovary surround each oocyte (Dumont, 1972; Van Gansen and Weber, 1972; Coggins, 1973), and by the early diplotene stage establish a large area of surface contact with the oocyte through numerous interdigitated microvilli. Shortly thereafter, deposits of material of the future vitelline membrane appear between the follicle cells and oocyte. Wischnitzer (1966) has suggested that cells of both types contribute material; the eventual vitelline membrane is known to comprise 11 kinds of glycoproteins (Wolf *et al.,* 1976) but the origin and time of appearance of each is unknown. Steinert *et al.* (1976) have identified glycoprotein deposits in the intercellular space of previtellogenic follicles, using fluorescent concanavalin A. If fluorescent antibodies against purified components of the vitelline membrane were available, one could answer the question of origin and schedule.

At the end of the previtellogenic period, blood vessels invade the follicle, establishing an arterial and venous circuit of flow through the stalk attaching the follicle to the continuous fabric of the ovary. This vascular pattern has been examined by Bellamy (1919) in an effort to evaluate the hypothesis of Child (1915) that egg polarity originates from the asymmetric blood supply of the follicle, one side of the isotropic oocyte being more exposed to oxygen and nutrients. Bellamy (1919) reported that 75–80% of *R. pipiens* oocytes are oriented with the equator within 20° of the stalk, whereas random orientation would predict only 34% of the oocytes within that 20° range. The equator was taken as the border of pigment, and these oocytes were scored at the vitellogenic stage. Also, Bellamy reported that the arterial capillaries always covered the animal hemisphere. Later (1921) he withdrew this claim, reporting that full-grown oocytes can also have the opposite orientation and that the relation of the oocyte polarity to blood supply was unclear at these late stages, perhaps because the capillary net changed its position. He proposed the possibility that a relation of blood supply and polarity existed in early stages of oogenesis. Child (1947) backed this suggestion, writing that oocytes are known to change their position in the follicle as they enlarge. Numerous authors subsequently examined oocyte orientation, all with the conclusion that follicle asymmetry is not related to oocyte asymmetry (Bronsted and Meyer, 1950; Wittek, 1952; Van Gansen and Weber, 1972). The last authors used transmission and scanning electron microscopy to observe that the previtellogenic oocyte has a marked asymmetry even before the capillary system is established. This asymmetry concerns the large mitochondrial cloud and germinal vesicle, the latter having a cap of nucleoli opposite the cloud, and abundant emissions of nuclear materials toward the cloud. They equated this early axis with the final yolk and pigment axis of the vitellogenic stages. Although this relationship is attractive, there is no evidence as yet that the early axis leads to the later one.

Even before yolk deposition begins, the stockpiling of other nutrient reserves is detectible. Lipid droplets or "lipochondria" appear about the time blood vessels arrive (Dumont, 1972; Balinsky and Devis, 1963). The source of lipids is unknown; transport from the liver might be a reasonable guess since this organ is the main site for fatty acid synthesis. The lipochondria possess a unit membrane of unknown origin. Lipid reserves are said to comprise 20–25% of the dry weight of the *R. pipiens* egg (Barth and Barth, 1954). This reserve provides not only fuel for energy production by oxidative metabolism but also fatty acids for membrane phospholipids. The earliest stage of glycogen synthesis is unknown; eventually the egg contains about 8% glycogen by dry weight, more than stored by liver or muscle. It is not known whether glycogen forms from glucose imported from the liver by the bloodstream or from glucose made in the oocyte from gluconeogenesis.

As evidence became conclusive for the role of germinal granules or polar granules in the determination of the germ line in anurans and insects, the question arose whether the granules themselves had a cytoplasmic continuity, whereby all stages of the germ line maintain their unique identity during multiplication and differentiation by synthesizing for themselves a steady supply of granules. This question was answered in the affirmative by Mahowald (1971) for *Drosophila* and Blackler (1970) for anurans. As mentioned above, the postvitellogenic oocyte contains granules identifiable in its vegetal cytoplasm by electron microscopy although they do not yet have the association with mitochondria and yolk-free areas, as is required for their identification by light microscopy, in the postfertilization stages (Mahowald and Hennen, 1971; Williams and Smith, 1971). When do presumptive primordial germ cells begin synthesizing granules for themselves to compensate for dilution during cell division, and when do oocytes stockpile granules for the eggs of the next generation? Coggins (1973), Kalt (1973, 1975), and Kerr and Dixon (1974) have obtained evidence from electron microscopy for the presence of a "nuage material," a fibrogranular complex in the nucleus of even the rapidly multiplying primordial germ cells of the undifferentiated male or female gonad, and the presence in primordial germ cells of similar material in the cytoplasm associated with mitochondria, looking very much like germinal granules except for the absence of a ring-shape. These nuage bodies are produced throughout the sequential divisions of gonial cells producing spermatocytes or oocytes. In the diplotene stage of the oocyte the rate of accumulation of nuage material increases and the bodies associate in large clusters with mitochondria. Later in the diplotene stage the complexes seem to disperse; Kerr and Dixon (1974) concur that nuage bodies go to the cortex of the oocyte, as originally suggested by Balinsky and Devis (1963). Thus, Kerr and Dixon (1974) and Kalt (1975) relate the future germinal granules of the egg to the nuage material present in all earlier stages of the germ line. In addition, there is a large fibroparticulate "nucleoluslike" or "chromatoid" body (Al-Muktar and Webb, 1971; Kalt, 1973) in the cytoplasm of both male and female primordial germ cells; its function is not known. Because of its fine structure and because it breaks into small vesicles in the diplotene oocyte, this chromatoid body has not been assigned a role in germinal granule formation. In conclusion, electron micrographs afford evidence that the nuclear and cytoplasmic "nuage" material is a precursor or alternate form of the germinal granules since it appears only in members of the germ line. Therefore the idea of a cytoplasmic continuity of the granules is supportable. However, until there is a direct assay for the material, such as the rescue of UV-irradiated eggs by injection of granule-containing cytoplasm (L. D. Smith, 1966), the evidence is not conclusive.

8.4.2 Vitellogenesis

In the full-grown oocyte of *X. laevis,* yolk platelets comprise 60–70% of the protein of the cell (Wallace and Bergink, 1974) and at least 45% of the dry weight (see Barth and Barth, 1954, for data for *R. pipiens*). The platelets are the reservoir of essential amino acids for embryonic protein synthesis prior to hatching and feeding. There is no other source. They are also the phosphate reservoir for nucleic acid synthesis, containing in each oocyte in the case of *X. laevis* (Coleman and Gadian, 1975) as much as 120 nmoles of phosphate, a large amount compared to the 6 nmoles in ATP and the 10 nmoles in RNA. Both protein and nucleic acid synthesis begin at high levels only after the neurula stage, drawing on the materials the oocyte has stockpiled long in advance. In addition, platelets contain large amounts of lipid (12% by weight of the platelet) and of carbohydrate (1.3%). They are probably not major energy reserves but rather nutrient reserves, whereas glycogen and lipid droplets provide the fuel reserves. The energy reserves of glycogen and fats distribute to most cells of the tadpole, but the yolk mass finally partitions mostly to the gut region where the largest yolky blastomeres of the endoderm disintegrate in the lumen of the gut and the digested materials circulate to the embryonic cells through the embryonic blood stream. Although the *X. laevis* tadpole hatches three days after fertilization, it does not feed for seven days, living off its yolk for the interim. Given the future long-term dependence of the embryo and tadpole on the yolk, it is easy to appreciate the adaptive value of a developmental strategy in which the fate map of the gastrula is organized around the position of the yolk, as seems to be the case: as discussed above, the fate map is inverted when the yolk mass is centrifuged to the opposite hemisphere, and the yolk mass acts as the inductor of mesoderm in the animal half. Nieuwkoop's results (1969*a,b*) indicate that the yolk mass is not only the carrier of the animal–vegetal axis but also the dorsal–ventral axis. Thus, as the major determinant of pattern in the amphibian egg, the yolk mass deserves our profound attention, despite its perhaps unexciting molecular nature.

Thanks to the recent work of R. A. Wallace and his colleagues, of Redshaw and Follett, and of Ansari and his colleagues, the nature and origin of yolk platelet proteins are now well understood, especially in *X. laevis.* Platelets contain three and perhaps four proteins: lipovitellin 1 and 2, and phosvitin 1 and 2, although the last may just be a breakdown product of the other phosvitin (Bergink and Wallace, 1974). The phosvitin molecules are highly phosphorylated with approximately every second residue a phosphoserine, making each a highly anionic protein. Lipovitellin is also phosphorylated to about one-fourth the level of the phosvitins. The single precursor of these proteins is vitellogenin (molecular mass approximately 200,000 daltons per chain) synthesized by the liver and transported by the bloodstream to the oocytes. It now appears that in *X. laevis* there are three different vitellogenins (Wiley and Wallace, 1978) and that there is a small family of vitellogenin genes the members of which are closely related, but quite definitely different (Wahli *et al.,* 1979). The hormonal stimulus to the liver for vitellogenin production is estrogen from the follicle cells, which in turn are stimulated by FSH from the pituitary. The pituitary receives stimulation from hypothalamic centers evaluating food intake, temperature, light, and unknown algal products in the water (Wallace and Bergink, 1974). Direct stimulation of the liver can be achieved by estrogen injections, even in the male which proceeds to accumulate vitellogenin to the level of 150 mg/ml in the blood, almost three times the level of serum albumin.

In the female, oocytes in the size range of 300–1200 μm in diameter remove vitellogenin actively from the blood by pinocytosis. Wallace has devised conditions to study this process *in vitro* with isolated follicles and radioactive vitellogenin. The rate of uptake is approximately constant per unit surface for vitellogenic oocytes (Wallace *et al.*, 1970). At diameters greater than 1200 μm diameter, oocytes become "postvitellogenic," showing much reduced uptake of the yolk precursor. In vitellogenic oocytes, other proteins such as serum albumin and hemoglobin are also taken up by pinocytosis but with much lower efficiency (1%). Even Nile blue enters pinocytotic vesicles of the oocyte after injection into the frog, and stains the vitellogenic oocytes selectively. DNA also enters slightly and without sequence selectivity; this uptake may explain the small amounts of DNA associated with platelets.

Radioactive vitellogenin is found shortly after uptake in the region beneath the plasma membrane where pinocytotic vesicles and membranous channels abound. The region of membrane *between* microvilli seems most directly involved, suggesting that the transport path is direct from the bloodstream to the oocyte without passage through the interior of the follicle cell. The pinocytotic vesicles appear as "coated vesicles" although high-resolution microscopy has not been done to identify the hexagonal and pentagonal arrays of clathrin found on other coated vesicles involved in protein uptake (Roth *et al.*, 1976). Deeper into the oocyte are found larger densely staining vesicles thought to represent transformed pinocytotic vesicles, perhaps several fused together and modified by internal reactions on the vitellogenin. Several hours later radioactive vitellogenin is found associated with platelets themselves. Somewhere along the path from pinocytosis to the platelet, the vitellogenin molecule is split, by an unknown but specific enzyme, to yield lipovitellin and phosvitin (Bergink and Wallace, 1974). One consequence of this proteolytic cleavage is the crystallization of the lipovitellin and phosvitin into a lattice of low solubility under cellular ionic conditions. This lattice has regular 70-Å spacings long recognized in yolk platelets by electron microscopists and recently studied by X-ray crystallography by Ohlendorf *et al.* (1978). The lattice occupies the major volume of the platelet and is called the "crystalline main body." Peripheral to the main body is a narrow amorphous layer, the "peripheral layer," thought to comprise a variety of proteins nonspecifically taken in by the pinocytotic vesicles, probably with modifications.

External to the amorphous layer of the platelet is the unit membrane formed by fusion of pinocytotic vesicles, probably with modifications during transit, such as removal of the "coated" surface of clathrin (Roth *et al.*, 1976). There are two points about this membrane that should be mentioned. First, it contributes a large amount of the total membrane of the mature oocyte; calculations based on the size, shape, and number of platelets (Robertson, 1977) suggest that the mature oocyte contains roughly 100 times as much membrane in its platelets as on its external surface. Yet this vast amount is originally derived from the surface membrane by pinocytosis, indicating the dynamic condition of the external plasma membrane during vitellogenesis. The second point concerns the similarity of the platelet membrane to the plasma membrane, and on this point there are few facts. The plasma membrane of the oocyte is an ion-pumping membrane, one well able to conduct nutrient uptake. To what extent do the platelets preserve these properties? Especially in question is whether the platelets retain ion-pumping properties when the external membrane undergoes its dramatic conversion to an impermeable state during meiotic maturation. Or does

the platelet membrane resemble the new ion-permeable furrow membranes of cleavage?
There is some recent evidence in favor of ion permeability for the platelet membranes, for
Ziegler and Morrill (1977) have reported that mature oocytes behave as if their large Na^+
reservoirs are completely exchangeable between compartments and the bulk cytosol, based
on high values for free Na^+ from selective electrodes. However, after fertilization the free
Na^+ levels are reported to be lower, as if the reservoirs have pumped the Na^+ into themselves. As mentioned above, it is the platelets which most likely contain the majority of the
sodium of the oocyte as the counterion of anionic groups of vitellogenin brought in originally from the sodium-rich bloodstream. The platelets thus represent a large sampling of
extracellular materials internalized by the oocyte in vitellogenesis. Furthermore, while
vitellogenin circulates in the bloodstream it carries calcium ion as the main counterion to
its phosphate groups, in a 1 : 1 ratio (Ansari *et al.,* 1971) and some of this calcium is
probably carried into the oocytes during pinocytosis. Calcium and sodium qualify as the
"extracellular" ions in contrast to the intracellular ions potassium and magnesium, and the
oocyte is unusual in its high internal levels of these external ions.

As a speculative aside, if one looks for a general mechanism for axis determination,
not just for amphibia and related taxonomic classes with yolky eggs, then one must include
in the discussion the mammalian egg where external positional signals seem to take the
place of cytoplasmic localizations (Hillman *et al.,* 1967; Tarkowski *et al.,* 1964). These
signals could plausibly include external ions and metabolites provided to the egg by its
uterine environment. If true, then the amphibian egg would merely have brought within
itself an internal reservoir of external ionic and metabolic signals in the form of its yolk
platelets, sequestered during oogenesis from the external medium of the maternal bloodstream. Along these lines can be mentioned the resemblance of the ion- and nutrient-rich
fluid cavity of the mammalian blastocyst to the yolk sac of the lower vertebrates; in Morgan's words, the mammalian embryo can be said to contain "a yolk sac without the yolk"
(Morgan, 1934). Perhaps the essential positional signals are still present in the blastocyst
cavity, despite the absence of platelets as such.

In the full-grown oocyte, the platelets range in size from approximately 3 μm to 14
μm in the long axis, in *X. laevis* (Nakatsuji, 1976; Robertson, 1978), a volume range of
approximately 100-fold. In the distribution of sizes, the large platelets and small platelets
abound, with a lower frequency of middle-size platelets (Bragg, 1939; Panijel, 1950; Nakatsuji, 1976). Some authors have suggested that two kinds of platelets exist in the oocyte.
Ubbels (1978) has detected differential staining of platelets with acidic Alcian blue in the
morula stage. Some amphibia such as *A. maculatum* have several kinds of yolk differing
in pigmentation, the darkest located at the vegetal pole (Harris, 1962). Panijel (1950) separated large and small platelets of *R. fusca* and the axolotl by velocity centrifugation and
reported that the large platelets contain higher amounts of phosphate and lower amounts
of phosphoprotein phosphatase, an enzyme thought to catalyze the removal of phosphate
from phosvitin and lipovitellin. These experiments would deserve repetition with modern
methods for purification, since the small platelets could have been contaminated with vesicles of the cell. Jared *et al.* (1973) have devised a centrifugal density separation of platelets
in 20–60% sucrose solutions; the platelets separate into two bands in the density range of
50% sucrose. These have not been analyzed thoroughly, although Jared *et al.* (1973)
reported that there is no distinct size correlation. The basis for the density difference is
unknown.

Until the existence of more than one kind of platelet has been well substantiated, the
question of dual origin or dual role in development must wait. But prior to the recent work
on vitellogenin, various authors, particularly Lanciavecchia, suggested dual origins for

platelets (reviewed in Wischnitzer, 1966), one path via vesicle coalescence presumably similar to the now-understood pinocytotic pathway, and a second path via mitochondria. The latter pathway was supported by electron micrographs of *Rana* oocytes in which mitochondria were found to contain crystalline lattice material. The role of these bodies remains unknown; they have not been found in urodele oocytes. Massover (1971) has measured the lattice spacings of crystals in *R. pipiens* mitochondria and finds them 15% smaller than those of the platelet crystalline main body. Thus, there is a question whether the mitochondrial crystals are even a yolk material. Wallace *et al.* (1972) have summarized their findings in relation to the hypothetical mitochondrial path by saying the mitochondria may be involved in stages of pinocytotic vesicle coalescence, perhaps adding or modifying materials, but oocyte mitochondria are clearly not involved in the synthesis of the yolk protein polypeptides or with the addition of phosphate groups to the vitellogenin, shown by radioactive labeling experiments to occur exclusively ($>$97%) in the liver before the protein is released to the bloodstream. There could, of course, be unknown modifications concerning vitellogenin phospholipid or carbohydrate groups, or concerning the platelet superficial layer or membrane.

8.5   Origins of the Animal–Vegetal Axis in Vitellogenesis

When platelets first appear in oocytes (diameter 300 μm in *X. laevis*; Dumont, 1972), they are located uniformly near the plasma membrane, that is, without a clear animal–vegetal difference. In fact, the oocyte at that time does not display a clear axis; the mitochondrial cloud has dispersed to the periphery (Balinsky and Devis, 1963; Billet and Adam, 1977) and the nucleoli have spread uniformly around the nucleus in a layer beneath the nuclear membrane. Furthermore, the earliest deposition of pigment, when the oocyte of *X. laevis* reaches a diameter of 450 μm, displays no animal–vegetal asymmetry, despite the fact that pigment eventually occupies only the animal hemisphere. The melanosome precursors or "premelanosomes" arise earlier as colorless organelles, perhaps from the Golgi system, and then darken progressively due to melanin deposition via tyrosinase activity, in a limited period while the oocyte grows from 450–600 μm in diameter. Then, pigment formation stops. At first, the premelanosomes are scattered throughout the cytoplasm and only gradually congregate at the plasma membrane. The picture for cortical granules is much the same. They first arise scattered in the cytoplasm when the oocyte is approximately 300 μm in diameter, probably as products of the Golgi complex (Wischnitzer, 1966; Balinsky and Devis, 1963), and then move slowly to the periphery. Wittek (1952) has interpreted stained sections in terms of a bidirectional flow of oocyte cytoplasmic contents at the early vitellogenic stages, with yolk granules moving inward and basophilic cytoplasm moving outward to the plasma membrane.

When the diameter of the oocyte of *X. laevis* reaches 600 μm (stage 4 of Dumont, 1972), the external signs of the vegetal pole of an axis are present as a nonpigmented patch appearing on the otherwise black surface. This patch continues to enlarge until the oocyte approaches the definitive pattern of one dark and one light hemisphere. The process of pigment segregation from the whole surface to half the surface is far from understood. Three hypothetical mechanisms can be suggested. First, Bellamy (1919) proposed that the side of the oocyte nearest the arterial capillaries would receive extra oxygen to complete the melanin oxidations and therefore become black. This is inconsistent with the pattern of pigment distribution which is at first uniform, and also inconsistent with the evidence for random oocyte orientation in the follicle. Second, there could be a localized expansion of the plasma membrane of the growing oocyte during the period after melanization has stopped, so that new membrane is pigmentless (J. Eppig, personal communication). It would require other assumptions to suggest why the pigmentless surface coincides with the vegetal hemisphere. A third hypothesis invokes a contraction of the pigment toward one area of the surface. Contraction is plausible in that the egg later displays the activation contraction toward the animal pole, the cortical contraction of gray crescent formation, and the cortical pigment expansion over the entire surface in epiboly. Whether the contraction would be active and actin-driven remains to be seen. It need not be, for Schultze (1894) and others have shown for fertilized eggs that the yolk mass often disrupts the integrity of the pigmented cortex in inverted eggs where gravity has forced yolk into contact with the animal surface. In the young oocyte, the enlarging yolk mass could plausibly just expel the pigment from the vegetal cortex—that is, a passive depigmentation of the vegetal hemisphere. In any event, the melanosomes become associated with only one-half the oocyte surface and do so in a large unified movement rather than as a localized movement to give a speckled pattern.

More attention has been given to the origin of the yolk distribution, particularly in the study of Wittek (1952). According to Wittek's light-microscopic, cytological studies of oogenesis in *R. temporaria, X. laevis,* and *T. alpestris,* vitellogenesis has a complex two-phase history. In the first phase, platelets arise near the periphery and move centripetally so that the largest and oldest platelets are innermost. Then, according to Wittek, when the platelets reach the germinal vesicle, a second phase of vitellogenesis begins in which other platelets start forming in the vicinity of the germinal vesicle and move outward, while the inward flow of new small platelets from the periphery continues. Small platelets appear in concentric zones at the center and the periphery of the oocyte with a zone of large platelets between. As the region of perinuclear platelets expands as a sphere, one region of it reaches the plasma membrane first, simply because the germinal vesicle is slightly excentric in its position. The zone of largest and oldest platelets is thereby interrupted where the zones of central and peripheral small platelets link up. Thereafter, the large platelets are more and more displaced from the region of the germinal vesicle and from the zone of small platelets, a region developing as the prospective animal hemisphere. Also, as the mass of largest platelets is displaced to one end, the production there of small platelets becomes slower and slower. This is the developing vegetal hemisphere. Wittek identifies the original zones in the full-grown oocyte of *R. temporaria* on the basis of platelet size. The peripheral platelets are smallest and this zone is thicker at the animal end; the central platelets near the germinal vesicle and equator are middle sized; and the vegetal platelets are largest and oldest. In Wittek's view, the animal–vegetal axis originates from the eccentric position of the germinal vesicle as it affects the countercurrent flows of platelets from their two sources, one at the surface and another at the center of the oocyte. These observations and interpretations would merit further study by electron microscopic methods as well as by light microscopy. Also, since platelets can be marked *in vivo* with Nile blue or radioactive vitellogenin injected into the frog's circulation, the relation of age and size could be verified. Very recently, Wallace and Misulovin (1978) have succeeded in growing *X. laevis* oocytes *in vitro* in a defined culture medium with vitellogenin from the pre- to postvitellogenic stages, while the animal–vegetal axis is developed. This achievement should make studies of the origin of the axis feasible for the first time. Wittek's idea of the perinuclear origin of some platelets requires verification in light of the current information about vitellogenin from the liver entering the oocyte only by pinocytosis at the surface. It is difficult to see how vitellogenin would reach the germinal vesicle without prior incorporation into platelets. Overall, though, Wittek considers the yolk distribution to result from the original asymmetric position of the germinal vesicle.

Whatever the mechanism for partitioning platelets to different cytoplasmic regions, the coherence of the body of large platelets is surprising. They remain as a mass, unmixed with small platelets. The oocyte cytoplasm is highly structured; in *X. laevis* oocytes it is essentially impossible to draw up yolky cytoplasm in a micropipette; also, the platelets show no settling by gravity as demonstrated by the fact that oocytes in the ovary have every orientation with respect to gravity. Thus, the platelets seem to stick together either directly or through some cytoskeletal network. Perry (1966) has commented on tubular elements associated with platelets in bottle cells at much later stages, but there is little information on the superstructure of the yolk mass in oogenesis.

At meiotic maturation, there is a decrease in cytoplasmic macroviscosity, and at fertilization a much larger decrease, at which time platelet distribution is definitely sensitive to gravity. Also, unfertilized eggs exposed to colchicine undergo gravitational redistribution of yolk (Manes *et al.*, 1978). But at the stages of oogenesis, the yolky regions remain unitary.

Some amphibian species such as *R. temporaria* and *T. alpestris* appear to have bilateral symmetry established long before grey crescent formation, as discussed above (Pasteels, 1938). Wittek (1952) observed precocious bilateral symmetry even in the early stages of vitellogenesis in these species, as far as the position of the germinal vesicle *vis à vis* the nascent zone of large platelets. As a result, the zone of medium-size platelets of perinuclear origin reached to the periphery on one side of the oocyte, displacing even the small peripheral platelets. At maturation, when the animal pole becomes defined as the spot of insertion of the meiotic spindle in the cortex, these eggs clearly had their animal pole well off-center in the pigmented animal hemisphere. To Wittek, these ever-clearer aspects of bilateral symmetry can be traced back to the lateral position of the nucleus and mitochondrial cloud at the previtellogenic stage.

In summary, there is a plausible cytological argument from Wittek's work that the axis defined by the nucleus and mitochondrial cloud presages the animal–vegetal axis of the full-grown oocyte, with the germinal vesicle eventually occupying the animal hemisphere. It is not known what factors affect the position of the nucleus at the beginning, but the original position of the centriole could well be important. Coggins (1973) noted an alignment in even the earliest oocytes (engaged in the premeiotic S period) of the division bridge, the centriole pair, and the nucleus. This same axis is present in the earliest spermatocytes (Kalt, 1973) and these cells, of course, build the obvious differentiated axis of the sperm directly from the nuclear–centriolar axis, with the centriole becoming the base of the flagellum. Perhaps the oocyte organizes itself on the same axis. These notions bring us close to the old conviction of Rabe and Van Beneden (cf. Wilson, 1896) that the true axis of any cell is defined by its centriole and nucleus, upon which other structures can be elaborated. Certainly these notions could be better defined nowadays for previtellogenic oocytes in experiments with fluorescent antibodies to visualize structural proteins such as tubulin, a technique that has demonstrated the unitary and unifying structure of the interphase aster of cultured cells (Connolly *et al.*, 1977; Schliwa, 1978; Kirschner, 1978).

8.6   Pattern Regulation

Pattern formation in the single-celled oocyte concerns the partitioning of the large yolk platelets to one hemisphere, and the germinal vesicle, pigment granules, and nonyolky cytoplasm to the other hemisphere. The partitioning mechanisms are poorly known, and in most cases even the description of the processes is incomplete. As a possible explanation for pattern formation in the oocyte, we have discussed the centriole and interphase aster of the oocyte as the organelle directing the polarization of materials, without resorting to extracellular means for orienting the axis. The early axis of the oocyte is apparent in the alignment of the cytoplasmic bridge, the centriole, and its nucleus with its basal end occupied by the bouquet stage chromosomes toward the centriole and its apical end by the cap of nascent nucleoli. This is a variation and elaboration of the basic centriolar–nuclear axis common to many cell types, including the sperm. In the oocyte a later elaboration of this same axis may be the localization of the mitochondrial cloud in the region of the centriole. Subsequently, however, the localization of yolk and pigment is difficult to relate to the original axis. Perhaps the centripetal movement of the yolk platelets from the periphery occurs under centriolar direction. And the germinal vesicle undoubtedly exerts a polarizing effect since it occupies the animal hemisphere, perhaps in connection with the output of ribosomal RNA and massive amounts of other transcription products in its vicinity, especially at the basal end.

Pattern regulation in oogenesis would concern the mechanisms insuring the formation of a single yolk mass of large platelets and its counterpart a single less-yolky region, with a single pigmented hemispheric cortex. It is perhaps appropriate to note that the cell carries only one nucleus and one centriole, so that their singularity may contribute to the singleness of the animal and vegetal hemispheres despite the large cytoplasmic volume. The yolk mass does not mix freely with small platelets or with nonyolky cytoplasm but remains coherent and close-packed. The cohesion of the platelets is not understood but could perhaps reflect the fact they were formed by fusion. In addition, the cytoplasm as a whole is coherent against the gravitational reorganization and presumably adheres to the cortex strongly.

Then, as a separate concern, the germinal vesicle arrives at a certain size characterized by a definite nucleocytoplasmic ratio. The previtellogenic oocyte starts with a ratio of about $1 : 4$ to $1 : 8$, a usual ratio for many cell types. After growth of the oocyte to full size the ratio is approximately $1 : 30$, but if a correction is made for the enormous amount of yolk, glycogen, and fat reserves, the ratio still falls in the range of $1 : 4$ to $1 : 8$ (Melton and Smorul, 1973; Peterson, 1971). We have no idea of the mechanism by which cells regulate their nucleocytoplasmic ratio. And finally, there must be many control mechanisms by which the oocyte produces definite amounts of certain materials stockpiled for use in the developmental stages at least until midblastula and in some processes at least until the neurula stage. These quantitative controls are unknown. There is the long period until the midblastula when gene expression is unused to accomplish pattern formation and regulation, and cytoplasmic circuitry of the oocyte will serve the purpose.

There has been almost no experimentation on pattern formation and pattern regulation in oogenesis, although techniques steadily improve for working with oocytes *in vivo* and *in vitro*. As cell biology provides more information on the basic organization of single cells it should be possible to return to the oocyte with questions about the cellular mechanisms by which the yolk, cytoplasm, germinal vesicle, and pigment are partitioned to establish the animal–vegetal axis of the egg.

9   Summary

We have reviewed amphibian development from early oogenesis to neurulation with respect to three points: (1) the anatomical and physiological pattern of each developmental stage, assuming each stage to have a distinct pattern of its own as opposed to an emerging, yet imperfect pattern of the adult; (2) the morphogenic mechanisms forming that pattern, or more accurately, transforming the previous pattern into the present one since these mechanisms originate within the previous pattern and operate with its geometrical limits (this is the "historical" component of development); and (3) the self-regulating properties of the morphogenic mechanisms. These properties are considered in this review as the source of pattern regulation during development.

In the following summary, pertinent sections of the review are indicated in parentheses. Literature references are found in the text of those sections. Although the summary is intended to apply to all amphibia, some conclusions are based almost entirely on experiments with urodeles, and others with anurans, especially *X. laevis*, and therefore generality remains to be demonstrated.

9.1   The Earliest Axis (8.1–8.4)

The primordial germ cell and the oogonium possess an axis based on the position of the nucleus and centriole, as do most animal cells. The oocyte even at the stage of premeiotic DNA synthesis elaborates this axis in terms of the position of an intercellular cytoplasmic bridge aligned with the nucleus and centriole after the last mitotic division from the oogonium. During the stages of leptotene, zygotene, and pachytene, this axis gains further structure as the nucleus displays condensed chromosomes at the basal side (toward the centriole) in the "bouquet stage" and a cap of nascent nucleoli toward the apical side. Later in the diplotene stage of anurans, a cloud of mitochondria and vesicles forms around the centriole, presumably because of the ability of the centriole and interphase aster to draw cell materials inward. At the same time, materials from the nucleus collect near its basal end, including a "nuage material" thought to contain precursors of the germinal plasm necessary for the establishment of germ cells in future embryonic development, still months away for the growing oocyte. Thus, the clear axial pattern during these early stages concerns organelle arrangement, most likely under the direction of the centriole and nucleus. Until the diplotene stage, the spermatocyte follows the same course of development as the oocyte; it then diverges and differentiates a head–flagellum axis based on the position of the nucleus and centriole. There is no evidence that extracellular agents orient the axis of internal components of the germ cells, although there is clear evidence that extracellular signals direct germ cells toward oogenesis or spermatogenesis. Pattern regulation at these early stages has not been studied but presumably involves at least those mechanisms used by all cells to restore their intracellular arrangement of contents each generation after cell division, perhaps via the interphase aster.

9.2   Disappearance of the Early Axis (8.4)

In the diplotene stage the mitochondrial cloud moves away from the nuclear surface and disperses. The nucleoli (1000–1500) move to positions in a uniform layer near the nuclear membrane. The cytoplasmic bridge to sister oocytes is severed as follicle cells cover the oocyte surface. Thus it is difficult to identify the oocyte axis at this stage. The state of the centriole and aster is not known. At some stage the centriole is inactivated for future use in the egg.

The period of vitellogenesis, in which the oocyte forms massive quantities of yolk platelets, begins after the follicle cells arrive. The major components of platelets are the proteins lipovitellin and phosvitin, derived by proteolytic cleavage from a precursor protein, vitellogenin. Vitellogenin is synthesized in the liver, transported by the blood stream, and sequestered in the oocyte by pinocytosis. Each platelet is surrounded by a unit membrane derived from the oocyte plasma membrane by way of numerous fusing pinocytotic vesicles. During vesicle fusion, proteolysis of vitellogenin leads to crystallization of the yolk proteins in an insoluble lattice in the center of the platelet. These proteins comprise the total reserve of essential amino acids for embryonic protein synthesis and of phosphate for nucleic acid synthesis. In addition, the platelet harbors phospholipid and various structural carbohydrates. By and large, the platelets contain all the nutrients a cell in an organism would normally obtain from the blood stream as processed by the liver. In that sense, the platelet is the adult liver's donation to embryogenesis. In addition, the platelet carries many sodium and calcium ions from the blood stream as counterions to the phosphate groups of phosvitin and thus comprises an intracellular reservoir of extracellular materials. The exposure of the oocyte and egg cytoplasm to these ions and nutrients depends in part on the permeability of the platelet membrane. Platelet membranes present approximately 100 times the surface of the oocyte plasma membrane, and may contribute significantly to the regulation of the composition of the cytosol of the cell.

9.4 Origin of the Animal–Vegetal Axis (8.5, 3.1)

Platelets become arranged in the oocyte according to a distinct pattern of size and number. The largest and probably oldest platelets collect with dense packing in one hemisphere, the vegetal hemisphere. In the opposite (animal) hemisphere the platelets are much smaller (approximately one-hundredth the volume) and only loosely packed. The animal hemisphere contains the large nucleus (the germinal vesicle), and the bulk of the cytosol and organelles of the oocyte. This pattern may be generated through the continued organizing action of the centriole and to the position of the nucleus as a source of newly synthesized materials. According to Wittek (1952) and others, young oocytes form small platelets uniformly around the periphery at first. These move inward, growing larger as they go. Their centripetal movement may be directed by the centriole and interphase aster although this point has not been established. The oldest and largest platelets reach the most interior position at the base of the germinal vesicle, where the centriole has previously been located. The germinal vesicle occupies a position toward one end of the oocyte, perhaps because of the centriole's position. As the oocyte grows in size, the germinal vesicle releases massive amounts of materials which collect around the nucleus in an enlarging sphere invaded only by small platelets. This becomes the animal hemisphere as the large platelets are displaced entirely to the other end, the vegetal hemisphere. Small platelets continue to form at the periphery, but more slowly at the vegetal end, perhaps because there is less active cytoplasm per unit of plasma membrane in that half. The vegetal half may be heterogeneous as far as its platelet composition, with intermediate platelets tending to occupy the equatorial level.

Early in vitellogenesis the cortical layer of the oocyte also differentiates according to the animal–vegetal axis. At first, melanosomes appear uniformly around the surface but gradually become attracted to the animal hemisphere (or excluded from the vegetal hemisphere) by an unknown mechanism. Cortical granules tend to remain uniformly distributed around the entire periphery, although there may be axial differences in the type of granule. While the animal–vegetal axis of the full-grown oocyte is plausibly an elaboration of the nuclear–centriolar axis of the oogonium, there are many unknown steps in this important polarization.

9.5   Pattern Regulation in Oogenesis (8.5, 8.6, 3.5)

While pattern formation in oogenesis concerns the pinocytosis of vitellogenin from the periphery, the centripetal movement of platelets under the direction of the centriole, and the centrifugal movement of materials from the eccentric germinal vesicle, pattern regulation at these stages may concern more the processes by which the oocyte unifies these movements and stabilizes the arrangements of organelles such as platelets. There may be two factors for the coherence of the vegetal hemisphere materials: one, the mutual affinity of large platelets; and two, the high viscosity of the cytoplasm. The affinity of platelets for one another may just reflect the fact that they formed by a series of fusions of vesicles mediated by close contact. The vegetal platelets also show affinity for the vegetal plasma membrane, again perhaps reflecting their origin from plasma membrane by pinocytosis. The high cytoplasmic viscosity precludes redistribution of materials by gravity during oogenesis when the oocytes are randomly oriented. This viscosity is presumably a reflection of the cytoskeletal system of the oocyte about which little is known. The coherence of the animal hemisphere may involve its own yolk-excluding properties or may result from the coherence of the vegetal yolk. The cortical specialization of the animal hemisphere may result from its cytoplasmic specialization.

According to Wittek (1952), the oocytes of some amphibian species display bilateral symmetry even in oogenesis, in terms of the lateral position of the germinal vesicle relative to the animal–vegetal axis of yolk platelets. She traces this arrangement back to the lateral position of the nucleus and mitochondrial cloud in the previtellogenic oocyte.

9.6   Oocyte Maturation (3.3, 3.4)

The full grown oocyte is a hormonally responsive cell, ready to mature to an unfertilized egg upon exposure to the steroid progesterone. Maturation requires the lifting of a G_2 or prophase arrest imposed on the oocyte since the last oogonial divisions months earlier. The transformation of the oocyte to an egg involves the nuclear and chromosomal changes of meiosis, alteration of the plasma membrane to a state of impermeability and preparedness for fertilization and activation, metabolic changes of the cytoplasm as the egg becomes nutritionally independent, and release from the follicle. The hormonal response is thought to be mediated entirely by cytoplasmic circuits, perhaps using cAMP and calcium ion as antagonists at an early step. Later, an autocatalytic cytoplasmic factor (maturation-promoting factor, MPF) triggers the nuclear and membrane changes, by processes thought to resemble mitotic steps, and still later a new cytoplasmic factor appears which arrests the egg at the second meiotic metaphase stage (cytostatic factor, CSF). The egg remains in this arrested state until fertilization. The oocyte has prepared a complex series of cytoplasmic changes executed without gene expression. This series runs until the midblastula stage when approximately 1000 cells are present. The full-grown oocyte and early developmental stages are suitable for the study of pattern formation and pattern regulation in the absence of gene expression.

9.7   Anisotropy of the Unfertilized Egg (2.6, 3.1, 4.1)

During maturation a new zone of cytoplasm originates in the animal hemisphere where the germinal vesicle breaks down and releases its sap. The meiotic spindle of second metaphase imbeds in the cortex at the apex of the animal hemisphere, defining the position of the animal pole. The vegetal pole is its geometric opposite but may have no discrete material basis to distinguish it from the surrounding vegetal regions. During maturation there are also changes in the arrangement of vegetal materials, such as a coalescence of islands of germinal granules. In the eggs of many amphibia the animal–vegetal axis defines the radial symmetry of the cell. In other species the egg already has bilateral symmetry, with its animal pole to one side. It is not known what determines the position of the animal pole in either case. Experiments on pattern regulation in oogenesis and maturation have become feasible in recent years due to progress in *in vitro* conditions, especially with the recent announcement by Wallace and Misulovin (1978) of success in growing oocytes from the previtellogenic stages, with development of the animal–vegetal axis.

9.8   Fertilization and Activation (4.2–4.4)

The unfertilized egg is activated by sperm entry which normally occurs in the animal hemisphere, or by the artificial stimulus of needle puncture or membrane-permeabilizing agents. Activation releases a series of prepared responses continuing past the midblastula stage. These include: (1) a rapid depolarization due to chloride efflux, thought to cause a rapid block to polyspermy; (2) an "activation wave" of discharge of cortical granules propogated from the point of sperm entry and perhaps spreading by way of a calcium ion flux; (3) a modification of the vitelline membrane and inner layer of the jelly associated with a slow block to polyspermy at the same time as egg interior repolarizes; (4) an inactivation of MPF and CSF, the cytoplasmic agents holding the egg in the second meiotic metaphase state, an inactivation perhaps dependent on a calcium ion flux; (5) the advancement of the egg through meiosis with extrusion of the second polar body and entry to interphase; and (6) the "activation contraction" in which the animal hemisphere cortex contracts toward the animal pole, perhaps mediated by a calcium ion flux. This contraction may mark the end of the meiotic period and the beginning of interphase.

During the first interphase of the egg, the sperm centriole organizes a large aster moving to the egg center followed by the sperm pronucleus. The egg is not able to mobilize its own fully functional centriole, which may have been inactivated sometime in oogenesis since it is absent from the meiotic spindle. The egg and sperm pronuclei meet in the center of the egg, near the centriole, at a time approximately half way from fertilization to first cleavage. As the sperm aster enlarges, a "postfertilization" wave is seen to travel across the egg surface from the site of sperm entry. Simultaneously, surface microvilli shorten and internally pigment granules associate more firmly with the cortex, the yolk-excluding cytoplasmic layer beneath the plasma membrane. Many of these surface changes occur in unfertilized activated eggs, but with an irregular schedule and orientation. In fertilized eggs these events are completed by the time of pronuclear pairing, the approximate time of a gray crescent formation.

9.9   The Gray Crescent (4.5, 4.6)

In many amphibian species a gray or clear crescent forms on one side of the animal hemisphere approximately halfway through the period from fertilization to first cleavage. The crescent is a surface region of lessened pigmentation. Although its mechanism of formation is not well studied, it seems likely that it arises in some species by a contraction of the animal hemisphere cortex toward the side of sperm entry, carrying pigment granules along. While the contractile machinery seems to be provided in part by the animal hemisphere, the signals for orienting the contraction may not be. Since unit gravity for short periods can easily determine the position of the crescent in horizontally positioned eggs, it is proposed that the orienting signal comes from a slight asymmetry in the easily displaceable yolk mass of the vegetal hemisphere. Normally, the gray crescent forms on the side opposite that of sperm entry—more explicitly, the side where the sperm aster originates. To reconcile this observation with that of gravitational reorientation, it is suggested that the aster normally affects slightly the contours of the vegetal yolk mass and thereby affects crescent position. In some species the crescent is not oriented by the sperm; there is evidence that in such cases the yolk mass already has an asymmetry before fertilization, one perhaps capable of fixing the crescent position. Since the crescent forms perfectly well in artificially activated eggs lacking the sperm aster, it is clear that the sperm aster is not a part of the crescent-forming machinery but of the orienting signal. The small aster of the activated egg may or may not function in crescent formation and orientation.

9.10   Internal Rearrangements (4.5, 4.6)

Internal movements accompanying crescent formation include: first, the shift of a yolk-free region of cytoplasm from the center of the animal hemisphere toward the crescent side; and second, the formation of a trail of large yolk platelets along the cortex on the crescent side (called the "vitelline wall"). The former occurs only in eggs that have been activated by sperm; the latter occurs even in eggs that have been artificially activated. The dorsal yolk-free cytoplasm seems to shift as a function of the aster whereas the vitelline wall forms as a function of the crescent. It is suggested that formation of the vitelline wall is the important function of the cortical contraction of crescent formation and that the gray crescent is just a record of the orientation of that asymmetric contraction. This proposal is made to bring the early events of crescent formation into line with the strong evidence of Nieuwkoop and his colleagues that the vegetal half is the site of the dorsalizing influences at the midblastula stage, not the animal half where the crescent resides.

9.11   Pattern Regulation before First Cleavage (4.6, 4.8)

Pattern regulation before the first cleavage concerns the means by which the vegetal yolk mass retains its coherence and the means by which the animal hemisphere cortex contracts as a unit so that only one prospective dorsal–ventral axis is formed. These aspects of the cell biology of the egg are unknown. As suggested before, the coherence of the yolk mass may reflect the history of the fusion of yolk granules in oogenesis, and the adherence of yolk granules to the vegetal membrane may reflect their origin from plasma membrane by pinocytosis.

9.12   The Prospective Dorsal–Ventral Axis (4.6)

The data of Dalcq and Pasteels (1937), Curtis (1960, 1962a,b), and Tompkins and Rodman (1971) for the uniqueness of the gray crescent cortex and for the cortical field have been reviewed in light of Nieuwkoop's evidence (1969a,b) for the exclusive role of the vegetal half as the carrier of the dorsal–ventral axis by the midblastula stage. It is suggested that many of the old data are equivocal, uncontrolled, or interpretable in terms of a vegetal as opposed to animal dorsalizing center. The Curtis experiments (1960, 1962a,b) on cortical transplantation, for example, require extensive control studies for twinning due to manipulations of the egg, such as rotation and surgery, at the time of first cleavage when the eggs of other species are known to be twinned by related treatments, without transplantation of the dorsal cortex.

The classical experiments of Schultze (1894, 1900), Born (1885), Penners and Schleip (1928a,b), and Pasteels (1938) on the rotation and inversion of eggs after gray crescent formation and even after first cleavage demonstrate that the position of the prospective dorsal–ventral axis can still be changed, or more explicitly the position of the vegetal dorsalizing center can be changed. Some of these effects may involve the shifting of an axis established at the time of crescent formation, but possibly some of the changes of position involve the elimination of an old dorsalizing center and creation of a new one. It is suggested that unit gravity acts by displacing the vegetal yolk mass, eliminating and creating a vitelline wall, a region of large vegetal yolk platelets exposed to the cytoplasm of the animal hemisphere. It is suggested that this region of hemispheric contact is the dorsalizing center. Normally, vitelline wall formation is a unique event, since it occurs by way of the one-time-only asymmetric contraction of the animal hemisphere cortex. However, gravity may simulate this mechanism at any time until cleavage furrows preclude the movement of internal materials—for example, with the horizontal cleavage of the 8-cell stage.

As found by Schultze (1894), and by Penners and Schleip (1928a,b), eggs inverted at the 2-cell stage can produce twin axes, the effect of twin vegetal dorsalizing centers. The cleavage furrow splits the vegetal yolk mass and perhaps allows its halves to move independently, perhaps forming independent vitelline walls. If vitelline wall formation is the primary event of dorsalization, the dorsal–ventral axis would be a modification of the original animal–vegetal axis shifted laterally in the vegetal hemisphere by the asymmetric cortical contraction of the animal hemisphere.

9.13   The Two Hemispheres of the Egg (4.6, 4.8)

As found by Motomura (1935) and Pasteels (1941), the fate map of the egg can be reversed 180° on the animal–vegetal axis by inverting the egg at the 1-cell stage and centrifuging it lightly to drive the vegetal yolk mass into the pigmented hemisphere, with the reciprocal displacement of the animal hemisphere cytoplasm. Also, the fate map can be shifted 90° in eggs centrifuged lightly at right angles to the original animal–vegetal axis.

Thus, any part of the cortex and surface can be used for any part of the fate map, including for the blastopore and for dorsal mesoderm, the organizer region of Spemann (1938). There is probably no autonomous irreversible cortical map provided to the egg from oogenesis. The cortex may assume regional differences in the two hemispheres as the result of its exposure to the yolk mass of one hemisphere and the much less yolky cytoplasm of the other hemisphere. As a general relationship, the position of the yolk mass always specifies the position of endoderm on the fate map, the position of the less yolky ("animal hemisphere") cytoplasm specifies the position of ectoderm, and the position of contact between the two is presumptive mesoderm. Thus the animal–vegetal organization of the fate map rests with the animal and vegetal internal contents, not on the cortex. As discussed later, the dorsal–ventral orientation of the map probably also depends on the positions of materials in the vegetal yolk mass, not the animal cortex.

When blastomeres of the animal hemisphere are isolated from the egg at early cleavage stages, or even at the 1-cell stage, by constriction, they divide and differentiate into ciliated epidermal cells as their single product. Vegetal blastomeres when isolated produce only large nutritive yolky cells similar to those found as the food source in the posterior gut lumen of the normal hatched but not yet feeding tadpole. Thus, cells of the two hemispheres have a very limited capacity for autonomous patterning, while the complex and reproducible fate map arises through interactions of the two hemispheres.

9.14   Cortical and Cytoplasmic Ingression (5.3)

As the first and second vertical cleavage furrows extend through the egg, materials of the cortex and cytoplasm stream inward along the newly formed plasma membrane in a process called "multipolar cytoplasmic ingression." This movement is prominent in the vegetal half where deep yolk platelets and islands of germinal plasm are carried upward toward the position of the future floor of the blastocoel. Vegetal ingression continues at least until the 64- to 128-cell stage when the first paratangential cleavages of the vegetal hemisphere cut off the path for further inward transport of materials. The function of ingression is unknown. It is suggested that it may affect the arrangement of materials of the vegetal hemisphere especially at the equatorial level where the two hemispheres meet, and therefore affect the approaching inductive interaction of the hemispheres. Thus, ingression might be classified as a morphogenic mechanism for the patterning of cytoplasmic localizations of the vegetal half, the same class in which the cortical contraction of gray crescent formation would belong. It is not known whether ingression contributes to pattern regulation at these early stages, perhaps by unifying the contents of the vegetal dorsalizing center.

9.15   Cleavage (4.7, 5.1)

The first 10 to 12 cycles of cell division occur at high speed and synchrony (or metachrony), without a G_1 or G_2 period. Furrow formation is preceded by a "surface contraction" wave propogated over the cell surface, coinciding with a wave of cortical tension and cortical competence for contractile ring formation. Surface contraction waves appear on schedule even if the nucleus has been removed and furrow formation is inhibited. The waves may reflect the activity of the fundamental cell cycle oscillator whose targets are the nucleus, spindle, centriole, and cortex. Synchronous or metachronous cleavage is preserved even by disaggregated blastomeres, indicating that synchrony reflects just the common origin of the cells and the absence of a "desynchronizing" mechanism. Division planes are displaced toward the less yolky end of the cell. This asymmetry of cleavage based on internal cell contents generates the animal–vegetal difference of cell size, an aspect of the pattern of the early blastula. Cleavage frequency is equivalent in the two hemispheres until shortly before gastrulation.

9.16   The Blastocoel (5.2)

An important element of pattern in the blastula stages is the blastocoel, a liquid-filled cavity in the interior of the animal hemisphere. It is generated by the following morphogenic mechanisms. First, the external egg membrane is impermeable to ions but slightly permeable to water, and seals itself during cleavages by tight junctions between blastomeres. Second, new plasma membrane, deposited between cleaving blastomeres, is permeable to ions and actively transports sodium ion from the cell interior to the intercellular spaces. The tight junctions between surface blastomeres are located at the border of the original egg membrane and the new plasma membrane, and thus ions are not lost from the intercellular space to the external medium of the egg. And third, there is more intercellular space in the animal hemisphere where the asymmetric cleavages produce smaller blastomeres, than in the vegetal hemisphere. As the sealed intercellular space becomes enriched in sodium ions, osmotic forces bring water in from the external pond water medium, across the semipermeable egg membrane and interblastomeric junctions. Thus, the blastocoel inflates. The yolk platelets are thought to be the intracellular reservoirs of sodium ion, replenishing the cytosol as the ion pumps expel sodium into the intercellular space. The permeability of the yolk platelet membrane may be a site of control of ion availability.

Gene expression is not required for blastocoel formation, nor for any aspect of pattern formation before the midblastula stage. The architecture of the midblastula is an elaboration of the pattern of the oocyte, transformed by the processes of cell division, tight junction formation, and ion pumping, acting within the geometrical context of the oocyte, particularly the position of the vegetal yolk mass and the ion-impermeable egg membrane. These morphogenic processes have limited properties of self-regulation in response to the displacement of internal materials. It is not known whether new plasma membrane can become impermeable when exposed to the external medium, a regulatory response suggested by Holtfreter (1943a,b).

9.17   Autonomy for Mesoderm Differentiation (5.5)

At the stage of 64–128 cells, the equatorial blastomeres acquire the ability, not expressed until much later, to differentiate mesoderm when removed from the egg and cultured in a balanced salts solution. Before this time, isolated blastomeres could form only ciliated epidermis (an ectodermal derivative), or nutritive yolk cells. This new autonomy presumably represents the completion of necessary cytoplasmic rearrangements or interactions between the hemispheres for later mesoderm development, with inductions and appropriate gene expression as intermediate steps. Gene expression is not detectable in amphibia until the 1000- to 2000-cell stage, the midblastula.

9.18   The Midblastula Transition (6.1, 6.3)

When the egg contains approximately 1000 cells from 10 metachronous cleavage cycles, a transition occurs in which the cell cycle lengthens, G_1 and G_2 periods first appear, RNA synthesis begins, and cleavages become asynchronous from blastomere to blastomere. This midblastula transition is thought to be triggered by an exhaustion of germinal vesicle materials supplied to the egg by the oocyte at maturation, since the summed volume of midblastula nuclei approximates that of the germinal vesicle. From the midblastula stage until well into gastrulation, nuclear number increases at the direct expense of nuclear volume. The importance of depletion of critical materials is further supported by the finding that the onset of the transition depends on ploidy, which affects nuclear volume. The identity of the critical materials is unknown, as is the sequence of events leading to RNA synthesis. The transition occurs at approximately the same time in all blastomeres, although perhaps slightly earlier in vegetal ones, and occurs as well in disaggregated blastomeres, indicating that the transition is cell-autonomous—that is, requiring no interactions in the egg. It is a major step in the sequence of temporal events played out by the cytoplasm over the long period starting from oogenesis and extending at least to gastrulation.

9.19   Mesoderm Induction (6.2)

Nieuwkoop and his colleagues (1969a,b, 1973, 1977) have identified intercellular effects that are necessary at the midblastula stage for the appearance of mesoderm in the embryo including the appearance of dorsal mesoderm, the "organizer" of the gastrula and neurula. In their experiments they surgically removed those equatorial blastomeres expected to be capable of autonomous mesoderm production by the midblastula stage, and checked their assumption by culturing the animal hemisphere fragment comprising the blastocoel roof and the vegetal fragment. The former produced ciliated epidermis, the apparent inherent differentiation of the animal hemisphere, whereas the latter produced nutritive yolk cells, the inherent differentiation of the vegetal hemisphere; no mesoderm appeared. When the two fragments were combined, mesoderm did appear and in fact the recombinates developed to remarkably coherent embryos with all regions of the fate map present in normal position. The animal fragment was identified as the source of the new-found mesoderm, and also of head endoderm and archenteron roof endoderm. It is possible that even the upper lip of the blastopore is derived from the animal fragment in these experiments. Thus the capacity for mesoderm production, including the organizer, is carried by the animal hemisphere, but is only evoked in the presence of the vegetal hemisphere, which presumably exerts an induction. Thus the positioning of mesoderm on the fate map at the zone of hemispheric contact is explained by the action of the vegetal hemisphere as an inductive source and the animal hemisphere as responsive target.

In addition, Nieuwkoop (1969b) rotated the animal fragment 180° relative to its vegetal partner and showed that the dorsal–ventral axis of the developing recombinate derived from the prospective axis of the vegetal piece, not of the animal piece. Thus, the vegetal hemisphere carries the prospective axis at the midblastula stage, including the capacity to determine the position of the future organizer.

At the midblastula stage the vegetal hemisphere behaves as a mosaic of inductive potencies, since a dorsal fragment in recombinates with a fragment of animal hemisphere induces dorsal mesoderm, whereas ventral or lateral fragments induce ventral mesoderm.

The activity of the dorsal fragment identifies it as the site of the "vegetal dorsalizing center," the inducer of the organizer (Boterenbrood and Nieuwkoop, 1973). Thus, the vegetal hemisphere at this stage has a definite pattern whereas the animal hemisphere does not. In the mid and late blastula stages, the pattern of the vegetal half is imprinted on the animal half by induction, organizing even the organizer. Thereafter the animal hemisphere takes over through the locomotor, inductive, and self-organizing properties of the organizer, the prospective dorsal mesoderm.

Since the pattern of the vegetal hemisphere is established in the period before gene expression begins, it is suggested that the pattern comprises one of cytoplasmic localizations, a pattern achieved by the rearrangement of vegetal materials received from the oocyte. Specifically, it is suggested that the vegetal dorsalizing center is the region derived from the vitelline wall, which may comprise a region of large vegetal yolk platelets brought into contact with animal hemisphere cytoplasm before first cleavage, and perhaps "activated" by this early contact. Also, it is suggested that these most inductive blastomeres must be located near the animal hemisphere to be effective. The vegetal dorsalizing center may owe its ultimate effectiveness both to its cytoplasmic composition and to its location.

The basis for the inductive patterning of the animal half by the vegetal half is not known. It is suggested that it is a quantitative pattern with the dorsal vegetal cells just emitting more inductor to their animal hemisphere targets than do the ventral vegetal cells. This proposal is consistent with certain experimental results itemized in Section 9.2. The induction is classified as a "vegetalizing induction" since animal blastomeres are induced to mesoderm and even endoderm derivatives.

The vegetal hemisphere may not impose on the animal hemisphere a pattern of high quantitative precision and detail, but rather one providing only rough threshold differences for the positioning of dorsal, lateral, and ventral mesoderm. The final accuracy of patterning of the animal hemisphere may result from the ability of its various regions of that hemisphere to self-organize when stimulated by induction. In this way, one could explain the approximately normal patterns of embryos in which artificial inductors have been used, or in which the dorsal–ventral axis has been reversed by temperature gradients or unilateral anaerobiosis, or in which the integrity of the vegetal hemisphere has been altered by surgery.

In the mid- and late blastula stages the blastocoel prevents the contact of prospective ectoderm with the inductively active vegetal blastomeres of the blastocoel floor. Thus, this cavity has an important role in the layout of the fate map.

9.20 Early Gene Expression (6.3, 6.5)

Nuclear synthesis of transfer and heterogeneous RNA is first detectable at the mid-blastula stage, as indicated by the incorporation of radioactive precursors. There is immediately a clear pattern of the regional differences in the extent of synthesis, presumably reflecting the response of the nuclei to preexisting regional cytoplasmic differences. Vegetal nuclei incorporate approximately 10 times more radioactivity than do animal hemisphere nuclei, and also dorsal vegetal nuclei incorporate almost twice as much as do ventral vegetal nuclei (Bachvarova and Davidson, 1966). Thus, the locus of highest incorporation corresponds topographically to the position of the vegetal dorsalizing center of Nieuwkoop. It is not known in what way RNA synthesis is required for mesoderm induction—that is, for the synthesis or release of inductor by the vegetal cells or for the reception or response to inductor by the animal hemisphere cells. Also it is not known whether there are regional differences in the kinds of RNA synthesized at this early stage. Ribosomal RNA is thought to comprise little or none of the transcripts until the gastrula and neurula stages.

9.21   Pattern Regulation in the Mid- and Late Blastula (6.5, 6.6)

Pattern regulation in the mid- and late blastula has been demonstrated by experiments on the constriction of blastulae into halves, either laterally or frontally, from which coherently patterned half-sized embryos result, and by experiments on the reversal of the prospective dorsal–ventral axis of blastulae by exposure to temperature gradients or to unilateral anaerobiosis. There may be regulative effects of two classes to distinguish. Those of the first class may comprise forced rearrangements of the animal–vegetal contact regions, even at these well cellularized stages, perhaps creating new vegetal dorsalizing centers and new formations of dorsal mesoderm. Constriction is a particularly effective means to force new contacts. Regulative effects of this class are related to the earlier gravitational effects, since both bring about new arrangements of materials of the egg—that is, altered cytoplasmic localizations. Effects of the second class may include temperature and oxygen gradients as modifiers of the pattern of inductor output by the vegetal half and inductor reception by the animal half. That is, cytoplasmic localizations are not changed but their consequences or expressions are. For example, in Penners' experiment (1936) of localized heating of the ventral vegetal region of the blastula, he may have intensified the output of inductor by the vegetal cells while prolonging the receptive period of the ventral animal blastomeres, thus changing the interhemispheric ventral balance to a dorsal one. Thus, effective axis reversal at the inductive blastula stages may involve changing the relative animal to vegetal rates of metabolism, but in opposite senses on the ventral and dorsal sides. The successful reversal of the axis by Lovtrup (1958; Landstrom and Lovtrup, 1975) with unilateral anaerobiosis is discussed as an effect of upsetting not only the dorsal–ventral balance but also the animal–vegetal one.

It is suggested that the vegetal half may have some properties of regulating the pattern of its inductor output, but that the main pattern regulation resides in the responsive animal blastomeres, perhaps as early as the late blastula stage and certainly by the gastrula stage when the prospective dorsal mesoderm takes over as the embryonic organizer.

The fundamental differences of prospective dorsal, lateral, and ventral mesoderm at the mid and late blastula stages are not known. They may be quantitative, with just more mesoderm cells on the dorsal side, or they may be qualitative.

9.22   Locomotion and Adhesion of Blastula Cells (6.4, 6.6)

Surface activity is observed in blastomeres immediately after the midblastula transition, even in blastomeres disaggregated from one another throughout cleavages. Thus, it is a cell-autonomous property, perhaps a function of the new G_1 or G_2 phase gained at the midblastula transition. Prospective mesodermal cells are thought to display lamellipodial motility earlier than do ectodermal and endodermal cells, especially in *X. laevis* where the mesoderm is internal. The tightly joined cells of the surface sheet may be inhibited from motility. Thus, there is a pattern imposed on this cell-autonomous property, based on the position of cells in the blastula. It is not known whether RNA synthesis is required for the acquisition of motility.

In the late blastula period there are also changes of cell surface adhesiveness and an increase in the production of extracellular materials by the blastomeres. Changes of adhesiveness may involve a substitution of one kind of affinity for another. Whereas the loss of preexising affinity may be independent of RNA synthesis, the gain of new affinity and the formation and release of extracellular polyanionic polymers requires gene expression, as indicated by studies of interspecies hybrids arresting at gastrulation. The production of extracellular materials occurs first among the prospective mesoderm cells. These cells also seem to gain a specific adhesiveness for both ectodermal and endodermal cells, appropriate to their intermediate position in the gastrula. Thus prospective mesoderm cells show early differences of behavior due to their exposure to the vegetal inductor and due to their origin in the animal hemisphere.

There are two pregastrulation movements of cell populations. One is the precocious migration of mesoderm in some amphibia, especially *X. laevis* where prospective mesoderm is internal and is halfway through with its movement around an internal blastopore lip before the external blastopore even appears. This movement is presumably an expression of the new lamellipodial motility and adhesiveness gained after induction. The other movement is epiboly, the expansion of the animal hemisphere surface with a compensatory contraction of the vegetal surface. During epiboly the original equatorial boundary of the hemispheres is driven toward the vegetal pole, positioning the future blastopore well below the geometrical equator. Epiboly may resemble the surface and cortical expansions of unfertilized eggs stored for prolonged periods ("pseudogastrulation"). According to this comparison, it would not require gene expression or the trigger of the midblastula transition, and would be scheduled by a cytoplasmic "clock" independent of the increasing nucleocytoplasmic ratio thought to time the midblastula transition. The timing of blastopore appearance and the susceptibility to grafts of the dorsal lip may also depend on this cytoplasmic clock. Thus gastrulation might represent the intersection of two lines of development, one based in the cytoplasmic and cortical circuitry set in the oocyte and the other including nuclear events, the increasing nucleocytoplasmic ratio, the exhaustion of materials, the midblastula transition, gene expression, and inductions.

9.23   The Blastopore (7.1)

Gastrulation formally begins with the blastopore lip appearing on the prospective dorsal side at the juncture of the large yolky vegetal blastomeres and the small less yolky animal blastomeres. The interface of yolky and less yolky cells is also the site of the multiple or displaced blastopores in rotated, inverted, or constricted eggs. This invariable location would be explained if the upper lip of the blastopore is the most vegetalized induction product from the animal hemisphere. The blastopore, especially the upper lip, contains bottle cells with long necks extending into the interior endodermal cell mass but retaining contact with the sheet of surface cells. Thus, they lead the invagination of surface cells, at least initially, although mesoderm migration probably controls and directs the continued movements of the expanding archenteron roof internally. The lining of the archenteron is derived entirely from surface cells of the egg which retain their specialized impermeable membrane and capacity for spreading as a tightly joined sheet.

As gastrulation proceeds, the blastopore extends laterally, forming bottle cells at its tips. Cells at the hemispheric interface gain the readiness or "competence" to be induced to form bottle cells well in advance of the time they actually do so; and in fact ventral cells are ready at the same time as are dorsal cells, as shown by the ability of ventral regions to add to the blastopore lip extending from a dorsal blastopore grafted into the ventral side. However, in the normal gastrula, the ventral cells do not form bottle cells until the dorsal lip has extended into their vicinity. Also, the ventral regions do not become autonomously able to form a blastopore until shortly before the lip reaches them, indicating that the actual induction occurs just shortly before bottle cell appearance. The basis of this bottle cell induction is not known.

9.24 Gastrulation (7.2)

Gastrulation involves the coordination of at least three morphogenic mechanisms: (1) the expansion of the surface sheet of blastomeres, (2) the migration of mesoderm across the roof of the blastocoel, and (3) the invagination of bottle cells. In the first, the uniform epibolic expansion of the animal hemisphere is supplemented by a strong expansion of the dorsal surface, both externally and internally, the latter blastomeres having invaginated to comprise the roof of the archenteron which stretches anteriorly. The surface expansion spreads laterally as gastrulation continues. In the second, the mesodermal cells migrate by lamellipodia across the blastocoel roof, a solid substrate. The population moves coherently, the individual loose cells governed by rules of contact inhibition and contact stimulation of movement. Although the blastocoel roof is the substrate, it does not provide directional information for migration. In the third mechanism, the bottle cells define the position of invagination of the surface cell sheet, and themselves are capable of invagination into the endoderm, but it is possible that the long-range internal movements of archenteron formation are less the result of the bottle cells than of the migrating mesoderm.

The means of coordination of these three mechanisms is unknown, but could plausibly reside in the mesodermal population located between the outer and internalized sides of the sheet of surface cells. As suggested by Holtfreter (1943b, 1944), the cell sheet of tightly connected surface blastomeres does provide unification since it moves as a unit and comprises an uninterrupted surface for mesodermal migration.

9.25 The Organizer (7.3, 7.4)

The organizer region of Spemann and Mangold (1924) comprises the dorsal mesoderm, including prospective head mesoderm, notochord, and somite as areas of the fate map. During gastrulation, this region takes over the dorsalizing role from the vegetal dorsalizing center, which simultaneously fades in effectiveness. In X. laevis the organizer derives from internal blastomeres at the dorsal wall and floor of the blastocoel, and not from surface cells of the gray crescent cortical region. In this species the organizer is not the topographical descendant of the crescent, further evidence of the separability of the two.

As a consequence of its own induction by the vegetal dorsalizing center, the dorsal mesoderm becomes inductive, perhaps first directing the migration of neighboring mesoderm from the lateral and ventral regions. This is indicated by the ability of dorsal lip transplants to attract to their domain a portion of the lateral and ventral mesoderm normally alligned with the host dorsal lip. As the migrating dorsal mesoderm population organizes itself as far as shape and position, it probably gains in its inductive effect on the overlying ectoderm and on the underlying endoderm, both of which are patterned in the anterior–posterior and mediolateral directions by the mesoderm. In the case of the induction of the neural plate in the ectoderm, there are inductions of at least two kinds at work. The first is the "neuralizing" induction by which the neural plate is outlined and set apart from the remaining ectoderm then destined for an epidermal fate. The neuralizing inductor is able to pass through nucleopore filters of the smallest pore size, suggesting it is a macromolecule or small molecule. The second induction defines the anterior–posterior regions of the plate through a caudalizing inductor not able to pass nucleopore filters. Nieuwkoop (1973) has suggested that the posterior mesoderm of the archenteron roof exerts the caudalizing influence. The caudalizing inductor may or may not be the same as the vegetalizing inductor acting at the mid- and late blastula stages. Once the regions of the neural plate are blocked out and oriented in the anterior–posterior and the mediolateral directions, pattern formation perhaps enters the phase of self-organization—that is, localized two-dimensional, intratissue patterning independent of the stimuli and quantitative information of inductors from neighboring tissues.

9.26   Self-Organization (7.3)

Self-organization becomes an increasingly demonstrable phenomenon in the gastrula and neurula stages, as stressed by Holtfreter (1951). First, ectodermal fragments removed from the early gastrula, dissociated, and reaggregated to induce neuralization and to destroy prior tissue organization, can nevertheless develop well-organized brain vesicles with eyes or nasal placodes—that is, under conditions in which the inductor is not likely to provide quantitative detail for patterning. Second, when an early gastrula receives in its ventral side an implant of the dorsal lip of the blastopore which has been disaggregated and reaggregated to destroy its structure, a coherent secondary head is still formed. And third, purified neuralizing and vegetalizing inductors from heterologous sources can induce head and tail structures, respectively, and these are well organized despite the abnormal spatial pattern of the inductors. In fact, if inductors of the two kinds are implanted as an oriented pair in solidified carriers, the embryo develops a complete secondary axis including trunk structures absent when either inductor is used singly. Thus the inductors may provide gross patterning on the basis of a double gradient, but at a more refined level the subregions of the neural plate may use self-organization. These results also indicate the continuing labile state of the ventral mesoderm and ectoderm at the gastrula stage, since they can be promoted to dorsal structures in the presence of inductors.

Self-organization belongs to the field phenomena discussed by Wolpert (1971) and Cooke (1975c) in terms of positional information, a patterning mechanism for two-dimensional sheets of cells. Inductions, on the other hand, may operate in the third dimension to coordinate in space the self-organization between two or more sheets of cells brought from distant parts of the embryo by morphogenic movements. Inductors might have a role in organ formation. Too little is known about self-organization in amphibia to discuss its mechanisms or its mechanistic relations to inductions.

9.27   Differential Cell Affinities (7.2–7.4)

The inside–outside axis of germ layers in the gastrula and neurula can be well explained on the basis of specific cell adhesion, as proposed by Townes and Holtfreter (1955) and by Steinberg (1963). Embryonic fragments dissociated to single cells and reaggregated at random sort out to restore the original layering of ectoderm, mesoderm, and endoderm. Thus, patterning during the cell movements of gastrulation and neurulation has a regulative capacity based in differential cell–cell affinities. In particular, mesoderm occupies an intermediate position because of its affinity for both ectoderm and endoderm; in the absence of ectoderm it sinks into the endoderm. The trilaminar structure of the embryo requires the interactions of the layers.

9.28 Pattern Regulation in the Gastrula and Neurula (7.3, 7.4)

Pattern regulation in the gastrula and neurula has been found in embryos receiving a secondary organizer region by transplantation. In this case, a double, single, or bifurcated axis develops depending on the angle between the host and graft organizers. The unification of slightly separated centers of dorsal mesoderm may occur during the stages of mesoderm migration when rules of contact stimulation and inhibition of movement and of cell sorting may govern the shaping of the mesodermal mantle. Lateral mesoderm can join the graft or host organizer in a period of apparent competition between organizers separated by a wide angle. The relative size of the two axes reflects this competition.

The dorsal mesoderm has the capacity to increase or decrease the size of its population during gastrulation. For example, embryos from which dorsal mesoderm has been ablated can replace it from neighboring mesoderm, and dorsal lip grafts can add more dorsal mesoderm to their axis by "assimilative induction." In the other direction, dorsal halves of gastrulae can reduce to half their final amount of dorsal mesoderm. Also, explants of the organizer can regulate to produce even ectodermal structures, essentially reversing their induced state. The homogeneity of the cell population of mesoderm is unknown. It appears that even the prospective dorsal mesoderm is a labile cell population at the gastrula stage, when the perpetuation of the mesodermal commitment for a cell depends upon the interactions it has with its neighbors. Thus the dorsal mesoderm is a strongly regulative cell population, completing extensive self-organization of its size and form before the induction of the neural plate begins. It is possible that much of the regulation of patterns disturbed at earlier stages occurs in gastrulation as the mesoderm organizes itself.

9.29 The Succession of Patterns (7.4)

The pattern of the neurula can be seen as the product of the morphogenic mechanisms of induction by the dorsal mesoderm and closure of the neural plate, operating within the geometry, or pattern, of the late gastrula. The gastrula can likewise be seen as the product of the mechanisms of cell locomotion of the mesoderm, of epibolic expansion of the ectoderm, and of specific cell adhesion operating within the pattern of the late blastula. And the arrangement of motile and adhesive cells in the late blastula depends on the pattern of the midblastula, particularly on the positions of the animal–vegetal contact regions where mesoderm induction occurs. Before this, the midblastula arises from the morphogenic mechanisms of ion-pumping, mitosis and cytokinesis, and tight junction formation acting on the architecture of the egg to form the blastocoel and thus limit the animal–vegetal contacts to the egg periphery at the equator. The spatial distribution of the blastocoel-forming mechanisms also depends on the architecture of the egg, especially its cytoplasmic pattern of yolk platelets and its surface position of impermeable membrane. Thus, we see how the pattern of each stage serves as the context for its own transformation into the next pattern, by defining not only the positions of the cells and cytoplasmic materials on which the morphogenic mechanisms will work, but also the positions of the morphogenic mechanisms themselves. Furthermore, we have seen that the morphogenic mechanisms have self-correcting properties related to their mode of action, although there is little information about these specific properties.

The amphibian embryo has such a great capacity for pattern regulation that it bears comparison more to the increasingly well-understood mammalian embryo than to the echinoderm or ascidian embryo to which it was compared years ago. This similarity will be still more apt if the amphibian gray crescent turns out to have only an indirect and transient role in axis determination and if the concept of the cortical field proves inappropriate. Whereas the mammalian embryo is considered to orient its axes according to the positions of external uterine signals, perhaps ions and metabolites, the amphibian embryo may do the same, but using external signals internalized months earlier by the oocyte by pinocytosis of materials from the maternal blood stream, deposited in the large yolk platelets of the vegetal hemisphere.

ACKNOWLEDGMENTS

The author wishes to thank Dr. P. D. Nieuwkoop for the opportunity to visit the Hubrecht Laboratory, Utrecht, The Netherlands, and use the excellent facilities and library, and to thank members of the laboratory for their hospitality, valuable discussions, and sharing of unpublished results, especially Drs. J. Bluemink, E. Boterenbrood, J. Faber, K. Hara, P. Nieuwkoop, and G. Ubbels. Also, the author thanks Drs. M. Thoman and J. Bonanno for their experimental contributions and suggestions, Dr. R. Keller for permission to use the material of Fig. 9 prior to publication, and Dr. Marc Kirschner for his perceptive questions and answers about amphibian embryology over the past seven years. The preparation of this review was facilitated by a fellowship from the Guggenheim Foundation and was supported by US PHS Grant GM 19363.

References

Adamson, E. D., and Woodland, H. R., 1974, Histone synthesis in early amphibian development: Histone and DNA synthesis are not co-ordinated, *J. Mol. Biol.* **88**:263.

Adamson, E. D., and Woodland, H. R., 1977, Changes in the rate of histone synthesis during oocyte maturation and very early development of *Xenopus laevis, Dev. Biol.* **57**:136.

Al-Mukhtar, K., and Webb, A. C., 1971, An ultrastructural study of the primordial germ cells, oogonia, and oocytes in *Xenopus laevis, J. Embryol. Exp. Morphol.* **26**:195.

Ancel, P., and Calame, S., 1959, Sur l'orientation du plan de symétrie bilatérale et de l'axe dorso-ventrale par la compression d'oeufs de *Rana fusca, C.R. Acad. Sci. Paris* **248**:893.

Ancel, P., and Vintemberger, P., 1948, Recherches sur le déterminisme de la symétrie bilatérale dans l'oeuf de Amphibiens, *Bull. Biol. Fr. Belg. (Suppl.)* **31**:1.

Ancel, P., and Vintemberger, P., 1949, La rotation de symétrisation, facteur de la polarisation dorso–ventrale des ébauches primordiales, dans l'oeuf des Amphibiens, *Arch. Anat. Microsc. Morphol. Exp.* **38**:167.

Anderson, D. M., and Smith, L. D., 1978, Patterns of synthesis and accumulation of heterogeneous RNA in lampbrush stage oocytes of *Xenopus laevis, Dev. Biol.* **67**:274.

Ansari, A. W., Dolphin, P. J., Lazier, C. B., Mundry, K. A., and Akhtar, M., 1971, Chemical composition of an oestrogen-induced calcium-binding glycolipophosphoprotein in *Xenopus laevis, Biochem. J.* **122**:107.

Arms, K., 1968, Cytonucleoproteins in cleaving eggs of *Xenopus laevis, J. Embryol. Exp. Morphol.* **20**:367.

Asashima, M., 1975, Inducing effects of the presumptive endoderm of successive stages in *Triturus alpestris*, *Roux Arch.* **177**:301.

Ave, K., Kawakami, I., and Sameshima, M., 1968, Studies on the heterogeneity of cell populations in amphibian presumptive epidermis, with reference to primary induction, *Dev. Biol.* **17**:617.

Bachvarova, R., and Davidson, E. H., 1966, Nuclear activation at the onset of amphibian gastrulation, *J. Exp. Zool.* **163**:285.

Baker, P., 1965, Fine structure and morphogenetic movements in the gastrula of the tree frog, *Hyla regilla*, *J. Cell Biol.* **24**:95.

Balinsky, B. I., 1966, Changes in the ultrastructure of amphibian eggs following fertilization, *Acta Embryol. Morphol. Exp.* **9**:132.

Balinsky, B. I., and Devis, R. J., 1963, Origin and differentiation of cytoplasmic structures in the oocytes of *Xenopus laevis*, *Acta Embryol. Morphol. Exp.* **6**:55.

Ballantine, J. E. M., Woodland, H. R., and Sturgess, E. A., 1979, Changes in protein synthesis during the development of *Xenopus laevis*, *J. Embryol. Exp. Morphol.* **51**:137.

Ballard, W., 1955, Cortical ingression during cleavage of amphibian eggs, studied by means of vital dyes, *J. Exp. Zool.* **129**:77.

Baltus, Ed., Brachet, J., Hanocq-Quertier, J., and Hubert, E., 1973, Cytochemical and biochemical studies on progesterone-induced maturation in amphibian oocytes, *Differentiation* **1**:127.

Banki, O., 1927, Die Lagebeziehungen der Spermium-Eintrittsstelle zur Medianebene und zur ersten Furche, nach Versuchen mit örtlicher Vitalfärbung am Axolotei, *Anat. Anz.* **63**:198.

Banki, O., 1929, Die Entstehung der ausseren Zeichen der bilateralen Symmetrie am Axolotlei, nach Versuchen mit örtlicher Vitalfärbung, *Verh. X Int. Zool. Kongr.*, Budapest, pp. 377–385.

Barth, L. G., and Barth, L. J., 1954, *The Energetics of Development*, Columbia University Press, New York.

Barth, L. G., and Barth, L. J., 1974, Ionic regulation of embryonic induction and cell differentiation in *Rana pipiens*, *Dev. Biol.* **39**:1.

Barth, L. J., 1956, Selective inhibition of cleavage in different regions of the frog egg by sulfhydryl inhibitors, *J. Embryol. Exp. Morphol.* **4**:73.

Bautzmann, H., 1927, Über Induction sekundärer Embryonalanlagen durch Implantation von Organisatoren in isolierte ventrale Gastrulahälften, *Roux Arch.* **110**:631.

Bautzmann, H., Holtfreter, J., Spemann, H., and Mangold, O., 1932, Versuche zur Analyse der Induktionsmittel in der Embryonalentwicklung, *Naturwissenschaften* **20**:972.

Beal, C. M., and Dixon, K. F., 1975, The effect of UV on cleavage of *Xenopus laevis* eggs, *J. Exp. Zool.* **192**:277.

Bellamy, A. W., 1919, Differential susceptibility as a basis for modification and control of early development in the frog, *Biol. Bull.* **37**:312.

Bellamy, A. W., 1921, Note concerning the origin of polarity in the frog's egg, *Biol. Bull.* **41**:351.

Benbow, R. M., and Ford, C. C., 1975, Cytoplasmic control of nuclear DNA synthesis during early development of *Xenopus laevis*: A cell-free assay, *Proc. Natl. Acad. Sci. U.S.A.* **72**:2437.

Benbow, R. M., Joenje, H., White, S. H., Breaux, C. B., Krauss, M. R., Ford, C. C., and Laskey, R. A., 1977, Cytoplasmic control of nuclear DNA replication in *Xenopus laevis,* in: *International Cell Biology 1976–1977* (B. P. Brinkley and K. R. Porter, eds.), pp. 453–463, Rockefeller University Press, New York.

Benford, H. H., and Namenwirth, M., 1974, Precocious appearance of the gray crescent in heat shocked *Axolotl* eggs, *Dev. Biol.* **39**:172.

Bergink, E. W., and Wallace, R. A., 1974, Precursor-product relationship between amphibian vitellogenin and the yolk proteins, lipovitellin and phosvitin, *J. Biol. Chem.* **249**:2897.

Bijtel, J. H., 1931, Uber die Entwicklung des Schwanzes bei Amphibien, *Roux Arch.* **125**:448.

Billet, F. S., and Adam, E., 1976, The structure of the mitochondrial cloud of *Xenopus laevis* oocytes, *J. Embryol. Exp. Morphol.* **36**:697.

Blackler, A. W., 1962, Transfer of primordial germ-cells between two subspecies of *Xenopus laevis*, *J. Embryol. Exp. Morphol.* **36**:697.

Blackler, A. W., 1965, Germ cell transfer and sex ratio in *Xenopus laevis*, *J. Embryol. Exp. Morphol.* **13**:51.

Blackler, A. W., 1970, The integrity of the reproductive cell line in the amphibia, *Curr. Top. Dev. Biol.* **5**:71.

Bluemink, J. G., 1971, Effects of cytochalasin B on surface contractility and cell junction formation during egg cleavage in *Xenopus laevis*, *Cytobiologie* **3**:1976.

Bluemink, J. G., 1972, Cortical wound healing in the amphibian egg: An electron microscopical study, *J. Ultrastruct. Res.* **41**:95.

Bluemink, J. G., and deLaat, S. W., 1973, New membrane formation during cytokinesis in normal and cytochalasin-B treated eggs of *Xenopus laevis* I. Electron microscopic observations, *J. Cell Biol.* **59**:89.

Bluemink, J. G., and deLaat, S. W., 1977, Plasma membrane assembly as related to cell division, in: *The Synthesis, Assembly, and Turnover of Cell Surface Components* (G. Poste and G. L. Nicolson, eds.), pp. 403–461, Elsevier/North-Holland Biomedical Press, Amsterdam.

Bluemink, J. G., and Tertoolen, L. G. J., 1978, The plasma membrane IMP pattern as related to animal/vegetal polarity in the amphibian egg, *Dev. Biol.* **62**:334.

Blumenthal, A. B., Kriegstein, H. J., and Hogness, D. S., 1973, The units of DNA replication in *Drosophila melanogaster* chromosomes, *Cold Spring Harbor Symp. Quant. Biol.* **38**:205.

Boell, E. J., 1947, Biochemical differentiation during amphibian development, *Ann. N.Y. Acad. Sci.* **49**:773.

Bonanno, J., 1977, Studies on dorsal–ventral polarity in embryos of Xenopus, Master's thesis, University of California, Berkeley, Calif.

Bonner, W. M., 1975a, Protein migration into nuclei. I. Frog oocyte nuclei *in vivo* accumulate microinjected histones, allow entry to small proteins, and exclude large proteins, *J. Cell Biol.* **64**:421.

Bonner, W. M., 1975b, Protein migration into nuclei. II. Frog oocytes nuclei accumulate a class of microinjected oocyte nuclear proteins and exclude a class of microinjected cytoplasmic proteins, *J. Cell Biol.* **64**:431.

Born, G., 1885, Über den Einfluss der Schwere auf das Froschei, *Arch. Mikrosk. Anat.* **24**:475.

Boterenbrood, E. C., 1962, On pattern formation in the prosencephalon: An investigation on disaggregated and reaggregated presumptive prosencephalic material of neurulae of *Triturus alpestris,* Ph.D. thesis, University of Utrecht, Utrecht.

Boterenbrood, E. C., and Nieuwkoop, P. D., 1973, The formation of the mesoderm in urodelian amphibians. V. Its regional induction by the endoderm, *Roux Arch.* **173**:319.

Brachet, J., 1968, *Chemical Embryology* (L. G. Barth, trans.), Hafner, New York.

Brachet, J., 1962, Effects of β-mercaptoethanol and lipoic acid on morphogenesis, *Nature (London)* **193**:87.

Brachet, J., 1972, Studies on nucleocytoplasmic interactions during early amphibian development. II. Cytochemical analysis of inversion and centrifugation experiments, *Arch. Biol. (Liège)* **83**:243.

Brachet, J., 1977, An old enigma: The gray crescent of amphibian eggs, *Curr. Top. Dev. Biol.* **11**:133.

Brachet, J., and Hubert, E., 1972, Studies on nucleocytoplasmic interactions during early amphibian development. I. Localized destruction of the egg cortex, *J. Embryol. Exp. Morphol.* **27**:121.

Brachet, J., Denis, H., and DeVitry, F., 1964, The effects of antinomycin D and puromycin on morphogenesis in amphibian eggs and *Acetabularia mediterranea, Dev. Biol.* **9**:398.

Brachet, J., Hanocq, F., and Van Gansen, P., 1970, A cytochemical and ultrastructural analysis of *in vitro* maturation in amphibian oocytes, *Dev. Biol.* **21**:157.

Brachet, J., Hubert, E., and Lievens, A., 1972, The effects of α-amanitin and rifampicins on amphibian egg development, *Rev. Suisse Zool.* **79**:47.

Brachet, J., Baltus, E., DeSchutter-Pays, A., Hancocq-Quertier, J., Hubert, E., and Steinert, G., 1975, Induction of maturation (meiosis) in *Xenopus laevis* oocytes by three organomercurials, *Proc. Natl. Acad. Sci. U.S.A.* **72**:1574.

Bragg, A. N., 1938, The organization of the early embryo of *Bufo cognatus* as revealed especially by the mitotic index, *Z. Zellforsch. Mikroskop. Anat.* **28**:154.

Bragg, A. N., 1939, Observations upon amphibian deutoplasm and its relation to embryonic and early larval development, *Biol. Bull.* **77**:268.

Brauns, A., 1940, Untersuchungen zur Ermittlung der Entstehung der roten Blutzellen in der Embryonalentwicklung der Urodelen, *Roux Arch.* **140**:741.

Bravo, R., and Knowland, J., 1979, Classes of proteins synthesized in oocytes, eggs, embryos, and different tissues of *Xenopus laevis, Differentiation* **13**:101.

Bravo, R., Otero, C., Allende, C., and Allende, J., 1978, Amphibian oocyte maturation and protein synthesis: Related inhibition by cAMP, theophylline, and papaverine, *Proc. Natl. Acad. Sci. U.S.A.* **75**:1242.

Brice, M. C., 1958, A reanalysis of the consequences of frontal and saggital constrictions of newt blastulae and gastrulae, *Arch. Biol. (Liège)* **69**:371.

Briggs, R., 1949, The influence of egg volume on the development of haploid and diploid embryos of the frog, *Rana pipiens, J. Exp. Zool.* **111**:255.

Briggs, R., and Cassens, G., 1966, Accumulation in the oocyte nucleus of a gene product essential for embryonic development beyond gastrulation, *Proc. Natl. Acad. Sci. U.S.A.* **55**:1103.

Briggs, R., and Justis, J. T., 1968, Partial characterization of the component from normal eggs which corrects the maternal effect of the gene *O* in the Mexican axolotl, *J. Exp. Zool.* **167**:105.

Briggs, R., and King, T. J., 1953, Factors affecting the transplantability of nuclei of frog embryonic cells, *J. Exp. Zool.* **122**:485.

Briggs, R., Green, E. U., and King, T. J., 1951, An investigation of the capacity for cleavage and differentiation in *Rana pipiens* eggs lacking "functional" chromosomes, *J. Exp. Zool.* **116**:455.

Bronsted, H. V., and Meyer, H., 1950, Does a correlation exist between the egg axis and the egg attachment in the ovary of *Rana temporaria? Vidensk. Medd. Dansk Naturhist. For. (Copenhagen)* **112**:253.

Brothers, A. J., 1976, Stable nuclear activation dependent on a protein synthesized during oogenesis, *Nature (London)* **260**:112.

Brown, D. D., and Blackler, A. W., 1972, Gene amplification proceeds by a chromosome copy mechanism, *J. Mol. Biol.* **63**:75.

Brown, R. D., and Brown, D. D., 1976, The nucleotide sequence adjoining the 3′ end of the genes coding for oocyte-type 5S ribosomal RNA in *Xenopus, J. Mol. Biol.* **102**:1.

Brown, D. D., and Dawid, I. D., 1968, Specific gene amplication in oocytes, *Science* **160**:272.

Brown, D. D., and Litna, E., 1964a, RNA synthesis during the development of *Xenopus laevis*, the South African clawed toad, *J. Mol. Biol.* **8**:669.

Brown, D. D., and Litna, E., 1964b, Variations in the synthesis of stable RNA's during oogenesis and development in *Xenopus laevis, J. Mol. Biol.* **8**:688.

Brown, D. D., and Litna, E., 1966, Synthesis and accumulation of low molecular weight RNA during embryogenesis of *Xenopus laevis, J. Mol. Biol.* **20**:95.

Brummett, A. R., and Dumont, J. N., 1976, Oogenesis in *Xenopus laevis* (Daudin). III. Localization of negative charges on the surface of developing oocytes, *J. Ultrastruct. Res.* **55**:4.

Brummett, A. R., and Dumont, J. N., 1977, Intracellular transport of vitellogenin in *Xenopus* oocytes: An autoradiographic study, *Dev. Biol.* **60**:482.

Brun, R. B., 1974, Studies on fertilization in *Xenopus laevis, Biol. Reprod.* **11**:513.

Brun, R., 1975, Oocytes maturation *in vitro:* Contribution of the oviduct to total maturation in *Xenopus laevis, Experentia* **31**:1275.

Bruns, E., 1931, Experiments über das Regulationsvermögen der Blastula von *Triton taeniatus* und *Bombinator pachypus, Roux Arch.* **123**:682.

Buehr, M. L., and Blackler, A. W., 1970, Sterility and partial sterility in the South African clawed toad following pricking of the egg, *J. Embryol. Exp. Morphol.* **23**:375.

Burnside, B., 1971, Microtubules and microfilaments in newt neurulation, *Dev. Biol.* **26**:416.

Burnside, M. B., and Jacobsen, A. G., 1968, Analysis of morphogenetic movements in the neural plate of the newt *Taricha torosa, Dev. Biol.* **18**:537.

Callan, H. G., 1963, The nature of lampbrush chromosomes, *Int. Rev. Cytol.* **15**:1.

Campanella, C., and Andreuccetti, P., 1977, Ultrastructural observations on cortical endoplasmic reticulum and on residual cortical granules in the egg of *Xenopus laevis, Dev. Biol.* **56**:1.

Caplan, A. I., and Rosenberg, M. J., 1975, Interrelationship between poly (ADP-Rib) synthesis, intracellular NAD levels, and muscle or cartilage differentiation from mesodermal cells of embryonic chick limb, *Proc. Natl. Acad. Sci. U.S.A.* **72**:1852.

Carroll, C. R., 1974, Comparative study of the early embryonic cytology and nucleic acid synthesis of *Ambystoma mexicanum* normal and *O* mutant embryos, *J. Exp. Zool.* **187**:409.

Carroll, E. J., Jr., and Hedrick, J. L., 1974, Hatching in the toad *Xenopus laevis:* Morphological events and evidence for a hatching enzyme, *Dev. Biol.* **38**:1.

Century, T. J., and Horowitz, S., 1974, Sodium exchange in the cytoplasm and nucleus of amphibian oocytes, *J. Cell Sci.* **16**:465.

Century, T. J., Fenichel, I. R., and Horowitz, S. B., 1970, The concentrations of water, sodium, and potassium in the nucleus and cytoplasm of amphibian oocytes, *J. Cell Sci.* **7**:51.

Chamberlin, M. E., Britten, R. J., and Davidson, E. H., 1975, Sequence organization in *Xenopus* DNA studied by the electron microscope, *J. Mol. Biol.* **96**:317.

Chang, C. Y., and Witschi, E., 1956, Genic control and hormonal reversal of sex differentiation in *Xenopus, Proc. Soc. Exp. Biol. N.Y.* **89**:150.

Chase, J. W., and Dawid, I. B., 1972, Biogenesis of mitochondria during *Xenopus laevis* development, *Dev. Biol.* **27**:504.

Child, C. M., 1915, *Individuality in Organisms,* University of Chicago Press, Chicago, Ill.

Child, C. M., 1928, The physiological gradients, *Protoplasma* **5**:447.

Child, C. M., 1929, Physiological dominance and physiological isolation in development and reconstitution, *Roux Arch.* **117**:21.

Child, C. M., 1941, *Patterns and Problems of Development,* University of Chicago Press, Chicago, Ill.

Child, C. M., 1946, Organizers in development and the organizer concept, *Physiol. Zool.* **19**:89.

Cho, W. K., Stern, S., and Biggers, J. D., 1974, Inhibitory effect of dibutyryl cyclic AMP on mouse oocyte maturation *in vitro, J. Exp. Zool.* **187**:383.

Chuang, H. H., 1940, Weitere Versuche über die Veränderung der Inductionsleistungen von gekochten Organ-teilen, *Roux Arch.* **140**:25.

Chulitskaia, E. V., 1970, Desynchronization of cell divisions in the course of egg cleavage and an attempt at experimental shift of its onset, *J. Embryol. Exp. Morphol.* **23**:359.

Chung, H. M., and Malachinski, G. M., 1975, Repair of ultraviolet irradiation damage to a cytoplasmic component required for neural induction in the amphibian egg, *Proc. Natl. Acad. Sci. U.S.A.* **72**:1235.

Citkowitz, E., 1971, The hyaline layer: Its isolation and role in echinoderm development, *Dev. Biol.* **24**:348.

Clark, T. G., and Merriam, R. W., 1977, Diffusible and bound actin in nuclei of *Xenopus laevis* oocytes, *Cell* **12**:883.

Clavert, J., 1962, Symmetrization of the egg of vertebrates, *Adv. Morphog.* **2**:27.

Cloud, J. G., and Schuetz, A. W., 1977, Interaction of progesterone with all or isolated portions of the amphibian *(Rana pipiens)* oocyte surface, *Dev. Biol.* **60**:359.

Coffman, G. K., Keem, K., and Smith, L. D., 1979, The progesterone receptor-like properties of *Xenopus laevis* oocyte melanosomes are probably due to eumelanin, *J. Exp. Zool.* **207**:375.

Coggins, L. W., 1973, An ultrastructural and autoradiographic study of early oogenesis in the toad, *Xenopus laevis*, *J. Cell Sci.* **12**:71.

Coggins, L. W., and Gall, J. G., 1972, The timing of meiosis and DNA synthesis during early oogenesis in the toad, *Xenopus laevis*, *J. Cell Biol.* **52**:569.

Colman, A., and Gadian, D. G., 1976, ^{31}P-nuclear-magnetic-resonance studies on the developing embryos of *Xenopus laevis*, *Eur. J. Biochem.* **61**:387.

Connolly, J. A., Kalnins, V. I., Cleveland, D. W., and Kirschner, M. W., 1977, Immunofluorescent staining of cytoplasmic and spindle microtubules in mouse fibroblasts with antibody to tau protein, *Proc. Natl. Acad. Sci. U.S.A.* **74**:2437.

Cooke, J., 1972a, Properties of the primary organization field in the embryo of *Xenopus laevis*. I. Autonomy of cell behavior at the site of initial organizer formation, *J. Embryol. Exp. Morphol.* **28**:13.

Cooke, J., 1972b, Properties of the primary organization field in the embryo of *Xenopus laevis*. II. Positional information for axial organization in embryos with two head organizers, *J. Embryol. Exp. Morphol.* **28**:27.

Cooke, J., 1972c, Properties of the primary organization field in the embryo of *Xenopus laevis*. III. Retention of polarity in cell groups excised from the region of the early organizer, *J. Embryol. Exp. Morphol.* **28**:47.

Cooke, J., 1973a, Properties of the primary organization field in the embryo of *Xenopus laevis*. IV. Pattern formation and regulation following early inhibition of mitosis, *J. Embryol. Exp. Morphol.* **30**:49.

Cooke, J., 1973b, Properties of the primary organization field in the embryo of *Xenopus laevis*. V. Regulation after removal of the head organizer in normal early gastrulae and in those already possessing a second implanted organizer, *J. Embryol. Exp. Morphol.* **30**:283.

Cooke, J., 1975a, Local autonomy of gastrulation movements after dorsal lip removal in two anuran amphibians, *J. Embryol. Exp. Morphol.* **33**:147.

Cooke, J., 1975b, Control of somite number during morphogenesis of a vertebrate, *Xenopus laevis*, *Nature (London)* **254**:196.

Cooke, J., 1975c, The emergence and regulation of spatial organization in early animal development, *Ann. Rev. Biophys. Bioeng.* **4**:185.

Cooke, J., 1979a, Cell number in relation to primary pattern formation in the embryo of *Xenopus laevis*. I. The cell cycle during new pattern formation in response to implanted organizers, *J. Embryol. Exp. Morphol.* **51**:165.

Cooke, J., 1979b, Cell number in relation to primary pattern formation in the embryo of *Xenopus laevis*. II. Sequential cell recruitment and control of the cell cycle, during mesoderm formation, *J. Embryol. Exp. Morphol.* **53**:269.

Cooke, J., and Zeeman, E. C., 1976, A clock and wavefront model for the control of the number of repeated structures during animal morphogenesis, *J. Theoret. Biol.* **58**:455.

Crippa, M., Davidson, E. N., and Mirsky, A. E., 1967, Persistence in early embryos of informational RNAs from the lampbrush chromosome stage in oogenesis, *Proc. Natl. Acad. Sci. U.S.A.* **57**:885.

Curtis, A. S. G., 1960, Cortical grafting in *Xenopus laevis*, *J. Embryol. Exp. Morphol.* **8**:163.

Curtis, A. S. G., 1962a, Morphogenetic interactions before gastrulation in the amphibian, *Xenopus laevis*—The cortical field, *J. Embryol. Exp. Morphol.* **10**:410.

Curtis, A. S. G., 1962b, Morphogenetic interactions before gastrulation in the amphibian, *Xenopus laevis*—Regulation in blastulae, *J. Embryol. Exp. Morphol.* **10**:451.

Curtis, A. S. G., 1963, The cell cortex, *Endeavor* **22**:134.

Curtis, A. S. G., 1965, Cortical inheritance in the amphibian *Xenopus laevis:* Preliminary results, *Arch. Biol. (Liège)* **76**:523.

Czolowska, R., 1969, Observations on the origin of the "germinal cytoplasm" in *Xenopus laevis, J. Embryol. Exp. Morphol.* **22**:229.

Czolowska, R., 1972, The fine structure of the "germinal cytoplasm" in the egg of *Xenopus laevis, Roux Arch.* **169**:335.

Dalcq, A. M., 1949, The concept of physiological competition (Spiegelman) and the interpretation of vertebrate morphogenesis, *Exp. Cell Res. (Suppl.)* **1**:483.

Dalcq, A., and Pasteels, J., 1937, Une conception nouvelle des bases physiologiques de la morphogénèse, *Arch. Biol.* **48**:669.

Daniel, J. F., and Yarwood, E. A., 1939, The early embryology of *Triturus torusus, Univ. Calif. Publ. Zool.* **43**:321.

Dasgupta, S., and Kung-Ho, C., 1971, Electrophoretic analysis of cell populations in presumptive epidermis of the frog, *Rana pipiens, Exp. Cell Res.* **65**:463.

Davidson, E. H., 1977, *Gene Activity in Early Development,* 2nd ed., Academic Press, New York.

Davidson, E. H., and Hough, B. B., 1969, High sequence diversity in the RNA synthesized at the lampbrush stage of oogenesis, *Proc. Natl. Acad. Sci. U.S.A.* **63**:342.

Davidson, E. H., Crippa, M., and Mirsky, A. E., 1968a, Genomic function during the lampbrush chromosome stage of amphibian oogenesis, *Proc. Natl. Acad. Sci. U.S.A.* **56**:856.

Davidson, E. H., Crippa, M., and Mirsky, A. E., 1968b, Evidence for the appearance of novel gene products during amphibian blastulation, *Proc. Natl. Acad. Sci. U.S.A.* **60**:152.

Dawid, I. B., 1965, Deoxyribonucleic acid in amphibian eggs. *J. Mol. Biol.* **12**:589.

deLaat, S. W., and Barts, P. W. J. A., 1976, New membrane formation and intercellular communication in the early *Xenopus* embryo. II. Theoretical analysis, *J. Membr. Biol.* **27**:131.

deLaat, S. W., Barts, P. W. J. A., and Bakker, M. I., 1976, New membrane formation and intercellular communication in the early *Xenopus* embryo. I. Electrophysiological analysis, *J. Membr. Biol.* **27**:109.

Del Pino, E. H., and Humphries, A. A., Jr., 1978, Multiple nuclei during oogenesis in *Plectonotus pygmaeus, Biol. Bull.* **154**:198.

Denis, H., 1966, Gene expression in amphibian development. II. Release of the genetic information in growing embryos, *J. Mol. Biol.* **22**:285.

Denis, H., 1971, Role of messenger ribonucleic acid in embryonic development, *Adv. Morphog.* **7**:115.

Denis, S., and Devlin, T. M., 1968, The effect of oligomycin on the development of amphibian eggs, *Exp. Cell Res.* **52**:308.

Denis, H., and Wegnez, M., 1977, Biochemical research on oogenesis: Oocytes of *Xenopus laevis* synthesize but do not accumulate 5S RNA of the somatic type, *Dev. Biol.* **58**:212.

DeRobertis, E. M., Longthorne, R. F., and Gurdon, J. B., 1978, Intracellular migration of nuclear proteins in *Xenopus* oocytes, *Nature (London)* **272**:254.

Dettlaff, T. A., 1962, Cortical changes in Acipenserid eggs during fertilization and artificial activation, *J. Embryol. Exp. Morphol.* **10**:1.

Dettlaff, T. A., 1964, Cell divisions, duration of interkinetic states, and differentiation in early stages of embryonic development, *Adv. Morphog.* **3**:323.

Dettlaff, T. A., Nikitina, L. A., and Stroeva, O. G., 1964, The role of the germinal vesicle in oocyte maturation in anurans as revealed by the removal and transplantation of nuclei, *J. Embryol. Exp. Morphol.* **12**:851.

Deuchar, E. M., 1956, Amino acids in developing tissues of *Xenopus laevis, J. Embryol. Exp. Morphol.* **4**:327.

Deuchar, E. M., 1958, Regional differences in catheptic activity in *Xenopus laevis* embryos, *J. Embryol. Exp. Morphol.* **6**:223.

Deuchar, E. M., 1972, *Xenopus laevis* and developmental biology, *Biol. Rev.* **47**:37.

Deuchar, E., 1975, *Xenopus, the South African Clawed Frog,* Wiley, New York.

DiCaprio, R. A., French, A. S., and Sanders, E. J., 1974, Dynamic properties of electrotonic coupling between cells of early *Xenopus* embryos, *Biophys. J.* **14**:387.

Dick, D. A. T., Fry, D. J., 1975, Sodium fluxes in single amphibian oocytes: Further studies and a new model, *J. Physiol.* **247**:91.

Dolecki, G. J., and Smith, L. D., 1979, Poly(A)$^+$RNA metabolism during oogenesis in *Xenopus laevis, Dev. Biol.* **69**:217.

Dollander, A., 1950, Etude des phénomènes de regulation consécutifs à la séparation des deux premiers blastomeres de l'oeuf de Triton, *Arch. Biol.* **61**:1.

Dollander, A., 1962, Organization corticale de l'oeuf d'amphibien, *Arch. Anat. Histol. Embryol.* **44**:93.

Dollander, A., and Melnotte, J. P., 1952, Variation topographique de la colorabilité du cortex de l'oeuf symé-trisé de *Triturus alpestris* au bleu de Nil et au rouge neutre, *C.R. Soc. Seances Biol. Paris* **146**:1614.

Doucet-de-Bruine, M. H. M., 1973, Blastopore formation in *Ambystoma mexicanum, Roux Arch.* **173**:136.

Driesch, H., 1892, Entwicklungsmechanische Studien. I. Der Werth der beiden ersten Furchungszellen in der Echinodermenentwicklung. Experimentelle Erzeugen von Theil- und Doppelbildung, *Z. Wiss. Zool.* **53**:160.

Drury, K. C., 1978, Method for the preparation of active maturation promoting factor (MPF) from *in vitro* matured oocytes of *Xenopus laevis, Differentiation* **10**:181.

Drury, K., and Schorderet-Slatkine, S., 1975, Effects of cycloheximide on the "autocatalytic" nature of the maturation promoting factor (MPF) in oocytes of *Xenopus laevis, Cell* **11**:269.

Dumont, J. N., 1972, Oogenesis in *Xenopus laevis* (Daudin). I. Stages of oocyte development in laboratory maintained animals, *J. Morphol.* **136**:153.

Duryee, W. R., 1950, Chromosomal physiology in relation to nuclear structure, *Ann. N.Y. Acad. Sci.* **50**:920.

Edwards, M. J., Mulley, R., Ring, S., and Wanner, R. A., 1974, Mitotic cell death and delay of mitotic activity in guinea-pig embryos following brief maternal hyperthermia, *J. Embryol. Exp. Morphol.* **32**:593.

Elinson, R. P., 1973, Fertilization of frog body cavity eggs enhanced by treatments affecting the vitelline coat, *J. Exp. Zool.* **183**:291.

Elinson, R. P., 1975, Site of sperm entry and a cortical contraction associated with egg activation in the frog *Rana pipiens, Dev. Biol.* **47**:257.

Elinson, R. P., 1977, Fertilization of immature frog eggs: Cleavage and development following subsequent activation, *J. Embryol. Exp. Morphol.* **37**:187.

Elinson, R. P., and Manes, M. E., 1978, Morphology of the site of sperm entry on the frog egg, *Dev. Biol.* **63**:67.

Elinson, R. P., and Manes, M. E., 1979, Gray crescent formation and its inhibition by ultraviolet light, *J. Cell Biol.* **83**:213a (abstract).

Epel, D., 1978, Mechanisms of activation of sperm and egg during fertilization of sea urchin gametes, *Curr. Top. Dev. Biol.* **12**:185.

Eppig, J. J., Jr., 1970, Melanogenesis in amphibians: The buoyant density of oocyte and larval *Xenopus laevis* melanosomes and the isolation of oocyte melanosomes from the eyes of PTU-treated larvae, *J. Exp. Zool.* **175**:467.

Fankhauser, G., 1930, Die Entwicklung diploid kerniger Hälften des ungefurchten Tritoneies, *Roux Arch.* **122**:671.

Fankhauser, G., 1932, Cytological studies on egg fragments of the salamander *Triton.* II. The history of the supernumerary sperm nuclei in normal fertilization and cleavage of fragments containing the egg nucleus, *J. Exp. Zool.* **62**:185.

Fankhauser, G., 1934a, Cytological studies on egg fragments of the salamander *Triton.* III. The early development of the sperm nuclei in egg fragments without the egg nucleus, *J. Exp. Zool.* **67**:159.

Fankhauser, G., 1934b, Cytological studies on egg fragments of the salamander *Triton.* IV. The cleavage of egg fragments without the egg nucleus, *J. Exp. Zool.* **67**:349.

Fankhauser, G., 1937, The development of fragments of the fertilized *Triton* egg with the egg nucleus alone ("gynomerogony"), *J. Exp. Zool.* **15**:413.

Fankhauser, G., 1945a, The effects of changes in chromosome numbers on amphibian development, *Q. Rev. Biol.* **20**:20.

Fankhauser, G., 1945b, Maintenance of normal structure in heteroploid salamander larvae through compensation of changes in cell size by adjustment of cell number and cell shape, *J. Exp. Zool.* **100**:445.

Fankhauser, G., 1948, The organization of the amphibian egg during fertilization and cleavage, *Ann. N.Y. Acad. Sci.* **49**:684.

Fankhauser, G., 1955, The role of nucleus and cytoplasm, in: *Analysis of Development* (B. H. Willier, P. A. Weiss, and V. Hamburger, eds.), pp. 126–150, W. B. Saunders, New York.

Fankhauser, G., and Moore, C., 1941a, Cytological and experimental studies of polyspermy in the newt, *Triturus viridescens.* I. Normal fertilization, *J. Morphol.* **68**:347.

Fankhauser, G., and Moore, C., 1941b, Cytological and experimental studies of polyspermy in the newt, *Triturus viridescens.* II. The behavior of the sperm nuclei in androgenetic eggs (in the absence of the egg nucleus), *J. Morphol.* **68**:387.

Faulhaber, I., 1970, Anreicherung des vegetalisierenden Inductions-factor aus der Gastrula des Krallenfrosches *(Xenopus laevis)* und Abgrenzung des Molekulargewichtsbereiches durch Gradientenzentrifugation, *Hoppe-Seyler's Physiol. Chem.* **351**:588.

Faulhaber, I., 1972, Die Inductionsleistung subzellularer Fractionen aus der Gastrula *Xenopus laevis, Roux Arch.* **171**:87.

Faulhaber, I., and Lyra, L., 1974, Ein Vergleich der Inductionsfähigkeit von Hüllenmaterial der Dotter-plättchen und der Mikrosomenfraktion aus Furchungs -sowie Gastrula und Neurulastadien des Krallen-frosches *Xenopus laevis, Roux Arch.* **176**:151.

Feldherr, C. M., 1975, The uptake of endogenous proteins by oocyte nuclei, *Exp. Cell Res.* **93**:411.

Feldherr, C. M., and Pomerantz, J., 1978, Mechanism for the selection of nuclear polypeptides in *Xenopus* oocytes, *J. Cell Biol.* **78**:168.

Ficq, A., and Brachet, J., 1971, RNA-dependent DNA polymerase: possible role in the amplification of ribo-somal DNA in *Xenopus* oocytes, *Proc. Natl. Acad. Sci. U.S.A.* **68**:2774.

Flickenger, R. A., Greene, R., Kohl, D. M., and Miyagi, M., 1966, Patterns of synthesis of DNA-like RNA in parts of developing frog embryos, *Proc. Natl. Acad. Sci. U.S.A.* **56**:1712.

Flickenger, R. A., Freedman, M. L., and Stambrook, P. J., 1967, Generation times and DNA replication patterns of cells of developing frog embryos, *Dev. Biol.* **16**:457.

Flickenger, R. A., Miyagi, M., Moser, C. R., and Rollins, E., 1967, The relation of DNA synthesis to RNA synthesis in developing frog embryos, *Dev. Biol.* **15**:414.

Flickenger, R. A., Lauth, M. R., and Stambrook, P. J., 1970, An inverse relation between the rate of cell division and RNA synthesis per cell in developing frog embryos, *J. Embryol. Exp. Morphol.* **23**:571.

Fortune, J. E., Concannon, P. W., and Hansel, W., 1975, Ovarian progesterone levels during in vitro oocyte maturation and ovulation in *Xenopus laevis, Biol. Reprod.* **13**:561.

Franke, W. W., Rathke, P. C., Seib, E., Trendelenburg, M. F., Osborn, M., and Weber, K., 1976, Distribution and mode of arrangement of the filamentous structures and actin in the cortex of amphibian oocytes, *Cytobiologie* **14**:111.

Fraser, B. R., and Zalik, S. E., 1977, Lectin-mediated agglutination of amphibian embryonic cells, *J. Cell Sci.* **27**:227.

Fraser, L. R., 1971, Physico–chemical properties of an agent that induces parthenogenesis in *Rana pipiens* eggs, *J. Exp. Zool.* **177**:153.

French, V., Bryant, P. J., and Bryant, S. V., 1976, Pattern regulation in epimorphic fields, *Science* **193**:969.

Galau, G. A., Lipson, E. D., Britten, R. J., and Davidson, E. H., 1977, Synthesis and turnover of polysomal mRNAs in sea urchin embryos, *Cell* **10**:415.

Garcia-Bellido, A., 1977, Homeotic and atavistic mutations in insects, *Am. Zool.* **17**:613.

Gebhardt, D. O. E., and Nieuwkoop, P. D., 1964, The influence of lithium on the competence of the ectoderm in *Ambystoma mexicanum, J. Embryol. Exp. Morphol.* **12**:317.

Geithe, H. P., Asashima, M., Born, J., Tiedemann, H., and Tiedemann, H., 1975, Isolation of a homogeneous morphogenetic factor inducing mesoderm and endoderm derived tissues in *Triturus* ectoderm, *Exp. Cell Res.* **94**:447.

Gilchrist, F., 1928, The effect of a horizontal temperature gradient on the development of the egg of the urodele, *Triturus torosus, Physiol. Zool.* **1**:231.

Gilchrist, F. G., 1933, The time relations of determination in early amphibian development, *J. Exp. Zool.* **66**:15.

Gingell, D., 1970, Contractile responses at the surface of an amphibian egg, *J. Embryol. Exp. Morphol.* **23**:583.

Glade, R. W., Burrill, E. M., and Falk, R. J., 1967, The influence of a temperature gradient on bilateral symmetry in *Rana pipiens, Growth* **31**:231.

Goerttler, K., 1926, Experimentell erzeugte "Spina Bifida" und "Ringembryo-Bildungen" und ihre Bedeutung für die Entwicklungsphysiologie der Urodeleneier, *Z. fur gesamte Anat.* **80**:283.

Goodwin, B. C., and Cohen, M. H., 1969, A phase-shift model for the spatial and temporal organization of developing systems, *J. Theoret. Biol.* **25**:49.

Graham, C. F., 1966, The regulation of DNA synthesis and mitosis in multinucleate frog eggs, *J. Cell Sci.* **1**:363.

Graham, C. F., and Morgan, R. W., 1966, Changes in the cell cycle during early amphibian development, *Dev. Biol.* **14**:439.

Grant, P., 1958, The synthesis of deoxyribonucleic acid during early embryonic development of *Rana pipiens, J. Cell. Comp. Physiol.* **52**:227.

Grant, P., 1960, The influence of folio acid analogs on development and nucleic acid metabolism in *Rana pipiens* embryos, *Dev. Biol.* **2**:197.

Grant, P., and Wacaster, J. F., 1972, The amphibian grey crescent—A site of developmental information? *Dev. Biol.* **28**:454.

Grant, P., and Youngdahl, P., 1974, Cell division and determination of the dorsal lip in *Rana pipiens* embryos, *J. Exp. Zool.* **190**:289.

Green, H. G., Goldberg, B., Schwartz, M., and Brown, D. D., 1968, The synthesis of collagen during the development of *Xenopus laevis, Dev. Biol.* **18**:391.

Greenfield, M. L., 1966, The oocyte of the domestic chicken shortly after hatching, studied by electron microscopy, *J. Embryol. Exp. Morphol.* **15**:297.

Grey, R. D., and Schertel, E. R., 1978, Ionic induction of polyspermy in *Xenopus:* Evidence for a fast block, *J. Cell. Biol.* **79**:164a (Abstract F919, *Am. Soc. Cell Biol. Mtg.*).

Grey, R. D., Wolf, D. P., and Hedrick, J. L., 1974, Formation and structure of the fertilization envelope in *Xenopus laevis, Dev. Biol.* **36**:44.

Grigliatti, T., and Suzuki, D. T., 1971, Temperature sensitive mutations in *Drosophila melanogaster*. VII. The homeotic mutant ss[a40a], *Proc. Natl. Acad. Sci. U.S.A.* **68**:1307.

Grunz, H., 1973, The ultrastructure of amphibian ectoderm treated with an inductor of actinomycin D, *Roux Arch.* **173**:283.

Grunz, H., Multier-Lajous, A.-M., Herbst, R., and Arkenberg, G., 1976, The differentiation of isolated amphibian ectoderm with or without treatment with an inductor; a scanning electron microscope study, *Roux Arch.* **178**:277.

Gruzova, M. N., and Parfenov, V. N., 1977, Ultrastructure of late oocyte nuclei in *Rana temporaria, J. Cell. Sci.* **28**:1.

Gurdon, J. B., 1960, The effects of ultraviolet irradiation on uncleaved eggs of *Xenopus laevis, J. Microscop. Sci.* **101**:299.

Gurdon, J. B., 1967, On the origin and persistence of a cytoplasmic state inducing nuclear DNA synthesis in frogs' eggs, *Proc. Natl. Acad. Sci. U.S.A.* **58**:545.

Gurdon, J. B., 1968, Changes in somatic cell nuclei inserted into growing and maturing amphibian oocytes, *J. Embryol. Exp. Morphol.* **20**:401.

Gurdon, J. B., 1974, *The Control of Gene Expression in Animal Development,* Oxford and Harvard University Press, Cambridge, Mass.

Gurdon, J. B., 1976, Injected nuclei in frog oocytes: Fate, enlargement, and chromatin dispersal, *J. Embryol. Exp. Morphol.* **36**:523.

Gurdon, J. B., 1977, Egg cytoplasm and gene control in development, *Proc. R. Soc. London, Ser. B* **198**:211.

Gurdon, J. B., and Woodland, H. R., 1969, The influence of the cytoplasm on the nucleus during cell differentiation, with special reference to RNA synthesis during amphibian cleavage, *Proc. R. Soc. London, Ser. B* **173**:99.

Gurdon, J. B., and Woodland, H. R., 1970, On the long-term control of nuclear activity during cell differentiation, *Curr. Top. Dev. Biol.* **5**:39.

Gurdon, J. B., Partington, G. A., and de Robertis, E. M., 1976, Injected nuclei in frog oocytes: RNA synthesis and protein exchange, *J. Embryol. Exp. Morphol.* **36**:541.

Gusseck, D. J., and Hedrick, J. L., 1971, A molecular approach to fertilization. I. Disulfide bonds in *Xenopus laevis* jelly coat and a molecular hypothesis for fertilization, *Dev. Biol.* **25**:337.

Guyer, M. F., 1907, The development of unfertilized frog eggs injected with blood, *Science* **25**:910.

Hamilton, L., 1969, The formation of somites in *Xenopus, J. Embryol. Exp. Morphol.* **22**:253.

Hamilton, L., and Tuft, P., 1972, The role of water-regulating mechanisms in the development of the haploid syndrome in *Xenopus laevis, J. Embryol. Exp. Morphol.* **28**:449.

Hanocq-Quertier, J., Baltus, E., and Brachet, J., 1976, Induction of maturation (meiosis) in small *Xenopus laevis* oocytes by injection of maturation promoting factor, *Proc. Natl. Acad. Sci. U.S.A.* **73**:2028.

Hara, K., 1977, The cleavage pattern of the axolotl egg studied by cinematography and cell counting, *Roux Arch.* **181**:73.

Hara, K., and Tydeman, P., 1979, Cinematographic observation of an "activation wave" on the locally inseminated egg of *Xenopus laevis, Roux Arch.* **186**:91.

Hara, K., Tydeman, P., and Hengst, R. T. M., 1977, Cinematographic observation of "post-fertilization waves" (PFW) on the zygote of *Xenopus laevis, Roux Arch.* **181**:189.

Hara, K., Tydeman, P., and Kirschner, M., 1980, A cytoplasmic clock with the same period as the division cycle in *Xenopus* eggs, *Proc. Natl. Acad. Sci. U.S.A.* **77**:462.

Harris, P., 1978, Triggers, trigger waves, and mitosis: A new model, in: *Cell Cycle Regulation* (J. R. Jeter, Jr., ed.), pp. 75–104, Academic Press, New York.

Harris, T. M., 1964, Pregastrular mechanisms in the morphogenesis of the salamander *Ambystoma maculatum, Dev. Biol.* **10**:247.

Harvey, E. N., and Fankhauser, G., 1933, The tension at the surface of the eggs of the salamander, *Triturus (Diemyctylus) viridescens, J. Cell. Compr. Physiol.* **3**:463.

Heasman, J., Mohun, T., and Wylie, C. C., 1977, Studies on the locomotion of primordial germ cells from *Xenopus laevis* in vitro, *J. Embryol. Exp. Morphol.* **42**:149.

Hebard, C. N., and Herold, R. C., 1967, The ultrastructure of the cortical cytoplasm in the unfertilized egg and first cleavage zygote of *Xenopus laevis, Exp. Cell. Res.* **46**:553.

Heidemann, S. R., and Kirschner, M. W., 1975, Aster formation in eggs of *Xenopus laevis:* Induction by isolated basal bodies, *J. Cell Biol.* **67**:105.

Herkovits, J., and Ubbels, G. A., 1979, The ultrastructure of the dorsal yolk-free cytoplasm and the immediately surrounding cytoplasm in the symmetrized egg of *Xenopus laevis, J. Embryol. Exp. Morphol.* **51**:155.

Hill, R. J., Maundrell, K., and Callan, H. G., 1974, Nonhistone proteins of the oocyte nucleus of the newt, *J. Cell Sci.* **15**:145.

Hillman, N., Sherman, M. I., and Graham, C., 1972, The effect of spatial arrangement on cell determination during mouse development, *J. Embryol. Exp. Morphol.* **28**:263.

Holland, C. A., and Dumont, J. N., 1975, Oogenesis in *Xenopus laevis* (Daudin). IV. Effects of gonadotropin, estrogen, and starvation on endocytosis in developing oocytes, *Cell Tiss. Res.* **162**:177.

Holliday, R., and Pugh, J. E., 1975, DNA modification mechanisms and gene activity during development, *Science* **187**:226.

Hollinger, T. G., and Schuetz, A. W., 1976, "Cleavage" and cortical granule breakdown in *Rana pipiens* oocytes, induced by direct microinjection of calcium, *J. Cell Biol.* **71**:395.

Holtfreter, J., 1938, Differenzierungspotenzen isolierter Teile der Urodelengastrula, *Roux Arch.* **138**:522.

Holtfreter, J., 1943a, Properties and functions of the surface coat in amphibian embryos, *J. Exp. Zool.* **93**:251.

Holtfreter, J., 1943b, A study of the mechanics of gastrulation. Part I, *J. Exp. Zool.* **94**:261.

Holtfreter, J., 1944, A study of the mechanics of gastrulation. Part II, *J. Exp. Zool.* **95**:171.

Holtfreter, J., 1946, Structure, motility, and locomotion in isolated embryonic amphibian cells, *J. Morphol.* **79**:27.

Holtfreter, J., 1947a, Observations on the migration, aggregation, and phagocytosis of embryonic cells, *J. Morphol.* **80**:25.

Holtfreter, J., 1947b, Neural induction in explants which have passed through a sublethal cytolysis, *J. Exp. Zool.* **106**:197.

Holtfreter, J., 1948a, Concepts on the mechanisms of embryonic induction and its relation to parthenogenesis and malignancy, *Symp. Soc. Exp. Biol.* **2**:17.

Holtfreter, J., 1948b, Significance of the cell membrane in embryonic processes, *Ann. N.Y. Acad. Sci.* **49**:709.

Holtfreter, J., 1949, Phenomena relating to the cell membrane in embryonic processes, *Exp. Cell Res. (Suppl.)* **1**:497.

Holtfreter, J., 1951, Some aspects of embryonic induction, *Growth Symp.* **10**:117.

Holtfreter, J., and Hamburger, V., 1955, Embryogenesis: Progressive differentiation—Amphibians, in: *Analysis of Development* (B. H. Willier, P. A. Weiss, and V. Hamburger, eds.), pp. 230–296, W. B. Saunders, Philadelphia.

Holtzer, H., Weintraub, H., Mayne, R., and Moohan, B., 1972, The cell cycle, cell lineages, and cell differentiation, *Curr. Top. Dev. Biol.* **7**:229.

Hough, B. R., and Davidson, E. H., 1972, Studies on the repetitive sequence transcripts of *Xenopus* oocytes, *J. Mol. Biol.* **70**:491.

Huff, R., 1962, The developmental role of material derived from the nucleus (germinal vesicle) of mature ovarian oocytes, *Dev. Biol.* **4**:398.

Huff, R. E., and Preston, J. T., 1965, The production of a cleavage-initiating factor by artificially activated eggs of *Rana pipiens, Texas J. Sci.* **27**:206.

Humphrey, R. R., 1960, A maternal effect of the gene (f) for a fluid imbalance in the Mexican axolotl, *Dev. Biol.* **2**:105.

Huxley, J. S., and DeBeer, G. A., 1934, *The Elements of Experimental Embryology,* Cambridge University Press, Cambridge.

Hyode, M., and Flickenger, R. A., 1973, Replicon growth rates during DNA replication in developing frog embryos, *Biochim. Biophys. Acta* **299**:29.

Ijiri, K.-I., 1977, Existence of ultraviolet-labile germ cell determinant in unfertilised eggs of *Xenopus laevis* and it sensitivity, *Dev. Biol.* **55**:206.

Ijiri, K.-I., and Egami, N., 1975, Mitotic activity of germ cells during normal development of *Xenopus laevis* tadpoles, *J. Embryol. Exp. Morphol.* **34**:687.

Ikenishi, K., and Kotani, M., 1975, Ultrastructure of the "germinal plasm" in *Xenopus* embryos after cleavage, *Dev., Growth Differ.* **17**:101.

Ikenishi, K., and Nieuwkoop, P. D., 1978, Location and ultrastructure of primordial germ cells (PGCs) in *Ambystoma mexicanum*, *Dev., Growth Differ.* **20**:1.

Imoh, H., and Sameshima, M., 1976, Possible significance of nuclear volume change during early development of newt embryos, *Dev., Growth Differ.* **18**:45.

Jaffe, L. A., 1976, Fast block to polyspermy in sea urchin eggs is electrically mediated, *Nature (London)* **261**:68.

Jared, D. W., Dumont, J. N., and Wallace, R. A., 1973, Distribution of incorporated and synthesized protein among cell fractions of *Xenopus* oocytes, *Dev. Biol.* **35**:19.

Johnson, K. E., 1969, Altered contact behavior of presumptive mesodermal cells from hybrid amphibian embryos arrested at gastrulation, *J. Exp. Zool.* **170**:325.

Johnson, K. E., 1970, The role of changes in cell contact behavior in amphibian gastrulation, *J. Exp. Zool.* **175**:391.

Johnson, K. E., 1976*a*, Ruffling and locomotion in *Rana pipiens* gastrula cells, *Exp. Cell Res.* **101**:71.

Johnson, K. E., 1976*b*, Circus movements and blebbing locomotion in dissociated embryonic cells of the amphibian, *Xenopus laevis*, *J. Cell Sci.* **22**:575.

Johnson, K. E., 1977*a*, Changes in the cell coat at the onset of gastrulation in *Xenopus laevis* embryos, *J. Exp. Zool.* **199**:137.

Johnson, K. E., 1977*b*, Extracellular matrix synthesis in blastula and gastrula stages of normal and hybrid frog embryos. I. Toluidine blue and lanthanum staining, *J. Cell Sci.* **25**:313.

Johnson, K. E., 1977*c*, Extracellular matrix synthesis in blastula and gastrula stages of normal and hybrid frog embryos. II. Autoradiographic observations on the sites of synthesis and mode of transport of galactose- and glucose-labeled material, *J. Cell Sci.* **25**:323.

Johnson, K. E., 1977*d*, Extracellular matrix synthesis in blastula and gastrula stages of normal and hybrid frog embryos. III. Characterization of galactose- and glucose-labelled materials, *J. Cell Sci.* **25**:335.

Johnson, K. E., 1978, Extracellular matrix synthesis in blastula and gastrula stages of normal and hybrid frog embryos. IV. Biochemical and autoradiographic observations on fucose-, glucose-, and mannose-labelled materials, *J. Cell Sci.* **32**:109.

Johnson, K. E., and Smith, E. P., 1976, The binding of concanavaline A to dissociated embryonic amphibian cells, *Exp. Cell Res.* **101**:63.

Johnson, R. T., and Rao, P. N., 1971, Nucleo-cytoplasmic interactions in the achievement of nuclear synchrony in DNA synthesis and mitosis in multinucleate cells, *Biol. Rev.* **46**:97.

Jones, K. W., and Elsdale, T. R., 1963, The culture of small aggregates of amphibian embryonic cells *in vitro*, *J. Embryol. Exp. Morphol.* **11**:135.

Jones, P., Jackson, H., and Whiting, M. H. S., 1975, Parthenogenetic development after chemical treatment of *Xenopus laevis* spermatozoa, *J. Exp. Zool.* **192**:73.

Kalt, M. R., 1971*a*, The relationship between cleavage and blastocoel formation in *Xenopus laevis*. I. Light microscopic observations, *J. Embryol. Exp. Morphol.* **26**:37.

Kalt, M. R., 1971*b*, The relationship between cleavage and blastocoel formation in *Xenopus laevis*. II. Electron microscopic observations, *J. Embryol. Exp. Morphol.* **26**:51.

Kalt, M. R., 1973, Ultrastructural observations on the germ line of *Xenopus laevis*, *Z. Zellforsch. Mikrosk. Anat.* **138**:41.

Kalt, M., 1976, Morphology and kinetics of spermatogenesis in *Xenopus laevis*, *J. Exp. Zool.* **195**:393.

Kalthoff, K., 1971, Photoreversion of UV induction of the malformation "double abdomen" in the egg of *Smittia* spec. (Diptera, Chironomidae), *Dev. Biol.* **25**:119.

Kane, R. E., 1973, Hyaline release during normal sea urchin development and its replacement after release at fertilization, *Exp. Cell Res.* **81**:301.

Kanno, T., Cochrane, D. E., and Douglas, W. W., 1973, Exocytosis (secretory granule extrusion) induced by injection of calcium into mast cells, *Can. J. Physiol. Pharmacol.* **51**:1001.

Karasaki, S., 1963, Studies on amphibian yolk. 5. Electron microscopic observations on the utilization of yolk platelets during embryogenesis, *J. Ultrastruct. Res.* **9**:225.

Karfunkel, P., 1971, The role of microtubules and microfilaments in neurulation in *Xenopus*, *Dev. Biol.* **25**:30.

Katagiri, C., 1975, Properties of the hatching enzyme from frog embryos, *J. Exp. Zool.* **193**:109.

Kauffman, S., 1973, Control circuits for determination and transdetermination, *Science* **181**:310.

Kauffman, S. A., and Wille, J. J., 1975, The mitotic oscillator of *Physarum polycephalum*, *J. Theoret. Biol.* **55**:47.

Kauffman, S. A., Shymko, R. M., and Trabert, K., 1978, Control of sequential compartment formation in *Drosophila, Science* **199**:259.

Kawakami, I., 1976, Fish swimbladder: An excellent mesodermal inductor in primary induction, *J. Embryol. Exp. Morphol.* **36**:315.

Keller, R. E., 1975, Vital dye mapping of the gastrula and neurula of *Xenopus laevis*. I. Prospective areas and morphogenetic movements of the superficial layer, *Dev. Biol.* **42**:222.

Keller, R. E., 1976, Vital dye mapping of the gastrula and neurula of *Xenopus laevis*. II. Prospective areas and morphogenetic movements of the deep layer, *Dev. Biol.* **51**:118.

Keller, R. E., 1978, Time-lapse cinematographic analysis of superficial cell behavior during and prior to gastrulation in *Xenopus laevis, J. Morphol.* **157**:223.

Keller, R. E., and Schoenwolf, G. C., 1977, An SEM study of cellular morphology, contact, and arrangement as related to gastrulation in *Xenopus laevis, Roux Arch.* **182**:165.

Kemp, N. E., and Istock, N. L., 1967, Cortical changes in growing oocytes and in fertilized or pricked eggs of *Rana pipiens, J. Cell Biol.* **14**:111.

Kerr, J. B., and Dixon, K. E., 1974, An ultrastructural study of germ plasm in spermatogenesis of *Xenopus laevis, J. Embryol. Exp. Morphol.* **32**:573.

Kinderman, N. B., and King, R. C., 1973, Oogenesis in *Drosophila virilis*. I. Interactions between the ring canal rims and the nucleus of the oocyte, *Biol. Bull.* **144**:331.

Kirschner, M. W., 1978, Microtubule assembly and nucleation, *Int. Rev. Cytol.* **54**:1.

Kirschner, M. W., Gerhart, J. C., Hara, K., and Ubbels, G. A., 1980, Initiation of the cell cycle and establishment of bilateral symmetry in *Xenopus* eggs, *Symp. Soc. Dev. Biol.* **38**:187.

Klag, J. J., and Ubbels, G. A., 1975, Regional morphological and cytological differentiation of the fertilized egg of *Discoglossus pictus* (Anura), *Differentiation* **3**:15.

Knowland, J. S., 1970, Polyacrylamide gel electrophoresis of nucleic acids synthesized during early development of *Xenopus laevis* Daudin, *Biochim. Biophys. Acta* **204**:416.

Kochav, S., and Eyal-Giladi, H., 1971, Bilateral symmetry in the chick embryo: Determination by gravity, *Science* **171**:1027.

Kocher-Becker, U., and Tiedemann, H., 1971, Induction of mesodermal and endodermal structures and primordial germ cells in *Triturus* ectoderm by a vegetalizing factor from chick embryos, *Nature (London)* **233**:65.

Kocher-Becker, U., Tiedemann, H., and Tiedemann, H., 1965, Exovagination of newt endoderm: Cell affinities altered by the mesodermal inducing factor, *Science* **147**:167.

Konishi, T., and Kosin, I. L., 1974, Morbidity of aging non-incubated chicken blastoderms: Further cytological evidence and interpretation, *J. Embryol. Exp. Morphol.* **32**:557.

Kosher, R. A., and Searles, R. L., 1973, Sulfated mucopolysaccharide synthesis during the development of *Rana pipiens, Dev. Biol.* **32**:50.

Kriegstein, H. J., and Hogness, D. S., 1974, Mechanism of DNA replication in *Drosophila* chromosomes: Structure of replication forks and evidence for bidirectionality, *Proc. Natl. Acad. Sci. U.S.A.* **71**:135.

Kubota, T., 1966, Studies of the cleavage in the frog egg. I. On the temporal relation between furrow determination and nuclear division, *J. Exp. Biol.* **44**:545.

Kubota, T., 1967, A regional change in the rigidity of the cortex of the egg of *Rana nigromaculata* following extrusion of the second polar body, *J. Embryol. Exp. Morphol.* **17**:331.

Kubota, T., 1969, Studies of cleavage in the frog egg. II. On the determination of the position of the furrow, *J. Embryol. Exp. Morphol.* **21**:119.

Kubota, H. Y., and Durston, A. J., 1978, Cinematographical study of cell migration in the opened gastrula of *Ambystoma mexicanum, J. Embryol. Exp. Morphol.* **44**:71.

Kusano, K., Miladi, R., and Stinnakre, Jr., 1977, Acetylcholine receptors in the oocyte membrane, *Nature (London)* **270**:739.

LaMarca, M. J., Smith, L. D., and Strobel, M. C., 1973, Quantitative and qualitative analysis of RNA synthesis in stage 6 and stage 4 oocytes of *Xenopus laevis, Dev. Biol.* **34**:106.

Landstrom, U., and Lovtrup, S., 1974, Oxygen consumption of normal and dwarf embryos of *Xenopus laevis, Roux Arch.* **196**:1.

Landstrom, U., and Lovtrup, S., 1975, On the determination of dorsal–ventral polarity in *Xenopus laevis* embryos, *J. Embryol. Exp. Morphol.* **33**:879.

Landstrom, U., Lovtrup-Rein, H., and Lovtrup, S., 1975, Control of cell division and cell differentiation by deoxynucleotides in the early embryo of *Xenopus laevis, Cell Differ.* **4**:313.

Landstrom, U., Lovtrup-Rein, H., and Lovtrup, S., 1976, On the determination of dorsal–ventral polarity in the amphibian embryo: Suppression by lactate of the formation of Ruffini's flask-cells, *J. Embryol. Exp. Morphol.* **36**:343.

Leder, A., and Leder, P., 1975, Butyric acid, a potent inducer of erythroid differentiation in cultured erythroleukemic cells, *Cell* **5**:319.

Legname, A. H., and Buhler, M. I., 1978, Metabolic behavior and cleavage capacity in the amphibian egg, *J. Embryol. Exp. Morphol.* **47**:161.

Legros, F., 1970, Effets de divers inhibiteurs de la synthèse des protéines sur le développement des oeufs d'amphibiens, *Arch. Biol. (Liège)* **81**:109.

Leikola, A., 1965, On the loss of mesodermal competence of the *Triturus* gastrula ectoderm *in vivo*, *Experientia* **21**:458.

Lewis, J., Slack, J. M. W., and Wolpert, L., 1977, Thresholds in development, *J. Theoret. Biol.* **65**:579.

Liskay, R. M., and Prescott, D. M., 1978, Genetic analysis of the G_1 period: Isolation of mutants (or variants) with a G_1 period from a Chinese hamster cell line lacking G_1, *Proc. Natl. Acad. Sci. U.S.A.* **75**:2873.

Loeffler, C. A., and Johnston, M. C., 1964, Permeability alterations in amphibian embryos caused by salt solutions and measured by tritiated thymidine uptake, *J. Embryol. Exp. Morphol.* **12**:407.

Lohmann, K., 1974, Analyse des Zellzyklus und der DNS-synthese in fruhembryonalen Geweben (Urodela, *Triturus vulgaris*), *Roux Arch.* **175**:135.

Lovtrup, S., 1958, A physiological interpretation of the mechanism involved in the determination of bilateral symmetry in amphibian embryos, *J. Embryol. Exp. Morphol.* **6**:15.

Lovtrup, S., 1962, Permeability changes in fertilized and activated amphibian eggs, *J. Exp. Zool.* **151**:79.

Lovtrup, S., 1965, Morphogenesis in the amphibian embryo: Fertilization and blastula formation, *Roux Arch.* **156**:204.

Lovtrup, S., 1974, *Epigenetics,* Wiley, London.

Lovtrup, S., 1975, Fate maps and gastrulation in amphibia—A critique of current views, *Can. J. Zool.* **53**:473.

Lovtrup, S., and Pigon, A., 1958, Inversion of the dorsal–ventral axis by unilateral restriction of oxygen supply, *J. Embryol. Exp. Morphol.* **6**:486.

Lovtrup, S., Landstrom, U., and Lovtrup-Rein, H., 1978, Polarities, cell differentiation, and primary induction in amphibian embryos, *Biol. Rev.* **53**:1.

Maéno, T., 1959, Electrical characteristics and activation potential of *Bufo* eggs, *J. Gen. Physiol.* **43**:139.

McClendon, J. F., 1910, The development of isolated blastomeres of the frog's egg, *Am. J. Anat.* **10**:425.

McKnight, S. L., and Miller, O. L., Jr., 1977, Electron microscopic analysis of chromatin replication in the cellular blastoderm of the *Drosophila melanogaster* embryo, *Cell* **12**:795.

McLaren, Anne, 1977, Mammalian chimeras, in: *Developmental and Cell Biology Series* (M. Abercrombie, D. R. Newth, and J. G. Torrey, eds.), pp. 1–154, Cambridge University Press, New York.

MacLennan, D. H., and Holland, P. C., 1975, Calcium transport in sarcoplasmic reticulum, *Ann. Rev. Biophys. Bioeng.* **4**:377.

McMahon, D., 1974, Chemical messengers in development: A hypothesis, *Science* **185**:1012.

MacMurdo-Harris, H., and Zalik, S. E., 1971, Microelectrophoresis of early amphibian embryonic cells, *Dev. Biol.* **24**:335.

Mahowald, A. P., 1971, Origin and continuity of polar granules, in: *Origin and Continuity of Cell Organelles* (J. Reinert and H. Ursprung, eds.), pp. 159–169, Springer-Verlag, New York.

Mahowald, A. P., and Hennen, S., 1971, Ultrastructure of the "germ plasm in eggs and embryos of *Rana pipiens, Dev. Biol.* **24**:37.

Malacinski, G. M., and Brothers, A. J., 1974, Mutant genes in the axolotl, *Science* **184**:1142.

Malacinski, G. M., Allis, C. D., and Chung, H. M., 1974, Correction of developmental abnormalities resulting from localized ultraviolet irradiation of an amphibian egg, *J. Exp. Zool.* **189**:249.

Malacinski, G. M., Benford, H., and Chung, H. M., 1975, Association of an ultraviolet-irradiation sensitive cytoplasmic localization with the future dorsal side of the amphibian egg, *J. Exp. Zool.* **191**:97.

Malacinski, G. M., Brothers, A. J., and Chung, H.-M., 1977, Destruction of components of the neural induction system of the amphibian egg with ultraviolet irradiation, *Dev. Biol.* **56**:24.

Malacinski, G. M., Ryan, B., and Chung, H.-M., 1978, Surface coat movements in unfertilized amphibian eggs, *Differentiation* **10**:101.

Malacinski, G. M., Chung, H.-M., and Asashima, M., 1980, The association of primary embryonic organizer activity with the future dorsal side of amphibian eggs and early embryos, *Dev. Biol.,* in press.

Maller, J. L., and Krebs, E. G., 1977, Progesterone-stimulated meiotic cell division in *Xenopus* oocytes. Inductions by regulatory subunit and inhibition by catalytic subunit of adenosine 3'5' monophosphate dependent protein kinase, *J. Biol. Chem.* **252**:1712.

Maller, J., and Krebs, E. G., 1980, Regulation of oocyte maturation, *Curr. Top. Cell. Regul.* **16**, in press.

Maller, J., Poccia, D., Nishioka, D., Kidd, P., Gerhart, J., and Hartman, H., 1976, Spindle formation and cleavage in Xenopus eggs injected with centriole-containing fractions from sperm, *Exp. Cell Res.* **99**:285.

Maller, J., Wu, M., and Gerhart, J., 1977, Changes in protein phosphorylation accompanying maturation of *Xenopus laevis* oocytes, *Dev. Biol.* **58**:295.

Malpoix, P., Quertier, J., and Brachet, J., 1963, The effects of β-mercaptoethanol on the morphogenetic movements of amphibian embryos, *J. Embryol. Exp. Morphol.* **11**:155.

Manes, M. E., and Barbieri, F. D., 1976, Symmetrization in the amphibian egg by disrupted sperm cells, *Dev. Biol.* **53**:138.

Manes, M. E., and Barbieri, F. D., 1977, On the possibility of sperm aster involvement in dorso–ventral polarization and pronuclear migration in the amphibian egg, *J. Embryol. Exp. Morphol.* **40**:187.

Manes, M. E., Elinson, R. P., and Barbieri, F. D., 1978, Formation of the amphibian grey crescent: Effects of colchicine and cytochalasin B, *Roux Arch.* **185**:99.

Mangold, O., and Seidel, F., 1927, Homoplastische und heteroplastische Verschmelzung ganzer *Triton* Keime, *Roux Arch.* **111**:594.

Martin, G. R., Wiley, L. M., and Damjanov, I., 1977, The development of cystic embryoid bodies *in vitro* from clonal teratocarcinoma stem cells, *Dev. Biol.* **61**:230.

Massover, W. H., 1971, Intramitochondrial yolk-crystals of frog oocytes. I. Formation of yolk-crystal inclusions by mitochondria during bullfrog oogenesis, *J. Cell Sci.* **48**:266.

Masui, Y., 1961, Mesodermal and endodermal differentiation of the presumptive ectoderm of the *Triturus* gastrula through the influence of lithium ion, *Experientia* **17**:458.

Masui, Y., 1966, pH-dependence of the inducing activity of lithium ion, *J. Embryol. Exp. Morphol.* **15**:371.

Masui, Y., 1967, Relative roles of the pituitary, follicle cells, and progesterone in the induction of oocyte maturation in *Rana pipiens*, *J. Exp. Zool.* **166**:365.

Masui, Y., and Clarke, H. J., 1979, Oocyte maturation, *Int. Rev. Cytol.* **57**:185.

Masui, Y., and Markert, C. I., 1971, Cytoplasmic control of nuclear behavior during meiotic maturation of frog oocytes, *J. Exp. Zool.* **177**:129.

Masui, Y., Meyerhof, P. G., Miller, M. A., and Wasserman, W. J., 1977, Roles of divalent cations in maturation and activation of vertebrate oocytes, *Differentiation* **9**:49.

Maufroid, J.-P., and Capuron, A., 1977, Induction du mesoderm et des cellules germinales primordiales par l'endoderm chez *Pleurodeles waltlii*, *C.R. Acad. Sci. (Paris)* **284**:1713.

Maundrell, K., 1975, Proteins of the newt oocyte nucleus: Analysis of the non-histone proteins from lampbrush chromosomes, nucleoli, and nuclear sap, *J. Cell Sci.* **17**:579.

Mayer, B., 1935, Über das Regulations-und Inductions-Vermogen der halbseitigen oberen Urmundlippe von *Triton*, *Roux Arch.* **133**:518.

Meinhardt, H., 1978, Space-dependent cell determination under the control of a morphogen gradient, *J. Theoret. Biol.* **74**:307.

Melton, C. G., 1965, Blastular arrest and partial cleavage of frog embryos injected with antinomycin D, mitomycin C, and homologous liver macromolecules, *Dev. Biol.* **12**:287.

Melton, C. G., Jr., and Smorul, R. P., 1973, Functional volume of frog eggs: Equivalence of metabolite diffusion space in chemically demembranated embryos and aqueous (non-yolk) volume, *J. Exp. Zool.* **187**:239.

Merriam, R. W., 1969, Movement of cytoplasmic proteins into nuclei induced to enlarge and initiate DNA or RNA synthesis, *J. Cell Sci.* **5**:333.

Merriam, R. W., 1971, Progesterone induced maturational events in oocytes of *Xenopus laevis*. II. Changes in intracellular calcium and magnesium distribution at germinal vesicle breakdown, *Exp. Cell Res.* **68**:81.

Meyerhof, P. G., and Masui, Y., 1977, Ca and Mg control of cytostatic factors from *Rana pipiens* oocytes which causes metaphase and cleavage arrest, *Dev. Biol.* **61**:214.

Miceli, D. C., del Pino, E. J., Barbieri, F. D., Mariano, M. L., and Raisman, J. S., 1977, The vitelline envelope-to-fertilization envelope transformation in the toad *Bufo arenarum*, *Dev. Biol.* **59**:101.

Mikamo, K., and Witschi, E., 1966, The mitotic chromosomes in *Xenopus laevis* (Daudin): Normal, sex reversed, and female WW, *Cytogenetics* **5**:1.

Moen, T. L., and Namenwirth, M., 1977, The distribution of soluble proteins along the animal–vegetal axis of frog eggs, *Dev. Biol.* **58**:1.

Monroy, A., and Bacetti, B., 1975, Morphological changes of the surface of the egg of *Xenopus laevis* in the course of development. I. Fertilization and early cleavage, *J. Ultrastruct. Res.* **50**:131.

Monroy, A., Baccetti, B., and Denis-Donini, S., 1976, Morphological changes in the surface of the egg of *Xenopus laevis* in the course of development. III. Scanning electron microscopy of gastrulation, *Dev. Biol.* **59**:250.

Mookerjee, S., Deuchar, E. M., and Waddington, C. H., 1953, The morphogenesis of the notochord in amphibia, *J. Embryol. Exp. Morphol.* **1**:399.

Moran, D., and Mouradian, W. E., 1975, A scanning electron microscopic study of the appearance and localization of cell surface material during amphibian gastrulation, *Dev. Biol.* **46**:422.

Morgan, T. H., 1895, Half embryos and whole embryos from one of the two first two blastomeres, *Anat. Anz.* **10**:623.

Morgan, T. H., 1934, *Embryology and Genetics,* Columbia University Press, New York.

Morgan, T. H., and Boring, A. M., 1903, The relation of the first plane of cleavage and the grey crescent to the median plane of the embryo of the frog, *Roux Arch.* **16**:680.

Morrill, G. A., and Kostellow, A., 1965, Phospholipid and nucleic acid gradients in the developing amphibian embryo, *J. Cell Biol.* **25**:21.

Morrill, G. A., Kostellow, A. B., and Murphy, J. B., 1975, Role of Na^+,K^+-ATPase in early embryonic development, *Ann. N.Y. Acad. Sci.* **242**:543.

Morrill, G. A., Schatz, F., Kostellow, A. B., and Poupko, J. M., 1977, Changes in cyclic AMP levels in the amphibian ovarian follicle following progesterone induction of meiotic maturation: Effect of phosphodiesterase inhibitors and exogenous calcium on germinal vesicle breakdown, *Differentiation* **8**:97.

Morrill, G. A., Ziegler, D., and Zabrenetsky, V. S., 1977, An analysis of transport, exchange, and binding of sodium and potassium in isolated amphibian follicles and denuded oocytes, *J. Cell Sci.* **26**:311.

Motomura, I., 1935, Determination of the embryonic axis in the eggs of amphibia and echinoderms, *Sci. Rep. Tohoku Univ., Ser. 4* **10**:211.

Müller, W. P., 1974, The lampbrush chromosomes of *Xenopus laevis, Chromosoma* **47**:283.

Murphy, D. B., and Tilney, L. G., 1974, The role of microtubules in the movement of pigment granules in teleost melanophores, *J. Cell Biol.* **61**:757.

Nakamura, O., and Kishiyama, K., 1971, Prospective fates of blastomeres at the 32 cell stage of *Xenopus laevis* embryos, *Proc. Jpn. Acad.* **47**:407.

Nakamura, O., and Takasaki, H., 1970, Further studies on the differentiation capacity of the dorsal marginal zone in the morula of *Triturus pyrrhogaster, Proc. Jpn. Acad.* **46**:546.

Nakamura, O., and Takasaki, H., 1971a, Effects of actinomycin on development of amphibian morulae and blastulae, with special reference to the organizer, *Proc. Jpn. Acad.* **47**:92.

Nakamura, O., and Takasaki, H., 1971b, Analysis of causal factors giving rise to the organizer. I. Removal of polar blastomeres from 32 cell embryos of *Xenopus laevis, Proc. Jpn. Acad.* **47**:499.

Nakamura, O., Hayashi, Y., and Asashima, M., 1978, A half-century from Spemann—Historical review of studies on the organizer, in: *Organizer: A Milestone of a Half-Century from Spemann* (O. Nakamura and S. Toivonen, eds.), pp. 1–48, Elsevier/North Holland Biomedical Press, N.Y.

Nakamura, P., Takasaki, H., and Mizohata, T., 1970, Differentiation during cleavage in *Xenopus laevis.* I. Acquisition of self-differentiation capacity of the dorsal marginal zone, *Proc. Jpn. Acad.* **46**:694.

Nakatsuji, N., 1974, Studies on the gastrulation of amphibian embryos: Pseudopodia in the gastrula of *Bufo bufo japonicus* and their significance to gastrulation, *J. Embryol. Exp. Morphol.* **32**:795.

Nakatsuji, N., 1975, Studies on the gastrulation of amphibian embryos: Light and electron microscopic observations of a urodele *Cynops pryyhogaster, J. Embryol. Exp. Morphol.* **34**:669.

Nakatsuji, N., 1976, Studies on the gastrulation of amphibian embryos: Ultrastructure of the migrating cells of anurans, *Roux Arch.* **180**:229.

Nass, M. M. K., 1962, Developmental changes in frog actomyosin characteristics, *Dev. Biol.* **2**:289.

Nicholas, J. S., 1945, Blastulation, its role in pregastrular organization in *Ambystoma punctatum, J. Exp. Zool.* **51**:159.

Nicholas, J. S., 1948, Form changes during pre-gastrular development, *Ann. N.Y. Acad. Sci.* **49**:801.

Nieuwkoop, P. D., 1956, Are there direct relationships between the cortical layer of the fertilized egg and the development of the future axial system in *Xenopus laevis* embryos? *Pubbl. Stz. Zool. Napoli* **28**:241.

Nieuwkoop, P. D., 1963, Pattern formation in artificially activated ectoderm *(Rana pipiens* and *Ambystoma punctatum), Dev. Biol.* **7**:255.

Nieuwkoop, P. D., 1967a, The "organization center." II. Field phenomena, their origin and significance, *Acta Biotheoret.* **17**:151.

Nieuwkoop, P. D., 1967b, The "organization center." III. Segregation and pattern formation in morphogenetic fields, *Acta Biotheoret.* **17**:178.

Nieuwkoop, P. D., 1969a, The formation of mesoderm in urodelean amphibians. I. Induction by the endoderm, *Roux Arch.* **162**:341.

Nieuwkoop, P. D., 1969b, The formation of mesoderm in urodelean amphibians. II. The origin of the dorso-ventral polarity of the mesoderm, *Roux Arch.* **163**:298.

Neiuwkoop, P. D., 1970, The formation of mesoderm in urodelean amphibians. III. The vegetalizing action of the Li ion, *Roux Arch.* **166**:105.

Nieuwkoop, P. D., 1973, The "organization center" of the amphibian embryo: Its origin, spatial organization, and morphogenetic action, *Adv. Morphog.* **10**:1.

Nieuwkoop, P. D., 1977, Origin and establishment of embryonic polar axes in amphibian development, *Curr. Top. Dev. Biol.* **11**:115.

Nieuwkoop, P. D., and Faber, J., 1975, *Normal Table of Xenopus laevis (Daudin)*, 2nd ed., 1st reprint, North-Holland, Amsterdam.

Nieuwkoop, P. D., and Florshutz, P., 1950, Quelques caractères spéciaux de la gastrulation et de la neurulation de l'oeuf de *Xenopus laevis,* Daud, et de quelques autres Anoures, *Arch. Biol. (Liège)* **61**:113.

Nieuwkoop, P. D., and Sutasurya, L. A., 1976, Embryological evidence for a possible polyphyletic origin of the recent amphibians, *J. Embryol. Exp. Morphol.* **35**:159.

Nieuwkoop, P. D., and Ubbels, G. A., 1972, The formation of mesoderm in urodelean amphibians. IV. Quantitative evidence for the purely "ectodermal" origin of the entire mesoderm and of the pharyngeal endoderm, *Roux Arch.* **169**:185.

Noda, S., and Kawakami, I., 1976, Cytological and microspectrophotometric analysis of mesodermalized explants of *Triturus* gastrula, *J. Embryol. Exp. Morphol.* **36**:55.

O'Connor, C. M., Robinson, K. R., and Smith, L. D., 1977, Calcium, potassium, and sodium exchange by full-grown and maturing *Xenopus laevis* oocytes, *Dev. Biol.* **61**:28.

O'Dell, D. S., Tencer, R., Monroy, A., and Brachet, J., 1974, The patterns of Concanavaline A binding sites during the early development of *Xenopus laevis, Cell Different.* **3**:193.

Ohlendorf, D. H., Wrenn, R. F., and Bonaszak, L. J., 1978, Three dimensional structure of the lipovitellin-phosvitin complex from amphibian oocytes, *Nature (London)* **272**:28.

Okada, T. S., 1957, The pluripotency of the pharyngeal primordium in the urodelan neurula, *J. Embryol. Exp. Morphol.* **5**:438.

Osborn, J. C., and Stanisstreet, M., 1977, Comparison of cell division and cell sizes in normal embryos and lithium-treated exogastrulae of *Xenopus laevis, Acta Embryol. Exp.* **3**:283.

Paleček, J., Ubbels, G. A., and Rzehak, K., 1978, Changes of the external and internal pigment pattern upon fertilization in the egg of *Xenopus laevis, J. Embryol. Exp. Morphol.* **45**:203.

Panijel, J., 1950, L'organisation du vitellus dans les oeufs d'Amphibiens, *Biochim. Biophys. Acta,* **5**:343.

Pasteels, J., 1937, Sur l'origine de la symétrie bilatérale des Amphibiens anoures, *Arch. Anat. Micros.* **33**:279.

Pasteels, J., 1938, Recherches sur les facteurs initiaux de la morphogénèse chez les Amphibiens anoures. I. Resultats de l'expérience de Schultze et leur interpretation, *Arch. Biol.* **49**:629.

Pasteels, J., 1940, Recherches sur les facteurs initiaux de la morphogénèse chez les Amphibiens anoures. III. Effets de la rotation de 135° sur l'oeuf insegmenté, muni de son croissant gris, *Arch. Biol.* **51**:335.

Pasteels, J., 1941, Recherches sur les facteurs initiaux de la morphogénèse chez les Amphibiens anoures. V. Les effets de la pesanteur sur l'oeuf de *Rana fusca* maintenu en position anormale avant le formation du croissant gris, *Arch. Biol.* **52**:321.

Pasteels, J., 1946, Sur la structure of l'oeuf insegmenté d'axolotl et l'origine des prodromes morphogénétiques, *Acta Anat.* **2**:1.

Pasteels, J., 1948, Les bases de la morphogénèse chez les vertébrés anamniotes au function de la structure de l'oeuf, *Folia Biotheor. (Leiden)* **3**:83.

Pasteels, J., 1964, The morphogenetic role of the cortex of the amphibian egg, *Adv. Morphog.* **3**:363.

Paterson, M. C., 1957, Animal–vegetal balance in amphibian development, *J. Exp. Zool.* **134**:183.

Penners, A., 1929, Schultzescher Umdrehungsversuch an ungefurchten Froscheieren, *Roux Arch.* **116**:53.

Penners, A., 1936, Neue Experimente zur Frage nach der Potenz der ventralen Keimhälfte von *Rana fusca, Z. Wiss. Zool.* **148**:189.

Penners, A., and Schleip, W., 1928a, Die Entwicklung der Schultzeschen Doppelbildungen aus dem Ei von *Rana fusca.* Teil I-IV, *Z. Wiss. Zool.* **130**:305.

Penners, A., and Schleip, W., 1928b, Die Entwicklung der Schultzeschen Doppelbildungen aus dem Ei von *Rana fusca.* Teil V and VI, *Z. Wiss. Zool.* **131**:1.

Perkowska, E., MacGregor, H. C., and Birnstiel, M. L., 1968, Gene amplification in the oocyte nucleus of mutant and wild-type *Xenopus laevis, Nature (London)* **217**:649.

Perlman, S. M., Ford, P. J., and Rosbash, M. M., 1977, Presence of tadpole and adult globin RNA sequences in oocytes of *Xenopus laevis, Proc. Natl. Acad. Sci. U.S.A.* **74**:3835.

Perry, M. M., 1966, Tubular elements associated with yolk platelets in *Triturus alpestris, J. Ultrastruct. Res.* **16**:376.

Perry, M., 1975, Microfilaments in the external surface layer of the early amphibian embryo, *J. Embryol. Exp. Morphol.* **33**:127.

Perry, M., and Waddington, C., 1966, Ultrastructure of the blastopore cells in the newt, *J. Embryol. Exp. Morphol.* **15**:317.

Perry, M. M., John, H. A., and Thomas, N. S. T., 1971, Actin-like filaments in the cleavage furrow of the newt egg, *Exp. Cell Res.* **65**:249.

Peterson, A. W., 1971, Relationships between germinal vesicle volume and cytoplasmic volume in amphibian oocytes, *Can. J. Genet. Cytol.* **13**:898.

Picard, J. J., 1975, *Xenopus laevis* cement gland as an experimental model for embryonic differentiation, *J. Embryol. Exp. Morphol.* **33**:957.

Porter, K., Prescott, D., and Frye, J., 1973, Changes in surface morphology of Chinese hamster ovary cells during the cell cycle, *J. Cell Biol.* **57**:815.

Pupko, J., Kostellow, A. B., and Morrill, G. A., 1977, Changes in histone patterns during amphibian embryonic development, *Differentiation* **8**:61.

Prescott, D., 1976, The cell cycle and the control of cellular reproduction, *Adv. Genet.* **18**:99.

Ramirez, S. A., and Huff, R. E., 1967, A cytological study of parthenogenetically activated eggs of *Rana pipiens, Texas J. Science* **19**:41.

Redshaw, M. R., 1972, The hormonal control of the amphibian ovary, *Am. Zool.* **12**:289.

Reed, S. C., and Stanley, H. P., 1972, Fine structure of spermatogenesis in the South African clawed toad, *Xenopus laevis* Daudin, *J. Ultrastruct. Res.* **41**:277.

Reynhout, J. K., and Smith, L. D., 1973, Evidence for steroid metabolism during the *in vitro* induction of maturation in oocytes of *Rana pipiens, Dev. Biol.* **30**:392.

Reynhout, J. K., and Smith, L. D., 1974, Studies on the appearance and nature of a maturation-inducing factor in the cytoplasm of amphibian oocytes exposed to progesterone, *Dev. Biol.* **38**:394.

Ridgeway, E. R., Gilky, J. C., and Jaffe, L. F., 1977, Free calcium increases explosively in activating medaka eggs, *Proc. Natl. Acad. Sci. U.S.A.* **74**:623.

Robertson, N., 1977, Structural carbohydrates of yolk platelets from early developmental stages of *Xenopus laevis,* Ph.D. thesis, University of California, Berkeley, Calif.

Robertson, N., 1978, Labilization of the superficial layer and reduction in size of yolk platelets during early development of *Xenopus laevis, Cell Different.* **7**:185.

Roeder, R. G., 1974a, Multiple forms of deoxyribonucleic acid-dependent ribonucleic acid polymerase in *Xenopus laevis.* Isolation and partial characterization, *J. Biol. Chem.* **249**:214.

Roeder, R. G., 1974b, Multiple forms of deoxynbonucleic acid-dependent ribonucleic acid polymerase in *Xenopus laevis.* Levels of activity during oocyte and embryonic development, *J. Biol. Chem.* **249**:249.

Rosbash, M., and Ford, P. J., 1974, Polyadenylic acid-containing RNA in *Xenopus laevis* oocytes, *J. Mol. Biol.* **85**:87.

Rosenbaum, R. M., 1960, Gastrular arrest and the control of autolytic activity in the egg of *Rana pipiens:* The comparative effects of oxygen, supramaximal temperature and dinitrophenol, *Dev. Biol.* **2**:427.

Rosenberg, M. J., and Caplan, A. I., 1974, Nicotinamide adenine dinucleotide levels in cells of developing chick limbs: Possible control of muscle and cartilage development, *Dev. Biol.* **38**:157.

Roth, T. F., Cutting, J. A., and Atlas, S. B., 1976, Protein transport: A selective membrane mechanism, *J. Supramol. Struct.* **4**:527.

Rothstein, A., 1970, Sulfhydryl groups in membrane structure and function, *Curr. Top. Membrs. Transp.* **1**:135.

Roux, W., 1888, Beiträge zur Entwicklungsmechanik des Embryo. Über die künstliche Hervorbringung halber Embryonen durch Zerstörung Einer der beiden ersten Furchungskugeln, sowie über die Nachentwicklung (Postgeneration) der fehlenden Körperhälfte, *Virchows Arch. Path. Anat.* **114**:113.

Ruderman, J. V., Woodland, H. R., and Sturgess, E. A., 1979, Modulations of histone messenger RNA during the early development of *Xenopus laevis, Dev. Biol.* **71**:71.

Rugh, R., 1951, *The Frog: Its Reproduction and Development,* McGraw-Hill, New York.

Ruud, G., 1925, Die Entwicklung isolierter Keimfragmente frühester Stadien von *Triton taeniatus, Roux Arch.* **105**:209.

Ruud, G., and Spemann, H., 1923, Die Entwicklung isölierter dorsaler und lateraler Gastrula Hälften von *Triton taeniatus* und *alpestris,* ihre Regulation und Postgeneration, *Roux Arch.* **52**:95.

Rzehak, K., 1972, Changes in the pigment pattern of eggs of *Xenopus laevis* following fertilization, *Folia Biologica* **20**:409.

Sanders, E. J., and DiCaprio, R. A., 1977, A freeze-fracture and concanavaline A-binding study of the membrane of cleaving *Xenopus* embryos, *Differentiation* **7**:13.

Sanders, E. J., and Singal, P. K., 1975, Furrow formation in *Xenopus* embryos. Involvement of the Golgi body as revealed by ultrastructural localization of thiamine pyrophosphatase activity, *Exp. Cell Res.* **93**:219.

Sanders, E. J., and Zalik, S. E., 1972, The blastomere periphery of *Xenopus laevis* with special reference to intercellular relationships, *Roux Arch.* **171**:181.

Satoh, N., 1977, Metachronous cleavage and initiation of gastrulation in amphibian embryos, *Dev., Growth Differ.* **19**:111.

Satoh, N., Kageyama, T., and Sirakami, K.-I., 1976, Motility of dissociated embryonic cells in *Xenopus laevis:* Its significance to morphogenetic movements, *Dev., Growth Differ.* **18**:55.

Sawai, T., 1972, Roles of cortical and subcortical components in cleavage furrow formation in amphibia, *J. Cell Sci.* **11**:543.

Sawai, T., 1974, Furrow formation on a piece of cortex transplanted to the cleavage plane of the newt egg, *J. Cell Sci.* **15**:259.

Sawai, T., 1979, Cyclic changes in the cortical layer of non-nucleated fragments of the newt's egg, *J. Embryol. Exp. Morphol.* **51**:183.

Sawai, T., and Yoneda, M., 1974, Wave of stiffness propagating along the surface of the newt's egg during cleavage, *J. Cell Biol.* **60**:1.

Saxen, L., and Toivonen, S., 1961, The two-gradient hypothesis in primary induction. The combined effect of two types of inductors in different ratio, *J. Embryol. Exp. Morphol.* **9**:514.

Saxen, L., and Toivonen, S., 1962, *Primary Embryonic Induction,* Logos Press, London.

Schaeffer, B. E., Schaeffer, H. E., and Brick, I., 1973, Cell electrophoresis of amphibian blastula and gastrula cells: The relationship of surface charge and morphogenetic movement, *Dev. Biol.* **34**:66.

Schaeffer, H. E., Schaeffer, B. E., and Brick, I., 1973, Electrophoretic mobility as a function of pH for disaggregated amphibian gastrula cells, *Dev. Biol.* **35**:376.

Scharf, S. R., and Gerhart, J. C., 1980, Determination of the dorsal–ventral axis in eggs of *Xenopus laevis.* Complete rescue of UV-impaired eggs by oblique orientation before first cleavage, *Dev. Biol.,* in press.

Schatten, G., and Thoman, M., 1978, Nuclear surface complex as observed with the high resolution scanning electron microscope. Visualization of the membrane surfaces of the nuclear envelope and the nuclear cortex from *Xenopus laevis* oocytes, *J. Cell Biol.* **77**:517.

Schechtman, A. M., 1934, Unipolar ingression in *Triturus torosus:* A hitherto undescribed movement in the pregastrula states of a urodele, *Univ. Calif. Publ. Zool.* **39**:303.

Schechtman, M., 1936, Relation between the grey crescent and the organizer center of a urodele egg *(Triturus torosus), Roux Arch.* **134**:207.

Scheer, U., 1973, Nuclear pore flow rate of ribosomal RNA and chain growth rate of its precursor during oogenesis of *Xenopus laevis, Dev. Biol.* **30**:13.

Schliwa, M., 1978, Microtubular apparatus of melanophores, *J. Cell Biol.* **76**:605.

Schliwa, M., and Bereiter-Hahn, J., 1975, Pigment movements in fish melanophores: Morphological and physiological studies. V. Evidence for a microtubule-independent contractile system, *Cell Tissue Res.* **158**:61.

Schmidt, G. A., 1933, Schnürungs-und Durchschneidungsversuche am Anurenkeim, *Roux Arch.* **129**:1.

Schönmann, W., 1938, Der diploide Bastard *Triton palmatus* ♀ × *Salamandra* ♂, *Roux Arch.* **138**:345.

Schorderet-Slatkine, S., and Drury, K. C., 1973, Progesterone induced maturation in oocytes of *Xenopus laevis.* Appearance of a "maturation promoting factor" in enucleated oocytes, *Cell Different.* **2**:247.

Schorderet-Slatkine, S., Schorderet, M., and Baulieu, E.-E., 1976, Initiation of meiotic maturation in *Xenopus laevis* oocytes by lanthanum, *Nature (London)* **262**:299.

Schorderet-Slatkine, S., Schorderet, M., and Baulieu, E.-E., 1977, Progesterone-induced meiotic reinitiation *in vitro* in *Xenopus laevis* oocytes. A role for the displacement of membrane-bound calcium, *Differentiation* **9**:67.

Schroeder, T. E., 1970, Neurulation in *Xenopus laevis.* An analysis and model based upon light and electron microscopy, *J. Embryol. Exp. Morphol.* **23**:427.

Schroeder, T. E., and Strickland, D. L., 1974, Ionophore A23187, calcium, and contractility in frog eggs, *Exp. Cell Res.* **83**:139.

Schuel, H., Kelly, W. J., Berger, E. R., and Wilson, W. L., 1974, Sulfated acid mucopolysaccharides in the cortical granules of eggs, *Exp. Cell Res.* **88**:24.

Schuetz, A. W., 1967, Effects of steroid on the germinal vesicle of oocytes of the frog *(Rana pipiens) in vitro,* *Proc. Soc. Exp. Biol. Med.* **124**:1307.

Schuetz, A. W., 1974, Role of hormones in oocyte maturation, *Biol. Reprod.* **10**:150.

Schuetz, A., Wallace, R. A., and Dumont, J. N., 1974, Steroid inhibition of protein incorporation by isolated amphibian oocytes, *J. Cell Biol.* **61**:26.

Schuetz, A. W., Hollinger, T. G., Wallace, R. A., and Samson, D. A., 1977, Inhibition of [³H]vitellogenin uptake by isolated amphibian oocytes injected with cytoplasm from progesterone-treated mature oocytes, *Dev. Biol.* **58**:528.

Schultze, O., 1894, Die künstliche Erzeugung von Doppelbildungen bei Froschlarven mit Hilfe abnormer Gravitation, *Roux Arch.* **1**:160.

Schultze, O., 1900, Über das erste Auftreten der Bilaterale Symmetrie in Verlauf der Entwicklung, *Arch. Mikr. Anat.* **15**:245.

Seilern-Aspang, F., 1959, Die Einwirkung von Mercaptoëthanol auf die Gastrulation von *Rana pipiens, Roux Arch.* **151**:159.

Selman, G. G., and Pawsey, G. J., 1965, The utilization of yolk platelets by tissues of *Xenopus* embryos studied by a safranin staining method, *J. Embryol. Exp. Morphol.* **14**:191.

Selman, G. G., and Perry, M. M., 1970, Ultrastructural changes in the surface layers of the newt's egg in relation to the mechanism of its cleavage, *J. Cell Sci.* **6**:207.

Selman, G. G., and Waddington, C. H., 1955, The mechanism of cell division in the cleavage of the newt's egg, *J. Exp. Biol.* **32**:700.

Selman, G. G., Jacob, J., and Perry, M. M., 1976, The permeability to cytochalasin B of the new unpigmented surface of the first cleavage furrow of the newt's egg, *J. Embryol. Exp. Morphol.* **36**:321.

Shaver, J. R., 1953, Studies on the initiation of cleavage in the frog egg, *J. Exp. Zool.* **122**:169.

Shields, R., 1978, Further evidence for a random transition in the cell cycle, *Nature (London)* **273**:755.

Shiokawa, K., and Yamana, K., 1967, Pattern of RNA synthesis in isolated cells of *Xenopus laevis* embryos, *Dev. Biol.* **16**:368.

Signoret, J., 1977, La cinétique cellulaire au cours de la segmentation du germe d'Axolotl: Proposition d'un modèle statistique, *J. Embryol. Exp. Morphol.* **42**:5.

Signoret, J., and Lefresne, J., 1969, Mise en evidence d'une modification chromosomique au cours de developpment chez l'axoltl, *Ann. Embryol. Morphog.* **2**:451.

Signoret, J., and Lefresne, J., 1971, Contribution a l'étude de la segmentation de l'oeuf d'axolotl. I. Definition de la transition blastuléenne, *Ann. Embryol. Morphog.* **4**:113.

Signoret, J., and Lefresne, J., 1973, Contribution a l'étude de la segmentation de l'oeuf d'axolotl. II. Influence de modification des noyau et du cytoplasme sur les modalitiés, de la segmentation, *Ann. Embryol. Morphog.* **6**:299.

Silverstein, S. C., Steinman, R. M., and Cohn, Z. A., 1977, Endocytosis, *Annu. Rev. Biochem.* **46**:669.

Singal, P. K., and Sanders, E. J., 1974a, Cytomembranes in first cleavage *Xenopus* embryos. Interrelationship between Golgi bodies, endoplasmic reticulum and lipid droplets, *Cell Tissue Res.* **154**:189.

Singal, P. K., and Sanders, E. J., 1974b, An ultrastructural study of the first cleavage of *Xenopus* embryos, *J. Ultrastruct. Res.* **47**:433.

Skoblina, M. N., 1976, The role of karyoplasm in the emergence of capacity of egg cytoplasm to induce DNA synthesis in transplanted sperm nuclei, *J. Embryol. Exp. Morphol.* **36**:67.

Slack, C., and Palmer, J. F., 1969, The permeability of intercellular junctions in the early embryo of *Xenopus laevis,* studied with a fluorescent tracer, *Exp. Cell Res.* **55**:416.

Slack, C., and Warner, A., 1973, Intracellular and intercellular potentials in the early amphibian embryo, *J. Physiol.* **232**:313.

Slack, C., Warner, A. E., and Warren, R. L., 1973, The distribution of sodium and potassium in amphibian embryos during early development, *J. Physiol.* **232**:297.

Smith, B. G., 1922, The origin of bilateral symmetry in the embryo of *Cryptobranchus allegheniensis, J. Morphol.* **36**:357.

Smith, B. G., 1929, The history of the chromosomal vesicles in the segmenting egg of *Cryptobranchus allegheniensis, J. Morphol. Physiol.* **47**:89.

Smith, J. A., and Martin, L., 1973, Do cells cycle? *Proc. Natl. Acad. Sci. U.S.A.* **70**:1263.

Smith, J. L., Osborn, J. C., and Stanistreet, M., 1976, Scanning electron microscopy of lithium-induced exogastrulae of *Xenopus laevis, J. Embryol. Exp. Morphol.* **36**:513.

Smith, L. D., 1966, The role of a "germinal plasm" in the formation of primordial germ cells in *Rana pipiens*, *Dev. Biol.* **14**:330.

Smith, L. D., 1975, Molecular events during oocyte maturation, in: *Biochemistry of Animal Development*, Vol. 3 (R. Weber, ed.), pp. 1–46, Academic Press, New York.

Smith, L. D., and Ecker, R. E., 1969, Role of the oocyte nucleus in physiological maturation in *Rana pipiens*, *Dev. Biol.* **19**:281.

Smith, L. D., and Ecker, R. E., 1970, Regulating processes in the maturation and early cleavage of amphibian eggs, *Curr. Top. Dev. Biol.* **5**:1.

Smith, L. D., and Ecker, R. E., 1971, The interaction of steroids with *Rana pipiens* oocytes in the induction of maturation, *Dev. Biol.* **25**:232.

Smith, L. D., Ecker, R. E., and Subtelny, S., 1968, *In vitro* induction of physiological maturation in *Rana pipiens* oocytes removed from their ovarian follicles, *Dev. Biol.* **17**:627.

Somerville, J., and Hill, R. J., 1973, Proteins associated with heterogeneous nuclear RNA of newt oocytes, *Nature (London), New Biol.* **245**:104.

Speaker, M. C., and Butcher, F. R., 1977, Cyclic nucleotide fluctuations during steroid-induced meiotic maturation of frog oocytes, *Nature (London)* **267**:848.

Spemann, H., 1902, Entwicklungsphysiologische Studien am Triton-Ei. II, *Roux Arch.* **15**:448.

Spemann, H., 1918, Uber die Determination der ersten Organanlagen des Amphibien Embryo. I–VI, *Roux Arch.* **43**:448.

Spemann, H., 1928, Die Entwicklung seitlicher und dorsalventraler Keimhälfte bei verzögerter Kernverzorgung, *Z. Wiss. Zool.* **132**:105.

Spemann, H., 1938, *Embryonic Development and Induction*, Yale University Press, New Haven.

Spemann, H., and Mangold, H., 1924, Über Induction von Embryonalanlagen durch Implantation artfremder Organisatoren, *Roux Arch.* **100**:599.

Spiegelman, S., 1948, Differentiation as the controlled production of unique enzyme patterns, *Symp. Soc. Exp. Biol.* **2**:286.

Stableford, L. T., 1948, The potency of the vegetal hemisphere of the *Ambystoma punctatum* embryo, *J. Exp. Zool.* **109**:385.

Stableford, L. T., 1949, The blastocoel fluid in amphibian gastrulation, *J. Exp. Zool.* **112**:529.

Stableford, L. T., 1967, A study of calcium in the early development of the amphibian embryo, *Dev. Biol.* **16**:303.

Stambrook, P. J., and Flickenger, P. A., 1970, Changes in chromosomal DNA replication patterns in developing frog embryos, *J. Exp. Zool.* **174**:101.

Steinberg, M. S., 1963, Tissue reconstruction by dissociated cells, *Science* **141**:401.

Steinberg, M. S., 1964, The problem of adhesive selectivity in cellular interactions, in: *Cell Membranes in Development*, Vol. 22, *Symposium of Social Development and Growth* (M. Locke, ed.), p. 321, Academic Press, New York.

Steinberg, M., 1970, Does differential adhesion govern self-assembly processes in histogenesis? Equilibrium configurations and the emergence of a hierarchy among populations of embryonic cells, *J. Exp. Zool.* **173**:395.

Steinert, K., Geuskens, M., Steinert, G., and Tencer, R., 1977, Concanavaline A binding to oocytes and eggs of *Xenopus laevis*, *Exp. Cell Res.* **105**:159.

Steinhardt, R., Zucker, R., and Schatten, G., 1977, Intracellular calcium release at fertilization in the sea urchin egg, *Dev. Biol.* **58**:185.

Stern, C. D., Goodwin, B. C., 1977, Waves and periodic events during primitive streak formation in the chick, *J. Embryol. Exp. Morphol.* **41**:15.

Streett, J. C., 1940, Experiments on the organization of the unsegmented egg of *Triturus pyrrhogaster*, *J. Exp. Zool.* **85**:383.

Subtelny, S., 1974, Nucleocytoplasmic interactions in development of amphibian hybrids, *Int. Rev. Cytol.* **39**:35.

Subtelny, S., and Bradt, C., 1961, Transplantation of blastula nuclei into activated eggs from the body cavity and the uterus of *Rana pipiens*, *Dev. Biol.* **3**:96.

Subtelny, S., and Bradt, C., 1963, Cytological observations on the early developmental stages of activated *Rana pipiens* eggs receiving a transplanted blastula nucleus, *J. Morphol.* **112**:45.

Subtelny, S., Smith, L. D., and Ecker, R. E., 1968, Maturation of ovarian frog eggs without ovulation, *J. Exp. Zool.* **168**:39.

Sudarwati, S., and Nieuwkoop, P. D., 1971, Mesoderm formation in the anuran *Xenopus laevis* (Daudin), *Roux Arch.* **166**:189.

Summerbell, D., Lewis, J. H., and Wolpert, L., 1973, Positional information in chick limb morphogenesis, *Nature (London)* **244**:492.

Sunkara, P. S., Wright, D. A., and Rao, P. N., 1979, Mitotic factors from mammalian cells induce germinal vesicle breakdown and chromosome condensation in amphibian oocytes, *Proc. Natl. Acad. Sci. U.S.A.* **76**:2799.

Sutasurya, L. A., and Nieuwkoop, P. D., 1974, The induction of primordial germ cells in the urodeles, *Roux Arch.* **175**:199.

Tahara, Y., and Nakamura, O., 1961, Topography of the presumptive rudiments in the endoderm of the anuran neurula, *J. Embryol. Exp. Morphol.* **9**:138.

Tarin, D., 1971*a*, Histological factors of neural induction in *Xenopus laevis*, *J. Embryol. Exp. Morphol.* **26**:543.

Tarin, D., 1971*b*, Scanning electron microscopical studies of the embryonic surface during gastrulation and neurulation in *Xenopus laevis*, *J. Anat.* **109**:535.

Tarin, D., 1973, Histochemical and enzyme digestion studies on neural induction in *Xenopus laevis*, *Differentiation* **1**:109.

Tarkowski, A. K., and Wroblewska, J., 1967, Development of blastomeres of mouse eggs isolated at the 4- and 8-cell steps, *J. Embryol. Exp. Morphol.* **18**:155.

Tchou-Su, 1950, Etude cytologique de la parthénogénèse experimente osmotique sur l'oeuf intraovarien mature *in vitro* chez le crapaud *(Bufo bufo asiaticus)*, *Chin. J. Exp. Biol.* **3**:1.

Thoman, M., and Gerhart, J. C., 1979, Absence of dorsal–ventral differences in energy metabolism in early embryos of *Xenopus laevis*, *Dev. Biol.* **68**:191.

Thornton, V. F., 1971, A bioassay for progesterone and gonadotropins based on the meiotic division of *Xenopus* oocytes *in vitro*, *Gen. Comp. Endocrinol.* **16**:599.

Tiedemann, H., 1976, Pattern formation in early developmental stages of amphibian embryos, *J. Embryol. Exp. Morphol.* **35**:437.

Tocchini-Valentini, G. P., and Crippa, M., 1970, Ribosomal RNA synthesis and RNA polymerase, *Nature (London)* **228**:993.

Toivonen, S., 1978, Regionalization of the embryo, in: *Organizer: A Milestone of a Half-Century from Spemann* (O. Nakamura and S. Toivonen, eds.), pp. 119–156, Elsevier/North Holland Biomedical Press, N.Y.

Toivonen, S., and Wartiovaara, J., 1976, Mechanisms of cell interaction during primary embryonic induction studied in transfilter experiments, *Differentiation* **5**:61.

Toivonen, S., Taris, D., Saxen, L., Tarin, P. J., and Wartiovaara, J., 1975, Transfilter studies on neural induction in the newt, *Differentiation* **4**:1.

Toivonen, S., Tarin, D., and Saxen, L., 1976, The transmission of morphogenetic signals from amphibian mesoderm to ectoderm in primary induction, *Differentiation* **5**:49.

Tompkins, R., and Rodman, W. P., 1971, The cortex of *Xenopus laevis* embryos: Regional differences in composition and biological activity, *Proc. Natl. Acad. Sci. U.S.A.* **68**:2921.

Townes, P. L., and Holtfreter, J., 1955, Directed movements and selective adhesion of embryonic amphibian cells, *J. Exp. Zool.* **128**:53.

Tuft, P., 1961*a*, Role of water-regulating mechanisms in amphibian morphogenesis: A quantitative hypothesis, *Nature (London)* **192**:1049.

Tuft, P., 1961*b*, The effect of a morphogenetic inhibitor (β-mercaptoethanol) on the uptake and distribution of water in the embryo of *Xenopus laevis*, *Proc. R. Phys. Soc. Edinburgh* **28**:123.

Tuft, P. H., 1962, The uptake and distribution of water in the embryo of *Xenopus laevis* (Daudin), *J. Exp. Biol.* **39**:1.

Tuft, P., 1965, The uptake and distribution of water in the developing amphibian embryo, *Soc. Exp. Biol. Symp.* **19**:385.

Tung, T. C., 1933, Recherches sur la détermination du plan médian dans l'oeuf de *Rana fusca*, *Arch. Biol.* **44**:809.

Turin, L., and Warner, A., 1977, Carbon dioxide reversibly abolishes ionic communication between cells of the early amphibian embryo, *Nature (London)* **270**:56.

Turing, A. M., 1952, The chemical based of morphogenesis, *Philos. Trans. R. Soc. London, Ser. B* **237**:37.

Ubbels, G. A., 1977, Symmetrization of the fertilized egg of *Xenopus laevis*, studied by cytological, cytochemical, and ultrastructural methods, in: *Progrès récents en biologie de développement des Amphibiens*, *Mém. de la Soc. Zool. France* **44**:103.

Ubbels, G. A., and Hengst, R. T. M., 1978, A cytochemical study of the distribution of glycogen and mucosubstances in the early embryo of *Ambystoma mexicanum*, *Differentiation* **10**:109.

Van Gansen, P., 1966a, Ultrastructure comparée du cytoplasm periphérique des oocytes murs et des oeufs vierges de *Xenopus laevis*, *J. Embryol. Exp. Morphol.* **15**:355.

Van Gansen, P., 1966b, Effect de la fecondation sur l'ultrastructure du cytoplasme periphérique de l'oeuf de *Xenopus laevis*, *J. Embryol. Exp. Morphol.* **15**:365.

Van Gansen, P., and Schram, A., 1972, Evolution of the nucleoli during oogenesis in *Xenopus laevis* studied by electron microscopy, *J. Cell Sci.* **10**:339.

Van Gansen, P., and Weber, A., 1972, Determinisme de la polarité de l'oocyte de *Xenopus laevis:* Etude aux microscopes électroniques a balayage et a transmission, *Arch. Biol. (Liège)* **83**:215.

Vintemberger, P., 1934, Resultats de l'autodifferentiation des quatre macromères isolés au stade de huit blastomères, dans l'oeuf d'un amphibien anoure, *C.R. Soc. Biol.* **117**:693.

Vintemberger, P., 1936, Sur le développement comparé des micromères de l'oeuf de *Rana fusca* divise en huit: (a) Après isolement, (b) après transplantation sur un socle de cellules vitellines, *C.R. Soc. Biol.* **122**:927.

Vitto, A., Jr., and Wallace, R. A., 1976, Maturation of *Xenopus* oocytes. I. Facilitation by ouabain, *Exp. Cell Res.* **97**:56.

Vogt, W., 1925, Gestaltungsanalyse am Amphibienkeim mit örtlicher Vitalfärbung, *Roux Arch.* **106**:542.

Vogt, W., 1929, Gestaltungsanalyse am Amphibienkeim mit örtlicher Vitalfärbung. II. Teil. Gastrulation und Mesodermbilding bei Urodelen und Anuren, *Roux Arch.* **120**:384.

Votquenne, J., 1933, La disposition générale des ébauches présomptives dans l'oeuf de Grenouille divisé en blastomères et les conséquences de la déstruction d'un micromère dorsal, *C.R. Soc. Biol. Paris* **113**:1531.

Waddington, C. H., and Perkowska, E., 1965, Synthesis of RNA by different regions of the early amphibian embryo, *Nature (London)* **207**:1244.

Waddington, C. H., and Perry, M. M., 1956, Teratogenic effects of trypan blue on amphibian embryos, *J. Embryol. Exp. Morphol.* **4**:110.

Wahli, W., Dawid, I. B., Wyles, T., Jaggi, R. B., Weber, R., and Ryffel, G. U., 1979, Vitellogenin in *Xenopus laevis* is encoded in a small family of genes, *Cell* **16**:535.

Wahn, H., Lightbody, L. T., and Tchen, T. T., 1976, Adenosine 3′,5′-monophosphate, morphogenetic movements and embryonic neural differentiation in *Pleurodeles waltlii*, *J. Exp. Zool.* **196**:125.

Wakahara, M., 1977, Partial characterization of "primordial germ-cell forming activity" localized in vegetal pole cytoplasm in anuran eggs, *J. Embryol. Exp. Morphol.* **39**:221.

Wall, R., and Faulhaber, I., 1976, Inducing activity of fractionated miscrosomal material from *Xenopus laevis* gastrula stage, *Roux Arch.* **180**:207.

Wallace, H., and Elsdale, T. R., 1963, Effects of actinomycin D on amphibian development, *Acta Embryol. Morphol. Exp.* **6**:275.

Wallace, H., Morray, J., and Langridge, W. H. R., 1971, Alternative model for gene amplification, *Nature (London), New Biol.* **230**:201.

Wallace, R. A., 1961, Enzymatic patterns in the developing frog embryo, *Dev. Biol.* **3**:486.

Wallace, R. A., and Bergink, E. W., 1974, Amphibian vitellogenin: Properties, hormonal regulation of hepatic synthesis and ovarian uptake, and conversion to yolk proteins, *Am. Zoologist* **14**:1159.

Wallace, R. A., and Ho, T., 1972, Protein incorporation by isolated amphibian oocytes. II. A survey of inhibitors, *J. Exp. Zool.* **181**:303.

Wallace, R. A., and Misulovin, Z., 1978, Growth and differentiation of *Xenopus* oocytes *in vitro*, *J. Cell Biol.* **79**:182a (abstract from *Am. Soc. Cell Biol.*).

Wallace, R. A., and Steinhardt, R. A., 1977, Maturation of *Xenopus* oocyte. II. Observations on the membrane potential, *Dev. Biol.* **57**:305.

Wallace, R. A., Jared, D. W., and Nelson, B. L., 1970, Protein incorporation by isolated amphibian oocytes. I. Preliminary studies, *J. Exp. Zool.* **175**:259.

Wallace, R. A., Nickol, J. M., Ho, T., and Jared, D. W., 1972, Studies on amphibian yolk. X. The relative roles of autosynthetic and heterosynthetic processes during yolk protein assembly by isolated oocytes, *Dev. Biol.* **29**:255.

Wartenburg, H., 1962, Elektronmikroskopische und histochemische Studien über den Oogenese der Amphibienzelle, *Z. Zellforsch. Mikroskop. Anat.* **58**:427.

Wartenburg, H., and Schmidt, W., 1961, Electronenmikroskopische Untersuchungen der struckturellen Veränderungen in Rindenbereich des Amphibieneies im Ovar und nach der Befruchtung, *Zeitschr. Zellforsch.* **54**:118.

Wassarman, P. M., Hollinger, T. G., and Smith, L. D., 1972, RNA polymerases in the general vesicle contents of *Rana pipiens* oocytes, *Nature (London), New Biol.* **240**:208.

Wasserman, W. J., and Masui, Y., 1975a, Effects of cycloheximide on a cytoplasmic factor initiating meiotic maturation in *Xenopus* oocytes, *Exp. Cell Res.* **91**:381.

Wasserman, W. J., and Masui, Y., 1975b, Initiation of meiotic maturation in *Xenopus laevis* oocytes by the combination of divalent cations and ionophore A23187, *J. Exp. Zool.* **193**:369.

Wasserman, W. J., and Masui, Y., 1976, A cytoplasmic factor promoting oocyte maturation: Its extraction and preliminary characteristics, *Science* **191**:1266.

Wasserman, W. J., and Smith, L. D., 1978, The cyclic behavior of a cytoplasmic factor controlling nuclear membrane breakdown, *J. Cell Biol.* **78**:R15.

Webb, A. C., and Smith, L. D., 1977, Accumulation of mitochondrial DNA during oogenesis in *Xenopus laevis*, *Dev. Biol.* **56**:219.

Weisz, P. B., 1945, The development and morphology of the larva of the South African clawed toad, *Xenopus laevis*. II. The hatching and the first and second-form tadpoles, *J. Morphol.* **77**:193.

Weyer, C. J., Nieuwkoop, P. D., and Lindenmayer, A., 1978, A diffusion model for mesoderm induction in amphibian embryos, *Acta Biotheoret.* **26**:164.

Whitington, P. McD., and Dixon, K. E., 1975, Quantitative studies of germ plasm and germ cells during early embryogenesis of *Xenopus laevis*, *J. Embryol. Exp. Morphol.* **33**:57.

Wiley, A. H., and DeRobertis, E. M., 1976, High tyrosinase activity in albino *Xenopus laevis* larvae, *J. Embryol. Exp. Morphol.* **36**:555.

Wiley, H. S., and Wallace, R. A., 1978, Three different molecular weight forms of the vitellogenin peptide from *Xenopus laevis*, *Biochim. Biophys. Res. Commun.* **85**:153.

Wilkins, A. S., 1976, Replicative patterning and determination, *Differentiation* **5**:15.

Williams, J., 1965, Chemical constitution and metabolic activities of animal eggs, in: *The Biochemistry of Animal Development*, Vol. 1 (R. Weber, ed.), pp. 13–71, Academic Press, New York.

Williams, M. A., and Smith, L. D., 1971, Ultrastructure of the "germinal plasm" during maturation and early cleavage in *Rana pipiens*, *Dev. Biol.* **25**:568.

Willmer, E. N., 1970, *Cytology and Evolution*, 2nd ed., Academic Press, New York.

Wilson, E. B., 1896, *The Cell in Development and Inheritance*, Johnson Reprint Corporation, New York, repr. 1966.

Wilson, E. B., 1925, *The Cell in Development and Heredity*, 3rd ed., Macmillan, New York.

Wintrebert, P., 1933, La mécanique du développement chez *Discoglossus pictus*. De l'ovogenèse a la segmentation, *Arch. Zool. Exp. Gen.* **75**:501.

Wischnitzer, S., 1965, The cytoplasmic inclusions of the salamander oocyte. I. Pigment granules, *Acta Embryol. Morph. Exp.* **8**:141.

Wischnitzer, S., 1966, The ultrastructure of the cytoplasm of the developing amphibian egg, *Adv. Morphog.* **5**:131.

Wischnitzer, S., 1976, The lampbrush chromosomes: Their morphology and physiological importance, *Endeavour* **35**:27.

Witschi, E., 1934, Appearance of accessory "organizers" in overripe eggs of the frog, *Proc. Soc. Exp. Biol. Med.* **31**:419.

Witschi, E., 1951, Gonad development and function. Embryogenesis of the adrenal and reproductive glands, *Recent Prog. Horm. Res.* **6**:1.

Witschi, E., 1952, Overripeness of the egg as a cause of twinning and teratogenesis: A review, *Cancer Res.* **12**:763.

Wittek, M., 1952, La vitellogénèse chez les Amphibiens, *Arch. Biol.* **63**:134.

Wolf, D. P., 1974a, The cortical granule reaction in living eggs of the toad, *Xenopus laevis*, *Dev. Biol.* **36**:62.

Wolf, D. P., 1974b, On the contents of the cortical granules from *Xenopus laevis* eggs, *Dev. Biol.* **38**:14.

Wolf, D. P., 1974c, The cortical response in *Xenopus laevis* ova, *Dev. Biol.* **40**:102.

Wolf, D. P., and Hedrick, J. L., 1971, A molecular approach to fertilization. II. Viability and artificial fertilization of *Xenopus laevis* gametes, *Dev. Biol.* **25**:348.

Wolf, D. P., Nishihara, T., West, D. M., Wyrick, R. E., and Hedrick, J. L., 1976, Isolation, physiochemical properties, and the macromolecular composition of the vitelline and fertilization envelopes from *Xenopus laevis* eggs, *Biochem.* **15**:3671.

Wolf, R., 1978, The cytaster, a colchicine-sensitive migration organelle of cleavage nuclei in an insect egg, *Dev. Biol.* **62**:464.

Wolpert, L., 1969, Positional information and the spatial pattern of cellular differentiation, *J. Theoret. Biol.* **25**:1.

Wolpert, L., 1971, Positional information and pattern formation, *Curr. Top. Dev. Biol.* **6**:183.

Woodland, H. R., 1974, Changes in the polysome content of developing *Xenopus laevis* embryos, *Dev. Biol.* **40**:90.

Woodland, H. R., and Adamson, E. D., 1977, The synthesis and storage of histones during oogenesis of *Xenopus laevis*, *Dev. Biol.* **57**:118–135.

Woodland, H. R., and Gurdon, J. B., 1968, The relative rates of synthesis of DNA, s-RNA, and r-RNA in the endodermal region and other parts of *Xenopus laevis* embryos, *J. Embryol. Exp. Morphol.* **19**:363.

Wu, M., and Gerhart, J. C., 1980, Partial purification and characterization of the maturation-promoting factor from eggs of *Xenopus laevis, Dev. Biol.,* in press.

Wylie, C. C., and Heasman, J., 1976, The formation of the gonadal ridge in *Xenopus laevis*. I. A light and transmission electron microscope study, *J. Embryol. Exp. Morphol.* **35**:125.

Wylie, C. C., and Roos, T. B., 1976, The formation of the gonadal ridge in *Xenopus laevis*. III. The behavior of isolated primordial germ cells in vitro, *J. Embryol. Exp. Morphol.* **35**:149.

Wyrick, R. E., Nishihara, T., and Hedrick, J. L., 1974, Agglutination of jelly coat and cortical granule components and the block to polyspermy in the amphibian *Xenopus laevis, Proc. Natl. Acad. Sci. U.S.A.* **71**:2067.

Yamada, T., 1950, Dorsalization of the ventral marginal zone of the *Triturus* gastrula. I. Ammonia treatment of the medio–ventral marginal zone, *Biol. Bull.* **98**:98.

Yamana, K., and Shiokawa, K., 1975, Inhibitor of ribosomal RNA synthesis in *Xenopus laevis* embryos. V. Inability of inhibitor to repress 5S RNA synthesis, *Dev. Biol.* **47**:461.

Young, R. S., Deal, P. H., Souza, K. A., and Whitfield, O., 1970, Altered gravitational field effects on the fertilized frog eggs, *Exp. Cell Res.* **59**:267.

Zalik, S. E., and Cook, G. M. W., 1976, Comparison of early embryonic and differentiating cell surfaces. Interaction of lectins with plasma membrane components, *Biochim. Biophys. Acta* **419**:119.

Ziegler, D., and Masui, Y., 1973, Control of chromosome behavior in amphibian oocytes. I. The activity of maturing oocytes inducing chromosome condensation in transplanted brain nuclei, *Dev. Biol.* **35**:283.

Ziegler, D., and Morrill, G. A., 1977, Regulation of the amphibian oocyte plasma membrane ion permeability by cytoplasmic factors during the first meiotic division, *Dev. Biol.* **60**:318.

Zimmerman, L., and Rugh, R., 1941, Effect of age on the development of the egg of the leopard frog, *Rana pipiens, J. Morphol.* **68**:329.

Zotin, A. I., 1964, The mechanism of cleavage in amphibian and sturgeon eggs, *J. Embryol. Exp. Morphol.* **12**:247.

Zotin, A. I., 1965, The uptake and movement of water in embryos, *Soc. Exp. Biol. Symp.* **19**:365.

Züst, B., and Dixon, K. E., 1975, The effect of U.V. irradiation of the vegetal pole of *Xenopus laevis* eggs on the presumptive primordial germ cells, *J. Embryol. Exp. Morphol.* **34**:209.

Züst, B., and Dixon, K. E., 1977, Events in the germ cell lineage after entry of the primordial germ cells into the germinal ridges in normal and U.V.-irradiated *Xenopus laevis, J. Embryol. Exp. Morphol.* **41**:33.

5

Biochemistry and Regulation of Nonmuscle Actins

Toward an Understanding of Cell Motility and Shape Determination

DENNIS G. UYEMURA and JAMES A. SPUDICH

1 Introduction

Actin is a component of virtually all eukaryotic cells. In cells of many types, it is the most abundant protein species. Moreover, many of its physical and chemical properties have been absolutely conserved throughout evolution. These observations underscore the critical nature of the contractile and cytoskeletal functions ascribed to actin. It is essential, therefore, in understanding cell motility and shape determination to address the regulation of actin polymerization, the intracellular distribution of actin, and the interaction of actin with myosin. It is the thesis of this chapter that substantial further insights into the regulation of actin in nonmuscle cells can be obtained only by a multifaceted approach involving biochemical, ultrastructural, and genetic techniques. Of course, the single common denominator in all such studies is actin itself. Hence, such studies can be built only on a solid foundation: a broad, quantitative characterization of actin purified from each organism selected for study.

The first actin to be isolated and characterized was that from vertebrate skeletal muscle. This was not surprising because of the availability of starting material, high actin con-

DENNIS G. UYEMURA ● Department of Biochemistry, State University of New York, Stony Brook, New York 11794   JAMES A. SPUDICH ● Department of Structural Biology, Sherman Fairchild Center, Stanford University School of Medicine, Stanford, California 94305

DENNIS G. UYEMURA
and JAMES A.
SPUDICH

tent of the tissue, and obvious significance of actin to muscle contraction. In fact, the molecular details of the skeletal muscle actin–myosin interaction and its regulation have served as a paradigm in most studies of actomyosin-based contractility in nonmuscle cells. Of primary importance was the finding that skeletal muscle actin is localized in thin filaments that are stably situated once myogenesis is completed. There is little turnover of actin itself, nor are there detectable changes in the intracellular distribution of the thin filament structures.

In contrast, actin in nonmuscle cells is appreciably more dynamic. That is, drastic alterations in its subcellular distribution, state of aggregation, and rate of turnover are the norm, particularly during growth and development.

In this chapter we shall examine recent progress (Spring, 1978) made in understanding the regulation of actin function in nonmuscle cells. Emphasis will be placed on those topics in which some biochemical insights have recently been made. These include the detailed characterizations of purified nonmuscle actins and investigations into the regulation of polymerization or aggregation. The interested reader is referred to excellent general reviews on actin and myosin in nonmuscle cells that address topics not covered here, such as the biochemistry of nonmuscle myosins and the regulation of the actin–myosin interaction (Pollard and Weihing, 1974; Clarke and Spudich, 1977; Hitchcock, 1977; Korn, 1978). In addition, detailed information can be obtained from *Cell Motility,* published by Cold Spring Harbor Laboratories (Goldman *et al.,* 1976), and the references cited therein.

2 Skeletal Muscle Actin: A Short Review of Pertinent Characteristics

Actin from vertebrate skeletal muscle is a globular single polypeptide chain with a mass of 42,000 daltons. It consists of 374 amino acid residues, including one residue of the rare amino acid 3-methylhistidine. The amino acid sequence of rabbit skeletal muscle actin has been determined by Collins and Elzinga (1975). The native monomer binds a single divalent cation (Ca^{2+} or Mg^{2+}) and a single ATP molecule.

The monomeric state (G-actin) is favored at low ionic strength. However, above about 25 mM KCl or 0.2 mM $MgCl_2$ actin polymerizes to yield filaments (F-actin). The structure of F-actin has been determined by electron microscopy and optical diffraction analysis. The filament is a one-start, left-handed helix with a genetic pitch of 59 Å. In negatively stained electron micrographs the filament gives the appearance of two strands of beads wound around each other into a right-handed helix with a pitch of 2×360 Å (Fig. 1) (Depue and Rice, 1965). During polymerization, each bound ATP of G-actin is hydrolyzed, resulting in one tightly bound ADP and the release of one inorganic phosphate per molecule polymerized.

It is important to note that under conditions that favor polymerization, there is always a characteristic critical concentration of monomeric actin in equilibrium with polymerized actin (for review, see Oosawa and Asakura, 1975). When the total actin concentration is lower than this critical concentration, all of the actin is monomeric. When the total actin concentration is higher than the critical concentration, only that amount of actin equal to it is monomeric; the rest is polymeric. Hence, actin polymerization is similar to the condensation of a gas to liquid or the crystallization of solute molecules.

When an actin molecule depolymerizes from F-actin, it will rapidly exchange its bound ADP for free ATP. That is, each complete cycle of polymerization and depolymerizalion results in the net hydrolysis of one ATP to free ADP and P_i. Hence, the polymerization and depolymerization reactions can be depicted as follows:

$$(\text{F-actin} \cdot \text{ADP})_n + \text{G-actin} \cdot \text{ATP} \underset{\text{ADP} \quad \text{ATP}}{\overset{P_i}{\rightleftarrows}} (\text{F-actin} \cdot \text{ADP})_{n+1}$$

Note that the hydrolysis of ATP is not directly coupled to polymerization. That is, depolymerization does not occur upon depolymerization. If it did, one would expect that the concentration of free P_i would affect the polymerization equilibrium, which it does not (Cooke, 1975).

A polymerization equilibrium may be approximated (Cooke, 1975) by the expression:

$$K_{eq} = \frac{[\text{F-actin}_{n+1}]}{[\text{G-actin}] [\text{F-actin}_n]} = \frac{[F_{n+1}]}{[G] [F_n]}$$

Since the concentration of F-actin does not change upon the addition of a subunit, $[F_{n+1}] = [Fn]$; and the expression simplifies to $K_{eq} = 1/[G]_{eq}$ where $[G]_{eq}$ = critical concentra-

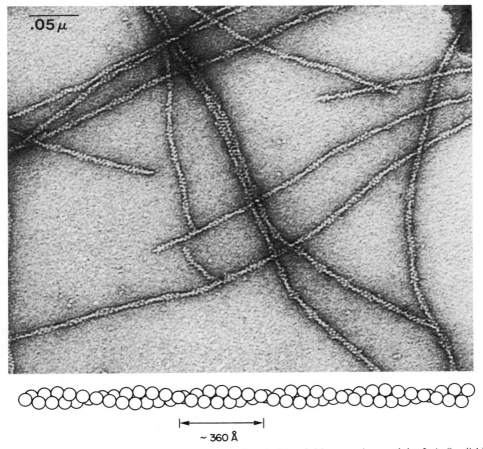

Figure 1. F-actin reconstituted from purified muscle G-actin (above) (electron micrograph by J. A. Spudich) and schematic drawing of F-actin (below).

DENNIS G. UYEMURA
and JAMES A.
SPUDICH

tion. Thus this approximated equilibrium constant for actin polymerization is equal to the inverse of the critical concentration.* Muscle actin at 0.1 M KCl shows a critical concentration of about 15 μg/ml. This value should be compared to the total concentration of actin in skeletal muscle, which is greater than 10 mg/ml. From the above, one would expect that, unless unknown factors are affecting polymerization, greater than 99% of the actin in skeletal muscle should be polymerized.

Although no physiological function has been ascribed to G-actin in skeletal muscle, F-actin must undertake several critical interactions. It must bind the regulatory proteins tropomyosin and troponin in the ratio of one of each of these molecules for every seven actins to form the thin filaments of the muscle sarcomere. These proteins can bind in one of two configurations, depending on the free calcium ion concentration (Spudich et al., 1972; Wakabayashi et al., 1975; Huxley, 1973). Below about 10^{-6} M Ca^{2+}, the regulatory proteins are thought to bind such that myosin would be sterically hindered from interacting with actin. When the Ca^{2+} concentration rises above that threshold, the tropomyosin molecules rotate about the long axis of the actin filament such that the myosin binding sites on the actin would be exposed. The actomyosin complex then forms, bound ADP and P_i are released from the myosin, and the myosin molecule is thought to change its configuration for its "power stroke" or "contraction" phase. Free ATP then binds, resulting in the release of actin, followed by hydrolysis of the ATP to complete the cycle. Note that purified F-actin will also bind myosin in an ATP dissociable linkage. When proteolytic subfragments of myosin are used to eliminate the complication of myosin thick filament formation, the binding of myosin "heads" to F-actin results in a distinctive arrowhead pattern, which is so characteristic that this phenomenon is currently used as a diagnostic test for the presence of actin filaments in cell extracts or permeabilized cell preparations (Fig. 2). All myosins bound to any particular filament are oriented in the same manner. Thus, there is a polarity to F-actin. Thin filaments are known to be anchored to structures within the muscle sarcomere termed Z lines. Only one end of the thin filament is anchored, such that decoration with myosin heads gives a pattern of arrowheads pointing away from the Z line. Thus, only one particular end of the thin filament interacts with the Z line. This is thought to be mediated at least in part by a protein called α-actinin.

In conclusion, muscle actin must be capable of interacting in a precise manner with many proteins: with itself to form F-actin, with tropomyosin and troponin to form thin filaments, with myosin for contraction, and probably with α-actinin. These numerous interactions take up a large part of the actin molecule surface and should place tight evolutionary constraints on the primary sequence of muscle actin.

Unfortunately, although large quantities of muscle actin can be obtained, the tendency of actin to polymerize and/or aggregate has hampered attempts to crystallize this protein. Hence, no fine-structure X-ray data are yet available.

3 Purification of Nonmuscle Actin: Some General Considerations

Muscle actin is distinctive for the large number of protein–protein interactive sites on its surface. For the purposes of studying actin in nonmuscle cells, two important classes of interactions can be defined: (1) actin–actin binding in 0.1 M KCl, imparting the ability to form F-actin, and (2) actin–myosin binding. These, along with the ability of actin to with-

---

*These considerations are only approximate since one end of the actin filament may be a primary assembly end and the other a primary disassembly end, each with a different equilibrium constant (Wegner, 1976; Engel et al., 1977).

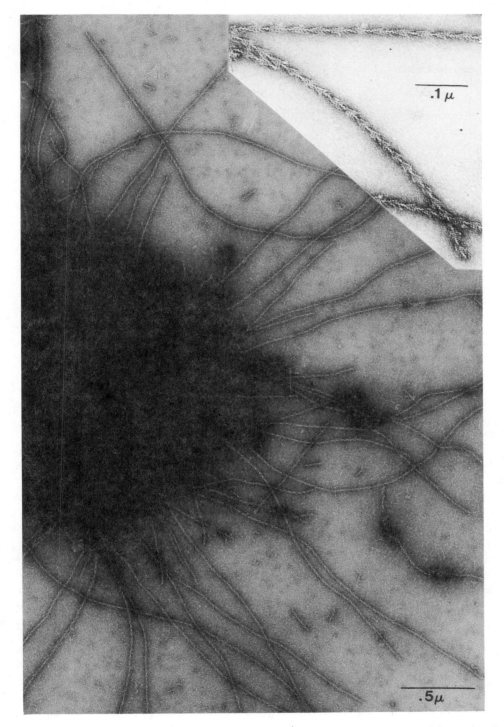

Figure 2. *Dictyostelium* F-actin grown off a polylysine-coated polystyrene bead (dark object left-center) and decorated with subfragment-1 (electron micrograph by S. S. Brown). Note that the beads initiate polar actin assembly (Brown and Spudich, 1979). Muscle F-actin decorated with subfragment-1 is shown at higher magnification in the upper right corner (electron micrograph by J. A. Spudich).

DENNIS G. UYEMURA
and JAMES A.
SPUDICH

stand dehydration on acetone, were adopted as steps for the purification of actin from numerous cell types prior to 1976. A common scheme would include acetone treatment of a crude actomyosin fraction, which was sometimes "seeded" by the addition of muscle myosin, followed by cycles of polymerization–depolymerization of the acetone extract. Some schemes also incorporated conventional steps such as gel filtration chromatography and ammonium sulfate precipitation. What is notable is that every purification scheme used to purify actin incorporated actomyosin precipitation and/or at least one complete cycle of polymerization–depolymerization.

A second notable feature was that the yields from these schemes were in all cases quite low. Actin in most of the cell types used comprises about 5% of the total cell protein. If we assume that 100 g wet weight of packed cells contains 5–8 g of protein, then about 200–400 mg of actin should be present. However, actual yields from such an amount of starting material typically fell in the range of 1–20 mg of purified actin. This indicates an overall yield of less than 10% in the best cases, with figures below 5% being routine.

Such low yields were of no great concern at the time. Indeed, the goal of such work was to obtain enough actin to undertake any analysis at all, and that was accomplished. However, in 1975 and 1976, two sets of results led to a necessary reexamination of purification schemes for actin from nonmuscle cells. They were (1) reports of microheterogeneous isoelectric species of vertebrate actins (Whalen *et al.,* 1976; Garrels and Gibson, 1976; Rubenstein and Spudich, 1977), and (2) reports of nonmuscle actins from vertebrate sources with polymerization properties qualitatively different from skeletal muscle actin (Bray and Thomas, 1976; Abramowitz *et al.,* 1975). This possibility for multiple chemical species of actin with functionally distinct properties raised a question as to whether the actins purified previously were a minor subpopulation of a large family of actins and were therefore not representative preparations. This concern was compounded by the combination of using functional characteristics (actomyosin interaction and actin polymerization) in the older purification schemes and the extremely low yields which ensued.

These problems were first addressed by Gordon *et al.* (1976a). These workers devised a new purification scheme for actin from *Acanthamoeba* involving DEAE-cellulose chromatography, polymerization–depolymerization of the actin, and Sephadex G-150 chromatography. Actin was assayed by dodecyl sulfate–polyacrylamide gel electrophoresis—that is, the band that migrated at the position of *Acanthamoeba* actin was followed. Quantitative densitometry of the Coomassie-blue-stained gels was used to calculate yields. This methodology recovered the majority of the actin at each step in the purification. The overall yield was in the range of 20–30%, a substantial increase over all previous preparations. This basic scheme has been used for the purification of actin from chicken embryonic brain, human platelets, and rat liver with comparable yields (Gordon *et al.,* 1977). This procedure is readily adaptable to a wide range of starting materials. Although the actin yield was greatly increased by this scheme, it should be noted that a cycle of polymerization–depolymerization was employed. Hence, any actin with intrinsically different polymerization properties would, in all probability, be lost. Nevertheless, it would indeed be extremely difficult to devise a purification scheme that was impervious to the polymerization state of actin altogether. The purification behavior of the G and F forms of actin are simply too different. During any particular step it is likely that one is greatly enriching for either the G or F form of actin.

A new step in actin purification schemes was introduced by Kane (1975). He found that upon warming an extract of sea urchin eggs, a gel formed that was highly enriched for actin, and from which actin was purified. Pollard *et al.* (1976) utilized this phenomenon to purify actin from *Acanthamoeba*. These and other recent purification schemes (Pardee and Bamburg, 1979; Adelman, 1977; Hatano and Owaribe, 1977) contain some original and useful steps, but none completely avoids actin polymerization–depolymerization, acto-myosin precipitation, or actin gel formation—all of which are highly regulated *in vivo.*

Recently, Uyemura *et al.* (1978) have devised a new purification scheme for actin from the cellular slime mold, *Dictyostelium discoideum.* This scheme is notable in that (1) it minimizes dependence on functional characteristics of actin, and (2) it optimizes for recovery of actin with no detectible loss due to proteolysis. The procedure incorporates Sephadex G-150 chromatography, DEAE-cellulose chromatography, ammonium sulfate fractionation, and differential centrifugation. As judged by quantitative densitometry of dodecyl sulfate–polyacrylamide gels, the overall yield was 20–30%. To our knowledge, this is the first procedure that does not utilize actomyosin precipitation, polymerization–depolymerization, acetone dehydration, or gel formation. That is, it does not rely on any functional characteristic that could result in a preparation of actin highly "selected" for some critical behavioral property.

What, then, are the ramifications of avoiding such "selective" purification steps? As will be discussed below, in several instances, the properties of the final actin fraction characterized directly reflect the strategy of the purification procedure used in its preparation. This is not surprising. Several important conclusions can be drawn from the studies discussed above, and the preliminary reports that led to them.

(1) It is critical to attempt to follow recoveries with quantitative assays for actin. This is particularly important if one desires to account for unusual losses in the course of "selective" purification steps. To date, the most reliable assay is densitometry of gels run under denaturing conditions. It is true that precise values cannot be obtained with crude fractions because of uncertainty as to where to draw a baseline for the actin peak. Nevertheless, one can assume an average baseline on the basis of the magnitude of general background staining in the vicinity of the actin peak.

Unfortunately, one cannot utilize the activation of the heavy meromyosin ATPase to assay for actin because of substantial ATPase activities present in crude extracts other than that of myosin. However, a potentially useful alternative assay procedure may be to utilize the phenomenon that actin specifically inactivates pancreatic deoxyribonuclease I (DNase I). This interaction will be discussed further in Section 6. Under appropriate conditions, it may be possible to follow "DNase I inhibitor activity" as a quantitation of actin (Blikstad *et al.,* 1978).

(2) Older purification schemes that utilize "selective" steps are generally far from being optimized for yields at individual steps or for controlling the extent of damage caused by proteolysis or denaturation. Such optimization can result in increased yields of about tenfold, but does not ensure that a representative sample of actin will be obtained.

(3) Actin can be purified by the conventional protein purification techniques of ion exchange chromatography, gel filtration, and differential precipitation. In such cases, it is mandatory to have a reliable assay procedure for following actin. This is probably the best strategy to follow if one is interested in obtaining a preparation of actin that is highly representative of the actin present *in vivo,* and in good yield.

4 Physicochemical Characteristics of Nonmuscle Actins

The vast majority of studies on the physical and chemical properties of nonmuscle actins have been performed with preparations of the older type. These studies have been reviewed by Pollard and Weihing (1974) and Clarke and Spudich (1977). In general, striking similarities between muscle and nonmuscle actins were found by all criteria. More recent studies on the chemical characteristics of nonmuscle actins are reviewed here. The most significant advances have been made in two areas: isoelectric focusing analyses and amino acid sequence determinations.

4.1 Isoelectric Focusing Microheterogeneity

Several groups have now reported that multiple microheterogeneous species of actins can be resolved on the basis of slight differences in charge density (Whalen *et al.*, 1976; Rubenstein and Spudich, 1977; Garrels and Gibson, 1976). Of the three species found from vertebrate sources, skeletal muscle contains the most acidic form, termed α-actin. Nonmuscle tissues contain the intermediate and basic forms, β and γ species, with a preponderance of β-actin. The questions of utmost significance are (1) can these species be correlated with distinct functional properties; and (2) are the differences due to posttranslational modifications or different genes, or both?

4.1.1 Functional Consequences of Multiple Isoelectric Species

As will be discussed in Section 5, highly purified preparations of vertebrate muscle (α), brain (β,γ), liver (β,γ), and platelet (β,γ) actins show no qualitative differences in polymerization or heavy meromyosin activation properties (Gordon *et al.*, 1977). Moreover, actin from *Dictyostelium* focuses at a yet more acidic position than does α-actin (Uyemura *et al.*, 1978). Nevertheless, this actin shows no qualitative differences in polymerization, heavy meromyosin activation, and filament structure properties compared to skeletal muscle actin. These results do not rule out the possibility that distinct isoelectric species have distinct functional properties *in vivo,* but they do suggest that differences in isoelectric point are not necessarily indicative of intrinsically distinct characteristics. In this regard, it should be noted that the two actins with the most divergent isoelectric points, those of *Acanthamoeba* and *Dictyostelium,* have essentially indistinguishable functional characteristics.

Whether the existence of multiple isoelectric species of actin in vertebrates is fortuitous and inconsequential or critical for normal growth and development is not known. However, the presence of only a single detectable species in *Acanthamoeba* and *Dictyostelium* suggests that one species may be sufficient for all the cellular functions required of actin in the course of the cell-division cycle, phagocytosis, and locomotion. Furthermore, the finding of but a single actin species in nematode (Schachat *et al.*, 1978) would further suggest that one species is sufficient even in a more complex organism, one that shows distinct myogenesis.

Further insight into the biological significance of multiple isoelectric species of actin will require substantial further work along several fronts: (1) Improved techniques for resolving microheterogeneous actins are needed to determine if multiple species really are present in amoebae or nematodes. Perhaps current methodology is not sufficient in this respect. (2) New and improved methods for testing the functional characteristics of purified actins are needed in order to detect possibly subtle, yet significant qualitative differences among the properties of various isoelectric species. In particular, methods to quantify actin–membrane interactions and actin supramolecular assembly (gel, cable, bundle, etc.) formation would be of great value. (3) Actin-binding proteins that modulate behavior must be purified and characterized. Whether any of these differentially bind particular isospecies of actin could then be addressed. (4) The application of genetic analysis is mandatory in order to substantiate that any postulated functional differences found *in vitro* are indeed operative *in vivo*.

It seems unlikely that, with current technology, one will be able to purify one iso-species from the others by conventional protein purification techniques. The physical differences between the various species are too small. This is in all probability one instance in which only a "selective" purification procedure, based on some possible specific functional characteristic, will succeed.

4.1.2 Molecular Basis for Multiple Isoelectric Species

The existence of multiple isoelectric species of actin in vertebrate organisms could be due to posttranslational modifications, distinct actin genes, or a combination of both. Recently, Vandekerckhove and Weber (1978) proposed that the isoelectric focusing microheterogeneity of actins is due to amino acid differences at the N-terminal three to four amino acids of the molecule. These workers found that the N-terminal sequences for several vertebrate actins were very acidic, with differences in composition which could account for the observed isoelectric focusing behavior (Table I). Since aspartic acid imparts a slightly lower pI than does glutamic acid, then of the two N-terminal sequences determined for brain actins, β-actin would be expected to have the three terminal Asp residues. γ-Actin would thus have three terminal Glu residues. Indeed, the major actin species of chicken gizzard, a γ species, has three terminal Glu residues. Since skeletal muscle actin contains two Asp and two Glu residues at its amino terminus, it would be expected to be the most acidic of the three vertebrate types.

4.2 Amino Acid Sequence Determinations

The reports of Elzinga *et al.* (1976), Elzinga and Lu (1976), and Lu and Elzinga (1977) established that all actins have closely related primary sequences. These analyses included actins from beef skeletal muscle (55), heart (72), and brain (157); human heart (78) and blood platelets (29); and *Acanthamoeba castellanii* (186). The numbers in parenthesis indicate the numbers of residues identified in relation to the known sequence of 374 residues in rabbit skeletal muscle actin. Since all the sequences show a high degree of similarity, it is more useful to consider only differences in sequences compared to that of rabbit skeletal muscle actin. All of the muscle actins (two skeletal and two cardiac) were identical in the identified residues, with the exception of a serine in cardiac actin at position 357 compared to a threonine in skeletal muscle actin. The two vertebrate nonmuscle actins were

TABLE I. Proposed N-Terminal Amino Acid Sequences of Vertebrate
Actins[a]

Tissues	Isoelectric species	Sequence[b]
Rabbit skeletal muscle	α	Ac-Asp-Glu-Asp-Glu-Thr-Thr-Ala-Leu
Bovine brain	(β)	x-Asp-Asp-Asp -Ile- Ala- Ala-Leu
Bovine brain	(γ)	x-Glu-Glu-Glu -Ile- Ala- Ala-Leu
Chicken gizzard	γ	x-Glu-Glu-Glu-Thr-Thr-Ala-Leu

[a]Taken from Vandekerckhove and Weber, 1978.
[b]The x- designates an N-terminal blockage of undetermined character.

DENNIS G. UYEMURA
and JAMES A.
SPUDICH

identical to one another in identified residues and showed 11 differences in 157 residues, compared to actin from skeletal muscle (93% homology). Actin from *Acanthamoeba* showed 11 differences in 186 residues, a 94% homology. Two important general conclusions were drawn from this work: (1) the vast majority of amino acid substitutions are conserved changes, and (2) tissue-specific differences within the same organism are greater than sequence variations among actins from the same tissues of different organisms. Indeed, certain positions may prove to be diagnostic for determining whether a particular actin is from muscle or from nonmuscle tissue. For example, the muscle actins from three different organisms contained a valine at position 129, whereas the cytoplasmic actins contained a threonine at that position. Also, two muscle actins had leucine at residue 16, and two nonmuscle actins had methionine. The extreme homology uncovered for actins from organisms as divergent as *Acanthamoeba* and humans is consistent with the constraints imposed by the numerous specific protein–protein interactions that actin must undertake, presumably over a large portion of its surface. The sequence differences that have been elucidated establish that there are at least three actin genes in mammals. The greater variability among different tissues than among different species suggests either that actin genes diverged earlier than did the various mammalian species, or that the few differences found between sequences from different tissues impart functional differences to these actins that are essential for proper actin function within each tissue. Such functional differences could be manifested as *intrinsically* distinct aggregative properties of the various actins and/or as differential binding properties of each actin to other cellular components with which proper interaction is essential.

Recently, Vandekerckhove and Weber (1978) have extended the work of Elzinga and co-workers. They have analyzed 62 to 97% of the amino acids in eight different actin preparations, including chicken gizzard, bovine thymus, and mouse SV101 and rat C6 cells. It should be noted that in this analysis, peptides were fingerprinted and analyzed for amino acid composition. Only those with differences in *composition* compared with rabbit skeletal muscle actin were sequenced. Hence, changes in the order of amino acid residues within a particular peptide may have been overlooked.

In general, this work further substantiated and generalized the major conclusions discussed above. In particular, certain amino acid positions can indeed be used to distinguish muscle from nonmuscle actins. Notable are positions 16 (muscle: leucine; nonmuscle: methionine), 76 (muscle: isoleucine; nonmuscle: valine), 106 (muscle: threonine; nonmuscle: valine), 201 (muscle: valine; nonmuscle: threonine), and 296 (muscle: asparagine; nonmuscle: threonine). These examples are cited because they have been identified in each of the nine vertebrate actins analyzed. Again, most substitutions are of a conserved nature. As discussed in Section 4.1.2, the only substitutions that could give rise to isoelectric point differences were found in the N-terminal three or four residues. Two other new findings of these workers were (1) a suggested different N-terminal sequence for skeletal muscle actin of Ac-Asp-Glu-Asp-Glu-Thr-Thr instead of the sequence Ac-Asp-Glu-Thr-Glu-Asp-Thr reported by Elzinga (see Elzinga and Lu, 1976); and (2) actin from rabbit skeletal muscle has four acidic N-terminal residues, whereas actins from bovine brain and chicken gizzard have three. Hence, most actins may have 373 residues, instead of the 374 found in skeletal muscle actin.

Finally, the conclusion that tissue variability is greater than species variability has been further strengthened. All skeletal muscle actins were identical to one another in the sequences analyzed. The cytoplasmic actins differ from skeletal muscle actins in at least 25 positions (representing 93% homology or less). The β and γ species of brain actins were assigned distinct N-terminal residues at positions 1, 2, 3, and 9. This is consistent with distinct genes coding for the two species. Hence, available data now suggest that there are at least five distinct actin genes in vertebrates: skeletal muscle, cardiac muscle, smooth muscle γ-, nonmuscle β-, and nonmuscle γ-actins.

4.3 Posttranslational Modifications

At least two posttranslational modifications have been suggested to occur on all actins: N-terminal blockage (acetylation in skeletal muscle actin), and histidine methylation (forming 3-methylhistidine at position 73 in skeletal muscle actin). Actin from *Acanthamoeba* is unique among all actins analyzed in that it also contains various forms of N^ϵ-methylated lysines. Whether actin–lysine methylation serves a biological function in this microorganism or is simply a fortuitous and innocuous side reaction is not known.

Much greater significance has been proposed for histidine methylation, as 3-methylhistidine is a rare amino acid (for review, see Paik and Kim, 1975) and because every actin analyzed shows about one 3-methylhistidine residue per molecule (see Pollard and Weihing, 1974). In skeletal muscle, histidine methylation is a posttranslational modification (Hardy *et al.,* 1970) and *S*-adenosylmethionine is the presumed methyl group donor (Hardy and Perry, 1969; Krzysik *et al.,* 1971). Moreover, one particular histidine residue, that at position 73, of the nine total histidines present in skeletal muscle actin is methylated. This sequence specificity, along with its conservation among all eukaryotic actins, is suggestive of an important biological role for this covalent modification. Two possibilities are that methylated and unmethylated actins show intrinsically distinct polymerization properties, and that methylation affects the interaction of actin with myosin or with any of the numerous other actin-binding proteins.

To address the biological significance of actin methylation it is important to quantitate the extent of methylation and to purify to homogeneity both the methylated and unmethylated forms for detailed biochemical and structural characterization. A further consideration is that methylation probably evolved originally in a primitive eukaryotic microorganism and certainly not in vertebrate skeletal muscle. Hence, one might be well-advised to investigate this topic by studying a eukaryotic microorganism or, at least, a nonmuscle tissue. The intracellular distribution and aggregation state of actin is relatively static in skeletal muscle compared to its various and ever changing structures and interactions in growing and developing cells (see Section 5). Therefore, the role of 3-methylhistidine may not be apparent in vertebrate skeletal muscle because of its relatively limited range of activities therein.

Unfortunately, little is known about actin methylation in nonmuscle cells. Methylhistidine has been demonstrated to exist in a cyanogen bromide fragment generated from *Acanthamoeba* actin, which is homologous to the CB-10 fragment from rabbit skeletal muscle actin (Weihing and Korn, 1972). In this amoeboid actin, its position in the primary sequence of amino acids is essentially the same as its position in muscle actin.

Certainly, additional investigations are warranted. An important first step will be to purify unmethylated and methylated forms of actin for comparative functional characterization. This could be accomplished by utilizing chicken embryos, which contain a large amount of unmethylated actin prior to 10 days of incubation, with methylation occurring gradually from the tenth to eighteenth day (Krzysik *et al.,* 1971). Such a characterization must include quantitative comparisons of polymerization, myosin activation, and interaction with all proteins known to modulate the behavior of actin *in vitro* (see Section 6). Furthermore, the actin methyltransferase must be purified and characterized particularly with respect to the regulation of its activity.

5 Functional Properties of Nonmuscle Actins

A striking feature of the behavior of actin in many nonmuscle cells, in contrast to its role in striated muscle, is that, during both cell division and locomotion, drastic changes occur in the subcellular distribution and state of aggregation of actin. To illustrate this point, consider the states of actin during the cell division cycle of a typical fibroblast cell.

During interphase, the cell is flattened as a result of substratum attachments, and much of the actin is localized in "stress fibers" or "bundles," which can reach several micrometers in length, and which generally are oriented along the long axis of the cell, parallel to the direction of movement. Immediately prior to mitosis, many of the substratum attachments are released and the cell rounds up. At this point the vast majority of stress fiber structures disappear altogether. Once the mitotic spindle apparatus is formed, some of the actin is found in association specifically with pole-to-chromosome spindle fibers, and not with pole-to-pole fibers. Upon completion of anaphase, septation of the two new daughter cells is accomplished by a "contractile ring," which is located just beneath the cytoplasmic membrane on the equator between the two newly formed nuclei. This ring constricts to cause septation. It is composed, in large part, of actin filaments. Once cytokinesis is complete, the new cells flatten and reattach to the substratum. This is accompanied by the reappearance of stress fibers and marks the completion of the mitotic cycle. Thus at various stages of the cycle, major changes occur in both the localization of actin within the cell and the structures in which actin can be identified. Clearly then, it is essential to address those regulatory mechanisms that underlie such changes. Two general classes of mechanisms should be considered:

(1) There may exist distinct chemical species of actins with different functional characteristics. Each particular species could be involved in a discrete set of activities, thereby simplifying the task of coordinating the diverse roles of actin. In this case, distinctions could arise from covalent posttranslational modifications of a common actin, and/or from the existence of multiple actin genes with different amino acid sequences. Although multiple chemical species of actin in vertebrates have been described (see Section 4.1), the critical point here is that *functional* distinctions must also be found before this mechanism is supported. This point will be discussed later in this section.

(2) Functional distinctions could arise from the reversible binding of other chemical species, probably proteins or small molecules such as divalent cations or nucleotides. Such binding could alter the behavior of some or all of a common pool of cellular actin according to the needs of the cell.

It is important to note that these are not mutually exclusive mechanisms. Indeed, because the roles of actin are so diverse, one might imagine that two or three distinct forms of actin might be further "differentiated" by small molecules or specific actin-binding proteins, resulting in a large array of functionally distinct actin populations.

Of course, for each cell type investigated, there are several steps that must be undertaken before the question of actin regulation can be properly addressed: purification to homogeneity of a representative sampling of the total cellular actin, followed by quantitative characterization of the actin in terms of polymerization properties, filament structure, and myosin ATPase activation. As discussed in Section 3, the majority of action preparations previously investigated were prepared by schemes that incorporated one or more "selective" purification steps. These include actin–myosin precipitation, polymerization–depolymerization cycles, and gel formation. Only recently have precautions been taken with respect to monitoring recoveries and checking side fractions for functionally different populations (Gordon *et al.*, 1976*a*, 1977) or avoiding such steps altogether (Uyemura *et al.*, 1978). These are also the only instances in which extensive *quantitative* characterization of the resulting preparations have been performed. Hence, this section will review the results of these recent studies. Some earlier reports of substantially different properties of nonmuscle and muscle actins are discussed in Section 6.

5.1 Polymerization Properties

As presented in Section 2, the polymerization equilibrium may be quantitated as the inverse of the concentration of G-actin present at equilibrium (critical concentration). The first quantitation of the polymerization of nonmuscle actin in this way was by Spudich and Cooke (1975). These workers found that actin from *Dictyostelium discoideum* showed an increase in absorbance at 232 nm upon polymerization, similar in extent to that of muscle actin. When the extent of this change was plotted against total actin concentration, identical profiles were obtained for both *Dictyostelium* and muscle actins. The profiles suggested that below a certain critical concentration of total actin, there was no polymerization. But above that concentration, polymerization was directly proportional to the total concentration of actin above the critical level. The extrapolated critical concentration found was approximately 1 μM.

Inasmuch as the *Dictyostelium* actin used in this study was purified by the older, conventional techniques, which included precipitation with myosin and a cycle of polymerization–depolymerization, Uyemura *et al.* (1978) have recently extended the characterization of *Dictyostelium* actin. In this study the actin was purified by "nonselective methods" (see Section 3), and a fivefold higher yield of actin was obtained. *Dictyostelium* actin prepared in this way also demonstrated a low critical concentration (about 0.4 μM) as determined by viscometric analyses. Such a low value for the critical concentration for polymerization leads to the interesting conclusion that, in the absence of factors that might perturb the polymerization equilibrium, greater than 99% of the total cellular actin would be expected to be polymerized *in vivo*.

The polymerization of *Acanthamoeba* actin has also been studied quantitatively. Gordon *et al.* (1976*b*) utilized viscometry to demonstrate that actin obtained in good yield from this organism polymerizes in a manner similar to that of muscle actin. They showed that the critical concentration and the reduced viscosity (indicative of the filament length distribution) are similar for both actins under conditions of ionic strength, pH, and temperature expected to be physiological. Moreover, by determining the effect of temperature on the critical concentration, they could calculate the enthalpy, entropy, and free energy of polymerization. *Acanthamoeba* actin polymerization was shown in this way to be strongly endothermic and entropy-driven, as was known for muscle actin.

The polymerization properties of vertebrate nonmuscle actins have been described by Gordon *et al.* (1977). Actins from rat liver, human platelets, and embryonic chick brains were purified by the procedure developed for obtaining *Acanthamoeba* actin; the overall

DENNIS G. UYEMURA
and JAMES A.
SPUDICH

yields were about the same in all cases. All of these actins showed critical concentrations and reduced viscosities similar to those of muscle actin. Hence, the general polymerization features of several nonmuscle actins from diverse sources are identical to those of skeletal muscle actin.

5.2 Myosin ATPase Activation

All of the actins described in the previous section have been analyzed with respect to their ability to activate the ATPase activity of heavy meromyosin (HMM), a proteolytic subfragment of skeletal muscle myosin. When the inverse of the activation is plotted against the inverse of the actin concentration, the values generated for each actin give straight lines which converge at the y axis, indicating that the extents of activation at extrapolated infinite actin concentrations are the same for each actin. However, the apparent Michaelis constants for all nonmuscle actins were greater than that of muscle actin. The vertebrate nonmuscle actins were less than 50% greater, whereas the amoeboid actins were two- to threefold greater. These findings suggest that the basic features of myosin activation are similar for all actins. That the amoeboid forms showed weaker binding interactions than the vertebrate forms is not surprising in view of the evolutionary distance between the organisms from which the actins were derived.

The identical maximal extents of activation are not unexpected. At low actin concentrations, some aspect of the actin–HMM interaction must be rate limiting for the ATPase reaction mechanism. But at infinite actin concentration, that interaction is accelerated to a point where it can no longer be rate limiting; and some other microscopic step of the overall mechanism *not* related to the actin interaction becomes limiting. This new rate-limiting step is presumably not affected by the type of actin present; hence, the maximal velocity at infinite actin should be the same for every actin.

5.3 Filament Structure

The general appearance of filaments of every actin is similar when visualized by negative staining in the electron microscope: it is a two-strand helix. However, the details of the filament parameters can only be obtained by more sophisticated techniques. One of the most useful of these is optical diffraction analysis, whereby the image obtained in the electron microscope is used to diffract a beam of coherent light. Although it is possible to diffract individual filaments, the signal-to-noise ratio is much improved with an array of filaments held in alignment. Skeletal muscle actin is known to form such an aggregate in the presence of magnesium ions (25–50 mM). This "magnesium paracrystal" is a single-layered, side-to-side aggregate of filaments with their helical crossover points in alignment. The optical diffraction pattern obtained with paracrystals of *Dictyostelium* actin reveals spots on the 1st, 6th, and 8th layer lines (Fig. 3) (Uyemura *et al.*, 1978). Such a layer line spacing is identical to that found for paracrystals of pure skeletal muscle actin and indicates that the helical parameters of *Dictyostelium* and muscle actins are essentially the same. This is in contrast to actin filament bundles from sea urchin eggs or sperm acrosomal processes (DeRosier *et al.*, 1977). Such bundles are multilayered aggregates formed by actin in the presence of other proteins. Optical diffraction analysis of these structures reveals a slightly different layer line spacing, consistent with a slight unwinding of the filament helix in bundles, compared to that found in muscle actin paracrystals. It is possible that this unwinding is the result of constraints imposed by the binding of the presumptive bundle-inducing protein(s).

From the results reviewed in this section, we must conclude that all purified actins from a variety of nonmuscle sources show functional properties remarkably similar to those of skeletal muscle actin. The general features of polymerization, myosin activation, and filament structure are identical in all cases. This conclusion points to the existence of other factors that must function to regulate the varied activities of actin in nonmuscle cells.

Moreover, these identical functional characteristics, coupled with the varied isoelectric focusing positions of these actins lead to one further conclusion. That is, differentiation of actin functions within a single organism, indeed, within a single cell, cannot be attributed to *intrinsic* differences in the discernible functional properties of the several actin isoelectric species thus far identified. Hence, not only is a single isoelectric species apparently sufficient for the needs of organisms such as amoebae and nematodes, but in organisms in which multiple species are present there is as yet no indication that those multiple species alone can account for differentiation of actin function. We would suggest that such multiple species are present rather to be differentially bound by "actin modulator proteins"—proteins that in turn serve to regulate and differentiate the diverse roles of actin in nonmuscle cells.

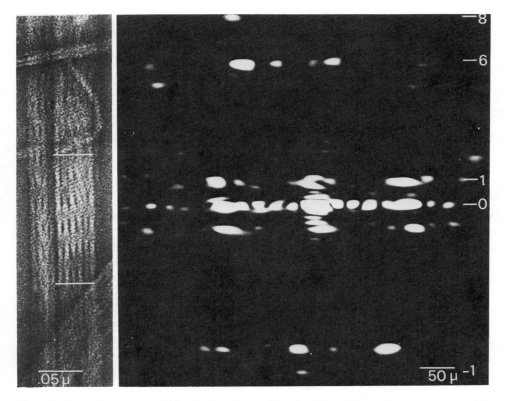

Figure 3. Magnesium paracrystal of actin from *Dictyostelium discoideum* (left panel: electron micrograph by S. S. Brown). The area denoted by the white lines was used to generate the optical diffraction pattern shown (right panel). Taken from Uyemura *et al.* (1978) with permission.

DENNIS G. UYEMURA
and JAMES A.
SPUDICH

It is tempting to speculate that a particular actin gene, the genes coding for its particular modulator proteins, as well as a particular myosin gene, and perhaps the respective myosin modulator gene(s) may be under coordinate regulation. If so, that particular actin species would evolve to optimize its regulatory interactions with its actin modulator protein(s); however, its basic filament structure, polymerization properties, and ability to activate myosin would nevertheless remain, of necessity, unchanged. This would lead to precisely those findings observed thus far, multiple chemical species of actins with identical intrinsic functional characteristics. We would predict, then, that particular modulator proteins will be able to distinguish between the multiple chemical species of actins. Clearly, the next major task to be confronted after a thorough characterization of nonmuscle actin is a search for such putative actin modulator proteins.

6   Actin Modulator Proteins

We will refer to all cell components that specifically and reversibly bind to actin, and thereby modify its functional characteristics, as *actin modulators*. It is assumed that most such species will be proteins, hence the term *modulator proteins*. These must be divided into two general classes: those which bind to G-actin and those which bind to F-actin. It is important to note that those that bind monomeric actin will probably bind stoichiometrically, and those which bind F-actin may act effectively on a large molar excess of actin molecules over modulator.

To date, several such modulator proteins have been identified and partially purified. Only one, however, binds reversibly to monomeric actin. This protein, called profilin, is present in bovine spleen, thymus, and brain, and in human lymphocytes and platelets (Carlsson *et al.*, 1977). It has not been detected in erythrocytes or skeletal muscle (Carlsson *et al.*, 1977). It shows the property of maintaining actin in a monomeric state under conditions that otherwise favor polymerization. Hence, its function may be to maintain a storage pool of monomeric actin in those cells in which rapid changes in the motility apparatus are necessary.

Profilin is a protein of 16,000 molecular weight. It forms a one-to-one complex with G-actin and was initially purified as such a complex (Carlsson *et al.*, 1976a). For those modulator proteins that affect a structural property of actin such as this, quantitative assays are generally unavailable. In this case, the profilin–actin complex, termed profilactin, showed the unique property of rapidly and stoichiometrically inhibiting the enzyme deoxyribonuclease I (DNase I) from bovine pancreas. Indeed, the identity of actin within this complex was not learned until several years after the purification of this inhibitor of DNase I (Lazarides and Lindberg, 1974).

The biochemical characterization of profilin and profilactin is still in a preliminary stage. Profilin may be recovered from the profilactin complex after exposure to 2 M urea followed by DEAE-cellulose chromatography. When mixed with skeletal muscle G-actin for 10 min, profilin retarded the rate of polymerization but had no effect on the extent of polymerization (Carlsson *et al.*, 1977). Recall that profilactin does not polymerize at all under standard conditions. Hence, profilactin was not properly reconstituted in this experiment. However, if skeletal muscle F-actin was depolymerized in the presence of profilin for 3 days, no repolymerization of actin was detected (Carlsson *et al.*, 1976b). Therefore profilactin can be successfully reconstituted from purified components.

These two experiments are consistent with three possible interpretations regarding profilactin reconstitution: (1) The kinetics of interaction between actin and profilin require times much longer than 10 min *in vitro*. (2) Filamentous, not monomeric, actin properly interacts with profilin. For example, perhaps only those actin molecules at the termini of depolymerizing filaments correctly bind profilin. (3) Monomeric actin with bound ADP (not ATP) will form profilactin. Hence, an actin · ADP molecule immediately after release from the filament and before nucleotide exchange with free ATP would react properly, whereas a solution of monomeric actin would have essentially all of its molecules bound with ATP. Alternatively, after release of an actin monomer from the filament and release of ADP, and yet before binding of free ATP, the monomer is nucleotide-free. Perhaps it is this nucleotide-free species that forms profilactin. Indeed, it is noteworthy that the profilactin complex has no bound nucleotide (Carlsson *et al.*, 1977). Either nucleotide-free actin interacts with profilin, or profilactin complex formation displaces bound nucleotide. Many further experiments are required, but these studies have already provided significant inroads pertaining to the methodology required for the purification and characterization of actin modulator proteins in general.

The biological role of profilin in specific contractile events should be addressed immunologically. Antibody to profilin could be prepared, and its effect on contractile events *in vitro*, such as gelation, could be assessed directly. Perhaps antibody could be microinjected into cells at specific stages of the cell division cycle to assess the possible role of profilin in the disappearance of stress fibers prior to mitosis or the disassembly of the contractile ring, which may be required to achieve cytokinesis.

The presence of profilin in spleen, brain, platelet, and lymphocyte tissues explains, at least in part, the persistence of an appreciable pool of monomeric actin in those and presumably other vertebrate tissues (Bray and Thomas, 1976), but the search for other modulator proteins that may bind monomeric actin must continue, along with attempts to investigate the biological role of profilin in specific contractile events. Monomeric actin binds to and inhibits DNase I much more rapidly than does filamentous (or otherwise aggregated) actin, and this can serve as a general assay for other proteins that affect the monomer $\rightleftharpoons$ polymer equilibrium in any cell type (Blikstad *et al.*, 1978). Of course, this assumes that the modulator protein does not in itself compete significantly with the DNase I–actin interaction.

The discovery of profilin, aside from providing the first reliable method for assaying a modulator protein, has also permitted one other breakthrough. The profilactin complex has been crystallized, with resultant crystals of sufficient size and quality for X-ray diffraction analysis (Carlsson *et al.*, 1976a). Hence, the potential exists for obtaining the first three-dimensional structure of the actin polypeptide backbone. Despite the fact that this will be a structure for actin in the profilactin complex, valuable information on the general polypeptide folding scheme of native actin will be obtained. This information will aid in the eventual elucidation of the structure of pure actin crystals. Similarly, the DNase I–actin complex has been crystallized (Mannherz *et al.*, 1977). This could also lead to a detailed X-ray structure for actin.

Within the second class of modulator proteins, those that bind filamentous actin, examples of only one type of modulator have been isolated and characterized. Those are the gelactins, or gel-forming factors.

A striking characteristic of amoeboid cell movement is the gel to sol transformation of the cytoplasm. Recently, several laboratories have described crude extract preparations from widely divergent cell types that gelate at 25 to 37°C. This topic has been comprehen-

DENNIS G. UYEMURA
and JAMES A.
SPUDICH

sively reviewed by Pollard (1976). Pertinent to this discussion is the isolation of several proteins that mediate such gelation. One direct attempt to purify such activities was performed by Maruta and Korn (1977), who were studying *Acanthamoeba castellanii*. Four different proteins (gelactins) were isolated that could account for essentially all the gelation activity present in the initial extract. Each acted on a great molar excess of actin (1 : 50 to 1 : 500), and Mg^{2+} or Ca^{2+} was required in the millimolar range, whereas ATP was not required. The subunit molecular weights of these proteins range from 23,000 to 38,000.

Two larger gelactins, which may be related, have been isolated. They are "actin-binding protein" from pulmonary macrophages (Hartwig and Stossel, 1975) and filamin from chicken gizzard (Wang *et al.,* 1975; Shizuta *et al.,* 1976; Wang, 1977). Both have subunits approximately 260,000 in molecular weight. Filamin has been shown to be a component of microfilament bundles (Heggeness *et al.,* 1977), although its ability to induce bundles in a reconstituted system has not yet been demonstrated. It exists as a dimer, which is consistent with its ability to crosslink actin filaments into gels (Wang and Singer, 1977). Although actin-binding protein (ABP) has not been as well characterized, antibody to ABP inhibits gelation in crude extracts of leukemic granulocytes (Boxer and Stossel, 1975), suggesting that ABP is the predominant gelating activity present in this higher cell type.

The current interest in actin gelation was initiated by the work of Kane (1975) on sea urchin egg extracts. In this system, two proteins of 58,000 and 220,000 subunit molecular weights were implicated (Kane, 1976). An interesting facet of the sea urchin egg gels was that the actin was aggregated into filament bundles with the filament helices in register. Hence, rather than a cross-linking of free filaments, these extracts showed gelation via cross-linking of bundles. Furthermore, the gels would not form in the presence of calcium ion.

The role of the 58 K and 220 K proteins in this process have now been elucidated (Bryan and Kane, 1978). The 58 K protein alone does not cause gelation when mixed with actin but does mediate bundle formation in a Ca^{2+}-insensitive manner. The 220 K protein does not mediate gelation with actin either, distinguishing it from actin-binding protein and filamin. In the presence of both 58 K protein and 220 K protein, gelation of actin occurs, but in a Ca^{2+}-insensitive manner. Hence, in the course of preparation of the isolated 58 K and 220 K proteins, a putative Ca^{2+}-sensitizing factor was separated and not recovered.

The above discussion dealt with those modulator proteins that have been separated from actin and reconstituted in active complexes. These recent efforts represent the beginning in the next major step toward the understanding of cellular contractile events. The particular lessons learned from the work accomplished thus far reemphasize the need for rapid, quantitative assays for each of the modulator activities. The most productive work has resulted from the fortuitous observation that actin monomers rapidly inhibit DNase I and from the temperature-dependent gelation of cytoplasmic extracts. Further progress must await the discovery of other, similarly useful, phenomena as the basis for new assays. The development of an *in vitro* assay for forms of movement (such as cytoplasmic streaming or directed motion of bundles of actin filaments when mixed with ATP, myosin, and other necessary factors) is of fundamental importance to understand the function of these constituents in cell motility.

Beyond the purification and characterization of modulator proteins, there remains the study of the regulation and biological role of such factors. As mentioned above, one possible approach is the microinjection of antibodies directed against individual effectors, which could bind to, and thus regulate, these factors. Finally, and most powerfully, is the genetic approach. Mutations in specific contractile events should correspond to defects in particular modulators if they indeed mediate the activity of actin as predicted. In this regard, the use of eukaryotic microorganisms that are amenable to genetic analysis may prove most profitable.

7 Concluding Remarks

From the preceding discussion, it is apparent that several topics must be addressed in order to further our understanding of the roles and regulation of actin in nonmuscle cells. These include (1) the detailed biochemical and structural characterization of highly purified actin(s) from *every* cell type utilized for study; (2) the separation of actin isoelectric species from vertebrate cells, along with studies on the regulation of expression of their respective genes; (3) the role of 3-methylhistidine in any actin; and (4) the development of quantitative assays for the detection, purification, and characterization of all actin modulators, most of which probably remain yet unidentified.

This last topic raises an important question. How does one purify and characterize proteins whose identities and functions are unknown? There is, indeed, one plausible answer: the application of the technique of genetics along with current biochemical technology.

This combination of biochemistry and genetics has proven to be extremely effective in the clarification of most biochemical pathways from glycolysis to phage morphogenesis. The feasibility of employing genetics depends on the amenability of the organism under study and the resourcefulness of the investigator in devising effective screening procedures for isolating pertinent mutant strains.

Two classes of mutations may arise: those defective in most or all contractile processes, and those defective in one or a few specific processes. Because general defects in cellular contractility may well prove lethal, it is essential to screen for conditional mutations. Such general contractility defectives could presumably be isolated by a succession of screening steps involving a variety of specific contractility defects. This approach has been employed by Clarke (1978) in an attempt to isolate motility mutants of *Dictyostelium discoideum*. Strains have been isolated that are defective in both phagocytosis and chemotaxis, as well as being temperature-sensitive for growth. Further screening is in progress.

Once a number of mutant strains are obtained, they can be analyzed for corresponding defects in contractile processes *in vitro* such as gelation, bundle formation, and myosin ATPase activation. Extracts from strains defective in any specific *in vitro* process may then be checked for complementation capability. For example, extracts from two strains that were unable to gelate alone under restrictive conditions might gain the ability to do so if they are mixed. This would suggest that the two extracts were defective in two different proteins necessary for gelation. Moreover, one could presumably purify each of these proteins based on those "complementation" activities. In this way, one can indeed identify and purify contractility proteins without knowing their specific functions beforehand.

Of course, contractility mutants would also be screened for discernible defects in known contractile components such as actin, myosin, gelactins, and others. Mutations in known proteins would serve to identify or confirm the precise *in vivo* roles of these proteins.

In summary, the incorporation of genetic technology into current biochemical studies will yield significant new advances into the understanding of cell motility and contractility at the molecular level. The organisms that will prove most immediately productive for study

DENNIS G. UYEMURA
and JAMES A.
SPUDICH

in this respect will be those that have already been studied with regard to contractility proteins and are currently amenable to genetic analyses. They are *Dictyostelium discoideum* and *Physarum polycephalum*. It is hoped that the considerations presented in this chapter will heighten interest in the use of such organisms by others working in this field.

References

Abramowitz, J. W., Stracher, A., and Detwiler, T. C., 1975, A second form of actin: Platelet microfilaments depolymerized by ATP and divalent cations, *Arch. Biochem. Biophys.* **167**:230–237.

Adelman, M. R., 1977, *Physarum* actin. Observation on its presence, stability, and assembly in plasmodial extracts and development of an improved purification procedure, *Biochemistry* **16**:4862–4871.

Blikstad, I., Markey, F., Carlson, L., Persson, T., and Lindberg, U., 1978, Selective assay of monomeric and filamentous actin in cell extracts, using inhibition of deoxyribonuclease I, *Cell* **15**:935–943.

Boxer, L. A., and Stossel, T. P., 1975, Isolation and interactions of contractile proteins from chronic myelogenous leukemia granulocytes (CMLG), *J. Cell Biol.* **67**:40a.

Bray, D., and Thomas, C., 1976, Unpolymerized actin in fibroblasts and brain, *J. Mol. Biol.* **105**:527–544.

Brown, S. S., and Spudich, J. A., 1979, Nucleation of polar actin filament assembly by a positively-charged surface, *J. Cell Biol.* **80**:499–504.

Bryan, J., and Kane, R. E., 1978, Separation and interaction of the components of sea urchin actin gel, *J. Mol. Biol.* **125**:207–224.

Carlsson, L., Nyström, L.-E., Lindberg, U., Kannan, K. K., Cid-Dresdner, H., Lövgren, S., and Jörnvall, H., 1976a, Crystallization of a non-muscle actin, *J. Mol. Biol.* **105**:353–366.

Carlsson, L., Nyström, L.-E., Sundkvist, I., Markey, F., and Lindberg, U., 1976b, Profilin, a low-molecular weight protein controlling actin polymerizability, in: *Contractile Systems in Nonmuscle Tissues* (S. V. Perry, A. Margreth, and R. S. Adelstein, eds.), pp. 39–49, Elsevier North-Holland Biomedical Press, Amsterdam.

Carlsson, L., Nyström, L.-E., Sundkvist, I., Markey, F., and Lindberg, U., 1977, Actin polymerizability is influenced by profilin, a low molecular weight protein in nonmuscle cells, *J. Mol. Biol.* **115**:465–483.

Clarke, M., 1978, A selection method for isolating motility mutants of *Dictyostelium discoideum,* in: *Cell Reproduction: Daniel Mazia Dedicatory Volume* (R. E. Dirkson, D. Prescott, and C. F. Fox, eds.), pp. 621–629, Academic Press, New York.

Clarke, M., and Spudich, J. A., 1977, Nonmuscle contractile proteins: The role of actin and myosin in cell motility and shape determination, *Annu. Rev. Biochem.* **46**:797–822.

Collins, J. H., and Elzinga, M., 1975, The primary structure of actin from rabbit skeletal muscle. Completion and analysis of the amino acid sequence, *J. Biol. Chem.* **250**:5915–5920.

Cooke, R., 1975, The role of the bound nucleotide in the polymerization of actin, *Biochemistry* **14**:3250–3256.

Depue, R. H., Jr., and Rice, R. V., 1965, F-actin is a right-handed helix, *J. Mol. Biol.* **12**:302–303.

DeRosier, D., Mandelkow, E., Silliman, A., Tilney, L., and Kane, R., 1977, Structure of actin-containing filaments from two types of nonmuscle cells, *J. Mol. Biol.* **113**:679–695.

Elzinga, M., and Lu, R. C., 1976, Comparative amino acid sequence studies on actins, in: *Contractile Systems in Nonmuscle Tissues* (S. V. Perry, A. Margreth, and R. S. Adelstein, eds.), pp. 29–37, Elsevier North-Holland Biomedical Press, Amsterdam.

Elzinga, M., Maron, B. J., and Adelstein, R. S., 1976, Human heart and platelet actins are products of different genes, *Science* **191**:94–95.

Engel, J., Fasold, H., Hulla, F. W., Waechter, F., and Wegner, A., 1977, The polymerization reaction of muscle actin, *Mol. Cell. Biochem.* **18**:3–14.

Garrels, J. I., and Gibson, W., 1976, Identification and characterization of multiple forms of actin, *Cell* **9**:793–805.

Goldman, R., Pollard, T., and Rosenbaum, J. (eds.), 1976, *Cell Motility,* Cold Spring Harbor Laboratory, Cold Spring Harbor, New York.

Gordon, D. J., Eisenberg, E., and Korn, E. D., 1976a, Characterization of cytoplasmic actin isolated from *Acanthamoeba castellanii* by a new method, *J. Biol. Chem.* **251**:4778–4786.

Gordon, D. J., Yang, Y. Z., and Korn, E. D., 1976b, Polymerization of *Acanthamoeba* actin. Kinetics, thermodynamics, and co-polymerization with muscle actin, *J. Biol. Chem.* **251**:7474–7479.

Gordon, D. J., Boyer, J., and Korn, E. D., 1977, Comparative biochemistry of nonmuscle actins, *J. Biol. Chem.* **252**:8300–8309.

Hardy, M. F., and Perry, S. V., 1969, *In vitro* methylation of muscle proteins, *Nature (London)* **223**:300–302.

Hardy, M. F., Harris, C. I., Perry, S. V., and Stone, D., 1970, Occurrence and formation of the N$^\epsilon$-methyllysines in myosin and the myofibrillar proteins, *Biochem. J.* **120**:653–660.

Hartwig, J. H., and Stossel, T. P., 1975, Isolation and properties of actin, myosin, and a new actin-binding protein in rabbit alveolar macrophages, *J. Biol. Chem.* **250**:5696–5705.

Hatano, S., and Owaribe, K., 1977, A simple method for the isolation of actin from myxomycete plasmodia, *J. Biochem.* **82**:201–205.

Heggeness, M. H., Wang, K., and Singer, S. J., 1977, Intracellular distributions of mechanochemical proteins in cultured fibroblasts, *Proc. Natl. Acad. Sci. U.S.A.* **74**:3883–3887.

Hitchcock, S. E., 1977, Regulation of motility in nonmuscle cells, *J. Cell Biol.* **74**:1–15.

Huxley, H. E., 1973, Structural changes in the actin- and myosin-containing filaments during contraction, *Cold Spring Harbor Symp. Quant. Biol.* **37**:361–376.

Kane, R. E., 1975, Preparation and purification of polymerized actin from sea urchin egg extracts, *J. Cell. Biol.* **66**:305–315.

Kane, R. E., 1976, Actin polymerization and interaction with other proteins in temperature-induced gelation of sea urchin egg extracts, *J. Cell Biol.* **71**:704–714.

Korn, E. D., 1978, Biochemistry of actomyosin-dependent cell motility (a review), *Proc. Natl. Acad. Sci. U.S.A.* **75**:588–599.

Krzysik, B., Vergnes, J. P., and McManus, I. R., 1971, Enzymatic methylation of skeletal muscle contractile proteins, *Arch. Biochem. Biophys.* **146**:34–45.

Lazarides, E., and Lindberg, U., 1974, Actin is the naturally occurring inhibitor of deoxyribonuclease I, *Proc. Natl. Acad. Sci. U.S.A.* **71**:4742–4746.

Lu, R. C., and Elzinga, M., 1977, Partial amino acid sequence of brain actin and its homology with muscle actin, *Biochemistry* **16**:5801–5806.

Mannherz, H. G., Kabsch, W., and Leberman, R., 1977, Crystals of skeletal muscle actin: Pancreatic DNAase I complex, *FEBS Lett.* **73**:141–143.

Maruta, H., and Korn, E. D., 1977, Purification from *Acanthamoeba castellanii* of proteins that induce gelation and syneresis of F-actin, *J. Biol. Chem.* **252**:399–402.

Oosawa, F., and Asakura, S., 1975, *Thermodynamics of the Polymerization of Protein*, Academic Press, London.

Paik, W. K., and Kim, S., 1975, Protein methylation: Chemical, enzymological, and biological significance, *Adv. Enzymol.* **42**:227–286.

Pardee, J. D., and Bamburg, J. R., 1979, Actin from embryonic chick brain. Isolation in high yield and comparison of biochemical properties with chicken muscle actin, *Biochemistry* **18**:2245–2252.

Pollard, T. D., 1976, Cytoskeletal functions of cytoplasmic contractile proteins, *J. Supramol. Struct.* **5**:317–334.

Pollard, T. D., and Weihing, R. R., 1974, Actin and myosin and cell movement, *CRC Crit. Rev. Biochem.* **2**:1–65.

Pollard, T. D., Fujiwara, K., Niederman, R., and Maupin-Szamier, P., 1976, Evidence for the role of cytoplasmic actin and myosin in cellular structure and motility, in: *Cell Motility* (R. Goldman, T. Pollard, and J. Rosenbaum, eds.), pp. 689–724, Cold Spring Harbor Laboratory, Cold Spring Harbor, N.Y.

Rubenstein, P. A., and Spudich, J. A., 1977, Actin microheterogeneity in chick embryo fibroblasts, *Proc. Natl. Acad. Sci. U.S.A.* **74**:120–123.

Schachat, F. H., Harris, H. E., and Epstein, H. F., 1978, Actin from the nematode, *Caenorhabditis elegans*, is a single electrofocusing species, *Biochim. Biophys. Acta* **493**:304–309.

Shizuta, Y., Shizuta, H., Gallo, M., Davies, P., Pastan, I., and Lewis, M. S., 1976, Purification and properties of filamin, an actin binding protein from chicken gizzard, *J. Biol. Chem.* **251**:6562–6567.

Spudich, J. A., and Cooke, R., 1975, Supramolecular forms of actin from amoebae of *Dictyostelium discoideum*, *J. Biol. Chem.* **250**:7485–7491.

Spudich, J. A., Huxley, H. E., and Finch, J. T., 1972, Regulation of skeletal muscle contraction. II. Structural studies of the interaction of the tropomyosin–troponin complex with actin, *J. Mol. Biol.* **72**:619–632.

Uyemura, D. G., Brown, S. S., and Spudich, J. A., 1978, Biochemical and structural characterization of actin from *Dictyostelium discoideum*, *J. Biol. Chem.* **253**:9088–9096.

Vandekerckhove, J., and Weber, K., 1978, Mammalian cytoplasmic actins are the product of at least two genes

338

DENNIS G. UYEMURA
and JAMES A.
SPUDICH

and differ in primary structure in at least 25 identified positions from skeletal muscle actins, *Proc. Natl. Acad. Sci. U.S.A.* **75**:1106–1110.

Wakabayashi, T., Huxley, H. E., Amos, L. A., and Klug, A., 1975, Three-dimensional image reconstruction of actin–tropomyosin complex and actin–tropomyosin–troponin T–troponin I complex, *J. Mol. Biol.* **93**:477–497.

Wang, K., 1977, Filamin, a new high-molecular-weight protein found in smooth muscle and nonmuscle cells. Purification and properties of chicken gizzard filamin, *Biochemistry* **16**:1857–1865.

Wang, K., and Singer, S. J., 1977, Interaction of filamin with F-actin in solution, *Proc. Natl. Acad. Sci. U.S.A.* **74**:2021–2025.

Wang, K., Ash, J. F., and Singer, S. J., 1975, Filamin, a new high-molecular-weight protein found in smooth muscle and non-muscle cells, *Proc. Natl. Acad. Sci. U.S.A.* **72**:4483–4486.

Wegner, A., 1976, Head to tail polymerization of actin, *J. Mol. Biol.* **108**:135–150.

Weihing, R. R., and Korn, E. D., 1972, *Acanthamoeba* actin. Composition of the peptide that contains 3-methylhistidine and a peptide that contains N$^\epsilon$-methyllysine, *Biochemistry* **11**:1538–1543.

Whalen, R. G., Butler-Browne, G. S., and Gros, F., 1976, Protein synthesis and actin heterogeneity in calf muscle cells in culture, *Proc. Natl. Acad. Sci. U.S.A.* **73**:2018–2022.

6

The Regulation of Cell Behavior by Cell Adhesion

PAUL C. LETOURNEAU, PETER N. RAY, and
MERTON R. BERNFIELD

1   Introduction

The complex morphogenesis of tissues and their successful participation in an organism's homeostasis depend on regulatory interactions of the tissue cells with their environment. The influence of soluble extrinsic molecules on genetic expression is best illustrated by the powerful effects of hormones. Less well appreciated but also with significant effects on genetic expression are the contacts and physical associations that nearly all cells make with other cells or with extracellular structures. The importance of these associations to cell function is poorly understood, because they are not easily manipulated or assessed *in situ,* and even when duplicated *in vitro* one must determine which aspects of cellular contact are regulatory.

Cell adhesion is commonly described as a basic cellular property of fundamental importance to multicellular organisms. In this chapter we examine closely the significance of cell adhesion to major aspects of cell behavior. Describing the role of cell adhesion in cell function and illustrating how genetic regulation may be mediated through adhesive interactions will be the purpose of this discussion rather than understanding the mechanisms of cell adhesion or the nature of the adhesive bond.

For each cell behavior discussed, we have considered three questions regarding cell adhesion. First, what is the significance to cell behavior of cell adhesion and of changes in

PAUL C. LETOURNEAU ● Department of Anatomy, University of Minnesota, Minneapolis, Minnesota 55455   PETER N. RAY ● Department of Medical Genetics, University of Toronto, Toronto, Canada M5S 1A8   MERTON R. BERNFIELD ● Department of Pediatrics, Stanford University School of Medicine, Stanford, California 94305 cell adhesion? A number of studies will be described demonstrating that cell adhesion is a controlling factor in cell movement, cell shape, cell division, and cell differentiation. Second, how does cell adhesion produce its effects on cell behavior? As with other cell surface functions transmembrane events may mediate the effects of cell adhesion, and we have emphasized the transmembrane influence of adhesion on the cytoskeleton, especially the arrangement of actin-containing microfilaments. The role of adhesion in cell movement and cell shape seems direct, but little is known about how adhesive contacts might influence nuclear events concerning cell division and cell differentiation. Third, how does adhesion regulate cell behavior? Cells *in vivo* make adhesive contacts with other cells and with elements of an elaborate but poorly understood extracellular matrix. Spatial and temporal changes and variations in these contacts may act as agents for genetic regulation of cell function. Such changes and variations could occur in the intrinsic adhesive properties of cells or may involve modifications in the distribution and abundance of extracellular materials.

Before proceeding, we will introduce cell adhesion with a definition, a description of cell adhesions *in vivo* and *in vitro,* a comment on the mechanism of cell adhesion, and a brief discussion of adhesive specificity.

1.1   How Is Cell Adhesion Defined?

Cell adhesions are noncovalent chemical bonds that cells make with other cells and with noncellular materials. The *in vitro* procedures employed for studying the nature and formation of these bonds measure cell adhesion as the resistance of cell contact to mechanical disruption. There is great diversity in the cell types used, the preparation of cells (potentially damaging proteolytic enzymes are often used to obtain single cells), and the means of initiating cell contacts and subsequently of applying disruptive forces (Cassiman and Bernfield, 1976a,b; Marchase *et al.,* 1976; McGuire and Burdick, 1976). Consequently, cell adhesion, as studied *in vitro,* is not a single phenomenon, but rather is defined operationally by the methods of each assay.

Among the molecules believed to participate in cellular adhesive bonds are components of the cell surface and of extracellular materials associated with cell surfaces: proteins, glycoproteins, proteoglycans, glycosaminoglycans, and gangliosides. The nature of the bonds is not clear, but may involve complementary molecular interactions such as those that occur in antibody–antigen complexes, enzyme–substrate interactions, or lectin–ligand binding. Curtis (1973) and others have discussed the possible bonding forces such as hydrogen bonds, calcium ion bridges, and electrostatic forces (Bell, 1978; Edwards, 1977; Trinkaus, 1969; Weiss, 1977).

1.2   Cell Adhesions in Vivo

The coherence and organization of all tissues involves the adhesive interactions of the component cells. In epithelial and central nervous tissues, cells are densely packed. Cell surfaces are often closely apposed, and intercellular adhesions are elaborately structured, as in the desmosomes and zonulae adherens of epithelia and synapses between neurons. These membrane specializations are characterized by accumulations of intracellular 10-nm filaments, 5- to 6-nm actin microfilaments, and dense material that is also present at the extracellular side of the membrane (Blomberg *et al.,* 1977; Weiss and Greep, 1977). The extremely close membrane appositions at gap junctions and the fusion of membranes at tight junctions are not characteristic of cellular adhesions. Probably because the outer boundary of a cell includes molecules that extend for various distances beyond the lipid bilayer (not usually resolvable by electron microscopy), most other adhesions between cells are morphologically simpler, appearing as limited areas of 10–20 nm separation between adjacent cellular projections. At the point of contact there may be an accumulation of microfilaments, but unlike microfilaments, 10-nm filaments and microtubules do not appear to contact the membrane (Brunk *et al.*, 1971; Heaysman, 1973; Spooner *et al.*, 1971). Intracellularly at adhesive sites, α-actinin and other proteins may act as anchors for microfilaments or may link filaments to membrane components (Ash and Singer, 1976; Bourguignon and Singer, 1977; Lazarides, 1976; Mooseker, 1976). These adhesive contacts may involve less than 1% of the cell surface.

In addition to intercellular adhesions, cells adhere to extracellular materials. As suggested above, clear morphological distinctions often cannot be made between extracellular materials and the cell surface. For example, epithelial tissues are always bound at their basal cell surfaces to a basal lamina, which by usual criteria consists of two layers: a 45- to 60-nm-thick electronlucent layer conforming closely to the contours of the basal cell membrane, possibly part of the cell surface, and an adjacent parallel denser layer of more variable thickness, which is clearly part of the extracellular matrix (Bernfield *et al.*, 1973; Trelstad *et al.*, 1974; Weiss and Greep, 1977). A similar basal lamina surrounds muscle fibers and lies beneath the endothelium lining blood vessels. Basal laminae consist of non-fibrillar types of collagen, glycoproteins, and in some cases contain proteoglycans and hyaluronic acid as well as fibronectin (Bernfield and Banerjee, 1972; Cohn *et al.*, 1977; Hay, 1973; Linder *et al.*, 1975). Fibronectin is a widely distributed, highly asymmetric, dimeric glycoprotein, which forms fibrillar aggregates and is very adhesive (L. B. Chen *et al.*, 1978; Yamada and Olden, 1978). Fibronectin binds to cell surfaces, collagen, heparen sulfate, gangliosides, and fibrinogen (Engvall and Ruoslahti, 1977; Hedman *et al.*, 1978; Stathakis and Mossesson, 1977; Yamada *et al.*, 1978). The components of the lamina may be highly organized into a regularly ordered scaffolding to which the cells attach (Cohn *et al.*, 1977). In epithelial tissues subject to abrasion, the cells adhere to the basal lamina by hemidesmosomes (Weiss and Greep, 1977).

In connective tissues, the large intercellular spaces are filled with a matrix containing varying proportions of fibrillar collagen, glycoproteins, fibronectin, proteoglycans, and hyaluronic acid (Greulich and Slavkin, 1975). Connective tissue cells, principally fibroblasts, have molecules at their cell surfaces that are very similar, and in some instances identical, to those in basal laminae (Bornstein and Ash, 1977; Cohn *et al.*, 1976; Goldberg *et al.*, 1979). One of these molecules, fibronectin, has been localized at sites of cell–cell and cell–substratum contact (Hedman *et al.*, 1978; Hynes and Destree, 1978; Mautner and Hynes, 1977). While these materials are not structurally arranged as is a basal lamina, they may be viewed as a primitive or rudimentary basal lamina. Molecules generally ascribed to the extracellular matrix are also components of the cell surface, and may be involved in cell adhesion to the matrix and possibly to other cells.

1.3 Cell Adhesions in Vitro

By far, the majority of cell adhesion studies have been done *in vitro* where cells are cultured on glass or plastic surfaces in liquid media. Although the relevance of studies involving cell contact with artificial surfaces can be questioned, recent work shows that molecules from the serum and from cells bind to tissue culture substrata (Culp and Buniel,

1976; Grinnell, 1978). The major molecules implicated in the coating of tissue culture substrata are the fibronectins, both cellular and plasma (Ali and Hynes, 1978; L. B. Chen *et al.,* 1978; Hedman *et al.,* 1978). Plasma fibronectin (previously known as cold insoluble globulin, cell attachment protein, etc.) is derived from the serum in culture medium and binds to *in vitro* substrata, promoting cell–substratum adhesion and cell spreading (Grinnell, 1978; Grinnell and Minter, 1978). It is very similar in molecular characteristics, but not identical to cellular fibronectin (Yamada and Kennedy, 1979), which is deposited on the substratum by cells via secretion, by desquamation of surface materials, or by portions of cells being torn away during migration across the substratum (L. B. Chen *et al.,* 1978; Culp, 1974). In addition to fibronectin, these deposits contain proteoglycans, hyaluronic acid, and, possibly from cell fragments, cytoskeletal components, including actin (Culp, 1974; Culp and Buniel, 1976; Rosen and Culp, 1977; Vaheri *et al.,* 1978). Thus, these substratum-associated materials, whether derived from the serum in the culture medium, or from the cells, mimic the substrata that the cells encounter *in vivo.*

Coating of *in vitro* culture surfaces with collagen or gelatin (denatured collagen) has been known for several years to enhance the survival and growth of epithelia and to promote nerve fiber formation and myoblast differentiation (Bornstein, 1958; Gospodarowicz *et al.,* 1978; Konigsberg and Hauschka, 1965; Luduena-Anderson, 1973). Some of these effects of collagenous substrata may be the result of fibronectin or fibronectinlike proteins, which bind specifically to collagen and mediate cell–substratum adhesion (Grinnell and Minter, 1978; Linsenmeyer *et al.,* 1978). In this regard, although exogenous fibronectin enhances the adhesion of fibroblasts to collagen, it does not affect the attachment of epidermal cells (Murray *et al.,* 1979). Collagen, however, may affect the *in vitro* formation of a basal lamina. When cultured on fibrillar collagen, but not on untreated plastic, mammary epithelial cells form a proteoglycan-rich basal lamina to which they attach, analogous to the *in vivo* situation (David and Bernfield, 1979; Emerman and Pitelka, 1977).

1.4 The Adhesive Process

The mechanism of cell adhesion remains elusive, in part because of differing interpretations of data from *in vitro* assays. A more basic impediment to understanding is that the formation of adhesions is a complex, multistep process. This complexity is illustrated by the diverse demonstrations that intercellular adhesion is inhibited by cold, metabolic inhibitors, proteolytic and glycolytic enzymes, cytochalasin B, colchicine, lectins, and the absence of Ca^{2+} (Cassiman and Bernfield, 1975, 1976a,b; Culp, 1978; Grinnell, 1978; Letourneau, 1979a; Marchase, 1977; McClay *et al.,* 1977; Roth, 1968; Steinberg *et al.,* 1973; Umbreit and Roseman, 1975). Current models generally propose that adhesion is initiated by a simple chemical interaction between components on opposing cell surfaces. Subsequent movements, clustering and changes in the mobility of membrane components, and transmembrane effects on the cytoskeleton, on ion fluxes or on enzymes may then stabilize and strengthen the intercellular bonds (Frazier and Glaser, 1979; Rees *et al.,* 1977; Umbreit and Roseman, 1975). These secondary events of the adhesive process are the subject of intense research and may be involved in some of the functional consequences of cell adhesion (Edelman, 1976).

1.5 Adhesive Specificity

An important concept, when considering the significance of cell adhesion, is adhesive specificity or cell recognition, meaning, in brief, that cell adhesions can differ quantitatively or qualitatively, depending on the identities of the cell interactants (Marchase *et al.*, 1976; Moscona, 1974). The idea of adhesive specificity springs from the "sorting out" of embryonic cells into homologous tissue masses, when cells from different organs are mixed and cultured together in a rotating suspension, and from demonstrations that cells adhere more rapidly or more firmly to preformed aggregates of homologous cells than of heterologous cells (Moscona, 1974; Moyer and Steinberg, 1976; Roth, 1968). An example of *in vivo* "sorting out" may be the exclusion of fibroblasts from clusters of myoblasts before fusion of the latter cells into myotubes.

Adhesive specificity applies not only to homologous cell interactions, but also to heterologous interactions, as in the affinity of Schwann cells for nerve axons and in interactions of T lymphocytes with B lymphocytes, of lymphocytes with macrophages, and of granulocytes with endothelial cells (Hoover *et al.*, 1978; Marchase *et al.*, 1976; Wood, 1976). Specificity may also exist in contacts between cells and elements of the extracellular matrix. For example, epidermal cells adhere more strongly to type IV collagen than to collagen of other types (Murray *et al.*, 1979).

A popular view is that adhesive specificity results from interactions between specific receptors and ligands on cell surfaces (Edelman, 1976; Moscona, 1974; Roth, 1973; Rutishauser *et al.*, 1976). In several models the ligand is a molecule containing carbohydrate, and the receptor is a molecule that binds carbohydrate (Roseman, 1974). Carbohydrates are also implicated in the adhesion of lower forms, such as the amoebae of cellular slime molds and dissociated cells of marine sponges, and there are receptors on sea urchin eggs for species-specific glycoproteins of sperm (Glabe and Vacquier, 1977; Henkart *et al.*, 1973; Muller and Gerisch, 1978; Turner and Burger, 1973). Algal mating, the survival of erythrocytes in the bloodstream, the binding of bacteria to mammalian cells, and the clearing of serum glycoproteins from the blood are other forms of biological recognition that involve molecules containing carbohydrate (Ashwell, 1977; Ashwell and Morell, 1974; Barondes and Rosen, 1976; Ofek *et al.*, 1977; Wiese and Wiese, 1975).

Adhesive specificity in *in vitro* assays is never all or none; rather, levels of affinity are obtained, often only after manipulation of the assay conditions (Cassiman and Bernfield, 1976*a,b*; McClay and Baker, 1975; McGuire and Burdick, 1976; Roth, 1968). Although distinct receptors and ligands may mediate adhesions of different cell types—for example, tissue-specific fibronectinlike molecules and aggregation-promoting factors (Balsamo and Lilien, 1974; Moscona, 1974)—the molecules may be structurally similar and able to interact with varying affinities.

In addition, other aspects of the adhesion process might generate adhesive selectivity. For example, the regional adhesive affinities of embryonic retinal cells for the optic tectum are believed to be due to topographic gradients in the abundance of complementary adhesive molecules, rather than the existence of regionally distinct molecules (Barbera, 1975; Gottlieb *et al.*, 1976; Marchase, 1977). Alternatively, differences in the distribution of adhesive molecules within the membrane or in membrane rearrangements and transmembrane events subsequent to initial molecular interactions might produce the differences in affinity implied by adhesive specificity and cell recognition (Cassiman and Bernfield, 1976*a,b*).

2 The Role of Cell Adhesion in Cell Movement and Cell Shape

Among the major events of organ morphogenesis are cell movements, crucial to the establishment of characteristic cell associations and cell shapes (Trinkaus, 1969, 1976). In adult organisms cell migration and changes in cell shape continue to be important in tissue repair and regeneration, and in granulocyte, macrophage, and lymphocyte activities (Marchase *et al.,* 1976; Trinkaus, 1976).

Cell movements and the maintenance of cell shape require adhesive contacts. This requirement is discussed here with emphasis on the basic relationship between adhesive contacts and the organization of cytoskeletal structures. In spite of tissue-specific characteristics, cell motility and the role of cell adhesion are very similar among fibroblasts, epithelial cells, and neurons. We discuss the role of cell adhesion in cell motility and cell shape in light of this generality, but also attempt to indicate how tissue-specific regulation of cytoskeletal and cell-surface properties is coupled with extrinsic cues from adhesive interactions with the environment to produce unique morphologies and characteristic cell movements.

2.1   Cell Movement in Vitro

Discussion begins with the attachment and spreading of fibroblastic cells onto a substratum. The involvement of cell–substratum adhesion is traced from the spreading of a freshly plated cell to the development of asymmetric cell shape and finally to cell movement. Fibroblastic cells are described first to provide a general statement of the role of cell adhesion in cell movement, and then epithelial cells and neurons are described.

2.1.1   Cell Spreading

All cells in suspension are rounded, although often many microvilli project from the cell periphery (Erickson and Trinkaus, 1976; Ukena and Karnovsky, 1977). In addition, the cytoskeleton is disorganized, as no microfilament bundles are seen. When cells are plated out, they settle onto the substratum, often making their initial contacts via the microvillar extensions.

The transformation of a rounded cell to the flattened spread shape of normal fibroblasts results from the protrusive activity of the cell margin. Once settled on a surface, the cell perimeter is protruded in two forms: cylindrical filopodia or microspikes and broader flattened lamellipodia (Abercrombie *et al.,* 1977; Albrecht-Buehler, 1976). These protrusions are extended independently, but a filopodium may also expand into a lamellipodium. The ultrastructure of these projections is similar to that of the cell cortex elsewhere, consisting of a lattice of actin-containing microfilaments, occasional vesicles, and the cell membrane (Abercrombie *et al.,* 1971; Spooner *et al.,* 1971).

2.1.1a   Cell Margin Activity.   The activity of the cell margin causes filopodia and lamellipodia to bend, wave, undulate, contact the substratum, and often to be resorbed or retracted into the cell margin (Abercrombie, 1961; Abercrombie *et al.,* 1977). Several phenomena suggest that protrusion is accompanied by activities directed backward toward the center of the cell (Harris, 1973*b*). Lamellipodia wrinkle and propagate undulations or ruffles back toward the nucleus. Particles attached to the front of these protrusions are transported backward along the extended process (Albrecht-Buehler and Goldman, 1976; Harris, 1973*b*). Concanavalin A binds to its receptors, forming complexes that move back from the cell margin and collect in centralized aggregates (Brown and Revel, 1976; Ukena and Karnovsky, 1977). Finally, the cell margin adjacent to intercellular adhesions often appears curved, as if undergoing local retraction (Abercrombie and Dunn, 1975).

Advance and withdrawal of the cell margin may represent opposing or alternating locomotory events in cell spreading (Abercrombie *et al.,* 1977). Extension may involve expansion of the cell surface through addition of components or by stretching of the membrane, and rearward movements may signify removal of membrane or recoil of the stretched portions. Microfilaments seem to be involved in all these activities, as cytochalasin B inhibits both extension of filopodia (and lamellipodia) and the rearward transport of surface-bound particles and of lectin–receptor complexes (Ash and Singer, 1976; Brown and Revel, 1976, Toh and Hard, 1977). How the filament network within the cell margin acts during extension and retraction is unclear, but filament assembly and disassembly, filament sliding, or changes in filament–membrane associations may occur.

2.1.1b Adhesive Contacts. On a nonadhesive substratum, usually a hydrophobic surface, protrusion occurs without cell spreading. But on a substratum to which the protrusions adhere, expansion of the cell margin begins (Albrecht-Buehler, 1976; Rajaraman *et al.*, 1974). Adhesion does not prevent the retractile events within a protrusion, because rearward transport of particles and lectin–receptor complexes continues; rather the extension is preserved in the presence of rearward movements or tensions. Should a contact persist until protrusion resumes ahead of and lateral to the adhesive site, a local spreading or expansion of the cell margin will result.

Adhesive contacts to glass coverslips can be visualized in the living state with interference reflection microscopy (Curtis, 1964; Izzard and Lochner, 1976). Adhesions appear as discrete spots or linear patches, separated from the substratum by approximately 10 nm. They may occur at one or more points beneath a filopodium and are spaced irregularly beneath lamellipodia. Occasionally, lamellipodia do not form discrete focal adhesions, but make broad adhesions with a separation of 30 nm from the substratum. Adhesions appear close to the front edge of the cell margin, but the edge itself is usually not adherent to the substratum (Ingram, 1969). Individual adhesions do not move; rather they disappear as the cell margin advances, forming new adhesions beyond the sites of previous ones (Izzard and Lochner, 1976).

2.1.1c Microfilament Organization. As a cell spreads, anchored by adhesive contacts near its margin, changes appear in the organization of cellular actin. Microfilaments in suspended cells are largely restricted to the filament network of the cell cortex and its microvilli. There are no bundles or cables of microfilaments in the cytoplasm of rounded cells, suggesting the existence of a pool of actin monomers (Hynes and Destree, 1978). When a cell settles onto an adhesive substratum, microfilament bundles appear first in filopodia and lamellipodia (Bragina *et al.*, 1976). Most strikingly, the bundles terminate or insert into the cell membrane at points of adhesive contact with a substratum or with other cells (Brunk *et al.*, 1971; Heath and Dunn, 1978; Heaysman, 1973; Izzard and Lochner, 1976). Although causality in this relationship is not clear, it is likely that the bundles form rapidly after establishment of adhesion.

It is not known how adhesion regulates microfilament organization. Transmembrane effects of adhesive contact may stimulate an isometric contraction of actomyosinlike filament proteins that is focused at the adhesive site and pulls filaments into bundles (Chen, 1979; Fleischer and Wolfarth-Botterman, 1975). Alternatively, contact-induced ion fluxes or formation of nucleation sites may stimulate filament assembly from actin monomers.

Studies of fibronectin further illustrate the relationship between adhesion, cell spreading, and the presence of microfilament bundles. As freshly plated cells spread onto a substratum, cellular actin and surface-associated fibronectin rearrange to become coincidently distributed (Hynes and Destree, 1978). Actin cables first appear when the margin begins to flatten, and at this time fibronectin becomes concentrated on the cell surface beneath the intracellular actin cables. The coincidence of actin and fibronectin is greatest at the ends of filament bundles where adhesive sites are located. Another example concerns transformed cells, which are often rounded and unspread *in vitro* and consequently have few micro-filament bundles. When purified fibronectin is added to the medium of established cell cultures, the cells spread, their adhesion to the substratum is increased, and many actin cables appear intracellularly (Ali *et al.,* 1977; Willingham *et al.,* 1977).

These studies with fibronectin raise interesting questions about the mechanism and consequences of cell adhesion. Does fibronectin move laterally in the membrane to accumulate at adhesive sites, and is there a transmembrane interaction between actin and fibronectin? Surface-associated fibronectin can be induced to form clusters with antibodies to fibronectin (Yamada, 1978), and fibronectin is released from cells by treatment with cytochalasin B (Kurikinen *et al.,* 1978). Other external surface proteins are reported to bind actin, also suggesting a transmembrane linkage (Koch and Smith, 1978).

The conclusions from these studies are that adhesion influences the organization of microfilaments in a localized fashion and that regulation of filament organization implies regulation of filament function. The interrelationships between microfilaments and their associations with the cell surface may be important in such filament activities as the extension and retraction of the cell margin, mobility and distribution of membrane molecules, and the positioning of intracellular elements. The assembly and redistribution of microfilaments, as produced by cell–substratum adhesion, could regulate these activites by altering the rates of filament functions or by establishing direction and polarity in microfilament functions (Small *et al.,* 1978).

2.1.2 Determination of Cell Shape

After spreading for some time, a fibroblast achieves a roughly circular outline with a uniformly flattened cell margin. Cell–substratum adhesions are located all around the cell and actin cables run diagonally within the spread cell margin, inserting at adhesive sites (Harris, 1973*c;* Heath and Dunn, 1978; Revel *et al.,* 1974). The commencement of cell migration involves the development of a polygonal, asymmetric cell shape. Large portions of the cell margin cease protrusive activity and become straight or slightly concave. These regions have few adhesive contacts, and lectin–receptor complexes are no longer transported centripetally from the cell margin in these areas (Brown and Revel, 1976; Vasiliev *et al.,* 1976). Microfilament bundles run parallel to the cell margin in these quiescent regions without inserting into the membrane (Luduena and Wessells, 1973; Spooner *et al.,* 1971).

The establishment of these quiescent regions of the cell margin is mandatory for net cell migration to occur. They may result from competition between areas of protrusion. Portions of the margin with more adhesions and more filament bundles may exert tension, possibly through microfilament activity, dislodging regions supported by fewer adhesions. Intact microtubules, though not involved in protrusive activity, are necessary for the restriction of protrusive activity and development of asymmetric cell shape (Domnina *et al.,* 1977; Vasiliev *et al.,* 1970). Microtubules apparently do not contact the cell membrane as do microfilaments, but it has been proposed that microtubules and microfilaments interact in a subplasmalemmal complex that controls membrane functions (Edelman, 1976; Puck, 1977).

In a migrating fibroblast, cell polarity is highly developed. The front of the cell is very broad, anchored by many adhesions, and actively extends lamellipodia and filopodia. Behind the leading edge the sides of the cell are usually straight, nonadherent, and quiescent. The cell tapers to a narrow tail, anchored by a few adhesions. The majority of the microfilament bundles run longitudinally along the axis of cell migration to insert at adhesive sites near the front edge of the cell (Albrecht-Buehler, 1977). The primary cilium, a microtubular structure located near the nucleus, also tends to be oriented along the axis of cell migration.

In one sense the shape of these cells is determined by the distribution of their adhesions (Abercrombie *et al.*, 1977). If a cell–substratum adhesion is broken, using a micromanipulator, the cell margin rapidly retracts, indicating that microfilament bundles do not remain in place without an adhesive anchor and may actually contract in response to being released (Harris, 1973*c*). The drug cytochalasin B is thought to alter microfilament organization, but does not disrupt cell–substratum adhesions. A cell treated with cytochalasin assumes an "arborized" shape as the cell margin collapses at many points, but adhesions remain, and when the drug is removed, the cell reassumes much the same shape as it formerly had (Sanger and Holtzer, 1972; Spooner *et al.*, 1971).

2.1.3 Migration of Tissue Cells

2.1.3a Fibroblasts. Fibroblast locomotion can be separated into three phases. The initial events, recalled from cell spreading, are extension and subsequent adhesion of filopodia and lamellipodia. These events advance the leading edge, and the locomotory cycle is completed by forward translocation of the cell and release of posterior adhesions. This movement, often an abrupt shortening or contraction of the cell forward, may result from actomyosinlike contraction of the microfilament bundles anchored at the front edge, or, alternatively, the movement signifies the release of tension developed against posterior adhesions by advance and spreading of the dominant leading edge (Chen, 1977; Fleischer and Wolfarth-Botterman, 1975; Huxley, 1973; Isenberg *et al.*, 1976; Luduena and Wessells, 1973). Posterior adhesions may decay slowly, as occurs at the base of lamellipodia, or they may detach rapidly during the forward shortening of the cell. In some cases, portions of the cell break off and remain attached to the substratum (Rosen and Culp, 1977).

Fibroblasts may migrate in a straight line for several hours, but over longer periods their paths approach a random walk due to changes in direction of movement (Gail, 1973). The direction of movement changes if protrusion activity becomes dominant at one side of the leading edge, or if protrusive activity begins elsewhere, forming a new leading edge, while the former leading edge becomes quiescent. The cell polarizes in the new direction, developing adhesive contacts beneath the new front edge, and there is reorientation of microfilament bundles and the primary cilium (Albrecht-Buehler, 1977). It is not known what prompts these changes. Alterations in adhesive contacts may promote protrusive activity at a new location or changes in cytoskeletal structures may redirect protrusive activity.

The most frequently observed interaction between fibroblasts is contact inhibition of locomotion (Abercrombie, 1967; Bell, 1977). Typically, the leading edge of a fibroblast encounters another cell, the cell margins cease to advance at the site of contact, and the cells remain immobile until a new leading edge forms to lead migration in a new direction (Trinkaus *et al.*, 1971). Firm cell–cell adhesions with associated actin cables are seen at the point(s) of cell–cell contact (Brunk *et al.*, 1971; Heaysman, 1973).

Contact inhibition has been explained by a proposal that protrusions do not adhere to the upper surface of the encountered cell, halting advance of the cell margin (DiPasquale and Bell, 1974). Protrusions also penetrate beneath the encountered cell, and it is suggested that their progress is impeded by the existing cell–substratum adhesions of the underlapped cell (Bell, 1977). This may explain why contact inhibition is more frequent in encounters between the leading edges of two cells than between a leading edge and the quiescent nonadherent sides of a cell, where one cell can underlap another (Abercrombie and Dunn, 1975; Bell, 1977). The reported absence of contact inhibition by transformed cells may occur because the cells have fewer adhesions to the substratum, making underlapping of a cell by other cells a common event (Bell, 1977; Erickson, 1978; Yamada *et al.*, 1976).

 2.1.3b Epithelial Cells. In adult organisms epithelial tissues form stable coverings over surfaces or line cavities and do not normally migrate. When an epithelial sheet is torn, the cells at the free margins spread and begin migratory activity. This wound healing and the spreading of epithelial sheets during morphogenesis within embryos occur by mechanisms similar to fibroblastic mobility, although *in vitro* studies reveal differences in cell–cell adhesion and the requirements for cell–substratum interaction.

 Freshly plated, isolated epithelial cells spread poorly on plastic or glass substrata and exhibit vigorous blebbing of the cell membrane (Middleton, 1977). Lamellipodia expand the cell margin only temporarily, and the cells do not remain spread. When two epithelial cells touch, they establish extensive intercellular contacts, surface blebbing ceases, and the free cell margins actively expand and spread over the substratum (Middleton, 1977). Apparently, adhesion between the cells influences the capacity of the free cell margins to extend and maintain stable adhesions with the culture substratum. The contacted cells do not separate as contact-inhibited fibroblasts do, but remain together, incorporating more cells into an enlarging cell sheet.

 As in polarized migrating fibroblasts, local differences exist in the cell margins within epithelial cell sheets. Intercellular adhesions within the sheet are very strong because of the numerous desmosomes (see Section 1.2), and in these areas, cell margins are unattached to the substratum and do not exhibit protrusive activity or surface movements (DiPasquale, 1975a; Domnina *et al.*, 1977; Vasiliev *et al.*, 1976). On the other hand, free cell margins at the edges of the sheet are adherent to the substratum and actively extend lamellipodia and filopodia.

 The free edges of an epithelial cell sheet resemble the leading edge of fibroblasts. A network of microfilaments extends into filopodia and lamellipodia, and microfilament bundles lie near the bottom of the cell (DiPasquale, 1975a,b). Unlike fibroblasts, microtubules are rarely seen in epithelial cells on *in vitro* surfaces (DiPasquale, 1975a,b). The fact that colchicine does not affect epithelial motility further indicates that microtubules are not involved (Domnina *et al.*, 1977; Dunlap, 1978).

 Intermediate filaments of 10 nm diameter are present in both fibroblasts and epithelial cells. In fibroblasts their function is unknown, but in epithelial cells they are grouped into bundles that insert into the cell membrane at desmosomes. That fibroblasts do not form desmosomes may explain why they do not form permanent intercellular contacts, but rather exhibit contact inhibition of cell movement. Epithelial cells from different organs make desmosomes with each other (Overton, 1977a), suggesting that the epithelial phenotype involves genetic expression of sufficiently common cytoskeletal and cell-surface features to form hybrid desmosomes.

 The role of the basal lamina in epithelial cell motility has not been well studied *in vitro*. Those epithelial cells which adhere to the basal lamina via hemidesmosomes form these adhesive structures at the spreading cell margins during wound healing *in vivo* (Krawczyk, 1971). Isolated epithelial cells on plastic or glass substrata may spread so poorly because they require more specific adhesive substrata for cell spreading and movement (that is, a basal lamina is needed) than do fibroblasts. This is supported by two recent *in vitro* studies, one showing that epidermal cell migration *in vitro* requires continual synthesis of type AB_2 collagen, a component of basal laminae (Stenn *et al.*, 1979), and the other demonstrating that epidermal cells adhere more to type IV collagen, another type of basal laminar collagen, than to other collagens, whereas fibroblasts adhere equally well to all collagen types tested (Murray *et al.*, 1979).

2.1.3c   Nerve Fiber Formation.   A third cell movement is the formation of axons and dendrites by embryonic and regenerating neurons. Unlike other cells, the nerve cell soma does not move; the outcome of nerve fiber formation is an elaborately extended cell shape. The regulation of this neuronal morphogenesis is especially important because hundreds of characteristic cell morphologies exist within the nervous system, and because the specificity of nervous function is due largely to the anatomy or circuitry of neuronal connections (Jacobson, 1970).

A freshly cultured neuron resembles other freshly plated cells, exhibiting protrusive activity all around the cell margin. The cell margin of neurons does not spread as extensively as in fibroblasts; rather, one or more nerve fibers is initiated and protrusive activity halts elsewhere around the cell (Collins, 1978*a*). The elongation of nerve fibers results from activities at the distal tip of the growing neurite, the growth cone, whose locomotory action and ultrastructure closely resemble the leading edge of fibroblasts (Luduena and Wessells, 1973; Yamada *et al.*, 1971). Filopodia and lamellipodia are extended from the growth cone margin, and small particles and lectin–receptor complexes are transported backward on the upper surface of these protrusions from the growth cone margin (Bray, 1970; Koda and Partlow, 1976; Letourneau, 1979*a*; Pfenninger and Pfenninger, 1975).

An additional necessary event in nerve fiber formation is the assembly of components into the microtubules, neurofilaments, and cell surface of the growing neurite. This may occur within the growth cone, coupled to extension of the neurite tip. A simple view, therefore, is that neurite elongation involves extension and adhesion of the cell margin, as occurs at the leading edge of a fibroblast, with the added aspect of assembly of neurite structures. The method of assembly is unknown, but probably involves polarity in the addition of components, transport of components for great distances, and construction of a cylindrical cell membrane around a core of microtubules and neurofilaments.

When neurons are cultured on unmodified tissue culture surfaces, nerve fibers usually adhere to the substratum only at the growth cone. The fibers appear taut, and if detached from the substratum, a nerve fiber rapidly shrivels into the cell body. In addition to being unattached to the substratum, the sides of nerve fibers lack protrusive activity. The surface of nerve fibers is not simply unadhesive, because growth cones and glial cells clearly adhere rapidly to the nerve fibers they encounter (Nakajima, 1965; Wessells *et al.*, 1980). Colchicine treatment induces protrusive activity and adhesive events along the sides of nerve fibers, suggesting that, as in fibroblasts, intact microtubules within nerve fibers restrict the protrusive and adhesive activities of the cell surface (Bray *et al.*, 1978).

The strength of neuron–substratum adhesion *in vitro* strongly affects the extension of nerve fibers (Collins, 1978*b*; Letourneau, 1975*a,b*). When cultured on a highly adhesive polylysine-coated surface, neurons initiate fibers sooner and more frequently, nerve fibers are more branched, and the rate of nerve fiber elongation is higher than on unmodified surfaces. If unadhesive surfaces such as bacteriological plastic or agar are used, few neurons extend fibers and the fiber growth that does occur is very slow (Strassman *et al.*, 1973).

In contrast to their behavior on untreated surfaces, nerve fibers adhere to a polylysine-treated substratum along great portions of their length, forming crooked neurites that trace the pathway taken by the growth cone. Growth cone morphology is also different on a polylysine-coated surface: the margin of growth cones is very spread, many filopodia project from the margin, and long filopodia, 30 μm and more, are seen.

Interference reflection microscopy can be used to examine the cell–substratum adhesion of cultured neurons (Letourneau, 1979*b*). In considering the role of adhesion in nerve fiber elongation, the adhesive interactions at the growth cone margin are most significant. Focal close contacts and linear adhesive contacts are seen beneath filopodia and beneath veillike lamellar expansions of the growth-cone margin. These linear adhesions frequently run inward toward the center of the growth cone. As proposed for fibroblasts, the microfilament lattice within growth cones is affected by adhesive contacts. Bundles of microfilaments within filopodia and veils project inward from the growth cone margins, just as do the linear adhesions described above. Vesicles are also seen oriented along these microfilament bundles, associated with cell–substratum adhesions (Letourneau, 1979*b*).

The enhancement of nerve fiber formation on highly adhesive substrata can be explained from these observations. The stabilization of filopodia and veils by strong cell–substratum adhesion leads to a highly spread cell margin and a large surface area of growth cones. Such net assembly of cell surface may be a key event in nerve fiber elongation (Bray, 1973; Letourneau, 1977). The arrangement of microfilaments into bundles at the adhesive sites may introduce direction or polarity into the microfilament functions, which may be involved in the positioning and assembly of precursors into nerve structures.

2.2 Regulation of Cell Movement and Shape

In considering the regulation of cell movements we want to know what controls the initiation and termination, as well as the directions and pathways of cell movement. Genetic control may be exerted through either cytoskeletal and cell-surface properties, or through the environment with which cells interact. Our discussion of how cell adhesion may regulate cell movement draws attention to the determining influences of variation in the environment, but intrinsic cellular properties also are regulated. In this regard cellular actin, protrusive activity, and adhesive properties have all been shown to change when embryonic cells begin periods of migration (Johnson, 1970; Santerre and Rich, 1976; Schaeffer *et al.*, 1973; Tickle and Trinkaus, 1973).

In vitro studies have shown that adhesion regulates cell movement. Measurements of cell motility on various surfaces reveal that locomotion is restricted to a range of cell–substratum adhesivity (Gail and Boone, 1972). If adhesion is too weak, cells do not spread and cannot actively migrate; when adhesion is very strong, motility is reduced because of the difficulty of detaching posterior adhesions. Although elongation of nerve fibers responds similarly to variations in growth cone–substratum adhesion, very strong adhesion may not be inhibitory, as adhesions need not break for neurite elongation to continue (Letourneau and Wessells, 1974).

The classic demonstration (Carter, 1965) that fibroblasts migrate up a gradient of increased adhesivity (palladium deposited on unadhesive cellulose acetate) showed that adhesion can influence the direction of cell movements. Harris (1973*a*) repeated these studies and also found that fibroblasts migrate onto the more adhesive areas of patterned substrata. Fibroblasts at the border between two substrata extend exploratory filopodia onto both surfaces, but contact with the more adhesive substratum leads to greater spreading of the cell margin, and eventually, the cell becomes polarized and moves onto the preferred, more adhesive substratum (Albrecht-Buehler, 1976).

Different areas of the cell margin may compete in their protrusive activity, perhaps through tensions developed by cell spreading and exerted on other adherent areas. Detachment of one side of a leading edge is rapidly followed by increased protrusion from the remaining active portions of the cell margin (Chen, 1978). Areas of stronger or more frequent adhesions may resist the tension exerted from other areas more successfully and may also exert more tension on other areas, possibly via the adhesion-associated microfilament bundles. These microfilament bundles may also be involved in the competition to establish cell direction by influencing the transport of materials to the cell margin or by aligning microtubules.

The shape of the substratum also appears to be important in producing spatial differences in cell–substratum adhesion (Weiss, 1945). Recent studies show that cell spreading and movement is restricted along ridges exceeding certain angles and by excessive curvature of the substratum (Dunn and Heath, 1976; Rovensky and Slavnaya, 1974; Rovensky et al., 1971). The proposal to explain this contact guidance states that microfilament bundles cannot assemble or operate in the bent states presumably imposed by spreading over sharp ridges or on highly curved surfaces (Dunn and Heath, 1976). A corollary to this statement is that protrusion and spreading are favored where more or larger microfilament bundles can be assembled (Dunn and Ebendal, 1978). This can explain guidance of fibroblasts and nerve fiber movements in oriented collagen matrices. Larger adhesive contacts and a greater density of contacts, with their consequent effects on filament bundle assembly, would be formed by filopodia extended along the axis of collagen fibrils than by those which contact fibrils from the side.

The pathways of nerve fiber elongation can also be controlled by adhesive events (Letourneau, 1975b). A growth cone at the boundary of two substrata extends microspikes onto both surfaces but spreads and elongates the nerve fiber on the more adhesive substratum. Thus adhesion influences the assembly of neurite structures (that is, microtubules, neurofilaments) in a localized manner. This is consistent with the very localized effects of adhesive contacts on the organization of microfilaments within the growth cone.

Nerve fiber elongation constitutes cell growth, because cell volume increases, as does cell-surface area, during the formation of axons and dendrites. This suggests that metabolism within the cell soma may change with the rate of nerve fiber extension, and in vitro studies show that protein synthesis increases when neurons are active in neurite elongation (Luduena, 1973; Zucco et al., 1975). Thus, modulation of the rate of nerve fiber elongation via adhesive events at the growth cone could act as a trophic factor on the metabolism of intracellular and cell-surface materials. How this might be accomplished is unclear, although changes in the extent of assembly of actin, tubulin, and neurofilament monomers might act as a regulatory signal. This is discussed further in Section 3.2.1c.

2.3 Cell Movements in Vivo

The in vitro studies discussed above suggest a scheme for the control of cell migration in vivo. Cell migrations could begin when contacts with extracellular features become available and are strong enough to allow prolonged protrusive activity and cell spreading or, conversely, when immobilizing adhesions to other cells or structures decrease, releasing cells to migrate. Cell movements might cease when adhesive contacts become so strong that cells are stuck in place or if adhesive events become too weak for spreading to continue. Direction of cell movement could be regulated by spatial variation in the distribution of extracellular adhesive molecules or by differences in the topography of the substrata.

In vivo observations indicate that the mechanism of cell locomotion is similar to that observed in vitro (Bard and Hay, 1975; Bard et al., 1975; Nelson and Revel, 1975; Overton, 1977b; Trinkaus, 1973). Extensions are formed at the front of a cell, apparent adhesions are made with other surfaces, and the cell body moves toward its front edge. Filopodia and lamellipodia are seen, as well as blunt protrusions called lobopodia. As occurs *in vitro,* these protrusions may transform into each other, possibly in response to adhesive events.

A major difference between cell migration *in vivo* and *in vitro* is the substrata involved. Disregarding for the moment molecular differences in the adhesive surfaces, the homogeneous planar substrata of tissue culture dishes do not exist in living tissues. Cells *in vivo* move in three-dimensional space and contact other cells, nerve fibers, basal laminae, other accumulations of amorphous extracellular materials, and fibrous elements within loose connective tissue spaces (Ebendal, 1977; Manasek, 1975; Trinkaus, 1976). Physical constraints imposed by the narrowness and discontinuity of *in vivo* surfaces may allow only limited opportunities for adhesive contacts, explaining why cell bodies often are rounded and why filopodia are more commonly observed than broad lamellipodia.

The influence of substratum shape on cell morphology is illustrated by the mesenchymal cells that form the corneal endothelium (Nelson and Revel, 1975). These cells are three dimensionally stellate when situated in loose connective tissue around the optic cup. When they enter the developing eye, they associate with the dense, more two-dimensional corneal stroma and lens capsule, spread and become flattened.

A series of elegant transplantation studies with the neural crest have established the importance of the tissue environment in controlling the extent and pathways of embryonic cell migrations (LeDouarin, 1976; LeDouarin and Teillet, 1974; Noden, 1975). Neural crest cells migrate within spaces rich in extracellular matrix materials (Ebendal, 1977; Pratt *et al.,* 1975). Elevation in hyaluronic acid content accompanies the opening of tissue spaces during neural crest migration, analogous to other morphogenetic cell movements (Pratt *et al.,* 1975; Derby, 1978; Toole *et al.,* 1972; Toole and Trelstad, 1971). Local removal of extracellular material accompanies the cessation of migration (Derby, 1978). Condensation of neural crest cells into the spinal and autonomic ganglia may involve the resultant increase in cell–cell contact. *In vitro* studies indicate that cell density, and presumably adhesive interactions, may determine whether neural crest cells differentiate into neurons or, alternatively, into pigment cells (Nichols and Weston, 1977; Nichols *et al.,* 1977).

Interpretation of cell–cell adhesive contacts *in vivo* is complicated by the association of extracellular materials with the cell surface. Cell-associated materials may significantly alter cell–cell interactions, for example, by acting as substrata for cells migrating across the surfaces of other cells, in a seeming violation of contact inhibition (Armstrong, 1978; Parkinson and Edwards, 1978). When *in vitro* studies are related to the *in vivo* situation, one should be aware that variations in the accumulation of extracellular materials on the substratum and around cells *in vitro* depend on culture conditions, such as the serum concentration and cell density (Mautner and Hynes, 1977).

The importance of cell–cell contacts in cell migration is most clear in the developing nervous system, where extracellular fibers and accumulations of matrix materials are sparse (Rakic, 1971). In several situations, migrating neuroblasts and elongating nerve fibers characteristically associate with certain cell types (Sidman and Wessells, 1975). *In vitro* demonstrations of regional differences in the adhesion between neural retinal cells and the optic tectum indicate that growing optic nerve fibers may follow adhesive gradients to synaptic targets in the tectum (Barbera, 1975; Marchase, 1977). The growth of nerve fibers and movements of cells along previously extended "pioneer fibers" has been frequently cited as a prime example of contact guidance (Lopresti *et al.,* 1973; Nornes and Das, 1972; Rakic, 1971; Weiss, 1941). Adhesive specificity may also operate, as Schwann cells preferentially migrate along nerve fibers *in vitro,* whereas fibroblasts in the same culture dish are not oriented by nerve fibers (Wood, 1976).

2.4 Summary

Cells must adhere in order to move and acquire distinctive shapes. Projections and extensions of the cell margin are unable to remain as extensions without adhesive anchors at many points. Cell adhesion also has a transmembrane effect on the organization or assembly of microfilaments, thereby influencing such microfilament functions in motility as contraction, extension, and transport and alignment of cytoplasmic components. Cells of different tissues have a similar locomotory apparatus but express specific adhesive and locomotory properties to produce tissue-specific motility and morphology. Finally, temporal and spatial variations in the adhesive interactions of a cell with other cells and with the extracellular matrix can control cell motility and shape.

3 The Role of Cell Adhesion in Cell Proliferation

The regulation of cell growth is complex and poorly understood at the molecular level, but because of its importance it has attracted substantial attention. A host of mechanisms has been proposed for this regulation, as a great number of effectors influence cell growth. Undoubtedly several mechanisms operate simultaneously on each cell, and these have been frequently reviewed (see, for example, Pardee, 1975; Holley, 1975; Fantes *et al.,* 1975). Our purpose in this discussion is to emphasize the role of cell adhesion as a regulator of cell growth.

3.1 Anchorage Dependence

The absence of growth regulation exhibited by certain cells is the basis for much of the evidence that adhesion is involved in the control of cell growth (MacPherson and Montagnier, 1964; Stoker *et al.,* 1968). When nonmalignant cells are cultured on the usual plastic or glass tissue culture substrata at a low density, they proliferate until no further free substratum is available, resulting in a confluent monolayer of nondividing cells. These cells are arrested in the G_1 phase of the cell cycle, and with adequate nutrition they will remain viable for weeks. If these cells are removed from the usual substratum and cultured under conditions in which they do not contact a surface (for example, suspension in liquid or in semisolid media), they lose the ability to proliferate and may even die, despite apparently adequate nutrition (Otkusa and Moskowitz, 1976). Normal cells must be bound to a surface in order to divide, a phenomenon termed anchorage dependence by Stoker *et al.* (1968).

On the other hand, neoplastic cells of many types, spontaneous or virally transformed, need not be anchored to a substratum for proliferation (MacPherson and Montagnier, 1964; Risser and Pollack, 1975). Transformed fibroblasts, for example, continue to proliferate when maintained in suspension culture or in semisolid media. Additionally, when grown on flat adhesive substrata, they continue to divide when the surface of the substratum is covered with cells, ultimately resulting in cells piling up on one another.

Anchorage dependence is further supported by observations that normal cells fail to proliferate when plated onto substrata to which they do not adhere, such as agar, or if they adhere poorly to a surface, such as bacteriologic plastic, proliferation rates are markedly reduced (Martin and Rubin, 1974; Folkman, 1977). The substratum apparently must also be within certain size and shape limits for growth to ensue. Growth of cells on glass beads or fibers is markedly reduced when the beads and fibers are sufficiently small so as to prevent spreading of the cell, although the cell remains attached (Maroudes, 1972, 1973).

This relationship between substratum adhesion and growth *in vitro* has correlates *in vivo*. For example, in the epidermis cell attachment to a suitable substratum is apparently required for DNA synthesis and mitosis. Relatively early in epidermal development, the capacity to synthesize DNA and to divide becomes restricted to the basal cells—those in contact with their natural substratum, the basal lamina (Sengel, 1976). Following wounding of the epidermis, the basal cells adjacent to the wound migrate over the exposed connective tissue as a single sheet of contiguous cells. These cells either do not divide or do so at a very low rate, but there is a high rate of DNA synthesis and proliferation of the cells, which remain adherent to the basal lamina at the wound margin.

This emphasis on cell–substratum adhesion in growth control does not diminish the possible regulatory role of direct cell–cell contacts. The most direct evidence that cell–cell contact *per se* influences cell growth is that the addition of plasma membrane-enriched preparations to cells mimics the effect of the cells from which the preparation was derived (Whittenberger and Glaser, 1977; Salzer *et al.*, 1977). For example, a preparation derived from confluent, growth-inhibited 3T3 cells will inhibit the growth of sparse cultures of 3T3 cells. The preparation binds to, but does not inhibit the growth of, SV40-transformed 3T3 cells (Whittenberger and Glaser, 1977). Some parameters of the growth inhibition resemble those that occur when normal cells reach confluency. Impressively, the specific mitogenic effect of dorsal root ganglion neurites on Schwann cells is duplicated by a membrane-enriched preparation from the neurites (Salzer *et al.*, 1977). The mechanism of these effects is unknown, but alteration in substratum adhesion as a consequence of the addition of the membrane-enriched preparations should be investigated.

3.2 Cell Shape and Growth Regulation

In considering anchorage dependence of fibroblast proliferation, the difference in cell shape between adherent and nonadherent cells is obvious. Nonadherent or suspended cells are rounded, have many microvilli, and lack microfilament bundles, whereas adherent cells are spread and flattened, with bundles of microfilaments spanning the cell (Revel *et al.*, 1975; Goldman *et al.*, 1976). The crucial role of cell–substratum adhesion in cell spreading and the formation of actin cables, and the correlation of cell shape and adhesion have been discussed in Sections 2.1.1 and 2.1.2.

Transformed cells are usually more rounded on tissue culture substrata than are their untransformed counterparts. In addition, they often lack actin cables and have reduced surface-associated fibronectin. These characteristics appear to be related to their reduced adhesivity to the substratum (Willingham *et al.*, 1977), because addition of fibronectin can cause them to resemble normal cells in every way but restoration of normal growth regulation (Ali *et al.*, 1977; Yamada and Pastan, 1976; Yamada *et al.*, 1977). Cell shape does not appear to be linked to growth control in transformed cells. This conclusion suggests that studies of growth regulation utilizing transformed cells may not be applicable to normal, untransformed cells.

The effect of adhesion on cell shape and on cell proliferation in normal cells has raised the issue of whether cell shape, rather than adhesion *per se*, is the primary growth-controlling element. The evidence to date is conflicting, but some generalities are becoming clear.

3.2.1a Reduced Substratum Adhesion Alters Cellular Metabolism and Growth. The idea that cell growth is controlled by changes in cell shape has been suggested on the basis of several indirect experiments (Foldman, 1977; Folkman and Greenspan, 1975). A more direct and quantitative assessment of this idea has recently been provided by studies in which cells were grown on tissue culture substrata whose adhesivity was modified over a wide range by treatment with a hydrophilic polymer (Folkman and Moscona, 1978). On untreated plastic, the cells are flattened, and with increasing thickness of the polymer film the cells become rounded, presumably because of decreased cell–substratum adhesion. When cells were made more spherical, fewer cells incorporated thymidine. A good correlation between the degree of cell flattening, measured by average cell height, and DNA synthesis was found for aortic endothelial cells, human diploid fibroblasts (WI-38) and BALB/c 3T3 cells (A-31). Impressively, a similar level of DNA synthesis occurred in sparse fibroblasts on a substratum whose adhesivity was modified to produce the same degree of flattening as in a population of confluent fibroblasts on untreated plastic. This result suggests that changes in cell shape, induced either by substratum adhesion or by crowding of adjacent cells, has similar effects on DNA synthesis.

It seems, from these studies, that flattened cells can synthesize DNA, but rounded cells cannot. This work utilized cells that normally have a flattened configuration (fibroblasts, endothelial and endothelioid cells), and it is unclear whether the cell shape or the adhesion of the cell to these substrata is the primary control. A simplistic view would suppose that enhanced cell adhesion and the resultant cell spreading reduces the concentration of elements that inhibit DNA synthesis. Flat, adherent cells have bundles of actin filaments, polymerized tubulin in microtubules, and accumulations of extracellular matrix materials at adhesive plaques. Reduced adhesion would make actin and tubulin monomers available and alter the distributions of cell-surface components, but although the interphase nucleus contains actin (Douvas *et al.*, 1975; Hitchcock *et al.*, 1976) and glycosaminoglycans (Furukawa and Terrayama, 1977), their function is unknown. Clearly, as recognized by Folkman and Moscona (1978), there are many potentially crucial alterations in cellular physiology introduced by converting flattened cells to rounded cells through alteration of their adhesion to the substratum.

The possibility that cell proliferation is controlled by adhesion or cell shape requires that information be transmitted from the cell periphery to the nucleus. Other than studies of cell proliferation and transport, the effects of adhesion (and related changes in cell shape) on cellular metabolism have not been well studied. Penman and his co-workers (Benecke *et al.*, 1978; Farmer *et al.*, 1978) have reported a sequence of changes in mRNA metabolism that accompany loss and resumption of cell–substratum adhesion by anchorage-dependent 3T6 cells. Loss of adhesion (that is, suspension in liquid media) results in a prompt reduction in the appearance of new cytoplasmic mRNA (without a change in the labeling of heterogeneous nuclear RNA), and the conversion of mRNA into a modified, inactive state. Additionally, protein synthesis slowly declines. The mRNA modification is unclear, but mRNA becomes no longer translatable. These effects are completely reversible when the cells are reattached to the substratum. The resumption of protein synthesis primarily involves the utilization of preexisting mRNA, which rapidly becomes translatable after reattachment. Not surprisingly, there is synthesis of actin mRNA following reattachment and elevated synthesis of actin and other cytoskeletal proteins.

The resumption of normal RNA metabolism on reattachment is reported to require only adhesion, not spreading, of the 3T6 cells on the substratum (Benecke *et al.*, 1978). Significantly, however, spreading is needed for DNA synthesis in these cells. The 3T6 cells are an established cell line and differ, therefore, from primary cell types. When mouse embryo or human diploid fibroblasts were similarly studied, suspension of the cells inhibited protein synthesis much more rapidly, and respreading of the cells onto a substratum was required for resumption of normal metabolism (Penman, 1979). This result raises the question of whether the behavior in response to adhesion of established cell lines, which are adapted for growth in a simplified *in vitro* environment, is relevant to the behavior of cells *in vivo*.

3.2.1b Protease Treatment Changes Cell Shape and Induces Cell Division. Treatment of resting fibroblasts in tissue culture with a variety of proteases, under defined conditions, causes the cells to round up without detaching from the substratum. Revel and his colleagues (Revel and Wolken, 1973; Revel *et al.*, 1975) showed that brief trypsin treatment modified the organization of actin microfilaments, resulting in the rounded cell shape, before the cells detached from the substratum. These cytoskeletal changes can be induced by treatment of fibroblast of several types with insolubilized trypsin, a variety of proteases including plasmin, and less efficiently, chymotrypsin and thrombin (Pollack and Rifkin, 1975). Using conditions similar to those that cause cell rounding, protease treatment will stimulate quiescent 3T3 cells and chick embryo cells to undergo a cycle of DNA synthesis and mitosis (reviewed in Noonan, 1978). Although conflicting results on the effect of proteases in influencing cell growth have been obtained using different cell lines, different growth conditions (for example, with or without prior serum deprivation), and differing number of prior cell passages (Carney and Cunningham, 1978; Carney *et al.*, 1978), it is well established that under optimal conditions certain proteases readily initiate the proliferation of resting fibroblasts.

The mitotic effect of trypsin and thrombin does not require entry of the protease into the cell. Because the action appears to be directly on the cell surface, it may involve a transmembrane event. In addition to changes in adhesion, instances in which the cytoskeleton appears to be modified by transmembrane effects include the binding of lectins (Ti and Nicolson, 1974; Albertini and Clark, 1975) and immunoglobulins to specific sites (Damsky *et al.*, 1979), as well as the removal of the basal lamina with hyaluronidase (Banerjee *et al.*, 1977).

Protease treatment also removes fibronectin very readily from cell surfaces. This evidence and observations that the amount of fibronectin is reduced on most transformed cells (Yamada and Olden, 1978) suggested the possibility that this adhesive glycoprotein was involved in growth regulation (Hynes, 1976). As previously mentioned, however, the reduced fibronectin on transformed cells is not the basis for their lack of growth control. Nor is fibronectin directly involved in the control of normal fibroblast proliferation, as papain or thermolysin can remove fibronectin without inducing cell division, and thrombin can induce cell division without removing fibronectin (Keski-Oja *et al.*, 1977; Teng and Chen, 1975).

In addition to these effects on fibronectin and on the cytoskeleton, protease treatment produces other cell-surface modifications, such as increased ease of lectin agglutinability, loss of certain glycopeptides, and increased precursor transport. None of these changes, however, has been causally linked to the initiation of cell division (Noonan, 1978).

In summary, it is not clear whether proteases induce cell division by their effect on cell adhesion. Fibroblasts in confluent layers already have a rounded shape before being stimulated to divide by protease treatment, suggesting that the protease effect involves something other than a change in cell shape.

   3.2.1c   Cell Shape and Growth of Transformed Cells and Mutants.   As previously noted, transformed cells characteristically are more rounded and lack anchorage dependence of growth. Several workers have suggested a direct relationship between the disappearance of actin microfilament bundles in these cells and their loss of growth control (Ash *et al.*, 1976; Edelman and Yahara, 1976; Pollack *et al.*, 1975; Rubin *et al.*, 1978). Several other characteristics have been proposed as *in vitro* phenotypic markers of the transformed state (for example, reduced cell–cell adhesion, reduced surface fibronectin, increased lectin agglutinability, and production of plasminogen activator). Exceptions (that is, cells with the normal phenotype) have been noted for each of these proposed markers of transformation, suggesting that transformation is associated with a spectrum of properties, but all transformed cells do not share the same properties. Moreover, although the *in vivo* tumorigenicity of cells seems most closely linked to loss of anchorage dependence for growth (Shin *et al.*, 1975), no marker is absolutely linked to the transformed state.

   The facts that transformed cells exhibit a general spectrum of altered behaviors and that no one trait is invariably linked to tumorigenicity may also explain why the behavior of certain induced mutants fails to correspond with some generally accepted correlations. For example, although rounded morphology and reduced anchorage dependence generally correlate with lack of contact inhibition of growth at high cell densities, a mutant of BALB/c 3T3 cells that shows these two properties grows to a saturation density similar to that of wild-type cells (Pouyssegur and Pastan, 1976).

   One perturbation of transformed cells may be a defect in the linkage between cell shape and growth control that acts to uncouple cell shape from DNA synthesis and cell proliferation and that also has pleiotropic effects on cell adhesion and other cell-surface properties. In this regard, the *src* gene of avian sarcoma virus may specify a protein kinase that modifies cellular proteins to produce the transformed state (Ash *et al.*, 1976; Collett and Erickson, 1978). Use of primary rather than established cell lines may provide more information on the relationship between cell shape and growth control because established lines are selected and adapted for growth on *in vitro* substrata where adhesive interactions may not allow normal expression of cell shape.

3.3   Extracellular Materials and Cell Proliferation

3.3.1   Fibroblasts

   Cell-surface glycosaminoglycans (GAG) have been implicated in growth control of fibroblasts. For example, increased cell surface GAG may be involved in enhanced cell–substratum adhesion and spreading, thereby reducing saturation densities (Roblin *et al.*, 1975). Many such studies, however, failed to account for cell-specific differences in GAG composition or did not control for the density of the cultures. A systematic examination of the cellular distribution of GAG, with procedures allowing direct comparisons among different GAG types, found that cell density modified the distribution and type of GAG (Cohn *et al.*, 1976). When dividing at similar rates at low densities, the relative proportions and amounts of the various GAG associated with the cell surface were similar for normal and transformed 3T3 cells. At high cell densities, the normal cells showed increased chondro-itin-6-sulfate and heparan sulfate, and reduced proportions of hyaluronic acid and chon-droitin, an undersulfated chondroitin sulfate.

Fibronectin, the major surface protein of fibroblastic cells, appears to be bound to GAG, probably heparan sulfate or its proteoglycan (Graham *et al.*, 1975; Rouslahti *et al.*, 1979; Yamada *et al.*, 1979). This complex interacts with collagen, and the ternary com-bination appears to be more stable than the complex formed by direct interaction of collagen and fibronectin. Heparan sulfate (Kraemer and Tobey, 1972) and fibronectin (Stenman *et al.*, 1977) are lost from the cell surface when cells round up prior to mitosis, suggesting that this surface matrix is involved in maintaining flattened cell shapes by enhancing adhe-sion. It has also been suggested that hyaluronic acid and undersulfated chondroitin sulfate destabilize the adhesive action of the heparan sulfate–fibronectin complex (Culp *et al.*, 1978). These less-sulfated GAGs may be involved in the detachment of cell–substratum adhesion, as during cell movement or mitosis. The reduction in the levels of these species in confluent cultures of normal cells and the maintenance of higher levels in transformed cells, mentioned above, may be related to the differences between these cell types in their mobility and proliferation at confluence. Matrix materials, by influencing cell adhesion and cell shape, may be involved in regulating proliferation of normal cells.

3.3.2 Epithelial Cells

The extracellular matrix has a role in the growth and differentiation of epithelial cells. Epithelial cell populations are continually and rapidly renewed on the exposed surfaces of the body and in the gut, endometrium, mucous membranes, and certain exocrine glands. This proliferation, in every instance, is from the basal layer of cells sitting on the basal lamina. Adhesion to the substratum also plays an important role in maintenance of epithe-lial cells in culture. On glass or plastic substrata, some primary cultures of epithelial cells flatten, do not proliferate, and rapidly become nonviable. Epithelial cells of other types form closely packed colonies that grow as circumferentially expanding monolayers of flat-tened cells, but despite this proliferation, these cells lose their polarity, normal shape, and functional characteristics.

Recently, gels of fibrillar (type I) collagen have been used as a culture substratum for epithelial cells of several types (Michaelopoulus and Pitot, 1975; Emerman and Pitelka, 1977). On collagen gels, unlike glass or plastic, these epithelia retain their polarity and their normal columnar or cuboidal configuration and certain functional characteristics. For example, mouse mammary epithelial cells cultured on collagen maintain their columnar shape and continue production of casein, but show low levels of DNA synthesis (Emerman *et al.*, 1977). Moreover, the cells produce a basal lamina on a collagen matrix, whereas no basal lamina is formed by these cells on the usual tissue culture substratum (David and Bernfield, 1979). Hence, the presence of a basal lamina *in vitro* is associated with the expression of normal *in vivo* behavior, be it cell proliferation, specific cell shape, or specific secretory activities. Whether or not DNA synthesis and proliferation occurs (and any phe-notype for that matter), depends on the genotype of the particular epithelial cells (for example, liver cells normally do not proliferate *in vivo* and similarly do not do so when cultured on collagen gels).

Adhesion and the mechanism by which it affects cell shape may also play a role in the proliferative response of epithelial cells to soluble growth factors. At least three purified proteins, nerve growth factor (NGF), epidermal growth factor (EGF), and fibroblast growth factor (FGF), are capable of initiating DNA synthesis in cultured cells. The specific response to these factors appears to depend not only on the cell type, but also on whether the cells are anchorage dependent and the nature of the substratum. For example, an anchorage-dependent line of rat embryonic myoblasts responds to FGF, but an anchorage-independent line does not (Schubert *et al.*, 1976). When bovine corneal epithelial cells are grown *in vitro* on a plastic substratum, the cells respond to FGF, but not to EGF, despite the presence of cell-surface EGF receptors. If these same cells are grown as organ cultures or, more importantly, on a collagen substratum, they respond to EGF and have a reduced response to FGF (Gospodarowicz *et al.*, 1978). The cells cultured on collagen are substantially more cuboidal, as they are *in vivo*, than are those on plastic. The response to hormonal mitogenic factors appears to require an appropriate substratum.

The crucial effect of the substratum may be brought about by its influence on cell shape. Both fibroblastic and epithelial cells produce materials that they deposit on the substratum and to which they attach. *In vitro* substrata, such as plastic, glass, and collagen, influence the cell's ability to make or degrade these extracellular materials (David and Bernfield, 1979), and may thereby alter adhesion and, hence, cell shape. If the influence of the substratum on cell proliferation is the result of this effect on extracellular materials involved in cell adhesion and shape, then control of the production and degradation of these extracellular materials would be a physiological regulator of cell proliferation.

3.4 Summary

Cells must adhere to proliferate. This anchorage dependence for proliferation of normal cells may actually be a requirement for a certain cell shape. For fibroblasts and endothelial cells, the ability to proliferate is correlated with a flattened, spread cell shape. The shape requirement for epithelial cell proliferation is less clear, but cell shape (columnar vs. flattened) is linked to the ability to respond to the mitogen, epidermal growth factor. Cell shape may be regulated by the adhesive associations of cells with extracellular materials. Nothing is known of the casual link between cell shape and either proliferation or alterations in transcription and translation, although changes in the concentrations of intracellular cytoskeletal monomeric units are a prime candidate for intracellular regulators. Transformed cells lack anchorage dependence of proliferation and do not show a correlation between cell shape and growth control. Transformation, certain induced mutations, and the induction of proliferation by proteases may involve perturbation of the link between cell shape and proliferation.

4 The Role of Cell Adhesion in Muscle Differentiation

4.1 Introduction

Interactions of the cell surface with extracellular factors and with other cells can elicit changes in genetic expression that direct cells along a path of differentiation. In this section we will discuss the involvement of cell–substratum adhesion and cell–cell adhesion in the differentiation of muscle cell precursors (myoblasts) to mature muscle fibers. Myogenesis is chosen as a model system because it offers many advantages for the *in vitro* study of cellular interactions during differentiation (for reviews, see Yaffe, 1969; Hauschka, 1972; Fischman, 1972; Merlie *et al.*, 1977).

Large numbers of myoblasts are relatively easily obtained from dissociated embryonic skeletal muscle. Such preparations consist of a high proportion of myoblasts, which can be morphologically distinguished and also purified from nonmyogenic cells. When cultured, myoblasts grow and can be induced to differentiate in an almost synchronous fashion. The process of myogenic differentiation is dependent on the substratum on which the cells are plated and on specific interactions at the level of both cell–cell contact and membrane fusion. The development of myoblast cell lines that can be propagated indefinitely and induced to differentiate in a nearly normal fashion has greatly facilitated understanding of many of the events of myogenesis.

4.2 The Process of Myogenesis

Morphological aspects of the differentiation of primary cultures of avian myoblasts have been studied *in vitro* by time-lapse cinematography (Capers, 1960; Cooper and Konigsberg, 1961; Fear, 1977). Trypsinized cell suspensions of muscle consist of two cell types, myoblasts and fibroblasts. When plated, both cell types adhere to the dish and undergo multiplication and migration. They can still be distinguished by their characteristic morphology and locomotion. Myoblasts are bipolar with a centrally located nucleus. This bipolar shape is maintained during cell locomotion, which generally occurs along the long axis of the cell. Fibroblasts, on the other hand, have a typical polygonal shape and are more mobile (Fear, 1977). During the first 2 days in culture, the myoblasts sort out from the fibroblasts and aggregate into broad multicellular strands, showing oriented side-by-side as well as end-to-end alignment. Usually between the second and third days in culture, the myoblasts fuse to form myotubes containing hundreds of nuclei. A detailed examination of this fusion has indicated that cells preferentially fuse through end-to-end encounters. Myoblasts align with their long axes parallel to the long axis of a myotube; however, they then migrate along and fuse near the end of the myotube (Fear, 1977). A polarized distribution of specific adhesion molecules in the myoblast plasma membrane may be important in aligning cells before fusion.

From approximately the third day in culture, the myotube matures into a myofiber. The nuclei migrate to the periphery, and typical cross-striations can be observed. At this point, the fiber is capable of spontaneous contraction and can be innervated if cocultured with neurons (Shimada *et al.*, 1969a,b; Fischbach, 1970; Harris *et al.*, 1971). This maturation process is accompanied by the deposition of a basal lamina on the surface of the myofiber (Bischoff and Ishikawa, quoted in Bischoff and Holtzer, 1969; Fischman, 1972). Because the appearance of the basal lamina correlated well with the loss of the ability of competent myoblasts to fuse with the myofiber, Fischman (1970) suggested that the two events may be causally related. The basal lamina may also be important in the formation and stabilization of myoneural junctions (Marshall *et al.*, 1977).

Concomitant with cell fusion is a dramatic change in genetic expression. DNA synthesis stops, and the synthesis of enzymes involved in replication is reduced. At the same time, the genes specifying "muscle-specific" proteins, such as creatine phosphokinase, actin, myosin, myokinase, and acetylcholine receptors, are activated (Merlie *et al.*, 1977).

The signal that prompts myoblasts to withdraw from the proliferative state and differentiate is not yet known. Myoblasts *in vitro* are capable of fusing only during the G_1

4.3 Cell–Substratum Adhesion

Konigsberg (1963) showed that muscle fibers could develop from clones of myoblasts arising from single cells. The formation of these fibers was dependent on the use of conditioned culture medium that had been exposed to dense fibroblastic cell populations for several days. In investigating this requirement for "conditioned medium," Konigsberg and Hauschka (1965) showed that it was not continually required during the differentiation process. In fact, if the culture dishes were preincubated with conditioned medium for 3 days and then seeded with myoblasts, differentiation proceeded in unconditioned medium. They concluded that the effect of the conditioned medium was to alter the substratum so as to permit differentiation. Because dense fibroblast cultures provided good conditioned medium, and because these cells had been shown to be producing collagen, Konigsberg and Hauschka (1965) proposed that a collagen substratum might be necessary for myoblast differentiation. They confirmed this by showing that dishes coated with purified collagen support myogenesis in the absence of conditioned medium (Hauschka and Konigsberg, 1966). A more detailed study (Konigsberg, 1970; Hauschka, 1972) showed that this substratum effect was specific to collagen, in that a number of molecules physically related to collagen did not stimulate myogenesis. In addition, several compounds known to promote attachment of cells to culture surfaces were shown to be ineffective. The native structure of collagen played no role, because both single-stranded collagen and gelatin (denatured collagen) were as effective as native triple-stranded collagen in promoting myogenesis.

While the stimulatory effect of collagen on myogenesis is generally accepted, the mechanism of this effect is controversial. Holtzer *et al.* (1974, 1975) suggested that the role of collagen is merely to permit myoblast adhesion and survival *in vitro*. Because myoblast fusion is preceded by a period of high cell mobility, when myoblasts are lining up in ordered arrays, and because cell mobility is sensitive to the substratum, one might reasonably assume that collagen acts primarily by providing an appropriately adhesive surface for cell migration. Hauschka and Konigsberg (1966) showed that the plating efficiency of primary chick myoblasts was the same with and without collagen, yet only cells on collagen differentiated into myofibers. Serum proteins and cellular molecules, notably fibronectin, are deposited generally onto culture substrata and onto collagen-coated substrata via specific interactions; therefore, a necessary factor for myogenesis may involve the binding and recognition by myoblasts of these molecules, which become associated with collagen during the culture period. In addition, the influence of extracellular materials on the synthesis and degradation of extracellular materials by cells may be important in the stimulation of myogenesis by collagen (David and Bernfield, 1979; Goldberg, 1979).

The reports of Yaffe (1969) that a rat myoblast line, L6, differentiates *in vitro* without added collagen, and of Konigsberg (1970, 1971) that single quail myoblasts can be cloned and will differentiate on plastic substrata without added collagen suggest that collagen is not absolutely required for myogenesis. Further examination indicates that these cells may provide their own collagenous substrata. Primary myoblasts do synthesize collagen, but it is not usually deposited on *in vitro* substrata (Lipton, 1977). The L6 line synthesizes substantially more collagen, possibly reaching sufficiently high levels to be deposited on the substratum. Alternatively, the L6 cells, unlike primary myoblasts, may process a collagen precursor so it precipitates on the culture substratum. In the case of quail cells, the clones become very dense and may form an environment that allows collagen to accumulate intercellularly and on the substratum.

The abundant extracellular matrix within developing muscle *in vivo* is largely produced by fibroblasts. In addition to the requirement that myoblasts must interact with extracellular materials to fuse, extracellular materials, produced by fibroblasts, may also be required for synthesis of the myofiber basal lamina (Lpiton, 1977). In pure cultures of myoblasts on collagen, myofibers lack a surrounding basal lamina, although sparse deposits of material are seen at the surface of myofibers. When whole dissociated embryonic muscle is cultured, basal laminae are seen around the differentiated myofibers, but the fibroblasts and the mononucleated myoblasts have no basal lamina associated with their surfaces. This specificity of basal lamina formation by myofibers may involve a specific differentiated sensitivity to factors that stimulate formation of the basal lamina or a specific ability to synthesize and accumulate components of the basal lamina.

4.4 Cell–Cell Adhesion

The fusion of mononucleated myoblasts into myotubes is the major phenotypic change associated with myogenesis and, as such, has been the subject of much study. The evidence to date suggests that this process is the culmination of a complex sequence of events involving cell–cell recognition, specific adhesion, and association of cells into linear aggregates—phenomena that can be shown to be distinct from membrane fusion and from each other by their differential ionic requirements, pH, and temperature optima, and sensitivities to various drugs and enzymes (Knudsen and Horwitz, 1977a,b).

The fusion of myoblasts shows a high degree of tissue specificity, which may be attributable to tissue-specific adhesive components. When myoblasts were mixed with cells of other types, such as kidney, heart fibroblasts, liver, or chondrocytes (Yaffe and Feldman, 1965; Okazaki and Holtzer, 1965), the cells sorted out according to type, and only myoblasts were incorporated into myotubes, although Yaffe and Feldman (1965) found that occasionally some cardiac myoblasts fused with skeletal myoblasts.

The ability of myoblasts to fuse also exhibits age dependency. Yaffe (1971) observed that primary rat myoblasts required approximately 52 hr in culture to become competent to fuse. When subcultures of "competent" cells were mixed with 24-hr cultures of "noncompetent" cells, only myoblasts from the older culture fused to form myotubes. Holtzer *et al.* (1972) found a similar age-dependent recognition process in avian myoblasts when they looked at the ability of myoblasts taken from 5-, 6-, 7-, and 8-day embryos to fuse with 10-day myoblasts in culture. Myoblasts from 8-day embryos fused with 10-day cells, but few 6-day myoblasts and no 5-day myoblasts fused. In another tissue, the neural retina, temporal changes in the adhesive properties of plasma membranes have been demonstrated (Gottlieb *et al.,* 1974).

In contrast to this tissue and age specificity, myogenesis shows little species specificity (Wilde, 1958, 1959; Yaffe and Feldman, 1965; Maslow, 1967, 1969). Myoblasts from species as diverse as rabbit, calf, mouse, rat, and chicken fuse with one another to form hybrid myotubes.

These recognition mechanisms could operate by two alternative mechanisms—one involving cell aggregation, in which only "competent" myoblasts could join elongated multicellular clusters, the other involving cell fusion, in which different cell types could aggregate, but fusion would occur only when "competent" myoblasts came in contact. To resolve this question Yaffe and Feldman (1965) examined aggregates of prefusion stage cells on confluent plates of mixed myoblasts and ^{3}H-labeled heart cells. They found that most of the cells in the elongated aggregates on the plates were skeletal myoblasts and concluded that cell–cell recognition occurred at the level of prefusion cell aggregation, not during the fusion process.

The biochemical nature of these surface changes has been probed by Knudson and Horwitz (1977a,b), who have operationally subdivided the intercellular interactions of myogenesis into recognition, adhesion, and fusion. To avoid ambiguities arising from cell–substratum interaction, they studied myoblast fusion in suspension. Briefly, myoblasts were grown as monolayers in a low calcium medium, which allowed the cells to become competent, but reversibly prevented cell fusion. The myoblasts were then harvested with EDTA and agitated in suspension. In the presence of calcium, the cells aggregated and eventually fused to form multinucleated myoballs. Under these conditions fusion of the suspended myoblasts showed the same tissue and age specificity, as well as the same pH, temperature, and calcium concentration optima as does fusion of myoblasts in monolayers, thus suggesting that the two processes were mechanistically similar.

Using this assay, Knudson and Horwitz (1977a,b) found that the initial intercellular interaction was a rapid event, labeled *recognition*, leading to a loosely adhesive aggregate of competent myoblasts. These aggregates could be dispersed by repeated pipetting or by incubation, with EDTA or trypsin. On further incubation, the cells became increasingly difficult to dissociate; by 30 min they were resistant to EDTA, and by 1–2 hr they were resistant to trypsin. The transition from recognition, or reversible adhesion, to irreversible adhesion could be separated by various inhibitors. For example, in the presence of 20 mM Mg^{2+}, myoblasts aggregated but did not become resistant to EDTA, whereas cells treated in suspension with cytochalasin B, colchicine, or concanavalin A formed aggregates that became EDTA resistant but did not become trypsin resistant. These data suggest events of the secondary phase of the adhesive process as described in Section 1.4.

Nameroff and his group have also been able to separate recognition from fusion (Nameroff and Munar, 1976; Leung et al., 1975). They found that myoblasts in monolayers could be reversibly prevented from fusing by treatment with phospholipase C. The enzyme effect was specific for fusion, as it did not inhibit growth or migration of the cells and did not prevent the cells from lining up, withdrawing from the cell cycle and expressing "differentiated functions." Leung et al. (1975) postulated that phospholipase C altered the lipids of the myoblast surface and thereby released a component necessary for fusion. Because recognition was not affected, they suggested that fusion and recognition are mediated by different surface molecules. In support of this hypothesis they found that phospholipase C released material from the surface of myoblasts, and that this material (tentatively identified as a glycoprotein) could reversibly block fusion of untreated myoblasts. This fusion-inhibiting material was developmentally regulated in that it could only be extracted from rapidly fusing myoblasts.

Further evidence that myoblast recognition and fusion involves a glycoprotein on the cell surface was provided by the finding of an α-D-galactoside-binding protein in embryonic chick muscle tissue (Teichberg et al., 1975). Two distinct lectin activities were subsequently found in cell extracts of a continuous rat myoblast line (Gartner and Podleski, 1975, 1976;

Nowak *et al.,* 1976) and of primary chick myoblast cultures (Den *et al.,* 1976; Nowak *et al.,* 1976; Nowak and Barondes, 1977). These activities were found to rise during periods of rapid fusion and to fall as fusion decreased. Gartner and Podleski (1975) suggested that at least one of these lectins was directly involved in cell fusion. They based this conclusion on the fact that they could show lectin activity on the external membrane of rat myoblasts and that thiodigalactoside, a disaccharide that blocked one of the lectin activities, was found to inhibit myoblast fusion at low concentrations. These results have been disputed by Den *et al.* (1976), who found lectin activity in cell extracts but found no evidence of lectin activity on the surface of embryonic chick myoblasts and saw no inhibition of fusion by thiodigalactoside. The system is complex, and it is likely that a disaccharide is too simple a ligand to simulate the specificity of a glycoprotein linked to the cell surface.

A different lectin isolated from myoblasts does not have specificity for galactoside, does bind to glycosaminoglycans, and is metabolized differently from the first lectin (Shadle *et al.,* 1979). This molecule may be present on myoblast cell surfaces and may interact with extracellular glycosaminoglycans, either in the matrix or on other cells.

4.5   Summary

Myogenesis involves distinct adhesive interactions of myoblasts—with the extracellular matrix and with other myoblasts. Interactions with collagen or a complex of collagen and other macromolecules may be required for sufficient cell migration during the formation of prefusion aggregates or may have transmembrane effects on cell metabolism and differentiation. Myoblast–myoblast adhesions may be important in the organization of properly aligned aggregates of myoblasts and in initiating membrane events that lead to fusion. The separate lectins isolated from myoblasts may be cell-surface molecules, which mediate these adhesive interactions, and further examination of the properties of these lectins may help us to understand the different roles of adhesive events in myogenesis.

5   Conclusions

As stated in the introduction, cell adhesion has been shown to be a major controlling factor in cell migration, cell shape, cell division, and cell differentiation. In concluding, we review how cell adhesion may produce its effects and speculate about how regulation by cell adhesion may occur within an organism.

It is not known which events of cell adhesion are most directly involved in controlling cellular behavior. Certain changes in membrane organization may be especially significant. For example, one can imagine that the distribution and mobility of the membrane molecules that participate in adhesive bonds is drastically altered by adhesion, and the best documented transmembrane event accompanying cell adhesion is the concentration of microfilaments into bundles that terminate at adhesive sites. It may be important in regulatory considerations that these changes in membrane structure are localized to particular areas of the cell surface, and may not be global membrane events.

Throughout this chapter we have emphasized the effects of cell adhesion on the cytoskeleton. The regulation of microfilament assembly and distribution seems to be a critical factor in the exertion of force during cell migration, cell shape changes, and the movement of intracellular components. But in addition to force generation and mechanical support, cytoskeletal assemblies and their subunits may influence cell physiology in other ways. For example, the free cytoplasmic polysomes may actually be bound to cytoskeletal structures (Lenk and Penman, 1979), monomeric actin inhibits deoxyribonuclease I (Hitchcock et al., 1976) and the γ subunit of phosphorylase kinase is similar to actin (Fischer et al., 1975). Cell adhesion may act through these nontraditional roles of the cytoskeleton to regulate such cell behavior as proliferation and differentiation.

Because cell adhesion affects cell behavior, it may be an important source of regulatory cues or signals from a cell's environment. Spatial differences in adhesion could determine directionality, as illustrated by the effects of local adhesive differences on cell migration. The local effects of cell adhesion on persistence of cytoplasmic extensions and on microfilament organization also allows the development of asymmetric cell shape and cell polarity. Cell polarity is often a feature of cell differentiation, for example, the apical glycocalyx and junctional complexes of intestinal epithelial cells and nonadhesiveness of the luminal surface of endothelial cells (Zetter et al., 1978). Adhesive interactions of these cells with their basal lamina may represent directional cues, which, through their effects on cytoskeletal distribution, are necessary for the establishment of cell polarity.

Adhesive interactions may provide other information about a cell's environment, such as cell density and the distribution and composition of the extracellular matrix. Myogenesis was chosen to illustrate the role of adhesion in cell differentiation because myoblast fusion is an easily recognized phenotype. Close cell–cell contacts seem to have general significance in cell differentiation (Adler et al., 1976; Medoff and Gross, 1971; Moscona, 1974). The amount of cell-surface fibronectin on myoblasts and chondroblasts decreases when the cells differentiate (L. B. Chen, 1977; Dessau et al., 1978; Pennypacker et al., 1978), suggesting that changes in adhesion are involved in the changes in cell shape and cell migration that accompany cytodifferentiation.

Cells may interact with and influence other cells, not only by direct cell–cell contact but also by modifying the extracellular matrix. The synthetic role of fibroblasts in connective tissues is well appreciated, but degradation of the matrix may be equally important in modifying cell adhesion. Cessation of cell movements in several embryonic organs is related to the hydrolysis of hyaluronate and shrinking of extracellular spaces. A novel proposal of Roth and Shur (Roth, 1973; Shur and Roth, 1973; Shur, 1977) is that migrating cells enzymically modify the pathways they follow, thereby influencing the subsequent migration of other cells along the same path.

The body tissues exhibit major differences in organization and morphology of their component cells. These differences may arise in part from differences between the cell types in their response to similar adhesive environments. For example, fibroblast migration becomes slower on highly adhesive substrata, whereas nerve fiber elongation, for which detachment of adhesions is unnecessary, goes faster under conditions of high adhesion. Contacts between epithelial cells are very strong and lead to the formation of cell sheets, while contacts between fibroblasts lead to contact inhibition and dispersal from areas of high cell density. We do not believe there are radical differences in the effects of cell adhesion among different cell types, but rather that tissue specificity in cell-surface receptors, cell-surface recognition of and interactions with extracellular materials, and organization of cytoskeletal structures operate in the context of common locomotory systems and adhesive properties to produce these unique tissue structures.

ACKNOWLEDGMENTS

This work was supported by NCI contract NO1-CB-53903 and NIH Grant HD 06763 awarded to M. R. B. and by NSF Grant PCM 77-21035 awarded to P. C. L. We thank the colleagues who sent us reprints and preprints of their work and are also very grateful to Kristie Blees and Paige Patch for skillfully typing numerous drafts of this paper.

References

Abercrombie, M., 1961, The bases of the locomotory behavior of fibroblasts, *Exp. Cell Res. Suppl.* **8**:188.

Abercrombie, M., 1967, Contact inhibition: The phenomenon and its biological implications, *Natl. Cancer Inst. Monogr.* **26**:249.

Abercrombie, M., and Dunn, G. A., 1975, Adhesions of fibroblasts to substratum during contact inhibition observed by interference reflection microscopy, *Exp. Cell Res.* **92**:57.

Abercrombie, M., Heaysman, J. E. M., and Pegnum, 1971, The locomotion of fibroblasts in culture. IV. Electron microscopy of the leading lamella, *Exp. Cell Res.* **67**:359.

Abercrombie, M., Dunn, G. A., and Heath, J. P., 1977, The shape and movement of fibroblasts in culture, in: *Cell and Tissue Interactions* (J. W. Lash and M. M. Burger, eds.), pp. 57–70, Raven Press, New York.

Adler, R., Teitelman, G., and Suburo, A. M., 1976, Cell interactions and the regulation of cholinergic enzymes during neural differentiation *in vitro, Dev. Biol.* **50**:48.

Albertini, D. F., and Clark, J. L., 1975, Membrane–microtubule interactions: Concanavalin A capping induced redistribution of cytoplasmic microtubules and colchicine binding proteins, *Proc. Natl. Acad. Sci. U.S.A* **72**:4976.

Albrecht-Buehler, G., 1976, Filopodia of spreading 3T3 cells, *J. Cell Biol.* **69**:275.

Albrecht-Buehler, G., 1977, Phagokinetic tracks of 3T3 cells: Parallels between the orientation of track segments and of cellular structures which contain actin or tubulin, *Cell* **12**:333.

Albrecht-Buehler, G., and Goldman, R. D., 1976, Microspike mediated particle transport towards the cell body during early spreading of 3T3 cells, *Exp. Cell Res.* **97**:329.

Ali, I. U., and Hynes, R. O., 1978, Effects of LETS glycoprotein on cell motility, *Cell* **14**:439.

Ali, I. U., Mautner, V., Lanza, R., and Hynes, R. O., 1977, Restoration of normal morphology, adhesion and cytoskeleton in transformed cells by addition of a transformation-sensitive surface protein, *Cell* **11**:115.

Armstrong, P. B., 1978, Modulation of tissue affinities of cardiac myocyte aggregates by mesenchyme, *Dev. Biol.* **64**:60.

Ash, J. F., and Singer, S. J., 1976, Concanavalin A-induced transmembrane linkage of concanavalin A surface receptors to intracellular myosin-containing filaments, *Proc. Natl. Acad. Sci. U.S.A.* **73**:4575.

Ash, J. F., Vogt, P. K., and Singer, S. J., 1976, Reversion from transformed to normal phenotype by inhibition of protein synthesis in rat kidney cells infected with a temperature-sensitive mutant of Rous sarcoma virus, *Proc. Natl. Acad. Sci. U.S.A* **73**:3603.

Ashwell, G., 1977, The role of cell-surface carbohydrates in binding phenomena, in: *Mammalian Cell Membranes: Membranes and Cellular Functions* (G. A. Jamieson and D. M. Robinson, eds.), Vol. 4, pp. 57–71, Butterworths, Boston, Mass.

Ashwell, G., and Morell, A. G., 1974, The role of surface carbohydrates in the hepatic recognition and transport of circulating glycoproteins, *Adv. Enzymol.* **41**:99.

Balsamo, J., and Lilien, J., 1974, The binding of tissue-specific adhesion molecules to the cell surface. A molecular basis for specificity, *Biochemistry* **14**:167.

Banerjee, S. D., Cohn, R. H., and Bernfield, M. R., 1977, The basal lamina of embryonic salivary epithelia: Production by the epithelium and role in maintaining lobular morphology, *J. Cell Biol.* **73**:445.

Barbera, A., 1975, Adhesive recognition between developing retinal cells and the optic tecta of the chick embryo, *Dev. Biol.* **46**:167.

Bard, J. B. L., and Hay, E. D., 1975, The behavior of fibroblasts from the developing avian cornea, *J. Cell Biol.* **67**:400.

Bard, J. B. L., Hay, E. D., and Meller, S. M., 1975, Formation of the endothelium of the avian cornea: A study of cell movement *in vivo, Dev. Biol.* **42**:334.

Barondes, S. H., and Rosen, S. D., 1976, Cell surface carbohydrate-binding proteins: Role in cell recognition, in: *Neuronal Recognition* (S. H. Barondes, ed.), pp. 331–356, Plenum Press, New York.

Bell, G. I., 1978, Models for the specific adhesion of cells to cells, *Science* **200**:618.

Bell, P. B., 1977, Locomotory behavior, contact inhibition, and pattern formation of 3T3 and polyoma virus-transformed 3T3 cells in culture, *J. Cell Biol.* **74**:963.

Benecke, B. J., Ben-Ze'ev, A., and Penman, S., 1978, The control of mRNA production, translation and turnover in suspended and reattached anchorage-dependent fibroblasts, *Cell* **14**:931.

Bernfield, M. R., and Banerjee, S. D., 1972, Acid mucopolysaccharide (glycosaminoglycan) at the epithelial–mesenchymal interface of mouse embryo salivary glands, *J. Cell Biol.* **52**:664.

Bernfield, M. R., Cohn, R. H., and Banerjee, S. D., 1973, Glycosaminoglycans and epithelial organ formation, *Am. Zool.* **13**:1067.

Bischoff, R., and Holtzer, H., 1969, Mitosis and the process of differentiation of myogenic cells *in vitro, J. Cell Biol.* **41**:188.

Blomberg, F., Cohen, R. S., and Siekevitz, P., 1977, The structure of postsynaptic densities isolated from dog cerebral cortex, *J. Cell Biol.* **74**:204.

Bornstein, M. B., 1958, Reconstituted rat-tail collagen used as a substrate for tissue cultures on coverslips in Maximow slides and roller tubes, *Lab. Invest.* **7**:134.

Bornstein, P., and Ash, J. F., 1977, Cell surface-associated structural proteins in connective tissue cells, *Proc. Natl. Acad. Sci. U.S.A* **74**:2480.

Bourguignon, L. Y. W., and Songer, S. J., 1977, Transmembrane interactions and the mechanism of capping of surface receptors by their specific ligands, *Proc. Natl. Acad. Sci. U.S.A* **74**:5031.

Bragina, E. E., Vasiliev, J. M., and Gelfand, I. M., 1976, Formation of bundles of microfilaments during spreading of fibroblasts on the substrate, *Exp. Cell Res.* **97**:241.

Bray, D., 1970, Surface movements during the growth of single explanted neurons, *Proc. Natl. Acad. Sci. U.S.A* **65**:905.

Bray, D., 1973, Model for membrane movements in the neural growth cone, *Nature (London)* **244**:93.

Bray, D., Thomas, C., and Shaw, G., 1978, Growth cone formation in cultures of sensory neurons, *Proc. Natl. Acad. Sci. U.S.A* **75**:5226.

Brown, S. S., and Revel, J. P., 1976, Reversibility of cell surface label rearrangement, *J. Cell Biol.* **68**:629.

Brunk, U., Ericsson, J. L. E., Ponten, J., and Westermark, B., 1971, Specialization of cell surfaces in contact-inhibited human glia-like cells *in vitro, Exp. Cell Res.* **67**:407.

Capers, C. R., 1960, Multinucleation of skeletal muscle *in vitro, J. Biophys. Biochem. Cytol.* **7**:559.

Carney, D. H., and Cunningham, D. D., 1978, Cell surface action of thrombin is sufficient to intiate division of chick cells, *Cell* **14**:811.

Carney, D. H., Glenn, K. S., and Cunningham, D. D., 1978, Conditions which affect initiation of animal cell division by trypsin and thrombin, *J. Cell. Physiol.* **95**:13.

Carter, S. B., 1965, Principles of cell motility and the directionality of cell movement and cancer invasion, *Nature (London)* **208**:1183.

Cassiman, J. J., and Bernfield, R. R., 1975, Transformation induced alterations in fibroblast adhesion: Masking by trypsin treatment, *Exp. Cell Res.* **91**:31.

Cassiman, J. J., and Bernfield, M. R., 1976a, Transformation-induced alterations in adhesion: Binding of preformed aggregates to call layers, *Exp. Cell Res.* **103**:311.

Cassiman, J. J., and Bernfield, M. R., 1976b, Intercellular adhesive recognition: Tissue specific differences and membrane site mobility, *Dev. Biol.* **52**:231.

Chen, L. B., 1977, Alterations in cell surface LETS protein during myogenesis, *Cell* **10**:393.

Chen, L. B., Murray, A., Segal, R. A., Bushnell, A., and Walsh, M. L., 1978, Studies on intercellular LETS glycoprotein matrices, *Cell* **14**:377.

Chen, W. T., 1977, Retraction of the trailing edge during fibroblast movement, *J. Cell Biol.* **75**:416a.

Chen, W. T., 1978, Induction of spreading during fibroblast movement, *J. Cell Biol.* **79**:83a.

Chen, W. T., 1979, Change in arrangement of cortical microfilaments associated with change in cell locomotor activity, *Biophys. J.* **25**:A32.

Cohn, R. H., Cassiman, J. J., and Bernfield, M. R. 1976, Relationship of tranformation, cell density, and growth control to the cellular distribution of newly synthesized glycosaminoglycan, *J. Cell Biol.* **71**:280.

Cohn, R. H., Banerjee, S. D. and Bernfield, M. R., 1977, Basal lamina of embryonic salivary epithelia, *J. Cell Biol.* **73**:464.

Collett, M. S., and Erikson, R. L., 1978, Protein kinase activity associated with the avian sarcoma virus *src* gene product, *Proc. Natl. Acad. Sci. U.S.A* **75**:2021.

Collins, F., 1978*a*, Axon initiation by ciliary neurons in culture, *Dev. Biol.* **65**:50.

Collins, F., 1978*b*, Induction of neurite outgrowth by a conditioned-medium factor bound to the culture substratum, *Proc. Natl. Acad. Sci. U.S.A* **75**:5210.

Cooper, W. G, and Konigsberg, R. I., 1961, Dynamics of myogenesis *in vitro, Anat. Rec.* **140**:195.

Culp, L. A., 1974, Substrate-attached glycoproteins mediating adhesion of normal and virus-transformed mouse fibroblasts, *J. Cell Biol.* **63**:71.

Culp, L. A., 1978, Biochemical determinants of cell adhesion, in: *Current Topics in Membranes and Transport: Cell Surface Glycoproteins* (R. L. Juliano and A. Rothstein, eds.), Vol XI, pp. 327–396, Academic Press, New York.

Culp, L. A., and Buniel, J. F., 1976, Substrate-attached serum and cell proteins in adhesion of mouse fibroblasts, *J. Cell. Physiol.* **88**:89.

Culp, L. A., Rollins, B. J., Buniel, J., and Hitri, S., 1978, Two functionally distinct pools of glycosaminoglycan in the substrate adhesion site of murine cells, *J. Cell Biol.* **79**:788.

Curtis, A. S. G., 1964, The mechanism of adhesion of cells to glass. A study by interference reflection microscopy, *J. Cell Biol.* **20**:199.

Curtis, A. S. G., 1973, Cell Adhesion, *Prog. Biophys. Mol. Biol.* **27**:315.

Damsky, C. H., Wylie, D. E., and Buck, C. A., 1979, Studies on the function of cell surface glycoproteins. II. Possible role of surface glycoproteins in the control of cytoskeletal organization and surface morphology, *J. Cell Biol.* **80**:403.

David, G., and Bernfield, M. R., 1979, Collagen reduces glycosaminoglycan degradation by cultured mammary epithelial cells: Possible mechanism for basal lamina formation, *Proc. Natl. Acad. Sci. U.S.A* **76**:786.

Den, H., Malinzak, D. A., and Rosenberg, A., 1976, Lack of evidence for the involvement of a β-D-galactosyl-specific lectin in the fusion of chick myoblasts, *Biochem. Biophys. Res. Commun.* **69**:621.

Derby, M. A., 1978, Analysis of glycosaminoglycans within the extracellular environments encountered by migrating neural crest cells, *Dev. Biol.* **66**:321.

Dessau, W., Sasse, J., Timpl, R., Jilek, F., von der Mark, K., 1978, Synthesis and extracellular deposition of fibronectin in chondrocyte cultures, *J. Cell Biol.* **79**:342.

DiPasquale, A., 1975*a*, Locomotory activity of epithelial cells in culture, *Exp. Cell Res.* **94**:191.

DiPasquale, A., 1975*b*, Locomotion of epithelial cells, *Exp. Cell Res.* **95**:425.

DiPasquale, A., and Bell, P. B., Jr., 1974, The upper cell surface: Its inability to support active cell movement in culture, *J. Cell Biol.* **62**:198.

Domnina, L. V., Pletyushkina, O. Y., Vasiliev, J. M., and Gelfand, I. M., 1977, Effects of antitubulins on the redistribution of crosslinked receptors on the surface of fibroblasts and epithelial cells, *Proc. Natl. Acad. Sci. U.S.A* **74**:2865.

Douvas, A. S., Harrington, C. A., and Bonner, J., 1975, Major nonhistone proteins of rat liver chromatin: Preliminary identification of myosin, actin, tubulin and tropomyosin, *Proc. Natl. Acad. Sci. U.S.A* **72**:3902.

Dulap, M. K., and Donaldson, D. J., 1978, Inability of colchicine to inhibit newt epidermal cell migration or prevent concanavalin A-mediated inhibition of migration, *Exp. Cell Res.* **116**:15.

Dunn, G. A., and Ebenbal, T., 1978, Some aspects of contact guidance, *Zoon* **6**:65.

Dunn, G. A., and Heath, J. P., 1976, A new hypothesis of contact guidance in tissue cells, *Exp. Cell Res.* **101**:1.

Ebendal, T., 1977, Extracellular matrix fibrils and cell contacts in the chick embryo, *Cell Tiss. Res.* **175**:439.

Edelman, G. M., 1976, Surface modulation in cell recognition and cell growth, *Science* **192**:218.

Edelman, G. M., and Yahara, I., 1976, Temperature-sensitive changes in surface modulating assemblies of fibroblasts transformed by mutants of Rous sarcoma virus, *Proc. Natl. Acad. Sci. U.S.A* **73**:2047.

Edwards, J. G., 1977, Cell adhesion, in: *Mammalian Cell Membranes: Membranes and Cellular Functions* (G. A. Jamieson and D. M. Robinson, eds.), Vol. 4, pp. 32–56, Butterworths, Boston, Mass.

Emerman, J. T., and Pitelka, D. R., 1977, Maintenance and induction of morphological differentiation in dissociated mammary epithelium on floating collagen membranes, *In Vitro* **13**:316.

Emerman, J. T., Enami, J., Pitelka, D. R., and Nandi, S., 1977, Hormonal effects on intracellular and secreted casein in cultures of mouse mammary epithelial cells on floating collagen membranes, *Proc. Natl. Acad. Sci. U.S.A* **74**:4466.

Engvall, E., and Ruoslahti, E., 1977, Binding of soluble form of fibroblast surface protein, fibronectin, to collagen, *Int. J. Cancer* **20**:1.

Erickson, C. A., 1978, Contact behavior and pattern formation of BHK and polyoma virus-transformed BHK fibroblasts in culture, *J. Cell Sci.* **33**:53.

Erickson, C. A., and Trinkaus, 1976, Microvilli and blebs as sources of reserve surface membrane during cell spreading, *Exp. Cell Res.* **99**:375.

Fantes, P. A., Grant, W. D., Pritchard, R. H., Sudbery, P. E., and Wheals, A. E., 1975, The regulation of cell size and the control of mitosis, *J. Theor. Biol.* **50**:213.

Farmer, S. R., Ben-Ze'ev, A., Benecke, B. J., and Penman, S., 1978, Altered translatability of messenger RNA from suspended anchorage-dependent fibroblasts: Reversal upon cell attachment to a surface, *Cell* **15**:627.

Fear, J., 1977, Observations on the fusion of chick embryo myoblasts in culture, *J. Anat.* **124**:437.

Fischbach, G. D., 1970, Synaptic potentials recorded in cell cultures of nerve and muscle, *Science* **169**:1331.

Fischer, E. H., Becker, J. U., Blum, H. E., Lehky, P., Malencik, D. A., and Pocinwong, S., 1975, Concerted regulation of carbohydrate metabolism and muscle contraction, *Hoppes-Seylers Z. Physiol. Chem.* **356**:381.

Fischman, D. A., 1972, in: *The Structure and Function of Muscle,* Vol. 1 (G. H. Bourne, ed.), Academic Press, New York.

Fleischer, M., and Wohlfarth-Botterman, K. E., 1975, Correlation between tension force generation, fibrillogenesis and ultrastructure of cytoplasmic actomyosin during isometric and isotonic contractions of protoplasmic strands, *Cytobiologie* **10**:339.

Folkman, J., 1977, Conformational control of cell and tumor growth, in: *Recent Advances in Cancer Research: Cell Biology, Molecular Biology and Tumor Virology,* Vol. 1 (R. C. Gallo, ed.), CRC, Cleveland, Ohio.

Folkman, J., and Greenspan, H. P., 1975, Influence of geometry on control of cell growth, *Biochem. Biophys. Acta* **417**:211.

Folkman, J., and Moscona, A., 1978, Role of cell shape in growth control, *Nature (London)* **273**:345.

Frazier, W., and Glaser, L., 1979, Cell surface components involved in cell recognition, *Annu. Rev. Biochem.* **48**:491.

Friedlander, M., and Fischman, D. A., 1975, Serological analysis of developing muscle cell surfaces, *J. Cell Biol.* **67**:124a.

Furcht, L. T., Mosher, D., and Wendelschafer-Crabb, G., 1977, Immunoelectron microscopic localization of fibronectin during myogenesis, *J. Cell Biol.* **75**:156a.

Furukawa, K., and Terayama, H., 1977, Isolation and identification of glycosaminoglycans associated with purified nuclei from rat liver, *Biochem. Biophys. Acta* **499**:278.

Gail, M., 1973, Time lapse studies on the motility of fibroblasts in tissue culture, *Ciba Found. Symp.* **14** (N. S.):287.

Gail, M. H., and Boone, C. W., 1972, Cell-substrate adhesivity, *Exp. Cell Res.* **70**:33.

Gartner, T. K., and Podleski, T. R., 1975, Evidence that a membrane bound lectin mediates fusion of L6 myoblasts, *Biochem. Biophys. Res. Commun.* **67**:972.

Gartner, T. K., and Podleski, T. R., 1976, Evidence that the types and specific activity of lectins control fusion of L6 myoblasts, *Biochem. Biophys. Res. Commun.* **70**:1142.

Glabe, C. G., and Vacquier, V. D., 1977, The surfaces of sea urchin eggs possess a species-specific receptor for binding, *J. Cell Biol.* **75**:56a.

Goldberg, B., 1979, Binding of soluble type I collagen molecules to the fibroblast plasma membrane, *Cell* **16**:265.

Goldman, R. D., Yerna M. J., and Schloss, J. A., 1976, Localization and organization of microfilaments and related proteins in normal and virus-transformed cells, *J. Supramol. Struct.* **5**:155.

Gospodarowicz, D., Greenberg, G., and Birdwell, C. R., 1978, Determination of cellular shape by the extracellular matrix and its correlation with the control of cellular growth, *Can. Res.* **38**:4155.

Gottlieb, D. I., Merrell, R., and Glaser, L., 1974, Temporal changes in embryonic cell surface recognition, *Proc. Natl. Acad. Sci. U.S.A* **71**:1800.

Gottlieb, D. I., Rock, K., and Glaser, L., 1976, A gradient of adhesive specificity in developing avian retina, *Proc. Natl. Acad. Sci. U.S.A* **73**:410.

Graham, J. M., Hynes, R. O., Davidson, E. A., and Bainkon, D. F., 1975, The location of proteins labeled by the ^{125}I-lactoperoxidase system in the NIL 8 hamster fibroblast, *Cell* **4**:353.

Greulich, R. C., and Slavkin, H. C. (eds.), 1975, *Extracellular Matrix Influences on Gene Expression,* Academic Press, New York.

Grinnell, F., 1978, Cellular adhesiveness and extracellular substrata, *Int. Rev. Cytol.* **29**:65.

Grinnell, F., and Minter, D., 1978, Attachment and spreading of baby hamster kidney cells to collagen substrata. Effects of cold-insoluble globulin, *Proc. Natl. Acad. Sci. U.S.A* **75**:4408.

Harris, A., 1973*a*, Behavior of cultured cells on substrata of variable adhesiveness, *Exp. Cell Res.* **77**:285.

Harris, A., 1973*b*, Cell surface movements related to cell locomotion, *Ciba Found. Symp.* **14** (N. S.):3.

Harris, A., 1973*c*, Location of cellular adhesions to solid substrata, *Dev. Biol.* **35**:83.

Harris, A. J., Heinemann, S., Schubert, C., and Tarakis, H., 1971, Trophic interactions between cloned tissue culture lines of nerve and muscle, *Nature (London)* **231**:296.

Hauschka, S. D., 1972, in: *Growth, Nutrition and Metabolism of Cells in Culture,* Vol. 2 (G. H. Rothblat and V. J., Cristofalo, eds.), pp. 67–130, Academic Press, New York.

Hauschka, S. D., and Konigsberg, I. R., 1966, The influence of collagen on developing muscle clones, *Proc. Natl. Acad. Sci. U.S.A* **55**:119.

Hay, E. D, 1973, Origin and role of collagen in the embryo, *Am. Zool.* **13**:1085.

Heath, J. P., and Dunn, G. A., 1978, Cell to substratum contacts of chick fibroblasts and their relation to the microfilament system, *J. Cell Sci.* **29**:197.

Heaysman, J., 1973, *Ciba Found. Symp.* **14** (N. S.):187–194.

Hedman, K., Vaheri, A., and Wartiovaara, J., 1978, External fibronectin of cultured human fibroblasts in predominantly a matrix protein, *J. Cell Biol.* **76**:748.

Henkart, P., Humphreys, S., Humphreys, T., 1973, Characterization of sponge aggregation factor: A unique proteoglycan complex, *Biochemistry* **12**:3045.

Hitchcock, S. E., Carlsson, L., and Lindberg, U., 1976, Depolymerization of F-actin by deoxyribonuclease I, *Cell* **7**:531.

Holley, R. W., 1975, Control of growth of mammalian cells in cell culture, *Nature (London)* **258**:487.

Holtzer, H., Weintraub, H., Mayne R., and Mochan, B., 1972, The cell cycle, cell lineages and cell differentiation, *Curr. Top. Dev. Biol.* **7**:229.

Holtzer, H., Rubinstein, N., Dienstman, S., Chi, J., Biehl, J., and Somlye, A., 1974, Perspectives in myogenesis, *Biochimie* **56**:1575.

Holtzer, H., Rubinstein, N., Fellini, S., Yeoh, G., Chi, J., Birnbaum, J., and Okayama, M., 1975, Lineages, quantal cell cycles, and the generation of cell diversity, *Q. Rev. Biophys.* **8**:523.

Hoover, R. L., Briggs, R. T., and Karnovsky, M. J., 1978, The adhesive interaction between polymorphonuclear leukocytes and endothelial cells *in vitro, Cell* **14**:423.

Huxley, H. E., 1973, Muscular contraction and cell motility, *Nature (London)* **243**:445.

Hynes, R. O., 1976, Cell surface proteins and malignant transformation, *Biochim. Biophys. Acta* **458**:73.

Hynes, R. O., and Destree, A. T., 1978, Relationships between fibronectin (LETS protein) and actin, *Cell* **15**:875.

Ingram, V. M., 1969, A side view of moving fibroblasts, *Nature (London)* **222**:641.

Isenberg, G., Rathke, P. C., Hulsmann, N., Franke, W. W., and Wohlfarth-Bottermann, K. E., 1976, Cytoplasmic actomyosin fibrils in tissue culture cells. Direct proof of contractility by visualization of ATP-induced contraction in fibrils isolated by laser microbeam dissection, *Cell Tissue Res.* **166**:427.

Izzard, C. S., and Lochner, L. R., 1976, Cell-to-substrate contacts in living fibroblasts: An interference reflection study with an evaluation of the technique, *J. Cell Sci.* **21**:129.

Jacobson, M., 1970, *Developmental Neurobiology,* Holt, Rinehart, and Winston, San Francisco, Calif.

Ji, T. H., and Nicolson, G., 1974, Lectin binding and perturbation of the outer surface of the cell membrane induces a transmembrane organization alteration at the inner surface, *Proc. Natl. Acad. Sci. U.S.A* **71**:2212.

Johnson, K. E., 1970, The role of changes in cell contact behavior in amphibian gastrulation, *J. Exp. Zool.* **175**:391.

Keski-Oja, J., Mosher, D. F., and Vaheri, A., 1977, Dimeric character of fibronectin, a major cell surface-associated glycoprotein, *Biochem. Biophys. Res. Commun.* **74**:699.

Knudson, K. A., and Horwitz, A. F., 1977*a*, Tandem events in myoblast fusion, *Dev. Biol.* **58**:328.

Knudson, K. A., and Horwitz, A. F., 1977*b*, Tandem events in myoblast fusion, *J. Supramol. Struct.* **1977**(S1):39.

Koch, G. L. E., and Smith, M. J., 1978, An association between actin and the major histocompatibility antigen H-2, *Nature (London)* **273**:274.

Koda, L. Y., and Partlow, L. M., 1976, Membrane marker movement on sympathetic axons in tissue culture, *J. Neurobiol.* **7**:157.

Konigsberg, I. R., 1963, Clonal analysis of myogenesis, *Science* **140**:1273.

Kongisberg, I. R., 1970, The relationship of collagen to the clonal development of embryonic skeletal muscle, in: *Chemistry and Molecular Biology of the Intercellular Matrix,* Vol. 3 (E. A. Balazs, ed.), p. 1779, Academic Press, New York.

Konigsberg, I. R., 1971, Diffusion-mediated control of myoblast fusion, *Dev. Biol.* **26**:133.

Konigsberg, I. R., and Hauschka, S. D., 1965, Cell and tissue interactions in the reproduction of cell type, *Symp. Soc. Devel. Biol.* **24**:243.

Kraemer, P. M., and Tobey, R. A., 1972, Cell-cycle dependent desquamation of heparan sulfate from the cell surface, *J. Cell Biol.* **55**:713.

Krawczyk, W. S., 1971, A pattern of epidermal cell migration during wound healing, *J. Cell Biol.* **49**:247.

Kurikenin, M., Wartiovaara, J., and Vaheri, A., 1978, Cytochalasin B releases a major surface-associated glycoprotein, fibronectin, from cultured fibroblasts, *Exp. Cell Res.* **111**:127.

Lazarides, E., 1976, Two general classes of cytoplasmic actin filaments in tissue culture cells: The role of tropmyosin, *J. Supramol. Struct.* **5**:531.

LeDouarin, N., 1976, Cell migration in early vertebrate development studied in interspecific chimeras, *Ciba Found. Symp.* **40**:71.

LeDouarin, N., and Teillet, M. A. M., 1974, Experimental analysis of the migration and differentiation of neuroblasts of the autonomic nervous system and of neuroectodermal mesenchymal derivatives, using a biological cell marking technique, *Dev. Biol.* **41**:162.

Lenk, R., and Penman, S., 1979, The cytoskeletal framework and polio virus metabolism, *Cell* **16**:289.

Letourneau, P. C., 1975*a*, Possible rolles for cell-to-cell substratum adhesion in neuronal morphogenesis, *Dev. Biol.* **44**:77.

Letourneau, P. C., 1975*b*, Cell-to-substratum adhesion and guidance of axonal elongation, *Dev. Biol.* **44**:92.

Letourneau, P. C., 1977, Regulation of neuronal morphogenesis by cell–substratum adhesion, *Soc. Neurosci. Symp.* **2**:67.

Letourneau, P. C., 1979*a*, Inhibition of cell adhesion by concanavalin A is associated with Con A-mediated rearrangements of surface receptors, *J. Cell Biol.* **80**:128.

Letourneau, P. C., 1979*b*, Cell-substratum adhesion of neurite growth areas, *Exp. Cell Res.* **124**:127.

Letourneau, P. C., and Wessells, N. K., 1974, Migratory cell locomotion versus nerve axon elongation. Differences based on the effects of lanthanum ion, *J. Cell Biol.* **61**:56.

Leung, J. P., Trotter, J. A., Munar, E., and Nameroff, M., 1975, Differentiation of the myogenic cell surface in: *ICN–UCLA Symposium on Molecular and Cellular Biology,* Vol. 2 (D. McMahon and D. F. Fox, eds.), W. A. Benjamin Co., Menlo Park, Calif.

Linder, E., Vaheri, A., Ruoslahti, E., and Wartiovaara, J., 1975, Distribution of fibroblast surface antigen in the developing chick embryo, *J. Exp. Med.* **142**:41.

Linsenmeyer, T., Gibney, E., Toole, B. P., and Gross, J., 1978, Cellular adhesion to collagen, *Exp. Cell Res.* **116**:470.

Lipton, B. H., 1977, Collagen synthesis by normal and bromodeoxyuridine-modulated cells in myogenic culture, *Dev. Biol.* **61**:153.

Lopresti, V., Macagno, E. R., and Levinthal, C., 1973, Structure and development of neural connections in isogenic organisms: Cellular interactions in the development of the optic lamina of *Daphnia, Proc. Natl. Acad. Sci. U.S.A* **70**:433.

Luduena-Anderson, M. A., 1973, Nerve cell differentiation *in vitro, Dev. Biol.* **33**:268.

Luduena-Anderson, M., and Wessells, N. K., 1973, Cell locomotion, nerve elongation, and microfilaments, *Dev. Biol.* **30**:427.

McClay, D., and Baker, S. R., 1975, A kinetic study of cell adhesion, *Dev. Biol.* **43**:109.

McClay, D. R., Gooding, L. R., and Fransen, M. E., 1977, A requirement for trypsin-sensitive cell-surface components for cell–cell interactions of embryonic neural retina cells, *J. Cell Biol.* **75**:56.

McEvoy, F. A., and Ellis, D. E., 1976, Incorporation of ^{14}C-palmitate into the lipids of cultured chick embryo myoblasts, *Biochem. Soc. Trans.* **4**:1065.

McGuire, E. J., and Burdick, C. L., 1976, Intercellular adhesive selectivity. I. An improved assay for the measurement of embryonic chick intercellular adhesion (liver and other tissues), *J. Cell Biol.* **68**:80.

MacPherson, I., and Montagnier, L., 1964, Agar suspension culture for the selective assay of cells transformed by polyoma virus, *Virology* **23**:291.

Manasek, F. J., 1975, The extracellular matrix: A dynamic component of the developing embryo, *Curr. Top. Dev. Biol.* **10**:35.

Marchase, R. B., 1977, Biochemical investigations of retino-tectal adhesive specificity, *J. Cell Biol.* **75**:237.

Marchase, R. B., Vosbeck, K., and Roth, S., 1976, Intercellular adhesive specificity, *Biochem. Biophys. Acta* **457**:385.

Maroudas, N. G., 1972, Anchorage dependence: Correlation between amount of growth and diameter of bead, for single cells grown on individual glass beads, *Exp. Cell Res.* **75**:337.

Maroudas, N. G., 1973, Chemical and mechanical requirements for fibroblast adhesion, *Nature (London)* **244**:353.

Marshall, L. M., Sanes, J. R., and McMahon, U. J., 1977, Reinnervation of original synaptic sites on muscle fiber basement membrane after disruption of the muscle cells, *Proc. Natl. Acad. Sci. U.S.A* **74**:3073.

Martin, G. R., and Rubin, H., 1974, Effects of cell adhesion of the substratum on the growth of chick embryo fibroblasts, *Exp. Cell Res.* **85**:319.

Maslow, D. E, 1967, The formation of multinucleated striated muscle, *Am. Zool.* **7**:751.

Maslow, D. E., 1969, Cell specificity in the formation of multinucleated striated muscle, *Exp. Cell Res.* **54**:381.

Mautner, V., and Hynes, R. O., 1977, Surface distribution of LETS protein in relation to the cytoskeleton of normal and transformed cells, *J. Cell Biol.* **75**:743.

Medoff, J., and Gross, J., 1971, *In vitro* aggregation of mixed embryonic kidney and nerve cells, *J. Cell Biol.* **50**:457.

Merlie, J. P., Buckingham, M. E., and Whalen, R. G., 1977, Molecular aspects of myogenesis, *Curr. Top. Dev. Biol.* **11**:61.

Michalopoulos, G., and Pitot, H. C., 1975, Primary culture of parenchymal liver cells on collagen membranes, *Exp. Cell Res.* **94**:70.

Middleton, C. A., 1977, The effects of cell–cell contact on the spreading of pigmented retina epithelial cells in culture, *Exp. Cell Res.* **109**:349.

Mooseker, M. S., 1976, Actin filament–membrane attachment in the microvilli of intestinal epithelial cells, in: *Cell Motility* (G. Goldman, T. Pollard and J. Rosenbaum, eds.), pp. 631–650, Cold Spring Harbor Laboratory, New York.

Moscona, A. A., 1974, Surface specification of embryonic cells: Lectin receptors, cell recognition, and specific cell ligands, in: *The Cell Surface in Development* (A. A. Moscona, ed.), pp. 67–100, Wiley, New York.

Moyer, W. A., and Steinberg, M. S., 1976, Do rates of intercellular adhesion measure the cell affinities reflected in cell-sorting and tissue-spreading configurations? *Dev. Biol.* **52**:246.

Muller, K., and Gerisch, G., 1978, A specific glycoprotein as the target site of adhesive blocking Fab in aggregating *Dictyostelium* cells, *Nature (London)* **274**:445.

Murray, J. C., Stingl, G., Kleinman, H. K., Martin, G. R., and Katz, S. I., 1979, Epidermal cells adhere preferentially to type IV (basement membrane) collagen, *J. Cell Biol.* **80**:197.

Nakajima, S., 1965, Selectivity in fasciculation of nerve fibers *in vitro*, *J. Comp. Neurol.* **125**:193.

Nameroff, M., and Munar, E., 1976, Inhibition of cellular differentiation by phospholipase C, *Dev. Biol.* **49**:288.

Nelson, G. A., and Revel, J. P., 1975, Scanning electron microscope study of cell movements in the corneal endothelium of the avian embryo, *Dev. Biol.* **42**:315.

Nichols, D. H., and Weston, J. A., 1977, Melanogenesis in cultures of peripheral nerve tissue. I. The origin and prospective fate of cells giving rise to melanocytes, *Dev. Biol.* **60**:217.

Nichols, D. H., Kaplan, R. A., and Weston, J. A., 1977, Melanogenesis in cultures of peripheral nerve tissue. II. Environmental factors determining the fate of pigment-forming cells, *Dev. Biol.* **60**:226.

Noden, D. M., 1975, An analysis of the migratory behavior of the avian cephalic neural crest cells, *Dev. Biol.* **42**:106.

Noonan, K. D., 1978, Proteolytic modification of cell surface macromolecules: Mode of action in stimulating cell growth, in: *Current Topics in Membranes and Transport: Cell Surface Glycoproteins,* Vol. XI (R. L. Juliano and A. Rothstein, eds.), pp. 397–461, Academic Press, New York.

Nornes, H. O., and Das, G. D., 1972, Temporal pattern of neurogenesis in spinal cord: Cytoarchitecture and directed growth of axons, *Proc. Natl. Acad. Sci. U.S.A* **69**:1962.

Nowak, T. P., and Barondes, S. H., 1977, Lectin activity in embryonic chick muscle: Developmental regulation and preliminary purification, *Prog. Clin. Biol. Res.* **15**:159.

Nowak, T. P., Haywood, P. L., and Barondes, S. H., 1976, Developmentally regulated lectin in embryonic chick muscle and myogenic cell line, *Biochem. Biophys. Res. Commun.* **68**:650.

Ofek, I., Beachey, E. H., Eyal, F., and Morrision, J. C., 1977, Postnatal development of binding of streptococci and lipoteichoic acid by oral mucosal cells of humans, *J. Infect. Dis.* **135**:267.

Okazaki, K., and Holtzer, H., 1965, An analysis of myogenesis *in vitro* using fluorescin-labelled antimyosin, *J. Histochem. Cytochem.* **13**:726.

Otsuka, H., and Moskowitz, M., 1976, Arrest of 3T3 cells in G1 phase in suspension culture, *J. Cell. Physiol.* **87**:213.

Overton, J., 1977*a*, Formation of junctions and cell sorting in aggregates of chick and mouse cells, *Dev. Biol.* **55**:103.

Overton, J., 1977b, Response of epithelial and mesenchymal cells to culture on basement lamella observed by scanning microscopy, *Exp. Cell Res.* **105**:313.

Pardee, A. B., 1975, The cell surface and fibroblast proliferation. Some current research trends, *Biochim. Biophy. Acta* **417**:153.

Parkinson, E. K., and Edwards, J. G., 1978, Non-reciprocal contact inhibition of locomotion of chick embryonic choroid fibroblasts by pigmented retina epithelial cells, *J. Cell Sci.* **33**:103.

Penman, S., 1979, Metabolic regulation is suspended and reattached anchorage-dependent fibroblasts, *J. Supramol. Struct. Suppl.* **3**:174.

Pennypacker, J. P., Hassell, J. R., Yamada, K. M., and Pratt, R. M., 1978, The influence of fibronectin on chondrogenic expression *in vitro, J. Cell Biol.* **79**:149a.

Pfenninger, K. H., and Pfenninger, M. F., 1975, Distribution and fate of lectin binding sites on the surface of growing neuronal processes, *J. Cell Biol.* **67**:332a.

Pollack, R., and Rifkin, D., 1975, Actin-containing cables within anchorage-dependent rat embryo cells are dissociated by plasmin and trypsin, *Cell* **6**:495.

Pollack, R., Osborn, M., and Weber, K., 1975, Patterns of organization of actin and myosin in normal and transformed cultured cells, *Proc. Natl. Acad. Sci. U.S.A* **72**:994.

Pouyssegur, J. M., and Pastan, I., 1976, Mutants of Balb/c 3T3 fibroblasts defective in adhesiveness to substratum: Evidence for alteration in cell surface protein, *Proc. Natl. Acad. Sci. U.S.A* **73**:544.

Pratt, R. M., Larson, M. A., and Johnston, M. C., 1975, Migration of cranial neural crest cells in a cell-free hyaluronic acid rich matrix, *Dev. Biol.* **44**:298.

Puck, T., 1977, Cyclic AMP, the microtubule-microfilament system, and cancer, *Proc. Natl. Acad. Sci. U.S.A* **74**:4491.

Rajaraman, R., Rounds, D. E., Yen, S. P. S., and Rembaum, A., 1974, A scanning electron microscope study of cell adhesion and spreading *in vitro, Exp. Cell Res.* **88**:327.

Rakic, P., 1971, Guidance of neurons migrating to the fetal monkey neocortex, *Brain Res.* **33**:471.

Rees, D. A., Lloyd, C. W., and Thom, D., 1977, Control of grip and stick in cell adhesion through lateral relationships of membrane glycoproteins, *Nature (London)* **267**:124.

Revel, J. P., and Wolken, K., 1973, Electron microscope investigations of the underside of cells in culture, *Exp. Cell Res.* **78**:1.

Revel, J. P., Hoch, P., and Ho, D., 1974, Adhesion of cultured cells to their substratum, *Exp. Cell Res.* **84**:207.

Risser, R., and Pollack, R., 1974, A nonselective analysis of SV40 transformation of mouse 3T3 cells, *Virology* **59**:477.

Roblin, R., Albert, S. O., Gelb, N. A., and Black, P. H., 1975, Cell surface changes correlated with density-dependent growth inhibition. Glycosaminoglycan metabolism in 3T3, SV3T3, and Con A selected revertant cells, *Biochemistry* **14**:347.

Roseman, S., 1974, Complex carbohydrates and intercellular adhesion, in: *The Cell Surface in Development* (A. A. Moscona, ed.), pp. 255–271, Wiley, New York.

Rosen, J. J., and Culp, L. A., 1977, Morphology and cellular origins of substrate-attached material from mouse fibroblasts, *Exp. Cell Res.* **107**:139.

Roth, S., 1968, Studies on intercellular adhesive selectivity, *Dev. Biol.* **18**:602.

Roth, S., 1973, A molecular model for cell interactions, *Q. Rev. Biol.* **48**:541.

Rovensky, Y. A., and Slavnaja, I. L., 1974, Spreading of fibroblast-like cells on grooved surfaces, *Exp. Cell Res.* **84**:199.

Rovensky, Y. A., Slavnaja, I. L., and Vasiliev, J. M., 1971, Behavior of fibroblast-like cells on grooved surfaces, *Exo. Cell Res.* **65**:193.

Rubin, R. W., Warren, R. H., Lukeman, D. S., and Clements, E., 1978, Actin content and organization in normal and transformed cells in culture, *J. Cell Biol.* **78**:28.

Ruoslahti, E., Hayman, E. G., and Engvall, E. 1979, Interaction of fibronectin with collagens, and its role in cell adhesion, *J. Supramol. Struct. Suppl.* **3**:173 (abstract).

Rutishauser, U., Thiery, J. P., Brackenbury, R., Sela, B. A., and Edelman, G. M., 1976, Mechanisms of adhesion among cells from neural tissues of the chick embryo, *Proc. Natl. Acad. Sci. U.S.A* **73**:577.

Salzer, J., Glaser, L., and Bunge, R. P., 1977, Stimulation of Schwann cell proliferation by a neurite membrane fraction, *J. Cell Biol.* **75**:118a (abstract).

Sanger, J. W., and Holtzer, H., 1972, Cytochalasin B: Effects on cell morphology, cell adhesion, and mucopolysaccharide synthesis, *Proc. Natl. Acad. Sci. U.S.A* **69**:253.

Schubert, D., Lacorbine, M., and Watson, J., 1976, Growth regulation of cells grown in suspension culture, *Nature (London)* **264**:266.

Santerre, R. F., and Rich, A., 1976, Actin accumulation in developing chick brain and other tissues, *Dev. Biol.* **54**:1.

Schaeffer, B. E., Schaeffer, H. E., and Brick, I., 1973, Cell electrophoresis of amphibian blastula and gastrula cells: The relationship of surface charge and morphogenetic movement, *Dev. Biol.* **34**:66.

Sengel, P., 1976, *Morphogenesis of Skin,* Cambridge University Press, Cambridge.

Shadle, P. J., Kobiler, D., and Barondes, S. H., 1979, Lectin-2 from embryonic chick muscle interacts with glycosaminoglycans from extracellular matrix, *J. Supramol. Struct. Suppl.* **3**:204.

Shimada, Y., Fischman, D. A., and Moscona, A. A., 1969a, Formation of neuromuscular junctions in embryonic cell cultures, *Proc. Natl. Acad. Sci. U.S.A* **62**:715.

Shimada, Y., Fischman, D. A., Moscona, A. A., 1969b, The development of nerve–muscle junctions in monolayer cultures of embryonic spinal cord and skeletal muscle cells, *J. Cell Biol.* **43**:382.

Shin, S. I., Freedman, V. H., Risser, R., and Pollack, R., 1975, Tumorigencity of virus-transformed cells in nude mice is correlated specifically with anchorage independent growth *in vitro, Proc. Natl. Acad. Sci. U.S.A* **72**:4435.

Shur, B. D., 1977, Cell-surface glycosyltransferases in gastrulating chick embryos, *Dev. Biol.* **58**:23.

Shur, B., and Roth, S., 1973, The localization and potential function of glycosyltransferases in chick embryos, *Am. Zool.* **13**:1129.

Sidman, R. L., and Wessells, N. K., 1975, Control of direction of growth during the elongation of neurites, *Exp. Neurol.* **48**:237.

Small, J. V., Isenberg, G., and Celis, J. E., 1978, Polarity of actin at the leading edge of cultured cells, *Nature (London)* **272**:638.

Spooner, B. S., Yamada, K. M., and Wessells, N. K., 1971, Microfilaments and cell locomotion, *J. Cell Biol.* **49**:593.

Stathakis, N. E., and Mossesson, M. W., 1977, Interactions among heparin, cold-insoluble globulin and fibronectin in formation of the heparin-precipitable fraction of serum, *J. Clin. Invest.* **60**:855.

Steinberg, M. S., Armstrong, P. B., and Granger, R. E., 1973, On the recovery of adhesiveness by trypsin-dissociated cells, *J. Membr. Biol.* **13**:97.

Stenman, S., Wartiovaara, J., and Vaheri, A., 1977, Changes in the distribution of a major fibroblast protein, fibronectin, during mitosis and interphase, *J. Cell Biol.* **74**:453.

Stenn, K. S., Madri, J. A., and Roll, F. J., 1979, Migrating epidermis produces AB₂ collagen and requires continual collagen synthesis for movement, *Nature (London)* **277**:229.

Stoker, M., O'Neill, C., Berryman, S., and Waxman, V., 1968, Anchorage and growth regulation in normal and virus transformed cells, *Int. J. Cancer* **3**:683.

Strassman, R. J., Letourneau, P. C., and Wessells, N. K., 1973, Elongation of axons in an agar matrix that does not support cell locomotion, *Exp. Cell Res.* **81**:482.

Teichberg, V. I., Silman, I., Beitsch, D. D., and Resheff, G., 1975, A β-ᴅ-galactoside binding protein from electric organ tissue of *Electrophorus electricus, Proc. Natl. Acad. Sci. U.S.A* **72**:1383.

Teng, N. N. H., and Chen, L. B., 1975, The role of surface proteins in cell proliferation as studied with thrombin and other proteases, *Proc. Natl. Acad. Sci. U.S.A* **72**:413.

Tickle, C. A., and Trinkaus, J. P., 1973, Change in surface extensibility of *Fundulus* deep cells during early development, *J. Cell Sci.* **13**:721.

Toh, B. H., and Hard, G. C., 1977, Actin co-caps with concanavalin A receptors, *Nature (London)* **269**:695.

Toole, B. P., and Trelstad, R. L., 1971, Hyaluronate production and removal during corneal development in the chick, *Dev. Biol.* **26**:28.

Toole, B. P., Jackson, G., and Gross, J., 1972, Hyaluronate in morphogenesis: Inhibition of chondrogenesis *in vitro, Proc. Natl. Acad. Sci. U.S.A* **69**:1384.

Trelstad, R. L., Hayashi, K., and Toole, B. P., 1974, Epithelial collagens and glycosaminoglycans in the embryonic cornea. Macromolecular order and morphogenesis in the basement membrane, *J. Cell Biol.* **62**:815.

Trinkaus, J. P., 1969, *Cells into Organs. The Forces That Shape the Embryo,* Prentice-Hall, Englewood Cliffs, N. J.

Trinkaus, J. P., 1973, Surface activity and locomotion of *Fundulus* deep cells during blastula and gastrula stages, *Dev. Biol.* **30**:68.

Trinkaus, J. P., 1976, On the mechanism of metazoan cell movements, in: *The Cell Surface in Animal Embryogenesis and Development* (G. Poste and G. L. Nicolson, eds.), pp. 225–329, North-Holland, Amsterdam.

Trinkaus, J. P., Betchaky, T., and Krulikowski, L. S., 1971, Local inhibition of ruffling during contact inhibition of cell movement, *Exp. Cell Res.* **64**:291.

Turner, R. S., and Burger, M. M., 1973, Involvement of a carbohydrate group in the active site for surface guided reassociation of animal cells, *Nature (London)* **244**:509.

Ukena, T. E., and Karnovsky, M. J., 1977, The role of microvilli in the agglutination of cells by concanavalin A, *Exp. Cell Res.* **106**:309.

Umbreit, J., and Roseman, S., 1975, A requirement for reversible binding between aggregating embryonic cells before stable adhesion, *J. Cell Biol.* **250**:9360.

Vaheri, A., Kurkinen, M., Lehto, V. P., Linder, E., and Timpl, R., 1978, Codistribution of pericellular matrix proteins in cultured fibroblasts and loss in transformation: Fibronectin and collagen, *Proc. Natl. Acad. Sci. U.S.A* **75**:4944.

Vasiliev, J. M., Gelfand, I. M., Domnina, L. V., Ivanova, O. Y., Komm, S. G., and Olshevskaja, L. V., 1970, Effect of colcemid on the locomotory behavior of fibroblasts, *J. Embryol. Exp. Morphol.* **24**:625.

Vasiliev, J. M., Gelfand, I. M., Domnina, L. V., Zacharova, O. S., and Ljubimov, A. V., 1975, Contact inhibition of phagocytosis in epithelial sheets: Alterations of cell surface properties induced by cell-cell contact, *Proc. Natl. Acad. Sci. U.S.A* **72**:719.

Vasiliev, J. M., Gelfand, I. M., Domnina, L. V., Dorfman, N. A., and Pletjushkina, O. J., 1976, Active cell edge and movements of concanavalin A receptors of the surface of epithelial and fibroblastic cells, *Proc. Natl. Acad. Sci. U.S.A* **73**:4085.

Weiss, L., 1977, Some biophysical aspects of cell contact, detachment and movement, in: *Cell Interactions in Differentiation* (M. Kartinen-Jaaskelainen, L. Saxen, and L. Weiss, eds.), pp. 279–289, Academic Press, New York.

Weiss, L., and Greep, R. O., 1977, *Histology*, McGraw-Hill, New York.

Weiss, P., 1941, Nerve patterns: The mechanics of nerve growth, *Growth (Suppl.)* **5**:163.

Weiss, P., 1945, Experiments on cell and axon orientation *in vitro:* The role of colloidal exudates in tissue organization, *J. Exp. Zool.* **100**:353.

Wessells, N. K., Letourneau, P. C., Nuttall, R. P., Luduena-Anderson, M. A., and Geiduschek, J. M., 1980, Reciprocal absence of contact paralysis when motile growth cones and glial cells meet, *J. Neurocytol.*, in press.

Whittenberger, B., and Glaser, L., 1977, Inhibition of DNA synthesis in cultures of 3T3 cells by isolated surface membranes, *Proc. Natl. Acad. Sci. U.S.A* **74**:2251.

Wiese, L., and Wiese, W., 1975, On sexual agglutination and mating type substances in isogamous dioecious *Chlamydomonads, Dev. Biol.* **43**:264.

Wilde, C. E., Jr., 1958, The fusion of myoblasts, a morphogenetic mechanism in striated muscle differentiation, *Anat. Rec.* **132**:517.

Wilde, C. E., Jr., 1959, Differentiation in response to the biochemical environment, in: *Cell, Organism and Milieu* (D. Rudnick, ed.), Ronald Press, New York.

Willingham, M. C., Yamada, K. M., Yamada, S. S., Pouyssegur, J., and Pastan, I., 1977, Microfilament bundles, and cell shape are related to adhesiveness to substratum and are dissociable from growth control in cultured fibroblasts, *Cell* **10**:375.

Wood, P. M., 1976, Separation of functional Schwann cells and neurons from normal peripheral nerve tissue, *Brain Res.* **115**:361.

Yaffe, D., 1969, Cellular aspects of muscle differentiation *in vitro, Curr. Top. Dev. Biol.* **4**:37.

Yaffe, D., 1971, Developmental changes preceding cell fusion during muscle differentiation *in vitro, Exp. Cell Res.* **66**:33.

Yaffe, D., and Feldman, M., 1965, The formation of multinucleated muscle fibers from myoblasts of different genetic origins, *Dev. Biol.* **11**:300.

Yamada, K., 1978, Immunological characterization of a major transformation-sensitive fibroblast cell surface glycoprotein, *J. Cell Biol.* **78**:520.

Yamada, K. M., and Kennedy, D. W., 1979, Fibroblast cellular and plasma fibronectins are similar but not identical, *J. Cell Biol.* **80**:492.

Yamada, K. M., and Olden, K., 1978, Fibronectins: Adhesive glycoproteins of cell surface and blood, *Nature (London)* **275**:179.

Yamada, K. M., and Pastan, I., 1976, Cell surface protein and neoplastic transformation, *Trends Biochem. Sci.* **1**:222.

Yamada, K. M., Spooner, B. S., and Wessells, N. K., 1971, Ultrastructure and function of growth cones and axons of cultured nerve cells, *J. Cell Biol.* **49**:614.

Yamada, K. M., Yamada, S. S., and Pastan, I., 1976, Cell surface protein partially restores morphology, adhesiveness, and contact inhibition of movement to transformed cells, *Proc. Natl. Acad. Sci. U.S.A.* **73**:1217.

Yamada, K. M., Olden, K., and Pastan, I., 1978, Transformation sensitive cell surface protein: Isolation, characterization, and role in cellular morphology and adhesion, *Ann. N.Y. Acad. Sci.* **312**:256.

Yamada, K. M., Hahn, L. E., and Olden, K., 1979, Structure and function of the fibronectins, *J. Supramol. Struct. Suppl.* **3**:187 (abstract).

Zetter, B. R., Chen, L. B., and Buchanan, J. M., 1976, Effects of protease treatment on growth, morphology, adhesion, and cell surface proteins of secondary chick embryo fibroblasts, *Cell* **7**:407.

Zetter, B. R., Johnson, L. K., Shuman, M. A., and Gospodarowicz, D., 1978, The isolation of vascular endothelial cell lines with altered cell surface and platelet-binding properties, *Cell* **14**:501.

Zucco, F., Persico, M., Felsani, A., Metafora, O. S., and Augusti-Tocco, G., 1975, Regulation of protein synthesis at the translational level in neuroblastoma cells, *Proc. Natl. Acad. Sci. U.S.A* **72**:2289.

7

Sensing the Environment
Bacterial Chemotaxis

ROBERT M. MACNAB

1 Introduction

Organisms at all levels of complexity are constantly subject to variations in factors imping-
ing on them from the external environment. It is obviously in the interest of the organism
to match as closely as possible its activity to these variations. Any mismatch can be remedied
by the organism according to one of three general strategies (Fig. 1), namely, (1) modification of the environment, (2) modification of the organism itself, or (3) migration to a
more favorable environment.

All three strategies are utilized extensively by higher organisms. For example, reduced
ambient temperature may stimulate a higher animal to modify its environment by building
shelter or (in the case of man) by wearing clothes or building a fire. Physiological responses
(self-modification) such as shivering, vigorous exercise, or growth of winter fur may also
be involved. And, of course, there is an option to migrate to warmer climates, in many
cases. Similar illustrations could be made of responses to other environmental challenges,
such as abundance or scarcity of certain foodstuffs, and so on.

The first strategy, that of *environmental modification*, becomes less and less feasible
with diminishing organismal size, and is virtually unavailable to single cells of a microbial
species. This is partly a reflection of simpler behavior, but a more important limitation is
probably that of sheer physical scale. Changes that a single cell such as a bacterium could
make on its environment by, for example, uptake or excretion of a chemical compound,
would be overwhelmed by diffusion processes. Concentrated populations of cells may, however, inflict appreciable change on the environment. Such change is usually adverse (deple-

ROBERT M. MACNAB ● Department of Molecular Biophysics and Biochemistry, Yale University, New
Haven, Connecticut 06511

tion of nutrient or accumulation of toxic excretion products), although there are examples of favorable environmental change accomplished by a cell population—for example, the maintenance of reduced concentrations of dissolved oxygen by obligate microaerophilic organisms. In this case, metabolic uptake depletes the oxygen supply faster than it can equilibrate with the air/water interface, and the resulting depleted levels are closer to the optimal environment for these cells.

The second strategy, *self-modification,* is universally used by organisms of all levels of complexity. Changes in the pattern of induction or repression of transport systems and enzymes, and of their activation or inhibition by allosteric regulators, occur in all cells in response to environmental change. Adaptation to the environment can of course also proceed by the longer-term process of mutation and natural selection. Thus, as an example, a hot-spring environment favors the evolution of enzymes that are unusually resistant to thermal denaturation.

The focus of this chapter will be the third strategy of response to the environment, that of *migration*—that is, active and directed motion from an unfavorable environment to a more favorable one. As was indicated above, this strategy is widely used by higher animals, but it is also well developed in the simplest of cells, the bacteria, in the form of migration toward or away from certain chemicals (the phenomenon of chemotaxis), toward oxygen (aerotaxis), and toward light (phototaxis). These phenomena were well known in the late 19th century (see Berg, 1975). Bacteria, because of their relative simplicity, are good subjects for the study of sensory reception and motor behavior at a molecular level, and have been extensively exploited in this regard in recent years. As a result bacterial chemotaxis is, at the present time, probably the best understood system of migratory responses to environmental stimulation, and will be described here as a model from which one may hope to gain insight into behavioral responses in more complex biological systems.

This chapter will attempt to use bacterial motility and chemotaxis as an illustration of an interlocking system of motor and sensory regulation. It does not aim to provide a detailed comprehensive coverage of the subject. The reader interested in more detailed information may consult any of a number of reviews which have appeared recently, for example, Hazelbauer and Parkinson (1977), Koshland (1977), Iino (1977), Parkinson (1977), Silverman and Simon (1977a), Macnab (1978a–c). To place the later aspects of

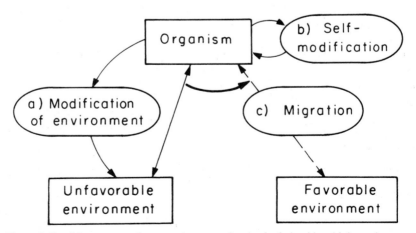

Figure 1. Possible responses of an organism to a suboptimal relationship with its environment.

the chapter in context a brief synopsis of current knowledge (July, 1978) regarding bacterial motility and chemotaxis follows.

1.1 Synopsis

Many bacterial species are motile, and demonstrate tactic responses to light, air, or chemicals. This chapter will only consider flagellar motility, although other types of motility, such as gliding, exist. The bacterial flagellum consists of a helical filament which assembles spontaneously—that is, without enzymic intervention—usually from a single protein. The helical filament is rotated by a motor at its base, using proton electrochemical potential across the membrane as the energy source. The motor is capable of rotation in both counterclockwise and clockwise senses, alternation between senses occurring randomly in the absence of environmental stimulation. Rotation in one sense enables translational motion, whereas rotation in the other sense gives either reversed translational motion, or chaotic tumbling, depending on the pattern of flagellation possessed by the species in question. In any case, a zigzag trajectory of some sort is executed by the cell (Fig. 2a). The effect of environmental stimulation is to modulate the probability of motor reversal so that more time is spent traveling in the favorable direction (Fig. 2b). This type of mechanism is distinct from, and much simpler than, a steering mechanism (see Section 2.2). Chemical stimuli influence motile behavior, not by general metabolic effects, but by binding to specific

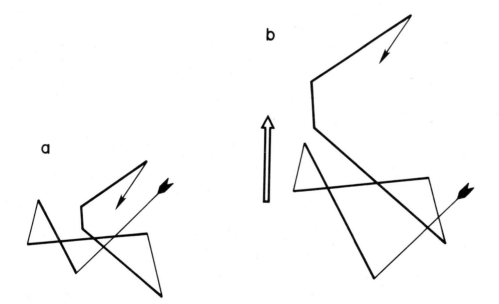

Figure 2. Bacterial motility in the absence and presence of a chemical gradient. (a) In the absence of a gradient, a cell alternately swims and randomly reorients by tumbling. The tumbles occur with constant probability. (b) In the presence of an attractant gradient (open arrow), the cell behaves similarly, but now the tumbles occur with a probability that varies with the current gradient stimulus, which is a function of the current direction of travel. A cell heading up-gradient is less likely to tumble and so a time bias is introduced into the trajectory— that is, the cell migrates. (From Macnab, 1978*b*.)

receptors, which transmit information regarding their state of occupancy to a central data processing system. Whenever the signal regarding current state of receptor occupancy differs from a slowly adapting signal regarding previous state of occupancy, a difference signal is generated, provising a basis for temporal gradient inference. The difference signal modulates motor reversals, and results in chemotactic migration. The slow adaptation component of the signal appears to be the extent of methylation of a set of membrane-bound proteins. The system shows many regulatory features, both with respect to synthesis and assembly, and also function.

There are two levels at which any biological topic may be discussed: the phenomenological level where one simply presents the observed facts, and the teleological level where one attempts to provide some sort of rationale for these facts in terms of survival value. There is no doubt that it is safer to stick to the first level of presentation, but only by consideration of the second level do general principles begin to emerge. The attempt will therefore be made in this chapter not only to describe, but also to consider significance. It is particularly important to consider the constraints under which a particular behavioral system may have evolved. Among the general questions which should be kept in mind while reading this chapter are the following:

1. What range of environments is the organism likely to encounter naturally?
2. What are the consequences of these environments in terms of nutritional benefit or toxicity?
3. What are probable time scales of variations in the environment?
4. What are probable spatial dimensions of these variations?
5. What chemicals are likely to provide useful clues to the availability of a more favorable environment elsewhere?
6. At what concentrations are chemical stimuli likely to provide useful information?
7. How should multiple stimuli be interpreted in a manner useful to the organism?
8. What motile and sensory mechanisms are feasible for an organism of given size and complexity?
9. What is the cost to the organism of the behavioral system, in terms of gene commitment and energy?
10. Are there circumstances in which possession of some or all of the response capability is unnecessary?
11. How does the cell take advantage of any existing machinery that has evolved for other purposes?
12. How does chemotactic signaling take place without introducing undesirable perturbations to other aspects of cell function?

Certain general considerations regarding the nature of the stimulus and the tactic response will be discussed first. This will be followed by an analysis of the underlying molecular events.

2 General Considerations

2.1 The Stimulus

The bacterium needs to obtain information about whether there are accessible environments that contain higher concentrations of nutrients and/or lower concentrations of toxic materials. In higher animals, remote sensing (sight or hearing) is possible, but it is hard to imagine a comparable system for bacteria, and indeed none has been found. Chemical transmission of information—by molecular diffusion—occurs widely in nature. Obvious examples include food aromas, insect pheromones, which attract males over great distances to females of the same species (Seabrook, 1978), and cyclic AMP as an aggregation signal in slime molds (Konijn, 1975). In these instances, the chemicals are relatively small molecules to permit rapid diffusion, and are acting in an informational capacity only, being themselves of no other significance to the organism. Returning now to the case of bacterial chemotaxis, the nutrients or toxic compounds of interest are themselves relatively small molecules, and therefore there is no need to rely on an independent set of molecules to carry out the signaling function. It is not surprising therefore that attractants include common sugars such as glucose, galactose, and ribose, and also amino acids such as serine, aspartate, and alanine (Mesibov and Adler, 1972; Adler *et al.*, 1973). It is particularly interesting to note that not all of the common amino acids are attractants, at least for *Escherichia coli* and *Salmonella*. Since the predominant source of amino acids in any natural context is likely to be from protein hydrolysis, the evolution of individual sensory systems for all the common amino acids would be redundant. The inclusion of serine in the repertoire may reflect a dual-purpose signal, for sensing availability of lipids (exemplified by phosphatidylserine) as well as of protein. Not all readily usable nutrients are attractants—indeed many of the principal intermediary metabolites such as pyruvate are not, possibly because free pools of these are less commonly available than are pools of polymeric breakdown products.

The picture regarding the evolution of repellent responses is far less clear (Tsang *et al.*, 1973; Tso and Adler, 1974). Adverse effects of known repellents such as phenol, indole, or valine can be demonstrated, but there is no satisfactory rationale at present as to why response systems have evolved for these particular compounds and not other toxic substances. Responses to repellents occur only when the repellents are present at high (but subtoxic) concentrations—a useful feature that prevents insignificant levels of these compounds from exerting control over cellular migration.

Presented with several chemotactic stimuli simultaneously, wild-type cells give responses that are roughly the algebraic sum of the responses expected from the individual stimuli (Tsang *et al.*, 1973; Adler and Tso, 1974; Berg and Tedesco, 1975; Spudich and Koshland, 1975), indicating that simultaneous access by several receptor systems to the "central processor" is possible. A slight desensitization is commonly observed, and in certain mutants substantial potentiation effects can be shown (Rubik and Koshland, 1978).

An important but unstudied question is whether responses are regulated by metabolic state—that is, whether the relative weights given to different stimuli are affected by their current importance to the cell. It would seem reasonable, for example, that, under nitrogen starvation conditions, the weight given to amino acid signals would be increased. (Interestingly, amino acids are in general stronger attractants than sugars, possibly an evolutionary consequence of their dual utility as carbon and nitrogen sources.)

How do chemotactic agents exert their effect? Consider first the possibility that agent concentration *per se* modifies motility—that is, that gradient information is unnecessary. In one mechanism of this sort, favorable environments would reduce swimming speed so that cells become trapped. There are, however, a number of difficulties with such a model. An unbounded environment would permit the futile situation of cells swimming faster and faster, as they traveled farther away from their optimum environment. Only if an abrupt change to essentially zero swimming speed in the optimum environment was allowed would there be an appreciable probability of the cells reaching this optimum environment and then being trapped. But how can the optimum environment be defined, inasmuch as it is relative to the other available environments? Arguments of this sort make such a speed-control mechanism seem implausible. The other class of nongradient model would involve modulation of turning frequency, higher concentrations of attractant resulting in a higher turning frequency. This would result in some tendency to stay in favorable environments, provided an integration of information is permitted, because a cell passing a given point from a favorable environment would have a higher probability of reversing than would a cell passing the same point from the opposite direction. This mechanism, however, has the disadvantage that in unfavorable environments the cell would seldom reverse, and therefore general dispersion would predominate over selected migration. Measurements of speed and turning frequency show that they are unaffected by various constant attractant levels, providing an experimental basis for rejecting the nongradient mechanisms discussed above (Berg and Brown, 1972; Macnab and Koshland, 1972).

Gradient-sensing, the next level of complexity above simple concentration sensing, provides a far more satisfactory basis for migrational response. Cell motility can now be made sensitive to the information of real importance to the cell, which is not how good or bad the current environment is, but whether better environments are accessible elsewhere.

Concentrations of chemicals in the environment will in general show both spatial and temporal variations. Temporal variations are irrelevant as direct sensory information to the cell, because it cannot elect to exist at a different point in time. Although a positive temporal gradient of attractant probably indicates that there is a newly available source nearby, this

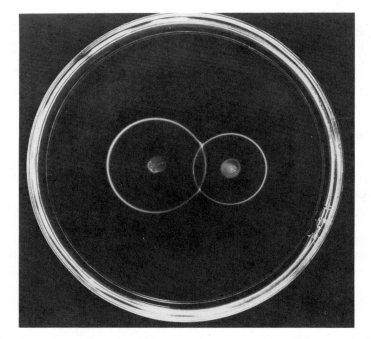

Figure 3. Migration in chemotactic bands, which occurs when bacteria exhaust the local supply of a particular attractant. The cells migrate in the narrow gradient-containing zone between the exhausted medium behind and the rich medium ahead. In this example (Adler, 1976), there are two attractants present, ribose and galactose, and two different *E. coli* strains have been inoculated at separate points on the agar. Because one strain is defective in ribose metabolism and the other in galactose metabolism they form bands that expand independently.

information is not usable because it does not indicate in which spatial direction the source lies. Spatial variations, in contrast, are highly relevant because motile cells *can* elect to exist at a different point in space. One is forced therefore to assume that the *usable* chemical information in the bacterial environment is spatial gradient information.

Spatial gradients may be imposed on the cells by diffusion of the relevant chemicals from a nearby source, or they may be generated by the cells themselves. In the latter case a localized population of cells in an initially uniform environment progressively reduces the concentration of some limiting substrate and then proceeds to migrate outward as a band confined between the depleted and untouched regions. This chemotactic band migration (Adler, 1966) is a very striking characteristic of bacterial behavior (Fig. 3) and constitutes an efficient use of the available resources.

How is spatial gradient information to be acquired by the cell? In the absence of a remote-sensing capability, spatial gradient information is confined to cell dimensions, say 1 μm. Gradients that are likely to be physiologically significant will be very shallow on this distance scale for two reasons: (1) steep gradients could only recruit bacteria within a very limited zone, outside which the receptors would be empty or saturated, and (2) diffusion would quickly collapse any appreciable concentration difference existing over such short distances. In support of these conclusions, it is found experimentally that bacteria have evolved responses to quite shallow gradients. Dahlquist *et al.* (1972) measured a migrational velocity of 2.8 μm sec^{-1} for *Salmonella* placed in a gradient of serine with a concentration change of only 3 parts in 10^4 over the cell body. This is quite a respectable response, considering that the maximum possible migrational velocity can be shown (see the following section) to be half of the instantaneous swimming velocity, or about 12 μm sec^{-1}.

The extreme shallowness of the gradients that can be sensed suggested that the mechanism responsible might be temporal rather than spatial in character. It was stated above that temporal gradient information is useless to the cell, but this statement was only intended to apply to temporal information in a stationary spatial frame of reference. In the case of a moving cell, it is evident that spatial gradient information can be received as temporal information. In fact, provided environmental temporal variations (other than very rapid stochastic fluctuations) are small, and swimming speed remains approximately constant, the temporal information received will be a linear function of the spatial gradient component in the direction of travel. In other words, a cell swimming at v μm sec^{-1} and receiving a temporal signal of dc/dt M sec^{-1} can infer that the spatial gradient in the direction of travel is $dc/v\,dt$ M μm^{-1}. The most direct evidence for temporal gradient sensing came from a series of rapid-mixing experiments (Macnab and Koshland, 1972; Tsang *et al.*, 1973) in which it was observed that "favorable" concentration jumps (increase of attractant or decrease of repellent) suppress tumbling (the smooth response*), whereas "unfavorable" concentration jumps (decrease of attractant or increase of repellent) enhance tumbling (the tumbling response*), with eventual adaptation to unstimulated behavior occurring in both cases (Fig. 4). These observations of response to temporal gradients have been substantiated by many subsequent experiments, such as those of Brown and Berg (1974). It should be noted that a temporal gradient-sensing mechanism does not alleviate the problem of signal size unless concentration difference information accumulates while the cell is swimming over a distance much greater than its size. Cells change direction on average every few seconds, so it would seem at first that the temporal device would need to operate on a time scale smaller than that, say less than 1 sec, to avoid excessive smoothing

*To be discussed in more detail in Section 2.2.

of information between successive segments of a cell's trajectory. With a swimming velocity of about 10 body lengths per second, a 1-sec accumulation time does not yield a particularly large signal.

The bacterium appears to have devised a very clever means of overcoming the apparently incompatible requirements of a signal that is both large and relevant. It was known a number of years ago that tumbling responses were of much shorter duration ($\sim$10 sec)

ATTRACTANTS   REPELLENTS

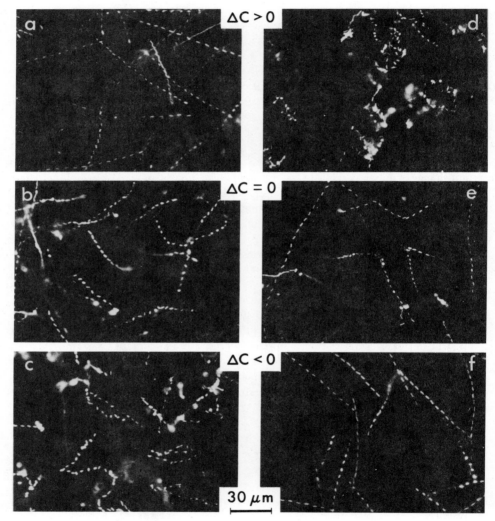

Figure 4. Motility tracks of *Salmonella* in the interval 2–7 sec after exposure to temporal gradients of attractants and repellents. Photographs taken in dark-field with stroboscopic illumination operating at 5 pulses/sec. (a–c) Attractants. (a) Serine increase from 0 to 7.5×10^{-4} M; (b) no concentration change (control); (c) serine decrease from 10^{-3} to 2.4×10^{-4} M. (d–f) Repellents. (d) Phenol increase from 0 to 7.5×10^{-4} M; (e) no concentration change (control); (f) phenol decrease from 3×10^{-4} to 7.5×10^{-5} M. The smooth motility response to "favorable" gradients (a and f) and the tumbling response to "unfavorable" gradients (c and d) eventually give way to the normal motility pattern (b and e). (From Tsang *et al.*, 1973.)

than smooth responses (up to several minutes), indicating substantial asymmetry in the mechanism (Macnab and Koshland, 1972; Tsang *et al.*, 1973)—indeed only the smooth response contributes significantly to the migration in spatial gradients (Berg and Brown, 1972). Berg and Tedesco (1975) subsequently demonstrated, in experiments utilizing a variety of concentration jump sequences, that a cell exposed to an elevated concentration of attractant and then briefly pulsed to a low concentration "forgets" the elevated concentration and gives the full duration of smooth response to the (restored) elevated concentration. Biochemical correlates of this slow adaptation to favorable stimuli and the rapid adaptation to unfavorable stimuli will be discussed in Section 3.4.9. The asymmetry permits a cell to accumulate information for as long as it happens to be traveling in a favorable direction, but to avoid the inappropriate carryover of this information into segments of travel in an unfavorable direction. Note also that there is a positive feedback aspect to the mechanism, in that the longer a cell travels in a favorable direction, the larger is the signal it acquires to promote its further travel in that direction. Cells in moderate gradients (with decay distances of the order of 10 mm) *persist*—that is, progress up-gradient—for mean intervals of the order of 20 sec (Macnab and Koshland, 1973, and unpublished data). It is not known how the effective integration time compares with this value, although Brown and Berg (1974) tested gradient smoothing functions in their analysis of tumbling frequency vs. gradient sensed, and found no improvement of fit, thus arguing against any extensive integration. Berg and Purcell (1977) have calculated that, in typical gradients studied experimentally, integration times of at least 1 sec or so would be needed to yield a measured concentration difference that exceeds the statistical variation expected from the measurement.

Berg and Purcell have shown theoretically another interesting feature regarding sensory reception, namely, that only a relatively small fraction (about 0.1%) of the cell surface need consist of receptors to provide half-maximal precision in measurement of concentration. This rather surprising result derives from the fact that the limiting step in reception is diffusion of molecules to the cell surface. After this, diffusion to a receptor (rather than escape to infinity) is highly probable, even when the number of receptors is relatively small. (It is important, however, that the receptors be well dispersed over the cell surface.) The argument made by Berg and Purcell only applies to the precision of measurement and not to the signal size which naturally will be dependent on the number of receptors. This may be an important factor in the elicitation of a motor response of suitable magnitude.

2.2   The Response

What type of motor response can be expected from the bacterium? The most efficient migrational response, of course, is one in which the organism continually reorients in the most favorable direction and therefore takes a route of steepest ascent up the gradient. Such a response, which occurs in higher cells, such as leukocytes (Zigmond and Hirsch, 1973), and also in many multicellular organisms, requires two sophisticated features that are lacking in bacteria. These are as follows: (1) A means of obtaining sensory information *normal* to the direction of travel, either by spatially differentiated receptors, or by excursions of the cell body in the normal direction. [The latter mechanism, in the form of head waving, is utilized by nematodes (Ward, 1973).] (2) A means of accomplishing specific changes of direction. Such a mechanism would require either a control of the orientation of the flagella or an independent control of the power delivered to each flagellum. A much simpler yet fairly effective response mechanism has actually evolved in the bacterium. It may be crudely expressed by the following "instructions" to the cell: "If the sensory information indicates that the current direction of travel is favorable, hold to this direction for a while longer. If the sensory information is unfavorable, execute a random reorientation maneuver (a tumble) and set off in the arbitrary new direction—the odds are high that it will be preferable to the previous direction." The bacterium, by biasing the duration of its runs according to their direction, accomplishes net migration (Fig. 2). The behavioral response is expressed by modulating the occurrence of just two motility modes, swimming and tumbling. As was indicated in the previous section, the actual stimulus responsible for modulating the motility modes is a temporal gradient of the chemotactic agent.

To obtain some idea of the efficiency of this response mechanism compared to a reorientation mechanism, consider the following: Let the swimming velocity be v. Then a perfect reorientation response would give a migration velocity $v_{mig} = v$. The time-biased random walk mechanism possessed by bacteria is less efficient than this, permitting a maximum response of $v_{mig} = \frac{1}{2}v$. This conclusion is reached by assuming that under maximum gradient stimulation, cells swimming with a positive velocity component in the gradient direction have totally suppressed their tumbling, whereas cells swimming with a negative component tumble immediately. Under these assumptions, all cells will be swimming with a positive velocity component with respect to the gradient direction, and the properly weighted mean velocity in the gradient direction is calculated (Fig. 5) to be $\frac{1}{2}v$. Although the bacterial cell will seldom be driven to this limit [for example, in the case discussed

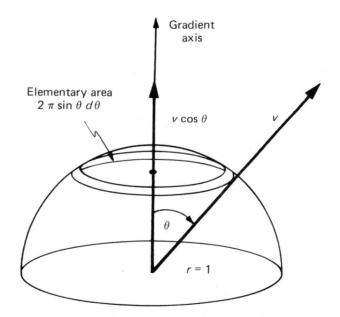

Figure 5. Maximum migrational velocity possible from a time-biased random walk. Any cell that attempts to swim down-gradient is immediately reoriented randomly, and so the population at any instant is uniformly distributed over all possible swimming directions in the up-gradient hemisphere. If the swimming velocity of each cell is v, the population-average migrational velocity in the gradient direction will be

$$\int_0^{\pi/2} v \cos \theta \cdot 2\pi \sin \theta \, d\theta \bigg/ \int_0^{\pi/2} 2\pi \sin \theta \, d\theta = \frac{1}{2} v$$

earlier (Dahlquist *et al.*, 1972), v_{mig} was actually $0.12 \times v$], the calculation shows that the simple mechanism of a time-biased random walk is not a hopelessly inefficient one. This conclusion is readily confirmed by visual observation in a microscope of the rapid accumulation of cells in the vicinity of an air bubble or a diffusing source of a chemotactic agent.

The basis of this response in terms of motor operation will be discussed in Sections 3.4.2 and 3.4.4.

The sole known function of motility in bacteria is to enable cells to reach environments that are optimal with respect to nutrition and toxicity. The chemotactic response is useful whenever the environment is suboptimal, and there is information that a better one is nearby. There can be other contexts in which control of motility would be useful. What, for example, would be useful motile behavior if the envirnment is already close to optimal?—presumably, cessation of motility. A long-term regulatory mechanism, which may well have evolved for this reason, is the repression of flagellar synthesis by elevated glucose concentration (the phenomenon of catabolite repression is discussed in more detail in Section 3.3.2), so that the complex flagellar apparatus is not synthesized when cells are in such a nutritionally rich environment.

As an example at the other extreme, what behavior would be appropriate to an environment that is unfavorable but contains no gradient information relating to better environments nearby? Evidence is beginning to accumulate that under these circumstances the cell displays maximum dispersive behavior, by suppressing the steady-state level of tumbling. Cells that are running out of energy to the motors, whether as a result of oxygen or substrate starvation, high-intensity light (Taylor and Koshland, 1975), or metabolic inhibition (Khan and Macnab, unpublished results) become smooth swimming. This, of course, does not result in selective migration, but by promoting random spreading presumably it enhances the probability that some subset of the population may reach an environment that will support life, whereas in the absence of dispersion the entire population might be annihilated.

Having discussed the chemical and gradient nature of the stimulus and the mechanism of the response at a phenomenological level, we will move on to consider the molecular basis of the chemotactic system.

3 The System at a Molecular Level

3.1 Genetic and Metabolic Investment

Since many, probably most, of the genes involved in motility and taxis in *E. coli* and *Salmonella* have been found and mapped (Fig. 6), perhaps the best way to introduce the topic of molecular mechanism is with a description of the genetics of the system. It should be emphasized, however, that in many instances only the general phenotype, and not the precise gene function, is known at the present time. More detailed reviews of genetic aspects of bacterial chemotaxis can be found in the references given in Section 1. Six broad mutant phenotypes are recognized in the literature: (1) Fla⁻ mutants, which are nonflagellate and therefore nonmotile and nonchemotactic. (2) Hag⁻(H⁻ in *Salmonella*) mutants, involving lesions in the structural gene(s) for flagellin, the protein of the flagellar filament. (3) Mot⁻ mutants, which possess flagella that have apparently normal morphology, yet do not rotate, and so the cells are paralyzed. (4) Che⁻ mutants, which are motile but have more or less

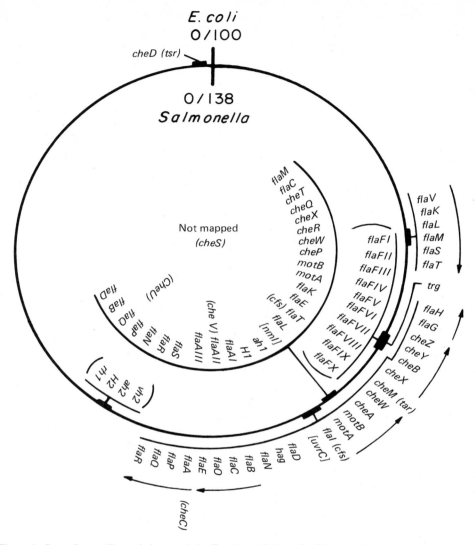

Figure 6. Genes for motility and chemotaxis in *E. coli* and *Salmonella*. The two circular genomes have been drawn coincidentally, with *E. coli* genes written outside and *Salmonella* genes written inside. There are distinct clusters of motility and chemotaxis genes at certain locations on the map. The *E. coli flaV → flaT* region is probably homologous to the *Salmonella flaF* region and is responsible for many of the basal body polypeptides. The *E. coli flaH → flaI* and *flaD → flaR* regions are homologous to the *Salmonella flaM → flaT* and *flaL → flaD* regions, respectively. Homology is not exact, in that some genes may be present in one species and absent in the other. The *Salmonella vh2 → rh1* region is responsible for phase 2 flagellar synthesis (see text) and has no counterpart in *E. coli*. Established operons are indicated by arrows. Undetermined orders are enclosed in round parentheses. *fla:* flagellar synthesis and assembly; *hag, H1, H2:* flagellin; *ah1, ah2, vh2, rh1:* regulation of flagellin phase; *nml:* methyltransferase for ε-*N*-methyllysine of flagellin; *mot:* motility; *che:* chemotaxis; *tsr:* taxis to serine and repellents [MCP(1)]; *tar:* taxis to aspartate and (other) repellents [MCP(2)]; *trg:* taxis to ribose and galactose; *uvr:* UV resistance (unrelated to motility). Receptor genes are not shown. [Based on data presented or reviewed in Iino (1977), Parkinson (1978), Silverman and Simon (1977a), Komeda *et al.* (1978), and DeFranco *et al.* (1979).]

general defects in tactic responses. In most cases the motility of Che⁻ mutants, though vigorous, is not normal; the incidence of tumbles is either unusually low (smooth phenotype) or unusually high (tumbling phenotype). From the discussion of the chemotactic response above, it will be apparent that either phenotype is likely to result in an impaired ability to suitably modulate tumbling in response to chemotactic stimuli. (5) Mutants with limited chemotactic deficiencies. [Certain Che⁻ mutants (Tar⁻ and Tsr⁻ phenotypes) fall into this category instead of category 4).] Mutants with limited chemotactic deficiencies are unable to respond to two or more classes of chemotactic agents—ribose and galactose, for example, or aspartate, maltose, and certain repellents. The lesion is presumed to lie in a secondary signaling component, rather than in the primary receptors. (6) Mutants with a specific chemotactic deficiency, the lesion being in the gene for a specific primary receptor.

So far a total of about 40–50 relevant genes have been identified for *E. coli* and *Salmonella:* 19 and 26 *fla* genes, respectively; 1 *hag* and 2 *H* + 4 regulatory genes for phase change (see below), respectively; 2 *mot* genes; 7 and 8 *che* genes, respectively; the *trg* gene; and about 9 receptor genes (Iino, 1977; Parkinson, 1977; Silverman and Simon, 1977*a*; and references cited therein). To this number should be added perhaps 16 more receptor genes and 3 or 4 secondary component genes, which have been characterized on the basis of specificity and competition, but for which the genes have not yet been found. Thus the bacterium has invested upward of 60 genes in order to be capable of versatile chemotactic responses. This number, though large, is still less than 2% of the estimated total gene complement, a small price the cell has been forced to pay because it may buy survival. It is not easy to make an intelligent *a priori* assessment of the survival value of chemotactic capability in a natural ecosystem, but perhaps the best assessment has been made for us in this way by the bacterial cell: chemotactic capability is worth almost 2% of its genetic capacity. In one recent study addressing this question of the importance of chemotaxis for survival, Allweiss *et al.* (1977) found that motile but nonchemotactic bacterial mutants had significantly impaired ability to invade mammalian mucosal tissue slices, and they suggest that chemotaxis may play an important part in the infectivity of pathogens such as *Salmonella* and *Vibrio,* as well as the maintenance of indigenous intestinal flora such as *E. coli.*

While on this subject of "cost/benefit analysis" we may note that although the vigorous motility of a bacterial culture is the most striking evidence of energy expenditure, and indeed of life, to an observer using the light microscope, the energy expenditure is actually quite small, about 10^{-17}–10^{-16} W (Coakley and Holwill, 1972) or less than 0.1% of the total useful energy expenditure of the cell (Macnab, 1978*c*). This calculation, however, refers only to energy of propulsion. Additional energy expenditures are involved in the construction of the motor and sensory apparatus as well as in sensory processing.

3.2   Overlap with Other Aspects of Cell Function

We will now consider whether components of the chemotactic machinery are shared with other aspects of cell function. Any such sharing would represent a genetic and biosynthetic economy and also would make the evolution of the system easier to contemplate (see Section 3.3.4). Sharing of components is likely to have a price associated with it, in terms of loss of exclusivity or compromise of suitability. An obvious indication that the chemotactic system is largely an autonomous entity within the cell is the existence of massive clusters of *fla, che,* and *mot* genes, with several operons (Fig. 6).

Synthesis and function of the chemotactic system of course utilizes many of the general biosynthetic and degradative pathways of the cell. Similarly, the energy source for motility (protonmotive force across the cytoplasmic membrane) does not serve this purpose exclusively, but drives other membrane-associated processes such as transport. These are illustrations of sharing in a very general sense, but there is in fact one example of sharing of quite specific components. Primary receptors for taxis toward sugars have in a number of cases been shown to be primary receptors for the corresponding transport systems (Hazelbauer and Parkinson, 1977). This is hardly surprising, since the compound that is acting as a signal to the cell to modulate its motility behavior is also a nutrient that must be transported before it can be utilized. Indeed, the transport and utilization of the compound is the whole point of the signaling system in the first place. It is known that only the recognition event, and not the transport itself or the subsequent metabolism, initiates chemotactic signaling. Wild-type cells presented with nonmetabolizable analogues, and transport-defective mutants, are both capable of normal tactic responses to the corresponding agent (Adler, 1969). The details of how a periplasmic protein, after specifically binding a given sugar (say ribose), conveys this information to the chemotactic signaling system as well as carrying the sugar molecule itself to the transport system are not known. The mechanisms must differ appreciably as receptor mutants may be defective with respect to taxis but not with respect to transport, and vice versa (Ordal and Adler, 1974). Hazelbauer (reported in Hazelbauer and Parkinson, 1977) has recently succeeded in forcing the "evolution" of a chemotaxis system for arabinose in an organism that already possessed an arabinose-binding protein that functioned in arabinose transport. The sharing of a chemoreceptor between the corresponding transport and taxis systems raises the ironic possibility that the two systems might be in competition for the chemoreceptor, in which case the capacity for transport would be diminished by the chemotaxis system it is supposed to serve. It seems unlikely, however, that competition resulting in drastically reduced transport would have evolved.

The nature of receptors for amino acids and repellents is still unclear because, although specificity of stimulus has been demonstrated, the molecular components responsible for this specificity have not been identified. Repellents present a peculiar problem in this regard because it seems highly unlikely that a specific binding protein for phenol, say, should exist for the purpose of transport. One therefore presumes that the corresponding receptor evolved for taxis only. Alternatively, repellents could be analogues of compounds that are (or once were) useful for the cell to transport.

As we have stated, the majority of the chemotactic apparatus appears to have evolved uniquely for the purpose of chemotaxis. Where signaling aspects are concerned, this may reflect a need to avoid "cross-talk" between different aspects of cell function. Suppose that a compound as centrally important as ATP had to act as a signal for the chemotaxis system. Then all ATP-dependent processes would be subject to fluctuations imposed on them by chemotactic signals in ways that would not necessarily make good metabolic sense. For example, a cell might receive a signal saying "aspartate concentration is increasing," yet the absolute concentration of aspartate might still be too low to justify a major shift in ATP-dependent metabolism. There are indications that one central parameter in the cell, namely, membrane potential, may participate in chemotactic signaling. Szmelcman and Adler (1976) found a transient membrane hyperpolarization associated with chemotactic stimulation. Others have failed to corroborate this (Miller and Koshland, 1977). If the results prove to be correct, it would seem essential that the fluctuations in membrane potential associated with chemotactic signaling should be small in magnitude or, if they are not, that other machinery in the cell that is dependent on membrane potential (such as ATP synthesis and catabolite transport) be effectively buffered from the fluctuations.

It should be noted that, although the role of membrane potential as a signal mediating specific chemotactic stimuli remains unsettled, it is clear that circumstances that cause fluctuations in the energy state of the cell do themselves act as tactic stimuli. This is almost certainly the basis of the aerotactic response, the phototactic response of photosynthetic organisms, and also responses artificially induced (for example, by proton ionophores, Mg^{2+}, and high-intensity blue light). A system capable of response to general cues regarding the metabolic well-being of the cell in addition to specific chemical information is obviously advantageous.

3.3 Regulation of Synthesis and Assembly

3.3.1 Sensory Components

Synthesis of a number of primary receptors of the chemotaxis system is inducible by the corresponding substrates (Adler et al., 1973). This inducibility permits the cell to utilize its limited biosynthetic resources and its surface area for reception to best advantage in terms of currently available molecules, where "current" refers to the time scale of one-half hour or so, which is characteristic of the induction process. Thus flexibility on this time scale is gained at the expense of ability to recognize transiently available molecular species. Receptor inducibility probably evolved in the context of transport anyway and so may not convey net advantage to the chemotaxis system.

Little is known yet regarding either the synthesis or assembly of other components of the sensory system, such as the various *che* gene products, although many of the sensory genes are under the same overall control (*flaI*) as the flagella (see Section 3.2.2). In at least two instances, expression of chemotaxis genes occurs in an operon (Silverman and Simon, 1976; Parkinson, 1978) and it seems likely that other examples of genetic regulation of this type will emerge as studies continue.

3.3.2 Motor Components

The motor system appears to be highly regulated, as might be expected for such an intricate structure (Fig. 7)—at least 11 structural genes [1 (2 in *Salmonella*) for flagellin, 1 for the hook protein, and 9 or more for different proteins in the basal body] are involved (Hilmen and Simon, 1976). A number of operons (for example, the entire *flaV–flaT* region of *E. coli*) participate in this regulation (Komeda et al., 1977). The process of flagellar synthesis appears to correlate with DNA replication, as temperature-sensitive replication mutants have defective flagellation at the restrictive temperature (Nishimura et al., 1975). A further overall control is provided by cyclic AMP, the virtually universal "second messenger" in biological systems (Bitensky and Gorman, 1973). Acting in complex with a cyclic AMP receptor protein (CRP, the product of the *crp* gene) it is thought to promote expression of the *E. coli flaI* (*Salmonella flaT*) gene (Yokota and Gots, 1970; Silverman and Simon, 1974a; Komeda et al., 1975). Certain alleles of *flaI* that are defective in a locus termed *cfs* obviate the cyclic AMP–CRP dependence, and result in constitutive flagellar synthesis. In wild-type cells cyclic AMP levels, and hence flagellation, fall off at elevated glucose concentrations—the phenomenon commonly known as catabolite repression.

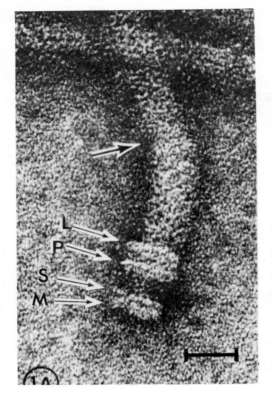

Figure 7. Intact isolated flagella from *E. coli*, negatively stained with uranyl acetate. Bar equals 20 nm. The basal body consists of 4 rings (M, S, P, L), which lie, respectively, in the cytoplasmic membrane, above the cytoplasmic membrane, in the peptidoglycan layer (cell wall), and in the lipopolysaccharide layer (outer membrane). The rings are connected by rods and cylinders. The M and S ring complex is likely to be the motor in which rotation originates. The hook between the basal body and the proximal end of the filament (unlabeled arrow) may be a flexible coupling. (From DePamphilis and Adler, 1971.)

No basal body structures are detected in the absence of *flaI* expression. This gene thus acts as a master locus for the synthesis of flagella and at least some signaling components of the chemotaxis system.

Until recently, it was believed that basal body assembly was so tightly regulated that assembly of defective structures was not possible. This has now been shown to be incorrect. Suzuki *et al.* (1978) demonstrated the existence of partial assembly mutants, but even for the simplest structures detected (lacking the outer rings, the hook, and the filament) some 14 functionally normal genes were still needed. From the additional features that appeared in strains possessing more and more of the normal gene complement, a first assembly scheme has been proposed (Fig. 8). The final addition of the external filament appears to be absolutely dependent on an intact hook/basal body complex. In this case, where mRNA for flagellin (the filament subunit protein, coded for by the *hag* or *H* gene) is absent in Fla⁻ strains, control is known to be at the synthetic level rather than at the assembly level (Suzuki and Iino, 1975). There is recent evidence from studies of ribonuclease mutants that posttranscriptional modification of mRNA may be important in motility (Apirion and Watson, 1978).

A remarkable example of regulation of flagellar assembly exists in the bacterium *Caulobacter* (Shapiro, 1976). This species grows and sheds a single polar flagellum at precise time points in the cell-division cycle, under the control of a cyclic GMP-dependent mechanism.

Among the questions that come to mind regarding the regulation of flagellar assembly are the following: (1) Does it require energy and/or enzymic mediation? The answer is not known for the hook or basal body, but the filament is known to readily self-assemble even in simple buffers and therefore to require neither enzymic control nor coupling to an exergonic process such as ATP hydrolysis (Asakura, 1970; Iino, 1974). (2) Is the stoichiometry of the components an intrinsic feature of the structure—that is, is it determined by structural constraints? If not, what determines it? Here again most is known about the filament. It is known that, provided monomer is not limiting, polymerization of filament *in vitro* occurs at a constant rate, with a constant probability of termination by faulty addition (Hotani and Asakura, 1974). Growth rate *in vivo* exponentially decreases as a function of length (Iino, 1969). It is known that addition of flagellin monomer occurs at the end distal to the cell (Iino, 1969; Emerson *et al.*, 1970), presumably by extrusion through the filament core. The exponentially decaying kinetics are most simply explained by a rate-limiting transport of monomer, without the need to invoke regulation of any more elaborate kind. Dark-field microscopy reveals a considerable number of detached filaments in the medium, probably as the result of simple mechanical breakage. This may be a factor ruling against extreme elongation. The filament is seen to be self-limiting, but of arbitrary length. The length of the hook, in contrast, is under genetic control (by the *flaE* and *flaR* genes of *E. coli* and *Salmonella*, respectively). When the responsible gene is defective, intermediate "polyhook" structures result (Silverman and Simon, 1972). Very little is known about the stoichiometry of the basal body, but as no morphological features of variable size or length have been detected so far one assumes that this stoichiometry is fixed. At least in the case of closed structures such as the rings, it seems likely that the number of subunits may be determined simply from the packing geometry.

3.3.3 Phase Variation in Salmonella

Salmonella exhibits a curious phenomenon in the regulation of its flagellar synthesis (Stocker, 1949). It has been called *phase variation* and is absent in the closely related enteric bacterium *E. coli*. It can be simply described at a superficial level—genetically homoge-

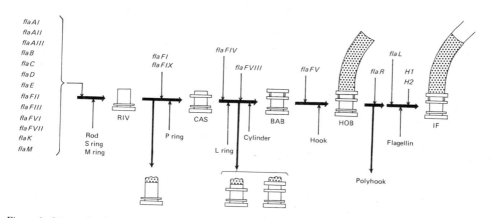

Figure 8. Scheme for flagellar assembly in *Salmonella*. The *fla* genes necessary to reach each successive stage are indicated. RIV ("rivet"), rod–inner ring complex; CAS ("candlestick"), RIV–P, ring complex; BAB, basal body; HOB, hook–basal body complex; IF, intact flagella. (From Iino, 1977, based on data of Suzuki *et al.*, 1978.)

neous cells can make either of two antigenically distinct flagellar filaments. A given cell at any given time makes filaments of only one type, but has a probability of about 10^{-3}/cell division of switching to synthesis of filaments of the other type. This switching is much higher than spontaneous mutational frequencies, but is attained in certain recombinational events such as the insertion of drug-resistance factors (Kleckner, 1977). The genetic basis of the phenomenon is partially understood. When the operon controlling the *H2* flagellar structural gene is being expressed, a repressor for the alternate (*H1*) gene is produced (Fig. 9). The gene responsible for turning on the *H2* operon is *vh2*, which exists in an oscillatory state, with a frequency of conversion between the "on" and "off" states of about 10^{-3}/cell division (Iino, 1977; Silverman and Silmon, 1977a). These two states have recently been shown (Zieg *et al.*, 1977) by heteroduplex analysis to differ by a recombinational inversion of the *vh2* gene, or a locus very close to it on the genome (Fig. 10). Strictly speaking, therefore, the cells in different phase are not genetically identical as was stated above. It is not known why *vh2* is particularly recombination-prone—even more puzzling is the question of why this elaborate regulatory machinery should have evolved in the first place. No difference between filaments of the two types has been noted as far as motor function is concerned, provided both genes are wild-type, yielding the normal waveform for the filament (see discussion of polymorphism in Section 3.4.4). The fact that the filaments are antigenically distinct raises the possibility of a counterdefense mechanism against the immune response of a host, but this seems plausible only in the unlikely context of an initial challenge of the host by only one of the antigenic types.

3.3.4   *The Puzzle of Evolution*

At this point, a digression to consider the process of evolution of organelles and organs seems appropriate. The flagellum constitutes a particularly clear illustration of the problem of how an organelle evolves. The usual "explanation" that the organelle evolves by a series of refinements, is readily acceptable where the entity under consideration is already a refined and sophisticated version (for example, the eye of a higher animal) and where simpler ancestral structures can therefore be imagined. But what of a structure such as the bacterial flagellum? It seems like quite an intricate structure (Figs. 7 and 8) but, consisting

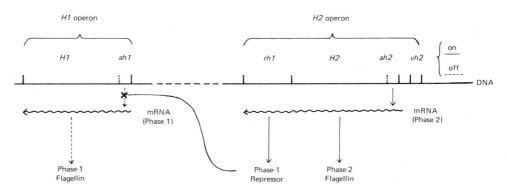

Figure 9. Regulation of flagellar phase in *Salmonella*. The *H1* and *H2* structural genes are alternately expressed with a frequency of alternation of about 10^{-3} per cell division. *vh2*, in the *H2* operon, oscillates between two states (see also Fig. 10). The "on" state activates the *H2* operon and represses the *H1* operon; the "off" state fails to activate the *H2* operon and fails to repress the *H1* operon. (From Iino, 1977.)

of a single rotary mechanism, an anchor in the cell surface, and an external helical screw to generate thrust, it would seem to approach a minimal design—in other words, it is hard to imagine any significant reduction in complexity that would still permit motor function of any sort. It is equally hard to think of any other useful function for an evolving, but not yet propulsive, flagellum that could confer an interim selective advantage on the cell.* How then do 10 or so structural genes (not to mention regulatory genes) coevolve to yield a flagellum? The question is even more vexing because the flagellum, although useful, does not approach other organelles or macromolecular assemblies such as ribosomes in terms of central importance to survival, and therefore one cannot invoke a massive evolutionary pressure for flagellar development.

Clearly there are large gaps in our present concepts of molecular evolution. One factor in evolution, which has perhaps been underestimated until recently, is the functional versatility of the sequences of polynucleotides, and the polypeptides for which they code. Examples of this versatility are found in viruses such as ϕX174 and G4, where a given nucleotide sequence can be read in all three frames and still give rise to functionally useful proteins (Godson *et al.*, 1978). Another quite different example is the extensive sequence homology that has been observed between certain apparently unrelated enzymes, such as

---

*Proton-driven transport systems might conceivably be candidates for such an earlier function.

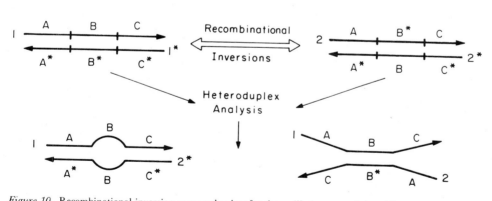

Figure 10. Recombinational inversion as a mechanism for the oscillating state of the *vh2* gene, which controls flagellar expression in *Salmonella* (see also Fig. 9). (a) Heteroduplex analysis of cloned DNA fragments containing the *H2* operon reveals nonannealed bubbles (left) and nonannealed flanks (right). (b) Reversible recombinational inversions in region B (presumed to be *vh2*) would alternately produce the duplexes 1–1* and 2–2*. A mixed population of these duplexes then subjected to denaturation and renaturation would yield heteroduplexes 1–2* and 1*–2, with nonhomology in the B region, and 1–2 and 1*–2*, with nonhomology in the flanking regions. (Based on data of Zieg *et al.*, 1977.)

β-galactosidase and dihydrofolate reductase (Hood *et al.,* 1978). It may be that in the case of the bacterial flagellum, the proteins from which it is constructed evolved from a variety of proteins fulfilling useful functions for the cell, but functions totally unconnected with motility.

3.4 Regulation of Function

As is evident from much of the foregoing discussion, the crucial regulatory feature of the entire chemotactic system in peritrichous* bacteria like *Salmonella* or *E. coli* is the regulation of the occurrence of swimming and tumbling modes, the means by which travel and reorientation are accomplished. At the molecular level some important questions to be answered are (1) How does the motor organelle function in each of the two modes? (2) What parameter places the motor in one or other mode? (3) How is the parameter regulated? (4) How is regulation influenced by sensory information from a wide variety of environmental stimuli?

3.4.1 Mechanism of the Flagellar Motor

The fundamental mechanism of the flagellum has now been established to be a rotation of the exterior filament by the basal body (Berg and Anderson, 1973; Silverman and Simon, 1974*b*; Larsen *et al.,* 1974*a*; Berg, 1974). Rotation has been demonstrated in a variety of ways, such as tethering of cells to glass by the filament and observing body rotation, or labeling with latex beads and observing rotation of these beads. The idea of a rotary motor initially met with considerable intuitive resistance on the part of some scientists, but it is unclear why they were uncomfortable about such a mechanism. There is no fundamental conceptual difference between the mutual sliding of elements in a linear fashion (as in muscle or eukaryotic cilia) and sliding in a closed circle (as in bacterial flagella).

The structure of the basal body (Fig. 7) has cylindrical symmetry, if we ignore for the moment the subunit structure. Both the apposed M and S rings, and the coaxial cylindrical surfaces of their vicinity, have a geometry that would be suitable for a device required to generate cyclic torque. The structure is embedded in the cell membrane, across which protonmotive force (electrochemical potential of protons) is developed by the energy-producing reactions of the cell—respiration and ATP hydrolysis (Harold, 1977). Motor function is dependent on the existence of an inwardly directed protonmotive force, but not on how it was derived; thus either respiration or the hydrolysis of ATP may (indirectly) drive the motor (Larsen *et al.,* 1974*b*). Motility can even occur in nonmetabolizing cells, as a result of artificially induced proton potential (Manson *et al.,* 1977; Matsuura *et al.,* 1977). It is not known how protons drive the motor (or for that matter whether they do so directly, although several other ionic species have been ruled out as intermediates), but obviously the mechanism cannot utilize acceleration of protons in an electrical field; the motor can run in the absence of an electrical field ($\Delta\psi$) across the membrane, provided an adequate proton chemical potential (ΔpH) exists. We are thus led to consider mechanisms in which the energy of the proton is conserved by means of chemical bonding. Binding at a site on the exterior of the cell membrane and subsequent release on the interior must proceed by a mechanism that is obligatorily linked to motor rotation. One way in which this might be

---

*Peritrichous: Possessing flagella, arbitrary in number and originating from random points on the cell surface.

accomplished, proposed by Laüger (1977), is by a proton-specific channel constructed from surfaces of two different structural elements in apposition. The geometry of the two surfaces would be such that mutual rotation is required to permit a proton to traverse the channel. An example of a suitable geometry is shown in Fig. 11. The specificity for protons could derive from a chain of acid–base or hydrogen-bonding sites. Because the geometry of the basal body is not, strictly speaking, cylindrical but does have subunit structure, it is reasonable to postulate that there might be definable rotational steps in the operation of the motor, and a search for such steps is being undertaken by Berg and co-workers (Berg, 1976).

3.4.2 Reversibility

The same series of experiments that demonstrated the rotational nature of the motor also demonstrated that it is reversible—that is, it can rotate in either a counterclockwise or a clockwise sense. Positive chemotactic stimuli enhance counterclockwise rotation relative to clockwise rotation; negative stimuli have the opposite effect (Larsen *et al.,* 1974a). How can we envisage this to happen in molecular terms?—not, presumably, by reversal of polarity of the protonmotive force, because this is used for purposes other than motility. The availability of channels of different handedness—for example, a left-handed channel for counterclockwise rotation and a right-handed channel for clockwise rotation, would permit reversal (Fig. 11). The left-handed and right-handed channels might be interconverted by conformational change (Laüger, 1977), or they might coexist but be subject to mutually exclusive inhibition.

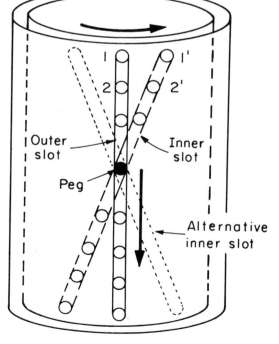

Figure 11. A model for proton translocation coupled to flagellar rotation. Consider first a mechanical model consisting of two coaxial cylinders, the outer with a slot parallel to the cylindrical axis, the inner with a slot tilted to the axis. A peg spanning both slots can move down only if the inner cylinder rotates with respect to the outer one. In the molecular case, discrete "divalent" sites 1–1′, 2–2′, etc., for protons would have the same effect. Reversal of rotation could be accomplished by utilizing an alternate inner "slot" tilted in the other direction. Laüger (1977) proposed a model of this sort with radial rather than cylindrical geometry.

ROBERT M. MACNAB

3.4.3 Role of mot Genes

The role of the *mot* gene products in motor function is still unclear—they are necessary for motility, but (although membrane-bound) are not part of the basal body, at least as currently defined. Some plausible suggestions for their function are: (1) energy-transducing devices, remote from the motor, needed to convert protonmotive force into another energy form that the motor actually uses, say the electrochemical potential of some ion other than H^+; (2) an "on/off" switch binding to the motor to enable it to utilize protonmotive force; (3) a labile element of the motor constituting part of the protonmotive force transducing apparatus; or (4) a structure surrounding the motor and needed for physical reasons to enable it to rotate or to transmit that rotation to the filament.

3.4.4 Flagellar Function in Swimming and Tumbling

Next we consider the consequence of counterclockwise and clockwise motor rotation on the behavior of the filaments and ultimately of the cell. It is necessary first of all to review some structural characteristics of the filaments themselves. They are helical, not because of active bending, but because that is their intrinsic structure—recall the *in vitro* reconstitution experiments mentioned in Section 3.3. The basic stacking pattern of the subunits makes a hollow tube rather like a brick smokestack (see O'Brien and Bennett, 1972), but this tube is not straight as one might have expected in view of the fact that all the subunits are the same. The fact that the tube is helical can only result if the subunits, although identical in primary sequence, are not quite identical in conformation or stacking geometry—the concept of quasi-equivalence in macromolecular assembles (Klug, 1967). It is found experimentally that flagellin, depending on conditions such as pH and ionic strength, is capable of polymerizing into a variety of discrete helical structures, each with its own characteristic parameters of wavelength, pitch angle, and handedness (Kamiya and Asakura, 1976). A plausible model for these structures has been developed by Calladine (1976, 1978). For our present purposes, it is only necessary to note that one way of converting a filament from one structure to another structure within the family is by subjecting it to torsion—for example, by rotation of the filament in a medium that provides viscous resistance (Macnab and Ornston, 1977; Hotani, unpublished). Two structures are found to predominate in motile cells: (1) "normal," a left-handed helix of wavelength 2.4 μm, and (2) "curly," a right-handed helix of wavelength 1.1 μm.

Counterclockwise rotation of the filaments, which are intrinsically left-handed helices (being in the normal waveform), causes them to operate as a bundle with a waveform traveling from base to tip so that the cell is pushed from behind—the behavior called *swimming* (Anderson, 1975; Macnab, 1977). The torsion from viscous resistance is left-handed, and so the structure remains stable. Clockwise rotation, on the other hand, applies right-handed torsion to the filaments, and initiates drastic changes in their quaternary structure, progressively converting them to the totally different right-handed curly helical form. It is the *transition* between structures that results in tumbling behavior of the cell (Fig. 12). Prolonged clockwise rotation as well as prolonged counterclockwise rotation should in fact give swimming behavior because as soon as all the filaments have been converted to the curly form they will be of the handedness appropriate for bundle formation in this sense of rotation. We may expect therefore that proper tactic responses will require quite precise regulation such that only brief periods of clockwise rotation are permitted. This expectation has recently been confirmed by the behavior of certain *Salmonella* mutants, which, in spite of "correct" sensory logic, swim toward repellents and away from attractants (Rubik and Koshland, 1978). The lesion in these mutants is found to be in the steady-state set point between counterclockwise and clockwise rotation. Whereas in wild-type cells this set point is predominantly counterclockwise, in the mutants it is predominantly clockwise (Khan *et al.*, 1978). The mutants are found to be swimming in the inverse mode, as a consequence of prolonged clockwise rotation, and therefore they tumble when a positive stimulus (attractant increase or repellent decrease) causes a change to counterclockwise rotation. Thus the unstimulated motility, although superficially similar to wild-type motility, is actually a mirror image of it and so chemotactic behavior becomes inverted (Fig. 13). This example serves to emphasize the subtlety of regulatory processes, even in so simple an organism as a bacterium.

The peritrichous species, which display this elaborate tumbling behavior with structural transitions in the flagella, are not the only ones that can migrate. Many other species do so by simple reversals of swimming direction modulated by chemotactic stimulation (Seymour and Doetsch, 1973; Taylor and Koshland, 1974). It is legitimate therefore to ask why the more elaborate behavior of swimming and tumbling has evolved. Two reasons, admittedly both speculative, come to mind: (1) Peritrichous flagellation appears to convey a definite advantage for motility in very viscous media or at solid–liquid interfaces (Henrichsen, 1972). It has been shown (Macnab, 1977) that simple reversal of peritrichous bundles, without structural transitions, would generally result in jamming and hence failure to reorient in a tactic response. Therefore the transitions appear to be necessary for such peritrichous species. (2) Reorientation by simple reversal of swimming direction is dependent on fortuitous asymmetries and Brownian rotational diffusion. Tumbling would seem to be a much more rapid and thorough means of exploring all cell orientations with respect to the environment.

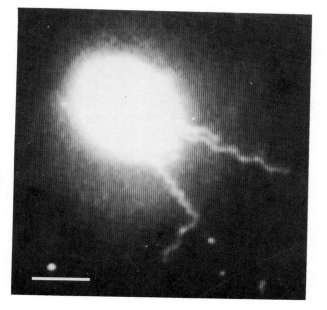

Figure 12. A cell of *Salmonella* showing a flagellum in the process of transition between the normal left-handed helical structure and a short wavelength "curly" right-handed structure. Such transitions, caused by viscous torsion during motor reversal, are important in generating tumbling behavior. Dark-field illumination using a 500-W xenon arc lamp; recorded with a silicon intensifying target Vidicon camera. (From Khan *et al.*, 1978.)

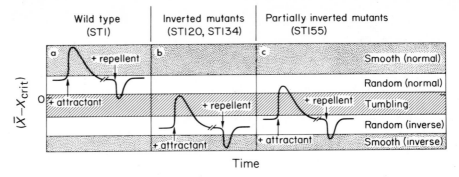

Figure 13. A "behavioral spectrum" of motility and chemotaxis in *Salmonella*. Cells have available to them two senses of motor rotation, counterclockwise (CCW) and clockwise (CW), and two senses of filament structure, left-handed (LH) and right-handed (RH) helicity. CCW rotation tends to stabilize the LH structure; CW rotation, the RH structure. Either CCW rotation of LH flagella or CW rotation of RH flagella can produce swimming. Reversal of rotation causes tumbling only during the brief period while the flagella are in transition between LH and RH forms. Therefore either predominantly CCW rotation with brief CW rotation or predominantly CW rotation with brief CCW rotation will give "random motility"—that is, swimming punctuated by tumbles. Frequent alternation between brief CCW and brief CW rotation will give endless tumbling. The steady-state CCW/CW distribution is determined by the mean level $\overline{X}$ of a response or rotation regulator relative to a reference value X_{crit}. Wild-type cells have random motility because they have a predominantly CCW distribution; "inverted" mutants have random motility because they have a predominantly CW distribution. Positive chemotactic stimuli drive the regulator up, causing a smooth response in wild-type cells but just the opposite response—that is, tumbling—in the mutants. Negative stimuli drive the regulator down, causing a tumbling response in wild-type cells but a smooth response in the mutants. (Based on Khan *et al.*, 1978.)

3.4.5 The Tumble, or Rotation Sense, Regulator

What parameter places the motor in counterclockwise or clockwise rotational mode, and what determines the state of this parameter? Counterclockwise rotation seems to be the default mode, since most mutations in the chemotaxis system, especially null ones, result in smooth swimming; double mutants constructed from parents of smooth and tumbling phenotype, respectively, are also found to be smooth (Parkinson, 1978). Blocking of the chemotactic methylation system (see Section 3.4.9), or energy deprivation have the same effect.

It is likely then that the motor's primary sense of rotation is counterclockwise and that something additional must happen for it to rotate clockwise. The factor or factors responsible for placing it in clockwise mode have not yet been determined, but one can think of various possible categories.

1. *Inversion of the polarity of the energy source.* As discussed earlier, this seems unlikely unless there is a compartmented energy source that is dedicated to motor function only. No evidence has been found for such compartmentation in the species we are considering.

2. *Transient membrane potential change.* The uncertain status of this possible candidate for the regulating factor was discussed earlier. Although clockwise rotation is normally transient, mutants have been described (Rubik and Koshland, 1978; Khan *et al.*, 1978) in which it is virtually continuous. The behavior of such mutants would be hard to reconcile with an intrinsically transient signaling mechanism. This statement does not of course preclude transient membrane potential change

(as might occur with a sudden pH shift) as one means of *affecting* the regulating factor.

3. *Direct macromolecular interaction between sensory components and the flagella.* Far too many copies of receptors and other sensory components exist for all of them to be in direct and permanent association with the flagella. For example, although there are only 5–10 flagella per cell, there are 6×10^4 copies/cell of the galactose chemoreceptor (Boos, 1974) and at least 700 copies/cell of the proteins of the methylation system (Kort *et al.*, 1975), but a subset of them might so interact, possibly employing some kind of antenna mechanism.

4. *Molecular species capable of diffusion to the motor and binding to it.* This is perhaps the most plausible candidate at the present time. Ordal (1977) finds that calcium ion is necessary for clockwise rotation in *Bacillus subtilis,* and he suggests that it may bind to the switching element of the motor. Another possibility would be a diffusible component of the methylation system. There is no information yet as to whether the tumble regulating factor, if it is a diffusible molecule, is cytoplasmic or is in permanent association with the membrane.

Whatever the factor is, it must be present in unstimulated cells at close to a critical value, and must be subject to appreciable fluctuation about that value, since frequent spontaneous alternation between counterclockwise and clockwise rotation occurs (Fig. 14). The alternation is stochastic, both counterclockwise and clockwise rotation intervals occurring with a Poissonian duration distribution (Berg and Tedesco, 1975). If the tumble-regulating factor is indeed a molecule that binds to the motor, a small pool size would result in appreciable stochastic fluctuations, and these fluctuations ought to occur independently at each motor. If the pool size is large, but fluctuates because it is determined by some other factor, the motors might be expected to show a high degree of synchrony. Clearly, when a cell is swimming, all the flagella in the bundle must be rotating in the same sense, but it is not

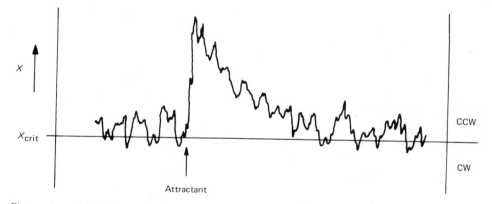

Figure 14. Model for the rotational response regulator of the flagellar motor. The motor is placed in CCW rotation if the level of the regulating parameter X exceeds a critical value X_{crit}, and in CW rotation if X falls below X_{crit}. In wild-type cells the mean steady-state value of X lies above X_{crit}, but stochastic fluctuations in X frequently take it below this level. The motor distributes between CCW and CW rotation with CCW predominating, and swimming punctuated by tumbling results. Attractant raises X to a level at which spontaneous fluctuations never result in its dropping below X_{crit}, and a smooth swimming response is displayed until adaptation occurs. The figure differs from Fig. 13 only in presentation. Figure 13 shows mean values of X and CCW/CW distribution, whereas here the instantaneous values of X and consequent CCW or CW states are given explicitly.

known whether or not flagella normally undergo synchronous reversal when they are freed from direct mechanical interactions. Nonsynchronous reversal has been seen in *Salmonella* (Macnab, unpublished observations), but the observations are technically difficult, and need to be quantitated.

Finally, it is worth reemphasizing that the motor is a simple two-state (counterclockwise/clockwise) device. Therefore, the observed graded responses (Dahlquist *et al.*, 1972) cannot result from a gradation in the state of the motor, but must result from a gradation in the distribution of time spent in the two discrete states of the motor. This may well be the most rudimentary class of mechanisms capable of producing a graded response.

3.4.6 The Sensory System—Excitation and Adaptation

Finally, we turn to the organization and molecular nature of the sensory transduction system, which takes information regarding changes in the external concentration of a variety of chemicals and processes it in such a way as to send a net controlling signal to the flagellar motors. What are the components of this system, where are they located, and how is information transmitted between them? At the time when the temporal-gradient-sensing character of the chemotaxis system was demonstrated, it was recognized that there must be a three-element device of some sort which would retain information about past concentration (slow element), monitor information about present concentration (fast element), and compare the two pieces of information (comparator). From the point of view of behavioral responses to a change in concentration of attractant or repellent we may speak of an excitation process (change in the fast element) followed by an adaptation process (change in the slow element).

The element responsible for monitoring present concentration is known to be quite fast, for cells display responses to concentration jumps of chemotactic agents within 200 msec of stimulation (Macnab and Koshland, 1972). This value is limited by the experimental protocol, so the actual excitation kinetics may be much faster. With the possible exception of membrane potential changes accompanying chemotactic stimulation (Szmelcman and Adler, 1976), no evidence regarding the character of the fast element of the gradient-sensing device is available.

Some idea of the time-scale of the memory device that stores information about past concentration may be gained from the duration of responses to a maximal single-agent stimulus—a stimulus that changes a receptor set from zero occupancy to full occupancy. For attractants, the range of the smooth swimming response is of the order of several minutes, whereas for repellents it is of the order of 10 sec. This difference between attractants and repellents (or, more strictly speaking, between positive and negative stimulation) is a characteristic feature of chemotaxis for which biochemical correlates in the methylation system (see Section 3.4.9) have now been found.

3.4.7 The Primary Chemoreceptors

The initial components of the information chain in chemotaxis are the primary receptors. On the basis of competition experiments it has been concluded that there is a limited number of receptors [about 20 have been identified so far in *E. coli* (Adler, 1975)] each with a fairly narrow range of closely related compounds it can recognize. The molecular nature of the receptors is only known in a few instances, all of them for sugars (Adler *et al.*, 1973; Adler and Epstein, 1974). They are either periplasmic—that is, located in the space between the inner and outer membranes—or they are membrane-bound. Examples of the two classes are the periplasmic ribose- and galactose-binding proteins, and the sugar-specific Enzymes II of the phosphoenolpyruvate-driven phosphotransferase systems, respectively.

Spectroscopic studies of the effects of substrate binding have led to the conclusion that the binding proteins undergo allosteric conformational changes. By the use of a double fluorescent label, this allosteric conformational change has been demonstrated to occur over a distance of at least 30 Å in the case of the galactose-binding protein (Zukin *et al.*, 1977*a*). This phenomenon may well reflect the activation by galactose of an allosteric site on the receptor for binding the next component in the information transfer mechanism. Although there have been a number of reports that a single receptor may have more than one sugar-binding site, these reports have since been refuted (Schwartz *et al.*, 1976; Zukin *et al.*, 1976*b*). Multiple-ligand binding, with differential kinetics of an associated conformational change, could theoretically serve as the basis for temporal gradient inference, but it now seems likely that the primary receptors transmit information about present concentration alone. The demonstration of competitive inhibition between receptors of different specificity, discussed in the following section, further reinforces this conclusion.

3.4.8 Secondary Components—Signal Focusing

A high level of occupancy of one receptor can inhibit responses to changes in occupancy of another receptor, presumably because both must interact with a common secondary component. The statement made earlier regarding algebraic additivity of stimuli is therefore restricted to cases in which the stimuli do not operate through a common secondary component. The *trg* gene product responsible for monitoring both ribose and galactose signals (via their respective periplasmic binding proteins) is an example of such a secondary component in the information chain (Ordal and Adler, 1974; Strange and Koshland, 1976). By this means, information from different receptor sets is focused into a smaller number of secondary component sets. When, as in the case of ribose and galactose, the effectors are chemically similar (and therefore presumably of comparable metabolic utility for the cell), the focusing may provide the cell with a means for rejecting gradient information about one chemical under conditions in which an ample supply of a similar one exists. Secondary components have also been found in the phosphotransferase system (Lengeler, 1975). In neither case has their location or mechanism of action been demonstrated.

3.4.9 Central Data Processing—The Chemotaxis Methylation System

There is another very important class of components that handle subsets of the total stream of sensory information. Components of this class are much more central in their function than is the *trg* gene product, and the signals they handle are not obviously related to one another in the way that ribose and galactose signals are. First described by Kort *et al.* (1975), they are proteins that are reversibly methylated in the process of handling chemotactic signals, and therefore they have been termed methyl-accepting chemotaxis proteins (MCPs). In *E. coli*, three genes, *cheD, cheM,* and *cheZ,* code for such proteins: MCP(1) MCP(2), and MCP(3), respectively (Springer *et al.*, 1977; Silverman and Simon 1977*b*). Both MCP(1) and MCP(2) are membrane-bound, and actually consist of more than one protein, possibly as a result of multiple-start transcription, or posttranscriptional modification.* [This feature is also found in other components of the chemotaxis system such as the *cheA* gene product (Silverman and Simon, 1977*c*).] From behavioral studies with mutants and from methylation patterns following chemotactic stimulation, it is known that the *cheD* gene product is responsible for handling signals for serine and a variety of repellents (hence, *t*axis to *s*erine and *r*epellents, or Tsr⁻ phenotype) whereas the protein specified by the *cheM* gene (Tar⁻ phenotype) is responsible for signals from aspartate, maltose, and a variety of repellents not handled by the *cheD* gene product. As no amino acid or repellent receptor has yet been characterized, it is conceivable that the MCPs themselves may serve this function as well as participating in the methylation system. Ironically, the chemotactic agents such as galactose, which are the best characterized at the level of their primary receptors, are least understood at the level of the MCPs because no reproducible methylation pattern has been observed for them. Even so, it is known that the MCP system is critically involved for sugars also, because Tar⁻ Tsr⁻ double mutants are nonchemotactic to all chemicals. The MCP(1) and MCP(2) systems appear to interact extensively because Tar⁻ mutants are hypersensitive to Tsr stimulation and vice versa.

If the bacterial cell is deprived of methionine (Adler and Dahl, 1967) or the active intermediate *s*-adenosylmethionine (Armstrong, 1972; Aswad and Koshland, 1975) tumbling is totally suppressed. One is tempted to conclude from this finding that clockwise rotation of the motor is dependent on methylation reactions, but actually the system must be more complicated than that. Studies on mutants defective in the methylating enzyme (Springer and Koshland, 1977) have demonstrated that such mutants can be made to tumble for very long periods when stimulated with the repellent phenol. Methylation therefore appears to be just one aspect of a more complicated determinant of tumbling behavior.

Study of the kinetics and equilibrium level of methylation of MCPs (1) and (2) has led to the following conclusions (Goy *et al.*, 1977): (1) The steady-state extent of methylation is a measure of the steady-state concentration of attractants and repellents in the environment. High serine concentration, for example, results in high methylation levels of MCP(1), and high repellent concentrations result in low methylation levels of MCP(1), compared with cells with no chemotactic agent in the environment. (2) On sudden addition of an attractant the extent of methylation increases at a roughly constant rate and then plateaus at the new steady-state level. The time interval between stimulation and reaching the plateau closely matches the time interval of the smooth swimming response before the cell adapts and reverts to its steady-state tumbling frequency. Upon addition of repellent, the methylation level drops rapidly, on a time scale comparable to the corresponding tumble response. Adaptation at the behavioral level corresponds with the methylation of MCP reaching a steady-state level. This makes it likely that the methylation level is the slow element of the gradient inference device. When the system is not at steady state, the methylation level is the cell's memory of "past concentration of attractant (or repellent)." The mismatch between the methylation level and the steady-state level that would pertain if the cell were fully adapted in its present environment reflects the degree of adaptation (Fig. 15). The methylation and demethylation reactions are controlled by different enzymes (Springer and Koshland, 1977; Stock and Koshland, 1978), which may explain the asym-

---

*Recent unpublished results (January, 1980) from the laboratories of Dahlquist, Hazelbauer, Koshland, and Simon all demonstrate that the above explanations are incorrect, and that the multiplicity is a multiplicity of methylation sites on a polypeptide such as MCP(1).

metry between methylation and demethylation kinetics, and also the corresponding asymmetry between behavioral responses to positive and negative chemotactic stimuli.

The above observations lead to two major (and so far unanswered) questions, the first regarding the link between the receptors and the methylation system, the second regarding the link between the methylation system and the motor. What causes methylation levels to change when attractants or repellents are added to the cell's environment? One possibility is that occupied receptors (or secondary components, such as the Trg protein) bind to the MCPs and alter their susceptibility to methylation, perhaps by exposing the esterifiable

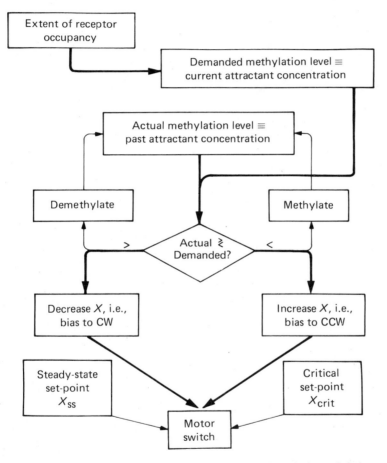

Figure 15. Schematic model for information flow from chemoreceptors through the methylation system to the flagellar motor. The current extent of occupancy of all receptors by their respective chemoeffectors "demands" a certain level of methylation of the methyl-accepting chemotaxis proteins (MCPs). Because both the methylation and demethylation reactions are slow (indicated by thin lines), the actual level of methylation under conditions of changing receptor occupancy will lag behind the level demanded. Any mismatch between actual and demanded levels acts on a response regulator X, which controls the sense of motor rotation. If X exceeds the critical value X_{crit} the motor rotates CCW; otherwise, it rotates CW. In the absence of gradient stimuli—that is, when actual and demanded methylation levels match—the cell is adapted, and X spontaneously fluctuates about a value X_{ss}, which is independent of the (constant) concentrations of chemoeffectors in the environment. Under these conditions the cell displays steady-state random motility.

residue, which is known to be a glutamyl γ-carboxyl (Van der Werf and Koshland, 1977; Kleene *et al.*, 1977). Alternatively, the susceptibility of MCPs to methylation might remain unchanged, but the availability or reactivity of enzyme or substrate might be affected by the extent of receptor occupancy. The fact that methylase mutants, but not MCP mutants, can be made to tumble suggests that MCP does undergo other changes in addition to reversible methylation.

How is motor function modulated by the methylation system, and why does the modulation operate only while methylation levels are changing—what is the comparator between the actual level of methylation and the level "demanded" by the current environment (Fig. 15)? There are a number of plausible mechanisms. The conformation of an MCP, unmethylated and not bound to an occupied attractant receptor, might be the same as that of a methylated MCP bound to occupied receptor (Fig. 16a). In a model of this type, methylation status and receptor occupancy status have compensatory effects on MCP, but, provided the kinetics of the two processes are different (receptor occupancy status effects being fast, methylation status effects being slow), the basis for a comparator exists. A specific example of such a model is given by Ordal and Fields (1977): An increase of attractant causes signal complexes—for example, (ribose) · (ribose binding protein) · (trg protein)—to bind to the MCPs which immediately increases their affinity for Ca^{2+} and also sets in motion the slow methylation reaction. Methylated MCP bound to signaler is presumed to have the same Ca^{2+} affinity as unmethylated MCP unbound to signaler. The overall effect is a rapid drop in free Ca^{2+} followed by a slow restoration. Ca^{2+} binding to the motor switch places it in clockwise rotation.

A model of a different type would have changes in the extent of methylation of MCP bring about depletion of the pool of another molecule (Fig. 16b). On the basis of data of Aswad and Koshland (1975) it seems unlikely that s-adenosylmethionine is limiting. Other possibilities would be (1) the methylase itself, (2) a regulator that binds to the methylase only when the latter has substrate (MCP) attached, or (3) a molecule requiring the methylase as enzyme. It is interesting in this context to note that there is evidence to suggest that methylation signals may directly interact with the flagella: MCP(3), unlike MCPs (1) and (2), is associated with the cytoplasm as well as the cytoplasmic membrane (Ridgway *et al.*, 1977) and certain mutations in the MCP(3) gene (*cheZ*) can complement defects in another chemotaxis gene (*cheC*) (Parkinson, 1977), that may code for a switch component of the flagellum. [Mutations in *cheC* can give rise to nonflagellate phenotype (Silverman and Simon, 1973).] Thus, it is conceivable that the information regarding extent of adaptation may be conveyed as a methylation signal from the membrane-bound MCPs to the cytoplasmic MCP and hence to the flagellum itself.

For technical reasons, the data published so far on the extent of methylation refer to pooled values for either MCP(1), MCP(2), or MCP(1) + MCP(2), not to values for the five or so individual proteins* (each encoded in a single gene) within, say, MCP(1). It is known, however, that not all components within a given MCP show the same methylation response to stimulation by attractant (Springer *et al.*, 1977). Thus, there may in fact be quite complex information processing even within the components of the MCP system, though this feature of the chemotaxis system has not yet been susceptible to detailed investigation.

---

*Recent unpublished results (January, 1980) from the laboratories of Dahlquist, Hazelbauer, Koshland, and Simon all demonstrate that the above explanations are incorrect, and that the multiplicity is a multiplicity of methylation sites on a polypeptide such as MCP(1).

A methylase of the *Salmonella* chemotactic system has been isolated and partially purified by Springer and Koshland (1977); it is a cytoplasmic protein encoded in the *cheR (E. coli cheX)* gene. The enzyme is capable of methylating both MCP(1) and MCP(2), suggesting that the specificity of methylation in response to a particular stimulus must stem from an activation of a particular MCP for methylation rather than specific activation of the methylase.

There are five *che* genes in addition to those for the MCPs and for the methylating enzymes. With the quite rapid demethylation observed on negative stimulation (Goy *et al.*, 1977) it was reasonable to assume that this process is enzymically mediated also. Recently,

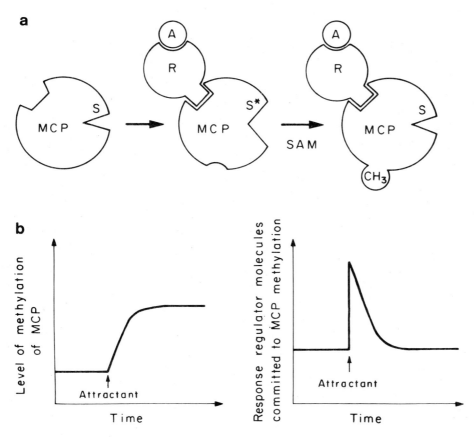

Figure 16. Two possible types of adaptation mechanisms involving the chemotaxis methylation system. (a) Binding of attractant A to receptor R enables R to bind (directly or through a secondary component) to the unmethylated methyl-accepting chemotaxis protein (MCP). This immediately makes MCP available for methylation and activates a site S to S* which results in an elevated level of the motor response regulator. Methylation, which is a slow process, restores S* to its inactive conformation S and the response regulator resumes its previous value. (b) As in (a), addition of attractant causes methylation of MCP to proceed and asymptotically approach a new steady-state value. The *process* of methylation ties up a limiting substrate, enzyme, or regulatory molecule (or, alternatively, releases a regulatory molecule), the molecule involved being part of the response regulator system. The effect is proportional to the net number of MCPs undergoing methylation, which is maximal immediately after subjecting cells to a sudden attractant increase, and decays asymptotically to zero thereafter.

a cytoplasmic demethylase activity has been correlated with the *Salmonella cheX (E. coli cheB)* gene and, perhaps, with the *Salmonella cheT (E. coli cheZ)* gene (Stock and Koshland, 1978). Of the remaining genes, one in the case of *E. coli* and two in the case of *Salmonella* appear to code for signal-receiving components of the flagella, hence mutations in them can result in nonflagellate phenotype as well as nonchemotactic phenotype. (Alternatively, they could be regulatory genes.) The other genes (*cheA, cheW,* and *cheY* in *E. coli*) code for cytoplasmic proteins with functions that are unknown at the present time.

4   Concluding Remarks

In this chapter we have chosen bacterial chemotaxis as an illustration of how an organism can regulate its environment by regulating its own movement. The bacterium, in order to accomplish this task, has evolved a sensory and motor system which, although rudimentary by comparison with those of higher animals, is still quite complex. Some 60 or so specific genes are involved, with considerable coordinate regulation of expression, as one might expect for a system with closely interlocking aspects, both at the structural and functional levels. Some of the more widespread aspects of genetic regulation, such as inducibility, regulation in trans, cyclic AMP control of expression and organization of groups of genes into operons, are found in the chemotaxis system along with mechanisms not yet described elsewhere, such as the "on–off" oscillator that controls flagellar phase.

The primary events in the sensing of chemicals by bacteria are remarkably similar to those in the sensing of neurotransmitters and hormones by the cells of higher organisms. The molecule constituting the stimulus is specifically bound to a protein receptor at the cell surface, and information regulating this event is communicated to the interior of the cell without the molecule itself having to cross the membrane. Many different types of attractants and repellents may initiate signals simultaneously. A hierarchy of access to the "central processor" has been developed, some chemicals having simultaneous access, some being mutual exclusive, some being capable of generating larger signals than others.

Because the cell requires gradient, or comparative, information regarding its environment, the chemotaxis system incorporates a primitive memory device capable of comparing past and present concentrations of chemoeffectors. This it does by playing a delicate kinetic game between fast consequences of effector binding and slow consequences, such as the process of methylation of specific proteins.

The target of all of this sensory transduction, the bacterial flagellar motor, is a remarkable device, which is capable of utilizing the proton electrochemical potential across the bacterial cell membrane to generate rotary motion, which is then transmitted to the external flagellar filament. Even the filament is an extraordinary structure, consisting of subunits of a single protein, yet possessing intrinsic helicity and sufficient rigidity for the provision of thrust, plus polymorphic structural versatility for randomizing that thrust when necessary.

One of the most instructive lessons to be learned from a consideration of this behavioral system is the manner in which a strategy of response evolves within the limitations of the organism concerned. The bacterium cannot in general alter its environment significantly, and therefore it evolves a migrational response. It requires information about other environments; a remote sensing system is beyond its capability, and so it develops a local gradient sensing mechanism and accepts an implied extrapolation regarding environments it has not yet reached. Its small physical size makes spatial comparison difficult, and so it develops a temporal comparison mechanism to increase signal size. The temporal comparison produces no information that would enable the cell to steer in the optimum direction, even if it had the requisite motor control mechanism, and so it evolves a mechanism that tries directions randomly and persists in ones that yield favorable signals. This strategy, put into effect by the molecular devices we have described, must surely constitute one of the first and simplest systems of sensory reception and motor response.

References

Adler, J., 1966, Chemotaxis in bacteria, *Science* **153**:708.

Adler, J., 1969, Chemoreceptors in bacteria, *Science* **166**:1588.

Adler, J., 1975, Chemotaxis in bacteria, *Annu. Rev. Biochem.* **44**:341.

Adler, J., 1976, The sensing of chemicals by bacteria, *Sci. Am.* **234**(4):40.

Adler, J., and Dahl, M. M., 1967, A method for measuring the motility of bacteria and for comparing random and non-random motility, *J. Gen. Microbiol.* **46**:161.

Adler, J., and Epstein, W., 1974, Phosphotransferase-system enzymes as chemoreceptors for certain sugars in *Escherichia coli* chemotaxis, *Proc. Natl. Acad. Sci. U.S.A.* **71**:2895.

Adler, J., and Tso, W-W., 1974, Decision-making in bacteria: Chemotactic response of *Escherichia coli* to conflicting stimuli, *Science* **184**:1292.

Adler, J., Hazelbauer, G. L., and Dahl, M. M., 1973, Chemotaxis toward sugars in *Escherichia coli*, *J. Bacteriol.* **115**:824.

Allweiss, B., Dostal, J., Carey, K. E., Edwards, T. F., and Freter, R., 1977, The role of chemotaxis in the ecology of bacterial pathogens of mucosal surfaces, *Nature (London)* **266**:448.

Anderson, R. A., 1975, Formation of the bacterial flagellar bundle, in: *Swimming and Flying in Nature* (T. Y.-T. Wu, C. J. Brokaw, and C. J. Brennen, eds.), Vol. 1, pp. 45–56, Plenum Press, New York.

Apirion, D., and Watson, N., 1978, Ribonuclease III is involved in motility of *Escherichia coli*, *J. Bacteriol.* **133**:1543.

Armstrong, J. B., 1972, An *S*-adenosylmethionine requirement for chemotaxis in *Escherichia coli*, *Can. J. Microbiol.* **18**:1695.

Asakura, S., 1970, Polymerization of flagellin and polymorphism of flagella, *Adv. Biophys.* **1**:99.

Aswad, D. W., and Koshland, D. E., Jr., 1975, Evidence for an *S*-adenosylmethionine requirement in the chemotactic behavior of *Salmonella typhimurium*, *J. Mol. Biol.* **97**:207.

Berg, H. C., 1974, Dynamic properties of bacterial flagellar motors, *Nature (London)* **249**:77.

Berg, H. C., 1975, Chemotaxis in bacteria, *Annu. Rev. Biophys. Bioeng.* **4**:119.

Berg, H. C., 1976, Does the flagellar rotary motor step? in: *Cell motility* (R. Goldman, T. Pollard, and J. Rosenbaum, eds.), pp. 47–56, Cold Spring Harbor Press, New York.

Berg, H. C., and Anderson, R. A., 1973, Bacteria swim by rotating their flagellar filaments, *Nature (London)* **245**:380.

Berg, H. C., and Brown, D. A., 1972, Chemotaxis in *Escherichia coli* analysed by three-dimensional tracking, *Nature (London)* **239**:500.

Berg, H. C., and Purcell, E. M., 1977, Physics of chemoreception, *Biophys. J.* **20**:193.

Berg, H. C., and Tedesco, P. M., 1975, Transient response to chemotactic stimuli in *Escherichia coli*, *Proc. Natl. Acad. Sci. U.S.A.* **72**:3235.

Bitensky, M. W., and Gorman, R. E., 1973, Cellular responses to cyclic AMP, *Prog. Biophys. Mol. Biol.* **26**:411.

Boos, W., 1974, The properties of the galactose-binding protein, the possible chemoreceptor for galactose chemotaxis in *Escherichia coli*, *Antibiot. Chemother.* **19**:21.

Brown, D. A., and Berg, H. C., 1974, Temporal stimulation of chemotaxis in *Escherichia coli*, *Proc. Natl. Acad. Sci. U.S.A.* **71**:1388.

Calladine, C. R., 1976, Design requirements for the construction of bacterial flagella, *J. Theor. Biol.* **57**:469.

Calladine, C. R., 1978, Change of waveform in bacterial flagella: The role of mechanics at the molecular level, *J. Mol. Biol.* **118**:457.

Coakley, C. J., and Holwill, M. E. J., 1972, Propulsion of micro-organisms by three-dimensional flagellar waves, *J. Theor. Biol.* **35**:525.

Dahlquist, F. W., Lovely, P., and Koshland, D. E., Jr., 1972, Quantitative analysis of bacterial migration in chemotaxis, *Nature (London), New Biol.* **236**:120.

DeFranco, A. L., Parkinson, J. S., and Koshland, D. E., Jr., 1979, Functional homology of chemotaxis genes in *Escherichia coli* and *Salmonella typhimurium, J. Bacteriol.* **139**:107.

DePamphilis, M. L., and Adler, J., 1971, Fine structure and isolation of the hook-basal body complex of flagella from *Escherichia coli* and *Bacillus subtilis, J. Bacteriol.* **105**:384.

Emerson, S. U., Tokuyasu, K., and Simon, M. I., 1970, Bacterial flagella: Polarity of elongation, *Science,* **169**:190.

Godson, G. N., 1978, A comparative DNA-sequence of the G4 and ϕX174 genomes, in: *The Single-Stranded DNA Phages (D. T. Denhardt, D. Dressler, and D. S. Ray, eds.), pp. 671–695, Cold Spring Harbor Laboratory, Cold Spring Harbor, N. Y.*

Goy, M. F., Springer, M. S., and Adler, J., 1977, Sensory transduction in *Escherichia coli:* Role of a protein methylation reaction in sensory adaptation, *Proc. Natl. Acad. Sci. U.S.A.* **74**:4964.

Harold, F. M., 1977, Membranes and energy transduction in bacteria, *Curr. Top. Bioenerget.* **6**:83.

Hazelbauer, G. L., and Parkinson, J. S., 1977, Bacterial chemotaxis, in: *Microbial Interactions: Receptors and Recognition,* Series B, Vol. 3 (J. L. Reissig, ed.), pp. 60–98, Chapman and Hall, London.

Henrichsen, J., 1972, Bacterial surface translocation: A survey and a classification, *Bacteriol. Rev.* **36**:478.

Hilmen, M., and Simon, M., 1976, Motility and the structure of bacterial flagella, in: *Cell Motility* (R. Goldman, T. Pollard, and J. Rosenbaum, eds.), pp. 35–45, Cold Spring Harbor Press, New York.

Hood, J. M., Fowler, A. V., and Zabin, I., 1978, On the evolution of β-galactosidase, *Proc. Natl. Acad. Sci. U.S.A.* **75**:113.

Hotani, H., and Asakura, S., 1974, Growth-saturation *in vitro* of *Salmonella* flagella, *J. Mol. Biol.* **86**:285.

Iino, T., 1969, Polarity of flagellar growth in *Salmonella, J. Gen. Microbiol.* **56**:227.

Iino, T., 1974, Assembly of *Salmonella* flagellin *in vitro* and *in vivo, J. Supramol. Struct.* **2**:372.

Iino, T., 1977, Genetics of structure and function of bacterial flagella, *Annu. Rev. Genet.* **11**:161.

Kamiya, R., and Asakura, S., 1976, Helical transformations of *Salmonella* flagella *in vitro, J. Mol. Biol.* **106**:167.

Khan, S., Macnab, R. M., DeFranco, A. L., and Koshland, D. E., Jr., 1978, The inversion of a behavioral response in bacterial chemotaxis: Explanation at the molecular level, *Proc. Natl. Acad. Sci. U.S.A.* **75**:4150.

Kleckner, N., 1977, Translocatable elements in procaryotes, *Cell* **11**:11.

Kleene, S. J., Toews, M. L., and Adler, J., 1977, Isolation of glutamic acid methyl ester from an *Escherichia coli* membrane protein involved in chemotaxis, *J. Biol. Chem.* **252**:3214.

Klug, A., 1967, The design of self-assembling systems of equal units, *Symp. Int. Soc. Cell. Biol.* **6**:1.

Komeda, Y., Suzuki, H., Ishidsu, J., and Iino, T., 1975, The role of cAMP in flagellation of *Salmonella typhimurium, Mol. Gen. Genet.* **142**:289.

Komeda, Y., Silverman, M., and Simon, M., 1977, Genetic analysis of *Escherichia coli* K-12 region I flagellar mutants, *J. Bacteriol.* **131**:801.

Komeda, Y., Silverman, M., and Simon, M., 1978, Identification of the structural gene for the hook subunit protein of *Escherichia coli* flagella, *J. Bacteriol.* **133**:364.

Konijn, T. M., 1975, Chemotaxis in the cellular slime moulds, in: *Primitive Sensory and Communication Systems* (M. J. Carlile, eds.), pp. 101–153, Academic Press, New York.

Kort, E. N., Goy, M. F., Larsen, S. H., and Adler, J., 1975, Methylation of a membrane protein involved in bacterial chemotaxis, *Proc. Natl. Acad. Sci. U.S.A.* **72**:3939.

Koshland, D. E., Jr., 1977, Sensory response in bacteria, in: *Advances in Neurochemistry* (B. W. Agranoff and M. H. Aprison, eds.), Vol. 2, pp. 277–341, Plenum Press, New York.

Larsen, S. H., Reader, R. W., Kort, E. N., Tso, W.-W., and Adler, J., 1974*a*, Change in direction of flagellar rotation is the basis of the chemotactic response in *Escherichia coli, Nature (London)* **249**:74.

Larsen, S. H., Adler, J., Gargus, J. J., and Hogg, R. W., 1974*b*, Chemomechanical coupling without ATP: The source of energy for motility and chemotaxis in bacteria, *Proc. Natl. Acad. Sci. U.S.A.* **71**:1239.

Läuger, P., 1977, Ion transport and rotation of bacterial flagella, *Nature (London)* **268**:360.

Lengeler, J., 1975, Mutations affecting transport of the hexitols D-mannitol, D-glucitol, and galactitol in *Escherichia coli* K-12: Isolation and mapping, *J. Bacteriol.* **124**:26.

Macnab, R. M., 1977, Bacterial flagella rotating in bundles: A study in helical geometry, *Proc. Natl. Acad. Sci. U.S.A.* **74**:221.

Macnab, R. M., 1978*a*, Bacterial flagella, in: *Encyclopedia of Plant Physiology, New Series* (W. Haupt and M. E. Feinleib, eds.), Vol, 7, pp. 207–223, Springer-Verlag, Heidelberg.

Macnab, R. M., 1978b, Chemotaxis in bacteria, in: *Encyclopedia of Plant Physiology, New Series* (W. Haupt and M. E. Feinleib, eds.), Vol, 7, pp. 310–334, Springer-Verlag, Heidelberg.

Macnab, R. M., 1978c, Bacterial motility and chemotaxis: The molecular biology of a behavioral system, *Crit. Rev. Biochem.* **5**:291.

Macnab, R. M., and Koshland, D. E., Jr., 1972, The gradient-sensing mechanism in bacterial chemotaxis, *Proc. Natl. Acad. Sci. U.S.A.* **69**:2509.

Macnab, R. M., and Koshland, D. E., Jr., 1973, Persistence as a concept in the motility of chemotactic bacteria, *J. Mechanochem. Cell Motility* **2**:141.

Macnab, R. M., and Ornston, M. K., 1977, Normal-to-curly flagellar transitions and their role in bacterial tumbling. Stabilization of an alternative quaternary structure by mechanical force, *J. Mol. Biol.* **112**:1.

Manson, M. D., Tedesco, P., Berg, H. C., Harold, F. M., and van der Drift, C., 1977, A protonmotive force drives bacterial flagella, *Proc. Natl. Acad. Sci. U.S.A.* **74**:3060.

Matsuura, S., Shioi, J., and Imae, Y., 1977, Motility in *Bacillus subtilis* driven by an artificial protonmotive force, *FEBS Lett.* **82**:187.

Mesibov, R., and Adler, J., 1972, Chemotaxis toward amino acids in *Escherichia coli, J. Bacteriol.* **112**:315.

Miller, J. B., and Koshland, D. E., Jr., 1977, Sensory electrophysiology of bacteria: Relationship of the membrane potential to motility and chemotaxis in *Bacillus subtilis, Proc. Natl. Acad. Sci. U.S.A.* **74**:4752.

Nishimura, A., Suzuki, H., and Hirota, Y., 1975, Flagellar formation in *E. coli* is coupled with cell division in regulatory mechanism, *Jpn. J. Hum. Genet.* **50**:484.

O'Brien, E. J., and Bennett, P. M., 1972, Structure of straight flagella from a mutant *Salmonella, J. Mol. Biol.* **70**:133.

Ordal, G. W., 1977, Calcium ion regulates chemotactic behaviour in bacteria, *Nature (London)* **270**:66.

Ordal, G. W., and Adler, J., 1974, Properties of mutants in galactose taxis and transport, *J. Bacteriol.* **117**:517.

Ordal, G. W., and Fields, R. B., 1977, A biochemical mechanism for bacterial chemotaxis, *J. Theoret. Biol.* **68**:491.

Parkinson, J. S., 1977, Behavioral genetics in bacteria, *Annu. Rev. Genet.* **11**:397.

Parkinson, J. S., 1978, Complementation analysis and deletion mapping of *Escherichia coli* mutants defective in chemotaxis, *J. Bacteriol.* **135**:45.

Ridgway, H. F., Silverman, M., and Simon, M. I., 1977, Localization of proteins controlling motility and chemotaxis in *Escherichia coli, J. Bacteriol.* **132**:657.

Rubik, B. A., and Koshland, D. E., Jr., 1978, Potentiation, desensitization, and inversion of response in bacterial sensing of chemical stimuli, *Proc. Natl. Acad. Sci. U.S.A.* **75**:2820.

Schwartz, M., Kellermann, O., Szelcman, S., and Hazelbauer, G. L., 1976, Further studies of the binding of maltose to the maltose-binding protein of *Escherichia coli, Eur. J. Biochem.* **71**:167.

Seabrook, W. D., 1978, Neurobiological contributions to understanding insect pheromone systems, *Annu. Rev. Entomol.* **23**:471.

Seymour, F. W. K., and Doetsch, R. N., 1973, Chemotactic responses by motile bacteria, *J. Gen. Microbiol.* **78**:287.

Shapiro, L., 1976, Differentiation in the *Caulobacter* cell cycle, *Annu. Rev. Microbiol.* **30**:377.

Silverman, M. R., and Simon, M. I., 1972, Flagellar assembly mutants in *Escherichia coli, J. Bacteriol.* **112**:986.

Silverman, M., and Simon, M., 1973, Genetic analysis of bacteriophage Mu-induced flagellar mutants in *Escherichia coli, J. Bacteriol.* **116**:114.

Silverman, M., and Simon, M., 1974a, Characterization of *Escherichia coli* flagellar mutants that are insensitive to catabolite repression, *J. Bacteriol.* **120**:1196.

Silverman, M., and Simon, M., 1974b, Flagellar rotation and the mechanism of bacterial motility, *Nature (London)* **249**:73.

Silverman, M., and Simon, M., 1976, Operon controlling motility and chemotaxis in *E. coli, Nature (London)* **264**:577.

Silverman, M., and Simon, M., 1977a, Bacterial flagella, *Annu. Rev. Microbiol.* **31**:397.

Silverman, M., and Simon, M., 1977b, Chemotaxis in *Escherichia coli:* Methylation of *che* gene products, *Proc. Natl. Acad. Sci. U.S.A.* **74**:3317.

Silverman, M., and Simon, M., 1977c, Identification of polypeptides necessary for chemotaxis in *Escherichia coli, J. Bacteriol.* **130**:1317.

Springer, M. S., Goy, M. F., and Adler, J., 1977, Sensory transduction in *Escherichia coli:* Two complementary pathways of information processing that involve methylated proteins, *Proc. Natl. Acad. Sci. U.S.A.* **74**:3312.

Springer, W. R., and Koshland, D. E., Jr., 1977, Identification of a protein methyltransferase as the *cheR* gene product in the bacterial sensing system, *Proc. Natl. Acad. Sci. U.S.A.* **74**:533.

Spudich, J. L., and Koshland, D. E., Jr., 1975, Quantitation of the sensory response in bacterial chemotaxis, *Proc. Natl. Acad. Sci. U.S.A.* **72**:710.

Stock, J. B., and Koshland, D. E., Jr., 1978, The identification of a protein demethylase in bacterial chemotaxis, *Proc. Natl. Acad. Sci. U.S.A.* **75**:3659.

Stocker, B. A. D., 1949, Measurements of rate of mutation of flagellar antigenic phase in *Salmonella typhimurium, J. Hyg.* **47**:398.

Strange, P. G., and Koshland, D. E., Jr., 1976, Receptor interactions in a signalling system: Competition between ribose receptor and galactose receptor in the chemotaxis response, *Proc. Natl. Acad. Sci. U.S.A.* **73**:762.

Suzuki, H., and Iino, T., 1975, Absence of messenger ribonucleic acid specific for flagellin in non-flagellate mutants of *Salmonella, J. Mol. Biol.* **95**:549.

Suzuki, T., Iino, T., Horiguchi, T., and Yamaguchi, S., 1978, Incomplete flagellar structures in nonflagellate mutants of *Salmonella typhimurium, J. Bacteriol.* **133**:904.

Szmelcman, S., and Adler, J., 1976, Change in membrane potential during bacterial chemotaxis, *Proc. Natl. Acad. Sci. U.S.A.* **73**:4387.

Taylor, B. L., and Koshland, D. E., Jr., 1974, Reversal of flagellar rotation in monotrichous and peritrichous bacteria: Generation of changes in direction, *J. Bacteriol.* **119**:640.

Taylor, B. L., and Koshland, D. E., Jr., 1975, Intrinsic and extrinsic light responses of *Salmonella typhimurium* and *Escherichia coli, J. Bacteriol.* **123**:557.

Tsang, N., Macnab, R., and Koshland, D. E., Jr., 1973, Common mechanism for repellents and attractants in bacterial chemotaxis, *Science* **181**:60.

Tso, W.-W., and Adler, J., 1974, Negative chemotaxis in *Escherichia coli, J. Bacteriol.* **118**:560.

Van der Werf, P., and Koshland, D. E., Jr., 1977, Identification of a γ-glutamyl methyl ester in bacterial membrane protein involved in chemotaxis, *J. Biol. Chem.* **252**:2793.

Ward, S., 1973, Chemotaxis by the nematode *Caenorhabditis elegans:* Identification of attractants and analysis of the response by use of mutants, *Proc. Natl. Acad. Sci. U.S.A.* **70**:817.

Yokota, T., and Gots, J. S., 1970, Requirement of adenosine 3′,5′-cyclic phosphate for flagella formation in *Escherichia coli* and *Salmonella typhimurium, J. Bacteriol.* **103**:513.

Zieg, J., Silverman, M., Hilman, M., and Simon, M., 1977, Recombinational switch for gene expression, *Science* **196**:170.

Zigmond, S. H., and Hirsch, J. G., 1973, Leukocyte locomotion and chemotaxis, *J. Exp. Med.* **137**:387.

Zukin, R. S., Hartig, P. R., and Koshland, D. E., Jr., 1977a, Use of a distant reporter group as evidence for a conformational change in a sensory receptor, *Proc. Natl. Acad. Sci. U.S.A.* **74**:1932.

Zukin, R. S., Strange, P. G., Heavey, L. R., and Koshland, D. E., Jr., 1977b, Properties of the galactose binding protein of *Salmonella typhimurium* and *Escherichia coli, Biochemistry* **16**:381.

8

Regulation of Membrane Transport

STEVEN C. QUAY and DALE L. OXENDER

1 Introduction

The intracellular levels of most nutrients are carefully controlled to meet the varying demands for nutrients presented by the normal growth cycle of the cell. The cell must balance the increases in nutrient levels (such as, *synthesis*) with the decreases in these nutrients (*metabolism* or macromolecular synthesis). Active transport is usually characterized as a process that will increase the cellular level of a nutrient. Most transport processes, however, are reversible to some extent and serve for exit as well as entry of nutrients. In a facilitated diffusion system the influx and efflux capacities are equal. In active transport systems the coupling of metabolic energy can lead to chemical gradients of transported solutes. Since transport activities alter the cellular levels of nutrients it is important that the cell have a way of regulating them. Regulation of the biosynthesis of various cellular nutrients has been extensively studied in the past, but regulation of transport systems has been largely ignored until recently.

Holley (1972) has suggested that regulation of nutrient transport has a primary effect in the growth regulation of mammalian cells. He has also proposed that transport-related membrane changes associated with malignancy may account for the loss of growth regulation and the uncontrolled proliferative nature of transformed cells.

This chapter is not a comprehensive review of the regulation of various transport systems. Instead, we will identify a few systems in which recent progress has been made in

STEVEN C. QUAY ● Department of Pathology, Harvard Medical School and Massachusetts General Hospital, Boston, Massachusetts 02114. *Present address:* Department of Pathology, Stanford University School of Medicine, Stanford, California 94305   DALE L. OXENDER ● Department of Biological Chemistry, The University of Michigan Medical School, Ann Arbor, Michigan 48109

STEVEN C. QUAY and
DALE L. OXENDER

the molecular basis of the regulation of transport. We have drawn heavily on studies in our laboratory on the regulation of branched-chain amino acid transport in the bacterium *Escherichia coli*, in which we have found a complex regulatory scheme similar to that found for the leucine biosynthetic pathway. The careful regulation of leucine transport appears to be important to the physiology of the *E. coli* cell.

It is hoped that the studies of transport regulation in procaryotic organisms will provide insight into the regulatory processes of eukaryotic organisms. As indicated in Section 5, there appear to be similarities in the regulation of amino acid transport in bacteria and in animal cells.

2 Regulation of Membrane Transport Activity

2.1 Cis Effects

Perhaps the simplest example of regulation of a transport system involves the inhibition of uptake of a compound when a structurally related compound is present in the medium surrounding the cell. Examples of interaction of this type are legion, especially among the amino acids (Oxender, 1972).

A quantitative treatment of the inhibition by a structural analogue of a substrate for a transport system often reveals a competitive mode of inhibition (Quay and Christensen, 1975). For an interaction of this type, the relationship of transport rate, V_{obs}, to substrate concentration, (S), given in Eq. (1) is altered to the form in Eq. (2), where (I) and K_I are the inhibitor level and the half-maximal inhibitor concentration, respectively.

$$V_{obs} = V_{max} \cdot (S)/K_m + (S) \tag{1}$$
$$V_{obs} = V_{max} \cdot (S)/K_m[1 + (K_I)] + (S) \tag{2}$$

Two types of cis-inhibitory interactions can be imagined: (1) the inhibitory compound both blocks the uptake of the substrate and undergoes the translocation reaction, to accumulate within the cell; and (2) the inhibitor has the necessary structural features to bind at the substrate recognition site but lacks the structural features needed for translocation. Inhibitors of both kinds have been found among the amino acid transport systems.

The amino acid transport system designated the LIV-I system, originally described as the uptake system for leucine, isoleucine, and valine (Piperno and Oxender, 1968), can also be shown to serve for the accumulation of threonine, serine, alanine, and glycine (Rahmanian *et al.*, 1973; Robbins and Oxender, 1973; Templeton and Savageau, 1974). For this system, all compounds that serve as inhibitors appear to be substrates.

In the case of the arginine-specific transport system (Quay and Christensen, 1974), although homoserine and the analogue, *trans*-hydroxyproline, were able to inhibit arginine transport completely, the lack of inhibition observed in the reciprocal experiment—that is, arginine did not inhibit homoserine transport—led to the conclusion that homoserine was accumulated by an uptake system that was not shared by arginine. We concluded that homoserine and the analogous *trans*-hydroxyproline do not provide all the structural features needed for recognition in transport, the deficiency presumably existing in the absence of a cationic structure represented by the two primary nitrogen atoms of the guanidinium group of arginine.

Trans effects refer to alterations in the transport activity of a test solute produced by another related solute that has accumulated on the opposite site of the membrane. Examples of both trans-stimulation and trans-inhibition phenomena have been reported for various organisms. Heinz and Walsh (1958) described the stimulation of the initial rate of glycine uptake into Ehrlich ascites mouse tumor cells produced by preloading the cells with high levels of glycine prior to uptake measurements. We have shown that the L-System amino acids, such as leucine and phenylalanine, are subject to strong trans-stimulation effects when Ehrlich cells are preloaded with another member of the L-System (Oxender and Christensen, 1963). In more recent studies we found that when the growth rate of 3T3 mouse cells grown in tissue culture is arrested by various methods such as serum removal, addition of protein synthesis inhibitors, or allowing the cells to reach confluent growth, the cellular levels of all the amino acids rise approximately twofold (Oxender *et al.*, 1976, 1977*a*). If transport measurements are made on these quiescent cells without first depleting the endogenous amino acids, strong trans effects can be observed. The initial rates of uptake of L-System amino acids are subject to trans-stimulation effects while A-System amino acid uptake is either not affected or for some amino acids trans inhibition is observed. The phenomenon of trans stimulation is believed to reflect a faster translocation process for the carrier–solute complex than for the empty or free carrier (Stein, 1967). Conversely, as Heinz and Durbin (1957) pointed out, if the empty carrier translocation process were faster than that of the carrier–solute complex, trans inhibition would result.

Trans inhibition of amino acid uptake has been observed in *Penicillium chrysogenum* by Benko *et al.* (1967), in yeast by Crabeel and Grenson (1970), in *Streptomyces hydrogenans* by Ring *et al.* (1970), and in *Neurospora crassa* by Wiley and Matchett (1966) and by Pall (1971), and in *E. coli* by Kadner (1975). Some negative interactions between transport solutes may also arise from a competition for a common source of energy.

Trans-inhibition effects can serve as a feedback mechanism to regulate transport activity and prevent excessive accumulation of amino acids. Trans-stimulation effects can serve to produce rapid exchange of one member of the L-System amino acid group for another, thereby maintaining adequate levels of all the amino acids required by the cell.

3   Regulation of Membrane Transport Protein Synthesis

3.1   Induction of Transport Systems

The current work on regulation of transport in bacteria by nutrients in the medium arose from concepts formulated during the late nineteenth century to explain the preferential utilization of one carbon source when two or more different carbon sources were present in the medium. The biphasic pattern of growth under these conditions has been termed *diauxie*. This phenomenon was initially investigated by Gale (1943) and Monod (1947), among others.

Examination of sugar transport systems reveals a variety of mechanisms for the regulation of these activities. For example, in the phosphoenolpyruvate phosphotransferase system (PTS) of bacteria, two of the proteins, HPr and Enzyme I, are commonly considered to be constitutive. On the other hand, most of the Enzyme II complexes, which provide for the specificity of the overall transport system, are clearly inducible. The latter situation pertains for fructose, mannitol, and sorbitol utilization (Hanson and Anderson, 1968; Fraenkel, 1968; Lengeler, 1975, 1977). However, these generalizations have been complicated by more recent work. For example, it has now been shown that the *pts* operon, which contains the structural genes for HPr (*ptsH*) and Enzyme I (*ptsI*), also contains a promoter-type region (Cordaro *et al.*, 1974). In addition, changes in the growth medium can result in fluctuations as great as threefold in the cellular level of these proteins (Saier *et al.*, 1970).

Although the Enzyme II complex for glucose, mannose, and fructose can be found in most membrane preparations regardless of the carbon source on which the cells were grown, there is some evidence that these uptake systems are inducible (Kornberg and Reeves, 1972). For these experiments, cells of *E. coli* B11 grown on glycerol were placed in medium containing fructose and the amount of fructose-1-phosphate formed was measured. Before growth recommenced the fructose-specific phosphotransferase was formed at a high rate; however, the cells did not synthesize glucose phosphotransferase. The reciprocal experiment indicated a preferential synthesis of glucose phosphotransferase when cells grown on glycerol were transferred to glucose. Under certain conditions these activities are clearly inducible.

An area of very active research and rapidly changing concepts is the role of the phosphotransferase system in the regulation of enzyme synthesis. While the PTS primarily serves for the accumulation of sugars in the form of their phosphate derivatives, other important roles of this protein complex include acting as sugar receptors for chemotaxis (Adler and Epstein, 1974) and modulating the activity of adenylate cyclase (Peterkofsky and Gazdar, 1974, 1975; Gonzalez and Peterkofsky, 1977; Saier *et al.*, 1976; Saier, 1977). The latter action has profound effects on the physiology of the cell.

The first evidence of a pleiotropic effect of the PTS system was the ability of glucose to inhibit the utilization of galactose, lactose, maltose, xylose, arabinose, and glycerol, all non-PTS sugars (McGinnes and Paigen, 1969). The mechanism of this inhibition, although incompletely understood, seems to involve inducer exclusion (Magasanik, 1970): glucose prevents the uptake of the inducer molecule and lowers the cAMP level, both by decreasing the activity of adenylate cyclase (Peterkofsky and Gazdar, 1974, 1975) and by stimulating the efflux of cAMP (Markman and Sutherland, 1965). The experimental evidence for these conclusions are reviewed by Postma and Roseman (1976).

In general, the amino acid transport systems are constitutive and regulated by repression–derepression systems. Tryptophan is, however, transported by three different systems in *E. coli* K12 (Oxender, 1972). One of these is the general aromatic transport system, which is regulated by a repression–derepression system (Whipp and Piffard, 1977). A second system specific for tryptophan is constitutive (Brown, 1970). A third system also specific for tryptophan is induced by growth on tryptophan and is subject to catabolite repression (Burrous and DeMoss, 1963).

3.2 Regulation of Membrane Transport by Derepression

In contrast to the induction of membrane transport observed for some sugar uptake systems, many of the amino acid transport systems are characterized by repression by their cognate amino acids (or some metabolically related derivative). For example, the transport systems for leucine, isoleucine, valine (Quay and Oxender, 1976), arginine, lysine, ornithine (Quay and Christensen, 1974), glutamate (Oxender, 1972), threonine, and serine (Templeton and Savageau, 1974) are all repressible.

In the case of branched-chain amino acid uptake systems of *E. coli*, early studies indicated a decrease in transport capacity on growth on complex media (Inui and Akedo, 1965). These early observations were the stimulus for more extensive studies on the regulation of this system. Since both the biosynthetic enzymes and the transport capacity are repressed by growth on medium containing leucine, it was important to examine if these two distinct cellular processes might be regulated in a concerted manner. For these studies, mutants of *E. coli* and *S. typhimurium* LT2 that were defective in either transport regulation or in the regulation of biosynthesis were examined (Quay *et al.*, 1975*a*). The *ilvB* gene product, acetohydroxy acid synthetase (AHAS), and the *leuB* gene product, 3-isopropylmalate dehydrogenase (IPMD), were used to estimate the level of derepression of the isoleucine–valine and leucine biosynthetic pathways, respectively. These biosynthetic enzymes were examined in *E. coli* K12 strains E0300 and E0312. Strain E0300 has wild-type transport regulation, whereas strain E0312 is a constitutively derepressed transport mutant isolated as a leucine auxotroph able to utilize D-leucine as a source of L-leucine (Rahmanian and Oxender, 1972). Subsequent genetic analysis indicated that this strain contains a genetic locus, *livR*, which maps at min 20 on the *E. coli* chromosome. Merodiploid analysis indicated that the *livR* locus codes for a diffusible, negative control element for leucine transport (Anderson *et al.*, 1976). The *ilv* operon in these two strains, as indicated by *ilvB* gene product activity, was studied under conditions of limiting and of excess branched-chain amino acids. The AHAS activity of the wild-type strain varied over a tenfold range under conditions in which the transport mutant showed a sixfold range of activity. Under all conditions of growth, the transport mutant showed a slightly lower activity of the *ilvB* gene. The leucine biosynthetic operon, as measured by the *leuB* gene activity, showed qualitatively the same pattern.

Reciprocal experiments were carried out in which strains with defined mutations in the operator–promoter region of the *leu* operon or with mutations leading to altered regulation of the *ilv* operon were tested for changes in the regulation of transport. In all cases studied, transport was found to be regulated in a normal fashion. These experiments allowed us to conclude that no portion of the leucine biosynthetic operon was required for branched-chain amino acid transport (Quay *et al.*, 1975*a*).

Two additional conclusions can be drawn from these experiments: (1) The *livR* locus, which leads to constitutively expressed transport, does not lead to constitutive synthesis of the biosynthetic enzymes and thus indicates a degree of independence in the regulation of these cellular processes; and (2) the mutation in transport regulation causes changes in expression of the biosynthetic operons that we believe result from changes in the ability of the mutant to maintain an adequate pool of internal amino acids for protein synthesis. That is, in order for a cell to grow at a normal growth rate, the transport and biosynthetic systems together must be maintained at some "total capacity." This capacity is set by a combination of the rates of amino acid synthesis and the ability of the cell to hold these pools of amino acids against their chemical gradients. If this is the case, some process (a mutation, for example) that raises the capacity for the transport system would be expected to lower the level of the biosynthetic system in a compensatory manner. The net result would be an unaltered branched-chain amino acid pool size with a shift upward in transport capacity and downward in biosynthetic capability. These predictions have, in fact, all been confirmed (Quay *et al.*, 1977).

Although it had been clearly shown that the bacterial transport system for leucine,

STEVEN C. QUAY and
DALE L. OXENDER

isoleucine, and valine is repressible by leucine (Inui and Akedo, 1965), a number of other amino acids, including isoleucine, valine (Piperno and Oxender, 1968), cysteine (Kanzaki and Anraku, 1971), methionine, and alanine (Guardiola *et al.*, 1974*a,b*) had been implicated in transport regulation. A regulatory system that involves both cognate (leucine, isoleucine, and valine) and noncognate (cysteine, methionine, and alanine) amino acids would be very unusual and seemed unduly complex. Therefore, we began a detailed study of the regulation of transport.

Initial experiments were undertaken to determine the kinetics of repression and derepression of branched-chain amino acid transport by leucine (Quay and Oxender, 1976). These experiments showed that derepression required both protein and RNA synthesis and that repression was the result of a reduction in the differential rate of synthesis of transport component(s) and probably not the result of a significant amount of protein degradation.

With these findings in hand, we wished to test if leucine, isoleucine, and valine together were necessary to lower transport capacity by multivalent repression (Burns *et al.*, 1966) in a manner analogous to that for regulation of the *ilvEDA* operon. In the latter case, the presence of all three branched-chain amino acids is required for repression. For these studies, a strain that requires leucine, isoleucine, and valine for growth was used, allowing us to limit selectively for a single branched-chain amino acid. When this strain was limited for leucine, a four- to sevenfold increase in the transport of leucine, isoleucine, and valine was observed. A similar derepression of the *ilvA* gene product, threonine deaminase, was observed. Because this enzyme derepresses in response to the limitation of any single branched-chain amino acid (Burns *et al.*, 1966), it served as an outside indicator of the level of these amino acids available to the cell. When limitation for isoleucine and valine was imposed, the transport capacity for all three branched-chain amino acids remained repressed, although the activity of threonine deaminase was increased. These results lead us to conclude that limitation for leucine alone is both necessary and sufficient for derepression of transport.

Since the LIV-I and LIV-II transport systems served for the accumulation of leucine, isoleucine, and valine, the regulation of these systems by leucine alone seemed paradoxical. At least two hypotheses could explain this situation. (1) The simple regulation pattern of the enteric organism *E. coli* has been adapted to the feast–famine conditions of the gut, where, it could be argued, the deprivation or surfeit of the three branched-chain amino acids is always coordinate. In such a situation the coupling of transport regulation to leucine may have been fortuitous. (2) The cell must regulate the leucine level independent of isoleucine and valine because of the central role played by leucine in various metabolic processes. The latter hypothesis is supported by an examination of some of the leucine-linked metabolic interactions in *E. coli* (see Table II). (1) Leucine represses membrane-bound reduced nicotinamide adenine dinucleotide phosphate (NADPH) : NADP$^+$ oxidoreductase (R. L. Hanson, personal communication), membrane-bound proline oxidase (Deutch and Stoffer, 1975), cystathionine synthetase (Greene and Radovich, 1975), and serine hydroxymethyltransferase (Greene and Radovich, 1975); inhibits aspartokinase III (Richaud *et al.*, 1974); causes growth inhibition after nutritional shiftdown (Alfoldin and Kerekes, 1964); and leads to the accumulation of unique isoaccepting species of leucine, histidine, arginine, valine, and phenylalanine tRNAs (Fournier and Peterkofsky, 1965; Yegian and Stent, 1969). (2) On the other hand, leucine stimulates the activity of lysyl-, methionyl-, and arginyl-tRNA synthetases (Hirschfield *et al.*, 1975), *S*-adenosylmethionine synthetase (Greene and Radovich, 1975), threonine and serine deaminases (Pardee and Prestidge,

1955), D-alanine, glycine *(dag)* transport (Robbins, 1973), and has a sparing effect on the utilization of glycine as a nitrogen source (Fraser and Newman, 1975) and the utilization of cyclic AMP during catabolite repression (Browman *et al.*, 1970). Fraser and Newman (1975) suggested that the intracellular leucine level may serve as a signal for nitrogen scavaging during periods of amino acid imbalance. This is analogous to a similar role played by ppGpp in amino acid imbalance or the function of cyclic AMP to signal a deficiency in energy supplies, and identifies leucine as a potential *alarmone* (Stephens *et al.*, 1975). This important role of leucine would require a careful regulation of leucine concentrations independent of fluctuations in the other branched-chain amino acids and dependent only on the relative rates of leucine supply and utilization in protein synthesis. The regulation, by leucine alone, of leucine biosynthesis (Burns *et al.*, 1966), tRNA aminoacylation (McGinnis and Williams, 1971), and transport is consistent with this hypothesis and contrasts with the multivalent nature of regulation of isoleucine and valine biosynthesis (Freundlich *et al.*, 1962) and tRNA aminoacylation (McGinnis and Williams, 1971). In addition, the leucine-specific transport system (Furlong and Weiner, 1970), which is present at low levels under normal conditions, but which can be genetically derepressed (Rahmanian *et al.*, 1973), may serve as yet another method to control leucine levels independent of isoleucine and valine.

To continue our exploration of the regulation of branched-chain amino acid transport, we drew on the extensive studies (Umbarger, 1973) of regulation in amino acid biosynthetic systems, which indicated that some aminoacyl-tRNA synthetases can serve both as components in protein synthesis and as a part of the regulatory machinery. In the latter capacity it has been shown that these enzymes serve to convert the appropriate amino acid to aminoacyl-tRNA, thereby producing the true corepressor.

To determine if leucine must be activated to an appropriate $tRNA^{Leu}$ to participate in the repression of transport and binding proteins, the expression of these activities was studied at growth-restricting temperatures in a strain with a temperature-sensitive leucyl-tRNA synthetase. A potential source of error in measuring transport in a temperature-sensitive tRNA synthetase mutant of a prototrophic strain is the large increase in the internal amino acid levels on derepression of the biosynthetic operons. For this reason, we chose a strain with a deletion of the leucine biosynthetic operon for our studies (Quay *et al.*, 1975*b*). Studies with another temperature-sensitive leucyl-tRNA synthetase mutant with an intact leucine biosynthetic operon produced qualitatively similar results to those reported here (Quay and Oxender, 1976).

Isogenic strains containing a deletion of the arabinose and leucine operons *(ara-leu Δ 1101)* and a temperature-sensitive leucyl-tRNA synthetase *(leuS^{ts})* were grown in a minimal medium with repressing levels of leucine at permissive temperatures (36°C) and transport and biosynthetic capacities measured. The two strains had nearly identical levels of leucine, isoleucine, valine, histidine, and proline transport. The regulation of histidine and proline transport is leucine-independent and served as a control in these experiments. When the cultures were shifted to a nonpermissive temperature for the mutant (42°C), growth of the *leuS^{ts}* mutant slowed and ceased after about one-half generation. During this period the differential rate of synthesis of the transport component(s) for leucine, isoleucine, and valine increased five- to tenfold in the *leuS^{ts}* strain. The temperature change had no effect on proline or histidine uptake in the mutant, nor did it have any effect on transport in the parental strain. In addition, threonine deaminase was greatly derepressed in the *leuS^{ts}* strain at nonpermissive temperatures, providing evidence that the activation of leucine to appropriate $tRNA^{Leu}$ had become the growth-rate-limiting step.

STEVEN C. QUAY and
DALE L. OXENDER

The possibility that the increase in branched-chain amino acid transport mediated by the leucyl-tRNA synthetase was due to activation of preexisting transport components was tested. Both RNA polymerase initiation and protein synthesis were found to be necessary for derepression of the transport system.

Measurement of the periplasmic binding protein activity for leucine showed a great increase at nonpermissive temperatures. These results are entirely consistent with the proposed role of this protein in the rate-limiting step for transport (Penrose *et al.*, 1968; Rahmanian and Oxender, 1972) and with the regulation of the rate of synthesis of this protein by some interaction of leucine with a functional leucyl-tRNA synthetase. The mechanism of this effect cannot be determined from these studies, although it presumably involves one of the following events: (1) the binding of leucine to the enzyme, (2) the formation of leucyladenylate, (3) the attachment of the leucyl residue to one (or more) of the specific tRNA(s) that functions in repression and/or protein synthesis, and (4) the attachment of the leucyl residue to some unknown aporepressor (Eidlic and Niedhardt, 1965). In addition, a role for uncharged $tRNA^{Leu}$ or the leucyl-tRNA synthetase as positive effectors of transport regulation is also consistent with these results.

In summary, our understanding of the regulation of leucine, isoleucine, and valine transport indicates that leucine, but not isoleucine or valine, interacts with $tRNA^{Leu}$ and the leucyl-tRNA synthetase, in effecting repression. This regulation has been established as primarily changing the differential rate of synthesis of transport components relative to total cellular proteins, without changing the rate of turnover of the rate-limiting component(s) for transport. While the studies described above were in progress, significant advances were being made in understanding regulation of the histidine and tryptophan biosynthetic operons. In addition to regulation by the classical mechanism of inhibition of RNA polymerase initiation by a corepressor–aporepressor complex (Jacob and Monod, 1961), there appear to be methods for changing the frequency of RNA transcriptional termination, apparently at an attenuator site, the "leader region," in the DNA proximal to the structural genes (Bertrand *et al.*, 1975). The protein factor rho, originally identified as important in relieving mutational polarity (Beckwith, 1963), seems to be involved in transcriptional termination at the attenuator site in the *trp* leader region (Korn and Yanofsky, 1976). We wished to determine if mutations of the rho factor affect leucine-dependent regulation of branched-chain amino acid transport.

Initial experiments involved growth of strain CU300 *(trpE ochre9851, leu amber 277)* and the isogenic strain CU2054 (containing *rho 120*) on repressing levels of leucine and then rapidly lowering the leucine level. This simple experiment provided a distinction between the strains: the wild-type strain (CV300) experienced a long lag in growth, whereas the *rho* mutant more quickly recovered from this shiftdown. Since the *leu*^am locus precluded derepression of leucine biosynthesis following this leucine limitation, we tentatively concluded that some other leucine-procuring mechanism, (that is, transport) was more rapidly derepressed in response to leucine limitation in the *rho 120* strain.

The uptake of a number of amino acids was examined in the wild-type strain and in the *rho 120* mutant grown on rich or on minimal media. Examining the cells grown on minimal medium first, we could see a number of differences in transport capacity as a result of the *rho* mutation. The uptake of arginine, histidine, leucine, isoleucine, and valine were increased from 30 to 90%; glutamine transport was decreased to one-third the wild-type capacity; and no change was seen in the uptake of proline or tryptophan. In contrast, growth in casamino acids had two effects: a tendency to negate the transport differences between the parental strain and the *rho 120* strain and a generalized decrease in all transport capacities.

The *leu*^{am} locus in these strains permitted a test for leucine-linked derepression of transport by limitation for leucine. The experiments consisted of transferring logarithmically growing cultures in 0.6 mM leucine to media containing from 0.05 to 0.6 mM leucine and measuring leucine transport after steady-state growth was established. In these experiments, the *rho*-associated increase in leucine transport was much more dramatic. While the wild-type strain underwent a 3.9-fold increase during leucine limitation, the *rho 120* strain showed a 15-fold derepression on shifting from 0.6 to 0.05 mM leucine. Kinetic experiments indicated that this 15-fold increase occurs primarily because of a greatly increased LIV-I transport system.

Analogous experiments were conducted to measure the specific activities of the various periplasmic binding proteins in these strains. For each binding protein examined (leucine, arginine, glutamine, and histidine), the transport activity and binding protein activity were regulated in a coordinate manner. This again supports the hypothesis that these amino acid-binding proteins are components of the corresponding amino acid transport systems.

Although the significance is not known, neither of the "membrane-bound" transport systems studied (proline and tryptophan) was derepressed. On the other hand, all systems involving binding proteins were altered by the *rho 120* mutation. In the branched-chain amino acid transport systems, although the membrane-bound LIV-II system was slightly increased, the major effect was on the binding protein-related LIV-I system.

The observation that the major effect of a mutation in *rho* was on the binding protein uptake systems must be correlated with the recent evidence that the β component of the membrane-bound Ca^{2+}, Mg^{2+}-adenosine triphosphatase is altered in *rho* mutants (S. Adhya, personal communication). Previous work had indicated that the binding protein uptake systems receive energy from ATP, while the membrane-bound systems use electron transport energy more directly (Berger and Heppell, 1974).

The effect of *rho* mutations on the glutamine transport system and binding protein may be related less to a direct regulatory effect of *rho* on that uptake system than to an indirect action on the intracellular level of energy or some other "effector" of glutamine transport capacity. Evidence has been presented that cAMP, NH_4^+, or some other compound may be a regulator of glutamine transport (Willis *et al.*, 1975).

The finding that growth in casamino acids negates the increase in transport components in the *rho* mutant indicates that additional regulatory controls exist under these growth conditions. One could imagine that the major regulatory control in this situation might be prevention of RNA polymerase initiation, as in the classic model of Jacob and Monod (1961).

The recent model of Travers (1976a), in which RNA polymerase can exist in two distinct forms, based on the presence or absence of ppGpp, is attractive in explaining certain aspects of promotor specificity during nutritional shifts. The indication that transport regulation involves some aspects other than transcriptional termination led us to investigate the question of whether or not ppGpp was required for promotor initiation in this system. Since the *relA* locus is required for ppGpp synthesis during nutritional step-down experiments, mutants of this locus could give us evidence about this aspect of regulation. In these experiments cells were grown in rich media to steady state and then shifted to minimal medium, allowed to grow for several hours, and then harvested and assayed for transport capacity. The results of these experiments indicate that both steady states of growth—that is, minimal media and rich media—lead to identical leucine transport capacities in the wild-type and in the *relA* strain. However, the shift-down growth condition demonstrated a regulatory abnormality in the *relA* strain, specifically, the inability to derepress transport under these growth conditions.

The kinetics of derepression of the wild-type strain were examined and indicated that the rate of derepression in these shift-down experiments is very similar to that seen earlier (Quay and Oxender, 1976) for leucine prototrophs, which were grown in minimal media during leucine limitation. In the latter case derepression was essentially complete in 90 min. In addition, this experiment again demonstrated the lack of derepression in the *relA* strain.

In order to determine the relative contributions of the LIV-I and LIV-II uptake systems to the derepression following the nutritional shift-down, the kinetics of leucine transport were determined under repressive conditions and after 3 hr in minimal medium. The results of these experiments indicated that the LIV-I system undergoes a fivefold derepression, while the LIV-II system increases only twofold. These values are reminiscent of earlier work in which the LIV-I system was always found to have a greater capacity for derepression.

As mentioned previously, a very attractive model for the effect of ppGpp on transport regulation is a modification of the specificity of the RNA polymerase. Both *in vitro* (Travers, 1976a) and *in vivo* (Travers, 1976b) studies indicate that the binding of ppGpp to the RNA polymerase induces a conformational change that alters the specificity of DNA sequences recognized as initiation codons. With this model one would predict that those operons necessary for growth under limiting conditions—that is, transport systems and biosynthetic operons—may have some sequence homology in the promotor region of the RNA. Clearly, the action of uncharged tRNA^{Leu} as a positive effector in regulation might simply be the production of ppGpp by tRNA^{Leu} and an idle ribosome (Haseltine and Block, 1973; Pedersen *et al.*, 1973), with ppGpp being the ultimate ("second messenger") effector. However, this cannot be the only action of tRNA^{Leu} in regulation, since ppGpp production under other conditions (such as threonine or arginine limitation) does not lead to derepression of leucine transport.

Recently, the maturation of tRNA^{Leu} has been shown to be incomplete in *relA* strains starved for leucine or phenylalanine (Kitchingman *et al.*, 1976; Kitchingman and Fournier, 1977). Specifically, leucine starvation leads to the accumulation of the major leucyl-tRNA species, tRNA$_1^{Leu}$, which is undermodified for the uridine at position 41 (Kitchingman and Fournier, 1977), the same alteration as that found in *hisT* mutants (Rizzino *et al.*, 1974). Thus the findings that both *relA* and *hisT* mutants are unable to produce tRNA^{Leu} with pseudouridine at position 41 and are also unable to derepress transport fully suggests a regulatory role for this key nucleotide and are consistent with the possibility that tRNA acts as a positive effector in regulation (Quay *et al.*, 1978).

In conclusion, it can be seen that the regulation of transport is extremely complex and requires the intricate interaction of leucine, tRNA^{Leu}, rho factor, and ppGpp and the various proteins responsible for the synthesis or recognition of these factors.

3.3   Processing by Proteases as a Regulatory Step

Many proteins are synthesized in precursor form and subsequently converted to active form by limited proteolysis. The role of limited proteolysis in physiological regulation has been the subject of several recent symposia (Ribbons and Brew, 1976; Reich *et al.*, 1975) and reviews (Goldberg and St. John, 1976). Examples of important processes that are triggered by specific proteases represent the generation of protein hormones (Steiner *et al.*, 1975), the activation of enzymes (Neurath, 1975), the assembly of viruses (Laemmli, 1975), blood coagulation (Davie *et al.*, 1975), and the turnover of various cellular proteins (Schimke and Bradley, 1975).

It has been well established that intracellular proteases are important for the break-down of abnormal proteins and the increased breakdown of normal proteins that occurs in starved cells, both bacterial and mammalian cells. Starvation of mammalian and bacterial cells for amino acids or glucose increases the activity of intracellular proteases. Goldberg and St. John (1976) have reported that the increased degradation of normal proteins during the starvation of bacteria responds to the level of charged tRNA and to the intracellular level of ppGpp. These findings suggest that the level of certain cellular proteolytic activities may be carefully regulated by the nutritional state of the cell. Alterations in the rates of certain proteolytic activities may also effect the processing and, therefore, the secretion of precursor proteins. Blobel and Sabbatini (1971) and Blobel and Dobberstein (1975a,b) have proposed the *signal hypothesis* for the transfer of proteins across cell membranes.

The signal sequence hypothesis states that a protein destined for crossing the membrane has an N-terminal amino acid sequence that triggers the attachment of the ribosome to the membrane, thus providing the topological conditions for transfer of the nascent chain across the membrane. While the protein is being transferred, the N-terminal signal sequence is cleaved by limited proteolysis to produce a processed protein.

3.3.1 Periplasmic Proteins

Recent evidence by Inouye and Beckwith (1977) and Inouye *et al.* (1977) has indicated that periplasmic proteins such as the enzyme alkaline phosphatase from *E. coli* may be transferred across the inner membrane by a mechanism similar to that involved in the signal hypothesis.

To reach these conclusions, Inouye and Beckwith (1977) established a cell-free system for synthesizing alkaline phosphatase. The alkaline phosphatase made *in vitro* had a higher molecular weight and was more hydrophobic in nature than authentic alkaline phosphatase. They found that the precursor form of alkaline phosphatase could be processed by limited proteolytic activity and, surprisingly, that this proteolytic activity was located in the outer membrane of *E. coli* (Inouye and Beckwith, 1977). Attempts are currently being made to obtain the amino acid sequence of the alkaline phosphatase precursor.

Are other periplasmic components such as the amino acid- and sugar-binding proteins made in precursor forms? As of this writing, a precursor form of the transport binding proteins has not been firmly established. The amino acid sequences of some of the binding proteins have been determined, and this should make it possible to look for precursor forms. Hogg and Hermodson (1977) have published the sequence of the arabinose-binding protein and Ovchinnikov *et al.* (1977) have published the complete sequence of the leucine-, iso-leucine-, and valine-binding protein. From these reports and preliminary data from other binding proteins, the early portions of the N-terminal sequences of periplasmic binding proteins have been made available to these authors (see Table I).

TABLE I. N-Terminal Sequences of Transport Binding Proteins of E. coli

1	2	3	4	5	6	7	8	9	Substrate	Reference
Glu	Asn	Leu	Lys	Leu	Phe	Leu	Val	Lys	Arabinose	Hogg and Hermodson, 1977
Glu	Asp	Ile	Lys	Val	Ala	Val	Val	Gly	Leu, Ile, and Val	Ovchinnikov *et al.*, 1977
Asp	Asp	Ile	Lys	Val	Ala	Val	Val	Gly	Leu	A. Y. Ovchinnikov, personal communication
Ala	Asp	Thr	Arg	Ile	Gly	Val	Thr	Ile	Galactose	R. L. Hogg, personal communication

Even though these N-terminal sequences were obtained from binding proteins that presumably have been processed, it is possible to see a number of common or similar amino acid residues in the early part of the sequence, such as Asp or Asn at position 2 and Lys or Arg at position 4. Leu, Ile, or Val also seem to be common residues at positions 5 and 7. It is not unlikely that processing for several of the binding proteins occurs by a common protease acting on the N-terminal signal sequence portion of the protein. The processed binding proteins would be expected to retain a portion of the common signal sequence. These findings are to be taken only as suggestive that the binding proteins for amino acids and sugars contain certain common N-terminal sequences that become cleaved during the secretion of the binding proteins into their periplasmic location. Further information will have to await *in vitro* studies of the synthesis of the binding proteins. This is an area of active investigation at the present time.

3.3.2 Hydrophobic Membrane Pencillinase

Yamamoto and Lampen (1976) have identified a hydrophobic form of the secretory enzyme, penicillinase, of *Bacillus licheniformis,* which is covalently attached to a phospholipid residue of the membrane. This membrane-bound penicillinase contains 24 extra amino acid residues that have been partially sequenced (Yamamoto and Lampen, 1976). The extra amino acids are mostly hydrophilic amino acids, but one of the serine residues in the N-terminal sequence is esterified to phosphatidic acid, giving rise to phosphatidyl-serine, which serves to attach the penicillinase to the membrane. Trypsin treatment of the hydrophobic membrane penicillinase yields a phospholipopeptide and a hydrophilic penicillinase differing from the exopenicillinase only by the absence of the NH_2-terminal lysine residue. Since membrane-bound penicillinase is formed during growth of the cells, it is proposed as a precursor or storage form of the exopenicillinase.

4   Role of Transport in Cellular Physiology

In this section we shall address the question of the role of transport systems in the overall physiology of bacterial cells. These discussions will be limited to the metabolism of leucine, isoleucine, and valine with which we are most familiar.

4.1   Relationship between Biosynthetic Systems and Transport

An examination of the quantitative aspects of the transport and biosynthetic systems for leucine, isoleucine, and valine in *E. coli* cultures growing under different growth conditions is quite revealing. In rich media both transport and biosynthesis are fully repressed. When the culture is shifted to minimal medium a period of unbalanced growth begins, which lasts 90–120 min, during which the level of the branched-chain amino acid transport system increases. This increase has been shown to represent an increase in the differential rate of synthesis of transport components with very little change in the rate of inactivation of these proteins (Quay and Oxender, 1976). This increase is in striking contrast to the level of the biosynthetic enzymes in prototrophic strains grown in minimal medium, which are 60–80% repressed (Umbarger, 1973). When the cells are further limited for leucine by starvation there is little increase in the transport capacity, and the biosynthetic enzymes undergo a large derepression. These considerations indicate that a prototrophic strain in minimal medium has a fully derepressed transport capacity. This relationship represents a hierarchy of regulation of transport and biosynthesis of branched-chain amino acids in *E. coli*. One could imagine that this difference in sensitivity to leucine limitation of transport and biosynthesis would allow cells to respond with maximum efficiency to a wider range of culture conditions.

4.2 Exchange-Linked Regulation of Biosynthesis

Early experiments on the response of *E. coli* to the transfer from rich to minimal medium (shift-down) indicated that regulatory mechanisms exist that control general RNA and protein synthesis (Borek *et al.*, 1955; Kjeldgaard *et al.*, 1958; Neidhardt, 1963). In addition, it was found that certain amino acids, such as leucine, could greatly lengthen the growth lag during a shiftdown experiment (Horvath and Gado, 1965; Neidhardt, 1963). The mechanism of growth inhibition by leucine was studied extensively, and suggestions included effects on RNA synthesis (Horvath and Gado, 1965), regulatory anomalies of acetohydroxy acid synthetase (Rogerson and Freundlich, 1970), and, more recently, inhibition of threonine deaminase (Calhoun, 1976; Vonder Haar and Umbarger, 1972). The finding that mutants consistently repressed for threonine deaminase were permanently leucine-sensitive pointed to this enzyme as the site of leucine sensitivity (Levinthal *et al.*, 1973).

Our interest in this area arose from the discovery that the derepressed LIV-I transport system in an *livR* mutant leads to enhanced leucine sensitivity in this strain (Quay *et al.*, 1977) and indicated that this transport system was important for this phenotype. One could imagine that the high LIV-I activity could increase sensitivity to leucine by increasing the size of the intracellular leucine pool. When direct pool size measurements indicated that the leucine pool in an *livR* and isogenic wild-type strain were very similar (Quay *et al.*, 1977) this explanation seemed inadequate. On the other hand, a marked reduction in intracellular isoleucine level during the transition period immediately following addition of leucine indicated that exchange of isoleucine might be involved in the mechanism of leucine toxicity. When direct measurements of leucine–isoleucine exchange transport indicated that the LIV-I system is primarily responsible for the exchange process, exchange became the most likely candidate for explaining the mechanism of leucine toxicity.

We envision the following sequence to occur in a wild-type strain during transfer to minimal medium supplemented with leucine. Immediately on transfer, extracellular leucine is taken up by the LIV-I system in exchange for intracellular isoleucine and valine. The biosynthetic enzymes for isoleucine and valine must increase their activity to compensate for the exchange-linked loss from the cell. Although transcription of the *ilv* operon can proceed at a normal rate, translation requires a supply of isoleucine and valine, which would be limited. Finally, after a period of restricted growth, a steady state is reached, in which the *ilv* operon is derepressed sufficiently to keep pace with the consumption of isoleucine and valine by protein synthesis and loss from the cell by exchange, and growth can proceed normally. During this transition period, the biosynthesis of transport proteins has been repressed by a process involving an interaction of leucine, tRNA^{Leu}, and the leucyl-tRNA synthetase.

In two special cases failure to repress the transport system leading to prolonged inhibition of growth can seriously interfere with this process. In one case, valine is substituted for leucine in the growth medium. The exchange by valine leads to depletion of isoleucine, causing a derepression of *ilv* biosynthetic enzymes, but failing to cause repression of the transport system. This is probably a minor aspect of valine inhibition, because mutants that have lost the LIV-I transport system are resistant to only low levels of valine (Anderson and Oxender, 1977); high-level valine resistance is the phenotype found only among biosynthetic mutants. The other case of prolonged inhibition was found in constitutively derepressed transport, as seen in the *livR* mutants. This strain could reach steady-state growth with leucine only after a substantial derepression of threonine deaminase. In addition, the derepressed LIV-I transport system is more effective in lowering the isoleucine levels, as indicated by direct measurement of pool size. Thus, under conditions in which threonine deaminase must derepress maximally, the rate of translation is most limited by the supply of isoleucine.

The transition from rich medium to leucine-supplemented medium differs from that from minimal medium to leucine-supplemented medium in that the initial level of the *ilv* enzymes is lower in the former case due to repression (in both the wild-type and *livR* strains). However, repression of the LIV-I transport system in the wild-type strain limits loss of amino acids from the pool during the transition period. The *livR* strain has two disadvantages in this situation: its LIV-I system is derepressed and able to rapidly catalyze exchange, and a reduced threonine deaminase level exists before addition of leucine, which must be increased if growth is to resume. The reduced threonine deaminase in this strain, when grown in complex medium, is similar to the lowered acetohydroxy acid synthetase and γ-isopropyl malate dehydrogenase activities in this strain relative to those of the wild type.

The observation that a constitutively derepressed transport mutant has more difficulty than the wild type in shifting from rich medium is in contrast to the advantage the former strain has in scavenging leucine from minimal medium (Quay *et al.,* 1976), and it represents a unique situation in which the ability to repress transport by excess solute provides a distinct advantage. Previous concepts about the role of transport systems in bacteria emphasized the ability to accumulate amino acids, a process which should be maximal under conditions when endogenous synthesis is low—that is, in rich medium. Conversely, our present work emphasizes those physiological conditions in which cells must be prepared to continue growth during periods of amino acid limitation, or even during exposure to unbalanced amino acid supplies. In this case, transport systems that could lead to loss of amino acids by exchange would be undesirable and should be repressible.

We now can understand the reason for the evolution of multiple uptake systems for leucine, isoleucine, and valine in *E. coli.* In rich medium, the branched-chain amino acid biosynthetic operons are maximally repressed, with the amino acids being supplied from the medium via the low-affinity LIV-II uptake system. As the exogenous supply of these amino acids dimishes, the LIV-I uptake system becomes derepressed via the leucyl-tRNA synthetase or its substrate, tRNA^{Leu}. This high-affinity system is well suited for scavenging branched-chain amino acids. When the cells are growing in minimal medium without supplemental amino acids, the transport systems are fully derepressed and are therefore able to maintain intracellular pools and to scavenge amino acids present in small amounts in the medium (it should be recalled that the K_m of the LIV-I system is 0.2 μM for leucine or 0.03 μg/ml). Under these conditions the biosynthetic enzymes are only 2–30% derepressed. It is only when auxotrophic strains reduce the pool amino acids still further that a full derepression of biosynthesis occurs. The role of the LIV-I system in scavenging amino acids from unsupplemented minimal medium may be difficult to envision but must be understood in the context of the evolutionary environment of *E. coli,* in which the total absence of exogenous amino acids was probably rare (and is probably only a laboratory phenomenon).

Under conditions of unbalanced growth—that is, conditions in which only one or a few of the potential substrates of a transport system are present in the medium—one can imagine that exchange-coupled uptake could result in a state of pseudoauxotrophy for those amino acids not present in the medium. The decrease in the intracellular concentrations of amino acids that occurs by this process could cause a derepression of appropriate biosynthetic pathways. In this way, exogenously supplied amino acids could appear to "derepress" biosynthetic pathways for amino acids of a shared transport system. These considerations may be related to the observation of "metabolic interlock" (Kane, 1975), in which histidine can derepress the tryptophan biosynthetic operon (Kane, 1975), as both are substrates of the *aroP* transport system (Ames, 1964).

4.3 Leucine-Linked Metabolic Interactions

As noted earlier (Section 3.2), leucine appears to be unique among amino acids in effecting a diverse group of metabolic events, primarily involving amino acid utilization. The possible reasons for this have been discussed elsewhere by Fraser and Newman (1975) and others (Quay and Oxender, 1976) and will not be elaborated here. Table II contains

TABLE II. *The Role of Leucine as a Regulator of Metabolic Interactions*

Effector	Action	Reference
Leucyl-tRNA	Represses NADP$^+$:NADPH oxidoreductase	R. L. Hanson, personal communication
Leucine	Represses membrane-bound proline oxidase	Deutch and Soffer, 1975
Leucine	Represses cystathionine synthetase	Greene and Radovich, 1975
Leucine	Represses serine hydroxymethyltransferase	Greene and Radovich, 1975
Leucine	Inhibits aspartokinase III	Richaud *et al.*, 1974
Leucine	Inhibits growth following nutritional shiftdown	Alfoldi and Kerkes, 1964
Leucine	Stimulates accumulation of unique tRNA species of leucine, histidine, arginine, valine, and phenylalanine	Fournier and Peterkofsky, 1975; Yegian and Stent, 1969
Leucyl-tRNA	Stimulates lysyl-, methionyl-, and arginyl-tRNA synthetases	Hirschfield *et al.*, 1975
Leucine	*S*-Adenosylmethionine synthetase	Greene and Radovich, 1975
Leucine	Stimulates threonine and serine deaminases via exchange transport	Pardee and Prestidge, 1955; Quay *et al.*, 1977
Leucine	D-Alanine and glycine (dag) transport	Robbins, 1973
Leucine	Spares glycine utilization as a nitrogen source	Fraser and Newman, 1975
tRNA$_1^{Leu}$	Most abundant tRNA^{Leu} isoaccepting species (In addition, tRNA^{Leu} species taken together are the most abundant tRNA)	von Ehrenstein, 1970
tRNA$_1^{Leu}$	Transcription of the genes for tRNA$_1^{Leu}$ is regulated in a unique manner	Ikemura and Dahlberg, 1973
tRNA$_1^{Leu}$	Although the most abundant tRNA^{Leu}, this species is used for protein synthesis to a much lower extent than other tRNA^{Leu} species (indicating a possible regulatory role)	Wettstein and Stent, 1968; Kano-Sueoka and Sueoka, 1969; Kitchingman and Fournier, 1975

a summary of these effects and, in addition, contains some recently identified properties of the $tRNA_1^{Leu}$, which are also unique. Taken together, these represent a wide range of effects. Our present state of ignorance precludes proposing a single unifying hypothesis for these phenomona, and our goal here is merely to define and enumerate these relationships with the hope that further research efforts can delineate the details of the role of leucine in regulating amino acid utilization.

5   Transport Regulation in Eukaryotic Organisms

It has been suggested that the growth of mammalian cells may be regulated by the availability of nutrients inside the cell (Pardee, 1964; Holley, 1972). Holley (1972) has proposed that crucial changes in the plasma membranes of cells may produce increased transport activity for certain critical nutrients. The altered growth rate resulting from the increased availability of critical nutrients is suggested by Holley (1972) as a possible primary cause of malignant transformation. These transport changes could result from structural changes in the membrane, the availability of energy supplies for transport, or a loss in the regulation of transport activity. Transport changes associated with viral transformation of mammalian cells have been reported for hexoses (Venuta and Rubin, 1973; Weber, 1973; Eckhart and Weber, 1973; Isselbacher, 1972; Kletzien and Perdue, 1974, 1976; Hatanaka, 1976; Weber et al., 1976), for amino acids (Isselbacher, 1972; Hillman and Otto, 1974; Foster and Pardee, 1969; Perdue, 1976; Kalckar et al., 1976), for inorganic phosphates (Cunningham and Pardee, 1969), and for K^+ (Kimelberg and Mayhew, 1975). It is not always clear whether increased uptake rates result from oncogenic transformation or from the increased growth rate that attends transformation. Weber et al. (1976) have made a distinction between growth-rate-dependent changes in transport activity and transformation-specific changes. They found that many transport systems of chick fibroblasts appear to be subject to growth-rate-contingent controls, but only the hexose transport system displayed a transformation-specific change in activity. The availability of a temperature-sensitive Rous sarcoma virus that transforms chick embryo fibroblasts has greatly aided such studies.

Studies such as those reported here have stimulated an interest in the examination of the regulation of transport activity in mammalian cells in the hope of obtaining information on the processes concerned with malignant transformation.

5.1   Amino Acid Transport Regulation

Riggs and Pan (1972) first showed that the incubation of immature rat uteri in buffer for 4 hr produced an increase in transport activity of α-aminoisobutyric acid. Guidotti and co-workers (Gazzola et al., 1972; Guidotti et al., 1975, 1976) greatly extended these studies to a variety of other animal tissues, showing that this enhancement of transport activity required RNA and protein synthesis. This increased transport occurred primarily for System A amino acids (Oxender and Christensen, 1963), such as proline, glycine, alanine, and aminoisobutyric acid (Guidotti et al., 1976). Other laboratories have reported enhanced uptake of amino acids following amino acid limitation (Reynolds and Segal, 1976; Peck et al., 1976; Heaton and Gelehrter, 1977). Heaton and Gelehrter (1977) found that threonine and phenylalanine were especially effective in preventing the enhanced uptake in cultured hepatoma cells. Other studies have shown that amino acid transport activity is altered in tissue culture cells when their rate of growth is changed by a variety of methods (Foster and Pardee, 1969; Oxender et al., 1976, 1977b). As 3T3 mouse cells approach confluence during growth in monolayer, the transport activity of System A amino acids decreases, whereas that of System L amino acids increases (Oxender et al., 1976, 1977b). Such behavior of the System L activity may be predicted from trans-stimulation effects of the endogenous amino acids, which are increased when cells become density-inhibited. System L amino acid transport is subject to trans-stimulation effects (Oxender and Christensen, 1963), as described in Section 2.2. However, the decrease in System A transport activity could not be explained by trans effects. From these studies it appeared that System A transport activity was increased during amino acid starvation and decreased when the intracellular levels of the amino acids were raised. If the cells are depleted of internal amino acids before uptake measurements to eliminate trans effects, it is often possible to observe similar behavior for the L System amino acids.

5.2   Role of Aminoacyl-tRNA Synthetases

Since the early work has shown that protein synthesis was required for the regulation of transport activity in animal cells, and we had found that leucyl-tRNA was important for regulation of leucine transport in E. coli, we examined the possible role of aminoacyl-tRNA synthetases in regulation of mammalian cell transport. The availability of several temperature-sensitive aminoacyl-tRNA synthetase mutants in Chinese hamster ovary cell lines (Thompson et al., 1973, 1975, 1977; Wasmuth and Caskey, 1976) provided us with the opportunity to test this possibility. When a temperature-sensitive leucyl-tRNA synthetase mutant, tsH1 (Thompson et al., 1973, 1975, 1977), was shifted from a normal growth temperature of 35°C to a marginally permissive temperature for growth (38°C), a significant enhancement in the initial rate of uptake of leucine and other L System amino acids was observed (Moore et al., 1977). In contrast, the uptake of A System amino acids showed no significant difference relative to the parental cell line. In a similar manner, a temperature-sensitive asparaginyl-tRNA synthetase mutant, RJK-4 (Wasmuth and Caskey, 1976), exhibited increased transport activity of System A amino acids when the growth temperature was shifted from 33° to 39.5°C.

Preliminary kinetic studies suggest that the V_{max} for transport is increased when the mutants are grown under conditions of amino acid limitation. It appears that when the Chinese hamster cells are starved for an A System amino acid they respond by increasing the transport activity of the A System, and when growth is limited by an L System amino acid the L System transport activity is increased. These results suggest that the aminoacyl-tRNA synthetases or their related products may play a role in regulation of amino acid transport in mammalian cells in a manner similar to that found for prokaryotic organisms.

5.3   Regulation of Hexose Transport in Eukaryotic Organisms

Hexose transport in animal cells occurs by a facilitated diffusion process followed by phosphorylation. 2-Deoxy-D-glucose is often used experimentally as an analogue of glucose, as it shares a common transport system with glucose and is phosphorylated but not further metabolized. The glucose analogue, 3-O-methyl-D-glucose, is also used. It has a lower affinity for the glucose system, although it allows the separation of the uptake and phosphorylation steps, because it is not phosphorylated. Extremely rapid kinetic analysis is required for studies with this analogue because it is not accumulated to levels above that of the external medium.

Variations in the transport activity of hexoses into animal cells have been correlated with the phases of the cell cycle, with stimulation by mitogenic agents, with oncogenic transformation, and with specific nutrient limitation. Enhanced rates of hexose uptake in *Neurospora crassa* (Scarborough, 1970; Neville *et al.*, 1971), in chick fibroblasts (Martineau *et al.*, 1972; Kletzien and Perdue, 1975, 1976), and in hamster cells (Kalckar and Ullrey, 1973; Christopher *et al.*, 1976, 1977) have been observed when these organisms are maintained on low levels of glucose. The inhibition of RNA and protein synthesis prevents the enhancement of hexose transport activity, suggesting that a regulatory control of the repression–derepression type may be operating in these tissues. Glucose-6-phosphate has been suggested as a possible repressor for the expression of hexose transport activity (Martineau *et al.*, 1972; Kletzien and Perdue, 1975). Support for this hypothesis comes from findings that when chick fibroblasts are grown in the presence of 2-deoxy-D-glucose, 2-deoxy-D-glucose-6-phosphate accumulates, and the hexose transport activity appears to be repressed. A more complicated pattern of regulation appears to be supported by the work of Ullrey *et al.* (1975) and Christopher *et al.* (1976, 1977). They showed that in hamster cells hexose uptake was increased fivefold to tenfold by either substituting D-fructose for glucose or by starving the cells for glucose. The high rates of uptake of glucose-starved cells could be decreased in about 6–8 hr by addition of glucose to the cells. The glucose analogue, 2-deoxy-D-glucose, did not cause hexose transport to decrease under the same conditions, suggesting that the 6-phosphoester of glucose does not serve as repressor for the hexose transport activity. Christopher *et al.* (1976, 1977) proposed that the regulation of hexose transport in animal cells is modulated partially by a repression mechanism and partially by variations in carrier turnover. Their model predicts the following: (1) inhibition of protein synthesis with no change in turnover would lead to a decrease in transport activity; (2) inhibition of turnover with continued protein synthesis would increase transport activity; and (3) when both protein synthesis and turnover are inhibited no change in transport rate would occur. This area of research in the regulation of transport is currently under active investigation.

ACKNOWLEDGMENTS

We wish to thank Dr. P. Moore and Ms. N. McGah for assistance in preparing this manuscript. Portions of this manuscript were written while S. C. Quay was a Research Associate in the laboratory of Dr. H. G. Khorana, Biology and Chemistry Department, Massachusetts Institute of Technology. The research from our laboratory which is described here was supported by Public Health Service Grants GM11024 and GM20737.

References

Adler, J., and Epstein, W., 1974, Phosphotransferase system enzymes as chemoreceptors for certain sugars in *Escherichia coli* chemotaxis, *Proc. Natl. Acad. Sci. U.S.A.* **71**:2895.
Alfoldi, L., and Kerekes, E., 1964, Neutralization of the amino acid sensitivity of RC[rel] *Escherichia coli,* *Biochim. Biophys. Acta* **91**:155.

Ames, G. F., 1964, Uptake of amino acids by *Salmonella typhimurium, Arch. Biochem. Biophys.* **104**:1.

Anderson, J. J., and Oxender, D. L., 1977, *E. coli* mutants lacking binding protein and other components of the branched-chain amino acid transport system, *J. Bacteriol.* **130**:384.

Anderson, J. J., Quay, S. C., and Oxender, D. L., 1976, Mapping of two loci affecting regulation of branched-chain amino acid transport in *Escherichia coli* K-12, *J. Bacteriol.* **126**:80.

Beckwith, J., 1963, Restoration of operon activity by suppressors, *Biochim. Biophys. Acta* **76**:162.

Benko, P. V., Wood, T. C., and Segel, I. H., 1967, Specificity and regulation of methionine transport in filamentous fungi, *Arch. Biochem. Biophys.* **122**:783.

Berger, E. A., and Heppel, L. A., 1974, Different mechanisms of energy coupling for the shock-sensitive and shock-resistant amino acid permeases of *Escherichia coli, J. Biol. Chem.* **249**:7747.

Bertrand, K., Korn, L., Lee, F., Platt, T., Squires, C. L., Squires, C., and Yanofsky, C., 1975, New features of the regulation of the tryptophan operon, *Science* **189**:22.

Blobel, G., and Dobberstein, B., 1975a, Transfer of proteins across membranes. I. Presence of proteolytically processed and unprocessed nascent immunoglobulin light chains on membrane-bound ribosomes of murine myeloma, *J. Cell Biol.* **67**:835.

Blobel, G., and Dobberstein, B., 1975b, Transfer of proteins across membranes. II. Reconstitution of functional rough microsomes from heterologous components, *J. Cell Biol.* **67**:852.

Blobel, G., and Sabatini, D. D., 1971, Ribosome-membrane interaction in eukaryotic cells, *Biomembranes* **2**:193.

Borek, E., Ryan, A., and Rockenbach, J., 1955, Nucleic acid metabolism in relation to the lysogenic phenomenon, *J. Bacteriol.* **69**:460.

Browman, R. L., Goldenbaum, P. E., and Dobrogosz, W. J., 1970, The effect of amino acids on the ability of cyclic AMP to reverse catabolite repression in *Escherichia coli, Biochem. Biophys. Res. Commun.* **39**:401.

Brown, K. D., 1970, Formation of aromatic amino acid pools in *Escherichia coli* K12, *J. Bacteriol.* **104**:177.

Burns, R. O., Calvo, J. M., Margolin, P., and Umbarger, H. E., 1966, Expression of the leucine operon, *J. Bacteriol.* **91**:5170.

Burrous, S. E., and DeMoss, R. D., 1963, Studies on tryptophan permease in *Escherichia coli, Biochim. Biophys. Acta* **73**:623.

Calhoun, D. H., 1976, Threonine deaminase from *Escherichia coli*. Feedback-hypersensitive enzyme from a genetic regulatory mutant, *J. Bacteriol.* **126**:56.

Christopher, C. W., Colby, W. W., and Ullrey, D., 1976, Derepression and carrier turnover: Evidence for two distinct mechanisms of hexose transport regulation in animal cells, *J. Cell. Physiol.* **89**:683.

Christopher, C. W., Colby, W., Ullrey, D., and Kalckar, H. M., 1977, Comparative studies of glucose-fed and glucose-starved hamster cell cultures: Responses in galactose metabolism, *J. Cell. Physiol.* **90**:387.

Cordaro, J. C., Anderson, R. P., Grogan, E. W., Wenzel, D., Engler, M., and Roseman, S., 1974, Promoter-like mutation affecting HPr and Enzyme I of the phosphoenolpyruvate:sugar phosphotransferase system in *Salmonella typhimurium, J. Bacteriol.* **120**:245.

Crabeel, M., and Grenson, M., 1970, Regulation of histidine uptake by specific feedback inhibition of two histidine permeases in *Saccaromyces cerevisiae, Eur. J. Biochem.* **14**:197.

Cunningham, D. D., and Pardee, A. B., 1969, Transport changes rapidly initiated by serum addition to contact-inhibited 3T3 cells, *Proc. Natl. Acad. Sci. U.S.A.* **64**:1049.

Davie, E. W., Fujikawa, K., Legaz, M. E., and Kato, H., 1975, Role of proteases in blood coagulation, in: *Proteases and Biological Control* (E. Reich, D. B. Rifkin, and E. Shaw, eds.), Vol. 2, pp. 65–78, Cold Spring Harbor Laboratory, Cold Spring Harbor, New York.

Deutch, C. E., and Soffer, R. L., 1975, Regulation of proline catabolism by leucyl-, phenyalanyl-tRNA protein transferase, *Proc. Natl. Acad. Sci. U.S.A.* **72**:405.

Eckhart, W., and Weber, M. J., 1973, Uptake of 2-deoxyglucose by BALB/3T3 cells: Changes after polyoma infection, *Virology* **61**:223.

Eidlic, T., and Neidhardt, F. C., 1965, Role of valyl-sRNA synthetase in enzyme repression, *Proc. Natl. Acad. Sci. U.S.A.* **53**:539.

Foster, D. D., and Pardee, A. B., 1969, Transport of amino acids by confluent and nonconfluent 3T3 and polyoma virus-transformed 3T3 cells growing on glass cover slips, *J. Biol. Chem.* **244**:2675.

Fournier, M. J., and Peterkofsky, A., 1975, Formation of chromatographically unique species of transfer ribonucleic acid during amino acid starvation of relaxed-control *Escherichia coli, J. Bacteriol.* **122**:538.

Fraenkel, D. G., 1968, The phosphoenolpyruvate-initiated pathway of fructose metabolism in *Escherichia coli, J. Biol. Chem.* **243**:6458.

Fraser, J., and Newman, E. M., 1975, Derivation of glycine from threonine in *Escherichia coli* K-12 mutants, *J. Bacteriol.* **122**:810.

Freundlich, M., Burns, R. O., and Umbarger, H. E., 1962, Control of isoleucine, valine, and leucine biosynthesis. I. Multivalent repression, *Proc. Natl. Acad. Sci. U.S.A.* **48**:1804.

Furlong, C. E., and Weiner, J. H., 1970, Purification of a leucine specific binding protein from *Escherichia coli, Biochem. Biophys. Res. Commun.* **38**:1076.

Gale, E. F., 1943, Factors influencing enzymic activities of bacteria, *Bacteriol. Rev.* **7**:139.

Gazzola, G. C., Franchi, R., Saibene, V., Ronchi, P., and Guidotti, G. G., 1972, Regulation of amino acid transport in chick embryo heart cells. I. Adaptive system of mediation for neutral amino acids, *Biochim. Biophys. Acta* **266**:407.

Goldberg, A. L., and St. John, A. C., 1976, Intracellular protein degradation in mammalian and bacterial cells: Part 2, *Annu. Rev. Biochem.* **45**:747.

Gonzalez, J. E., and Peterkofsky, A., 1977, The mechanism of sugar-dependent-repression of synthesis of catabolic enzymes in *Escherichia coli, J. Supramol. Struct.* **6**:495.

Greene, R. C., and Radovich, C., 1975, Role of methionine in the regulation of serine hydroxymethyl-transferase in *Escherichia coli., J. Bacteriol.* **124**:269.

Guardiola, J., DeFelice, M., Klopotowski, T., and Iaccarino, M., 1974*a*, Multiplicity of isoleucine, leucine, and valine transport systems in *Escherichia coli* K12, *J. Bacteriol.* **117**:383.

Guardiola, J., DeFelice, M., Klopotowski, T., and Iaccarino, M., 1974*b*, Mutations affecting the different transport systems for isoleucine, leucine, and valine in *Escherichia coli* K-12, *J. Bacteriol.* **117**:393.

Guidotti, G. G., Gazzola, G. C., Borghetti, A. F., and Franchi-Gazzola, R., 1975, Adaptive regulation of amino acid transport across the cell membrane in avian and mammalian tissues, *Biochim. Biophys. Acta* **406**:264.

Guidotti, G. G., Borghetti, A. F., Gazzola, G. C., Tramacere, M., and Dall'asta, V., 1976, Insulin regulation of amino acid transport in mesenchymal cells from avian and mammalian tissues, *Biochem. J.* **160**:281.

Hanson, T. E., and Anderson, R. L., 1968, Phosphoenolpyruvate-dependent formation of D-fructose-1-phosphate by a four-component phosphotransferase system, *Proc. Natl. Acad. Sci. U.S.A.* **61**:269.

Haseltine, W. A., and Block, R., 1973, Synthesis of guanosine tetra- and pentaphosphate requires the presence of a codon specific, uncharged transfer ribonucleic acid in the acceptor site of ribosomes. *Proc. Natl. Acad. Sci. U.S.A.* **70**:1564.

Hatanaka, M., 1976, Saturable and nonsaturable process of sugar uptake: Effect of oncogenic transformation on transport and uptake of nutrients, *J. Cell. Physiol.* **89**:745.

Heaton, J. H., and Gelehrter, T. D., 1977, Derepression of amino acid transport by amino acid starvation in rat hepatoma cells, *J. Biol. Chem.* **252**:2900.

Heinz, E., and Durbin, R. P., 1957, Studies of the chloride transport in the gastric mucosa of the frog, *J. Gen. Physiol.* **41**:101.

Heinz, E., and Walsh, P. O., 1958, Exchange diffusion, transport, and intracellular level of amino acids in Ehrlich carcinoma cells, *J. Biol. Chem.* **233**:1488.

Hillman, R. E., and Otto, E. F., 1974, Transport of L-isoleucine by cultured human fibroblasts, *J. Biol. Chem.* **249**:3430.

Hirshfield, I. N., Yeh, F. M., and Sawyer, L. E., 1975, Metabolites influence control of lysine transfer ribonucleic acid synthetase formation in *Escherichia coli* K12, *Proc. Natl. Acad. Sci. U.S.A.* **72**:1364.

Hogg, R. L., and Hermodson, M. A., 1977, Amino acid sequence of the L-arabinose-binding protein from *Escherichia coli* B/r, *J. Biol. Chem.* **252**:5135.

Holley, R. W., 1972, A unifying hypothesis concerning the nature of malignant growth, *Proc. Natl. Acad. Sci. U.S.A.* **69**:2840.

Horvath, I., and Gado, I., 1965, Possible causes of leucine inhibition in *Escherichia coli* K₁₂ λ-28, *Acta Microbiol. Acad. Sci. Hung.* **12**:103.

Ikemura, T., and Dahlberg, J. E., 1973, Small ribonucleic acids of *Escherichia coli*. II. Noncoordinate accumulation during stringent control, *J. Biol. Chem.* **248**:5033.

Inouye, H., and Beckwith, J., 1977, Synthesis and processing of an *Escherichia coli* alkaline phosphatase precursor *in vitro, Proc. Natl. Acad. Sci. U.S.A.* **74**:1440.

Inouye, H., Pratt, C., Beckwith, J., and Torriani, A., 1977, Alkaline phosphatase synthesis in a cell-free system using DNA and RNA templates, *J. Mol. Biol.* **110**:75.

Inui, Y., and Akedo, H., 1965, Amino acid uptake by *Escherichia coli* grown in presence of amino acids. Evidence for repressibility of amino acid uptake, *Biochim. Biophys. Acta* **94**:143.

Isselbacher, K. J., 1972, Increased uptake of amino acids and 2-deoxy-D-glucose by virus-transformed cells in culture, *Proc. Natl. Acad. Sci. U.S.A.* **69**:585.

Jacob, F., and Monod, J., 1961, Genetic regulatory mechanisms in the synthesis of proteins, *J. Mol. Biol.* **3**:318.

Kadner, R. J., 1975, Regulation of methionine transport activity in *Escherichia coli, J. Bacteriol.* **122**:110.

Kalckar, H. M., and Ullrey, D., 1973, Two distinct types of enhancement of galactose uptake into hamster cells: Tumor virus transformation and hexose starvation, *Proc. Natl. Acad. Sci. U.S.A.* **70**:2502.

Kalckar, H. M., Christopher, C. W., and Ullrey, D., 1976, Neoplastic potentials and regulation of uptake of nutrients. II. Inverse regulation of uptake of hexose and amino acid analogues in the neoplastic GIV line, *J. Cell. Physiol.* **89**:765.

Kane, J. F., 1975, Metabolic interlock: Mediation of interpathway regulation by divalent cations, *Arch. Biochem. Biophys.* **170**:452.

Kano-Sueoka, T., and Sueoka, N., 1969, Leucine tRNA and cessation of *Escherichia coli* protein synthesis upon phage T2 infection, *Proc. Natl. Acad. Sci. U.S.A.* **62**:1229.

Kanzaki, S., and Anraku, Y., 1971, Transport of sugars and amino acids in bacteria. IV. Regulation of valine transport activity by valine and cysteine, *J. Biochem. (Tokyo)* **70**:215.

Kimelberg, H. K., and Mayhew, E., 1975, Increased ouabain-sensitive ^{86}Rb$^+$ uptake and sodium and potassium ion-activated adenosine triphosphatase activity in transformed cell lines, *J. Biol. Chem.* **250**:100.

Kitchingman, G. R., and Fournier, M. J., 1975, Unbalanced growth and the production of unique transfer ribonucleic acids in relaxed-control *Escherichia coli*, *J. Bacteriol.* **124**:1382.

Kitchingman, G. R., and Fournier, M. J., 1977, Modification-deficient transfer ribonucleic acids from relaxed control *Escherichia coli:* Structures of the major undermodified phenylalanine and leucine transfer RNAs produced during leucine starvation, *Biochemistry* **16**:2013.

Kitchingman, G. R., Webb, E., and Fournier, M. J., 1976, Unique phenylalanine transfer ribonucleic acids in relaxed control *Escherichia coli:* Genetic origin and some functional properties, *Biochemistry* **15**:1848.

Kjeldgaard, N. O., Maaløe, O., and Schaechter, M., 1958, The transition between different physiological states during balanced growth of *Salmonella typhimurium*, *J. Gen. Microbiol.* **19**:607.

Kletzien, R. F., and Perdue, J. F., 1974, Sugar transport in chick embryo fibroblasts. II. Alterations in transport following transformation by a temperature-sensitive mutant of the Rous-sarcoma virus, *J. Biol. Chem.* **249**:3375.

Kletzien, R. F., and Perdue, J. F., 1975, Induction of sugar transport in chick embryo fibroblasts by hexose starvation: Evidence for transcriptional regulation of transport, *J. Biol. Chem.* **250**:593.

Kletzien, R. F., and Perdue, J. F., 1976, Regulation of sugar transport in chick embryo fibroblasts and in fibroblasts transformed by a temperature-sensitive mutant of the Rous sarcoma virus, *J. Cell. Physiol.* **89**:723.

Korn, L. J., and Yanofsky, C., 1976, Polarity suppressors defective in transcription termination at the attenuation of the tryptophan operon of *Escherichia coli* have altered rho factor, *J. Mol. Biol.* **106**:231.

Kornberg, H. L., and Reeves, R. E., 1972, Inducible phosphoenolpyruvate dependent hexose phosphotransferase activities in *Escherichia coli*, *Biochem. J.* **128**:1339.

Laemmli, U. K., 1975, Cleavage associated with the maturation of the head of bacteriophage T4, in: *Proteases and Biological Control* (E. Reich, D. B. Rifkin, and E. Shaw, eds.), Vol. 2, pp. 661–687, Cold Spring Harbor Laboratory, Cold Spring Harbor, New York.

Lengeler, J., 1975, Mutations affecting transport of the hexitols D-mannitol, D-glucitol, and galactitol in *Escherichia coli* K12: Isolation and mapping, *J. Bacteriol.* **124**:26.

Lengeler, J., 1977, Analysis of mutations affecting the dissimilation of galactitol (Dulcitol) in *Escherichia coli* K12, *Mol. Gen. Genet.* **152**:83.

Levinthal, M., Williams, L. S., Levinthal, M., and Umbarger, H. E., 1973, Role of threonine deaminase in the regulation of isoleucine and valine biosynthesis, *Nature (London) New Biol.* **246**:65.

McGinnis, J. F., and Paigen, K., 1969, Catabolite inhibition: A general phenomenon in the control of carbohydrate utilization, *J. Bacteriol.* **100**:902.

McGinnis, E., and Williams, L. S., 1971, Regulation of synthesis of the aminoacyl-transfer ribonucleic acid synthetases for the branched-chain amino acids of *Escherichia coli*, *J. Bacteriol.* **108**:254.

Magasanik, B., 1970, Catabolite repression, in: *The Lactose Operon* (J. R. Beckwith and D. Zipser, eds.), pp. 189–219, Cold Spring Harbor Laboratory, Cold Spring Harbor, New York.

Markman, R. S., and Sutherland, E. W., 1965, Adenosine 3′,5′-phosphate in *Escherichia coli*, *J. Biol. Chem.* **240**:1309.

Martineau, R., Kohlbacher, M., and Shaw, S. N., 1972, Enhancement of hexose entry into chick fibroblasts by starvation: Differential effect on galactose and glucose, *Proc. Natl. Acad. Sci. U.S.A.* **69**:3407.

Monod, J., 1947, Phenomenon of enzymatic adaption and its bearings on problems of genetics and cellular differentiation, *Growth* **11**:223.

Moore, P. A., Jayme, D. W., and Oxender, D. L., 1977, A role for aminoacyl-tRNA synthetases in the regulation of amino acid transport in mammalian cell lines, *J. Biol. Chem.* **252**:7427.

Neidhardt, F. C., 1963, Properties of a bacterial mutant lacking amino acid control of RNA synthesis, *Biochim. Biophys. Acta* **68**:365.

Neurath, H., 1975, Limited proteolysis and zymogen activation, in: *Proteases and Biological Control* (E. Reich, D. B. Rifkin, and E. Shaw, Eds.), Vol. 2, pp. 51–64, Cold Spring Harbor Laboratory, Cold Spring Harbor, New York.

Neville, M. M., Suskind, S. R., and Roseman, S., 1971, A derepressible active transport system for glucose in *Neurospora crassa*, *J. Biol. Chem.* **246**:1294.

Ovchinnikov, A. Y., Aldanova, N. A., Grinkevich, V. A., Arzamazova, N. M., Moroz, I. N., and Nazimov, I. V., 1977, The primary structure of LIV-binding protein from *E. coli*, *Bioorg. Chem.* **3**:564.

Oxender, D. L., 1972, Membrane transport, *Annu. Rev. Biochem.* **41**:777.

Oxender, D. L., and Christensen, H. N., 1963, Distinct mediating systems for the transport of neutral amino acids by the Ehrlich cell, *J. Biol. Chem.* **238**:3686.

Oxender, D. L., and Quay, S. C., 1976a, Regulation of leucine transport and binding proteins in *Escherichia coli*, *J. Cell. Physiol.* **89**:517.

Oxender, D. L., and Quay, S. C., 1976b, Isolation and characterization of membrane binding proteins, in: *Methods in Membrane Biology* (E. D. Korn, ed.), Vol, 6, pp. 183–242, Plenum Press, New York.

Oxender, D. L., Lee, M., and Ceccini, G., 1976, Regulation of transport in mammalian cell culture, in: *Progress in Clinical and Biological Research, Membranes and Neoplasia: New Approaches and Strategies* (V. T. Marchesi, ed.), Vol. 9, pp. 41–47, Alan R. Liss, New York.

Oxender, D. L., Lee, M., Moore, P. A., and Cecchini, G., 1977a, Neutral amino acid transport systems of tissue culture cells, *J. Biol. Chem.* **252**:2675.

Oxender, D. L., Lee, M., and Cecchini, G., 1977b, Regulation of amino acid transport activity and growth rate of animal cells in culture, *J. Biol. Chem.* **252**:2680.

Pall, M. L., 1971, Amino acid transport in *Neurospora crassa*. IV. Properties and regulation of methionine transport, *Biochim. Biophys. Acta* **233**:201.

Pardee, A. B., 1964, Cell division and a hypothesis of cancer, *Natl. Cancer Inst. Monogr.* **14**:7.

Pardee, A. B., and Prestidge, L. S., 1955, Induced formation of serine and threonine deaminase by *Escherichia coli*, *J. Bacteriol.* **70**:667.

Peck, W. A., Rockwell, L. H., and Lichtman, M. A., 1976, Adaptive enhancement of amino acid uptake and exodus by thymic lymphocytes: Influence of pH, *J. Cell. Physiol.* **89**:417.

Pedersen, F. S., Lund, E., and Kjeldgaard, N. O., 1973, Codon specific, tRNA dependent *in vitro* synthesis of ppGpp and ppGpp, *Nature (London), New Biol.* **243**:13.

Penrose, W. R., Nicholalds, G. E., Piperno, J. R., and Oxender, D. L., 1968, Purification and properties of a leucine-binding protein from *Escherichia coli*, *J. Biol. Chem.* **243**:5921.

Perdue, J. F., 1976, Loss of the post-translational control of nutrient transport in in vitro and in vivo virus-transformed chicken cells, *J. Cell. Physiol.* **89**:729.

Peterkofsky, A., and Gazdar, C., 1974, Glucose inhibition of adenylate cyclase in intact cells of *Escherichia coli* B, *Proc. Natl. Sci. U.S.A.* **71**:2324.

Peterkofsky, A., and Gazdar, C., 1975, Interaction of enzyme I of the phosphoenolpyruvate: Sugar phosphotransferase system with adenylate cyclase of *Escherichia coli*, *Proc. Natl. Acad. Sci. U.S.A.* **72**:2920.

Piperno, J. R., and Oxender, D. L., 1968, Amino acid transport systems in *Escherichia coli* K12, *J. Biol. Chem.* **243**:5914.

Postma, P. W., and Roseman, S., 1976, The bacterial phosphoenolpyruvate: Sugar phosphotransferase system, *Biochim. Biophys. Acta* **457**:213.

Quay, S. C., and Christensen, H. N., 1974, Basis of transport discrimination of arginine from other basic amino acids in *Salmonella typhimurium*, *J. Biol. Chem.* **249**:7011.

Quay, S. C., and Oxender, D. L., 1976, Regulation of branched-chain amino acid transport in *Escherichia coli*, *J. Bacteriol.* **127**:1225.

Quay, S. C., and Oxender, D. L., 1977, Regulation of amino acid transport in *Escherichia coli* by transcriptional termination factor rho, *J. Bacteriol.* **130**:1024.

Quay, S. C., Oxender, D. L., Tsuyumu, S., and Umbarger, H. E., 1975a, Separate regulation of transport and biosynthesis of leucine, isoleucine, and valine in bacteria, *J. Bacteriol.* **122**:994.

Quay, S. C., Kline, E. L., and Oxender, D. L., 1975b, Role of the leucyl-tRNA synthetase in regulation of transport, *Proc. Natl. Acad. Sci. U.S.A.* **72**:3921.

Quay, S. C., Dick, T. E., and Oxender, D. L., 1977, Role of transport systems in amino acid metabolism: Leucine toxicity and the branched-chain amino acid transport systems, *J. Bacteriol.* **129**:1257.

Quay, S. C., Lawther, R. P., Hatfield, G. W., and Oxender, D. L., 1978, Branched-chain amino acid transport regulation in mutants blocked in tRNA maturation and transcriptional termination, *J. Bacteriol.* **134**:683.

Rahmanian, M., and Oxender, D. L., 1972, Derepressed leucine transport activity in *Escherichia coli, J. Supramol. Struct.* **1**:55.

Rahmanian, M., Claus, D. R., and Oxender, D. L., 1973, Multiplicity of leucine transport systems in *Escherichia coli* K-12, *J. Bacteriol.* **116**:1258.

Reich, E., Rifkin, D. B., and Shaw, E. (eds.), 1975, *Proteases and Biological Control,* Vol. 2, Cold Spring Harbor Laboratory, Cold Spring Harbor, New York.

Reynolds, R. A., and Segal, S., 1976, Regulatory characteristics of amino acid transport in newborn rat renal cortex cells, *Biochim. Biophys. Acta* **426**:513.

Ribbons, D. W., and Brew, K. (eds.), 1976, *Proteolysis and Physiological Regulation, Miami Winter Symposia,* Vol. II, Academic Press, New York.

Richaud, C., Mazat, J.-P., Felenbok, B., and Patte, J.-C., 1974, The role of lysine and leucine binding on the catalytic and structural properties of aspartokinase III of *Escherichia coli* K12, *Eur. J. Biochem.* **48**:147.

Riggs, T. R., and Pan, M. W., 1972, Transport of amino acids into the oestrogen-primed uterus: Enhancement of uptake by a preliminary incubation, *Biochem. J.* **128**:19.

Ring, K., Gross, W., and Heinz, E., 1970, Negative feedback regulation of amino acid transport in *Streptomyces hydrogenans, Arch. Biochem. Biophys.* **137**:243.

Rizzino, A. A., Bresalier, R. S., and Freundlich, M., 1974, Derepressed levels of the isoleucine-valine and leucine enzymes in *hisT 1504,* a strain of *Salmonella typhimurium* with altered leucine transfer ribonucleic acid, *J. Bacteriol.* **117**:449.

Robbins, J. C., 1973, Ph.D. thesis, Transport systems for alanine, serine, and glycine in *Escherichia coli* K-12, The University of Michigan, Ann Arbor.

Robbins, J. C., and Oxender, D. L., 1973, Transport systems for alanine, serine and glycine in *Escherichia coli* K-12, *J. Bacteriol.* **116**:12.

Rogerson, A. C., and Freundlich, M., 1970, Control of isoleucine, valine, and leucine biosynthesis. VIII. Mechanism of growth inhibition by leucine in relaxed and stringent strains of *Escherichia coli* K12, *Biochim. Biophys. Acta* **208**:87.

Roon, R. J., Larimore, F., and Levy, J. S., 1975, Inhibition of amino acid transport by ammonium ion in *Saccaromyces cerevisiae, J. Bacteriol* **124**:325.

Roon, R. J., Levy, J. S., and Larimore, F., 1977, Negative interactions between amino acid and methylamine/ammonia transport systems of *Saccaromyces cerevisiae, J. Biol. Chem.* **252**:3599.

Saier, M. H., Jr., 1977, Bacterial phosphoenolpyruvate:sugar phosphotransferase systems: Structural, functional, and evolutionary interrelationships, *Bacteriol. Rev.* **41**:856.

Saier, M. H., Jr., Simoni, R. D., and Roseman, S., 1970, The physiological behavior of enzyme I and heat-stable protein mutants of a bacterial phosphotransferase system, *J. Biol. Chem.* **245**:5870.

Saier, M. H., Jr., Feucht, B. U., and Hofstadter, L. J., 1976, Regulation of carbohydrate uptake and adenylate cyclase activity mediated by the enzyme II of the phosphoenolypruvate:sugar phosphotransferase system in *Escherichia coli, J. Biol. Chem.* **251**:883.

Scarborough, G. A., 1970, Sugar transport in *Neurospora crassa, J. Biol. Chem.* **245**:1694.

Schimke, R. T., and Bradley, M. O., 1975, Properties of protein turnover in animal cells and a possible role for turnover in "quality" control of proteins, in: *Protease and Biological Control* (E. Reich, D. B. Rifkin, and E. Shaw, eds.), Vol. 2, pp. 515–530, Cold Spring Harbor Laboratory, Cold Spring Harbor, New York.

Stein, W. D., 1967, in: *The Movement of Molecules across Cell Membranes,* pp. 52–61, Academic Press, New York.

Steiner, D. F., Kemmler, W., Tager, H. S., Rubenstein, A. H., Lernmark, A., and Zühlke, H., 1975, Proteolytic mechanisms in the biosynthesis of polypeptide hormones, in: *Proteases and Biological Control* (E. Reich, D. B. Rifkin, and E. Shaw, eds.), Vol. 2, pp. 531–549, Cold Spring Harbor Laboratory, Cold Spring Harbor, New York.

Stephens, J. C., Artz, S. W., and Ames, B. N., 1975, Guanosine-5′-diphosphate-3′-diphosphate (ppGpp): Positive effector for histidine operon transcription and general signal for amino acid deficiency, *Proc. Natl. Acad. Sci. U.S.A.* **72**:4389.

Templeton, B. A., and Savageau, M. A., 1974, Transport of biosynthetic intermediates: Regulation of homoserine and threonine uptake in *Escherichia coli, J. Bacteriol.* **120**:114.

Thompson, L. H., Harkins, J. L., and Stanners, C. P., 1973, A mammalian cell mutant with a temperature-sensitive leucyl-transfer RNA synthetase, *Proc. Natl. Acad. Sci. U.S.A.* **70**:3094.

Thompson, L. H., Stanners, C. P., and Siminovitch, L., 1975, Selection by (^{3}H) amino acids of CHO-cell mutants with altered leucyl- and asparaginyl-transfer RNA synthetases, *Som. Cell Genet.* **1**:187.

Thompson, L. H., Lofgren, D. J., and Adair, B. M., 1977, CHO cell mutants for arginyl-, asparaginyl-, glutaminyl-, and methionyl-transfer RNA synthetases: Identification and initial characterization, *Cell* **11**:157.

Travers, A., 1976a, RNA polymerase specificity and the control of growth, *Nature (London)* **263**:641.

Travers, A., 1976b, Template selection by *E. coli* RNA polymerase holoenzyme, *FEBS Lett.* **69**:195.

Ullrey, D., Gammon, M. T., and Kalckar, H. M., 1975, Uptake patterns and transport enhancements in cultures of hamster cells deprived of carbohydrates, *Arch. Biochem. Biophys.* **167**:410.

Umbarger, H. E., 1973, Genetic and physiological regulation of isoleucine, valine, and leucine formation in the *Enterobacteriaceae,* in: *Genetics of Industrial Organisms* (Z. Vanek, Z. Hostalek, and J. Culdin, eds.), pp. 195–218, Academic Publishing House, Prague.

Venuta, S., and Rubin, H., 1973, Sugar transport in normal and Rous-sarcoma virus-transformed chick embryo fibroblasts, *Proc. Natl. Acad. Sci. U.S.A.* **70**:653.

Vonder Haar, R. A., and Umbarger, H. E., 1972, Isoleucine and valine metabolism in *Escherichia coli.* XIX. Inhibition of isoleucine biosynthesis by glycyl-leucine, *J. Bacteriol.* **112**:142.

von Ehrenstein, G., 1970, Transfer RNA and amino acid activation, in: *Aspects of Protein Biosynthesis* (C. B. Anfinsen, Jr., ed.), Part A, pp. 139–214, Academic Press, New York.

Wasmuth, J. J., and Caskey, C. T., 1976, Selection of temperature-sensitive CHO asparaginyl-tRNA synthetase mutants using the toxic lysine analog, S-2-aminoethyl-L-cysteine, *Cell* **9**:655.

Weber, M. J., 1973, Hexose transport in normal and in Rous sarcoma virus-transformed cells, *J. Biol. Chem.* **248**:2978.

Weber, M. J., Hale, A. H., Yau, T. M., Buckman, T., Johnson, M., Brady, T. M., and Larossa, D. D., 1976, Transport changes associated with growth control and malignant transformation, *J. Cell. Physiol.* **89**:711.

Wettstein, F. O., and Stent, G., 1968, Physiologically induced changes in the properties of phenylalanine tRNA in *Escherichia coli, J. Mol. Biol.* **38**:25.

Whipp, M. J., and Piffard, A. J., 1977, Regulation of aromatic amino acid transport systems in *Escherichia coli* K-12, *J. Bacteriol.* **132**:453.

Wiley, W. R., and Matchett, N. H., 1966, Tryptophan transport in *Neurospora crassa.* I. Specificity and kinetics, *J. Bacteriol.* **92**:1698.

Willis, R. C., Iwata, K. K., and Furlong, C. E., 1975, Regulation of glutamine transport in *Escherichia coli, J. Bacteriol.* **122**:1032.

Yamamoto, S., and Lampen, J. O., 1976, The hydrophobic membrane penecillinase of *Bacillus lichenformis* 749/C. Characterization of the hydrophilic enzyme and phospholipopeptide produced by trypsin cleavage, *J. Biol. Chem.* **251**:4102.

Yegian, C. D., and Stent, G. S., 1969, An unusual condition of leucine transfer RNA appearing during leucine starvation of *Escherichia coli, J. Mol. Biol.* **39**:45.

9

Immunoglobulin Diversity
Regulation of Expression of Immunoglobulin Genes during Primary Development of B Cells

ALEXANDER R. LAWTON, JOHN F. KEARNEY, and
MAX D. COOPER

1 Introduction

Since it was perceived that the capacity of immunoglobulin molecules to bind specifically to antigenic determinants was determined by their amino acid sequences, and therefore represented genetic information, there have been impressive attempts to clarify the problem of how the immense diversity of genes coding for antibody molecules of different specificities might be generated. Approaches to the question of generation of diversity have been largely based on comparative analysis of the amino acid sequences of the immunoglobulin molecules derived from plasma cell tumors of mice and humans. These studies have provided fundamental information on the nature of the structural genes that code for immunoglobulin molecules, have given insights into their evolution, and have defined constraints within which models for the generation of diversity must be fitted (Hood *et al.*, 1977*b*).

In this chapter we will attempt to treat the problem of immunoglobulin diversity from a different perspective. Rather than ask how diversity is generated, we will focus on the regulation of expression of immunoglobulin genes during the primary development of lymphoid cells (B lymphocytes), which utilize these molecules as receptors, and whose progeny

ALEXANDER R. LAWTON, JOHN F. KEARNEY, and MAX D. COOPER ● The Cellular Immunology Unit of the Tumor Institute, Departments of Pediatrics and Microbiology, and the Comprehensive Cancer Center, University of Alabama in Birmingham, Birmingham, Alabama 35294 (plasma cells) secrete specific antibodies. It is now apparent that very complex genetic rearrangements must occur during differentiation of B lymphocytes. Understanding the mechanisms involved may be relevant to the more general problem of genetic regulation of differentiation in eukaryotic cells.

B lymphocytes represent one major division of immunocompetent cells. T lymphocytes, whose name is derived from their developmental dependence on the thymus, represent the other. T lymphocytes carry out a variety of functions collectively called cell-mediated immunity. They may use molecules similar to immunoglobulins as a part of their specific antigen receptors, but do not secrete antibodies. Because precise information on the nature of the antigen-specific T cell receptor has been lacking, analysis of the early development of this cell line has been much more difficult. We have therefore chosen to limit this review to B cell development.

1.1 Information Handling and the Immune System

We begin with a brief statement of the dogma of immunology within which schemes for lymphoid differentiation must be accommodated. The central characteristics of immunity are diversity, specificity, and memory. With proper conditions of immunization, higher vertebrates have the capacity to respond to an almost limitless array of organic compounds, whether naturally occurring or synthetic. The antibodies produced and secreted following immunization are highly specific, able to distinguish very minor chemical or conformational differences among haptens. (Haptens are the chemical groupings on macromolecules that can be accommodated within the binding site of antibodies.) Initial contact with a particular antigen induces both a response and a state of memory. Reintroduction of the same, or a closely related, antigen results in an increase in the quantity of antibody produced and in the average affinity with which antibody molecules bind the antigen.

In the early part of this century Erlich (1900) proposed that cells might bear preformed antibodies on their surfaces and that interaction with antigen resulted in their more rapid production and secretion. This germinal idea was restated and expanded by Burnet (1959) as the *clonal selection hypothesis*. The essence of clonal selection is that the specificity of antibodies is genetically determined. Each individual possesses families of lymphocytes (clones) defined by the expression of antibody receptors of a unique predetermined specificity. Antigens select from among this clonal repertoire those cells having complementary receptors and induce their proliferation and differentiation to cells secreting antibodies of the original specificity. The contemporary opposing theories, called *instructionist*, held that antigen served as a template to direct either primary synthesis or secondary folding of antibody molecules. The basic tenants of the clonal selection hypothesis have not been seriously challenged, as virtually all of the information gained in both molecular and cellular immunology during the past two decades have fulfilled its predictions.

The cellular requirements of clonal selection are basically two. First, the diversity of genetic information for antibody specificity must be expressed on B lymphocyte precursors of secretory plasma cells. Second, in order to account for specificity and memory, each lymphocyte (or clone) must bear receptors of a single unique specificity. The process of lymphoid differentiation is therefore divisible into two separate phases, clonal development and clonal selection. Clonal development encompasses the mechanism through which genetic information coding for specific antibody molecules becomes expressed as functional receptors on separate clones of lymphocytes. Clonal selection constitutes the differentiative events that follow selective stimulation of clones bearing receptors of predetermined specificity by antigens.

The distinction between these stages of lymphoid development is to some extent dependent on the mechanisms for generation of antibody diversity. Two general theories have been proposed. According to the *germ-line hypothesis,* the entire specificity repertoire of each individual organism is represented in inherited germ-line genes. The mutations and other genetic events involved in development of antibody diversity have occurred during evolution. *Somatic mutational theories,* on the other hand, hold that only a small proportion of the diversity present in a mature animal is inherited; the bulk of antibody diversity is generated by mutational events during the life of each individual.

In its purest form, somatic generation of diversity would obviate the distinction we have made between clonal development and clonal selection, as the only requirement of primary development would be clonal expression of a minimum number of germ-line specificity genes. Available information on the sequences of the specificity regions of antibody molecules indicates that the minimum number of different germ-line genes may be on the order of 100s (Cohn *et al.,* 1974; Hood *et al.,* 1977a,b). Moreover, ontogenetic evidence strongly suggests that substantial diversity is generated and expressed independently of exogenous antigenic stimulation. Even if somatic mutations account for expansion of the clonal specificity repertoire by several orders of magnitude, understanding the mechanisms involved in initial expression of hundreds of inherited genes by different clones of lymphocytes remains a problem of considerable significance and complexity.

1.2   Two Genes, One Polypeptide

Early studies on the primary structure of immunoglobulin light chains revealed a paradox. Approximately half the peptides generated by proteolysis of light chains from different myeloma tumors proved to be identical by peptide mapping. The remaining peptides appeared to be unique for each individual kappa chain (Putnam, 1962; Potter *et al.,* 1964). On the basis of these observations, Dreyer and Bennett (1965) proposed that immunoglobulin light and heavy chains might each be the product of two separate genes. All the data since generated on amino acid sequences of both heavy and light chains have been consistent with this formulation. The variable regions of light and heavy chains, products of V genes, consist of the NH_2-terminal residues, approximately 110 amino acids in length. The remainder of each chain is coded by a constant (C) gene. The V genes contain information that determines the specificity of antibodies, whereas C genes determine the type of light chain (kappa or lambda) and the class of heavy chain (IgM, μ chain, IgG, γ chain, etc.).

1.3   Variable Region Structure

Present concepts of the nature of immunoglobulin genes depend heavily on recognition that sequence variation among V regions is limited. For example, amino acids present at certain positions within the N-terminal sequence distinguish a particular sequence as belonging to a kappa or lambda light chain or to a heavy chain. By similar comparative sequence analysis, subgroups of V_κ, V_λ, and V_H sequences have been established (See Hood *et al.,* 1077a). The sequence variation within a subgroup is considerably less than that occurring between different subgroups. A major advance in our understanding was made by Wu and Kabat (1970) who identified the specificity-determining residues of V regions. By aligning a series of V-region sequences to obtain maximum homology, it was determined that substitutions were much more frequent at the approximate positions 25–35, 50–60, and 95–100 than at other positions in light chains. V_H sequences have hypervariable regions at roughly homologous positions, and an additional area at positions 80–87. Analysis of the three-dimensional structure of antibodies by X-ray crystallography have confirmed the prediction that the contact residues forming the combining site are within the hypervariable regions (Amzel *et al.*, 1974). There are variations of two types within V regions: substitutions within the hypervariable regions, which determine specificity, and substitutions in framework residues, some of which distinguish V regions as belonging to different species or to groups and subgroups within species.

1.4 Gene Families

Immunoglobulin genes comprise three families, each consisting of a set of V genes and corresponding C genes. One family encodes kappa light chains; a second, lambda light chains; and the third, heavy chains. The chromosomal location of any family is not known with certainty, although there is recent evidence that the kappa family may be in chromosome 6 of the mouse (Gottlieb and Durda, 1977). There is no evidence for linkage between families, so they may well be on three separate autosomes. On the other hand, studies with rabbits and with mice have clearly established linked expression of V and C genes within the heavy-chain family (Mage, 1971; Weigert and Riblet, 1977).

Immunoglobulin molecules in different rabbits possess structural variants coded by allelic genes. Allotypes of the *a* locus are associated with multiple amino acid differences in V_H genes. The *d* and *e* locus allotypes represent single amino acid substitutions in the constant region of the γ chains. Two sorts of linkage relationships have been established using these markers in doubly heterozygous rabbits. First, the V and C gene allotypes are closely linked in terms of their transmission to progeny, although recombinants have been observed at low frequency. Second, the expression of V_H and C_H genes is linked at the cellular and molecular level. Thus the individual IgG molecules from an animal of the genotype *a1, d_{12}/a3, d_{11}* have the corresponding phenotypes; less than 2% represent recombinants having the *a*-locus marker from one parental chromosome and the *d*-locus marker from another (that is, *a1, d11*). This phenomenon is called allelic exclusion. Also, individual antibody-forming cells in heterozygous animals secrete immunoglobulin of only one of alternative allotypes (Pernis *et al.*, 1965).

Similar linkage relationships have been established in inbred mice. In this case, the V_H markers have been idiotypes, or antigenic markers distinguishing the binding site of one or a closely related set of V regions. Breeding studies have provided a tentative map of the locations of different idiotypes, and have also demonstrated linkage of idiotypes to C gene allotypic markers.

There is strong evidence that individual V_H genes may be expressed in combination with any C_H gene, at least within the confines of the principle of allelic exclusion. The *a*-locus V_H markers of rabbit immunoglobulin are found on IgG, IgM, and IgA immunoglobulins (see Kim and Dray, 1973). In mice, idiotypic specificities defining individual or related families of V_H genes are also found in molecules of different immunoglobulin classes (Gearhart *et al.*, 1975a). Plasma cell tumors may synthesize immunoglobulins having identical light chains and identical N-terminal (V_H) sequences but different C_H regions (Wang *et al.*, 1970).

Before beginning our discussion of B cell differentiation we will briefly summarize the dimensions of heterogeneity of antibody molecules and the characteristics of the terminally differentiated plasma cells which secrete them. Estimates of the size of the specificity repertoire of mature animals range from 10^5 to 10^8 antibody molecules with unique combining sites for antigens. Not all this information need be encoded in separate genes. Assuming that V regions of both light and heavy chains contribute to the structure and specificity of the binding site, and that any V_L and V_H can be combined, then the repertoire may be expanded to the product of numbers of V_H and V_L genes ($nV_H \times nV_L$) (see Perlmutter *et al.*, 1977). Another molecular mechanism for expanding specificity is degeneracy of the combining site such that a single site may accommodate unrelated antigenic determinants (see Hood *et al.*, 1977*b*).

The total diversity represented by specific combining sites is further amplified by the fact that each V_H region may be expressed in conjunction with any of the C_H regions that determine immunoglobulin class, or isotype. This fact has considerable biological significance, because the effector functions of immunoglobulins are largely, if not entirely, determined by C-region structure. Pentameric IgM molecules, for example, are highly efficient effectors of agglutination and complement fixation. A single IgM pentamer affixed to the surface of an erythrocyte can activate the complement cascade sufficiently to cause lysis of that cell. Some subclasses of IgG molecules also fix complement, while others do not. IgG molecules of primates and rodents have the unique capacity to cross the placenta and provide passive immunity to the fetus and newborn. IgA molecules appear to be especially designed to function on the external mucus surfaces of the body. IgE has the property, conferred by its C-region structure, of attaching to specific receptors of mast cells and basophils and causing release of vasoactive amines when triggered by antigen. IgD, a very minor class of serum immunoglobulin, is a major surface receptor molecule for B lymphocytes, and is believed to play some critical function in cell triggering. Through the mechanism of association of each V_H region with all C_H regions, the entire range of antibody specificities can be expressed among immunoglobulin molecules having a wide range of specific biologic attributes, each presumably conferring some survival advantage.

The four chain subunits of immunoglobulin molecules are invariably symmetrical. That is, the paired light chains and paired heavy chains are identical. As indicated above, each of the chains is apparently coded by separate V genes and C genes on one of the diploid set of chromosomes. Individual plasma cells, with rare exceptions, secrete only one molecular species of antibody.

We can now make the following broad generalizations concerning expression of immunoglobulin genes during primary development of B cells: (1) The process results in generation of a large number, perhaps 10^7, clones of lymphocytes, each defined by expression of a single $V_L V_H$ gene pair conferring a unique specificity for antigen. (2) Commitment of individual clones is accomplished by a dual process of gene selection and repression. Selection of lambda light chain for expression in a given cell, for example, results in permanent repression of the kappa gene family by that cell and its progeny. By a mechanism apparently unique among autosomal genes, selection of $V_H C_H$ or $V_L C_L$ genes present on one chromosome leads to repression of all corresponding genes on the homologous chromosome (allelic exclusion). (3) A critical event in commitment is the process by which V and C genes are joined. Hozumi and Tonegawa (1976) used light-chain mRNA probes for only the C region or both V and C regions to study the arrangement of V and C genes in embryonic DNA and in DNA derived from differentiated plasma cell tumors. The respective DNAs were fragmented with restriction endonucleases and the fragments separated according to size by polyacrylamide gel electrophoresis. In embryo DNA, hybridizable V and C sequences were found on separate fragments, whereas both probes recognized a single fragment in DNA from the homologous myeloma. One of the problems raised by these results was the failure to identify separate V- and C-containing fragments together with the integrated V–C gene in myeloma DNA. If V–C joining serves as the mechanism for initiating transcription of mRNA, then one would predict the absence of V–C joining on the homologous, repressed, chromosomal segment. Similar studies with a different, but closely related, plasma cell tumor gave results consistent with the existence of both integrated V–C segments and unjoined V and C genes in the myeloma DNA (Tonegawa *et al.*, 1977). In other experiments the existence of a nonintegrated V_κ gene was demonstrated in DNA from a myeloma-producing immunoglobulin of a different V_κ subgroup. Finally, Tonegawa's group have successfully cloned in a bacteriophage a segment of mouse embryonic DNA that contains all or a portion of a V_λ gene. Partial sequencing of the distal

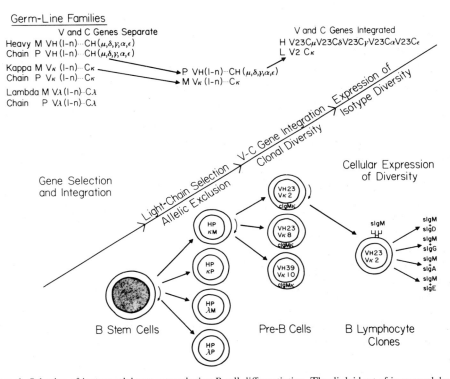

Figure 1. Selection of immunoglobulin genes during B cell differentiation. The diploid set of immunoglobulin gene familes, heavy, kappa, and lambda, are depicted in the upper left. Each family consists of a series of V genes linked to genes encoding the constant region of each chain. In the example shown, the maternal kappa family and the paternal heavy family are selected for expression; corresponding genes on homologous chromosomes will not be expressed by this clone (allelic exclusion). Alternative selection and exclusion (lower left) is one factor that contributes to diversity. A particular V_H and V_L gene is then selected for each clone and becomes integrated with C_H genes. The lower part of the diagram indicates the stages of B cell development during which these events are presumed to occur. The last cell depicted (lower right) corresponds to the immature B lymphocyte depicted in Figs. 2 and 3. This diagram is taken from Cooper *et al.* (1978) with permission of Excerpta Medica, Amsterdam.

portion of the λ gene and adjacent nucleotides suggests that there is not a C_λ gene contiguous to the V_λ gene (Tonegawa *et al.*, 1978). These results strongly suggest the two-gene, one-polypeptide hypothesis of Dreyer and Bennett, and indicate that V–C joining is a major mechanism in the regulation of immunoglobulin synthesis.

Some of the principal genetic events that occur during primary B cell differentiation are shown in the upper portion of Fig. 1. The lower part of this figure outlines some characteristics of the cells in which these events appear to occur.

2 Morphologic and Functional Aspects of Primary B Cell Differentiation

In the sections that follow, we will describe the stages of B cell differentiation. Our goal is to correlate morphologic, physiologic, and functional changes occurring during development with the genetic events involved in expression of clonal diversity of immunoglobulins.

2.1 Stem Cells

Lymphocytes, erythrocytes, granulocytes, monocytes, and megakaryocytes share a common stem cell precursor. Hemopoietic stem cells are first found within the blood islands of the primitive yolk sac, and later migrate to the fetal liver. During adult life they reside primarily in the bone marrow. Pleuripotential stem cells generate, in turn, a number of self-renewing stem cells of more restricted potential, including a common precursor for both T and B lymphocyte lineages (Adamson *et al.*, 1976).

The subsequent determination of differentiation pathways almost certainly depends on interaction between stem cells of mesodermal origin and specialized epithelia derived from entoderm. This relationship, with regard to lymphocytes, has been most clearly demonstrated by studies in birds. In both birds and mammals, development of T lymphocytes is dependent on the thymus, the epithelium of which originates in the third and fourth branchial pouches. Birds have a second central lymphoid organ, the bursa of Fabricius, which is formed as a dorsal evagination of gut epithelium and is solely responsible for the primary development of the B cell lineage (Cooper *et al.*, 1966). The homologue of the bursa in mammals has long been controversial; there is now substantial evidence that fetal liver and adult bone marrow serve this function (Owen *et al.*, 1974; Gathings *et al.*, 1977). The ability to remove the bursa, even in young embryos, has greatly facilitated investigation of the early events of B cell development.

Moore and Owen (1966) demonstrated that the stem-cell precursors of bursal lymphocytes entered the bursa via the circulation. Le Dourain and her colleagues (1975) have used an elegant system of interspecies grafting to study factors controlling stem-cell influx. Bursal rudiments from quail embryos, grafted onto chicken embryos, were populated by either host or donor lymphocytes, distinguishable by nuclear morphology, depending on the age of the donor. The results indicated that the bursa becomes receptive to circulating stem cells at a specific time during development and remains so for a limited period of only a few days. The presence of circulating stem cells overlapped the time at which invasion occurred. A particularly important observation derived from experiments in which the transplanted bursal rudiments were obtained from 8- to 9-day quail and already contained some donor-type stem cells. Nearly 50% of the lymphoid follicles examined after a 10-day residence in the chicken contained either donor or host lymphocytes exclusively, and the remaining 50% had mixed populations. These results strongly suggest that signals generated by the bursal rudiment, and expressed during a limited period of development, regulate the ingress of stem cells. Stem-cell invasion is restricted in terms of both time and numbers; a significant proportion of lymphoid follicles appear to be derived from one or a very few stem cells.

2.2 Identification of B Cells

B lymphocytes are not readily distinguishable on morphological grounds from T lymphocytes and other mononuclear cell types. Tracing their life history depends on the use of a variety of techniques for identifying cell line-specific markers. These markers have been called differentiation antigens because they distinguish different cell lineages and are frequently recognized using immunological tools. Qualitative and quantitative changes in the expression of several differentiation antigens occur during maturation of lymphocytes and provide the means of identifying cells at a particular point in their developmental sequence. Among the variety of B cell differentiation antigens that have been described (see Cooper *et al.*, 1978), the most useful in tracing B cell development has been immunoglobulin.

B lymphocytes synthesize immunoglobulin molecules in relatively small amounts and insert them into the membrane, where they function as antigen-specific receptors. Plasma cells, the terminal stage of the B lineage, synthesize and secrete immunoglobulin at rates on the order of 10^4 molecules/sec and may or may not bear membrane-bound immunoglobulin. T lymphocytes may utilize molecules similar or identical to the V_H regions of immunoglobulin as their specific antigen receptors (Binz and Wigzell, 1977; Krawinkel *et al.*, 1977). However, the quantity and type of immunoglobulin or immunoglobulinlike molecules that may be expressed by T cells at any point in their life history is insufficient to be identified by conventional techniques. The question of T cell immunoglobulin is further complicated by the fact that distinct T cell subpopulations bear receptors for IgM and IgG antibodies, and thus may utilize B cell products as a part of their receptor (see Moretta *et al.*, 1978).

The expression of immunoglobulins by individual cells can be conveniently studied by immunofluorescence techniques. Antibodies prepared in one species to immunoglobulins, or fragments of immunoglobulins, are covalently coupled to the fluorescent dyes fluorescine (green) or rhodamine (red). With appropriate absorptions such reagents can be made specific for different immunoglobulin classes or light-chain types, for different allotypes, or even for V-region-determined antigen-combining sites. The integral cell membrane immunoglobulin receptors of B lymphocytes are identified by reacting viable cells with fluorochrome-labeled antiimmunoglobulins. As long as the cells are viable, antiimmunoglobulin is excluded from the cytoplasmic compartment and stains only membrane-associated molecules. Cytoplasmic immunoglobulin is identified after the integrity of the membrane is disrupted by fixation. By use of pre- and postfixation staining with antibodies conjugated to fluorochromes of different colors, it is possible to classify individual cells on the basis of localization of immunoglobulin. Most B lymphocytes contain too little intracytoplasmic immunoglobulin to be detected by immunofluorescence and therefore have the phenotype: surface immunoglobulin positive (sIg^+) and cytoplasmic Ig negative (cIg^-). The mature plasma-cell progeny of B lymphocytes have an extensive rough endoplasmic reticulum within which immunoglobulin is synthesized at a rapid rate. These cells exhibit bright cytoplasmic fluorescence (cIg⁺) but may or may not retain membrane-associated immunoglobulin (sIg⁺). A representation of the phenotypic characteristics of the major stages in B cell differentiation discussed in the following sections is shown in Fig. 2.

2.3   Pre-B Cells

A series of complementary observations made in different laboratories during the past several years have resulted in definition of the direct precursor of B lymphocytes (pre-B cell) and in identification of the mammalian equivalent of the avian bursa. Characterization of this previously unrecognized cell type represents an important forward step in analysis of the generation of clonal diversity, because it is in this cell that initial expression of immunoglobulin genes occurs.

Owen *et al.* (1974) demonstrated *de novo* generation of sIg⁺ B lymphocytes in organ cultures of mouse embryo liver obtained as early as the eleventh day of gestation, 6 days

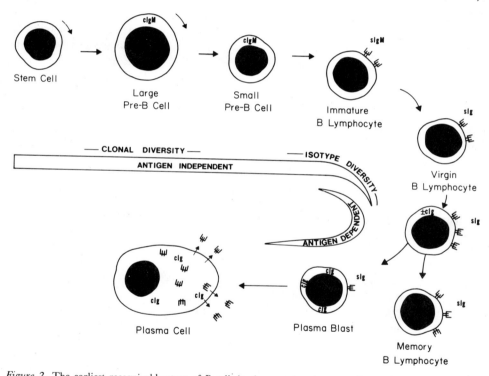

Figure 2. The earliest recognizable stage of B cell development is a large pre-B cell that synthesizes small amounts of intracellular IgM but lacks easily detectable membrane-bound immunoglobulin. These rapidly replicating cells are probably the major site of clonal diversification. Large pre-B cells generate small cIgM⁺ sIg⁻ pre-B cells, which can express sIgM without further division. The first sIgM⁺ B lymphocytes are designated immature because of their unique susceptibility to specific inactivation by multivalent antigen. Isotype diversity is expressed among clonal progeny of these cells (see Fig. 3). The differentiation of any cell expressing sIg receptors may be modified by contact with specific antigen; however, clonal diversity and isotype diversity develop independently of exogenous antigen stimulation. Clonal selection operates primarily at the level of the mature virgin B lymphocyte. Cells stimulated to proliferate via specific antigen binding and signals derived from T cells give rise to memory cells and develop into terminally differentiated plasma cells.

prior to the appearance of these cells *in vivo*. In studying the effects of antiimmunoglobulin antibodies on the development of B lymphocytes *in vitro*, Raff *et al.* (1975) made a serendipitous discovery. Explants cultured in the presence of anti-μ contained no cells bearing detectable sIg when stained as viable cell suspensions. The same cultures contained medium to large lymphoid cells with cytoplasmic IgM (cIgM) when stained with the same reagent after fixation. It was then demonstrated that cIgM$^+$ · sIg$^-$ cells could be found in mouse fetal liver as early as the 12th gestational day and were also present in fetal spleen and bone marrow. In adult animals these cells were found only in bone marrow. Pre-B cells detected by immunofluorescence contained μ heavy chains, but were not stained by antibodies to γ or α heavy chains (Raff *et al.*, 1976). Melchers *et al.* (1975) demonstrated biosynthesis of 8 S IgM in rapidly sedimenting (that is, large) lymphoid cells obtained from mouse embryos before the thirteenth day of gestation. Using the technique of lactoperoxidase-catalyzed iodination of cell-surface proteins, they detected sIgM on cells in the same velocity gradient fractions. We will return to the question of surface immunoglublin expression on pre-B cells, as it has some significance in understanding the regulation of clonal diversification.

Pre-B cells are readily detected by staining viable cell suspensions with fluorescein anti-μ (green) followed by fixation and counterstaining with rhodamine anti-μ (red). B lymphocytes stain with superimposible red and green spots representing sIgM but generally do not contain diffuse cytoplasmic IgM. Pre-B cells are not stained by the green antibody (sIg$^-$) but contain small amounts of red cytoplasmic antibody (cIgM$^+$) stain. The intensity of cytoplasmic fluorescence of morphologically distinct plasma cells containing IgM is far greater than that of pre-B cells. Plasmablasts more closely resemble pre-B cells in morphology, but they too stain very brightly for cytoplasmic Ig and also express surface Ig.

In mice, rabbits, and humans, pre-B cells develop first in fetal liver and invariably precede development of sIgM$^+$ B lymphocytes (Raff *et al.*, 1976; Hayward *et al.*, 1978; Gathings *et al.*, 1977). These cells are heterogeneous in size, but on average are larger than mature B lymphocytes. They have a high nuclear to cytoplasmic ratio and a highly convoluted nuclear contour. The larger pre-B cells incorporate DNA precursors at a very high rate both *in vivo* and *in vitro*: 80–90% of large pre-B cells in cultures of human fetal liver or adult bone marrow are labeled following exposure to [^{3}H]thymidine for 45 min. Under these conditions, fewer than 1% of sIg$^+$ B lymphocytes take up the label. Large pre-B cells pulse-labeled with [^{3}H]thymidine in culture form smaller pre-B cells (Okos and Gathings, 1977). Similar results were obtained in mice by Owen *et al.* (1977), using both *in vivo* and *in vitro* labeling. By analogy to studies on the genesis of B lymphocytes in bone marrow of adult mice (Osmond and Nossal, 1974; Ryser and Vassali, 1974), it is likely that smaller pre-B cells begin to express sIgM and thus become B lymphocytes without a requirement for further cell division.

Light chains as well as μ heavy chains have been demonstrated in the cytoplasm of rabbit pre-B cells, although all of these cells may not synthesize light chains. In contrast, the great majority of pre-B cells from mice and humans apparently do not synthesize light chains (Burrows *et al.*, 1979; Levitt and Cooper, 1980). In rabbits heterozygous for *b* locus light-chain allotypes (markers for C$_\kappa$ genes), individual pre-B cells express only one of alternative alleles (Hayward *et al.*, 1978). This indicates that allelic exclusion, or repression of one of the parentally derived κ gene families, has occurred at this stage. The question of whether pre-B cell IgM contains V$_H$ determinants has been more difficult to resolve, as there is as yet no methodology for obtaining purified populations of cells. Our approach has been to use antibodies to idiotypic determinants (that is, antigenic determinants associated with specific V_H–V_L combining sites for antigen) of myeloma proteins of humans to stain normal pre-B cells. Pre-B cells staining with two different antiidiotypic antisera have been detected in a human bone marrow sample containing an unusually high frequency of pre-B cells. Idiotype-positive cells have also been observed among pre-B cells of myeloma patients (Kubagawa, 1978, 1979). Both observations suggest that V genes are expressed at this early stage in differentiation.

Pre-B cells are found in liver of mice, rabbits, and man prior to development of hemopoiesis in spleen and bone marrow. As the latter organs become hemopoietic sites, they also contain pre-B cells. During fetal life, pre-B cells are also occasionally seen in blood and lymph nodes. As noted earlier, in adult animals or humans pre-B cells are found almost exclusively in bone marrow. The association of pre-B cells with sites of hemopoiesis is not invariant. Adult mice made anemic by treatment with phenylhydrazine develop extramedullary hematopoiesis, particularly in spleen, but their spleens contain no pre-B cells (Burrows et al., 1978). It seems quite certain that fetal liver is a primary site at which the interactions between mesodermal stem cells and entodermally derived epithelium that initiate B-cell differentiation occur. It is possible that similar interactions may occur elsewhere, but it is equally possible that the pre-B cells found in other organs are self-renewing populations originally derived from liver.

In describing the functional attributes of pre-B cells it is necessary to take some liberties with the phenotypic definition (cIgM$^+$ · sIg$^-$) we have used in previous sections. It has not yet been possible to isolate pure populations of pre-B cells; therefore, their functional capabilities can only be inferred from studies of the properties of lymphoid populations known to contain pre-B cells and to lack substantial numbers of mature sIg$^+$ B lymphocytes. The term *pre-B cell* was initially used in this setting by LaFleur et al. (1972). They demonstrated a population of relatively large cells in mouse bone marrow, which was not capable of being immediately triggered by antigen when transferred to a lethally irradiated host. After a period of residence in the adaptive host, these cells, or their progeny, could respond to antigen and differentiate into antibody-secreting cells. Osmond and Nossal (1974) and Ryser and Vassalli (1974) showed that a large, sIg$^-$, rapidly dividing cell in bone marrow of the adult mouse was the precursor of sIg$^+$ B lymphocytes. Organ cultures of fetal liver, spleen, and bone marrow, all of which contain morphologically identifiable pre-B cells, generate sIgM$^+$ B lymphocytes, whereas culture of thymus and yolk sac, which lack pre-B cells, fail to produce sIgM$^+$ lymphocytes (Owen et al., 1974, 1975). Chronic treatment of mice from birth with anti-μ antibodies prevents development of sIg$^+$ B lymphocytes and plasma cells, but does not affect pre-B cells. Cultured bone marrow from these mice generates sIg$^+$ B lymphocytes that can be triggered by bacterial lipopolysaccharide (LPS) to differentiate to mature plasma cells; cultured spleen cells from the same animals, which do not contain identifiable pre-B cells, do not give rise to LPS-responsive cells (Burrows et al., 1978). Melchers (1977a) found that liver cells from mouse embryos established in culture with LPS at 13- to 19-days gestation generated plasma cell responses all of which peaked at the equivalent of 5–7 days after birth. These observations suggest that liver contains precursors, which are themselves unresponsive to LPS, but which give rise to LPS-responsive cells at a fixed time of development. Using a limiting dilution assay, Melchers (1977b) demonstrated that the frequency of these precursors rose from ~1/106 liver cells at day 13 to ~1/30 on the day of birth (day 19) and then declined rapidly. Considering the errors involved in direct counting of cells present in low frequency, these estimates are in reasonable accord with those based on enumeration of cIgM$^+$ · sIg$^-$ pre-B cells in fetal liver (Raff et al., 1976).

On the basis of these observations, we can give a general summary of the properties of pre-B cells. They are rapidly replicating cells that synthesize μ chains. Although they may transiently display IgM on their surfaces, these molecules do not seem to function as stable receptors for antigen (see Raff, 1977; Melchers *et al.*, 1977). If functional receptors for antigen are lacking, it is likely that the replication and differentiation of these cells is neither dependent on antigens nor modifiable by antigens. Pre-B cells give rise to a progeny of sIgM$^+$ B lymphocytes whose further differentiation can be markedly influenced by interaction with specific antigens, as will be discussed in the following section.

2.4   Immature B Lymphocytes

B lymphocytes with 8 S IgM molecules integrated into their surface membranes appear in the liver of fetal mice at 16- to 17-days gestation (Owen *et al.*, 1974); in man they are found in the same site at about 8-weeks gestation. IgM is the only class of immunoglobulin expressed by these early cells (Gathings *et al.*, 1977; Vitetta *et al.*, 1977; Abney *et al.*, 1978). Immature B lymphocytes are distinguished as a separate stage of B cell development by a unique functional characteristic: their susceptibility to inactivation when their IgM receptors are cross-linked by multivalent antigens or bivalent antibodies to IgM.

The development of the B cell line, excluding pre-B cells, can be aborted by treatment of mice from birth with antibodies to μ chains. This treatment regimen is effective only when begun on the day of birth or shortly thereafter; initiation of injections at 1 week of age may enhance rather than suppress B cell development (Lawton and Cooper, 1974). Exposure of B lymphocytes to bivalent anti-Ig antibodies *in vitro* results in an energy-dependent migration of cell membrane immunoglobulin molecules to a single pole of the cell *(capping)*, which is followed by ingestion of receptor–antibody complexes (Taylor *et al.*, 1971). The stripping of cell-surface antigens by this mechanism is called *antigenic modulation*. Adult B lymphocytes, stripped of their receptors by anti-μ, reexpress their IgM receptors within a period of a few hours. The sIgM of B lymphocytes from neonatal mice is modulated by much lower concentrations of anti-μ antibodies, and removal of antibodies from the environment is not followed by reexpression of sIgM (Raff *et al.*, 1975; Sidman and Unanue, 1975). Probably through a similar mechanism, low concentrations of anti-μ antibodies inhibit the LPS-induced differentiation of neonatal mouse B lymphocytes to plasma cells. Equivalent suppression of the response of adult splenic B lymphocytes requires 100 to 1000 times as much antibody (Kearney *et al.*, 1976a). Adult B lymphocytes activated by LPS in the presence of suppressive concentrations of anti-μ proliferate and continue to express B-cell differentiation antigens other than sIgM. The effect of the antibody appears to be a selective disruption of the organization of cytoplasmic organelles involved in high-rate antibody synthesis. The suppressed B cells lack the well-developed rough endoplasmic reticulum characteristic of mature plasma cells, although they continue to synthesize small amounts of IgM. Lymphocytes from newborn mice treated similarly with LPS and anti-μ are suppressed by a totally different mechanism; they neither proliferate nor differentiate (Kearney *et al.*, 1978). Physiological changes in immature B cells occur rapidly following birth, so that by 7–10 days of life the B cells in spleen behave as adult cells (Kearney *et al.*, 1976a).

Experimental results of Metcalf and Klinman (1976, 1977) indicate that multivalent antigens may inactivate immature B lymphocytes in much the same way as do bivalent antiimmunoglobulin antibodies. Their rather complex experimental system permits enumeration of individual B lymphocyte clonal precursors by identifying the antibodies their differentiated progeny produce in culture, and allows analysis of the requirements for B cell triggering. Both adult and neonatal B cell precursors require the help of carrier-primed

T cells to generate an antibody response to a hapten when stimulated by a hapten–carrier complex. As long as this requirement is met, the frequency of clonal precursors for the haptens dinitrophenol and trinitrophenol is similar in the newborn and adult. Exposure of adult clonal precursors to the hapten on the wrong carrier (that is, a protein to which T cells have not been primed) has no effect on the ability of these cells to respond at a later time to the same hapten on the correct carrier. In contrast, neonatal B cells exposed to a hapten on the wrong carrier are rendered incapable of later responding to the correct hapten–carrier conjugate. The major requirement for induction of tolerance among immature B cells is that the hapten–carrier conjugate be multivalent and thus capable of cross-linking multiple receptors on the target B lymphocyte. The same requirement holds for modulation of sIgM receptors on B cells by antiimmunoglobulin. The fraction of clonal precursors in spleen that is easily tolerized is over 90% at birth and diminishes rapidly thereafter; by 1 week of age, susceptibility to tolerance is lost (Metcalf and Klinman, 1976). Bone marrow, however, contains easily tolerizable precursors, even in adult life (Nossal and Pike, 1975; Metcalf and Klinman, 1977). Similarly, adult bone marrow B lymphocytes are more susceptible to receptor modulation by anti-μ than are spleen cells (Raff *et al.*, 1975). The latter observations provide additional evidence that immature B lymphocytes are generated in bone marrow throughout life.

One of the requirements of the clonal selection theory is a mechanism for elimination of self-reactive clones of cells. These findings suggest a basis for one such mechanism: each developing clone of B lymphocytes passes through a maturational stage during which it is subject to permanent inactivation through contact with its specific antigen. The existence of such a stage of B cell development raises some theoretical considerations with regard to the expression of diversity. There are two major reasons for proposing that significant diversification occurs prior to this developmental stage. First, the mechanism for eliminating self-reactive cells would be ineffective for any specificities generated at later stages of B cell development. For example, autoreactive clones arising from mutational events during antigen-driven clonal proliferation would not be easily tolerized. This may not be a critical argument; there are clearly other mechanisms for suppressing such potentially damaging clones (Basten *et al.*, 1977). The second reason for proposing expression of diversity at an earlier stage of differentiation is the possibility that the clonal abortion mechanism might severely limit later expansion of the repertoire. Let us assume for this argument that clonal diversity is generated by a series of somatic mutations of a relatively small number of germ-line genes, and that each clone expressing germ-line specificities will eventually give rise to 10^4 useful mutants. If most of this diversification process were to follow the stage at which contact with specific antigen results in clonal abortion, then early exposure to antigens might be expected to eliminate a substantial amount of potential diversity. Similarly, early expression of specificities to self-determinants could limit development of derivative clones having needed specificities.

3 Ontogenetic Expression of Clonal Diversity

The potential contribution of ontogenetic studies to unraveling the mechanisms for generation of diversity can be distilled to a few questions. What is the time frame during which diversity is generated? To what extent does environmental manipulation (that is, exposure to antigens) modify the development of diversity? Is there evidence for ordered development of different clonotypes?

It is interesting that the favored answers to some of these questions have changed considerably as knowledge of regulatory mechanisms in immune responses has accumulated. Newborn animals have a very limited capacity to respond to antigenic challenge; for many years, the neonate was considered "immunologically null." From this perspective it was perfectly logical to assume that the development of immunologic capacity represents a learning process in which experience with environmental antigens rapidly expands the store of immunological information. From this assumption, in turn, it seemed logical to conclude that diversification is generated through somatic mutations occurring in the course of antigen-driven clonal proliferation.

During the past decade it has become apparent that the elicitation of immune responses is an extremely complex process, requiring collaborative interactions among at least three cell types, which are required by a variety of nonimmunoglobulin genes (see Chapter 10). Therefore, the absence of an immune response in a fetal or neonatal animal cannot be taken as evidence for the absence of the relevant specific clones of B lymphocytes (Mosier and Johnson, 1975; Sherwin and Rowlands, 1975). Reexamination of the ontogeny of B cell clonal diversity, using experimental methods that circumvent these regulatory factors, has indicated that substantial diversification occurs within a short time period, and that this process is probably not driven by environmental antigens.

The direct demonstration that rare lymphocytes from nonimmune animals bear surface receptors for specific antigens was first made approximately 10 years ago (Naor and Sulitzeanu, 1967; Byrt and Ada, 1969; Davie and Paul, 1971). The experiments of Davie and Paul (reviewed in 1974a) firmly established the feasibility of using antigen-binding techniques to assess both numbers of precursor B lymphocytes and the fine-specificity of their immunoglobulin receptors, and provided a means for directly examining the ontogeny of B lymphocyte clones.

Lydyard *et al.* (1976) examined the ontogeny of chicken bursal lymphocytes binding to each of several complex antigens. The first sIgM-positive lymphocytes appear in the bursa on the 12th day of embryonation; they increase at an exponential rate consistent with a generation time of 10 hr until near the time of hatching (days 20–21). Beyond this time the numbers of sIgM$^+$ cells per bursa continues to increase, but at a lower rate following a straight line on a linear-logarithmic plot. This change may be explained on the basis of emigration of cells from the bursa beginning several days before hatching.

Bursal lymphocytes binding keyhole limpet hemocyanin (KLH) and the synthetic polypeptide, poly-L (Tyr, Glu) poly-D,L-Ala-poly-L-Lys (TGAL), respectively, were first detected at day 16, 4 days following initiation of IgM synthesis. Sheep erythrocyte antigen-binding cells (SE-ABC) did not appear until day 18. From this point until 14 days after hatching, the frequency of SE-ABC increased in log-linear fashion, but at a rate significantly lower than the initial rate of increase of sIgM$^+$ bursal lymphocytes. The frequencies of SE-ABC and of sIgM$^+$ B lymphocytes, respectively, in bursas and spleens of birds raised for 2 weeks in a germ-free environment were identical to those of conventionally raised controls. Deliberate immunization with SE caused a 300-fold increase in frequency of SE-ABC in peripheral blood but did not alter the frequency of these cells in the bursa. These results suggest the following conclusions:

(1) Generation of specific ABC in the bursa occurs in a sequential fashion, such that different specificities appear at different times. This interpretation is supported not only by the differences in times of appearance of cells binding KLH or TGAL and of SE-ABC, but by the observation that the total population of sIgM$^+$ cells in the bursa initially increased much more rapidly than did the specific population of SE-ABC.

(2) Environmental antigens do not influence the emergence of specific antigen-binding cells in the bursa. Although proliferation of bursal lymphocytes is influenced by envi-

ronmental antigens or mitogens, the frequency of specific populations of antigen-binding cells is not. By contrast, the frequency of postbursal ABC can be dramatically increased by exposure to specific antigens.

In other experiments, autoradiographic techniques were used to study the anatomic localization of specific ABC in the bursa. Cells binding [^{125}I]-KLH or [^{125}I]-TGAL were present in multiple follicles (0–5 positive follicles in sections containing 30–50 follicles). In each case, the ABC existed as singlets or as clusters of 2–4 contiguous cells; the great majority of sIgM$^+$ cells within positive follicles did not bind the radiolabeled antigen. As was discussed earlier, the experiments of LeDourain and her colleagues (1975) suggest that the lymphocytes of individual bursal follicles are the progeny of one or a very few stem cells. It therefore follows that a single stem cell gives rise to diverse clones of B cells within each follicle, and that separate stem cells, localized in different follicles, generate progeny having similar or identical specificities for antigen.

Edelman and his colleagues have analyzed the ontogeny of specific ABC in mice (Spear *et al.*, 1973; D'Eustachio and Edelman, 1975; D'Eustachio *et al.*, 1976; Cohen *et al.*, 1977). Their results indicate that the generation of diverse clones of B lymphocytes occurs very rapidly during the later part of gestation (days 15–19). Cells binding to 11 different antigens were present in 18-day fetal spleen. The numbers of antigen-binding cells of different specificities increased in parallel with age, whether expressed as the fraction of nucleated cells or on the basis of numbers per spleen. The range of avidities with which fetal cells bound specific antigen was found to be similar to that determined for ABC in adult spleen (D'Eustachio and Edelman, 1975). This pattern of development is clearly inconsistent with the idea that initial generation of diversity is regulated by random exposure to environmental antigens. Also inconsistent is the evidence for genetic control of the ontogenetic expression of diversity. Cohen *et al.* (1977) compared frequencies of ABCs specific for four different antigens in individual fetal mice of different inbred strains. Among individuals of one strain, there was minimal variation in the frequency of specific ABCs per nucleated spleen cell and in ratios of ABCs of different specificities. However, the absolute number of ABCs per spleen varied for the two strains. This difference was correlated with the number of sIg$^+$ B cells in the spleens of 18-day fetuses of the two strains; breeding experiments suggested that B lymphocyte numbers were regulated by one or a few closely linked autosomal genes. Similar studies in outbred Swiss mice revealed significant individual difference in the total frequency of ABCs. The ratios of ABCs of different specificity were relatively constant among different individuals, but differed numerically from the ratios of cells binding the same antigens in the two inbred strains. It was concluded that the size of the B cell repertoire (indicated by the absolute frequency of ABCs) and the composition of the repertoire (indicated by ratios of ABCs of different specificity) were regulated by independent genes.

The only major difference between the studies with mice and those with chickens concerns the question of sequential emergence of different clones. This is a particularly critical point in understanding the mechanisms involved in generation of diversity, but is difficult to analyze at the level of antigen-binding cells. Many related clones may bind a given antigenic determinant. Measurements of heterogeneity of binding avidities (see D'Eustachio and Edelman, 1975) may distinguish one clone from two or three with similar specificity but cannot distinguish a few clones from hundreds. Also, the detection of sequential emergence of different specificities is dependent to a major extent on luck in the selection of a few antigens for study from among the millions to which the animal might respond.

Klinman and his colleagues have brought both luck in selection of antigens and an assay capable of detecting the products of single clonal precursors to bear on this problem. In their assay, limiting numbers of B cell precursors are injected into a mouse that has previously been primed with a protein carrier and then lethally irradiated. The spleen of the recipient mouse is removed 1 day later and cut into fragments, which are distributed in individual culture dishes. The cultures are stimulated with hapten–carrier conjugates, and supernatant fluids analyzed 1 week or more later for specific antihapten antibodies. By a variety of analytical techniques, it has been demonstrated that the antibodies produced by a single fragment represent the products of a clone derived from a single precursor. Specific clonotypes can be identified by isoelectric focusing or by use of antiidiotypic antibodies (Klinman and Press, 1975a).

By determining the frequencies of clonal precursors reactive with a number of different antigens in neonatal and adult BALB/c mice, Klinman and his colleagues have reached the conclusion that a repertoire of approximately 10^4 clonotypes present at or shortly after birth is expanded to more than 10^7 clonotypes in the adult (see Klinman *et al.*, 1977). The diversification process is highly ordered and occurs independently of exogenous antigenic stimulation. The key observations supporting these conclusions may be briefly summarized as follows.

The frequency of precursors reactive with the haptens dinitrophenol (DNP) and trinitrophenol (TNP) in neonatal animals, expressed as a fraction of splenic sIg^+ B lymphocytes, is very similar to that in adults. In the neonate, approximately 90% of the precursors for DNP and TNP, respectively, belong to one of three clonotypes, identified by their isoelectric focusing patterns. The same predominant clonotypes appear regularly in different neonatal BALB/c mice. It was estimated that each clonotype was represented by 50–200 cells. By 9–10 days of life the original clonotypes can still be identified, but they constitute only a small minority of DNP- and TNP-specific precursors. Cells reactive to another hapten, fluorescein, are present in very low frequency at birth, but in adult mice are found in frequencies similar to DNP and TNP precursors (Press and Klinman, 1974; Klinman and Press, 1975b). Neither the time of appearance nor the initial frequency of clones reactive to a particular hapten is closely correlated with the eventual frequency of B lymphocytes with that specificity in adults (Klinman *et al.*, 1977).

In BALB/c mice the great majority of antibodies formed to phosphorylcholine (PC) are products of a single clonotype. This clonotype can be defined by use of antibodies specific for the antigen-binding site of a PC-reactive BALB/c plasma cell tumor, TEPC-15 (Claflin *et al.*, 1974; Lee *et al.*, 1974). Although other PC-reactive precursors can be detected, more than 70% of B lymphocytes in BALB/c mice that can be stimulated by PC bear the T15 idiotype (Gearhart *et al.*, 1975b). The extremely high frequency of this clone argues strongly that its specificity is encoded in germ-line genes. This germ-line clonotype is not detectable in mouse spleen until 5–7 days after birth (Sigal *et al.*, 1976). When PC-reactive cells do appear, they have the characteristic susceptibility to tolerance induction of immature B cells, whereas DNP or TNP precursors at this age are already nontolerizable (Klinman *et al.*, 1977). Finally, the frequency of PC precursors, and, in fact, the frequency of precursors for all haptenic determinants that have been analyzed, is as high in germfree as in conventionally reared mice (Sigal *et al.*, 1975; Press and Klinman, 1974).

Goidl and Siskind (1974) have examined the ontogeny of DNP-reactive cells by transferring fetal or neonatal cells together with adult T cells into irradiated adult recipients that were then challenged with antigen. They found that the heterogeneity of avidities (an indirect measure of numbers of different clones responding to this antigen) was greatly restricted with respect to adult precursors. Peripheral B cells from 2-week-old mice generated a fully heterogeneous response, whereas bone marrow precursors remained restricted. The heterogeneity of precursors obtained from germ-free adults was indistinguishable from that of normal animals.

Although a few areas of controversy remain, all of these observations coincide on a few very important points. Generation of diverse clones of B lymphocytes begins well before birth and proceeds rapidly. By the time of birth, and preceding exposure to a variety of environmental antigens, a considerable repertoire of different clones is already present. Maintenance of animals to maturity in an antigen-sheltered environment does not substantially alter the extent of B cell diversity. There is also evidence that both the size and composition of the repertoire are genetically regulated. These observations effectively eliminate positive selection by environmental antigens as a major factor in initial generation of diversity. They do not, however, exclude the possibility that negative selection mediated by "self" antigens may play a role in this process (Jerne, 1971), nor do they argue against the possibility that useful mutants may arise and be selected during the course of an immune response.

We believe the weight of evidence favors the notion that specific clones, at least those expressing germ-line specificities, emerge according to a predetermined, genetically regulated program. We have not discussed rather extensive evidence that fetal animals of several species acquire the ability to respond to different antigens at fixed times (Silverstein et al., 1963; Rowlands et al., 1974; Sherwin and Rowlands, 1975) because it is not known that the observed hierarchies of antibody responses reflect the sequential emergence of specific B cell clones. Nevertheless, the existence of these reproducible developmental patterns adds to the evidence that ontogenetic expression of diversity is not a random process.

What might sequential expression of distinct clonotypes mean with regard to mechanisms of generation of diversity? First, this phenomenon seems incompatible with diversification models based on random somatic alterations (mutations, chromosomal recombinations) of a very small number of germ-line genes. However, there is no reason to discount the possibility that somatic mutations serve to expand a germ-line repertoire sufficiently large to account for the predominant neonatal clones that Klinman and his colleagues have defined (see Klinman et al., 1977).

Through the use of antiidiotypic antibodies, which recognize the combining sites of other antibodies, it has been possible to begin to map the location of mouse V_H genes. As was mentioned earlier, there is evidence that V and C genes are linked, and that V_H genes are arranged in tandem on a single (haploid) chromosome (Weigert and Riblet, 1977). It is obviously tempting to relate sequential appearance during ontogeny of putative germline specificities to the linkage relationships of the relevant V genes. At the present time there are no data to support or refute this possibility.

4   Development of Isotype Diversity

The proposition that the variable and constant regions of immunoglobulin chains were coded by separate genes raised the possibility that initial expression of particular genes might be regulated by V–C integration (Dreyer and Bennett, 1965). This novel idea, now directly supported by the experiments of Hozumi and Tonegawa (1976), provides a rationale for approaching the problem of antibody diversity by studying the cellular expression of C_H gene products (Cooper et al., 1972). Analysis of the timing, sequence, and regulation of C_H gene expression has not elucidated the mechanism of V–C joining, but it has limited the theoretical possibilities and generated considerable information on the factors that regulate this process.

The concept that expression of class or isotype diversity is related to V–C gene integration rests heavily on evidence that B cells synthesizing antibodies of different classes but the same specificity originate from a common clonal precursor. From this perspective, generation of isotype diversity is an intraclonal differentiation event, which follows cellular commitment to expression of a unique combining site. This suggested the possibility of a repetitive "switch" mechanism, whereby a selected V_H gene might be sequentially integrated with a series of different C_H genes, giving rise to subclones of B cells synthesizing antibodies of the same specificity but of different classes (Cooper et al., 1972; Cooper and Lawton, 1974).

It was shown that injections of antibodies to μ heavy chains into chickens (Kincade et al., 1970) and mice (Lawton et al., 1972a; Manning and Jutila, 1972; Murgita et al., 1973) inhibits development of cells secreting all classes of immunoglobulin. Wang et al. (1969, 1970) demonstrated, in a patient with a biclonal myeloma, that IgG and IgM myeloma immunoglobulins had identical light chains and V-region amino acid sequences. A proportion of IgG-secreting plasma cells in rabbits were found to bear IgM antibodies on their surfaces (Pernis et al., 1971). The development of cells secreting IgG and IgA antibodies in response to in vitro challenge with sheep erythrocytes was shown to be inhibited by antibodies to IgM (Pierce et al., 1972). A small population of antibody-forming cells simultaneously secreting IgM and IgG antibodies of the same specificity was demonstrated (Nossal et al., 1971). Antibodies with identical idiotypic determinants but of different classes were shown to originate from single precursors (Press and Klinman, 1973; Gearhart et al., 1975a). These observations, supported by studies on the ontogeny of isotype expression, established the principle that isotype diversity is generated by an intraclonal switch mechanism through which a selected V_H gene initially expressed with C_μ subsequently is expressed in association with other C_H genes.

As indicated earlier in discussing the development of pre-B cells and immature B cells, IgM is the first immunoglobulin class to be expressed during ontogeny. The pivotal cell in development of isotype diversity is the immature $sIgM^+$ B lymphocyte, from which B cells destined to secrete all other classes are derived. That pre-B cells cannot directly give rise to cells expressing other isotypes is demonstrated by the suppression of synthesis of all Ig classes in anti-μ treated mice, despite persistence of normal numbers of functional pre-B cells in their marrow (Burrows et al., 1978).

B lymphocytes bearing sIgM become detectable in mouse fetal liver at 15–17 days gestation (Spear et al., 1973; Owen et al., 1974) and in human fetal liver by about the ninth week (Lawton et al., 1972b; Gathings et al., 1977). Because the subsequent sequence of expression of different isotypes by these cells appears to be identical in mouse and human except with respect to timing, we will present a composite description based on immunofluorescence observations of Gathings et al. (1977) and Abney et al. (1978). Cells bearing isotypes other than IgM (that is, IgD, IgG, IgA) begin to appear at about the time of birth in the mouse and between 10 and 12 weeks gestation in humans. Invariably, expression of these isotypes occurs on cells that also bear sIgM. The sequence of appearance of these different classes is difficult to establish because of their rarity: in humans, expression of sIgG probably occurs before sIgD and clearly precedes development of sIgA. Experiments utilizing the double-staining technique with cells from neonatal mice suggest that each of these isotypes is initially expressed on a different population of $sIgM^+$ cells. While all cells bearing sIgD, sIgG, or sIgA also have sIgM, few if any doubles are detected by staining simultaneously for sIgD and sIgG or for sIgD and sIgA. This situation changes dramatically by about 7 days in the mouse and by birth in humans. At this point the majority of B cells that bear sIgG or sIgA stain for sIgM and also for sIgD. On the basis of the numbers of doubles observed for each combination, the conclusion that cells originally expressing sIgG + sIgM or sIgA + sIgM have begun to synthesize a third receptor isotype, sIgD, is inescapable. The expression of different isotypes by individual cells is not totally unrestricted, however. Cells bearing two IgG subclasses are rare; cells bearing both sIgG and sIgA are observed more frequently, particularly in neonatal human blood.

Further maturation of B cells is accompanied by restriction of isotype expression. The great majority of cells bearing sIgA in adult lymphoid tissues do not express any other isotype. The frequency with which sIgG-positive cells stain for sIgM or sIgD decreases greatly, although doubles of these types are still easily detectable.* It is important to note that the pattern of isotype distribution on B cells from pathogen-free or athymic mice is similar to that in normals, indicating that neither environmental antigens nor T cell factors are required for initial expression of isotype diversity. A model describing these events is shown in Fig. 3. (Also see Cooper et al., 1976; Abney et al., 1978; Cooper et al., 1978.)

The significance of these findings for the problem of regulation of expression of immunoglobulin genes may be most easily appreciated by considering the evolution of concepts of isotype switching. Originally, the idea of switching was applied to events that followed antigen stimulation and that might explain the earlier appearance of IgM than IgG during the immune response, or in the ontogeny of immunoglobulin synthesis (Nossal et al., 1964). The attractiveness of this model was diminished by the rarity with which antibody-secreting cells could be caught in the act of switching, that is, synthesizing both IgM and IgG antibodies (Nordin et al., 1970; Nossal et al., 1971). The discovery by Raff et al. (1970) and Pernis et al. (1970) that B lymphocytes expressed cell-surface immunoglobulin led to a series of experiments collectively indicating that switching was accomplished at an earlier stage of B cell differentiation (see Cooper et al., 1971, 1972; Lawton and Cooper, 1974; Lawton et al., 1975; Pernis et al., 1977). Key observations were that isotype diversity is generated by a switch mechanism during bursal lymphopoiesis in chickens (Kincade and Cooper, 1971; Kincade et al., 1970) and is expressed early during fetal life in humans (Lawton et al., 1972b). It was also found that removal of the chicken bursa at different times during embryonation selectively interfered with subsequent production of different immunoglobulin classes. Embryonic bursectomy could result in complete agammaglobulinemia or in synthesis of only IgM; bursectomy at hatching frequently led to permanent IgA deficiency in birds synthesizing normal or elevated levels of IgM and IgG (Cooper et al., 1969; Kincade and Cooper, 1973; Kincade et al., 1973). These results, together with less direct evidence from studies on other species (see Lawton and Cooper, 1974; Lawton et al., 1975) supported the hypothesis that intraclonal isotype diversity was generated through sequential association of a selected V gene with C_μ, C_γ, and C_α. It was argued that commitment to isotype, like commitment to clonotype, is regulated within the bursa or bursa equivalent and not by contact with environmental antigens. A single genetic mechanism for translocating V genes could thus account for both the development of clonal diversity (by integrating different V genes with C_μ) and of isotype diversity (by sequentially integrating a selected V gene with different C genes) in the progeny of a single stem cell (Cooper and Lawton, 1974).

---

*An exception to this rule occurs with the IgG3 subclass in the mouse. Even in adults, the great majority of sIgG3[+] B lymphocytes are also stained for both sIgM and sIgD (Abney et al., 1978). The significance of this observation is not apparent.

This sequential integration model could accommodate early evidence that precursors of IgG plasma cells have IgM on their surfaces (Pernis *et al.*, 1971; Pierce *et al.*, 1972) by proposing that sIgM is the product of a long-lived mRNA. The discovery that the majority of circulating B lymphocytes simultaneously express both sIgM and sIgD (Knapp *et al.*, 1973; Rowe *et al.*, 1973) and that IgM and IgD molecules on single cells have identical idiotype and specificity for antigens (Pernis *et al.*, 1974; Fu *et al.*, 1975; Goding and Layton, 1976) created two major problems for this simple model of switching. First, it became clear that IgD must play some central role as a cell receptor, as immunoglobulin of this class is expressed on the majority of B lymphocytes but is either absent (mouse) or rare (primates) in serum. Second, the persistent expression of two Ig classes by single cells implied simultaneous transcription of two integrated V–C genes. It was these observations that prompted reexamination of the ontogenetic expression of isotype diversity in mice and humans and in turn led to the model shown in Fig. 3.

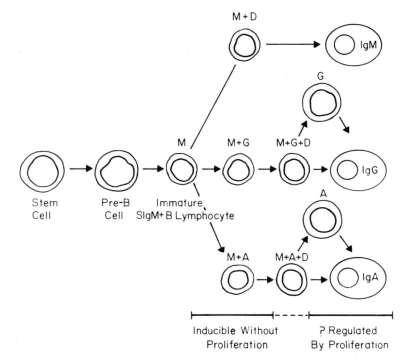

Figure 3. Development of isotype diversity. Expression of isotype diversity is an intraclonal event, which follows cellular commitment to synthesis of a particular antibody specificity. Although commitment to different isotypes might be programmed earlier, the pivotal cell in expression of class diversity is the sIgM$^+$ immature B lymphocytes. During ontogenetic development, separate sublines of sIgM$^+$ cells begin to express a second isotype— sIgD, sIgG (or its subclasses), sIgA or presumably sIgE (not shown). Cells committed to IgG or IgA synthesis later begin to express sIgD as a third isotype. Following antigen stimulation, sIgD expression ceases rapidly and sIgM expression diminishes at a lower rate.

The factors regulating expression of isotype are not well understood. Expression of a second isotype on sIgM$^+$ cells can be induced without cellular proliferation. By contrast, eventual commitment to IgA or IgG synthesis appears to be dependent upon several cycles of proliferation (Section 5). The scheme shown indicates that some B lymphocytes express only sIgG or sIgA; these are presumably memory cells, but IgG or IgA memory cells may also express sIgM or sIgD. This model is more fully discussed by Abney *et al.* (1978) and Cooper *et al.* (1978).

Several conclusions relevant to regulation of expression of immunoglobulin genes can be drawn from these results:

(1) At one stage in their life history, individual B lymphocytes simultaneously express immunoglobulins of at least three classes, all presumably having identical light chains and V_H regions. If transcription occurs only with integrated V–C genes (Hozumi and Tonegawa, 1976), this observation implies that copies of a selected V_H gene are inserted near each C_H gene. Sequential expression of different isotypes (that is, sIgM $\rightarrow$ sIgM + sIgG $\rightarrow$ sIgM + sIgG + sIgD) would then occur by derepression of already integrated V–C gene complexes.

Simultaneous integration of multiple V_H copies into different C_H genes has important implications with regard to theories of somatic diversification. If this model is correct, then it follows that somatic mutations of V_H genes must be generated at a stage of development that precedes the expression of multiple isotypes on single cells. Mutations occurring in a single V_H gene at a later stage of differentiation would presumably be expressed on only one of the cell's receptor isotypes. Although multispecificity of receptors on single cells can probably never be formally excluded as a possibility, the available data suggest that sIgM and sIgD on individual cells have identical idiotypes and antigen-binding specificities (Pernis et al., 1974; Goding and Layton, 1976). These considerations reemphasize the idea that the pre-B cell is the most likely site of generation of diversity. They further suggest that antigens cannot play a major role in driving somatic diversification.

(2) Whatever the genetic mechanism controlling isotype switching, it is apparent that isotypes are expressed in sequence on different sublines of B lymphocytes at a particular time in development. There is no evidence that IgD expression is an intermediate step in eventual commitment of IgM-bearing lymphocytes to synthesis of IgG or IgA. On the contrary, expression of sIgG or sIgA on $sIgM^+$ B lymphocytes precedes expression of sIgD on the same cells. These observations imply that eventual commitment to synthesis of a given isotype is established early in development, is not dependent on external antigens, and is determined by the action of still undefined regulatory genes.

5 Regulation of Isotype Expression

In our original model for generation of isotype diversity it was proposed that a single V_H was sequentially integrated into different C_H genes during successive proliferative cycles (Cooper et al., 1972; Cooper and Lawton, 1974). This single mechanism could account for (1) programmed expression of different germ-line specificities, by sequential expression of V_H genes according to their linkage order on the chromosome; (2) development of isotype diversity, by sequential integration of one V gene with different C genes ($C_\mu \rightarrow C_\gamma \rightarrow C_\alpha$, etc.); and (3) restriction of plasma cells to synthesis of a single species of antibody, by assuming that only the single integrated V–C gene pair in each cell could be expressed. As indicated in the preceding section, observations that single B cells may simultaneously produce two or more immunoglobulins of different class but the same specificity have made this model untenable (see Conclusion, Section 6). More complex regulatory mechanisms must be invoked to explain the sequential expression of different isotypes and the eventual commitment to synthesis of a single isotype. In this section we will discuss some of the factors that may play a role in this process.

The ontogenetic data indicate that initial expression of isotype diversity occurs in a programmed manner and independently of influences provided by exogenous antigens and T cells. There is also considerable evidence suggesting that antigens and/or T cell factors do modify expression of isotype, both on the surface of B cells and with regard to the class of immunoglobulin produced by mature plasma cells (Pierce *et al.*, 1972; Davie and Paul, 1974*b*; Pierce and Klinman 1975). These observations raise an important question: Is eventual commitment to synthesis of a given isotype predetermined prior to activation of individual B lymphocytes, or do environmental influences determine commitment?

The effects of embryonic bursectomy in chickens clearly indicate that the first B cells seeded from the bursa can make IgM only, and that subsequent waves of cells can be triggered to synthesize IgM and IgG, but not IgA (Cooper *et al.*, 1969; Kincade and Cooper, 1973; Kincade *et al.*, 1973). Kearney and Lawton (1975*a*) observed that precursors for IgM and IgG2 responses are present in mouse fetal liver earlier than those capable of generating IgG1 and IgA responses. Goidl *et al.* (1976) observed a similar restriction to IgM production with B cells derived from 14-day mouse embryo liver; in contrast to the chicken experiments, however, they found that repeated antigen boosting of recipients of fetal liver cells did result in an indirect (presumably IgG) antibody response. Press and Klinman (1974) found a distinct predominance of IgM-producing precursors in fetal and neonatal mice.

We have used an *in vitro* culture system, employing bacterial lipopolysaccharide (LPS) to trigger B cell differentiation as a means of studying the switching process directly. Plasma cells synthesizing each of the major Ig classes—IgM, IgG, and IgA—are generated independently of the presence of T lymphocytes (Kearney and Lawton, 1975*b*). All of the LPS-inducible precursors for IgM, IgG, and IgA responses bear cell-surface IgM, as incorporation of anti-μ chain antibodies into cultures abolishes the response to all classes (Kearney *et al.*, 1976*a*). Kinetic studies of changes in cell-surface and cytoplasmic immunoglobulins during the course of LPS-induced differentiation have provided several insights into the regulation of isotype expression (Kearney *et al.*, 1976*b*). Within a period of 6–12 hr after LPS stimulation, a population of $sIgM^+$ cells are induced to express a second isotype (Pernis *et al.*, 1977, Kearney *et al.*, 1977). This induction event does not require DNA synthesis, but is blocked by inhibitors of RNA and protein synthesis, respectively (Kearney *et al.*, 1977). Induction is clearly restricted to distinct subpopulations of $sIgM^+$ cells. Only a fraction of such cells can be induced to express a second class, and more importantly, different induced isotypes appear on different cells. For example, sIgG2, sIgG3, and sIgA appear on mutually exclusive populations of $sIgM^+$ cells (Pernis *et al.*, 1977; Kearney *et al.*, 1977). As the cells proliferate, sIgM is lost from the surface of cells differentiating to IgG synthesis. After approximately 4 days in culture, the IgG2 response can no longer be inhibited by anti-μ antibodies (Kearney *et al.*, 1976*b*). These observations suggest that the precursors for IgG or IgA responses may have several phenotypes. One population bears both sIgM and the additional isotype (and perhaps sIgD as well) and is present prior to stimulation by LPS (Abney *et al.*, 1978). A second population bears only sIgM and is already programmed to express either sIgG or sIgA, but not both.

We mentioned earlier, that plasma cells simultaneously secreting both IgM and IgG, although detectable, are present in very low frequency (Nossal *et al.*, 1971). In our initial studies with LPS-induced responses, we failed to detect a significant frequency of *cytoplasmic* IgM–IgG doubles, as opposed to cells bearing both *surface* IgG and IgM (Kearney *et al.*, 1976*b*). Pernis *et al.* (1977), using an identical system, did find a relatively high frequency of IgM–IgG doubles early in the response. We have recently found an explanation for this discrepancy, which has considerable implications for the regulation of isotype switching. A variety of inhibitors of cell proliferation, including thymidine, sodium butyrate, and suboptimal concentrations of fetal calf serum in growth medium, were found to inhibit induction of IgG responses while having a much smaller effect on development of IgM producers. Virtually all the IgG-containing cells in blocked cultures were also stained for IgM. By examining the effect of thymidine or sodium butyrate added to cultures at various times after initiation and in several concentrations, it was possible to substantiate an inverse relationship between cellular proliferation and the switch from IgM to IgG synthesis (Kearney *et al.*, 1977; Lawton *et al.*, 1978). Double producers were restricted to IgM and one additional class. For example, blocked cultures of newborn spleen or liver in which 100% of IgG- and IgA-positive cells were also stained for IgM had less than a 10% frequency of IgG–IgA doubles.

Taken together, these observations suggest that most B lymphocytes are committed with respect to isotype prior to their activation by polyclonal mitogens—and presumably by antigens. Appropriate activation signals may result in expression of a new isotype in addition to sIgM without DNA synthesis; however, different populations of sIgM$^+$ cells are induced to express different isotypes. Several cycles may be required to complete the transition from sIgM$^+$ lymphocyte to IgG-secreting plasma cell (Askonas and North, 1977). Under normal circumstances, the loss of IgM synthetic capacity is rapid, and cytoplasmic doubles are accordingly rare.

This interpretation must be reconciled with data that suggest that antigens, or T-cell factors, or both, are critical determinants in isotype commitment. There is no doubt that T cells may modify expression of both cell-surface Ig (Davie and Paul, 1974*b*; Pierce *et al.*, 1972) and development of plasma cells secreting various isotypes (Pierce and Klinman, 1975). The critical question is whether these factors activate expression of predetermined sequence of differentiation or direct B cells which are multipotential with respect to isotype, to differentiate along a particular pathway.

Klinman and associates have shown that individual B cell precursors may give rise to clones of plasma cells secreting antibodies of the same specificity and idiotype but of different classes (Press and Klinman, 1973; Gearhart *et al.*, 1975*a*). Pierce and Klinman (1975) found that substitution of allogeneic helper T cells for syngeneic T cells did not change the frequency of detectable B cell precursors but dramatically altered the characteristics of the antibodies produced. Individual clones generated with allogeneic help produced less IgM and virtually no detectable IgG antibodies (see Section 3 for a description of this experimental technique). These results suggest that (1) single precursor B cells may give rise to progeny secreting several different classes, and (2) T cells determine whether a given precursor will give rise to IgG-secreting progeny.

The latter result may still be interpreted in a manner consistent with predetermined B-cell commitment. It is reasonable to suggest that help from allogeneic T cells may be sufficient to initiate differentiation but may fail to provide an adequate stimulus for proliferation. By analogy to LPS-activated cultures in which proliferation was blocked, this might result in selective inhibition of IgG responses due to failure of IgG-committed cells to complete the switch.

The observation that single precursors may give rise to clones secreting antibodies of two or three classes in an apparently random fashion cannot be reconciled with the concept that all B lymphocytes have a predetermined commitment to generate progeny synthesizing a *single* isotype. Gearhart (1977) has analyzed the isotypes of antibodies to phosphorylcholine produced by the progeny of single precursors at various times after stimulation. In clones producing multiple isotypes the onset of synthesis of IgM, IgG, and IgA occurred in an apparently random manner; there was no indication of sequential expression of IgM → IgG → IgA. Lawton and Klinman (unpublished) stained cells from similar clones with fluorescent antibodies to different heavy chains at 10 days after stimulation. Clones known to be producing two Ig classes contained separate populations of cells staining for each class. No doubly producing cells were found. These observations indicate that some single precursors may give rise to sublines of cells secreting IgM, IgG, and IgA; others, stimulated under identical conditions generate progeny producing a single isotype.

The complex factors regulating isotype expression obviously defy any simple explanation. Nevertheless, we believe that most B cells do have a restricted potential, which is determined prior to their stimulation by antigen. Beyond the ontogenetic evidence discussed earlier, the strongest support for this statement has come from comparisons of the immunoglobulin classes produced by cells from different lymphoid tissue. For example, precursors derived from Peyer's patches (lymphoid nodules of the intestine) of both mice and rabbit are highly enriched for IgA precursors, whereas spleen or lymph node cells produce primarily IgG and IgM (Craig and Cebra, 1971, 1975; Cebra *et al.*, 1977). Whether cells home to these sites because of their restrictions or acquire restrictions subsequent to homing has not been determined.

6   Conclusions

With the hope that we have at least indicated areas of controversy, we will close by summarizing our own views of the contributions of studies on development of B lymphocytes to the general problem of cellular differentiation and the specific problems of the generation of antibody diversity.

Primary differentiation of lymphoid cells, like that of other somatic cells, is regulated genetically through interactions of mesenchymal precursor cells with inductive microenvironments of epithelial origin. Once set in motion, the early events of B cell differentiation seem to follow a predetermined sequence in which clonal (specificity) diversification occurs first and is followed by intraclonal development of cells committed to synthesis of the various immunoglobulin isotypes. The basic processes of development of clonal and isotype diversity are not dependent on the presence of external antigens; however, antigens may modify development both negatively, by elimination of specific clonotypes of a critical early stage of development, and positively, by regulation of isotype expression at later stages of B cell maturation.

Ontogenetic observations have not resolved the question of the relative importance of somatic mutations in generation of V-region diversity. They have suggested that the bulk of somatic diversification must occur during the very early stages of B cell development, probably within the population of rapidly dividing pre-B cells. Since pre-B cells apparently lack functional receptors for specific antigens, it is unlikely that antigens play any direct role in driving somatic diversification.

Since this manuscript was submitted (late 1977) there has been an explosion of new information on the structure and organization of immunoglobulin genes in germ-line DNA and the rearrangements which occur during differentiation. We have chosen not to attempt to revise this chapter to incorporate all of this information, since most of the problems we have discussed remain unresolved.

Very recently several groups of molecular geneticists have turned their attention to the phenomenon of isotype switching. Two major discoveries have been made: (1) Davis *et al.* (1980) and Maki *et al.* (1980) have clearly demonstrated that at least two rearrangements are involved in constructing complete genes for α chains and γ2b chains, respectively. The first involves translocation of a selected V_H gene from its embryonic location into apposition to one of a family of minigenes (J genes) which are located approximately one kilobase to the 3' side of the C_μ gene. (The newly discovered J genes code for a part of the third hypervariable region and the carboxy-terminal end of the variable region.) This is followed by a "switch-recombination" event in which the V–J segment and several hundred bases to its 3' side are integrated into sites a few hundred bases to the 5' side of C_α or $C_\gamma 2b$ genes, respectively. Thus in the mature C_α gene, the intron separating VJ and C_α structural gene segments is divisible into two parts. The left end is identical to the left end of the intron between J and C_μ in the germ line, while the right end is adjacent to germ line C_α.

(2) Current evidence favors deletion of intervening DNA as a mechanism for both V–J joining and heavy-chain switching (Sakano et al., 1979; Honjo and Kataoka, 1978; Coleclough et al., 1980; Cory et al., 1980). This supposition is based on observations that IgA-secreting tumors lack one or more copies of C_γ and C_μ genes, IgG secreting tumors lack C_μ but have C_α genes, while IgM secreting tumors contain the whole family of C_H genes. These data have suggested that the murine C_H gene family is linked in the order C_μ, (?C_δ), $C_\gamma 3$, $C_\gamma 1$, $C_\gamma 2b$, $C_\gamma 2a$, C_α, (C_ϵ).

A simple deletional mechanism for V–J joining and subsequent isotype switching is both logical and attractive. In order to explain the sequence of expression of multiple isotypes on single cells outlined in Section 4, it seems necessary to postulate rather complex mechanisms for processing and storing of messenger RNA. There is much truth in the cliché that new information creates more questions than answers.

ACKNOWLEDGMENTS

We dedicate this review to the memory of Mrs. Mary Huckabee, whose irrepressible humor, loyalty, and hard work played a major role in the function of our laboratories for many years. Original work from our laboratories was supported by grants from the Department of Health, Education and Welfare, National Institutes of Health (AI 11502; CA 16673; CA 13148) and The National Foundation, March of Dimes 1-354. Alexander R. Lawton is recipient of an NIH Research Career Development Award, AI 70780. We are greatly indebted to Mrs. Summer King and Mrs. Martha Dagg for help in preparation of the manuscript, and to Drs. Larry Vogler and Elliott Pearl for their critical review.

References

Abney, E. R., Cooper, M. D., Kearney, J. F., Lawton, A. R., and Parkhouse, R. M. E., 1978, Sequential expression of immunoglobulin on developing mouse B lymphocytes. A systemic survey which suggests a model for the generation of immunoglobulin isotype diversity, *J. Immunol.* **120**:2041.

Adamson, J. W., Fialkow, P. J., Murphy, S., Prahl, J. R., and Steinmann, L., 1976, Polycythemia vera: Stem-cell and probable clonal origin of the disease, *N. Engl. J. Med.* **295**:913.

Amzel, L. M., Poljak, R. J., Saul, F., Varga, J. M., and Richards, F. F., 1974, The three dimensional structure of a combining region-ligand complex of immunoglobulin NEW at 3.5-Å resolution, *Proc. Natl. Acad. Sci. U.S.A.* **71**:1427.

Askonas, B. A., and North, J. R., 1977, The life style of B cells—Cellular proliferation and the invariancy of IgG, *Cold Spring Harbor Symp. Quant. Biol.* **41**:749.

Basten, A., Loblay, R., Chia, E., Collard, R., and Pritchard-Briscoe, H., 1977, Suppressor T cells intolerance to non-self and self antigens, *Cold Spring Harbor Symp. Quant. Biol.* **41**:93.

Binz, H., and Wigzell, H., 1977, Antigen-binding, idiotypic receptors from T lymphocytes: An analysis of their biochemistry, genetics and use of immunogens to produce specific immune tolerance, *Cold Spring Harbor Symp. Quant. Biol.* **41**:275.

Burnet, F. M., 1959, *The Clonal Selection Theory of Immunity,* Vanderbilt University Press, Nashville, Tenn.

Burrows, P. B., Kearney, J. F., Lawton, A. R., and Cooper, M. D., 1978, Pre-B cells: Bone marrow persistence in anti-μ suppressed mice, conversion to B lymphocytes, and recovery following destruction by cyclophosphamide, *J. Immunol.* **120**:1526.

Burrows, P. B., LeJeune, M., and Kearney, J. F., 1979, Evidence that murine pre-B cells synthesize μ chains but no light chains, *Nature (London)* **280**:838.

Byrt, P., and Ada, G. L., 1969, An *in vitro* reaction between labeled flagellin or haemocyanin and lymphocyte-like cells from normal animals, *Immunology,* **17**:503.

Cebra, J. J., Gearhart, P. J., Kamat, R., Robertson, S. M., and Tseng, J., 1977, Origin and differentiation of lymphocytes involved in the secretory IgA response, *Cold Spring Harbor Symp. Quant. Biol.* **41**:201.

Claflin, J. L., Lieberman, R., and Davie, J. M., 1974, Clonal nature of the immune response to phosphorylcholine. I. Specificity, class, and idiotype of phosphorylcholine-binding receptors on lymphoid cells, *J. Exp. Med.* **139**:58.

Cohen, J. E., D'Eustachio, P., and Edelman, G. M., 1977, The specific antigen-binding cell populations of individual fetal mouse spleens: Repertoire composition, size, and genetic control, *J. Exp. Med.* **146**:394.

Cohn, M., Blomberg, B., Geckeler, W., Raschke, W., Riblet, R., and Weigert, M., 1974, First order considerations in analyzing the generation of diversity, in: *The Immune System: Genes, Receptors, Signals* (E. Sercarz, L. A. Herzenberg, and C. F. Fox, eds.), pp. 89–117, Academic Press, New York.

Coleclough, C., Cooper, D., and Perry, R. P., 1980, Rearrangement of immunoglobulin heavy chain genes during B lymphocyte development as revealed by studies of mouse plasmacytoma cells, *Proc. Natl. Acad. Sci. U.S.A.* **77**:1422.

Cooper, M. D., and Lawton, A. R., 1974, The development of the immune system, *Sci. Am.* **231**:58.

Cooper, M. D., Peterson, R. D. A., South, M. A., and Good, R. A., 1966, The functions of the thymus system and the bursa system in the chicken, *J. Exp. Med.* **123**:75.

Cooper, M. D., Cain, W. A., Van Alten, P. J., and Good, R. A., 1969, Development and function of the immunoglobulin-producing system. I. Effect of Bursectomy at different stages of development on germinal centers, plasma cells, immunoglobulins and antibody production, *Int. Arch. Allergy Appl. Immunol.* **35**:242.

Cooper, M. D., Kincade, P. W., and Lawton, A. R., 1971, Thymus and bursal function in immunologic development: A new theoretical model of plasma cell differentiation, in: *Immunologic Incompetence* (B. M. Kagan and E. R. Stiehm, eds.), pp. 81–104, Year Book Medical Publishers, Chicago, Ill.

Cooper, M. D., Lawton, A. R., and Kincade, P. W., 1972, A developmental approach to the biological basis for antibody diversity, in: *Contemporary Topics in Immunobiology,* Vol. 1 (M. G. Hanna and M. D. Cooper, eds.), pp. 33–47, Plenum Press, New York.

Cooper, M. D., Kearney, J. F., Lawton, A. R., Abney, E. R., Parkhouse, R. M. E., Preud'homme, J. L., and Seligmann, M., 1976, Generation of immunoglobulin class diversity in B cells: A discussion with emphasis on IgD development, *Ann. Immunol. (Inst. Pasteur)* **127**:573.

Cooper, M. D., Kearney, J. F., and Lawton, A. R., 1978, The life history of antibody producing B cell cells, in: *Proceedings of the Fifth International Congress* (J. W. Rittlefield and J. de Grouchy, eds.), pp. 178–195, *Excerpta Medica,* Amsterdam and Oxford.

Cory, S., Jackson, J., and Adams, J. M., 1980, Deletions in the constant region locus can account for switches in immunoglobulin heavy chain expression, *Nature (London)* **285**:450.

Craig, S. W., and Cebra, J. J., 1971, Peyer's patches: An enriched source of precursors for IgA-producing immunocytes in the rabbit, *J. Exp. Med.* **134**:188.

Craig, S. W., and Cebra, J. J., 1975, Rabbit Peyer's patches, appendix, and popliteal lymph node B lymphocytes: A comparative analysis of their membrane immunoglobulin components and plasma cell precursor potential, *J. Immunol.* **114**:492.

Davie, J. M., and Paul, W. E., 1971, Receptors on immunocompetent cells. II. Specificity and nature of receptors on dinitrophenylated guinea pig albumin-^{125}I-binding lymphocytes of normal guinea pigs, *J. Exp. Med.* **134**:495.

Davie, J. M., and Paul, W. E., 1974a, Antigen-binding receptors on lymphocytes, in: *Contemporary Topics in Immunobiology,* Vol. 3 (M. D. Cooper and N. L. Warner, eds.), pp. 171–192, Plenum Press, New York.

Davie, J. M., and Paul, W. E., 1974b, Role of T lymphocytes in the humoral immune response. I. Proliferation of B lymphocytes in thymus-deprived mice, *J. Immunol.* **113**:1438.

Davis, M. M., Calame, K., Early, P. W., Livant, D. L., Joho, R., Weissman, I. L., and Hood, L., 1980, An immunoglobulin gene is formed by at least two recombinational events, *Nature (London)* **283**:733.

D'Eustachio, P., and Edelman, G. M., 1975, Frequency and avidity of specific antigen-binding cells in developing mice, *J. Exp. Med.* **142**:1078.

D'Eustachio, P., Cohen, J. E., and Edelman, G. M., 1976, Variation and control of specific antigen-binding cell populations in individual fetal mice, *J. Exp. Med.* **144**:259.

Dreyer, W. J., and Bennett, J. C., 1965, The molecular basis of antibody formation: A paradox, *Proc. Natl. Acad. Sci. U.S.A.* **54**:864.

Ehrlich, P., 1900, On immunity with special reference to cell life, *Proc. R. Soc.* **66**:424.

Fu, S. M., Winchester, R. J., and Kunkel, H. G., 1975, Similar idiotypic specificity for the membrane IgD and IgM of human B lymphocytes, *J. Immunol.* **114**:250.

Gathings, W. E., Lawton, A. R., and Cooper, M. D., 1977, Immunofluorescent studies of the development of pre-B cells, B lymphocytes, and immunoglobulin isotype diversity in humans, *Eur. J. Immunol.* **7**:804.

Gearhart, P. J., 1977, Non-sequential expression of multiple immunoglobulin classes by isolated B-cell clones, *Nature (London)* **269**:812.

Gearhart, P. J., Sigal, N. H., and Klinman, N. R., 1975a, Production of antibodies of identical idiotype but diverse immunoglobulin classes by cells derived from a single stimulated B cell, *Proc. Natl. Acad. Sci. U.S.A.* **72**:1707.

Gearhart, P. J., Sigal, N. H., and Klinman, N. R., 1975b, Heterogeneity of the BALB/c antiphosphorylcholine antibody response at the precursor cell level, *J. Exp. Med.* **141**:56.

Goding, J. W., and Layton, J. E., 1976, Antigen-induced co-capping of IgM and IgD-like receptors on murine B cells, *J. Exp. Med.* **144**:852.

Goidl, E. A., and Siskind, G. W., 1974, Ontogeny of B-lymphocyte function. I. Restricted heterogeneity of the antibody response of B lymphocytes from neonatal and fetal mice, *J. Exp. Med.* **140**:1285.

Goidl, E. A., Klass, J., and Siskind, G. W., 1976, Ontogeny of B-lymphocyte function. II. Ability of endotoxin to increase the heterogeneity of affinity of the immune response of B lymphocytes from fetal mice, *J. Exp. Med.* **143**:1503.

Gottlieb, P. D., and Durda, P. J., 1977, The I_B-peptide marker and the L_{4-3} surface alloantigen: Structural studies of a V_κ-region polymorphism and a T-cell marker determined by linked genes, *Cold Spring Harbor Symp. Quant. Biol.* **41**:805.

Hayward, A. R., Simons, M., Lawton, A. R., Mage, R. G., and Cooper, M. D., 1978, Ontogeny of B lymphocytes in rabbits: Allotype exclusion and suppression at the b locus, in: *Developmental Immunobiology* (J. B. Soloman, ed.), pp. 181–188, Elsevier/North-Holland Biomedical Press, Amsterdam.

Hood, L., Loh, E., Hubert, J., Barstad, P., Eaton, G., Early, P., Fuhrman, J., Johnson, N., Kronenberg, M., and Schilling, J., 1977a, The structure and genetics of mouse immunoglobulins: An analysis of NZB myeloma proteins and sets of BALB/c myeloma proteins binding particular haptens, *Cold Spring Harbor Symp. Quant. Biol.* **41**:817.

Hood, L., Kronenberg, M., Early, P., and Johnson, N., 1977b, Nucleic acid chemistry and the antibody problem, in: *ICN–UCLA Symposia on Molecular and Cellular Biology,* Vol. VI, *The Immune System: Genetics and Regulation* (E. E. Sercarz, L. A. Herzenberg, and C. F. Fox, eds.), pp. 1–27, Academic Press, New York.

Honjo, T., and Kataoha, T., 1978, Organization of immunoglobulin heavy chain genes and allelic deletion model, *Proc. Natl. Acad. Sci. U.S.A.* **75**:2140.

Hozumi, N., and Tonegawa, S., 1976, Evidence for somatic rearrangement of immunoglobulin genes coding for variable and constant regions, *Proc. Natl. Acad. Sci. U.S.A.* **73**:3628.

Jerne, N. K., 1971, The somatic generation of immune recognition, *Eur. J. Immunol.* **1**:1.

Kearney, J. F., and Lawton, A. R., 1975a, B-lymphocyte differentiation induced by lipopolysaccharide. II. Response of fetal lymphocytes, *J. Immunol.* **115**:677.

Kearney, J. F., and Lawton, A. R., 1975b, B-lymphocyte differentiation induced by lipopolysaccharide. I. Generation of cells synthesizing four major immunoglobulin classes, *J. Immunol.* **115**:671.

Kearney, J. F., Cooper, M. D., and Lawton, A. R., 1976a, B-lymphocyte differentiation induced by lipopolysaccharide. III. Suppression of B cell maturation by anti-mouse immunoglobulin antibodies, *J. Immunol.* **116**:1664.

Kearney, J. F., Cooper, M. D., and Lawton, A. R., 1976b, B cell differentiation induced by lipopolysaccharide. IV. Development of immunoglobulin class restriction in precursors of IgG-synthesizing cells, *J. Immunol.* **117**:1567.

Kearney, J. F., Lawton, A. R., and Cooper, M. D., 1977, Multiple immunoglobulin heavy chain expression by LPS stimulated murine B lymphocytes, in: *ICN–UCLA Symposia on Molecular and Cellular Biology,* Vol. VI, *Immune System: Genetics and Regulation* (E. Sercarz, L. A. Herzenberg, and C. F. Fox, eds.), pp. 313–320, Academic Press, New York.

Kearney, J. F., Klein, J., Bockman, D. E., Cooper, M. D., and Lawton, A. R., 1978, B cell differentiation induced by lipopolysaccharide. V. Suppression of plasma cell maturation by anti-μ: Mode of action and characteristics of suppressed cells, *J. Immunol.* **120**:158.

Kim, B. S., and Dray, S., 1973, Expression of the *a, x,* and *y* variable region genes of heavy chains among IgG, IgM, and IgA molecules of normal and *a* locus allotype-suppressed rabbits, *J. Immunol.* **111**:750.

Kincade, P. W., and Cooper, M. D., 1971, Development and distribution of immunoglobulin-containing cells in the chicken: An immunofluorescent analysis using purified antibodies to μ, γ and light chains, *J. Immunol.* **106**:371.

Kincade, P. W., and Cooper, M. D., 1973, Immunoglobulin A: Site and sequence of expression in developing chicks, *Science* **179**:398.

Kincade, P. W., Lawton, A. R., Bockman, D. E., and Cooper, M. D., 1970, Suppression of immunoglobulin G synthesis as a result of antibody-mediated suppression of immunoglobulin M synthesis in chickens, *Proc. Natl. Acad. Sci. U.S.A.* **67**:1918.

Kincade, P. W., Self, K. S., and Cooper, M. D., 1973, Survival and function of bursa-derived cells in bursectomized chickens, *Cell. Immunol.* **8**:93.

Klinman, N. R., and Press, J. L., 1975*a*, The B cell specificity repertoire: Its relationship to definable subpopulations, *Transplant. Rev.* **24**:41.

Klinman, N. R., and Press, J. L., 1975*b*, The characterization of the B-cell repertoire specific for the 2,4-dinitrophenyl and 2,4,6-trinitrophenyl determinants in neonatal BALB/c mice, *J. Exp. Med.* **141**:113.

Klinman, N. R., Sigal, N. H., Metcalf, E. S., Pierce, S. K., and Gearhart, P. J., 1977, The interplay of evolution and environment in B-cell diversification, *Cold Spring Harbor Symp. Quant. Biol.* **41**:165.

Knapp, W., Boluis, R. L. H., Radl, J., and Jijmans, W., 1973, Independent movement of IgD and IgM molecules on the surface of individual lymphocytes, *J. Immunol.* **111**:1295.

Krawinkel, U., Cramer, M., Berek, C., Hämmerling, G., Black, S. J., Rajewsky, K., and Eichmann, K., 1977, On the structure of the T-cell receptor for antigen, *Cold Spring Harbor Symp. Quant. Biol.* **41**:285.

Kubagawa, H., Vogler, L., Lawton, A., and Cooper, M., 1978, Use of idiotypic antibodies to explore development and differentiation of human B cell clones, *Clin. Res.* **26**:517A.

Kubagawa, H., Vogler, L. B., Capra, J. D., Conrad, M. E., Lawton, A. R., and Cooper, M. D., 1979*a*, Studies on the clonal origin of multiple myeloma: Use of individually specific (idiotype) antibodies to have the oncogenic event to its earliest point in B cell differentiation, *J. Exp. Med.* **150**:792.

LaFleur, L., Miller, R. G., and Phillips, R. A., 1972, A quantitative assay for progenitors of bone marrow associated lymphocytes, *J. Exp. Med.* **135**:1363.

Lawton, A. R., and Cooper, M. D., 1974, Modification of B lymphocyte differentiation by anti-immunoglobulins, in: *Contemporary Topics in Immunobiology,* Vol. 3 (M. D. Cooper and N. L. Warner, eds.), pp. 193–225, Plenum Press, New York.

Lawton, A. R., Asofsky, R., Hylton, M. B., and Cooper, M. D., 1972*a*, Suppression of immunoglobulin class synthesis in mice: I. Effects of treatment with antibody to μ chain, *J. Exp. Med.* **135**:277.

Lawton, A. R., Self, K. S., Royal, S. A., and Cooper, M. D., 1972*b*, Ontogeny of B lymphocytes in the human fetus, *Clin. Immunol. Immunopathol.* **1**:104.

Lawton, A. R., Kincade, P. W., and Cooper, M. D., 1975, Sequential expression of germ line genes in development of immunoglobulin class diversity, *Fed. Proc.* **34**:33.

Lawton, A. R., Kearney, J. F., and Cooper, M. D., 1978, Control of expression of C region genes during development of B cells, in: *Progress in Immunology, III. Proceedings of the Third International Congress of Immunology* (T. E. Mandel, C. Cheers, C. S. Hoshing, I. F. C., McKenzie, and G. J. V. Nossel, eds.), pp. 171–182, Elsevier/North Holland, Inc., New York.

LeDouarin, N. M., Houssaint, E., Jotereau, F. V., and Belo, M., 1975, Origin of haemopoetic stem cells in the embryonic bursa of Fabricius and bone marrow studied through intraspecific chimaeras, *Proc. Natl. Acad. Sci. U.S.A.* **72**:2701.

Lee, W., Cosenza, H., and Köhler, H., 1974, Clonal restriction of the immune response to phosphorylcholine, *Nature (London)* **247**:55.

Levitt, D., and Cooper, M. D., 1980, Mouse pre-B cells synthesize and secrete μ heavy chains but not light chains, *Cell* **19**:617.

Lydyard, P. M., Grossi, C. E., and Cooper, M. D., 1976, Ontogeny of B cells in the chicken: I. Sequential development of clonal diversity in the bursa, *J. Exp. Med.* **144**:79.

Mage, R. G., 1971, Structural localization, allelic exclusion, and linkage relationships of rabbit allotypes, *Progr. Immunol.* **1**:47.

Maki, R., Traunecker, A., Sakano, H., Roeder, W., and Tonegawa, S., 1980, Exon shuffling generates an immunoglobulin heavy chain gene, *Proc. Natl. Acad. Sci. U.S.A.* **77**:2138.

Manning, D. D., and Jutila, J. W., 1972, Immunosuppression of mice injected with heterologous anti-immunoglobulin heavy chain antisera, *J. Exp. Med.* **135**:1316.

Melchers, F., 1977*a*, B lymphocyte development in fetal liver. I. Development of reactivities to B cell mitogens "in vivo" and "in vitro," *Eur. J. Immunol.* **7**:476.

Melchers, F., 1977*b*, B lymphocyte development in fetal liver. II. Frequencies of precursor B cells during gestation, *Eur. J. Immunol.* **7**:482.

Melchers, F., von Boehmer, H., and Phillips, R. A., 1975, B-lymphocyte subpopulations in the mouse, *Transplant. Rev.* **25**:26.

Melchers, F., Andersson, J., and Phillips, R. A., 1977, Ontogeny of murine B lymphocytes: Development of Ig synthesis and of reactivities to mitogens and to anti-Ig-antibodies, *Cold Spring Harbor Symp. Quant. Biol.* **41**:147.

Metcalf, E. S., and Klinman, N. R., 1976, *In vitro* tolerance induction of neonatal murine B cells, *J. Exp. Med.* **143**:1327.

Metcalf, E. S., and Klinman, N. R., 1977, *In vitro* tolerance induction of bone marrow cells: A marker for B cell maturation, *J. Immunol.* **118**:2111.

Moore, M. A. S., and Owen, J. J. T., 1966, Experimental studies on the development of the bursa of Fabricius, *Dev. Biol.* **14**:40.

Moretta, L., Ferrarini, M., and Cooper, M. D., 1978, Characterization of human T cell subpopulations as defined by specific receptors for immunoglobulins, in: *Contemporary Topics in Immunobiology,* Vol. 8 (M. D. Cooper, and N. L. Warner, eds.), pp. 19–53, Plenum Press, New York.

Mosier, D. E., and Johnson, B. M., 1975, Ontogeny of mouse lymphocyte function. II. Development of the ability to produce antibody is modulated by T lymphocytes, *J. Exp. Med.* **141**:216.

Murgita, R., Mattioli, C., and Tomasi, T. B., 1973, Production of a runting syndrome and selective IgA deficiency in mice by the administration of anti-heavy chain antisera, *J. Exp. Med.* **138**:209.

Naor, D., and Sulitzeanu, D., 1967, Binding of radioiodinated bovine serum albumin to mouse spleen cells, *Nature (London)* **214**:687.

Nordin, A. A., Cosenza, H., and Sell, S., 1970, Immunoglobulin classes of antibody-forming cells in mice. II. Class restriction of plaque-forming cells demonstrated by replica plating, *J. Immunol.* **104**:495.

Nossal, G. J. V., and Pike, B. L., 1975, Evidence for the clonal abortion theory of B lymphocyte tolerance, *J. Exp. Med.* **141**:904.

Nossal, G. J. V., Szenberg, A., Ada, G. L., and Austin, G. M., 1964, Single cell studies on 19S antibody production, *J. Exp. Med.* **119**:485.

Nossal, G. J. V., Warner, N. L., and Lewis, H., 1971, Incidence of cells simultaneously secreting IgM and IgG antibody to sheep erythrocytes, *Cell. Immunol.* **2**:41.

Okos, A. J., and Gathings, W. E., 1977, Characterization of precursor B cells in human bone marrow, *Fed. Proc.* **36**:1294.

Osmond, D. G., and Nossal, G. J. V., 1974, Differentiation of lymphocytes in mouse bone marrow. II. Kinetics of maturation and renewal of antiglobulin-binding cells studied by double labeling, *Cell. Immunol.* **13**:132.

Owen, J. J. T., Cooper, M. D., and Raff, M. C., 1974, *In vitro* generation of B lymphocytes in mouse foetal liver—A mammalian "bursa equivalent," *Nature (London)* **249**:361.

Owen, J. J. T., Raff, M. C., and Cooper, M. D., 1975, Studies on the generation of B lymphocytes in the mouse embryo, *Eur. J. Immunol.* **5**:468.

Owen, J. J. T., Wright, D. E., Habu, S., Raff, M. C., and Cooper, M. D., 1977, Studies on the generation of B lymphocytes in fetal liver and bone marrow, *J. Immunol.* **118**:2067.

Perlmutter, R. M., Briles, D. E., and Davie, J. M., 1977, Complete sharing of light chain spectrotypes by murine IgM and IgG anti-streptococcal antibodies, *J. Immunol.* **118**:2161.

Pernis, B., Chiappino, G., Kelus, A. S., and Gell, P. G. H., 1965, Cellular localization of immunoglobulins with different allotypic specificities in rabbit lymphoid tissue, *J. Exp. Med.* **122**:853.

Pernis, B., Forni, L., and Amante, L., 1970, Immunoglobulin spots on the surface of rabbit lymphocytes, *J. Exp. Med.* **132**:1001.

Pernis, B., Forni, L., and Amante, L., 1971, Immunoglobulins as cell receptors, *Ann. N.Y. Acad. Sci.* **190**:420.

Pernis, B., Brouet, J. C., and Selgimann, M., 1974, IgD and IgM on the membrane of lymphoid cells in macroglobulinemia. Evidence for identity of membrane IgD and IgM antibody activity in a case with anti-IgG receptors, *Eur. J. Immunol.* **4**:776.

Pernis, B., Forni, L., and Luzzati, A. L., 1977, Synthesis of multiple immunoglobulin classes by single lymphocytes, *Cold Spring Harbor Symp. Quant. Biol.* **41**:175.

Pierce, C. W., Solliday, S. M., and Asofsky, R., 1972, Immune responses *in vitro*. IV. Suppression of primary γM, γG, and γA plaque-forming cell responses in mouse spleen cultures by class-specific antibody to mouse immunoglobulins, *J. Exp. Med.* **135**:675.

Pierce, S. K., and Klinman, N. R., 1975, The allogeneic bisection of carrier-specific enhancement of monoclonal B-cell responses, *J. Exp. Med.* **142**:1165.

Potter, M., Dreyer, W. J., Kuff, E. L., and McIntire, K. R., 1964, Heritable variation in Bence–Jones protein structure in an inbred strain of mouse, *J. Mol. Biol.* **8**:814.

Press, J. L., and Klinman, N. R., 1973, Monoclonal production of both IgM and IgG_1 anti-hapten antibody, *J. Exp. Med.* **138**:300.

Press, J. L., and Klinman, N. R., 1974, Frequency of hapten-specific B cells in neonatal and adult murine spleen cells, *Eur. J. Immunol.* **4**:155.

Putnam, F. W., 1962, Structural relationships among normal human γ-globulin, myeloma globulins, and Bence–Jones proteins, *Biochim. Biophys. Acta* **63**:539.

Raff, M. C., 1977, Development and modulation of B lymphocytes: Studies on newly formed B cells and their putative precursors in the hemopoietic tissues of mice, *Cold Spring Harbor Symp. Quant. Biol.* **41**: 159.

Raff, M. C., Sternberg, M., and Taylor, R. B., 1970, Immunoglobulin determinants on the surface of mouse lymphoid cells, *Nature (London)* **225**:553.

Raff, M. C., Owens, J. J. T., Cooper, M. D., Lawton, A. R., Megson, M., and Gathings, W. E., 1975, Differences in susceptibility of mature and immature mouse B lymphocytes to anti-immunoglobulin-induced immunoglobulin suppression *in vitro*: Possible implications for B cell tolerance to self, *J. Exp. Med.* **142**:1052.

Raff, M. C., Megson, M., Owen, J. J. T., and Cooper, M. D., 1976, Early production of intracellular IgM by B-lymphocyte precursors in mouse, *Nature (London)* **259**:224.

Rowe, D. S., Hug, K., Forni, L., and Pernis, B., 1973, Immunoglobulin D as a lymphocyte receptor, *J. Exp. Med.* **138**:965.

Rowlands, D. T., Blakeslee, P. Angala, E., 1974, Acquired immunity in opossum (Didelphis virginiana) embryos, *J. Immunol.* **112**:2148.

Ryser, J-E., and Vassalli, P., 1974, Mouse bone marrow lymphocytes and their differentiation, *J. Immunol.* **113**:719.

Sakano, H., Hiippi, K., Heinrich, G., and Tonegawa, S., 1979, Sequences at the somatic recombination sites of immunoglobulin light-chain genes, *Nature (London)* **280**:288.

Sherwin, W. K., and Rowlands, D. T., Jr., 1975, Determinants of the hierarchy of humoral immune responsiveness during ontogeny, *J. Immunol.* **115**:1549.

Sidman, C. L., and Unanue, E. R., 1975, Receptor-mediated inactivation of early B lymphocytes, *Nature (London)* **257**:149.

Sigal, N. H, Gearhart, P. J., and Klinman, N. R., 1975, The frequency of phosphorylcholine-specific B cells in conventional and germfree BALB/c mice, *J. Immunol.* **114**:1354.

Sigal, N. H., Gearhart, P. J., Press, J. L., and Klinman, N. R., 1976, The late adquisition of a "germline" antibody specificity, *Nature (London)* **259**:57.

Silverstein, A. M., Uhr, J. W., Kramer, K. L., and Lukes, R. J., 1963, Fetal response to antigenic stimulus. II. Antibody production by the fetal lamb, *J. Exp. Med.* **117**:799.

Spear, P. G., Wang, A., Rutishauser, U., and Edelman, G. M., 1973, Characterization of splenic lymphoid cells in fetal and newborn mice, *J. Exp. Med.* **138**:557.

Taylor, R. B., Duffus, W. P. H., Raff, M. C., and De Petris, S., 1971, Redistribution and pinocytosis of lymphocyte surface immunoglobulin molecules induced by anti-immunoglobulin antibody, *Nature (London) New Biol.* **233**:225.

Tonegawa, S., Brach, C., Hozumi, N., and Pirrotta, V., 1977, Organization of immunoglobulin genes, *Cold Spring Harbor Symp. Quant. Biol.* **42**(Part I):921.

Vitetta, E. S., Cambier, J., Forman, J., Kettman, J. R., Yuan, D., and Uhr, J. W., 1977, Immunoglobulin receptors on murine B lymphocytes, *Cold Spring Harbor Symp. Quant. Biol.* **41**:185.

Wang, A. C., Wang, I. Y. F., McCormick, J. N., and Fudenberg, H. H., 1969, The identity of light chains of monoclonal IgG and monoclonal IgM in one patient, *Immunochemistry* **6**:451.

Wang, A. C., Wilson, S. K., Hopper, J. E., Fudenberg, H. H., and Nisonoff, A., 1970, Evidence for control of synthesis of the variable regions of the heavy chains of immunoglobulins G and M by the same gene, *Proc. Natl. Acad. Sci. U.S.A.* **66**:337.

Weigert, M., and Riblet, R., 1977, Genetic control of antibody variable regions, *Cold Spring Harbor Symp. Quant. Biol.* **41**:837.

Wu, T. T., and Kabat, E. A., 1970, An analysis of the sequences of the variable regions of Bence–Jones proteins and myeloma light chains and their implications for antibody complementarity, *J. Exp. Med.* **132**:211.

10

Immune Response Genes in the Regulation of Mammalian Immunity

JAY A. BERZOFSKY

1 Introduction

1.1 Definition and Perspective

The continuous and growing excitement elicited by the concept of immune response (Ir)* genes since their discovery in the mid-1960s (McDevitt and Benacerraf, 1969; Benacerraf and McDevitt, 1972; Benacerraf and Katz, 1975; Benacerraf and Germain, 1978) can be understood as soon as that concept is fully defined. Immune response genes are operationally defined as antigen-specific genes that control the ability of an animal to raise an immune response, humoral or cellular, to a particular antigen. The antigen specificity is a

---

JAY A. BERZOFSKY ● The Metabolism Branch, National Cancer Institute, National Institutes of Health, Bethesda, Maryland 20205.

Abbreviations used in this chapter: BSA, Bovine serum albumin; BUdR, bromodeoxyuridine, CFA, complete Freund's adjuvant; C_H, constant portion of immunoglobulin heavy chain; CI, cellular interaction; cM, centi-Morgans or recombinational map units; DNP, dinitrophenyl; DTH, delayed-type hypersensitivity; GAT, random linear copolymer $Glu^{60},Ala^{30},Tyr^{10}$; GLLeu, poly-(Glu,Lys,Leu); GLPhe, random copolymer of Glu^{53}, Lys^{36},Phe^{11}; GLT^5, poly-$(Glu^{57},Lys^{38},Tyr^5)$; *H-2,* The major histocompatibility complex of the mouse; (H,G)-A–L, poly(His,Glu)-poly-D,L-Ala-poly-L-Lys; *Ir,* immune response; KLH, keyhole limpet hemocyanin; LPS, lipopolysaccharide from bacterial cell wall; Mb, myoglobin; MBSA, methylated bovine serum albumin; MHC, major histocompatibility complex; MLR, allogenic mixed lymphocyte reaction; NIP, (4-hydroxy-5-iodo-3-nitrophenyl) acetyl; NP, (4-hydroxy-3-nitrophenyl) acetyl; (Phe,G)-A–L, poly(Phe,Glu)-poly-D,L-Ala-poly-L-Lys; PLL, poly-L-lysine; (T,G)-A–L, poly(Tyr,Glu)-poly-D,L-Ala–poly-L-Lys; (T,G)-Pro-L, poly(Tyr,Glu)-poly-L-Pro–poly-L-Lys; TNP, trinitrophenyl; V_H, variable portion of immunoglobulin heavy chain; V_L, variable portion of immunoglobulin light chain.

crucial aspect of the definition. Thus genes that lead to broad immune deficiency diseases, such as Wiscott–Aldrich syndrome or ataxia telangiectasia in man (Waldmann *et al.,* 1980) and the CBA/N defect in the mouse (Mosier *et al.,* 1977), are excluded from the concept. However, the antigen specificity is also what leads to all the excitement. The hallmark of immunology has always been the exquisite, fine specificity of antigen recognition combined with the seemingly endless diversity of specificities that could be elicited. Despite many investigator years of research invested in trying to explain this diversity, primarily with regard to immunoglobulins, providing us with some understanding of the molecular bases of specificity (Kabat, 1978; Berzofsky and Schechter, 1980, and reviews cited therein), the mechanisms of generation of antibody diversity are still the subject of continuing controversy (Cunningham, 1976; Kabat *et al.,* 1979; Seidman *et al.,* 1979, and references cited therein). The discovery of *Ir* genes introduced an apparently new level of antigen recognition whose diversity and specificity had to be explained in addition to those of familiar immunoglobulins. The distinctness of this new level of diversity was emphasized when it was found that most *Ir* genes are not genetically linked to the structural genes for immunoglobulins (McDevitt and Benacerraf, 1969). Thus most *Ir* gene defects are not merely deficiencies in the antibody structural gene repertoire or alterations in regulatory genes that might form part of an immunoglobulin operon (see Goldberger, Volume 1 of this series). Moreover, no macromolecule has yet been identified that can be demonstrated with certainty to be the product of an *Ir* gene, although many contenders for this title exist (see Sections 3, 6, 7, and 9). Therefore, *Ir* genes are still defined solely on a functional basis. Furthermore, *Ir* gene "defects" do not arise out of simple self-tolerance. For instance, if a series of mice were immunized with immunoglobulin from one of them, the strain from which the immunoglobulin was derived would be tolerant, whereas some of the other strains would produce antibodies to determinants not present on their own immunoglobulin. This result would mimic the action of an *Ir* gene. However, the F_1 hybrid between a tolerant stain and a responding strain would be tolerant, and therefore a low responder. In contrast, in the case of most *Ir* genes we shall describe, high responsiveness is dominant. Therefore, no obvious trivial explanation was adequate to account for *Ir* genes. From the very beginning, then, it has been apparent that interpreting the riddle of *Ir* gene function would likely lead to a new level of understanding of not only the regulation of immune responses but also, *pari passu,* the mechanism of induction of immunity in general.

It should be pointed out that while some *Ir* gene control appears to be all or none in that no response can be detected in nonresponders even by the most sensitive assays available, most *Ir* gene "defects" are "leaky." The strain differences in responsiveness are quantitative, but not absolute, so that one must refer to high and low responders, but not to nonresponders. In addition, some low responders can produce a primary IgM response indistinguishable from that of high responders, but these low responders can be recognized by their failure to produce a secondary IgG response. In other cases, *Ir* gene control of the primary IgM response is seen as well. Furthermore, some *Ir* gene "defects" can be overcome by providing an additional nonspecific stimulus, such as an ongoing graft-versus-host reaction, or by immunizing low responders with the antigen attached to an immunogenic "carrier" molecule. Some differences may also be observed depending on the mode of immunization (route, adjuvant, number of immunizations) and the type of response studied (cellular or humoral). Despite this evidence, and that discussed below, that most *Ir* genes do not exert absolute control, all *Ir* genes are characterized by three findings: (1) that a particular group of animals (or inbred strain) produces a lower immune response to a specific antigen, over a reasonable range of antigen dose, than do other animals of the species; (2) that these low-responder animals can respond as well as the others to a variety of other antigens of the same type; and (3) that the low or high responsiveness is a hereditary trait.

469

IMMUNE RESPONSE
GENES IN THE
REGULATION OF
MAMMALIAN
IMMUNITY

The field of immune response genes has changed enormously over the past 6 or 7 years. Some popular theories, thought at the beginning of the 1970s to be all but proven have now been essentially discarded. In their place, other theories, in some sense variants on theories considered initially but then rejected, have risen to ascendancy. While at the level of popular models ideas may have reversed, the fundamental experimental findings on which each theory was based have not been disproven. Indeed, the kernels of these several theories may all turn out to be true and ultimately reconcilable. We are all still very much like the proverbial blind men groping to describe the elephant. Some descriptions are relevant to the legs, some to the trunk, others to the tail. None might be incorrect. Yet none might be the complete description of the whole animal that is so eagerly sought. In fact, there may not even be a whole animal, but rather many separate ones. There may be several disparate phenomena that have been lumped together as *Ir* genes because their effects, if not their fundamental mechanisms, are similar. Attempts to unify these may, in fact, impose artificial and misleading constraints on all the theories. Nevertheless, most investigators would like to find a unified theory to explain all observations by some single central underlying mechanism. Through these attempts, the field of immune response genetics has grown to be both exciting and crucial to all of immunology, as well as perhaps to our understanding of more general mechanisms of biological regulation and development.

Considerations of space and the interest of the general reader of this volume do not permit a comprehensive review of either immune response genes or the regulation of mammalian immunity in this chapter. Rather, this chapter attempts to highlight the major themes that have intertwined and led to the current state of understanding. While some ideas may be emphasized, no ultimate truth has yet been found that can be documented or analyzed in these pages. For this reason, to give the reader a fairer perspective of the various experiments and hypotheses I find so exciting, I shall take a semihistorical approach in describing them. Many other areas, such as the nature of T cell* receptors, major histocompatibility gene products, genetic restriction phenomena, and immune regulatory circuits, must be covered as well, since the concepts of immune response genes have grown up hand in hand with these areas. I have therefore tried to maintain a broad enough scope of coverage to be of interest to the general reader, and yet sufficient depth in certain areas to be of value to the immunologist as well.

1.2 Early History and Speculations on Mechanisms

Throughout the history of immunology, some individuals of a species have been found to respond strongly to a given immunogen, while other individuals responded less well or not at all. When the goal was to raise antibodies to a certain antigen, this variability and unpredictability were regarded as more troublesome than interesting. Animals that failed

---

*T lymphocytes or T cells are a "thymus-dependent" or "thymus-derived" class of lymphocytes responsible for most cellular immune reactions, such as delayed hypersensitivity, cellular cytotoxicity, and graft-versus-host disease, and for both positive and negative regulation of the immune response. B lymphocytes or B cells (originally standing for "bursa-derived," since they arise in the gut-associated lymphoid organ called the Bursa of Fabricius in chickens and later extended to stand for "bone marrow-derived" in mammals) are the precursors of antibody-producing cells.

to produce good antisera were discarded. The solution was simply to immunize enough animals that at least one would be likely to produce the desired antiserum. It is to the great credit of Benacerraf and his colleagues (Kantor *et al.,* 1963; Levine *et al.,* 1963*a,b*; Levine and Benacerraf, 1965) and McDevitt and Sela (1965) that they converted what was originally a nuisance to most investigators into what is now one of the central themes of cellular and molecular immunology. Moreover, the groundwork was laid without the benefit of inbred strains of animals. Benacerraf and co-workers (references above) found that immunization of outbred Hartley guinea pigs with a dinitrophenyl conjugate of poly-L-lysine (DNP-PLL) in complete Freund's adjuvant (CFA) resulted in a bimodal distribution of responsiveness: some animals produced high titers of serum anti-DNP antibodies as well as a delayed-type hypersensitivity (DTH) reaction to DNP-PLL, while others produced neither of these. The difference in responsiveness was antigen-specific and inherited as a single Mendelian autosomal gene, with responsiveness dominant to nonresponsiveness. However, the nonresponders could produce antibodies to DNP, albeit not a DTH reaction, when DNP-PLL was injected bound to an immunogenic "carrier"* protein such as bovine serum albumin (BSA) (Green *et al.,* 1966). Thus the defect was not caused by lack of the appropriate immunoglobulin structural gene. This last conclusion has turned out to be true for most but not for all *Ir* genes (see Section 2.1).

Shortly after these initial discoveries in the guinea pig, a similar *Ir* gene phenomenon was discovered by McDevitt and Sela (1965) in the mouse for the branched amino acid polymers† poly (Tyr,Glu)-poly-D,L-Ala–poly-L-Lys [abbreviated (T,G)-A–L], poly(His, Glu)-poly-D,L-Ala–poly-L-Lys [abbreviated (H,G)-A–L], and poly(Phe,Glu)-poly-D,L-Ala-poly-L-Lys [abbreviated (Phe,G)-A–L]. These studies, along with the availability of inbred strains of mice, led to two more important discoveries about immune response genes: the genetic locus controlling responsiveness to (T,G)-A–L, (Phe,G)-A–L, and (H,G)-A–L was tightly linked to the strong transplantation antigens or major histocompatibility complex (MHC) of the mouse, known as *H-2* (McDevitt and Tyan, 1968; McDevitt and Chinitz, 1969).‡ Moreover, with the availability of intra-*H-2* recombinant strains of mice, these *Ir* genes were found to map between the serologically defined *H-2K* and *H-2D* loci of *H-2* (McDevitt *et al.,* 1972). Thus a new region of *H-2* was defined as that region in which *Ir* genes mapped, and was accordingly dubbed the *Ir,* or simply *I,* region. Similarly, the *Ir* gene for PLL in the guinea pig was found to be linked to the MHC of that species (Ellman *et al.,* 1970), an indication of the generality of this association.

With the discovery of MHC-linkage came another simple explanation of *Ir* genes that was considered but ruled out (Benacerraf and McDevitt, 1972). Since animals were tolerant to their own major histocompatibility antigens, perhaps certain antigens simply mimicked the MHC antigens of low-responder animals and led to tolerance rather than immu-

---

*The use of the term "hapten" to refer to a molecule which was not immunogenic itself but which could be made immunogenic by attachment to an immunogenic "carrier" molecule was first introduced by Landsteiner (1921; see also Landsteiner and Simms, 1923).

†These polymers consist of a poly-L-lysine backbone with poly-D,L-alanine chains extending from ϵ-NH$_2$ groups of lysines, and terminating in short random sequences of two other amino acids, such as Tyr and Glu, at the amino-terminus of the polyalanine.

‡It is interesting to note that this critical discovery, like so many other major discoveries in science, was made serendipitously. *H-2* linkage was encountered unexpectedly when McDevitt and Tyan (1968) were studying the transfer of responsiveness by spleen cells and needed strains which were *H-2*-identical (but different in the rest of the genome) to avoid graft-versus-host reactions.

471

IMMUNE RESPONSE
GENES IN THE
REGULATION OF
MAMMALIAN
IMMUNITY

nity. However, the dominance of responsiveness in both the guinea pig and the mouse*
ruled against this explanation, since the hypothesis in its simplest form would predict that
an F_1 hybrid between a responder and a nonresponder, bearing the MHC antigens of both
parents, would also be tolerant and thus a nonresponder, rather than a responder. Other
evidence against this tolerance hypothesis has been reviewed by Benacerraf and McDevitt
(1972), including the failure to detect cross-reactivity between (T,G)-A–L and MHC products and the failure to produce low responsiveness by inducing MHC tolerance in chimeras.† However, more complex versions of this hypothesis have now been reconsidered
(see below).

With the simplest explanations excluded, the major effort in the field was directed
toward the mechanism of *Ir* gene function, and in particular which lymphoid cell type was
responsible for *Ir* gene expression. After brief digressions to discuss Ia antigens and non-
MHC-linked *Ir* genes, the remainder of this chapter is devoted to a systematic analysis of
this question with respect to MHC-linked *Ir* genes.

1.3   The Major Histocompatibility Complex and Ia Antigens

In order to understand the studies of MHC-linked *Ir* genes, one must have some
background knowledge of the MHC itself. There are many excellent reviews and texts on
the major histocompatibility complex in a variety of species (Klein, 1975; Snell *et al.*, 1976;
Shreffler and David, 1975; Bodmer, 1978) and of Ia antigens in the mouse (Sachs, 1976;
Davies and Staines, 1976; Hess, 1976; Hämmerling, 1976; Klein and Hauptfeld,
1976; Niederhuber and Frelinger, 1976; McDevitt *et al.*, 1976; Cullen *et al.*, 1976; David,
1976), the guinea pig (Schwartz *et al.*, 1976), and man (Van Rood *et al.*, 1976; Wernet,
1976; Mann, 1977; Mann and Murray, 1979). This section therefore only briefly outlines
the major facts that bear on the remainder of the chapter. For the sake of simplicity and
brevity, we shall concentrate on the mouse, but most of the observations hold true for the
several other species studied as well.

1.3.1   H-2, The Mouse MHC

The MHC of the mouse, termed *H-2* for historical reasons, has been mapped to chromosome 17 (linkage group IX) (Fig. 1). It includes some of the most polymorphic genes of
the species. The complex itself is only 0.5 cM wide (Shreffler and David, 1975), bounded
by the two loci *H-2K* and *H-2D*, which encode the serologically determined transplantation antigens present on the surfaces of all cells in the organism. A third locus, *H-2L,* with
many similar properties and mapping very close to *H-2D*, has recently been discovered
(Lemonnier *et al.*, 1975; Hansen *et al.*, 1977; Hansen and Levy, 1978; Levy *et al.*, 1978).

---

*There are only rare examples of both *H-2*-linked (Urba and Hildeman, 1978) and non-*H-2*-linked (Rotman,
1978) *Ir* genes for which responsiveness is recessive. The mechanism of action of these is unknown.
†A chimera, named after the mythological creature consisting of parts from several different animals, is an
organism in which the cells of two or more donor individuals coexist side-by-side in a relatively physiological
situation. These are most commonly made by fusing embryos at an early stage (called an allophenic chimera)
or by irradiating an adult animal and reconstituting it with bone marrow stem cells, thymus, or some other
tissue from an appropriate donor animal. In some cases, chimeras arise naturally in utero, as in Freemartin
cattle. The mutually coexisting cells are tolerant to one another.

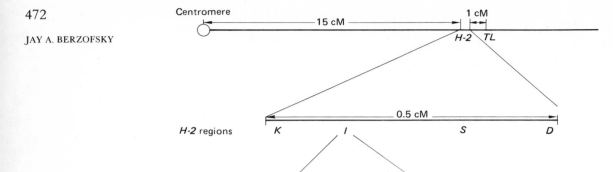

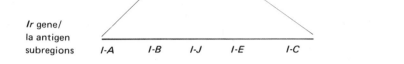

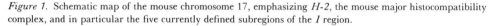

Figure 1. Schematic map of the mouse chromosome 17, emphasizing *H-2*, the mouse major histocompatibility complex, and in particular the five currently defined subregions of the *I* region.

These antigens correspond in structure and function to HLA-A, B, C in man (on human chromosome 6) and GPLA-B, S in the guinea pig. They have a molecular weight of about 45,000 (Schwartz *et al.*, 1973) and are associated with β_2-microglobulin, of molecular weight 12,000, which is not polymorphic. Their major activity in experimental cellular immunology appears to be in graft rejection and as targets for cellular cytotoxicity of allo-geneic cells (Klein, 1975) and of syngeneic cells infected with virus or chemically modified (Doherty *et al.*, 1976*a;* Shearer *et al.*, 1976). In the last respect especially they will be of interest in this chapter, primarily Sections 8 and 10, where they serve as models for antigen-plus-Ia-associative phenomena. However, their major physiological functions in the normal animal remain uncertain.

To the left of the *H-2D* region is the *S* region, which encodes a component of com-plement (Carroll and Capra, 1978), but no known cell-surface antigens, and which does not appear to influence *Ir* gene phenomena. One aspect, however, that would be of interest to molecular geneticists is the evidence that sex-limited expression of the Slp allotype on the serum protein encoded by this region fits an operon model very nicely (see Volume I in this series) in which testosterone acts as an inducer (derepressor) (Hansen and Shreffler, 1976). Noninducible as well as constitutive mutants are known that are compatible with this hypothesis.

H-2 haplotypes, defined by the whole *H-2* region of prototype inbred strains, are denoted by a lower-case superscript—for example, $H-2^b$ for the haplotype of the C57BL/10 strain. Alleles at individual subregions are designated similarly, such as $H-2K^b$ or $I-A^b$ for the alleles derived from the $H-2^b$ haplotype.

1.3.2 The I Region and Ia Antigens

The *I* region of the MHC was defined originally as that region to which *Ir* genes mapped, between the *K* and *S* regions of *H-2* (Fig. 1). The first was *Ir-1,* for (T,G)-A–L

(McDevitt *et al.*, 1972). When another gene, *Ir-IgG*, controlling the response to an IgG myeloma protein (Lieberman *et al.*, 1972), was mapped to the right of *Ir-1* on the basis of known intra-*H-2* recombinant strains, the *I* region was divided into two subregions, *Ir-1A* and *Ir-1B*, or simply *I-A* and *I-B*, respectively.

473

IMMUNE RESPONSE
GENES IN THE
REGULATION OF
MAMMALIAN
IMMUNITY

During 1973–1974, several groups concurrently discovered cell-surface antigens, detectable by alloantisera on a subpopulation of spleen cells, which mapped in this *I* region (Sachs and Cone, 1973; Hauptfeld *et al.*, 1973, 1974; David *et al.*, 1973; Hämmerling *et al.*, 1974). Most of the original Ia specificities mapped in *I-A* on the basis of known recombinants. A unified nomenclature of Ia, for "*Ir*-associated," was devised for these antigens (Shreffler *et al.*, 1974). It should be stressed that the primary definition of an Ia antigen is any cell-surface antigen mapping within the *I*-region of the MHC, as defined by *Ir* genes. Correlative criteria such as tissue distribution, molecular weight, and association with allogeneic mixed lymphocyte reactions (the proliferative response of T cells to allogeneic lymphocytes), are all secondary, not defining, characteristics.

On the basis of Ia antigens as markers, several more subregions of the *I* region were defined by known and newly discovered intra-*H-2* recombinant strains. The *I-C* region, originally defined primarily on the basis of Ia specificities Ia.6 and Ia.7 (David *et al.*, 1975), mapped to the right of *I-B* and is the farthest right of the known *I* subregions. However, with the discovery of the *I-E* subregion (encoding specificities Ia.22 and Ia.23) between *I-B* and *I-C* (Colombani *et al.*, 1976; Sachs, 1978), Ia.7 was remapped into the *I-E* subregion (David and Cullen, 1978) because it was found to be on the same molecule as Ia.22, and with it one of the complementing *Ir* genes for GLPhe (see below) was remapped to *I-E*. Therefore, the only Ia antigen specificity left to mark the *I-C* subregion was Ia.6, for which the defining antiserum had been completely consumed and could not be reproduced. The resultant confusion about whether a distinct *I-C* subregion existed led to the now widespread notation *I-E/C* to denote a single region or group of two that were poorly resolved from one another. However, despite the lack of Ia antigens mapping in *I-C*, a distinct *I-C* subregion is defined by several functional genes that clearly map to the right of *I-E*: one of the *Ir* genes for myoglobin (Berzofsky, 1978), one of the suppressor genes for poly-(Glu,Tyr) (Benacerraf and Dorf, 1976; Debré *et al.*, 1976a), a suppressor of the mixed lymphocyte reaction (MLR) (Rich and Rich, 1976), a target antigen for antibodies blocking T cell receptors for the Fc region of immunoglobulin (Stout *et al.*, 1977), and a mixed lymphocyte reaction-stimulating antigen (Okuda and David, 1978), and one of the complementing *Ir* genes for phage fd (Kölsch and Falkenberg, 1978). In this chapter, *I-E* and *I-C* are used to denote the distinct subregions as just described, separated by the crossover position in *H-2*ᵃ haplotype.* Thus both the *I-B* and *I-C* subregions are currently defined by functional markers rather than by Ia antigens. A fifth subregion, *I-J*, was defined by Murphy *et al.* (1976), between *I-B* and *I-E*, resulting in the current map shown in Fig. 1. The *I-J* subregion is distinctive in that it encodes Ia antigens primarily on suppressor T cells—for example, for allotype-specific suppression (Murphy *et al.*, 1976), antigen-specific suppression (Tada *et al.*, 1976) and concanavalin A-induced suppression

---

*The *H-2*ᵃ haplotype, present in A/J and B10.A mouse strains, behaves in all tests, cellular and serologic, as though it were a recombinant between the *H-2*ᵏ haplotype on the left and the *H-2*ᵈ haplotype on the right, with the crossover between the *I-E* and *I-C* subregions (Klein, 1975). However, because it did not arise from a documented laboratory recombination event, exceptions to this view of the *H-2*ᵃ haplotype as a simple recombinant could be discovered at any time.

(Frelinger *et al.*, 1976), but also on some types of helper cells (Tada *et al.*, 1978) and macrophages (Niederhuber, 1978). Antigens encoded in this region have not been found on B cells.

Ia antigens, in contrast to *H-2K* and *D* antigens, have a very limited tissue distribution: primarily B cells (Unanue *et al.*, 1974; Sachs, 1976), some T cells (Fathman *et al.*, 1975; Goding *et al.*, 1975; Sachs, 1976), splenic and peritoneal macrophages (Unanue *et al.*, 1974; R. H. Schwartz *et al.*, 1976*d*; Hämmerling *et al.*, 1975; Cowing *et al.*, 1978; Niederhuber, 1978), Langerhans cells of the epidermis (Stingl *et al.*, 1978), liver Kupffer cells (Richman *et al.*, 1979), and spermatozoa (Hämmerling *et al.*, 1975), but not fibroblasts, liver or kidney parenchymal cells, muscle cells or other tissues, which do bear *H-2K/D* antigens. Thus, aside from sperm cells, the distribution of Ia antigens is largely limited to cells of the lymphoid and macrophage–monocyte lineages.

Ia antigens have been shown to be 2-chain glycoproteins, noncovalently joined, with molecular weights of about 25,000 for the β chain and 33,000 for the α chain (Cullen *et al.*, 1976). The sequence polymorphism appears to be greater in the smaller β chain than in the α chain, on the basis of both the partial sequence data available and peptide maps (Cook *et al.*, 1979*a,b*).

Although Ia antigens were initially hoped to be the long-sought structural products of *Ir* genes, their presence primarily on B cells, rather than T cells, and their lack of demonstrable antigen binding or specificity left serious doubts. One very exciting recent result that suggests that they may play a direct role in *Ir* gene function is the discovery by Jones *et al.* (1978) of intrasubregion Ia complementation. They found that an additional polypeptide chain was encoded in the *I-A* subregion but was not expressed on the cell surface unless a complementary chain was encoded by the *I-E* subregion. This result was extended by Cook *et al.* (1979*c*), who demonstrated that only the α chain of the *I-E* molecule was actually encoded in that subregion and that the β chain for this molecule was encoded in the *I-A* subregion. Thus both appropriate genes had to be expressed to get a single Ia molecule. This mechanism is consistent with the observation of an MLR-stimulating antigen expressed uniquely in some F_1 hybrids but in neither parent (Fathman and Nabholz, 1977; Fathman *et al.*, 1978). In addition, the genetics of this molecular complementation exactly paralleled the genetics of *Ir* gene complementation described by Dorf *et al.* (1975) and R. H. Schwartz *et al.* (1979*c*) for the response to the random copolymer $Glu^{53}Lys^{36}Phe^{11}$ (GLPhe) (see Section 4.2). As shall be seen in Sections 9 and 10, this finding, in fact, helps solidify the notion that Ia antigens on antigen-presenting cells (macrophages?) determine *Ir* gene function.

Ia antigens appear to be the main stimulatory determinants in a mixed lymphocyte reaction (Meo *et al.*, 1973; Lozner *et al.*, 1974*a*) and it is in this regard that the *D* region of HLA in man (which is also closely associated with the *DR* locus encoding alloantigens expressed primarily on B cells and monocytes) was felt to be the human analogue of the *I* region (Bodmer, 1978; Mann and Murray, 1979). Indeed, proliferation in the MLR can be blocked by anti-Ia alloantisera (Meo *et al.*, 1975; Schwartz *et al.*, 1976; Kano *et al.*, 1976). In these regards as well as in chemistry and association with *Ir* genes, the strain 2 and 13 guinea pig alloantigens also appear to be Ia-like (Schwartz *et al.*, 1976). Such anti-Ia antisera also block the binding of aggregated IgG to the receptors for the Fc portion of these molecules (that is, Fc receptors) on both B cells (Dickler and Sachs, 1974) and T cells (Dickler *et al.*, 1976; Stout *et al.*, 1977), although there is evidence that Fc receptors and Ia determinants are not covalently linked. However, it is their role in restricting macrophage–T cell–B cell interactions and, probably as part of the same process, in T cell recognition of antigen (see Sections 8 and 9), that will be of paramount concern in unraveling the mechanism of action of *Ir* genes.

475

IMMUNE RESPONSE
GENES IN THE
REGULATION OF
MAMMALIAN
IMMUNITY

2   *Ir Genes Not Linked to the Major Histocompatibility Complex*

While the greatest excitement about *Ir* genes has been directed toward the MHC-linked genes, probably because of the large interest in the function of the MHC itself, a number of *Ir* genes have been described which are not linked to the MHC. They may be classified according to their linkage to heavy chain allotype, the X chromosome, or other known genetic markers. A sampling of each type is given in Table I. Some of these, in retrospect, would no longer be classified as *Ir* genes by the definition given in Section 1.1, since they are not antigen-specific genes, but rather affect broader classes of antigens. However, they were historically included, and the immune defects attributable to them have sometimes led to important new insights into the mechanisms of the immune response.

2.1   *Heavy-Chain Allotype-Linked Ir Genes*

The classic example of an allotype-linked *Ir* gene is that controlling the response to α-1,3-dextran in mice (Blomberg *et al.,* 1972). High responsiveness is genetically linked to

TABLE I.   *Ir Genes Not Linked to the Major Histocompatibility Complex*

Linkage	Antigen	Prototype high-responder strain	Prototype low-responder strain	Reference
Ig heavy chain Allotype	α-1,3-Dextran	BALB/c	C57BL/6	Blomberg *et al.,* 1972
X chromosome	Type III pneumococcal polysaccharide	BALB/c	CBA/HN	Amsbaugh *et al.,* 1972
	Poly I:poly C RNA	BALB/c, DBA/2, ALN	CBA/HN	Scher *et al.,* 1973
	DNP-Ficoll		CBA/N	Cohen *et al.,* 1976; Mosier *et al.,* 1977
	Denatured DNA–mBSA	SJL	DBA/2	Mozes and Fuchs, 1974
Agouti coat color (chromosome 2)	Ea-1 Mouse erythrocyte antigen	YBR	BALB/c, CBA/J	Gasser, 1969; Gasser and Shreffler, 1972
Other	(T,G)-Pro–L	SJL	DBA/1	Mozes *et al.,* 1969
	Lysozyme loop	DBA/1	SJL	Maron *et al.,* 1973
	Lipopolysaccharide	C3HeB/FeJ	C3H/HeJ	Watson and Riblet, 1974
	Sheep erythrocytes	A/J	C57BL/10J	Silver *et al.,* 1972
	GAT	A/J, A.BY	B10.A, B10	Dorf *et al.,* 1974
	Staphylococcal nuclease	A/J, A.BY	B10.A, B10	Berzofsky *et al.,* 1977a; Pisetsky *et al.,* 1978
	Sperm whale myoglobin	A.SW, A.BY	B10.S, B10	Berzofsky, 1978
	(D-Tyr, D-Glu)-D-Pro–D-Lys	AKR/Cu	C57BL/6	Schmitt-Verhulst *et al.,* 1974

the BALB/c immunoglobulin heavy chain allotype* marker. However, responsiveness also depends on the presence of the gene for a specific λ light chain, and leads to the production of antibodies bearing the same idiotype* as the BALB/c myeloma protein, J558, which has the same specificity and same λ light chain. The *Ir* gene linkage has also been found to be closer to the immunoglobulin heavy chain variable region genes, V_H, than to the constant region genes, C_H; the BAB-14 recombinant strain with the low responder C57BL/6 C_H allotype but most of its V_H genes from the BALB/c behaves as does the BALB/c strain in this response (Blomberg *et al.*, 1972). Thus the linkage and the idiotype studies strongly suggest that this *Ir* gene is the structural gene for the appropriate immunoglobulin variable region. Moreover, unlike MHC-linked *Ir* genes, its action cannot be overcome by attachment of dextran to an immunogenic thymus-dependent "carrier" (see Section 3), such as sheep erythrocytes (Blomberg *et al.*, 1972). The *Ir* gene for dextran is therefore a case in point for the mechanism considered in Section 1.1, the simple absence of an appropriate Ig structural gene which leads to unresponsiveness. The fact that such *Ir* genes are relatively rare is probably due to the heterogeneity of most antibody responses; many different structural genes can be used to respond to each of the vast majority of antigens.

2.2 X-Chromosome-Linked Ir Genes

X-chromosome-linked *Ir* genes are of two types. The most common are those for so-called T-independent† antigens, such as pneumococcal SSSIII polysaccharide (Amsbaugh *et al.*, 1972), double stranded RNA poly I:poly C (Scher *et al.*, 1973), and DNP-Ficoll (Cohen *et al.*, 1976; Mosier *et al.*, 1977), for which the CBA/N strain is a low responder. These findings were important because they led to the discovery of a broad X-linked immune deficit (Mond *et al.*, 1978) in a certain class of B lymphocytes in the CBA/N strain, which has been useful in defining subclasses of B lymphocytes and surface markers on these lymphocytes (Ahmed *et al.*, 1977; Huber *et al.*, 1977), but for the same reason, although they were historically classified as *Ir* genes, they actually define an immune deficiency disease in the CBA/N mouse, not a group of antigen-specific *Ir* genes.

The X-linked *Ir* gene of the second type is that described for the antibody response to denatured DNA coupled to methylated bovine serum albumin (Mozes and Fuchs, 1974). In this case the low responder is the DBA/2 strain, which does not carry the CBA/N defect. Whether or not this is truly antigen specific remains to be determined, and its mechanism of action is unknown. One relevant finding is its association with a newly defined

---

*Idiotype is defined as the set of antigenic determinants (idiotopes) on an immunoglobulin which allow unique serological identification of a particular antibody molecule with a given specificity made in an individual animal. As such, it is usually taken as a marker for the variable region of the immunoglobulin, and may even represent determinants in the antibody combining site, since some antiidiotypic antibodies can be inhibited by blocking the combining site of their target antibody with its antigen. In contrast, allotype is defined as the set of antigenic determinants common to all or most immunoglobulin molecules of a given class (such as IgG1, IgG2a, IgM) in a given strain. In the mouse, these are generally markers for the heavy-chain constant region genes of a particular strain.

†T-independent antigens are those for which thymus-derived ("T") helper lymphocytes are not necessary to obtain an antibody response from B lymphocytes. Conversely, T-dependent antigens require these "T helper cells" in addition to B lymphocytes (and macrophages) to elicit an antibody response.

477

IMMUNE RESPONSE
GENES IN THE
REGULATION OF
MAMMALIAN
IMMUNITY

X-linked lymphocyte surface alloantigen (Zeicher *et al.*, 1977). The existence of this gene suggests that there is more than one aspect of the immune response controlled by X-chromosome genes.

2.3   Other Non-MHC-Linked Ir Genes

A number of other *Ir* genes have been described which are not linked to the MHC, allotype, or sex chromosomes (Table I). Some, such as the *Ir-3* gene controlling the response to poly(Tyr,Glu)-poly-L-Pro–poly-L-Lys [(T,G)-Pro–L] (Mozes *et al.*, 1969), the *Ir* gene for the "loop" of lysozyme (Maron *et al.*, 1973), and the *Ir-2* gene for agglutinating antibodies to Ea-1 mouse erythrocyte antigen (Gasser, 1969; Gasser and Shreffler, 1972), may be antigen specific. The latter response is also influenced by an *H-2* linked gene. Another response under non-H-2-linked genetic control, the antibody response to lipopolysaccharide (LPS), which correlates with a generalized deficit in C3H/HeJ B cells in the mitogenic response to LPS (Watson and Riblet, 1974), may be specific for LPS, but involves a receptor defect on all B cells of the strain. Therefore, it does not represent an antigen-specific *Ir* gene in the usual sense.

An interesting non-*H-2*-, nonallotype-, non-X-linked *Ir* gene controls a response to the all-D-amino acid analogue of (T,G)-Pro–L which is T-independent but almost entirely of the IgG class, unusual for T-independent responses (Schmitt-Verhulst *et al.*, 1974).

There is a group of non-MHC-linked *Ir* genes that is defined by a higher antibody response in congenic strains on the A background* than in the corresponding strains of like *H-2* type on the B10 background. However, several differences between the traits they control suggest that they may not represent a single gene. The difference in antibody response to sheep erythrocytes between A/J mice and B10 or B10.A mice was primarily in the switch from IgM to IgG in that the A/J mice produced IgG after repeated immunization, whereas the B10 and B10.A mice continued to produce primarily IgM (Silver *et al.*, 1972). The non-*H-2* differences between strains A/J and B10.A (both *H-2*ª), and also between A.BY and B10 (both *H-2*ᵇ), in antibody response to the random linear copolymer $Glu^{60}Ala^{30}Tyr^{10}$ (GAT), was multigenic. At least one gene was possibly linked to heavy chain immunoglobulin allotype, but no difference in affinity of antibodies produced was noted (Dorf *et al.*, 1974). A similar non-*H-2*-linked difference between the same four strains was observed for the plateau level of antibody response after hyperimmunization to staphylococcal nuclease (Berzofsky *et al.*, 1977a; Pisetsky *et al.*, 1978). However, in this case, a backcross analysis was consistent with the action of a single gene that was not allotype-linked, and a difference in affinity for a major fragment of the antigen was observed.†

---

*Congenic strains are ones which have been bred to be identical except for a single gene or gene complex. Such strains, denoted B.A, are produced by repeated backcrosses and intercrosses, selecting at each intercross for the desired trait from inbred strain A, but always backcrossing to inbred strain B. The result is a strain whose genome is identical to that of B except for the single gene (and closely linked ones) that have been preserved from the original A strain parent. Conventionally, one refers to the common genome of such congenic strains as the "background" genes, as distinguished from the gene complex at which they differ—for example, *H-2*. Collections of congenic strains are extremely useful because differences between them should be linked to the gene complex at which they differ, with all else held constant.

†The higher affinity antibodies in the mice producing the lower plateau level of antibody concentration suggested a possible feedback inhibition in which the binding of antigen by antibody shut off the response.

Finally, both high and low responder congenic strains of mice on the A background produced about six-fold more antibody to sperm whale myoglobin than did corresponding *H-2*-identical strains on the B10 background (Berzofsky, 1978). Although the antibody responses to all three of the last mentioned antigens are under *H-2*-linked *Ir* gene control as well, the interaction between *H-2* and non-*H-2*-linked genes differed.* Therefore, while it is tempting to consider all these as aspects of the same genetic difference between the A and B10 non-*H-2* background genomes, the functional differences discussed above indicate that they may not be. No linkage study has been performed to see if the non-*H-2*-linked controls for these several antigens segregate together or independently.

Once MHC-linked *Ir* genes are better understood, the interaction between these genes and non-MHC-linked *Ir* genes for the same antigen may provide a fruitful approach to unraveling the complex interrelated regulatory mechanisms of the immune response.

3   Cellular Site of Expression: I. Initial Evidence for Expression in T Cells versus B Cells and Macrophages

Since most *Ir* genes with a high degree of antigen specificity were found to be MHC-linked, and since the function of the MHC was a major puzzle in itself, most of the mechanistic studies concentrated on MHC-linked *Ir* genes. Of the three cell types involved in the immune response, T lymphocytes and B lymphocytes† are known to consist of populations of many distinct antigen-specific clones, whereas no evidence exists to date for either antigen specificity or clonal diversity in macrophages. Therefore, the site of expression of antigen-specific *Ir* genes was from the beginning felt to be either T cells or B cells or both, but not macrophages. While the recent evidence we shall discuss in Sections 8–10 suggests a major role for macrophages (or "accessory cells"‡) in *Ir* gene control, the antigen specificity of the control still resides in T cells and B cells. Adding macrophages to the schema introduces a new twist to the interpretations of earlier evidence for *Ir* gene expression in T and B cells, but does not invalidate the basic notions of many of them. This section will serve to summarize the evidence (see Mozes, 1975) and consider briefly how macrophages could play a role.

3.1   Evidence for T Cell Involvement in Ir Gene Expression

The following major points have been marshalled to suggest expression of MHC-linked *Ir* genes in T lymphocytes:

*For GAT, *H-2*-linked low responsiveness overrode the non-*H-2*-linked effect. The same was true for the early response to nuclease, but after hyperimmunization the non-*H-2*-linked control overrode the *H-2*-linked gene and *H-2*-low responders made as much antibody as high responders. For myoglobin, both genetic effects were apparent at all stages of immunization, superimposed on one another.

†See footnote on page 469.

‡The exact cell type among the macrophagelike populations which serves the function of antigen presentation is still not firmly established. For simplicity, I shall refer to it as a macrophage, with the caveat that the true "accessory cell" may not be a classical macrophage.

(1) No known MHC-linked *Ir* gene has been found which does not involve a T-dependent response* (Benacerraf, 1975), whereas several non-MHC-linked *Ir* genes control T-independent responses. This generalization suggests that T cells play an obligate role in MHC-linked *Ir* gene-controlled responses. This much is not in dispute.

(2) Delayed-type hypersensitivity (Levine and Benacerraf, 1965) and T cell proliferation, which are both independent of B cells, appear to be controlled by the same *Ir* genes as antibody production (Lonai and McDevitt, 1974; Schwartz and Paul, 1976; Schwartz *et al.*, 1978; Berzofsky *et al.*, 1979). In fact, in a system in which the responses to different determinants on the same antigen are controlled by distinct *H-2*-linked genes, the gene which controls the production of antibodies specific for a given region of the molecule also controls the T cell proliferative response to that same region of the molecule (Berzofsky *et al.*, 1979). Therefore, there is no question that responses involving only T cells and macrophages can manifest virtually all the *Ir* genetic phenomena seen for B cell responses. These results do not distinguish between the roles of T cells and macrophages. While T cells and macrophages are involved in B cell responses as well, the lack of a requirement for B cells in some *Ir* gene-controlled responses does not imply that B cells as well cannot express *Ir* genes in responses in which they *are* involved. In fact, a type of symmetry between T-macrophage and T–B cell interactions may tie these results together (see Sections 8, 9, 12).

(3) *Ir* genes have been found to control the T cell-dependent switch from IgM to IgG. When adjuvant was not used for immunization with (T,G)-A–L, both high and low responders produced comparable primary antibody responses, consisting exclusively of IgM (Grumet, 1972). However, with repeated immunization, the high responder strains made a secondary IgG response, while the low responders continued to make only IgM. Since the bulk of evidence suggested that the switch from IgM to IgG is helper T cell-dependent (Taylor and Wortis, 1968), this result suggested that the *Ir* gene-controlled failure to switch from IgM to IgG was due to lack of T cell help in the low responder. This direct conclusion has not been seriously challenged by later experiments. However, the observation does not definitively localize the cellular site of *Ir* gene expression. The lack of T cell help in the low responder could be caused by an intrinsic T cell *Ir* gene defect, but it could also be caused by a failure of low responder macrophages to present antigen to the antigen-specific T cell clones, or to a failure of the relevant antigen-specific B cell clones to receive the help. Another problem with generalizing from these experiments is that *Ir* genes have been found to control IgM responses as well as IgG responses (Cheung *et al.*, 1977; Singer *et al.*, 1977; Smith *et al.*, 1977; Waltenbaugh *et al.*, 1979), although in these cases, the IgM response was T cell dependent. Thus while MHC-linked *Ir* genes may always be manifested in the activation or expression of T cell helpers (or other T cells), the switch to IgG may not be obligatory for the manifestation of humoral *Ir* gene effects.

(4) Another observation in the series of experiments just described (Mitchell *et al.*, 1972) was that thymectomy of high-responder mice converted them to phenotypic low responders, whereas thymectomy of low responders did not affect their response. Therefore, low responders behaved (with respect to the *Ir* gene-controlled response) as if they had been thymectomized. This result shows directly that thymus function is necessary for high responsiveness, and suggests less directly that T cell help is necessary. However, as in point

*See second footnote (†) on page 476.

479

IMMUNE RESPONSE
GENES IN THE
REGULATION OF
MAMMALIAN
IMMUNITY

(3) above, it does not rule out *Ir* gene expression in the cell activating the T cell, or in the cell receiving the T cell signal.

(5) Another important experiment by this same group (Ordal and Grumet, 1972) demonstrated that an ongoing graft-versus-host reaction could bypass the *Ir* gene control. $(H-2^k \times H-2^q)F_1$ hybrids between mice of two low-responder haplotypes for (T,G)-A–L were injected with homozygous parental $H-2^k$ spleen and lymph node cells and were challenged with the antigen on the same day. The parental lymphocytes, which recognized the $H-2^q$ antigens from the other parent as foreign, reacted against the cells of the F_1 host, whereas the host F_1 lymphocytes did not recognize the $H-2^k$ grafted lymphocytes as foreign. Concomitant with this ongoing graft-vs.-host reaction, the low-responder F_1 recipient produced a higher anti-(T,G)-A–L response accompanied by a switch from IgM to IgG antibodies. Therefore, an antigen-nonspecific (allogeneic) stimulus could substitute for antigen-specific T cell help and allow low responder B cells to produce antibody. One conclusion was that low-responder B cells were competent to produce anti-(T,G)-A–L. A corollary was that the low responder defect was at the level of T cell help which could be bypassed by the allogeneic stimulus. Katz *et al.* (1971) had shown that a graft-vs.-host reaction could bypass the requirement for specific help, and they (Armerding and Katz, 1974) also showed that an "allogeneic effect factor" obtained from supernatant fluids of an *in vitro* allogeneic reaction in tissue culture ("mixed lymphocyte reaction" or MLR) could substitute for helper T cells in sustaining some B cell responses. The requirements for such "positive allogeneic effects" have been analyzed by Panfili and Dutton (1978). However, the ability to bypass antigen-specific T cell help by substitution of other stimuli does not distinguish between an intrinsic defect in antigen-specific clones of helper T cells and a failure of antigen-presenting macrophages to activate these clones. Moreover, the ability of B cells to make anti-(T,G)-A–L antibodies, while proving that (T,G)-A–L-specific B cells exist in the low-responder mice, does not rule out the possibility that an *Ir* gene defect in these B cells prevents them from receiving antigen-specific T cell help but leaves intact their ability to be activated by other stimuli. Thus, this result does not clearly rule out *Ir* gene expression in any of the three cell types.

(6) Other evidence taken to suggest *Ir* gene expression in T rather than B cells was the ability to overcome the *Ir* gene defect by attachment of the antigen in question to an immunogenic "carrier" molecule. Green *et al.* (1966) originally showed that the nonresponder strain 13 guinea pigs could make anti-PLL antibodies if immunized with PLL attached to bovine serum albumin. In similar experiments, McDevitt (1968) found that immunization with (T,G)-A–L attached noncovalently to methylated bovine serum albumin (made very basic by esterification of the carboxyl groups) could induce production of anti-(T,G)-A–L antibodies in low-responder mice. More recently, M. Schwartz *et al.* (1978) followed up these results with the important finding that the idiotypes of anti-(T,G)-A–L antibodies made in congenic high-responder [to (T,G)-A–L] and low-responder [immunized with (T,G)-A–L-methylated BSA] mice were the same. Thus the B cell clones producing the antibodies in the low-responder mice immunized with (T,G)-A–L attached to a carrier were apparently the same as those in the congenic high responder mice immunized with (T,G)-A–L alone. It is hard to avoid the conclusion that low responders have B cells capable of making antibodies to the *Ir* gene-controlled antigen, and that they are even the same B cells as in the high responder. However, in the low responder, they do not receive the help of (T,G)-A–L-specific (or PLL-specific) helper T cells. Helper

cells specific for a carrier can bypass this problem since the helper cells need no longer be specific for the antigen under genetic control. Again, these results could be interpreted as a failure of the (T,G)-A-L-specific T helper cells to be activated, an intrinsic defect in these cells, or an inability of B cells to respond to these cells even though they can be helped by T cells with other specificities.

(7) Another hapten-carrier phenomenon has also been taken to suggest *Ir* gene expression in T cells. Strain 2 guinea pigs, which are high responders to PLL, made anti-DNP antibodies to DNP-PLL (Kantor *et al.*, 1963; Levine *et al.*, 1963*b*) but not to DNP-guinea pig albumin (Bluestein *et al.*, 1971*a*). Conversely, the same authors showed that strain 13 guinea pigs made anti-DNP antibodies to DNP-guinea pig albumin but not to DNP-PLL. Thus, although both strains could make DNP-specific antibodies, whether or not they did so depended on the *Ir* gene control of the response to the carrier molecule. Comparable findings for other haptens such as benzylpenicilloyl, tosyl, and dimethylaminonaphthalene sulfonyl, each attached to PLL (Levine *et al.*, 1963*a*), demonstrated that the response to any hapten would probably depend on the *Ir* gene control of recognition of the carrier molecule. Similarly, Mozes and McDevitt (1969) found that in mice high responders to (T,G)-A-L made anti-DNP antibodies to DNP-(T,G)-A-L, whereas low responders did not, even though both strains made antibodies equally well to the DNP group on other carriers such as BSA. The additional observation that the helper T cells were specific for the carrier moiety rather than the hapten moiety in many such conjugates suggested that the carrier-specific *Ir* genes were also expressed in the helper T cells. While the juxtaposition of these findings was certainly suggestive, the syllogism is not valid. If A = *Ir* gene, B = carrier, and C = helper T cell, one cannot rigorously deduce from the premises (1) that A is specific for B and (2) that C is specific for B, the conclusion that A is mediated by C. In addition, the alternative explanation implicating macrophage presentation of antigen to helper T cells could be applied here as well.

(8) Two studies were done by the same laboratory using tetraparental chimeric mice to test the ability of low-responder B cells to (T,G)-A-L to make antibody to this antigen, with apparently conflicting results. In the first study the tetraparental chimeras were "allophenic" mice, made by fusing early blastula stage embryos from high- and low-responder mice, differing in heavy chain immunoglobulin allotype as well as *H-2* (Bechtol *et al.*, 1974; Bechtol and McDevitt, 1976). T cells from these mice were tolerant to cells from either parental strain, so that cooperation between high-responder T cells and low-responder B cells could be investigated. Antibodies to (T,G)-A-L made by some of these chimeras bore the allotype of the low-responder strain, an indication that they were made by B cells from the low-responder parent. This result was taken as evidence that low-responder B cells could respond to (T,G)-A-L, without an additional carrier, under conditions in which they could receive help from high-responder T cells. However, when the frozen sera from these experiments were retested years later, the results could not be reproduced (Press and McDevitt, 1977). The second study employed tetraparental chimeras made by reconstituting lethally irradiated F_1 mice with bone marrow from the two parents (Press and McDevitt, 1977). This time, no anti-(T,G)-A-L of the low-responder allotype was found, consistent with a possible *Ir* gene defect in B cells as well. A similar result was obtained by Warner *et al.* (1978) in allophenic mice. As we shall see in Sections 8–10, the latter results are consistent with some subsequent studies on bone marrow chimeras and antigen presentation (Kappler and Marrack, 1977; Marrack and Kappler, 1978; Yamash-

ita and Shevach, 1978; Sprent, 1978*a,b*) but not others (Singer *et al.*, 1979; Hodes *et al.*, 1979). In any case, one cannot definitively exclude *Ir* gene expression in B cells from these studies.

3.2 Evidence for Ir Gene Expression in B Cells and T Cells

While the evidence described in the previous section clearly implicated T cells as necessary for MHC-linked *Ir* gene controlled responses, there was also evidence that another non-T cell, found in the bone marrow, was sometimes the determining factor in *Ir* gene expression. For the most part this cell was felt to be a B cell, although the experiments generally could not explicitly distinguish between a B cell and a macrophage.

3.2.1 Adoptive Transfer and Limiting Dilution

In 1968, McDevitt and Tyan (1968) showed that the *Ir* phenotype of high-responder (C3H $\times$ C57BL/6)F$_1$ mice could be transferred to C3H low-responder mice by adoptive transfer of spleen cells into lethally irradiated recipients. Similar ability to adoptively transfer responsiveness was shown in the guinea pig by Foerster *et al.* (1969) and Ellman *et al.* (1970*b*). Thus lymphoid cells clearly expressed *Ir* gene phenotypes.

To determine whether low responsiveness was due to a limiting number of antigen-specific precursor cells, Mozes *et al.* (1970) and Shearer *et al.* (1971) titrated limiting numbers of spleen cells into lethally irradiated *syngeneic* recipients, then immunized, and, by applying Poisson statistics to the numbers of recipients making detectable antibody 14 days later, calculated the frequency of antigen-specific precursor cells in each strain. The use of syngeneic transfers avoided allogeneic effects. In the case of (T,G)-Pro–L under control of the *Ir-3* gene not linked to *H-2*, they found that the low responders had fewer precursors by this method. Similar results were obtained for two determinants of (Phe,G)-Pro–L. Since both B and T cells were necessary for the antibody response, Mozes and Shearer (1971) then sought to distinguish whether thymus-derived or bone marrow-derived precursors, or both, were limiting in the low responders. They found, to the surprise of many, that bone marrow cells, but *not* thymocytes were lower in precursor frequency for (Phe,G)-Pro–L in low compared to high responders (when the other cell type was transferred in excess). These studies were extended by Shearer *et al.* (1972) to compare the limiting cell types in the non-*H-2*-linked *Ir-3*-controlled response to (T,G)-Pro–L and the *H-2*-linked *Ir-1*-controlled response to (Phe,G)-A–L. For the non-*H-2*-linked response, only the bone marrow precursors (which include B cell and macrophage precursors) were reduced in the low-responder strain. The syngeneic thymocyte titrations were superimposable for the two strains. In contrast, for the *H-2*-linked response, both cell types of precursors were less frequent in the low-responder strain, although strict Poisson statistics did not hold for the thymocyte titration in the high-responder strain. It was concluded that the *Ir* gene control was reflected in (mediated by?) the number of antigen-specific precursor cells in unimmunized mice, and that for some responses, the limiting precursor cell was bone marrow-derived, not thymus derived. This approach was extended to the widely studied *H-2*-linked *Ir* gene-controlled response to (T,G)-A–L by Lichtenberg *et al.* (1974). For this antigen, *H-2*^{b} mice were high responders while *H-2*^{k} and *H-2*^{s} mice were low responders. However, the low responders differed in that *H-2*^{k} mice could respond to other antigens with the A–

L backbone [for example, (H,G)-A–L], whereas *H-2*[s] mice failed to respond to any of these polypeptides. Interestingly, it was found that the defect in *H-2*[k] mice was reflected in only the bone marrow precursor number, while that of the *H-2*[s] mice was reflected in both thymocyte and bone marrow cell numbers. It was suggested that the thymocyte precursor deficiency correlated with the defect in recognition of the A–L backbone (carrier?) in *H-2*[s] mice, while the bone marrow precursor frequency correlated with recognition of the poly(Tyr,Glu) moiety, to which the antibodies were made (haptenlike?). The results for the *H-2*[k] mice were most important, however, because this haplotype was the prototype low responder type for (T,G)-A–L used in most studies cited above and below, which represent a significant fraction of the mechanistic studies on *Ir* genes in mice. These studies have been criticized because (1) the fits to Poisson statistics were not always very good (although the qualitative differences were generally clear cut); (2) the bone marrow was not treated with anti-thy antisera and complement to eliminate mature T cells; and (3) no distinction was made between B cells and macrophages or other cells as the limiting cell in the bone marrow. In addition, they are puzzling since they are the only studies which attribute *Ir* gene control to differences in precursor cell frequency. Studies designed for determining the difference in the numbers of specific antigen-binding cells between high- and low-responder unimmunized mice never detected any difference (Dunham *et al.*, 1972; Hämmerling *et al.*, 1973). Nevertheless, these limiting dilution studies were among the first to strongly suggest that *Ir* genes were expressed in cells other than T cells. The general results were corroborated by studies of T cell helper factors (see Sections 3.2.2 and 7.1) and of *in vitro* cell cooperation (Marrack and Kappler, 1978; see Section 9).

3.2.2 T Cell Helper Factors

The other major line of evidence before 1976 for an *Ir* gene defect in B cells came from the study of cell-free antigen-specific "helper factors" produced by T cells. Although the major discussion of these factors will be deferred to Section 7, we shall consider them here only as they pertain to this tissue. This helper factor was made by "education" of thymocytes to antigen *in vivo* in a syngeneic adoptive transfer and then incubation of the resulting spleen cells of the recipient with antigen *in vitro* to induce release of the factor into the supernatant fluid (Taussig, 1974). Its presence was detected by injection of this (supernatant) factor with antigen and bone marrow cells into a syngeneic irradiated recipient, whose spleen was then tested for specific antibody-producing cells. The factor replaced the requirement for thymocytes or T cells to obtain a response, in an antigen-specific fashion. Conveniently, the factor acted across allogeneic barriers so that experiments could be performed with allogeneic bone marrow cells without the complication of the positive or negative allogeneic effects mentioned earlier. Such experiments were performed using (T,G)-A–L by Taussig *et al.* (1974), who found that thymocytes from congenic high responder (C3H.SW, *H-2*[b]) or low-responder (C3H, *H-2*[k]) mice made the factor equally well. In contrast, bone marrow cells only from the responder *H-2*[b] strain were receptive to the action of factor (made in either strain), while bone marrow cells from the low-responder *H-2*[k] strain failed to respond to either factor. This result, like that of the limiting dilution studies, placed the *Ir* gene defect of *H-2*[k] mice for (T,G)-A–L squarely in the bone marrow-derived cell population, not the thymocyte population. Also in accordance with the limiting dilution studies, the SJL (*H-2*[s]) low-responder mice were found to have defects in both factor production and reception (Munro and Taussig, 1975). The defect in SJL bone

483

IMMUNE RESPONSE
GENES IN THE
REGULATION OF
MAMMALIAN
IMMUNITY

marrow receptor for the factor turned out to be non-*H-2*-linked, so that only the inability of thymocytes to make the factor was linked to *H-2*^{s}, as witnessed by the A.SW (*H-2*^{s}) strain on the A/WySn background, which could respond to the factor but not make it (Munro *et al.*, 1978). Similarly, the B10.M (*H-2*^{f}) low-responder strain was found to have a defect in T cell factor production to (T,G)-A–L but not in bone marrow cell acceptance of the factor. The existence of two types of defects in different low responders suggested that complementation might be observed in F$_1$ hybrids between low responders of different types. In fact, several examples of such complementation have been reported, in further support of the distinction between the two types of defects (Munro and Taussig, 1975; Munro *et al.*, 1978). One of these examples (that between *H-2*^{f} and *H-2*^{k} low responders) has been difficult to reproduce (Munro and Taussig, 1977; Rüde *et al.*, 1977; Deak *et al.*, 1978; Marrack and Kappler, 1978), but the other [complementation between *H-2*^{s} and *H-2*^{a} (like H-2^k) low responders] has not been questioned. Moreover, the difference between *H-2*^{k} and *H-2*^{s} defects has been confirmed by Erb *et al.* (1979a) and Howie and Feldmann (1977).

An objection could be raised that the bone marrow contained macrophages and other cell types besides B cells. Perhaps the acceptor for the T cell factor was on a macrophage, which then transmitted the signal to a B cell as suggested by the data of Feldmann *et al.* (1979), of McDougal and Gordon (1977), and of Howie and Feldmann (1977, 1978). To counter this alternative interpretation, Munro and Taussig (1975) used two approaches. First, they showed that the factor activity could be removed from supernatant fluids by adsorption to bone marrow cell populations, and that the ability of bone marrow to remove the factor exactly paralleled in genetic strain distribution its ability to respond to the factor. Importantly, the same ability was found in preliminary studies in putative populations of peripheral B cells, made by treatment of lymph node cells with anti-thy antiserum plus complement to remove T cells and then depletion of macrophages. Second, they used irradiated (high responder × low responder) F$_1$ recipients to test the ability of *H-2*^{a} (like *H-2*^{k}) bone marrow cells to respond to the helper factor. If the defect in the *H-2*^{a} bone marrow cells was solely in its macrophage, not B cell, population, then radioresistant* high responder macrophages from the F$_1$ hybrid recipient should have filled this role and allowed the *H-2*^{a} B cells to respond. The fact that they did not do so suggested that there was at least a B cell defect, whether or not there was also a macrophage defect. The controversy between B cell and macrophage defects still remains unresolved (see Section 9), but the helper factor and limiting dilution studies clearly implicated a cell type other than the T cell as one site of *Ir* gene expression.

The recent evidence on the nature of these other cells and their roles in *Ir* gene expression will be examined in Sections 8–10, but first we must consider the variety of types of MHC-linked *Ir* genes, and the possible roles of T cell and macrophage factors, T cell receptors, and suppressor T cells in *Ir* gene control.

4 Complex H-2-Linked Ir Genes: Natural Protein Antigens and Complementing Systems

A number of *Ir* gene systems have been discovered which are more complex than the original ones of PLL and (T,G)-A–L. Some systems probably involve multiple genes con-

---

*Radioresistance has been used to distinguish macrophage functions, which are generally radioresistant, from B cell functions, which are largely radiosensitive.

trolling the response to multiple determinants on the same antigen molecule (Berzofsky *et al.*, 1979), while others may involve complementation of two or more genes for what may be the response to a single determinant (Dorf *et al.*, 1975; Schwartz *et al.*, 1979). These complexities have led to new insights into *Ir* gene function. While we cannot cite all *Ir* genes described to date, we shall try to highlight examples representative of the different types. Complementation of immune suppressor genes will be deferred to Section 5.

485

IMMUNE RESPONSE
GENES IN THE
REGULATION OF
MAMMALIAN
IMMUNITY

4.1 Determinant-Specific Ir Gene Control

The first *Ir* genes discovered controlled responses to simple synthetic antigens with only one or a few determinants. Besides the cases already discussed, a striking example which illustrates the exquisite specificity of *Ir* gene control can be seen by comparing the known sequence polymers consisting of repeating tetramers $(Tyr-Glu-Ala-Gly)_n$ and $(Tyr-Ala-Glu-Gly)_n$ (Maurer *et al.*, 1973; Zeiger and Maurer, 1976). Both polymers are α-helices, have similar molecular weights (53,000 and 33,000, respectively), and have identical amino acid compositions. Yet the former was immunogenic in strain 13 but not strain 2 guinea pigs, while the latter was immunogenic in strain 2 but not strain 13 guinea pigs, whether the response measured was antibody production, delayed hypersensitivity, or *in vitro* T cell proliferation. Thus, reversal of the order of two of the four amino acids in these repeating, presumably single-determinant polymers completely reversed the *Ir* gene control. In addition to illustrating the fine resolving power of *Ir* gene specificity, this example points out the types of stringent controls first observed with synthetic polypeptides with relatively few distinct antigenic determinants.

When protein antigens were studied, the ones found to be under *Ir* gene control were either the simpler, smaller ones, such as staphylococcal nuclease (Lozner *et al.*, 1974; Berzofsky *et al.*, 1978; Sachs *et al.*, 1978), insulin (Arquilla and Finn, 1965; Keck, 1975; Barcinski and Rosenthal, 1977), lysozyme (Hill and Sercarz, 1975; Sercarz *et al.*, 1978), myoglobin (Berzofsky, 1978; Berzofsky *et al.*, 1979), and cytochrome *c* (Schwartz *et al.*, 1978; Solinger *et al.*, 1979), or ones that differed in only a small way from the homologous protein of the host, for example, DNP-guinea pig albumin in guinea pigs (Benacerraf and Germain, 1978), or IgG myeloma protein (Lieberman *et al.*, 1972). One possible explanation for the failure to see *Ir* gene control of more complex antigens is that each determinant of a complex antigen is under the control of a different MHC-linked gene. Thus, the likelihood that any single strain would be a low responder to all of the determinants, and therefore to the whole complex antigen, should be exceedingly small unless the total number of independent determinants is very small. While fundamentally appealing in some aspects, this hypothesis was at odds with the view based on experience with hapten-carrier systems (see Section 3.1), that *Ir* gene control was mediated through helper T cell recognition of a "carrier" which could then provide help for stimulation of B cells to make antibodies to any hapten on that carrier. If this latter hypothesis could be applied in its simplest form to natural protein antigens, which would be viewed as having some determinants analogous to haptens to which antibodies were made, and other determinants recognized by helper T cells, then *Ir* gene-controlled responsiveness would depend on whether or not a strain had helper T cells that could recognize a "carrier" determinant. A strain that had helper T cells of such a specificity could then make antibodies to any "haptenic" determinant on the molecule, provided it had appropriate immunoglobulin structural genes. By contrast, a strain which did not have such a clone of helper T cells would not make antibodies to any determinants on the molecule. This hypothesis would predict that the levels of antibodies to different determinants would be regulated in concert—all high or all low, whereas the former hypothesis of independent control of the response to different determinants by different genes would predict independent variation in antibody levels to the several determinants.

In 1974, we set out to test these questions (Berzofsky *et al.*, 1977*a,b;* Berzofsky *et al.*, 1978), using a protein antigen, staphylococcal nuclease, to which the antibody response had just been found to be under *H-2*-linked *Ir* gene control (Lozner *et al.*, 1974*b*), and whose primary sequence (Anfinsen *et al.*, 1971), three-dimensional structure (Cotton and Hazen, 1971), and conformational equilibrium in solution (Anfinsen, 1972, 1973) had been extensively described. High-responder (B10.A, *H-2*ᵃ) and low-responder (B10, *H-2*ᵇ) congenic strains of mice were used since, differing only in *H-2*, they had identical immunoglobulin structural gene repertoires (on a different chromosome from *H-2*) (see footnote (*), p. 477). We immunized groups of both strains multiple times with whole nuclease and monitored levels of nuclease-specific serum antibody. The first surprise was that, although the initial (3-week) sera after one immunization showed the high–low response difference described by Lozner *et al.* (1974*b*), after three immunizations, the total antinuclease made by the two strains was identical (Berzofsky *et al.*, 1977*a*). This suggested that the *Ir* gene control was directed toward an immunodominant* determinant, but that after hyperimmunization, other weaker determinants elicited antibodies, and that the ceiling on total antibodies to nuclease in aggregate was controlled by another mechanism, which turned out to be a non-*H-2*-linked *Ir* gene described in Section 2. To test this possibility, the specificity of antibodies raised against nuclease in these strains was assessed by binding to radiolabeled fragments of nuclease using a method which measured antibody concentration independent of affinity (above a certain threshold) (Berzofsky *et al.*, 1977*b*). This approach was possible despite the exquisite conformational sensitivity of the antinative nuclease antibodies because the fragments, although random in conformation, were in equilibrium with the native conformation (Sachs *et al.*, 1972). The result was that after three immunizations, when the levels of total antinuclease and antibody to the fragment (1–126) were quite comparable in the two strains, antibodies to fragment (99–149), which represented a significant fraction of the antibody in the high responder strain (B10.A), were undetectable in the low responder strain (B10).†

Thus the levels of antibodies of the various determinants were not controlled in concert. This result was incompatible with the simple form of *Ir* gene-helper T cell carrier model described above (see Fig. 2). Either *Ir* genes could function to directly control which B cell clones responded without the services of the carrier–specific–helper T cell pathway, or else multiple "carrier"-type determinants exist, none of which can provide T cell help for all the "haptenic" determinants on the molecule. In fact, a full pendulum swing away from the idea of a single sufficient carrier determinant is the concept that every "haptenic" determinant can serve as its own individual "carrier" determinant, so that a natural protein

---

*An immunodominant determinant is one toward which the majority of the response is directed, and which may preempt the response that could potentially occur to other determinants.

†The possibility that the results were due to genetic drift in the immunoglobulin structural genes between the ostensibly congenic strains could be ruled out on the basis of the following considerations: (1) the same results were found when mice of these two *H-2* types were compared in congenic strains on a different background genome (A/J vs A.BY) (Pisetsky *et al.*, 1978); and (2) progeny of an F₂ cross between A/J and B10, selected to have the *H-2*ᵃ type of the A/J parent but to be homozygous for the immunoglobulin allotype of the B10 parent, behaved like the B10.A (Berzofsky, Killion, and Sachs, unpublished observations).

487

IMMUNE RESPONSE
GENES IN THE
REGULATION OF
MAMMALIAN
IMMUNITY

antigen may not fit the mold fashioned on the example of a carrier to which multiple copies of the same hapten are attached. This latter extreme is supported by the findings that (1) when each fragment was injected as immunogen, only the fragment (99–149) was under the same apparent *Ir* gene control as the whole nuclease (Berzofsky *et al.*, 1977a); and (2) when T cells from nuclease-immunized mice were stimulated *in vitro* with individual fragments of nuclease, *Ir* gene control of the T cell proliferative response appeared to be directed at the same fragment (99–149) (Schwartz *et al.*, 1978). Perhaps one reason for the difference between these results and those for hapten-carrier models (see Fig. 2) is that multiple copies of the same hapten are attached to many parts of the carrier molecule— that is, perhaps to many different carrier determinants, which cannot be distinguished in such studies.

In retrospect, results suggesting the same phenomena had been seen earlier in the guinea pig by Bluestein *et al.* (1972). Both strain 2 and strain 13 guinea pigs made antibodies to the random terpolymer poly-$(Glu^{60},Ala^{30},Tyr^{10})$ (GAT), but of these, antibodies which bound to poly-(Glu,Ala) (GA) were found only in the strain 2 animals, which could respond when immunized directly with GA. This result anticipated that found for nuclease. However, both strains made anti-GAT antibodies which bound to poly-(Glu,Tyr) (GT), even though only strain 13 could respond when immunized directly with GT. Thus, the findings with the two strains were not symmetrical, and the lack of congenicity in the guinea pigs made it impossible to attribute the differences to MHC genes with any certainty. Results such as these for the antinuclease and anti-GAT antibody responses, as well as the observation in a T cell proliferative response for nuclease and fragments that recombinant strains behaved unlike either parent (Schwartz *et al.*, 1978), suggested that multiple genes might be involved. However, distinct MHC-linked genes could not be mapped.

The discovery that the antibody response to sperm whale myoglobin was under control

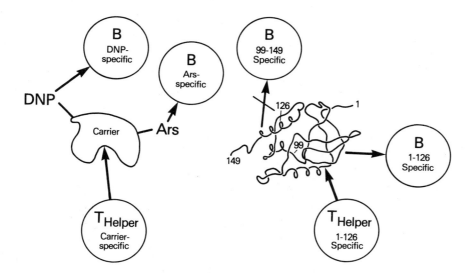

Figure 2. Comparison of Hapten-carrier model with antibody response to nuclease in low-responder B10 mice. (Left) Carrier-specific helper T cell provides help for B cells specific for any hapten on the carrier—for example, DNP and Ars (Azobenzene arsonate). (Right) Low responder has helper T cells (which appear to be specific for determinants in the region 1–126) which provide help for B cells specific for 1–126. However, B cells specific for 99–149 are not helped, even though they must be present, as discussed in the text.

of two *H-2*-linked genes mapping in distinct subregions of the *I*-region (*Ir-Mb-1* in *I-A* and *Ir-Mb-2* in *I-C*) (Berzofsky, 1978a) allowed further pursuit of these questions in a situation in which distinct genes could be defined by virtue of available recombinant strains. Strains with the high-responder *H-2*^{d} allele in *I-C* (*Ir-Mb-2*) and the low-responder *H-2*^{k} or *H-2*^{b} alleles in *I-A* made only intermediate levels of antibody compared to the congenic *H-2*^{d} high-responder and *H-2*^{k} or *H-2*^{b} low-responder strains. This result raised the possibility that the intermediate response was due to production of antibodies to only a subset of determinants—that is, to fewer determinants than recognized by high-responder strains with both *Ir* genes. This hypothesis was a corollary to the more general hypothesis that different genes were responsible for the control of responses to different determinants (Berzofsky, 1978b). To test this hypothesis, we examined the specificity of antibodies raised against whole myoglobin in various strains for binding to different CNBr-cleaved peptide fragments of myoglobin (Berzofsky, 1978b; Berzofsky *et al.*, 1979). Antimyoglobin sera from strains of mice with only *Ir-Mb-2*, while clearly binding to some parts of myoglobin (for example, in the amino-terminal region) showed no more binding to the carboxyl-terminal fragment (132–153) than did sera from corresponding low-responder strains (Fig. 3). In contrast, strains with both genes or with only *Ir-Mb-1* produced antimyoglobin antibodies with specificity for this carboxyl-terminal fragment, as well as others. These results were extended to an *in vitro* T cell proliferative response in which it was found that high, intermediate and low proliferative responses exactly paralleled the antibody responses of the several strains.

Moreover, the strains that failed to make antibodies to the carboxyl-terminal fragment also lacked a T cell proliferative response to this fragment (Berzofsky *et al.*, 1979). In contrast, they did generate a proliferative response to the amino-terminal fragment (1–55), while the strains with only *Ir-Mb-1* often (though not always) gave a reduced response to the latter fragment (Fig. 3). Two major conclusions could be drawn from these experi-

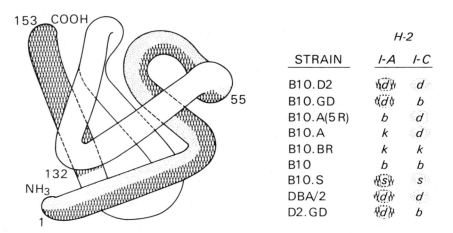

Figure 3. Determinant specificity of *Ir* genes for sperm whale myoglobin. The model at left is an artist's representation of the three-dimensional structure of myoglobin. The cross-hatched alleles in *I-A* allow responses to determinants in both the carboxyl-terminal and amino-terminal regions of the molecule (also cross-hatched). The stippled alleles in *I-C* allow responses to the amino-terminal region only (also stippled) (the effect of *I-C*^{s} is uncertain). The 1–55 segment is bisected longitudinally in shading to indicate the uncertainty of the relative positions of different determinants in this region. The alleles indicated by unshaded letters appear to have no effect. (Based on Berzofsky, 1978b, and Berzofsky *et al.*, 1979.)

489

IMMUNE RESPONSE
GENES IN THE
REGULATION OF
MAMMALIAN
IMMUNITY

ments: (1) The hypothesis that the immune responses to different antigenic determinants on the same antigen molecule could be controlled by separate MHC-linked *Ir* genes was correct, and applied at least in the case of myoglobin. It probably also explained the observations for staphylococcal nuclease in mice and for GAT in guinea pigs. (2) The gene controlling the T cell proliferative response to a given region of myoglobin was the same gene which controlled production of antibodies specific for that same region of myoglobin. This conclusion supports the notion that if there are haptenlike and carrierlike determinants on natural protein antigens, the determinant which serves the carrierlike role is either the same as the haptenlike determinant it serves or is closely associated to it.* Of broader import is the suggestion from this conclusion that a common regulatory mechanism, at the level of each determinant, controls both T cell and B cell responses to that determinant. There is now evidence that this common mechanism may operate in the recognition of each antigenic determinant of myoglobin in association with a particular Ia antigen on the antigen-presenting macrophage (Richman *et al.*, 1980). Full discussion of this and related results with other antigens will be deferred to Section 9.

It should be pointed out that by coupling a synthetic polypeptide for which the response was controlled by a non-*H-2*-linked gene (*Ir-3*) to one for which the response was under *H-2*-linked control, Mozes *et al.* (1969) obtained an antigen for which different types of genetic control applied to different determinants on the polymer. These experiments proved at least for an *H-2* and non-*H-2*-linked combination of genes, that one could construct a situation in which different parts of the antigen were controlled separately. The observations with myoglobin extend this point in two important ways: They show that two genes both within *H-2* can also act apparently independently to control responses to different determinants on the same molecule, and that this mechanism can explain the complex genetic control observed for natural protein antigens which have multiple antigenic determinants. They also therefore support the idea expressed at the beginning of this Section that the reason that MHC-linked *Ir* gene control is not seen for responses to very complex multideterminant antigens like keyhole limpet hemocyanin (KLH) is that the product of the independent probabilities of being a nonresponder to each determinant is so small that no single strain fails to respond to all of them. If one could assess the response to individual determinants, presumably *Ir* gene control would be found.

The observations made for myoglobin may also explain some of the cases of *Ir* gene complementation previously observed, such as that of lactic dehydrogenase (LDH$_B$) (Melchers and Rajewsky, 1978; Melchers and Rajewsky, 1975). In this case, the hybrid between two partial responders was a high responder. The complementation may be due to the additive effect of genes controlling responses to different sets of determinants inherited from the two parents.

Similar mechanisms may apply to *Ir* gene control of both antibody and T cell-proliferative responses to insulin in both mice and guinea pigs. In mice, the amino acid residues A8, 9, 10 of the A chain, which are tightly constrained in a loop formed by a disulfide bridge between residues A6 and A11, and also constrained by an interchain disulfide bridge between residues A7 and B7, have been shown to constitute the crucial determinant which distinguishes between the immunogenicity of bovine insulin and nonimmunogenicity of porcine insulin in B10 mice (Keck, 1975). This determinant was considered to play a "car-

*The hole in this argument is the uncertainty that the proliferating T cells are helper T cells. However, preliminary evidence suggests that they are (Berzofsky and Richman, unpublished observations; J. Chiller, personal communication).

rier" role since it controlled whether or not antibodies would be made to DNP attached to lysine B29, and also because antibodies made in B10 mice against bovine insulin would mostly bind to porcine insulin. Thus the ability to discriminate was at the *Ir* gene level, not the antibody level [although evidence for some antibodies specific for beef insulin was found (Schroer *et al.*, 1979)]. In this respect, these results differ from those described for nuclease and myoglobin. However, the evidence for multigene interactions in the response to insulin could be interpreted in terms of the hypothesis that different genes control the response to different determinants on the molecule, as we have described for myoglobin. For instance, Keck (1977) demonstrated complementation between the low-responder *H-2^b* and *H-2^k* haplotypes, as long as one of them was on the C3H background. He interpreted this finding with the suggestion that each haplotype provides *Ir* gene-controlled recognition for a single carrier determinant and that recognition of both determinants was required for an immune response. While this interpretation would be consistent with the myoglobin story, the only evidence for this multiple-carrier recognition mechanism in insulin was the finding that low-responder mice could be made to respond to fluorescein hapten conjugated to porcine insulin if immunogenic sheep insulin was injected simultaneously. However, this type of experiment is open to other interpretations.

Another experiment consistent with the idea that different genes control the responses to different determinants was recently described by Rosenwasser *et al.* (1979). The *Ir* genes of *H-2^b* mice for response to beef insulin and of *H-2^d* mice for pork insulin both mapped in the *K* or *I-A* region of *H-2*. However, the *H-2^b* mice were responding to the A-chain loop of beef insulin (consistent with the results of Keck, 1975) whereas the *H-2^d* mice were responding to the B chain of pork insulin. Thus, *Ir* genes for insulin, while mapping in the same subregion in different *H-2* haplotypes, need not have the same determinant specificity. This conclusion would have been even stronger if mice of both haplotypes had been studied with the same insulin (beef, to which they both could respond), rather than two different ones. It remains possible that *H-2^d* mice respond to both determinants of beef insulin, utilizing two different *Ir* genes. This result resembles the findings of Bluestein *et al.* (1972) for GAT in the guinea pig even more than do the observations on nuclease described above, since in the guinea pig example, two strains which could respond to the same antigen were actually responding to different determinants. In fact, a somewhat analogous result was observed by Barcinski and Rosenthal (1977) in the guinea pig response to insulins, in that strains 2 and 13 were found to be responsive to different parts of the insulin molecule.

Another related, but not identical, situation has been observed also for staphylococcal nuclease. Among *H-2^k*, *H-2^a* and *H-2^d* mice, all high responders to nuclease, the *Ir* gene in *H-2^k* and *H-2^a* mice appears to specify responsiveness toward a determinant between residues 99 and 149 (Berzofsky *et al.*, 1977a,b; Schwartz *et al.*, 1978). In contrast the high-responder *H-2^d* mice respond poorly to this determinant, but recognize instead a determinant more toward the amino terminus of the molecule (Schwartz *et al.*, 1978). In addition, whereas the *Ir* gene for the antinuclease antibody response of *H-2^k* and *H-2^a* mice had been mapped in the *I-B* subregion (Lozner *et al.*, 1974b), we now have tentatively mapped the *Ir* gene for antinuclease antibody response of *H-2^d* mice to the *I-A* (or *H-2K*) subregion (Berzofsky, Pisetsky, Killion and Sachs, unpublished observations). Thus, *Ir* genes for the same antigen (but different determinants) need not map in the same *I* subregion in different responder *H-2* haplotypes.

The overall message of all of these studies in that *Ir* genes are highly determinant specific, and that the "*Ir* gene" for a complex antigen is actually a set of independent *Ir* genes for different determinants on that antigen, which may map in the same or in different *I* subregions in a single strain or in different responder strains. Some examples of apparent complementation may actually be due to this antigenic fine structure rather than true genetic complementation. One internally consistent explanation for all of these phenomena relates to antigen presentation by macrophages, to be discussed in Section 9.

491

IMMUNE RESPONSE
GENES IN THE
REGULATION OF
MAMMALIAN
IMMUNITY

4.2 True Complementing Ir Gene Systems

In addition to the pseudocomplementation described in Section 4.1, several cases of apparent true complementation among *Ir* genes have been described. One possible type, complementation between genes expressed in different cell types required for the antibody response to (T,G)-A-L, was described in Section 3.2.2. However, there are also cases of complementation between genes which operate at the same cellular level, and perhaps even the same molecular level, in the response. Examples of these include the responses to poly-(Glu^{53},Lys^{36},Phe^{11}) (GLPhe) (Dorf *et al.*, 1975; Dorf and Benacerraf, 1975), poly-(Glu^{55}, Lys^{35},Leu^{10}) (GLLeu) (Dorf *et al.*, 1976) poly-(Glu^{57},Lys^{38},Tyr^{5}) (GLT^{5}) (Schwartz *et al.*, 1979*b*), the H-2.2 alloantigen (Dorf and Stimpfling, 1977), and pigeon cytochrome *c* (Schwartz *et al.*, 1979*c*; Solinger *et al.*, 1979). Other complementing *Ir* genes, such as those in *I-A* and *I-C* for the antibody response to bacteriophage fd (Kölsch and Falkenberg, 1978), show even more complex behavior and their mechanisms of action are less well understood.

The original and most thoroughly studied case is that of GLPhe. The first evidence that this complementation was not of the two-cell type described for (T,G)-A-L (Section 3.2) was that the same complementation was seen for a T-cell-proliferative response, which did not require B cells, as had been observed for the antibody response (R. H. Schwartz *et al.*, 1976*a*). Of course, that result alone could have been explained by postulating two interacting subpopulations of T cells necessary to obtain proliferation, one helping the other. However, for the antibody response as well, it was shown in adoptive transfer that responder F$_1$ T cells would not help nonresponder parental B cells to make anti-GLPhe; conversely, responder F$_1$ B cells were not helped by nonresponder parental T cells (Katz *et al.*, 1976). The definitive proof that both genes must be expressed in the same cell came from experiments in allophenic chimeras by Warner *et al.* (1977), and experiments in irradiation bone marrow chimeras by Schwartz *et al.* (1979*a*). In the former study, allophenic tetraparental mice made from the two low responder complementation types failed to show complementation, even though F$_1$ hybrids did. In the latter study double irradiation bone marrow chimeras (reconstituted with bone marrow stem cells from both low responder parental strains) responded to antigens that each parental strain alone could respond to, indicating intact function of cells from both donors, but did not respond to GLPhe. Only F$_1$ cells bearing both genes complementing in the same cell could respond to GLPhe. In the same study it was shown that one such cell, which must be F$_1$ in origin, was the antigen-presenting cell (macrophage?), as will be discussed in Sections 9 and 10. These results, and the genetics of the other complementing systems listed at the beginning of this section, are consistent with a mechanism in which expression of the *Ir* gene phenotype depends on expression of a "hybrid" Ia molecule consisting of one chain encoded in *I-A* and the other chain encoded in *I-E,* as described by Jones *et al.* (1978) and Cook *et al.* (1979*c*), and suggested by the MLR data of Fathman and Nabholz (1977) and Fathman *et al.* (1978) (see Section 1.3). Since both chains would have to be synthesized in the same cell to obtain the hybrid molecule, mixtures of cells of the complementing H-2 types, even in chimeras in which alloreactivity is not a problem, would not produce a response. That products of both genes must, in fact, be expressed on the cell surface has been shown by the ability to block the response to GLPhe with anti-Ia antisera directed against either subregion product (Schwartz *et al.*, 1978*b*). The mechanism involved will be discussed further in Sections 9 and 10.

While the other examples listed follow the general behavior of GLPhe as far as they have been tested, each of them adds some new wrinkle. The pattern of response to the alloantigen H-2.2 (Dorf and Stimpfling 1977) differs from that to GLPhe in that in the latter case H-2^k mice respond but H-2^d and H-2^a mice do not. This result necessitated mapping one of the complementing genes in *I-C*—that is, just to the right of the recombination event between H-2^k and H-2^d in the recombinant H-2^a haplotype*—rather than *I-E*, and postulating a system of "coupled complementation." Coupled complementation is defined as restricted complementation in which certain *I-A* genes complement with certain *I-C* genes but not others, although these others complement with different *I-A* genes. (In this case, I-A^k complemented with I-C^k but not I-C^d, even though the latter did complement with I-A^b.)† Even more complex restricted sets of interactions occur in the case of the response to GLLeu (Dorf *et al.*, 1978), which can be described only by a coupled complementation mechanism. Even such complex coupled complementation is compatible with the general mechanism of hybrid Ia molecules, although not with the known hybrid Ia molecules so far described.

Another potentially important principle discovered in these complementing *Ir* gene systems is that *Ir* gene phenotypes may be subject to gene dosage effects, suggestive of codominance rather than complete dominance. In both the case of GLPhe (Dorf *et al.*, 1979) and the case of GLT[5] (Schwartz *et al.*, 1979b), complementation of *I-A* and *I-E* genes in F_1 hybrids led to lower responses than did complementation in strains which were recombinant between *I-A* and *I-E*. This difference was shown in both studies to be due to the fact that the F_1 had only one copy of each high-responder allele, whereas the homozygous recombinant strain had two copies of each of the complementing genes. The alternative explanation of cis–trans effects was ruled out by diluting the genes of the recombinant by crossing it with a low-responder strain which had neither of the complementing genes. This F_1, with the 2 genes on the same chromosome (cis) behaved as did the F_1 with the genes on opposite chromosomes (trans). However, it is difficult from such studies to distinguish gene dosage effects from yet another alternative explanation—namely, that the

---

*See footnote on page 473.

†A possible alternative explanation is also interesting to consider. The gene which was mapped in *I-C* could be remapped in *I-E*, consistent with the Ia hybrid molecules described by Jones *et al.* (1978) and Cook *et al.* (1979*c*), and the need to postulate coupled complementation in this case could be eliminated if the responsiveness of the H-2^k mice could be attributed to other mechanisms. One possibility is suggested by the fact that the antigen, in this case a cell surface alloantigen of H-$2D^b$, could not be used in a "pure" form, since whole cells were injected. Genetic matching of recipient and donor (immunogen) strain was carried out to the greatest extent possible with existing strains so that the only difference was in the H-$2D$ region. However, the B10.BR (H-2^k) was immunized with cells from a B10.A(2R) recombinant, which differs in *I-C* as well as H-$2D$. Since a mixed lymphocyte reaction can be stimulated by this difference in *I-C* (for example, in B10.AM vs. B10.A(2R) and A.AL vs. A/J: Okuda and David, 1978), the response of the B10.BR may have been due to the added help of this stimulation. Then the complementation of *Ir* genes in this response could be explained by a mechanism identical to that for GLPhe. This example illustrates how tricky the interpretations of these complex interactions can be. Of course, this idea is only speculation and coupled complementation may still apply to the response to the H-2.2 alloantigen.

nonresponder alleles, present in both types of F_1 hybrid but not in the homozygous inbred recombinant, actively induced suppression of the response. While precedent exists for suppressive mechanisms of Ir gene control (see Section 5), gene dosage effects would be more consistent with a mechanism based on the known formation of hybrid Ia antigen molecules.

493

IMMUNE RESPONSE
GENES IN THE
REGULATION OF
MAMMALIAN
IMMUNITY

Finally, in the case of complementation for the response to the globular protein antigen cytochrome c, the complementation of two Ir genes was necessary for the response to a single antigenic determinant (Solinger $et\ al.$, 1979), in contrast to the antigens described in Section 4.1. While complementation for the synthetic polypeptide antigens such as GLPhe was presumed to involve a single determinant, no direct evidence for this contention exists.

In summary, complementation can occur between Ir genes (for a single determinant) at at least two levels: (1) Both Ir genes may be expressed in a single cell type in which complementation may occur between comparable types of molecules (Ia antigen subunits); and (2) complementation may be observed between Ir genes which function at different levels in the immune response (for example, in collaborating T and B cells). The study of complementing systems has contributed much to the understanding of Ir gene function. They enormously reinforce the notion that Ir gene function may be mediated by the cell surface Ia antigen molecules themselves—that is, because of the tremendous correlation between hybrid Ia molecules and complementing Ir genes, it now appears much more likely that Ia antigens and Ir genes bear a relationship that is closer than that suggested by the mere proximity of genes in a small segment of the genome.

5 Suppressive Mechanisms and Regulatory Circuits in Ir Gene Function

Several examples of Ir genes have been discovered for which active suppression of responsiveness is an important amplification mechanism, if not the primary mechanism of action. Some of these have been appropriately termed "Is (immune suppression)" rather than "Ir" genes (Benacerraf and Dorf, 1976), but for purposes of the present discussion we shall combine them all as types of Ir genes. Only a few examples can be discussed, and these are chosen to illustrate certain points from which generally applicable conclusions may be drawn.

5.1 The Random Polymers GAT and GT

The first example of suppressive activity found selectively in low-responder animals for an Ir-gene-controlled response was the case of poly-$(Glu^{60},Ala^{30},Tyr^{10})$ (GAT) in the mouse (Kapp $et\ al.$, 1974). Nonresponder strains of mice ($H\text{-}2^{p,\ q,\ or\ s}$) could not respond to GAT itself (Martin $et\ al.$, 1971; Merryman and Maurer, 1972), but could make anti-GAT antibodies when immunized with GAT attached to the immunogenic carrier molecule methylated bovine serum albumin (MBSA). However, preimmunization with GAT suppressed the ability of the nonresponder mice to respond to subsequent challenge with GAT-MBSA (Kapp $et\ al.$, 1974). Mixing splenic lymphocytes from suppressed low responders with spleen cells from normal syngeneic mice suppressed the $in\ vitro$ antibody response of the latter to GAT-MBSA. The cells mediating the transferrable suppression were shown to be suppressor T lymphocytes.

Subsequent studies revealed that sonication of the suppressor cells from GAT-immunized nonresponder mice released a "suppressor factor" into the cell-free supernatant which could substitute for live suppressor cells in mediating suppression (Kapp et al., 1976). Similar extracts from responder mice did not suppress in vitro responses to GAT-MBSA. Cell fractionation experiments demonstrated that the cell which produced the "factor" was probably a T cell, in that it did not adhere to plastic, did not bind to an anti-mouse Ig column, and was enriched in lymph node-free thymus relative to spleen cell populations (Kapp et al., 1977). The suppression was antigen specific and dependent on the dose of factor added to the cultures. Extracts from GAT-immunized animals of any nonresponder strain suppressed the response to GAT-MBSA of any other nonresponder strain, but not of any responder strain (Kapp et al., 1977; Kapp, 1978). Thus, while there was restriction in terms of Ir responsiveness, the factor could act across major histocompatibility barriers, in contrast to the very similar factor described by Tada and Taniguchi (1975) (see Section 7).

The mechanism of action of this suppressor factor proved to be complex. Rather than act directly on the B cell, or even indirectly on the helper T cell, the factor was found to induce a secondary "suppressor effector" cell, also a T cell, which mediated the actual suppression. Thus Waltenbaugh et al. (1977) found that injection of GAT-specific suppressor factor into immunized nonresponder animals induced the production of a second type of suppressor T cell in the recipients' spleens which did not produce the factor but which suppressed the response to GAT-MBSA when these cells were adoptively transferred into a third animal. The suppressor effector cells, like the factor itself, were antigen specific. In an in vitro analog of this approach, Germain et al. (1978) showed that normal nonresponder spleen cells, cultured for two days in the presence of GAT-suppressor factor, were induced to become suppressor effectors which, after washing and injection into recipient nonresponder mice, could suppress the subsequent response to GAT-MBSA. The responsible effector cells were also shown to be T cells, not adherent to nylon, and sensitive to treatment with anti-thy 1.2 serum plus complement. When purified factor was used (see Section 7) instead of crude extracts, the ability of the factor to induce suppressor effectors in vitro was found to require the presence of nanogram quantities of GAT as well. The chemical characterization of this factor will be considered further in Section 7.

With regard to the present subject of Ir genes, it is important to distinguish whether the suppressive activity seen in low responders for GAT is the primary mechanism of Ir gene control in this case or whether supression is seen secondarily because a positive response is absent, weak, or delayed due to some other Ir gene mechanism. In this regard, it is important to note that a suppressor factor made in nonresponder mice, while not suppressing the response of responder mice to antigen given simultaneously, would suppress the response if given 7 days previously (Germain and Benacerraf, 1978a). In addition, the same authors showed that in vitro culture of responder spleen cells with nonresponder suppressor factor induced secondary suppressor effector T cells in these populations which could suppress the response of normal responder cells both in vitro and in vivo. Thus, although suppressor effectors could be generated in responders, the fact that they were not generated if immunization was simultaneous with treatment with factor suggested that the helper response in responders was strong enough and rapid enough to preempt and override any suppressive mechanisms. The possibility that the Ir gene action was in the production of suppressor factor in low responders, rather than in its ability to induce suppressor effects, also appears unlikely in view of the evidence that even responder spleen cells can produce suppressor factor in vitro under appropriate conditions of macrophage scarcity (Pierres and Germain, 1978) (see also Ishizaka and Adachi, 1976; Feldmann and Kontiainen, 1976). In addition, Kontiainen et al. (1979) have described a suppressor factor for GAT, secreted into the culture medium (rather than extracted by sonication) by spleen cells immunized *in vitro* (rather than *in vivo*), which is made by responder as well as nonresponder strains and which acts on responder as well as nonresponder strains. However, this factor may be the product of the secondary effector itself, so that its production would be analogous to the demonstration of suppressor effectors in responder strains by Germain and Benacerraf (1978a). The latter authors conclude that the difference between responders and nonresponders may not be in whether the former are defective in any step of the suppressor pathway, but rather in the balance between help and suppression in the various strains. It was further suggested that suppression may be preferentially induced by soluble GAT whereas helper cell activation may require macrophage-bound GAT (Pierres and Germain, 1978; Germain and Benacerraf, 1978a,b; Pierce and Kapp, 1978a,b). Therefore, even in systems in which suppression appears to be the overwhelming process in low responders, the deciding factor in *Ir* gene control may be the ability of macrophages to present antigen to appropriate clones of T cells. For this reason, the story of GAT will be taken up again in Sections 8 and 9.

A related antigen, the random copolymer poly-(Glu^{50},Tyr^{50}) (GT), induces an almost identical circuit of suppressor cells and factors to those described for GAT, and the two antigens have largely been studied in parallel in this regard (Debré *et al.*, 1976b; Thèze *et al.*, 1977a,b; Waltenbaugh *et al.*, 1977a,b, 1979). Two differences, both genetic, make the GT-response a unique and interesting case. The first is that none of 19 inbred strains of mice tested produced anti-GT antibodies after direct immunization with GT, but some of these strains would produce anti-GT when immunized with GT noncovalently complexed to the carrier methylated bovine serum albumin (MBSA) (Debré *et al.*, 1975a). Of those which did respond to GT-MBSA, some strains were susceptible to suppression of the response by prior immunization with GT, whereas others were not, and this susceptibility to suppression was inherited as a dominant trait, linked to *H-2* (Debré *et al.*, 1975a,b). Therefore, this type of gene was dubbed an *Is* gene by these investigators, to distinguish it from most *Ir* genes for which responsiveness is dominant (or at least codominant). The suppression could be transferred to normal syngeneic recipients with suppressor T cells from the suppressed mice. Thus, there were at least two categories of nonresponders to GT: those strains for which nonresponsiveness was accompanied by (and possibly but not necessarily attributable to) the generation of suppressor T cells, and those strains for which other unknown mechanisms had to be invoked.*

*The genetic control was quite distinct from that for GAT in that some low responders to GAT (*H-2^s*), as well as some high responders to GAT (*H-2^d*), were in the suppressible category for GT, whereas other low responders for GAT (*H-2^q*) were not suppressed by GT. Moreover, although some cross-reactions between the two antigens were found, they were not universal and in one case were unidirectional. For instance, in an *H-2^s* strain (suppressible to GT), both GAT and GT could suppress a subsequent response to GAT-MBSA, whereas only GT, not GAT, could suppress a subsequent response to GT-MBSA. In the case of an *H-2^q* strain, not suppressible to GT-MBSA by prior injection of either GT or GAT, the response to GAT-MBSA was also not suppressible by GT, although it was suppressible by GAT (Debré *et al.*, 1975b; Waltenbaugh *et al.*, 1976). This unidirectionality was not completely explicable on the basis of the fine specificity of the suppressor factors extracted from the suppressor cells induced by the two antigens. The GAT-suppressor factor from *H-2^q* was only partially removed by columns of GA-Sepharose or GT-Sepharose, but completely removed by GAT-Sepharose (Thèze *et al.*, 1977a). This result suggests that there may have been two components in the GAT factor, one specific for GA determinants and one for GT determinants, although this possibility was not tested explicitly and the alternative explanation of a single factor with lower affinity for GA and GT than for GAT remains possible. The GT suppressor factor (made in *H-2^d* mice, responders to GAT), bound to both GT and GAT-Sepharose but not GA-Sepharose. Unfortunately for the interpretation of the unidirectional cross induction of suppression, the fine binding specificity of *H-2^s* suppressor factors was not tested. Also,

The second unique feature of the immune-suppressor genetic control of the response to GT is that two complementing *Is* genes could be mapped to different *I* subregions of *H-2*, one in *I-A* and one in *I-C*, and the interaction between these is so complex as to require for its explanation the postulate of coupled complementation (Debré *et al.*, 1976a; Benacerraf and Dorf, 1976). This phenomenon of coupled complementation (as described for the *Ir* genes for GLLeu and the H-2.2 alloantigen in Section 4.2) is defined by a complex set of complementation groups such that an allele at *I-A* which complements with one allele at *I-C* fails to complement with another allele at *I-C* even though the latter allele is capable of complementation with other alleles at *I-A*. For instance, *H-2^k*, *H-2^d*, and *H-2^s* mice are all suppressible by GT, but the intra-*H-2* recombinants (with crossovers between *I-A* and *I-C*), *k/d*, *s/d*, and *s/k*, are not. In addition, while *H-2^b* mice and *d/b* recombinants are not suppressible, the reciprocal recombinant, *b/d*, is suppressible. Whether these data are analyzed in terms of complementation for suppression or complementation for lack of suppression, coupled complementation must be invoked. The mechanism remains to be explained.

5.2 Suppressor Determinants on Protein and Polypeptide Antigens

A very exciting concept demonstrated in the responses to the several copolymers and terpolymers of glutamic acid, alanine, and tyrosine is the idea that one determinant on a molecule can induce supression of responses to other determinants on the same molecule. We shall refer to the former type of determinant as a supressor determinant. It was noted that the random polymers GAT and GT induced suppressor T cells in *H-2^s* mice, whereas GA did not. Also, *H-2^s* mice in particular were responders to GA but not GAT. The latter differs from the former by only 10% tyrosine, most of which may polymerize in oligotyrosine sequences at the amino terminus because of the slower rate of polymerization of the N-carboxy anhydride of L-Tyr compared to the equivalent derivatives of the other amino acids involved. To test the hypothesis that the oligotyrosine sequences functioned as suppressor determinants, M. Schwartz *et al.* (1976) synthesized a copolymer of 65% glutamic acid and 31% alanine to which an oligotyrosine sequence, representing only 4% of the total molecular weight of 14,000, was attached at the carboxyl terminus [denoted (G,A)-T]. This polypeptide behaved as did the random polymer GAT in all strains tested, and in particular elicited a low response in *H-2^s* mice, in contrast to the same polymer without the oligotyrosine. Moreover, preimmunization of *H-2^s* mice with this synthetic (G,A)-T induced suppression of the response to a subsequent challenge with normally immunogenic GA (without tyrosine) as well as to a challenge with GAT-MBSA or (G,A)-T-MBSA. It was concluded that the oligotyrosine was a suppressor determinant which could induce suppression of responses to other determinants on the same molecule, even when the rechallenge with the other determinants was in the absence of the oligotyrosine moiety. Thus, the mechanism of this suppression cannot result from the antigen's serving as a bridge between the

the secondary suppressor effector cells induced in *H-2^d* mice, at least, by GAT factor or GT factor were shown to be completely specific for the antigen which was used to generate the factor, with no detectable cross-reactive suppression at the effector phase of the process (Germain and Benacerraf, 1978a). At this phase also it would be interesting to examine the *H-2^s* factors and suppressor effectors to try to explain the unidirectionality noted above.

suppressor T cell binding to one determinant and the helper T cell or B cell binding to another determinant, as has been suggested for the case of lysozyme (see below).*

497

IMMUNE RESPONSE
GENES IN THE
REGULATION OF
MAMMALIAN
IMMUNITY

There are several similar examples of determinants on fragments of protein antigens which can induce suppression of an immune response to subsequent challenge with the whole protein antigen, although *Ir* gene control has not been demonstrated for most of these. They include the antigens myelin basic protein (Swanborg, 1975; Swierkosz and Swanborg, 1975; Hashim *et al.*, 1976), β-galactosidase (Eardley and Sercarz, 1976; Turkin and Sercarz, 1977), and bovine serum albumin (Muckerheide *et al.*, 1977). In all three cases, evidence at least suggested the participation of suppressor T cells.

One protein antigen for which a suppressor determinant has been reported and which is also under *H-2*-linked *Ir* gene control is hen egg white lysozyme (Sercarz *et al.*, 1978). Lysozymes from four species of birds elicit an antibody and/or T cell proliferative response in *H-2*[b] mice, whereas hen lysozyme and five other species of lysozyme fail to elicit either type of response in mice of this *H-2* type (although they do in mice of other *H-2* types). All four lysozymes which are immunogenic in *H-2*[b] mice have a tyrosine residue at position 3, whereas all six nonimmunogenic ones have a phenylalanine residue at this position. This region and that between residues 99 and 103 are the only ones that consistently differ between immunogenic and nonimmunogenic (for *H-2*[b] mice) lysozymes (Hill and Sercarz, 1975). Furthermore, whereas neither hen lysozyme nor its reduced and carboxymethylated derivative elicits a T cell proliferative response in *H-2*[b] mice, the reduced and carboxy-methylated cyanogen–bromide cleavage fragment from residues 13 to 105 (called L$_{II}$) elicits a good response. In contrast, another peptide called "N-C," isolated after mild acid hydrolysis of hen lysozyme and consisting of amino-terminal residues 1–17 and carboxyl terminal residues 120–129 covalently coupled by a disulfide bridge between residues 6 and 127, induces suppression of the anti-TNP response to TNP-guanidinylated-hen lysozyme, just as does preimmunization with whole hen lysozyme. In addition, preimmunization with hen lysozyme or the N-C fragment has been found to induce suppressor T cells which suppress the antibody response to lysozyme attached to erythrocytes (Adorini *et al.*, 1979*a*,*b*). The suppressor cells did not affect the antibody response to ring-necked pheasant lysozyme, and the latter lysozyme did not induce suppressors of the antihen lysozyme erythrocyte response. Thus the suppressor T cells showed a fine specificity not reflected in the bulk antibody produced, which reacted with both lysozymes.

In contrast, intraperitoneal immunization with the L$_{II}$ fragment alone elicited helper T cells in spleens of nonresponder *H-2*[b] mice for an *in vitro* response to lysozyme bound to erythrocytes (Adorini *et al.*, 1979*b*). Another method of demonstrating intact helper cell function in nonresponder *H-2*[b] mice was to use foot-pad immunization and study the response in the draining popliteal lymph nodes (Araneo *et al.*, 1979). Ten days after immunization with hen lysozyme there was at least as much helper activity in the popliteal lymph nodes as after immunization with the immunogenic ring-necked pheasant lysozyme, but the activity fell off sharply with time. Mixture of one part of popliteal lymph node T cells taken 21 days after immunization with nine parts of 10-day lymph node cells revealed potent suppressor activity in the 21-day T cell population. The suppressor cells were sen-

*It was subsequently found that the random terpolymer, poly-(Glu[58],Ala[38],Tyr[4])(GAT[4]), containing the same percentage of tyrosine residues as (G,A)-T but randomly distributed, was immunogenic in *H-2*[a] mice in contrast to (G,A)-T (Maurer *et al.*, 1978). This result is further evidence that the "suppressor determinant" must include a stretch of several contiguous tyrosine residues, clearly present in (G,A)-T, probably common in GAT molecules, but relatively rare in GAT[4].

sitive to treatment with anti-thy-1 and complement. However, whereas the suppression by spleen T cells from mice immunized intraperitoneally with lysozyme was sensitive to treatment with antisera against $I\text{-}J^b$-encoded Ia antigens plus complement (Adorini et al., 1979b), the suppression occurring in these mixtures of 10-day and 21-day lymph node T cell populations was sensitive to such treatment of the 10-day helper T cell population but not the 21-day suppressor T cell population (Araneo et al., 1979). Putting these several observations together, it was suggested that a suppressor determinant on the N-C part of hen lysozyme induces suppressor T cells which block the response of helper or proliferating T cells which recognize predominantly the L_{II} region of lysozyme (residues 13–105). The suppression requires the interaction of at least two T cells, one bearing antigens encoded in the $I\text{-}J$ subregion which is present in the spleen and lymph node both early and late, and one not bearing $I\text{-}J$-encoded antigens which appear early in the spleen but only late in the popliteal lymph node. It was suggested that the late-appearing N-C-specific cell is the suppressor precursor, whereas the early-appearing $I\text{-}J^+$ cell in the helper T cell population is a suppressor inducer, analogous either to the $Ly1^+2^+3^+$ auxiliary cell described by Tada (1977) or to the $Ly1^+2^-3^-$ inducer of feedback suppression described by Eardley et al. (1978, 1979), Cantor et al. (1978), and McDougal et al. (1979).

To test the hypothesis that the antigen itself serves as an intercellular bridge to direct the N-C-specific suppressor cell to the L_{II}-specific helper cell, the proliferative response of T cells from $H\text{-}2^b$ mice immunized with hen L_{II} fragment was tested. These proliferated in response to L_{II} and to the whole lysozyme, whereas T cells from $H\text{-}2^b$ mice immunized with whole hen lysozyme responded to neither antigen. As little as 2% of these latter T cells suppressed the response of the remaining 98% L_{II}-immune T cells to whole lysozyme but not to L_{II} (Yowell et al., 1979). If a mixture of all the fragments was used in vitro to stimulate the mixture of L_{II}-immune and lysozyme-immune T cells, a good proliferative response was seen—a response comparable to that with the L_{II} fragment alone (E. Sercarz, personal communication). The requirement that the N-C and L_{II} parts of the molecule be linked in order to elicit suppression was taken as strong evidence that the antigen served as a bridge between suppressor and proliferating T cells which, because of their different specificities, had no other specific mechanism by which to interact. While this proposed mechanism is certainly appealing, it cannot easily explain all cases of suppressor-determinant action, as pointed out in the case of suppression of the GA response by prior immunization with (G,A)-T [especially if the latter situation involves transferrable suppressor cells which act at the time of challenge with GA, rather than elimination of GA-specific helper cells at the time of preimmunization with (G,A)-T]. Similarly, it does not explain why lysozyme-immune T cells do not respond to L_{II} alone. It is also worth noting that although the putative suppressor determinant of lysozyme does not involve tyrosine, it is the difference between the aromatic residues tyrosine and phenylalanine that appears to be crucial in generating the suppressor determinant. It is striking that both of these identified suppressor determinants involve aromatic amino acid residues.

Finally, in the high-responder strain to hen lysozyme, B10.A, two types of helper T cells have been discovered (Adorini et al., 1979c; Harvey et al., 1979). Incubation of helper T cells on plastic plates coated either with lysozyme or with purified antilysozyme (idiotype-bearing) antibodies removed helper activity from the nonadherent population, while incubation on control plates coated with ribonuclease or normal mouse serum did not.*

---

*This result is somewhat surprising since most attempts to remove antigen-specific helper T cells on antigen-coated plates or beads have failed. When helper T cells have been specifically removed, it is usually on monolayers of macrophages pretreated with antigen.

Mixing the two inactive nonadherent populations from the antigen-coated and idiotype-coated plates restored full activity. This result suggested that helper T cells of two types were necessary, those of one type specific for antigen and interacting with B cells via an antigen bridge, and those of the other type specific for the common idiotype of the majority of antilysozyme antibodies (which were specific for the N-C portion of the molecule) and interacting with B cells via the idiotype of their surface immunoglobulin. These results and conclusions are completely analogous to those of Woodland and Cantor (1978), who demonstrated idiotype-specific and carrier-specific helper T cells in a hapten-carrier model. Since suppressor T cells in the H-2^b low responder were also shown to bear this common idiotype and specificity for the N-C region (Adorini *et al.*, 1979*c*), it was postulated that they could interact with L_{II} antigen-specific helper T cells, as suggested above for proliferating T cells, via an antigen bridge, and with idiotype-specific helper T cells via the

499

IMMUNE RESPONSE
GENES IN THE
REGULATION OF
MAMMALIAN
IMMUNITY

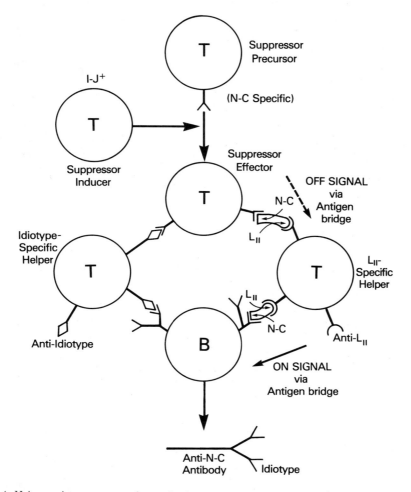

Figure 4. Helper and suppressor regulatory circuits suggested to modulate the response to hen lysozyme. The cell interactions are suggested to be via idiotype–antiidiotype binding or via an antigen bridge. N-C and L_{II} are peptides which correspond to two regions of the lysozyme molecule, as described in the text. The model is modified from Sercarz *et al.* (1978) and Harvey *et al.* (1979), with permission.

idiotype of their receptors. Therefore, N-C-specific idiotype-positive suppressor T cells and the N-C-specific, idiotype-positive, B cells would occupy symmetrical positions with respect to the two types of helper T cells.

The several suppressor and helper regulatory circuits postulated to be involved in the modulation of the antibody response to hen lysozyme are depicted in Fig. 4. It should be noted that this "model" is still quite speculative, especially since some parts were demonstrated in high-responder B10.A mice and other parts in low-responder B10 mice. However, it is consistent with regulatory circuits suggested in other immune systems not under *Ir* gene control (Eardley *et al.,* 1978, 1979; Cantor *et al.,* 1978; Tada, 1977; Woodland and Cantor, 1978; McDougal *et al.,* 1979). Therefore, the response to lysozyme may serve as an excellent model system in which to explore the interaction of *Ir* genes with the other complex regulatory machinery of the immune system. In particular, the immune suppression *(Is)* genes may be equivalent to other *Ir* genes but marked by an imbalance of suppression, perhaps induced in a feedback circuit by suboptimal amounts of helper activity.

In conclusion, suppressor mechanisms have been found to dominate the response of low-responder strains in several systems regulated by MHC-linked *Ir* genes. Whether the suppressor mechanism represents the primary determining event, or rather is a secondary phenomenon arising in a nonresponding animal, remains to be determined. The interaction of *Ir* genes and the other regulatory mechanisms of the immune network remains a fruitful area for research (see Section 12).

6   The Nature and Specificity of T Cell Receptors

Since T lymphocytes play so prominent a role in the function of all known MHC-linked *Ir* genes, it was long considered possible that the products of *Ir* genes were the T cell receptors themselves. The nature of T cell receptors still remains very poorly understood, in contrast to B cell receptors for antigen which, being immunolobulin, have been more thoroughly studied. If *Ir* gene products were another type of antigen receptor encoded separately from immunoglobulin (that is, in the MHC), and if T cells also had receptors which were not immunoglobulin in nature, might not they be one and the same? Recent evidence strongly suggests that at least the variable portion or combining site region of T cell receptors is the same as or closely related to, that of humoral antibodies, encoded with the immunoglobulin structural genes, not in the MHC. In this Section I will summarize this evidence, and then discuss whether *Ir* genes may be related to T cells in terms of their specificity.

6.1   The Evidence That V_H of Immunoglobulin Is the Combining Region of T Cell Receptors

Four basic lines of evidence have been taken to demonstrate that the combining site of T cell receptors includes V_H, the variable region of the immunoglobulin heavy chain with the same specificity. All of them depend on serological methods of identification, with all the pitfalls inherent in characterization of specificity of the antibody used as defining reagent and in the possibility of cross-reactivities, and three of them depend on idiotype, as defined in the first footnote (*) on p. 476. Also relevant is the evidence discussed in Section 7.1 for V_H idiotype on T cell-derived helper and suppressor factors (Mozes, 1978; Germain et al., 1979; Bach et al., 1979; M. Feldman, personal communication).

501

IMMUNE RESPONSE
GENES IN THE
REGULATION OF
MAMMALIAN
IMMUNITY

6.1.1 Receptors on Alloantigen-Specific T Cells

One approach to the nature of T cell receptors, uniquely possible for alloantigen-specific T cell receptors, was that proposed by Ramseier and Lindenmann (1969, 1971, 1972). The concept was that an F_1 hybrid animal should have all the receptors in the repertoires of both parents *except* for the receptors that one parental strain would have against the alloantigens (such as MHC antigens) of the other parent, since the F_1 should be tolerant to these. Therefore, immunization of an F_1 hybrid with cells of one parent might result in antibodies specific for the receptors on that subpopulation of parental cells which recognizes the other parent as foreign. The evidence produced by this conceptually rather elegant approach was reviewed recently by Ramseier et al. (1977). The original experiments employed a rather indirect assay system in which the end point was the number of granuloycytes chemotactically attracted to the site at which material was injected into the skin of hamsters.

The conceptual approach was taken up subsequently in a different system by Binz and Wigzell (1975a–c). The key experiments were to raise antisera in (Lewis × DA) F_1 rats by immunization with T cells from the Lewis parental strain and to show that these putative (Lewis × DA) F_1 anti-(Lewis anti-DA) antibodies were specific both for idiotypic determinants on Lewis anti-DA serum alloantibodies and for determinants on Lewis cells which were alloreactive against DA (Binz and Wigzell, 1975a). The important demonstration that it was the same antibodies in F_1 anti-Lewis serum, which reacted with both humoral alloantibodies and alloreactive T cells (those mediating graft-vs.-host disease and MLR) was accomplished by reciprocal absorption studies: Columns of Lewis anti-DA alloantibodies attached to Sepharose removed the antibodies which reacted specifically with T cells mediating Lewis anti-DA MLR and graft-vs.-host disease; conversely, absorption with Lewis T cells removed the antibodies binding to Lewis anti-DA alloantibodies (Binz and Wigzell, 1975a). The F_1 anti-Lewis serum was specific for Lewis antibodies and cells reactive against DA alloantigens, not for Lewis antibodies or T cells against third party strains. Therefore it recognized the idiotype (or antigenic determinants unique to the combining region) of the Lewis antibodies and T cell receptors with this particular specificity (anti-DA). This demonstration was the strongest direct evidence at the time for sharing of idiotypes between T cell receptors and antibodies with the same specificity. As further confirmation of this conclusion, it was shown that affinity columns of Sepharose-bound antibodies could be used to separate Lewis T cells reactive against various other strains such as DA, BN, and AU (Binz and Wigzell, 1975c), and that columns of the same type could be used to purify shed T cell receptor material. When the latter material was cross-linked with gluteraldehyde and injected back into Lewis rats, it could elicit autoantiidiotypic antibodies against Lewis receptors for DA and induce specific transplantation tolerance toward skin grafts from DA but not third party rats (Binz and Wigzell 1975d, 1976a,b). Similar experiments were carried out in the mouse (Binz and Askonas, 1975).

Crucial to the concerns of this review, the idiotype of the Lewis anti-DA T cell receptor was found to have no genetic linkage to the rat MHC, tested in a backcross of (Lewis × BN) F_1 × BN (Binz et al., 1976), or to the light chain of immunoglobulin. However, tight linkage was found to the Lewis heavy chain immunoglobulin allotype in (Lewis ×

DA)F_2 rats which were homozygous for the Lewis MHC (Binz *et al.*, 1976; Andersson *et al.*, 1978). Thus, the idiotypic determinants on Lewis T cells specific for DA were not only shared with Lewis anti-DA antibodies, but also linked genetically to the immunoglobulin heavy chain structural genes, consistent with the notion that T cells employed V_H, the heavy chain variable region of immunoglobulin, as at least the variable or combining site portion of their receptors. Lack of linkage to the MHC meant that at least the portion of the receptor bearing the idiotype was not encoded in the MHC as, for example, an *Ir* gene product.

A related approach tried in mice was to purify T lymphoblasts of strain A proliferating in an MLR against strain B and to use these enriched populations to immunize an (A × B) F_1 hybrid in analogy to the above studies (Kramer, 1978) or to immunize a syngeneic strain A mouse to produce an autoantiidiotype and specific transplanation tolerance (Andersson *et al.*, 1976, 1977; Aguet *et al.*, 1978; Binz and Wigzell, 1978). The results so far have been similar to those in the rat, with one major exception. In addition to linkage to the heavy chain Ig structural genes, the idiotypes on T cells detected by this method appear to be linked to the MHC as well (Kramer and Eichmann, 1977). Whether this linkage is due to direct involvement of an MHC gene product in the receptor, or to an indirect influence of MHC in selection of T cell repertoire specific for alloantigens, is not yet clear. Also the idiotypes detected by this method differ for different subclasses of T cells (Binz *et al.*, 1979).

6.1.2 Receptors on Exogenous Antigen-Specific T Cells

In contrast to the approach used in the experiments discussed in Section 6.1.1, in which antiidiotype raised against T cell receptors was shown to cross-react with serum antibody of the same specificity, other investigators showed that antiidiotype raised against serum antibodies could react with antigen-specific T cells and either trigger them or block their response. The first such demonstration for exogenous antigen-specific antibodies and T cells was for the case of the A5A idiotype of antistreptococcal Group A carbohydrate described in A/J mice by Eichmann (1972). A significant fraction of antibody against this antigen in most A/J mice bears this idiotype, defined by guinea pig antisera raised against a purified antibody from a clone of cells from hyperimmune mouse A5A whose whole serum showed restricted heterogeneity by isoelectric focusing even before the cloning. Guinea pig IgG2 class anti-A5A idiotype, when injected *in vivo*, induced suppressor T cells which specifically suppressed the production of anti-Group A carbohydrate bearing the A5A idiotype (Eichmann, 1974, 1975). The mechanism of induction of suppressor cells is unknown and does not necessarily imply idiotype-positive T cells. However, more important for the question of receptors on antigen-specific T cells, the IgG1 class of guinea pig antiidiotype, rather than inducing suppression, was found to actually "prime" or sensitize clones of idiotype-positive B cells as well as helper T cells, so that subsequent challenge with streptococcal A carbohydrate resulted in a secondary type of response (Eichmann and Rajewsky, 1975). The B cell secondary response after sensitizing with antiidiotype was specific for A carbohydrate and was idiotype positive. The helper T cell activation, measured by help provided for a B cell antihapten response to hapten attached to A carbohydrate, was therefore taken as evidence that the antigen-specific helper T cells also bore the same (or a cross-reactive) idiotype.

Two types of evidence suggested that this idiotype apparently shared by antigen-specific B and helper T cells was a marker for the variable region of the heavy chain of immunoglobulin (V_H). One was that the A5A idiotype was genetically linked to the immunoglobulin heavy chain allotype locus, and could be mapped relative to other putative V_H marker idiotypes, such as that of antibodies common to most A/J mice immunized with the p-azophenylarsonate hapten (Eichmann et al., 1974). The second piece of evidence was that only antiidiotype antibodies specific for the V_H, not the V_L, portion of the A5A anti-A-carbohydrate antibodies were capable of sensitizing helper T cells (Eichmann, 1977). Finally, to avoid possible artifacts incurred by raising antiidiotype against an induced antibody which could be contaminated by antigen (A-carbohydrate) or by T cell receptors shed into serum, Black et al. (1976) reproduced the same findings in BALB/c mice using antiidiotypic antibodies raised against the BALB/c myeloma protein (IgA, Kappa) S117, which had specificity for streptococcal A carbohydrate, but had never been in contact with antigen. Not only could the same type of helper T cell sensitization be demonstrated, but also Hämmerling et al. (1976) showed that the ability of the two antiidiotypes, anti-A5A and anti-S117, to sensitize helper T cells specific for streptococcal A carbohydrate was inherited strictly in accordance with the respective A/J and BALB/c heavy chain immunoglobulin allotype markers. Even finer mapping was provided by the BAB14 strain which inherited the heavy chain constant region (C_H) from C57BL/6 but the rest of the genome, including much of the heavy chain variable region (V_H) genes, from the BALB/c. This strain also inherited the S117 idiotype and the ability to undergo T cell priming with anti-S117. Thus, the ability of T cells to be primed by antiidiotype was linked to V_H, not merely C_H. No linkage to *H-2* was seen for any of these effects. Thus, these results, like those of Binz and Wigzell (see Section 6.1.1), suggested that the combining or variable portion of the antigen receptor on T cells was similar or identical to V_H of the corresponding immunoglobulins.

Another approach was taken by Geczy et al. (1976), who raised autoantiidiotypic antibodies in syngeneic guinea pigs against antipenicilloyl-bovine gamma globulin made in the same strain. These antiidiotypic sera inhibited the T cell proliferative response *in vitro* to the same antigen. The inhibition occurred only in syngeneic animals; that is, strain 13 anti-(strain 13 antipenicilloyl) inhibited the response of strain 13 T cells but not strain 2 T cells and conversely for strain 2 antibodies, despite the fact that strains 2 and 13 differ only for Ia antigens and not, as far as is known, for immunoglobulin genes. Since the sera were always syngeneic, no antibodies to MHC antigens should have been involved. Again the conclusion reached was that T cells, this time ones which can be induced to proliferate in response to antigen stimulation *in vitro*, bore the same idiotype as antibodies of the same specificity. In addition, antigen binding by hapten-specific T cells has been shown to be blocked by antiidiotype raised against humoral antibodies (Prange et al., 1977). It should be noted that the appearance of V_H idiotypes on T cells has therefore been demonstrated for T cells functioning in MLR, graft-vs.-host reactions, help for antibody synthesis, and antigen-stimulated proliferation, as well as in direct antigen binding studies (references cited above).

6.1.3 *Isolated Hapten-Binding Receptors from T Lymphocytes*

In the process of following up a method described by Kiefer (1973, 1975) for isolating hapten-specific cells by binding to hapten-coupled nylon mesh, Krawinkel and Rajewsky (1976) serendipitously made the exciting discovery that after the cells were released from the mesh by a temperature shift from 4 to 25°C, something remained behind blocking the hapten on the mesh (U. Krawinkel, personal communication). This blocking material,

503

IMMUNE RESPONSE
GENES IN THE
REGULATION OF
MAMMALIAN
IMMUNITY

which could be eluted from the nylon at pH 3.2, was found to bind specifically to hapten. Some of it was apparently immunoglobulin, as judged by binding to affinity columns of anti-Ig (the anti-Ig$^+$ fraction). However, some of it, with lower affinity for hapten, did not bind to polyvalent rabbit antimouse immunoglobulin with specificity for IgG, IgM, and IgA heavy chains and κ light chains. This latter (anti-Ig$^-$) fraction also increased with increasing proportions of T cells in the T-enriched, nylon wool-nonadherent spleen cell population from which it was derived. Therefore, the nonimmunoglobulin fraction was postulated to consist of isolated T cell receptors.

When the hapten used was (4-hydroxy-3-nitrophenyl) acetyl (NP), the nonimmuno-globulin fraction of "receptor" from C57BL/6 mouse spleen cells displayed two character-istics of the variable region of the heavy chain of antibodies from the same strain with the same specificity. One was the presence of the so-called NP^b idiotype characteristically pres-ent on anti-NP antibodies from the C57BL/6 strain (Mäkelä and Karjalainen, 1977; Imanishi-Kari et al., 1979), and the other was a fine specificity property, heteroclicity (Krawinkel et al., 1977a). The latter property is the manifestation of a higher affinity for NIP ([4-hydroxy-5-iodo-3-nitrophenyl] acetyl) than for the immunizing hapten NP and is characteristic of anti-NP antibodies made by C57BL/6 mice. In backcross and F_2 studies, the idiotype on both anti-Ig$^+$ and anti-Ig$^-$ fractions of receptors was found to be linked to the heavy chain allotype locus of the C57BL/6 (Krawinkel et al., 1977a; Reth et al., 1977). Therefore it was suggested that these putative T cell receptors shared the V_H portion with the corresponding antibodies, even though they lacked immunoglobulin C_H markers. More-over, whereas the heteroclitic antibodies to NP all bore the λ light chain obligatorily (Imanishi-Kari et al., 1979), the putative T cell receptors were not bound to anti-λ light chain (Reth et al., 1977; Cramer et al., 1979). This result further supported the notion (see Sections 6.1.1 and 6.1.2) that the combining site of T cell receptors consisted of V_H but not V_L, and that the single-chain nature of the site might contribute to the lower affinity. Additional strong support for the V_H nature of these isolated putative T cell receptors came from studies in the rabbit, in which there is an allotype marker on the V_H as well as C_H region of immunoglobulin (Mage et al., 1973; Kindt, 1975). The receptors isolated from rabbit nylon-nonadherent lymphocytes that were negative for all C_H immunoglobulin determinants nevertheless carried the V_H marker, the a allotype (Krawinkel et al., 1977b).

The certainty of the interpretation of all of these data, however, rested on the premise that the receptors which were not bound to antiimmunoglobulin antisera were actually derived from T cells. Besides the enrichment with increasing proportions of T cells, two further observations strongly supported this contention. One was that even after hyperim-munization of C57BL/6 mice with NP, when most of the B cells and antibodies were negative for the NP^b idiotype (which characterizes the primary response), the Ig$^-$ fraction of receptors still bore the idiotype (Krawinkel et al., 1978). This finding suggested that the Ig$^-$ receptors did not come from B cells, and furthermore that perhaps the T cell receptor, not diversifying with repeated immunization as did the B cell receptor, was restricted to major idiotypes common to all mice of a strain and that may correspond to germ-line genes. The second, stronger, piece of evidence came from experiments in which receptors were isolated from mixed populations of lymphocytes from two congenic strains which differed only in heavy chain allotype. Chimeras were made by injecting thymocytes from one strain into congenitally athymic nu/nu mice of the other strain, which served as the B cell source. When T cells were derived from one allotype-congenic strain and B cells were derived from the other strain, or vice versa, the Ig$^+$ receptors always had the idiotype and fine specificity characteristic of the B cell donor, whereas the Ig$^-$ receptors always had both markers char-acteristic of the T cell donor (Krawinkel et al., 1979). Also, the SJL strain, which does not

505

IMMUNE RESPONSE
GENES IN THE
REGULATION OF
MAMMALIAN
IMMUNITY

express the NP^b idiotype on B cells because of lack of λ light chains, does express the NP^b idiotype on Ig^- receptors for NP (Cramer *et al.*, 1979). The likelihood is extremely high, therefore, that the Ig^- fraction of receptors was of T cell origin. These data then become some of the strongest direct evidence that the V_H region of immunoglobulin is also employed as all or part of the combining region of antigen-specific receptors on T lymphocytes.*

6.1.4 Receptors on T Cells Detected by Antibodies to Immunoglobulin Variable Region Framework Determinants

An approach that is older and rather different from the ones described above is to use antisera raised against broad classes of immunoglobulin, rather than specific idiotypic determinants, to search for the presence of immunoglobulinlike molecules on T cells. Attempts using classical antiimmunoglobulin antisera, usually directed at class-specific determinants on the Fc portion of the molecule, or against light chain determinants, were negative or at best controversial (Warner, 1974; Ada and Ey, 1975). This approach was modified by Szenberg *et al.* (1977), who used antisera to purified mouse $(Fab')_2$ fragments of IgG. In order to detect determinants shared by IgG of many mammals, these investigators raised their anti-$(Fab')_2$ antisera in chickens, possible descendents of the dinosaurs. The antibodies were then purified by affinity chromatography. With this antibody reagent, they found Ig-like molecules on the surface of T cells, which could be shed and resynthesized by the T cells. The binding capacity of the antibodies for T cells could be abolished by absorption with κ light chains attached to Sepharose. While this result is consistent with the notion that the variable region of immunoglobulin is found on T cells, even if the constant region is not, it does not fit with the findings of the studies described in Sections 6.1.1–6.1.3, which demonstrated V_H, but not V_L, on the surface of T cells.

Another variant of this approach has recently been reported by Lonai *et al.*, (1978), who used antibodies raised in mammals, but raised them against the isolated V_H, V_λ and V_κ dissociated parts of Fab fragments of myeloma proteins. These antisera were not idiotype-specific, but appeared to react with V-region "framework" (as opposed to the hypervariable or complementarity-determining) sequences common to many V regions of that class. Using microscopic autoradiography, these investigators found that the anti-V_H serum, but not the others, inhibited antigen binding by T cells bearing surface markers characteristic of helper T cells ($Ly1^+2^-$).† In contrast, the anti-V_λ serum, but not the others, inhibited antigen binding by T cells bearing the surface markers of killer and/or suppressor T lymphocytes ($Ly1^-2^+$). Furthermore, the anti-V_H serum, but not anti-V_λ or anti-V_K, was found to prevent the radiation-suicide of fowl gamma globulin-specific helper T cells caused by $[^{125}I]$fowl gamma globulin, as measured by their ability to provide carrier-specific help in an *in vitro* antibody response to hapten coupled to fowl gamma globulin (P. Lonai,

---

*Support for the functional nature of the idiotype-bearing NP receptor on T cells comes from the recent work of Weinberger *et al.* (1979a,b) who found the fine specificity marker (heteroclicity) characteristic of the NP^b idiotype on both delayed hypersensitivity and suppressor T cells, and the serologically detectable NP^b idiotype at least on suppressor cells. These markers were genetically linked to the immunoglobulin V_H structural gene locus.

†The cell-surface antigens Ly 1, 2, and 3 found on T cells have been shown to be useful in defining subpopulations of T cells. The use of anti-Ly antisera plus complement has allowed three populations to be defined: $Ly1^+2^-3^-$, $Ly1^-2^+3^+$, and $Ly1^+2^+3^+$, of which helper T cells appear to be $Ly1^+2^-3^-$ and killer and suppressor T cells $Ly1^-2^+3^+$ (Cantor and Boyse, 1977). Although recent studies using a fluorescence-activated cell sorter suggest that all T cells bear some Ly1 on their surfaces (B. Mathieson, personal communication), not all of them are killed by this antiserum and complement, so that the operational definition of subpopulations remains useful.

personal communication). In addition, the isolated T cell receptors specific for the NP hapten described in Section 6.1.3 react with the anti-V_H serum (Cramer *et al.*, 1979). Other studies using anti-Fab sera to inhibit T cell binding of antigen have been reviewed by Warner (1974) and Greaves (1970).

These studies using anti-Ig V region or Fab-specific antisera which are not idiotype specific complement those using anti-idiotypic antisera described in Sections 6.1.1–6.1.3. The conclusion of all these studies in Section 6.1, taken in the aggregate, is that at least some part of the antigen-specific receptor of T cells is shared with the corresponding receptor of B cells (immunoglobulin), and that minimally, this consists of all or part of the variable or combining-site portion of these receptors. The bulk of the evidence suggests that the immunoglobulin heavy chain variable region (V_H) is the part shared by T cells, either alone or in association with a light chain counterpart that may or may not, according to different studies, bear any resemblance to V_κ or V_λ.

6.2 The Specificity of T Cell Receptors

Some information about the relationship of T and B cell receptors for antigen can be gleaned from a comparison of their fine specificity. In addition, some evidence for or against the role of T cell receptors in *Ir* gene control can be obtained from a comparison of T cell and *Ir* gene specificity. The large body of data comparing T and B cell specificity has been reviewed extensively elsewhere (Paul, 1970; Schlossman, 1972; Warner, 1974; Ada and Ey, 1975; Goodman, 1975) and is beyond the scope of the present review. This Section will serve to highlight a few selected observations which may be relevant to the specificity of *Ir* genes, and to the role of T cell receptors in *Ir* gene specificity.

There are some examples suggestive of similarity of specificity between T and B cell receptors for the same antigen (Rajewsky and Pohlit, 1971; Taylor and Iverson, 1971; Rajewsky and Mohr, 1974; Schwartz *et al.*, 1978; Berzofsky *et al.*, 1978; Berzofsky *et al.*, 1979). Even small haptens, the classic example of determinants recognized primarily by B cells, can elicit hapten-specific helper and suppressor T cells as well (Alkan *et al.*, 1971, 1972; Bush *et al.*, 1972; Hanna *et al.*, 1973; Rubin and Wigzell, 1973; Goodman, 1975; Yamamoto *et al.*, 1977; Weinberger *et al.*, 1979b), and also T cells mediating delayed hypersensitivity (Weinberger *et al.*, 1979a,c). However, for some hapten-specific T cell reactions there is evidence that the carrier plays a role (Davie and Paul, 1970), while in other cases part of the specificity may be directed toward conformational changes in the carrier specifically induced by a particular hapten, rather than toward the hapten itself (Ray and Ben-Sasson, 1979).

However, the bulk of studies have aimed at contrasting the fine specificities of T and B cells, and for our present purposes it is important to correlate these differences with the specificities of *Ir* genes to see which, if either, cell receptor may play a role in *Ir* gene function. Two major categories of differences have been observed between the specificities of T and B cell receptors, one relating to recognition of distinct determinants on a single antigen, and the other relating to the sensitivity with which the receptors distinguish conformational changes in a single determinant. In some cases both differences may apply. It must be pointed out from the outset that neither of these differences proves that T and B cells use fundamentally different molecules as receptors, a conclusion which would be in contrast to the findings discussed in Section 6.1. Both types of receptor may draw their combining regions from the same V_H pool, and the differences may all be attributable to different activation pathways which select which clones are stimulated. The mode of antigen presentation and degree of antigen processing at the time of first antigen contact may be critical. In addition, T and B cell clones may be subject to different selective pressures in the organism independent of antigen stimulation—for example, during maturation of T cells in the thymus (see Section 10).

The classic example of the first type of difference between T and B cell specificities is the hapten-carrier complex. As described in Section 3.1, if the helper T cell is carrier specific and the B cell is hapten specific, the *Ir* gene control shows carrier specificity. This result suggests that the specificity of the *Ir* gene control is determined by the T cell receptor. There are several illustrative examples in the case of defined protein antigens in which T cells and B cells show different specificities. Using tryptic peptides to delineate the antigenic determinants on the 29-residue polypeptide hormone, bovine glucagon, Senyk *et al.* (1971) found that antibodies from over two dozen guinea pigs bound primarily to the amino-terminal peptide (1–17), not to the carboxyl-terminal peptide (18–29). However, only the intact hormone and the carboxyl-terminal peptide elicited a T cell-proliferative response. Thus the determinants recognized by B cells and by proliferating T cells were on different parts of the molecule. Along similar lines, in studies of delayed hypersensitivity and antibody production to the flagellar H antigens of *Salmonella* in mice (Cooper, 1972; Cooper and Ada, 1972), it was found that B cells preferentially recognized the H_V determinants that were unique for each strain of *Salmonella*, whereas delayed hypersensitivity T cells recognized primarily the Hc determinants which were common to all *Salmonella* strains. It thus appeared that T cells showed more cross-reactivity (and therefore less specificity) than B cells, but in fact the cross-reactivity may have been caused by T cell specificity for these shared determinants rather than to a T cell receptor with less resolving power. In a similar fashion, helper T cells and B cells of rabbits appeared to recognize different determinants on gallinaceous lysozymes (Cecka *et al.*, 1976).

Similar examples have been seen for antigens under *Ir* gene control. In an autoradiographic study of [^{125}I]-(T,G)A–L binding to cells of both immunized and unimmunized mice, binding to B cells was inhibited by (H,G)A–L and (Phe,G)-A–L as well as by (T,G)-A–L, whereas binding to T cells was inhibited only by the homologous (T,G)-A–L (Hämmerling and McDevitt, 1974). Similarly, Lonai and McDevitt (1974) found that antigen-stimulated *in vitro* T cell proliferation in mice in response to (T,G)-A–L, (H,G)-A–L and (Phe,G)-A–L was specific for the antigen used to immunize the mice, whereas antibodies crossreacted with all three. In experiments of both types, the T cells showed the differences in fine specificity that had been observed for the *Ir* genes regulating responses to these antigens (for instance, H-2^b high and H-2^k low responders to (T,G)-A–L are reciprocally low and high responders to (H,G)-A–L), whereas the B cells did not. Of course, the difference in specificity may have been due to recognition of distinct determinants by T and B cells rather than differences in sensitivity of B cells to substitution of Phe or His for Tyr (that is, the opposite of the *Salmonella* flagellar antigen case above), but the important finding for our purposes was that the *Ir* gene specificity resembled the T cell specificity rather than the B cell specificity. Similarly, the *Ir* gene control of the response to insulin in guinea pigs was found to parallel the specificity of proliferating T cells, even though the antibodies made crossreacted with nonimmunogenic insulins (Barcinski and Rosenthal, 1977). These results all suggest that T cell receptors, rather than B cell receptors, determine the antigen specificity of *Ir* genes.*

The second category of specificity difference seen between T cells and B cells involves

---

*One possible exception to this generalization is the case illustrated by poly-L-lysine and poly-L-(Glu,Lys), which show little or no crossreactivity with each other for either antibody or delayed hypersensitivity T cell responses, and yet appear to be under identical *Ir* gene control in guinea pigs (Kantor *et al.*, 1963). The *Ir* genes controlling these responses were never found genetically separated in studies involving large numbers of outbred guinea pigs.

507

IMMUNE RESPONSE
GENES IN THE
REGULATION OF
MAMMALIAN
IMMUNITY

the ability to discriminate between different three-dimensional conformations of the same molecule. In this case, the difference has always been found to be in the same direction: B cells and antibodies can discriminate exquisitely between native and denatured or otherwise altered conformations of a protein, whereas T cells show virtually total cross-reactivity between these forms. This property of the two cell types was demonstrated twenty years ago by Gell and Benacerraf (1959), comparing delayed hypersensitivity and antibody production in guinea pigs. The heat-denatured and native forms of several proteins, such as bovine gamma globulin, ovalbumin, and bovine and human serum albumins, were equally effective in inducing delayed hypersensitivity regardless of which of the pair was the original immunogen, whereas antibodies showed little cross-reactivity between members of each pair. Similarly, Thompson, *et al.* (1972), found T cell cross-reactivity (in delayed hypersensitivity reactions and in production of migration inhibitory factor) between native lysozyme and the reduced and carboxymethylated form, whereas B cells distinguished readily between the two. Addressing the same question for helper T cells and B cells, Ishizaka *et al.* (1974, 1975) found that urea-denatured ragweed antigen E and its isolated α and β polypeptide chains, although not reacting with antibodies to the native protein, were quite capable of inducing helper T cells which could enhance the B cell response to native antigen E or to DNP coupled to antigen E.

A related line of studies showed that coupling haptens to protein carriers had similar effects, perhaps by modifying the charge and/or conformation of the carrier. Benacerraf and Gell (1959), in the companion study to that mentioned earlier, found that hapten–protein conjugates could elicit antibodies to the hapten or delayed hypersensitivity to the carrier, but not antibodies to the carrier. Subsequently, Parish (1971) showed that increasing degrees of acetoacetylation of *Salmonella* flagellin led to increases in ability to induce delayed hypersensitivity and reciprocal decreases in ability to elicit antibody in rats. Schirrmacher and Wigzell (1972, 1974), using an adoptive cell transfer approach, found that spleen and lymph node cells from mice immunized with methylated, acetylated, or succinylated albumins could transfer T cell help for an antihapten antibody response to hapten-conjugated albumin, but did not transfer to the recipient the ability to make anti-albumin antibodies. Presumably, although albumin-specific helper T cells were primed by the modified albumin, B cells specific for native albumin were not primed.

In at least three systems under known *Ir* gene control, similar distinctions between T and B cell sensitivity to conformation prevail. Antibodies to staphylococcal nuclease discriminate between native and random conformations of the protein with a difference in affinity of 10^3 to 10^4, whether the antibodies are raised against the native protein (Sachs *et al.*, 1972) or against random conformation fragments (Furie *et al.*, 1975). In contrast, T cells from mice immunized with native nuclease proliferate at least as well when stimulated *in vitro* with the fragments as when stimulated with native nuclease (Schwartz *et al.*, 1978; Berzofsky *et al.*, 1978). Similar findings have been made for sperm whale myoglobin (Berzofsky *et al.*, 1979) and cytochrome *c* (Corradin and Chiller, 1979; Solinger *et al.*, 1979), both also under *Ir* gene control. As for *Ir* genes, the same *Ir* genes appear to control the T cell proliferative response as control the antibody response to nuclease and myoglobin (Schwartz *et al.*, 1978; Berzofsky *et al.*, 1978, 1979). Also when mice are immunized with fragments of nuclease, the fragment (99–149) bearing the dominant determinant under *Ir* gene control is the only one against which the anti-random-conformation fragment antibody response follows the same *Ir* gene control as the anti-native-conformation antibody response to the whole nuclease (Berzofsky *et al.*, 1977a). Likewise, the proliferative response of T cells from mice immunized with this fragment appears to be under the same genetic control (R. H. Schwartz, J. A. Berzofsky, and D. H. Sachs, unpublished observations). All of these results point to the same conclusion—namely, that *Ir* gene recognition of determinants, like T cell recognition of the same determinants, is not sensitive to changes in conformation between the native protein and random conformation fragments, whereas the corresponding antibodies are highly sensitive.* The specificity of *Ir* genes again parallels the T cell, not the B cell, specificity *even when all three* (at least in the cases of nuclease and myoglobin) *appear to be directed at the same determinants* (or at least, regions) *of the molecule.*†

509

IMMUNE RESPONSE
GENES IN THE
REGULATION OF
MAMMALIAN
IMMUNITY

At least two explanations are consistent with these data, both of which are compatible with the notion that the antigen specificity of *Ir* genes is derived from that of T cell receptors, not B cell receptors. One is that T cell receptors, while sharing V_H combining sites with the corresponding antibodies, lack the light chain counterpart and therefore have a less rigid binding site than antibodies and a correspondingly greater ability to accommodate conformational changes in a determinant. The second is that T cell receptors are identical in their variable regions with antibodies, but are selected out to be specific for determinants which have more conformational flexibility than those commonly recognized by antibodies. One possible mechanism for such selection, which would also be compatible with T cell, B cell, and *Ir* gene specificity for the same determinant (in terms of sequence), might be suggested. The idea is that while B cells must be able to recognize the native protein, since they make antibodies exquisitely specific for this conformation, perhaps some types of T cells never see antigen in the native form. The classes of T cells for which the above data have been amassed, namely those mediating help for antibody synthesis, *in vitro* antigen-stimulated proliferation, and delayed hypersensitivity (all of which bear the same $Ly1^+2^-3^-$ surface marker phenotype), may never see antigen before it is degraded by macrophages and presented on the macrophage surface. Thus they may never see whole native protein antigen, either when it is used as immunogen or when it is used for subsequent challenge to test for help, proliferation, or delayed hypersensitivity, all of which require macrophages as well. The arguments against this hypothesis are that helper as well as suppressor T lymphocytes have been reported to be enriched by binding to plates coated with ostensibly native antigens (Maoz *et al.*, 1976), and that if helper T cells recognized only random nonnative conformations, then it would be hard to use the native antigen as a bridge between the helper T cell and the B cell specific for the native antigen. Of course, if the B and helper T cells are specific for the same nonrepeating sequence of amino acid residues on a protein antigen (Berzofsky *et al.*, 1979) then the antigen bridge concept is subject to problems of steric hindrance as well. Both of these problems might be avoided if there were a macrophage or soluble factor intermediary rather than direct contact between helper cell and B cell.

Another mechanism may allow both B and T cells to draw their receptor combining sites from the same V gene pool and yet have different specificity repertoires. This mech-

---

*One possible exceptional case is the T cell proliferative response to pigeon cytochrome *c* in B10.A mice (Solinger *et al.*, 1979). There is suggestive evidence that some of the T cells recognize a determinant involving residues 100 and 104 plus residue 3, which is brought into close proximity to the other two residues by the folding of the native protein. However, even in this case, the proliferative response to the isolated fragment (81–104) is greater than to the native protein, despite the loss of residue 3. Another possible exception may be the suppressor T cell recognition of the N-C portion of hen lysozyme described in Section 5.2 (Sercarz *et al.*, 1978; Adorini *et al.*, 1979*b*). In this case, the amino-terminal and carboxyl-terminal portions of the molecule are held together by a disulfide bridge, which would not be disrupted by many forms of denaturation or proteolysis. Therefore, this case may not be a true example of T cell recognition of a "conformational" determinant.

†A major caveat to this whole issue is that B cell receptor specificity is generally taken to be that of soluble antibody, whereas T cell receptor specificity is based on activation of whole cells. If B cell specificity were measured by cell activation, some of the differences between T and B cell specificities might evaporate.

anism was suggested in general terms by Warner (1974), who postulated that "in one of these cell series, an additional cell surface component plays a role in the binding and cell triggering by antigen." At the time, it was suggested that this additional component was the *Ir* gene product on the surface of T cells which acted as a second receptor for antigen and imposed restrictions on specificity, while B cells did not require this second receptor. More recent evidence suggesting that T cells must recognize antigen in association with MHC-encoded antigens on the antigen-presenting macrophage or sensitizing cell and on the B cell or target cell (Paul and Benacerraf, 1977) would fit into an altered version of this hypothesis. The additional receptor on the T cell would not be antigen specific, but specific for *MHC* (*H-2* or *I*-region)-encoded antigens. The latter, which in some senses might turn out to be *Ir* gene products, would be expressed on B cells and macrophages, and stimulators and targets, rather than the responding T cells. Antigen restrictions under this mechanism could occur both during T cell maturation in the thymus and on first exposure to antigen on a macrophage or sensitizing cell. This dual specificity of T cells may be the most important difference between B and T cell recognition of antigen, and may explain at least one class of *Ir* gene phenomena (Paul and Benacerraf, 1977; Zinkernagel, 1978; Benacerraf, 1978). As such, it will be the subject of three parts of this review, Sections 8–10. Whatever the mechanisms involved, it seems safe to conclude that the *antigen* specificity of *Ir* gene control depends on the specificity of the T cell receptor.

7 Antigen-Specific Helper and Suppressor Factors

Since their discovery in the mid 1970s, the antigen-specific cell-free factors have been one of the major candidates for an *Ir* gene product. Their special appeal derives from the fact that they are the only molecules or molecular complexes which are known to bear Ia antigenic determinants and yet bind antigen. Factors derived from T cells and from macrophages have both been described.

7.1 Antigen-Specific T Cell Helper and Suppressor Factors

A number of antigen-specific soluble "factors" secreted by or extracted from T cells have been described which substitute either for live helper cells (Taussig, 1974; Munro *et al.*, 1974; Mozes, 1978; Howie and Feldmann, 1977; Feldmann *et al.*, 1979; McDougal and Gordon, 1977; Shiozawa *et al.*, 1977) or for live suppressor cells (Takemori and Tada, 1975; Tada and Okumura, 1979; Kapp *et al.*, 1976; Thèze *et al.*, 1977*a*; Waltenbaugh *et al.*, 1977*a,b*, 1979; Kontiainen and Feldman, 1977, 1978; Feldmann *et al.*, 1979; Moorhead, 1977*a,b*; Zembala and Asherson, 1974), in the *in vitro* or *in vivo* enhancement or suppression of both B cell and T cell responses. A recent comprehensive review is available (Tada and Okumura, 1979). What is striking about all of these is their remarkable similarity, although each individual factor has its own unique features as well. We have already discussed the method of production, the genetics and the cellular aspects of some helper and suppressor factors in Sections 3.2 and 5, respectively. One aspect worth reemphasizing is the genetic restriction between source of factor and its target. Most of the helper factors will act across MHC differences or even species differences (Taussig *et al.*, 1975; Luzzati *et al.*, 1976; Feldmann *et al.*, 1979). A notable exception is the factor of Shiozawa *et al.*

(1977), which is prepared as a cell extract rather than a product released into the supernatant fluid during cell culture, and which acts only on syngeneic cells. This difference between extracted and secreted factors applies to some extent for suppressor factors as well, in that some prepared by sonication show genetic restriction in their action, either a strict requirement for identity at the *I-J* subregion between producer and acceptor of the factor (Taniguchi *et al.*, 1976a; Tada *et al.*, 1976) or a requirement for nonresponder *Ir* status (Kapp, 1978), whereas some secreted into the supernatant show no genetic restriction (Kontiainen *et al*, 1979). However, there are other differences in the way these factors are generated (*in vivo* vs. *in vitro* immunization) and in the cells on which they act (see Section 5), and another secreted suppressor factor does show genetic restriction, albeit to the *H-2K/D* regions, not the *I* region (Moorhead, 1977a,b). Therefore, while it is tempting to speculate that the factors extracted from whole cells have a component which produces the genetic restriction and which is not present on the factors which are shed from the cell membrane, the correlation is not good enough to draw any conclusions.

A second generalization worth reemphasizing is the complexity of the regulatory circuits involving many cell types, of which these factors are only a small part. As described in Section 5, some of the suppressor factors have been shown to act by inducing a secondary suppressor effector cell which executes the actual suppression (Tada and Okumura, 1979; Waltenbaugh *et al.*, 1977b; Germain *et al.*, 1978b; Germain and Benacerraf, 1978a). This places them as part of a complex multicellular suppressor regulatory mechanism which involves not only these two types of T cells and amplifier T cells (Tada, 1977; Feldmann *et al.*, 1977) but also macrophages and in addition a cell that resembles in cell surface markers the helper T cell, which can induce a "feedback" type of suppression (Eardley *et al.*, 1978, 1979).

The requirement for macrophages for production and expression differs for suppressor versus helper cells and factors. While suppression is not so macrophage-dependent for its induction as is help (Feldmann and Kontiainen, 1976; Pierres and Germain, 1978; Feldmann *et al.*, 1979; McDougal and Gordon, 1977a), both types of factors, once produced, appear to be transmitted via macrophages, at least in the cases in which this was studied (Feldmann *et al.*, 1973; McDougal and Gordon, 1977b; Howie and Feldmann, 1978; Ptak *et al.*, 1978).

Since the other relevant genetic and cellular aspects of helper and suppressor T cell factors in *Ir* gene-controlled systems have been discussed in Sections 3.2 and 5.1, the present Section will concentrate on the biochemical properties of the factors and their possible relationship to products of *Ir* genes. All of the antigen-specific T cell factors, whether helper or suppressor, are extremely similar in their biochemical properties. In general, their molecular weights are in the range of 30,000–50,000; they are protein in nature; they have specific binding sites for antigen (that is, they can be removed and eluted from antigen affinity columns); and they are bound by anti-Ia antisera generally specific for *I-A* encoded determinants in the case of helper factors and *I-J* encoded determinants in the case suppressor factors) but not by antisera raised against any immunoglobulin heavy or light chain constant region determinant (Taussig and Munro, 1974; Munro *et al.*, 1974; Feldmann *et al.*, 1979; McDougal and Gordon, 1977b; Takemori and Tada, 1975; Taniguchi *et al.*, 1976a,b; Thèze *et al.*, 1977a; Tada and Okumura, 1979). The Ia antigenic determinants appear to be the same as those found on B cells and macrophages, since the same monoclonal hybridoma antibody reacts with all three sources of Ia (M. Feldmann, personal communication). The binding site for antigen appears to be on the same molecule as, or tightly complexed to, the determinant(s) recognized by anti-Ia alloantisera, as determined by suc-

511

IMMUNE RESPONSE
GENES IN THE
REGULATION OF
MAMMALIAN
IMMUNITY

cessive affinity columns or by mixing the eluates from each type of affinity column (Thèze *et al.*, 1977*b*; Bach *et al.*, 1979). This finding is crucial to the interpretation and significance of these factors because they are therefore the only direct evidence for an association between a receptor or binding site for antigen and an Ia antigen—that is, a molecule coded in the same region as *Ir* genes. It was therefore essential to determine whether this binding site on the factor was the same as the antigen receptor on the corresponding T cell. It will be recalled from Section 6.1. that T cell receptors appear to share V_H idiotypes with the corresponding immunoglobulins, but not C_H or C_L determinants, and also not Ia antigenic determinants. Several recent studies have now demonstrated that at least some of the antigen-specific helper and suppressor factors also share the V_H idiotype corresponding to antibodies of the same specificity and that the idiotypic and Ia determinants are not separable by successive affinity columns—that is, they are on a single molecule or complex (Mozes, 1978; Germain *et al.*, 1979; Bach *et al.*, 1979; M. Feldmann, personal communication). This finding is consistent with the hypothesis that the antigen-specific binding site of the T cell factors is, in fact, the T cell receptor for antigen. One contradictory finding is that the fine specificity of at least the (T,G)-A–L helper factor, measured both by its binding specificity and by its helper-functional specificity using related antigens (Phe,G)-A–L and (H,G)-A–L, was found to be more like that of the B cell receptor (immunoglobulin) than like that of the T cell receptor (Isac and Mozes, 1977; see Section 6.2). However, the corresponding helper factor produced *in vitro* did have fine specificity more like that of the T cell (Howie and Feldmann, 1977). Despite this possibly lower order of specificity of the helper T cell factor compared to the T cell receptor on the cell membrane, it is still tempting to speculate that these extracted or secreted or shed factors are, at least in part, T cell receptors. In that case, even if the T cell receptors released onto hapten-coated nylon discs did not bear Ia determinants (Section 6.1.4), one must wonder whether the intact T cell receptor contains or is tightly associated with Ia determinants. One possibility is a single-chain structure in which the variable region is identical to immunoglobulin V_H but the constant region is Ia rather than C_H (Fig. 5A). Although it could be more difficult to

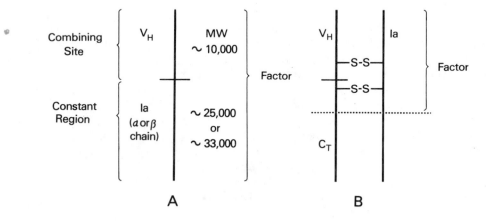

Figure 5. Some possible models of T cell antigen-specific receptors involving V_H (immunoglobulin heavy chain variable region) combining sites and Ia antigenic determinants. These models are based primarily on the soluble T cell factors discussed in this section (see text for references). Model A would account for the whole molecular weight of the receptor as identical to the factors. Model B would require cleavage between V_H and C_T (T cell constant region) in the process of factor production to account for the molecular weight. The disulfide bonds are suggested by analogy with immunoglobulin, but are not required.

513

IMMUNE RESPONSE
GENES IN THE
REGULATION OF
MAMMALIAN
IMMUNITY

join two genes on different chromosomes (V_H and Ia) to make a single polypeptide chain, instead of two genes on the same chronomosome (V_H and C_H) it might not be impossible. This possibility would be consistent with the molecular weights of single Ia chains (25,000 and 33,000), V_H (about 10,000), and factors (35,000–50,000). A second possibility would be a two-chain structure of V_H $C_?$ and Ia, perhaps joined by disulfide bonds in analogy with heavy and light chains of immunoglobulin, or even associated without covalent bonds (Fig. 5B). The latter possibility would be more consistent with the loss of the Ia-bearing moiety in the receptor isolated by Krawinkel and Rajewsky (Cramer *et al.*, 1978). It would also be consistent with the evidence for one T cell suppressor factor with a molecular weight of approximately 200,000, consisting of two noncovalently coupled chains, one with a molecular weight of approximately 85,000 which binds to antigen specifically, and one with a molecular weight of approximately 25,000 which is necessary for the factor to be removed by anti-*H-2* antisera (Taussig *et al.*, 1979; Taussig and Holliman, 1979). The latter molecular weight is consistent with the β chain of Ia. Either possibility might nicely tie together T cell receptors, T cell factors, and *Ir* genes. However, other evidence we shall discuss in Sections 9 and 10 suggests that the Ia antigens relevant to *Ir* gene function are those on macrophages not T cells, and that it is the Ia recognized by the T cell, not the Ia borne on its surface, which determines *Ir* gene phenotype.

In view of the latter consideration, and the Ia-bearing antigen-carrying factor released from macrophages to be discussed in Section 7.2., it becomes of interest to consider whether the Ia antigenic determinants on these T cell factors are actually derived from the T cells or whether they come originally from the macrophages used to present the antigen. [A related hypothesis has also been suggested by McDougal and Gordon (1979*b*).] The possibility is made less likely for the suppressor factors by the facts that their production is less macrophage dependent, and more importantly that several suppressor factors bearing Ia determinants mapping in the *I-J* subregion have recently been obtained from single clones of hybridoma tumor lines made by hybridizing suppressor T cells with T lymphoma cells (Taniguchi and Miller, 1978; Kontiainen *et al.*, 1978; Taniguchi *et al.*, 1979; Taussig *et al.*, 1979; Taussig and Holliman, 1979). The latter evidence is particularly strong for cases in which the tumor was grown as a single clone *in vitro* rather than *in vivo*. For the helper factors, I know of no proof that the Ia determinants are not derived from macrophages rather than the T cells themselves. Because of the requirements for syngeneity of macrophages in the presentation of antigen to T cells (see Section 8), even in studies using F_1 cells one cannot prove genetically whether the Ia antigen was derived from the T cell or the macrophage, since they share the same Ia. The most definitive way to test this hypothesis for helper factors is to use tolerant T cells from a chimeric animal, which are able to recognize antigen on a completely allogeneic macrophage without the complications of allogeneic effects. For instance, using purified H-2^k T cells from an H-$2^k \longleftrightarrow H$-$2^b$ chimera, and antigen presenting macrophages from an H-2^b mouse, one could determine in an unambiguous experiment whether the helper factor obtained bore Ia^k determinants from the T cells or Ia^b determinants from the macrophages. Such an experiment needs to be done, as it would either prove beyond doubt that the Ia came from the T cell and thus leave viable the hypotheses described above for association of Ia and antigen-specific receptor on the T cell itself, or it would open up a whole new interpretation for the derivation of these so-called T cell factors. If the Ia on the factor derived from the antigen-presenting macrophage, it might itself be bound to an Ia-specific binding site associated with the antigen-specific T cell receptor, as postulated in both the one- and two-receptor models for T cell recognition of antigen in association with Ia (see Sections 8-10). It would also be consistent with the macrophage-derived factor to be described in Section 7.2. Therefore, the appeal of this hypothesis comes from its ability to tie together diverse findings from many experimental systems into a single unified mechanism. It might also explain the failure of some *Ir* low-responder strains to make an antigen-specific factor (Munro and Taussig, 1975; Munro *et al.*, 1978; Howie and Feldmann, 1977) on the same basis as one might explain the failure of other strains to respond to a particular antigen determinant on a particular macrophage—namely, that there were no clones of T cells with the appropriate combination of binding sites for that Ia antigen and that particular exogenous antigen (see Section 9).

Finally, we must consider whether these Ia-bearing antigen-specific T cell factors are the physical embodiment of the long-sought, always elusive, *Ir* gene products. On genetic grounds alone, the suppressor factors probably are not. The antigen-specific suppressor factors in several systems have been shown to bear Ia determinants which map in the *I-J* subregion (Tada *et al.*, 1976; Thèze *et al.*, 1977*b;* Greene *et al.*, 1977; Feldmann *et al.*, 1979), the same region as the Ia determinants on the suppressor cells themselves. In contrast, the *Ir* functional behavior, in the two cases of GAT and GT where the production of suppressor cells appears to be under *Ir* gene control, maps in the *I-A,* and *I-A* and *I-C,* regions, respectively. Therefore, the *I-J* encoded product on the factor cannot be the *Ir* gene product. In the case of helper factors, the *Ir* gene control and the Ia determinants on the factors map in the *I-A* subregion in all cases studied (Taussig *et al.*, 1975; McDougal *et al.*, 1977). Thus, whether the Ia on the factor derives from the T cell or the macrophage, it could still be an *Ir* gene product, on genetic grounds. The ability of T cells to change their *Ir* phenotype when maturing in a chimeric environment (see Section 10) argues against the T cell factor as an *Ir* gene product unless the Ia moiety derives from the macrophage, as suggested above. The fine specificity of one of the helper factors for (T,G)-A–L, which resembles more closely the specificity of B cells than that of T cells or of the *Ir* gene control itself (Isac and Mozes, 1977) also argues strongly against the identity of helper factor as *Ir* gene product. Therefore, for the present, all we can say is that while the production of factors and the susceptibility to the action of factors may both be under *Ir* gene control (see Section 3.2), the identity of the factor itself with the elusive *Ir* gene product is questionable at best.

7.2 Antigen-Specific Macrophage Factors

In addition to factors produced by T cells, both antigen-specific and nonspecific factors derived from macrophages have been described. The most relevant to our discussion of *Ir* genes is the antigen-specific "genetically related factor" described by Erb and Feldmann (1975*c*). These authors found that although the induction of keyhole limpet hemocyanin (KLH)-specific helper T cells *in vitro* required syngeneic macrophages (Erb and Feldmann, 1975*b*), the macrophages could be separated from the T cells by a Nucleopore membrane (Erb and Feldmann, 1975*c*). The action across the Nucleopore membrane required syngeneity between the T cells and macrophages. To analyze this action at a distance, these authors investigated the effects of cell-free supernatant fluids obtained from peritoneal exudate macrophages, pretreated with anti-T and anti-B cell antisera and complement and also with 2000R irradiation to eliminate effects of lymphocytes, and cultured for four days with antigens, KLH or (T,G)-A–L. The supernatant fluids so obtained from syngeneic but not allogeneic macrophages substituted for live macrophages in the induction of helper T cells and were specific for the antigen used in culturing the macrophages. Only if the antigen was particulate (such as KLH-Sepharose) did supernatants from allogeneic macrophages work. For nonparticulate antigens, the antigen present in the macrophage supernatant factor was sufficient, even at a final dilution of 1 : 10 or 1 : 100 of supernatant fluid, to induce helper cells without additional antigen. Mapping studies with recombinant strains revealed that the requirement for genetic identity between macrophages producing the factor and helper T cell induced was limited to the *I-A* subregion of *H-2*—that is, that identity at the *I-A* subregion was both necessary and sufficient (Erb and Feldmann, 1975*c*). The mechanism of action of the factor was found to be activation of shortlived "T_1" cells (sensitive to adult thymectomy) which were involved in the induction of long-lived "T_2" cells (sensitive to antilymphocyte serum *in vivo* but not to thymectomy), which were the actual precursors of the helper T cells (Erb *et al.*, 1977).

This "genetically related factor" was found to be a heat-labile protein with an approximate molecular weight of 55,000, bearing Ia determinants (mapping in the *I-A* subregion) but not *H-2K* or *H-2D* determinants, as determined by adsorption to and elution from affinity columns of various anti-Ia and anti-*H-2* alloantisera (Erb *et al.*, 1976). The active principle in the supernatant fluids also bound to affinity columns of anti-antigen, but not of antigen itself. Thus, the factor bore a piece of the antigen (not the whole antigen, which was larger than the factor), but did not have a (free) binding site for native antigen. In addition, it was not bound by antibodies specific for mouse Ig, kappa light chain, β_2-microglobulin, or the C3 component of complement. Thus, the serological and physicochemical properties of the factor closely resembled those of the T cell factors described in Section 7.1., except for the presence of an antigen fragment and absence of a demonstrable binding site for antigen. Unless the antigen affinity were enormous (virtually irreversible interaction), the antigen on the affinity column would be expected to compete effectively with the antigen already bound to the factor, for binding to an antigen-specific binding site. The failure to bind to an antigen column suggested that the antigen was bound essentially irreversibly, and not to an antigen-specific receptor resembling that of T cells or antibody. In addition, the ability to adsorb to *and elute* the activity from either an anti-antigen or anti-Ia column suggested that the Ia determinants and antigen fragment were tightly associated, and not on independent molecules in the macrophage supernatant fluids. To test this important point further, mixing experiments were performed using KLH-factor and the effluent from passing (T,G)-A–L factor over an anti-Ia column. The resulting mixture induced KLH-specific but not (T,G)-A–L-specific helper T cells, in contrast to the expectation if the (T,G)-A–L and KLH moieties were separate from the Ia moieties. The same result was obtained when the roles of the two antigens were reversed. Also, the effluent from an anti-(T,G)-A–L column of (T,G)-A–L factor did not provide the Ia moiety necessary to allow an allogeneic (T,G)-A–L factor to work. All these experiments indicated that the antigen moiety (presumably the part conferring antigen specificity) and the Ia moiety (presumably the part conferring genetic restriction of action) were not free to dissociate and reassociate in random fashion. This observation is extremely important because it remains *the only demonstration of a physical association between antigen and Ia antigen on a macrophage or macrophage product*. The theories which we shall discuss in Sections 8–10 all hinge on the associative *recognition* of antigen and Ia on an antigen-presenting cell (macrophage), but the distinctions between them and the viability of some theories of *Ir* gene control rest on whether or not there is a direct physical interaction between the antigen and the Ia molecule on the macrophage as opposed to the need for simultaneous recognition of both independent molecules on the surface of the macrophage by the responding T cell.

515

IMMUNE RESPONSE
GENES IN THE
REGULATION OF
MAMMALIAN
IMMUNITY

This physical association between antigen and Ia in the genetically restricted and antigen-specific macrophage factor, in addition to the evidence for a role of *Ir* genes in macrophage presentation of antigen, led to the hypothesis that this factor derived from macrophages was itself the long-sought *Ir* gene product (Erb *et al.*, 1979). To test this hypothesis, (T,G)-A–L-specific helper T cells were induced in T cell populations derived from F_1 hybrids between high responder *H-2*^{b} mice and low responder *H-2*^{k} or *H-2*^{s} mice. These helper T cells were tested using syngeneic F_1 B cells, either unprimed or DNP-primed using a different carrier, for the *in vitro* anti-DNP antibody respnse to DNP-(T,G)-A–L. In the case of the (*H-2*^{b} × *H-2*^{k}) F_1 T cells, comparable helper T cells were induced using F_1, *H-2*^{b} or *H-2*^{k} macrophages, or using the supernatant factor derived from these three types of macrophages. This result confirmed the observation (Taussig *et al.*, 1974; Howie and Feldmann, 1977) that the defect in *H-2*^{k} low responders was not in the induction of helper T cells or helper T cell factors. However, when (*H-2*^{b} × *H-2*^{s}) F_1 T cells were used, only F_1 or *H-2*^{b} macrophages or their factors could induce helper T cells. The *H-2*^{s} (B10.S) macrophages and *H-2*^{s} macrophage supernatant fluids failed to induce (T,G)-A–L-specific help, but did not induce suppressor cells which would prevent the action of helper T cells stimulated with *H-2*^{b} macrophages or their factors, as judged by mixing the T cells from the two types of induction cultures. Also, the *H-2*^{s} defect was specific for (T,G)-A–L, in that B10.S macrophages or factor worked well for KLH. This result confirms the finding (Munro *et al.*, 1978; Howie and Feldmann, 1977), that the *H-2*^{s} mice have an *Ir* gene defect at the level of production of helper T cell factor, but extends this result to place the "defect" in the macrophage or macrophage factor's ability to induce helper T cells. These experiments strongly support the notion that this "genetically related factor" from macrophages, containing both an antigen fragment and Ia tightly bound together (or more strictly the endogenously synthesized Ia portion of this complex) is a soluble form of the elusive *Ir* gene product. Of course, it is still not clear whether the *Ir* gene defect is in the production of the *H-2*^{s} genetically related (T,G)-A–L factor, which would place the defect solely in the macrophage, or whether the defect is in the failure of *H-2*^{s} or F_1 T cells to recognize or respond to this factor. This latter interpretation, placing the defect in the lack of appropriate clones of T cells in the repertoire, is still a viable alternative. These ideas will be discussed more extensively in Sections 9, 10, and 12.

In summary, this "genetically related factor" from macrophages, which induces helper T cells, which manifests both antigen specificity and genetic restriction requiring *I-A* subregion identity in the T cells on which it acts, and which bears both Ia determinants and a fragment of antigen but not Ig determinants, may very well be a soluble *Ir* gene product. The weight of evidence is that the antigen moiety is bound to the Ia nonspecifically rather than to an antigen-specific receptor, but this question warrants more detailed investigation. Also, it would be extremely interesting to determine the nature of the presumably small fragment of the antigen. This fragment must be small enough to be accommodated in a complex of 55,000 molecular weight, which also includes Ia determinants (which are usually found on chains with molecular weights of 25,000 and 33,000) even though KLH itself has a molecular weight of several million. Nevertheless, these antigen fragments are bound by antibody to native KLH, even though antibody to native antigen does not block macrophage presentation of antigen (Ben-Sasson *et al.*, 1977; Ellner *et al.*, 1977; Thomas *et al.*, 1978; see Section 8.2.2). Also, using a better-defined antigen (such as myoglobin), it would be of great interest to explore whether the choice of antigen fragment bound is random or selective for key antigenic determinants under *Ir* gene control (Berzofsky *et al.*, 1979). Finally, it is crucial to determine whether the failure of supernatant fluids from certain

517

IMMUNE RESPONSE
GENES IN THE
REGULATION OF
MAMMALIAN
IMMUNITY

low-responder macrophages to induce helper T cells for an *Ir* gene controlled response is due to the failure of these macrophages to synthesize the appropriate factor or secrete it into the medium, or whether the factor's structure and production are perfectly normal, but no T cells can respond to it.

8 Genetic Restrictions on Immunological Cell–Cell Interactions

In the early 1970s, a number of studies detected requirements for some degree of histocompatibility between different classes of cells for successful interactions leading to an immune response. As these so-called genetic restriction phenomena were discovered in more and more types of cellular interactions, it became apparent that they reflected some fundamental mechanisms in the immune response. Moreover, since these restrictions generally mapped to the MHC, and in many cases the *I* region, the question arose as to what relationship these restrictions, seen for antigens not under known *Ir* gene control, bore to *Ir* gene phenomena. In the last few years it has become apparent that these two phenomena, genetic restrictions on cellular interactions and *Ir* gene control, are so intimately intertwined that it is hard to believe that they are not mechanistically related. In fact, a rather successful set of internally consistent theories to explain *Ir* genes is based almost completely on mechanisms related to these genetic restrictions. In order to elaborate these theories, we must first review the several types of genetic restriction phenomena observed. Since these have all been extensively reviewed elsewhere (see references below), the present summary will be limited primarily to those facts which are most directly relevant to *Ir* genes. The parallel mechanisms suggested for *Ir* genes will then be explored in Sections 9 and 10.

Three major types of interactions have been described which involve genetic restriction: interactions between T cells and B cells, interactions between macrophages and T (and possible also B) lymphocytes, and interactions between cytotoxic T lymphocytes and the cells serving as stimulators and targets of the cytotoxicity. Although this last type of genetic restriction was the most recent to be discovered, the concepts that have arisen from it have greatly contributed to the interpretation of the others. Therefore, it will be discussed first.

8.1 Restrictions on Interactions in T Cell-Mediated Cytotoxicity

Several groups demonstrated that T cell-mediated killing, which was specific for antigens other than MHC antigens (that is, other than classic allogeneic cell-mediated lympholysis), was nevertheless restricted by MHC antigens. This was true whether the specificity of killing was for virus (Zinkernagel and Doherty, 1974*a,b;* Koszinowski and Ertl, 1975), for chemical hapten (Shearer, 1974), for the male antigen H-Y (Gordon *et al.,* 1975), or even for minor (non-MHC) histocompatibility antigens (Bevan, 1975*a,b*). If the initial stimulation was *in vivo* (such as virus-infection of the whole organism), then the stimulator cells were syngeneic. If the stimulation was *in vitro,* then stimulator cells had to be syngeneic (in the case of hapten or H-Y) or at least *H-2*-compatible (in the case of minor histocompatibility antigens) in order to avoid reactions directed at the MHC which might overwhelm or cloud the interpretations of the reactions against the non-MHC antigens. The genetic restriction observed was that for any of the above modes of stimulation, the target cells also had to be histocompatible with the cells used for stimulation (Shearer *et al.*, 1977). Moreover, the restriction, unlike that which we shall discuss for T cell–B cell and T cell–macrophage interactions, was mapped to the *H-2K* and *H-2D* regions rather than the *I* region (Shearer *et al.*, 1975; Blanden *et al.*, 1975; Gordon *et al.*, 1975; Bevan, 1975*b*).

8.1.1 Altered Self versus Dual Recognition

Three types of hypotheses were proposed to explain these genetic restrictions. The first, termed the *intimacy hypothesis,* suggested that the histocompatibility requirement had nothing to do with specificity, but rather was necessary for the cells to come together to interact. This possibility was quickly ruled out by experiments using F_1 mice as sources of responder cells (Zinkernagel and Doherty, 1974*b*, 1975; Shearer *et al.*, 1975; Forman, 1975; Bevan, 1975*b*; Gordon *et al.*, 1976; Blank *et al.*, 1976). The rationale of these experiments was that (A × B) F_1 mice stimulated with antigen on parent A stimulator cells should be able to recognize the same antigen on parent B target cells if the specificity were solely for the antigen, and sharing of MHC antigens between responder and target were necessary only to allow cellular interaction. The finding for all four types of antigens, however, was that (A × B) F_1 responder cells stimulated with antigen on parent A stimulator cells could kill only parent A antigen targets, not parent B targets bearing the same antigen. Thus the intimacy hypothesis was excluded, and *pari passu* it was shown that the histocompatibility restriction was for MHC sharing between stimulator and target, not necessarily responder and target. In addition, this type of experiment suggested that the MHC antigens were part of the specificity recognized, not just the virus, hapten, or other cell surface antigen. Moreover, it suggested that there may be two populations of responder T cells in the F_1, one specific for antigen in association with parent A and one specific for the same antigen in association with parent B. Only if there were allelic exclusion of MHC antigens on the surface of these two populations of F_1 responder T cells could the intimacy hypothesis be resurrected, and all available evidence is that no allelic exclusion occurs for MHC antigens (Cullen *et al.*, 1976; Klein, 1975).

The alternative hypothesis, that the target antigen involved both self-MHC and the nominal antigen such as virus, hapten, or minor histocompatibility antigen, has been split into at least two forms. These have been termed *"altered self"* and *"dual recognition,"* and their distinctions and the experiments to attempt to distinguish them have been extensively reviewed (Shearer *et al.*, 1976, 1977; Doherty *et al.*, 1976*a,b*; Simpson and Gordon, 1977). One extreme is the strict altered-self concept that what is recognized is a neoantigenic determinant formed by the interaction of *H-2* and virus, hapten, or other antigen, so that neither self-*H-2* nor other antigen in its native form is the target (Figure 6). The new determinant is either *H-2* modified by the action of virus, hapten, etc., or virus, hapten, or minor histocompatibility antigen altered by the catalytic action of *H-2,* or some interaction product of the two. The opposite extreme is the purest form of dual recognition, that the responder T cell has two independent receptors, one for self-*H-2* and one for the other antigen, and that the only association of the two receptors to one another, and of the two antigens to one another, need be that they be on the same responder cell or same target cell, respectively (Fig. 6). The difference between the intimacy hypothesis and the dual recognition hypothesis is that the latter requires separate clones of T cells (with dual specificity) for each *combination* of *H-2* type and other antigen, and thus is compatible with the F_1 experiments. In between these extremes is a whole spectrum of hypotheses in which two

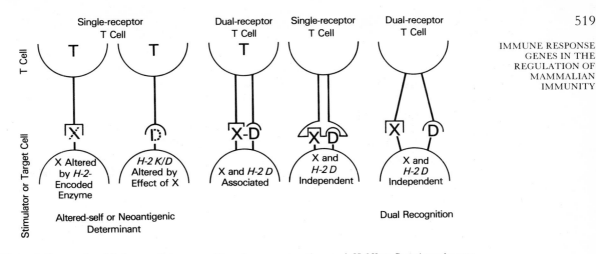

519

IMMUNE RESPONSE
GENES IN THE
REGULATION OF
MAMMALIAN
IMMUNITY

Figure 6. Range of models for associative recognition of exogenous antigen and *H-2K* or *D* antigens by cytotoxic T cells. At left is the neoantigenic determinant form of the altered-self hypothesis. At right is the most extreme form of the dual recognition hypothesis. In between are two of several possible intermediate versions. X, exogenous antigen; x, altered form of X; D, *H-2D* (or *K*)-encoded antigen; $\mathbb{D}$, altered form of D.

independent receptors, or two separate receptors in an interacting complex, or a single receptor with a dual combining site, recognize antigen plus self-H-2. The antigen plus *H-2* antigen may be viewed as either diffusing independently in the cell membrane or somehow physically associated, and as either still in their native forms or slightly altered (Fig. 6). Needless to say, it is much more difficult to distinguish experimentally among the overlapping possibilities in this spectrum of hypotheses than it was to rule out the intimacy hypothesis. Indeed, despite many experiments in the last several years and strong advocates for each hypothesis, I think it is safe to conclude that no functional experiment (as opposed to chemical isolation and characterization of the molecules involved) has yet been devised which can rigorously distinguish among these alternatives. For instance, although the responder T cells show exquisite specificity for both *H-2* and other antigen, be it chemical (Rehn *et al.*, 1976*a*) or virus (Ennis *et al.*, 1977), cold targets (that is, unlabeled target cells) bearing the same *H-2* without the other antigen, or the antigen and a different *H-2*, fail to inhibit the killing competitively, even when mixed. To compete, cold targets must bear both the antigen and the appropriate *H-2* antigens on the same cell (Zinkernagel and Doherty, 1975; Shearer *et al.*, 1975; Bevan, 1975*b*). However, these results do not disprove dual recognition since the affinity of either receptor alone might be too low to result in effective blocking or killing. Conversely, although antibodies to *H-2* or TNP or virus can block the killing of appropriate targets (Koszinowski and Ertl, 1975; Koszinowski and Thomssen, 1975; Schmitt-Verhulst *et al.*, 1976; Burakoff *et al.*, 1976; Forman *et al.*, 1977*a*), the possibility that antibodies to a determinant adjacent but not identical to that recognized by cytotoxic T cells could sterically block the response leaves at least most forms of the altered self hypothesis still viable. Similarly, although there is evidence that *H-2* antigens are among the cell-surface proteins to which TNP groups are bound when cells are treated with trinitrobenzene sulfonate, and that a positive correlation exists between the extent of TNP coupling to *H-2* and the ability of these cells to serve as stimulators or targets (Forman *et al.*, 1977*a,b*), such evidence does not prove that TNP on other cell surface proteins works any less effectively. In fact, TNP-modified soluble proteins such as BSA or bovine gamma globulin, after incubation with unmodified stimulator or target cells, can bind to the cell surface and convert the cells into stimulators or targets which function exactly like trinitrobenzene sulfonate-modified cells (Schmitt-Verhulst *et al.,* 1978). While is is conceivable that TNP was attached to *H-2* either via chemical exchange or via macrophage digestion and presentation of the TNP-BSA, experiments in which macrophages were depleted or in which the TNP on the cell surface was shown to be primarily still attached to BSA make these explanations less likely. On the other hand, the opposite result was found when TNP groups were put on the cell surface attached to a dextran-lipid compound (Henkart *et al.,* 1977). In this case, the cells did not function as do trinitrobenzene sulfonate-modified stimulators or targets. Thus, depending on what molecule carries the TNP, one can obtain results consistent with either altered self or dual recognition hypotheses. To reconcile the findings, one can suggest either that TNP on protein becomes associated with *H-2,* whereas TNP on dextran does not (compatible with altered self) or that neither associates physically with *H-2* but that the specificity is for TNP-lysine, not just TNP alone, so that the TNP-dextran is not recognized by the responder T cells (compatible with dual recognition). Consistent with the latter explanation is the finding that TNP attached to cells via an alanyl-gylcyl-glycyl spacer arm rather than attached directly to lysine does not produce crossreactive targets (Rehn *et al.,* 1976*b*). While these results can be most easily reconciled overall by a hypothesis which does not require modified *H-2* neoantigenic determinants, an intriguing observation compatible with altered-self is the finding that Friend leukemia virus particles preferentially bud off the cell membrane carrying those *H-2* antigens (such as *H-2D*^{b} rather than *H-2K*^{b}) that determine the genetic restriction of Friend virus-specific T cell-mediated cytotoxicity (Bubbers and Lilly, 1977; Bubbers *et al.,* 1976, 1978; Blank *et al.,* 1977).

8.1.2 *The Nature of Self: Experiments in Chimeras*

A series of experiments with bone marrow chimeras and thymus transplants, while having only indirect implications on the number and type of receptors involved, have been seminal in the subsequent development of the field (Bevan, 1977; Zinkernagel *et al.,* 1978*a,b*; Fink and Bevan, 1978). The use of long-term chimeras allowed circumvention of the problem mentioned earlier that responders and stimulators had to be MHC compatible to avoid anti-MHC responses that would mask or alter the responses specific for virus, hapten, or weak histocompatibility antigens, since T cells from the chimera were tolerant to other MHC types present in the chimera. Therefore, questions could be addressed regarding what constitutes self and when the recognition of self-*H-2* is learned—prior to or in the process of primary exposure to antigen. When lethally irradiated mice are reconstituted with fetal liver or bone marrow cells depleted of mature T cells, after several months all lymphocytes and macrophage as well can be shown to be of donor origin, but tolerant to the host. The initial experiments employed responder cells whose primary exposure to vaccinia virus was in the chimeric host. When the chimeras were made by injecting T cell-depleted bone marrow from parental A strain into (A × B) F$_1$ lethally irradiated recipients, designated A → (A × B), and these were stimulated several months later with either virus or TNP coupled to strain B cells, the resulting cytotoxic T cells could kill infected or TNP-modified strain B targets (Zinkernagel, 1976; von Boehmer and Haas, 1976; Pfizenmaier *et al.,* 1976). The result was originally interpreted to mean that strain A T cells made tolerant in the F$_1$ host to strain B alloantigens, could now be stimulated by strain B plus virus or TNP, as if lack of tolerance were the cause of the restriction.

521

IMMUNE RESPONSE
GENES IN THE
REGULATION OF
MAMMALIAN
IMMUNITY

However, when the chimeras were made by injecting T cell-depleted bone marrow of (A × B) F$_1$ mice into lethally irradiated parental A recipients, designated (A × B) → A, and after several months infected in situ with vaccinia (or primed against minor histocompatibility antigens), the (A × B) chimeric T cells, shown themselves to be virtually all (A × B) F$_1$ in origin, would lyse only vaccinia-infected (or specific minor histocompatibility antigen-bearing) strain A target cells, not B target cells (Bevan, 1977; Zinkernagel et al., 1978a–c). Nevertheless, these T cells were still tolerant to parent B MHC as well as parent A MHC. Donor (A × B) F$_1$ T cells, immunized in the F$_1$ mouse from which they derived, could lyse infected targets from either parent. Two explanations were considered: (1) the narrowing of the genetic restriction was caused by some occurrence during the maturation of the F$_1$ stem cells in the A host, or (2) it was caused merely by the exposure to virus in an A "environment,"—that is, in a situation in which most infected cells that could stimulate were cells not arising from bone marrow and thus of recipient A origin. The latter explanation would be compatible with the F$_1$ experiments described earlier in which F$_1$ responder cells stimulated with virus-infected parent A cells could lyse only infected parent A targets. To distinguish these alternatives, chimeras were made by reconstituting irradiated A recipients with untreated adult (A × B) F$_1$ spleen cells (Zinkernagel et al., 1978a). The only difference between these and the first set of chimeras was that these contained mature F$_1$ T cells that had matured in the F$_1$ donor, whereas the former chimeras had only T cells which had differentiated from stem cells in the parent A environment. In this case, the chimeric F$_1$ T cells, first exposed to vaccinia in the chimera, killed both parent A and parent B infected targets. Therefore, it was the environment in which the stem cells matured into T cells that determined the genetic restriction. [Exceptions to this phenomenon have been noted by Forman et al. (1977c), Wilson et al. (1977), Blanden and Andrew (1979), and Doherty and Bennink (1979).]

To determine what part of the chimeric host environment was responsible for imposing the genetic restriction, thymic transplants were employed (Zinkernagel et al., 1978a,c; Fink and Bevan, 1978). Lethally irradiated and thymectomized (A × B) F$_1$ recipients were reconstituted with syngeneic (A × B) F$_1$ fetal liver stem cells plus irradiated thymus grafts from parent A or parent B. Since the thymus grafts were irradiated, they contributed only the radioresistant "thymic epithelium," not T cells, which after several months all derived from (A × B) F$_1$ stem cells. However, if the thymus was from parent A, the F$_1$ chimeric T cells lysed infected targets from parent A only, not parent B; and conversely, if the thymus was from parent B, the chimeric F$_1$ T cells lysed only parent B infected targets (Zinkernagel et al., 1978a). Similar findings for cytotoxic responses to minor histocompatibility antigens were obtained concurrently by Fink and Bevan (1978). These landmark experiments showed that it was the radioresistant portion of the thymus (the thymic epithelium?) that imposed the genetic restriction for self, prior to and independent of exposure to antigen, and that what was "learned" to be recognized as self were the MHC antigens of the thymus in which the stem cells matured.

Several important questions remained—namely, was there also a restriction imposed by antigen-presenting cells during the primary exposure to antigen, and could the repertoire of self-recognition be expanded as well as contracted during T cell maturation in the thymus of chimeras? To answer the first question, responder T cells were sensitized to virus in an acutely irradiated adoptive recipient. In the long-term chimeras, the macrophages surviving after several months were all of donor origin, whereas in this acute adoptive transfer the macrophages were still primarily of irradiated recipient origin. When (A × B) F$_1$ T cells were sensitized in an irradiated, infected parent A (or B) recipient, they lysed only parent A (or B) infected targets, respectively (Zinkernagel et al., 1978b). Thus even though these F_1 T cells could recognize both parent A and parent B as self, an additional restriction was imposed at the time of primary exposure to antigen, probably because only a subpopulation of the F_1 T cells was stimulated, namely those clones that recognize virus plus (or in association with) the MHC antigens of the recipient cells on which the virus was first seen.

To address the second question, the reverse type of chimera was made: parent A stem cells into irradiated (A × B) F_1, designated A → (A × B) (Zinkernagel et al., 1978b). If the long-term chimeras were infected directly with virus, their T cells lysed only parent A infected targets. However, if the chimeric A → (A × B) T cells were removed and adoptively transferred to an acutely irradiated (A × B) F_1 recipient infected with virus, they now lysed both parent A and parent B infected targets. The simplest interpretation, consistent with the earlier results, was that the repertoire of recognition of self was, in fact, expanded when the A stem cells matured in the (A × B) F_1 thymus. However, since the long-term chimera had "lymphoreticular" cells (macrophages?) derived from the donor parent A, not the F_1 recipient, the initial exposure to virus in the chimera was effectively in a parent A environment, so only the A-specific clones were sensitized. Sensitization in an F_1 acutely irradiated adoptive recipient, providing infected macrophages of F_1 origin, allowed stimulation of both A- and B-specific clones. Implicit in this interpretation is the assumption that only cells derived from bone marrow stem cells (such as macrophages and lymphocytes) are effective in the primary presentation of virus for sensitization, since in both cases the majority of cells in the animal (long-term chimera or acutely irradiated recipient) were of F_1 origin. This assumption is internally consistent within the present set of experiments, but not consistent with Zinkernagel's (1976) earlier study using A → (A × B) chimeras, which did not require priming in an acutely irradiated secondary recipient in order to detect the expanded repertoire. The disparity is probably due to a lower dose of irradiation used in the earlier experiments, which left residual recipient macrophages (Zinkernagel, 1978). However, the earlier explanation, that all that was required in the chimera was tolerance to B, was ruled out by testing T cells from an A mouse, neonatally tolerized to B, sensitized in an (A × B) F_1 acutely irradiated recipient. Even though these T cells were tolerant to B, they failed to lyse infected B targets. (For exceptions, see below.)

As further proof that expansion and contraction of the repertoire of self-MHC recognition can occur simultaneously, (A × B) → (A × C) chimeras were made (Zinkernagel et al., 1978a,b). The results were completely as predicted from the principles demonstrated above. If these chimeras were sensitized directly in the chimera, they lysed only A infected targets, not B or C, presumably because the self-recognition repertoire of the donor (A × B) F_1 stem cells was expanded to include C but contracted to exclude B in the (A × C) F_1 thymus, and the sensitization was by (A × B) F_1 lymphoreticular cells in the long-term chimera, so that the C-specific clones were not activated. If the chimeric T cells were sensitized in an acutely irradiated infected (A × C) F_1 recipient, the expanded ability to lyse infected C targets, as well as infected A targets, was revealed. However, B targets were still not lysed. Finally, if the genotypically (A × B) F_1 chimeric T cells were sensitized in an acutely irradiated infected (A × B) F_1 recipient, they lysed only infected A, not B or C, targets. Therefore, even though the chimeric T cells were themselves genotypically (A × B) F_1, they had lost the ability to recognize B as self when they matured from stem cells in the (A × C) F_1 thymus, so no "virus + B"-specific clones of T cells were available to be sensitized in the (A × B) F_1-infected recipient.

The conclusion of these elegant experiments is that what is recognized as self is a function not of the genotype of the T cells themselves, but of the thymic environment in which they mature from stem cells. This learned recognition of self should probably not be

523

IMMUNE RESPONSE
GENES IN THE
REGULATION OF
MAMMALIAN
IMMUNITY

thought of as learning by individual cells as by some instructional mechanism (analogous to the old instructional theory of antibody specificity), but rather as selection and deletion of clones in the thymus, analogous to the clonal selection theory of antibody formation (Burnet, 1959). Thus, only the population as a whole learns (has its clonal repertoire altered) because only clones with certain specificity are ever released from the thymus into the periphery (see Jerne, 1971). In addition, a second level of genetic restriction is imposed in the periphery at the time of initial antigen exposure, when only certain clones are sensitized (expanded?). This second level of restriction during "priming" with antigen can only further contract, not expand, the repertoire, whereas in the thymus, either contraction or expansion can occur.

One hole in this otherwise elegant and internally consistent theory is the failure of (A × B) → A chimeric T cells to recognize parent B as foreign. They can be sensitized only to virus-infected A cells, so that in that sense their repertoire of self-recognition has been restricted to parent A, but they remain completely tolerant to parent B, so that B is not seen as allogeneic (or nonself) either. The recognition of MHC antigens of parent B in any fashion is simply a *hole* in the repertoire which is not explained by any part of the theory (see Singer *et al.,* 1979). Evidently, there is another meaning of self beyond what is learned in the thymus.

Other exceptions to the absoluteness of restrictions imposed in the thymus should be noted. Blanden and Andrew (1979) found preference for but not absolute restriction to parent A virus-specific lysis in F_1 → parent A chimeras. Three groups found that elimination of alloreactivity allowed virus or TNP-specific cellular cytotoxicity against allogeneic stimulators and target cells, implying that no absolute restriction had been imposed in the autologous thymus, but rather that alloreactivity masked the potential responsiveness. In one case the alloreactivity was eliminated by neonatal tolerance (Forman *et al.,* 1977*c*), and in the other two cases it was eliminated by short-term passage of parent A T cells through (A × B) F_1 recipients and collection of thoracic duct lymph, leaving donor cells reactive to parent B trapped in the recipient spleen (Wilson *et al.,* 1977; Doherty and Bennink, 1979). All three results conflict with the neonatal tolerance experiment of Zinkernagel *et al.* (1978*b*; see above). Analogous results to those of Forman *et al.* (1977*c*) and Doherty and Bennink (1979) were obtained by Thomas and Shevach (1977*b*) in the case of antigen-specific T cell proliferation (see Section 8.2.1). Therefore, the question of how absolute the restriction is which occurs during T cell maturation in the thymus still remains controversial.*

In the same studies (Zinkernagel *et al.,* 1978*a,b*), another series of experiments was performed which remains more controversial. These experiments involved completely allo-

---

*One possible explanation for the conflicting results obtained in similar experiments but using different strain combinations by Bennink and Doherty (1978) and Doherty and Bennink (1979) is that specific *H-2* determinants must be shared between the thymus of the donor and the infected stimulator cell for eliciting cytotoxicity specific for a particular virus. If such determinants are not shared, no responder cells are induced. If such determinants are shared, but other unshared *H-2* determinants induce an allogeneic response, the ability to stimulate virus-specific responder cytotoxic T cells is masked or blocked by the allogeneic reaction unless alloreactive T cells are first removed. This explanation might reconcile this exception (Doherty and Bennink, 1979) with the neonatal tolerance results of Zinkernagel *et al.* (1978*b*), since the latter also used vaccinia virus and either the same *H-2* combination as Bennink and Doherty (1978), with whom they agree, or a totally different *H-2* combination. This explanation is not in conflict with the lack of restriction of TNP-specific killing of *H-2*^k responders tolerant to *H-2*^b (the same *H-2* combination used by Bennink and Doherty, 1978) (Wilson *et al.,* 1977), since different specific determinants may have to be shared between *H-2* molecules to allow killing specific for different antigens, such as different viruses or haptens.

geneic chimeras of the general type A → B. T cells from such chimeras, which also had a lower survival rate, failed to lyse any vaccinia-infected targets no matter what environment was used for sensitization. The minimal requirement necessary and sufficient for immunologic competence in these chimeras was matching of the left part of the *H-2* complex (including the *I* region) between stem cell donor and recipient. The hypothesis was suggested that the need for matching, presumably in the *I* region, reflected a need for helper T cells in generating a cytotoxic response. In a chimera of the sort $I^A → I^B$ (even if *H-2K* and *H-2D* were shared), according to the principles of the theory above, helper T cells would be restricted in the I^B thymus to recognize cells bearing I^B, not I^A. However, the cytotoxic T cells in the chimera would still be genotypically I^A and would bear only I^A on their surfaces, so that they could not be helped by the I^B-specific T cells in the same chimeric population. Thus, the population as a whole would be impotent, due to failure of internal collaboration. However, this hypothesis is further removed from the data than is the more general theory elaborated above, and requires many more ad hoc assumptions. In addition, other workers have obtained completely allogeneic chimeras which are immunologically competent, for instance in T cell cytotoxicity specific for minor histocompatibility antigens and restricted by *H-2* (Matzinger and Mirkwood, 1978). Thus, for two reasons this hypothesis regarding helper T cells in chimeras remains less firm than the general theory above. Nevertheless, it will be a useful hypothesis in explaining some intriguing observations of *Ir* gene phenomena in chimeras to be discussed in Section 10.

One additional conclusion drawn from the above studies was that the restriction for self-MHC was imposed long before and independent of exposure to antigen, and that therefore these results were more consistent with the dual recognition than with the "altered self" hypotheses. A consistent result was obtained also by negative selection procedures, using bromodeoxyuridine incorporation into reactive cells, followed by exposure to light, to induce suicide of clones of cells with particular specificity (Janeway *et al.*, 1978). This study also confirmed the existence of separate clones of T cells in the F_1 specific for the different parental *H-2K* and *H-2D* antigens in association with TNP. All these results are more easily interpreted in terms of two receptors, one for self-*H-2* and one for the non-MHC antigen, on each T cell. However, it is also possible to envision the selection in the thymus of a repertoire of clones of T cells with single receptors reacting with an interaction product formed by the association of self-*H-2* antigen and other antigen. As one approaches the more extreme neoantigenic determinant hypotheses, it becomes more difficult to reconcile these results. Therefore, they support but do not prove the notion of dual recognition.

8.2   Restrictions on Macrophage–T Cell Interactions

A number of studies have demonstrated the importance of macrophages or macrophagelike cells (glass or plastic adherent, radioresistant, non-T, non-B, probably phagocytic) in immune responses by T and B cells, at least *in vitro* (Mosier, 1967; Roseman, 1969; Unanue and Cerottini, 1970; Hartmann *et al.*, 1970; Haskill *et al.*, 1970; Mishell *et al.*, 1970; Cosenza *et al.*, 1971; Leserman *et al.*, 1972; Sjöberg *et al.*, 1972).

The first demonstration of a requirement for histocompatibility between macrophages and T cells for successful interaction in initiating an immune response was in the classic work of Rosenthal and Shevach (1973) in the guinea pig. These investigators showed that macrophages, preincubated with antigen and then washed, could stimulte the *in vitro* proliferation of T cells immune to that antigen, without additional soluble antigen, only if the macrophages were histocompatible with the T cells. F_1 hybrid T cells could be stimulated by antigen bound to macrophages of either parent. Moreover, the cell surface alloantigens themselves appeared to be involved in the interaction since alloantisera directed against these blocked the response (Shevach *et al.*, 1972). An elegant experiment suggested that the blocking required anti-MHC antibodies against determinants shared by the macrophage and the T cell (Rosenthal and Shevach, 1973). If strain (2×13) F_1 T cells were stimulated with antigen on strain 2 macrophages, anti-2 serum blocked the response but anti-13 serum did not, even though the latter would equally well react with F_1 T cell. The reciprocal result was obtained using strain 13 macrophages. On the other hand, if F_1 macrophage-bound antigen was used to stimulate parental T cells, only the antiserum directed against the MHC antigens shared by the T cells and macrophages could block the response. Thus, any anti-MHC antibody reacting with the macrophage or T cell was not sufficient. The action of the alloantisera in blocking was localized to the MHC antigens held in common by the T cells and macrophages.

Similar results on genetic restriction of macrophage presentation of antigen for T cell proliferation, and on blocking an antigen-induced proliferation by anti-MHC antisera were found in the mouse (Yano *et al.*, 1977; R. H. Schwartz *et al.*, 1978; R. H. Schwartz *et al.*, 1976c). In this case, the restriction could be mapped to the *I-A* subregion of *H-2*, and the blocking was by anti-Ia, not anti-*H-2K/D* antibodies. This distinction held for the guinea pig also when it was shown that the main difference between strain 2 and 13 guinea pigs was in the guinea pig analog of the *I* region (B. D. Schwartz *et al.*, 1976). Also, a requirement for histocompatible macrophages was demonstrated for the *in vitro* induction of helper T cells, and the restriction was mapped to the *I-A* subregion of *H-2* (Erb and Feldmann, 1975b).

8.2.1 Genetic Mechanisms in Restriction

The same types of mechanisms considered for genetic restrictions between cytotoxic T lymphocytes and their stimulators and targets could be considered for these macrophage-T cell restrictions. The first to be considered was the analogue of the intimacy hypothesis, namely that T cells and macrophages had to share Ia antigens in order to interact for successful presentation of antigen. This simple requirement for sharing was ruled out in exactly the same way as the intimacy hypothesis above, when Thomas and Shevach (1976) developed the method to immunize nonimmune T cells *in vitro* and then restimulate in a second culture *in vitro*. F_1 hybrid T cells "primed" with antigen associated with macrophages of one parent could be restimulated by the same antigen associated with macrophages of that parent but not with the same antigen when it was associated with macrophages of the other parent, even though the T cells were equally histocompatible with either set of macrophages. (Related results in antibody-producing systems are discussed in Section 8.3.)

Similar results were obtained concurrently *in vivo* for delayed-type hypersensitivity, for which T cell proliferation is thought to be the *in vitro* correlate, and which requires *I-A* subregion compatibility for successful adoptive transfer of the response (Miller *et al.*, 1975). Miller *et al.* (1976) found that T cells from $(A \times B)$ F_1 mice, sensitized either in parental A recipients or in the F_1 with antigen bound to parent A macrophages, could transfer delayed-type hypersensitivity only to parent A, not parent B, recipients, even though both recipients were equally histocompatible.

The exclusion of this intimacy hypothesis and the ongoing discussion of altered-self

525

IMMUNE RESPONSE
GENES IN THE
REGULATION OF
MAMMALIAN
IMMUNITY

recognition in T cell-mediated cytotoxicity (see Section 8.1.) led to consideration of mechanisms involving T cell recognition of antigen only "in association" with Ia antigens on the antigen-presenting cell (for reviews, see Thomas *et al.*, 1977b; R. H. Schwartz *et al.*, 1978; Rosenthal, 1978). The above experiments with F_1 T cells suggested that what determined the restriction was the first exposure to antigen (priming). Two important questions thus arose: (1) Was the same population of T cells in the F_1 being restricted to antigen in association with a particular parental Ia at the time of priming, or did two different populations exist in the F_1, already restricted to recognize antigen plus only one of the parental Ia types, such that only one of these populations was expanded during priming? (2) Was the restriction truly to self-Ia, or only apparently so because allogeneic macrophages could not be used for priming and restimulation without an overwhelming allogeneic effect (Thomas and Shevach, 1976)? To answer the first question, Paul *et al.* (1977) employed two elegant methods of positive (Ben-Sasson *et al.*, 1975) and negative (Janeway and Paul, 1976) selection to separate the two putative populations. Positive selection of F_1 T cells on monolayers of parental macrophages pretreated with antigen revealed two populations, each enriched for T cells specific for antigen plus a particular parental MHC type. [A similar positive selection procedure was employed for helper T cells in an antibody-producing system with similar results (Swierkosz *et al.*, 1978) and other positive selection procedures led to the same conclusion (Sprent, 1978a,b); see Section 8.3.] Perhaps even more convincing were the results of the negative selection procedure (Paul *et al.*, 1977). (2 × 13) F_1 guinea pig T cells were first exposed to antigen X on parent strain 2 macrophages, and the cells that proliferated were allowed to incorporate bromodeoxyuridine (BUdR) into their DNA. These were prevented from further proliferation by exposure to UV light, which induced crosslinking of DNA containing the BUdR. The resulting population was recultured with antigen X or Y associated with either parental strain 2 or 13 macrophages. No proliferation was seen to antigen X on strain 2 macrophages, but proiferative responses to antigen X on strain 13 macrophages, and to antigen Y on macrophages of either strain, were unaffected. Comparable results were found when the other three possible combinations of antigen plus macrophage were used for the first culture. Thus, in the F_1 animal, different subpopulations of T cells exist which are specific for each possible combination of antigen plus Ia (Fig. 7).

The second question was answered on one level *in vivo* for delayed-type hypersensi-

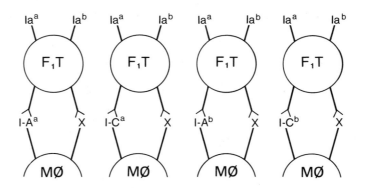

Figure 7. Four clones of T cells specific for each exogenous antigen in (a × b) F_1 hybrid. All the T cells share the same F_1 surface Ia antigen phenotype—that is, there is no allelic exclusion of Ia on the cell surface. However, each clone has a different combination of specificities for X plus one of the several possible Ia^a or Ia^b antigens on the macrophage. The T cells are depicted with 2 receptors for simplicity, but a single receptor (or neoantigenic determinant) model is also compatible.

527

IMMUNE RESPONSE
GENES IN THE
REGULATION OF
MAMMALIAN
IMMUNITY

tivity by testing T cells from tetraparental bone marrow chimeras (Miller *et al.*, 1976). If (A × B) F_1 lethally irradiated recipients were reconstituted with bone marrow stem cells from *both* parents A and B, and the chimeras sensitized two months later, pure homozygous parent A T cells from the chimera (after elimination of parent B T cells with alloantisera and complement), could transfer delayed-type hypersensitivity to recipients of the allogeneic strain B as well as to recipients of the syngeneic strain A. At this time, no distinction was made between tolerance and priming in the presence of both types of macrophages, on the one hand, and differentiation of T cells in the chimeric environment, on the other.

To answer the second question *in vitro* without the use of chimeras, Thomas and Shevach (1977*b*) employed the technique using BUdR and light to inactivate alloreactive T cells that would proliferate in a mixed lymphocyte reaction (MLR). They could then look for a population of T cells that would respond to antigen associated with histoincompatible macrophages. When strain 13 guinea pig T cells were precultured with strain 2 macrophages and treated with BUdR and light, the allogeneic MLR was sufficiently reduced to expose a TNP-specific proliferative response to TNP-coupled strain 2 macrophages when the T cells were primed with TNP-strain 2 macrophages in a second culture and then tested with TNP- on strain 2 macrophages in a third culture. The response was not only TNP-specific, but also specific for strain 2 (that is, allogeneic!) macrophages [see the related results of Pierce *et al.* (1976) in an antibody system, Section 8.3.2]. This result demonstrated not only that antigen could be presented on allogeneic macrophages, but also that the population of T cells specific for antigen on allogeneic macrophages was not a subset of those specific for unmodified allogeneic macrophages, since the latter cells were inactivated (fivefold to sevenfold) in the preculture. This conclusion about the relation of these T cells to alloantigen-specific T cells is important for theories of dual recognition vs. modified self, but suffers from the same loopholes, such as affinity, as the cold-target blocking studies described in Section 8.1.1, and also from the fact that alloreactivity was not completely eliminated. The former conclusion, that allogeneic macrophages could present antigen, is important because it suggests that priming is the all-important event in this T cell–macrophage genetic restriction and that no limitation exists prior to exposure to antigen. In contrast, the results for cytotoxicity suggest a restriction to self during T cell maturation in the thymus and only a secondary restriction imposed during priming. The experiments on which this conclusion was based used $F_1 \rightarrow$ parent chimeras (see Section 8.1.2), but have not been done in a study of T cell proliferation. However, several studies provide evidence that helper T cell–macrophage interactions are MHC restricted by maturation of F_1 T cells in a parental chimeric recipient (Sprent, 1978*c;* Kappler and Marrack, 1978; Erb *et al.*, 1979*b*; see Section 8.3.3). Thus, it is likely that the same result would be found for T cell proliferation. Also, in experiments analogous to those of Thomas and Shevach (1977*b*), but involving T cell-mediated cytotoxicity, Bennink and Doherty (1978) obtained the opposite result from that of Thomas and Shevach (1977*b*). They found that after acute depletion of strain A T cells alloreactive to strain B, there were no T cell clones left that react with virus-infected strain B cells. Therefore, they concluded that either there were no strain A clones specific for virus plus strain B alloantigens (in agreement with Zinkernagel *et al.*, 1978*a,b*, above) or that if there were any, they were a subset of the alloreactive clones against unmodified strain B. On the other hand, even in the cytotoxicity studies, results comparable to those of Thomas and Shevach (1977*b*) were obtained, using several methods to remove alloreactive cells (Forman *et al.*, 1977*c*; Doherty and Bennink, 1979; see Section 8.1.2).* Therefore, it still remains controversial how much or how abso-

*See footnote on page 523.

lute is the restriction that occurs during T cell maturation in the thymus. It is not clear whether the genetic restriction that occurs during priming with antigen is a lower order restriction than that occurring in the thymus (that is, it can further contract, but not expand the repertoire) or whether it can in fact lead to a restriction to recognition of antigen on allogeneic presenting cells.

8.2.2 Molecular and Cellular Mechanisms in Restriction

The nature of the macrophage–T cell interaction in the presentation of antigen has been investigated by directly studying the binding of immune T lymphocytes to antigen-bearing macrophages in clusters or rosettes. It has been found that antigen-specific binding can occur (Lipsky and Rosenthal, 1975; Lipscomb et al., 1977; Ellner et al., 1977; Ben-Sasson et al., 1978), and that the binding is genetically restricted to syngeneic T cells and macrophages (Lipsky and Rosenthal, 1975; Braendstrup et al., 1979), although the restriction applies only to the central lymphocytes in contact with the macrophage, not to the peripheral lymphocytes of the cluster which make contact with other lymphocytes (but not with the macrophage). The specific binding can be blocked by anti-Ia antisera against Ia determinants shared by the lymphocyte and macrophage (Lyons et al., 1979).

One intriguing and potentially extremely important finding with regard to antigen-specific T lymphocyte binding to macrophages is that although binding is antigen dependent, it cannot be blocked by stripping of antigen from the macrophage surface with trypsin or pronase, or by competition with excess antibody directed against the antigen (Ellner et al., 1977; Ben-Sasson et al., 1977). These results suggested that the antigen on the macrophage that is recognized by T cells is not merely native antigen displayed on the surface, but rather is antigen or antigen fragments which are "sequestered" so as to be protected from proteolytic enzymes or antibody but not from T cells. (An exception was noted in earlier studies of the binding of the protein antigen KLH to macrophages, in which native antigen was found bound to the macrophages and proteolytic enzymes or anti-KLH antibodies partially diminished the immunogenicity of such bound antigen (Unanue and Cerottini, 1970).) Furthermore, antibody to antigen could not be made to block the binding of T cells to antigen associated with macrophages even when further covered with a second layer of heterologous antiimmunoglobulin, or when the original antigen was multivalently coupled with hapten and antibodies to hapten were used (Ben-Sasson et al., 1977). In addition, a large excess of soluble or Sepharose-bound antigen did not compete for T cell binding to the antigen-treated macrophages (Ben-Sasson et al.. 1977).

As for effects on function, antibody to antigen did not block the ability of macrophage-bound antigen to induce antigen-specific T cell proliferation (Ellner et al., 1977). The only exception to this generalization is antibody to TNP in the case of Ia-restricted TNP-specific proliferation (Thomas et al., 1978). Even in that case, if cells were cultured to allow the majority of TNP-coupled proteins to be shed from the cell surface, the macrophages still stimulated a TNP-specific proliferative response, but it was no longer blockable by anti-TNP (Thomas et al., 1978). Therefore, this apparent exception may have been due to antibodies reacting with many cell surface proteins (including possibly Ia and mimicking anti-Ia blocking). In order to test the possibility that the T cell-stimulatory antigen was exposed on the macrophage surface (not sequestered) but was no longer native and therefore not recognized by antibody to native antigen, we tested the effect of antibodies raised to a random-conformation fragment of staphylococcal nuclease on the proliferative response to that peptide fragment (R. H. Schwartz, J. A. Berzofsky, and D. H. Sachs, unpublished

529

IMMUNE RESPONSE
GENES IN THE
REGULATION OF
MAMMALIAN
IMMUNITY

observations). We also did not see blocking of the response by these antibodies, but the negative result could always be attributable to too low a concentration of specific antibodies in the serum. However, it might be predicted from the failure of anti-hapten antibody to block the response to heavily hapten-coupled human gamma globulin (Ben-Sasson et al., 1977) that lack of a native conformation of antigen bound to macrophages was not the problem. Thus, the nature of antigen associated with macrophages, in particular how it can be sequestered from attack by antibodies or by proteolytic enzymes and yet be available to stimulate T cells, remains a mystery.

If it is so difficult to block antigen-specific proliferation by antibody to the antigen itself, why is it comparatively easy to block with anti-Ia antisera? How certain is it that the anti-Ia antibodies are actually reacting with Ia on the macrophage, not the T cell? The results of Rosenthal and Shevach (1973) discussed above, while suggestive that the macrophage Ia was the target, were actually symmetrical for blocking of F_1 T cells and parental macrophages or F_1 macrophages and parental T cells, so that the only firm conclusion was that the anti-Ia antibodies had to be directed against Ia shared by the strains donating the T cells and the macrophages. Several other pieces of evidence, taken together, suggest strongly that the macrophage, not the T cell, is the target of the anti-Ia. First of all, the Ia-positive population of macrophages is necessary for antigen presentation, as shown by killing with anti-Ia plus complement (R. H. Schwartz et al., 1978c; Niederhuber et al., 1979; S. Habu, personal communication), whereas the Ia-negative subpopulation of T cells proliferates as well as the untreated T cell population (Thomas et al., 1977a,b). Second, anti-Ia present only during the pretreatment of macrophages with antigen and then washed away blocks the ability of those macrophages to present the antigen to primed T cells (J. A. Berzofsky and L. K. Richman, unpublished observations; Geha et al., 1979). Finally, using BUdR and light to inactivate alloreactive cells and thereby examine antigen presentation by completely allogeneic macrophages, Thomas et al. (1977a) found that anti-Ia directed at the macrophage blocked the antigen-specific proliferative response whereas anti-Ia specific for the strain donating the T cells had no effect. This last important result was confirmed by D. Longo and R. H. Schwartz (personal communication), who used (A + B) → (A × B) F_1 long term chimeras to generate more complete tolerance of the donor A T cell to parent B macrophages, and found that anti-Ia directed against the parent B macrophage blocked proliferation, whereas anti-Ia against the parent A-responding T cell did not.

These results suggest that Ia on the macrophage is directly involved in antigen presentation to the T cell, but they do not elucidate the mode of association, if any, between antigen and Ia on the macrophage surface. Just as in the case of altered-self-H-2 cytotoxicity (see Section 8.1.1), few chemical data are available, and these are still controversial. Two interesting and opposite (but not contradictory) results are worth reporting. First, Thomas and Shevach (1977a) used anti-Ia antibodies as the *antigen* and elicited an antigen-specific, genetically restricted response. In this case, the antigen, being itself an antibody specific for Ia, was clearly bound to Ia molecules on the macrophage surface, and possibly only to Ia molecules (although binding to Fc receptors cannot be ruled out). In contrast, in the same study it was shown that antibodies directed against the B.1 MHC alloantigen shared by strain 2 and 13 guinea pigs could serve as an antigen under Ia restriction. Since these antibodies should not have reacted with Ia, the two results together suggest that for an I region-genetically restricted response, the antigen may be directly bound to Ia molecules on the macrophage but need not necessarily be so bound. The other study is that of Thomas et al. (1978) who could not detect TNP groups on molecules precipitated by anti-Ia, or Ia molecules precipitated by anti-TNP antibodies, in material extracted from TNP-coupled macrophages which could present antigen in a TNP-specific, Ia-restricted proliferative response that could be blocked by anti-Ia antibodies. Therefore, it was concluded that Ia restriction of the anti-TNP response did not require covalent coupling TNP-groups to the Ia molecules themselves. This conclusion contrasts with the finding of TNP coupled to *H-2* by Forman *et al.* (1977*a,b*) in the TNP-specific cytotoxic response, but agrees with the results of Schmitt-Verhulst *et al.* (1978) in the same type of response.

In conclusion, the facts that macrophage-associated antigen is more highly stimulatory to T cells than soluble antigen (Katz and Unanue, 1973; Waldron *et al.*, 1973), that this stimulation is *I* region genetically restricted and involves Ia antigens on the macrophage surface, and that different clones of T cells in an F_1 are specific for each combination of antigen plus parental Ia antigen, are well established. However, the nature of this macrophage-associated antigen, its physical relationship to surface Ia molecules, and the number of T cell receptors necessary to recognize antigen plus Ia (in analogy with the various versions of altered-self vs. dual recognition) remain to be elucidated.

8.3 Restrictions on T Cell–B Cell Interactions and the Role of Macrophages Therein

Histocompatibility requirements for functional cooperation between T and B lymphocytes were the first type of genetic restriction on cell–cell interaction to be described, and yet remain the most controversial to date.

There is considerable disagreement as to whether they reflect solely a T cell–macrophage restriction or whether there are also true genetic restrictions on T cell–B cell and/or B cell–macrophage interactions. The studies of these apparent restrictions on T cell–B cell cooperation to produce antibody, and the theories adduced to explain them, have evolved through extremely similar stages to those of the studies of T cell–macrophage and cytotoxic T cell–stimulator–target cell genetic restrictions (for reviews, see Benacerraf and Katz, 1975; Katz, 1977).

8.3.1 Demonstration and Mapping of Genetic Restrictions on T Cell–B Cell Cooperation

Genetic restrictions on the ability of T and B lymphocytes to cooperate to produce antibody were first demonstrated in two different systems. Kindred and Shreffler (1972) found that reconstitution of congenitally athymic nude mice with thymocytes to allow an antibody response to sheep erythrocytes was effective if, and only if, the thymus cells shared the *H-2* complex with the athymic nude recipients. In a more complex adoptive transfer scheme, Katz *et al.* (1973*a*) used carrier-immunized helper T cells from one strain (A) transferred into an (A × B) F_1 recipient which was then irradiated with 600 R a day later. This dose of irradiation was not lethal and did not abrogate the helper function of the carrier-primed T cells, but prevented a positive or negative allogeneic effect when allogeneic B cells (spleen cells treated with antitheta serum to kill T cells) from a hapten-immunized mouse of strain B were transferred to the same recipient. The recipient was then immunized with the hapten on the carrier to which the T cells had been immunized and serum antibody was measured seven days later. It was found that cooperation occurred (that is, serum antihapten antibody was produced) only when the T and B cells were syngeneic or semisyngeneic (F_1) to one another. This genetic restriction was mapped to the *H-2* complex (Katz *et al.*, 1973*b*), and subsequently to the *I* region of *H-2* (Katz *et al.*, 1975). It is important to note that cooperation occurred between F_1 helper T cells and parental B cells as well as between parental T cells and F_1 B cells, so that the presence of allogeneic Ia antigens on either the T cell or the B cell did not interfere as long as some Ia antigens were shared between them (Katz *et al.*, 1973*a*; Katz, 1977).

In addition, Katz *et al.* (1976) showed that mixing allogeneic carrier-primed helper T cells with syngeneic carrier-primed helper T cells did not prevent the latter from cooperating with hapten-specific B cells. Thus, the failure of cooperation across a histocompatibility barrier did not appear to be due to active suppression.

The requirement for *I* region sharing, rather than any block by the presence of non-shared Ia antigens or allogeneic cells, led to a theory of "cellular interaction" or "CI" structures which mapped in the *I* region of *H-2* and were required to be identical for physiological cell interaction to occur, presumably by a type of like–like interaction. The alternative, a complementary interaction between different molecules, would require other assumptions to explain how the complementary structures always stayed together in recombinants and mutants (Katz and Benacerraf, 1975; Katz, 1977). The latter problem is one basis for Katz's proposal of "adaptive differentiation," which suggests that during differentiation of T and B lymphocytes, appropriate matching (like or complementary) interaction structures are produced or at least selected (Katz, 1976; Katz, 1977) (see chimera studies below). It should be noted that the cellular interaction structure (CI) hypothesis was fundamentally a form of the intimacy hypothesis, and suffered from the same potential problems (see Section 8.1.1). However, the necessary experiments using F_1 T cells primed in parental recipients or with antigen on parental macrophages had not been done at that time to rule it out. In conjunction with the hypothesis of adaptive differentiation, it could be converted into a dual recognition theory more compatible with the results to be described below.

8.3.2 *Cooperation across Histocompatibility Barriers and the Role of Priming*

Several studies raised questions as to whether the above genetic restrictions on T cell–B cell cooperation were due to an absolute requirement for *I* region identity, or were due to subtle allogeneic effects that were not detected in the adoptive transfer scheme, despite the F_1 experiments to the contrary. Heber-Katz and Wilson (1975) eliminated alloreactive Lewis strain T cells from rat thoracic duct lymphocytes by passing them through (Lewis × AUG) F_1 rats and collecting the thoracic duct lymph which was depleted of Lewis T cells specific for AUG antigens (which had migrated to other organs). These negatively selected Lewis T cells were found to cooperate with AUG B cells in a primary *in vitro* anti-sheep erythrocyte antibody response, with a cellular dose–response curve identical to that of syngeneic AUG T cells. Thus allogeneic cooperation was possible.

Similarly, von Boehmer *et al.* (1975) eliminated alloreactivity by inducing tolerance in tetraparental bone marrow irradiation chimeras, in which lethally irradiated F_1 recipients were reconstituted with T cell-depleted marrow from both parental strains. T cells were immunized against sheep erythrocytes *in situ* in the chimeras, and then primed chimeric T cells of one parental type were isolated by killing cells of the other parental type with alloantisera and complement. These tolerant primed helper T cells cooperated with primed allogeneic B cells of the opposite parental type as well as they did with syngeneic B cells, in an *in vivo* secondary anti-sheep erythrocyte antibody response in an adoptive

531

IMMUNE RESPONSE
GENES IN THE
REGULATION OF
MAMMALIAN
IMMUNITY

irradiated F_1 recipient. Again, allogeneic cooperation occurred in the presence of tolerance to the particular alloantigens. This result was reminiscent of the apparent allogeneic cooperation in allophenic chimeras seen by Bechtol *et al.* (1974).

Similar experiments with tetraparental bone marrow chimeras were carried out by Waldmann *et al.* (1975), with identical results. The important additional points were (1) that the soluble carrier-hapten complex TNP–KLH was the antigen, the same one used by Katz *et al.* (1973*a*), rather than the particulate antigen sheep erythrocytes, and the T cells were primed in the chimera to KLH only, while the B cells were primed to TNP on a different carrier; and (2) that the response with *H-2*^{k} B cells was proven to be due to allogeneic *H-2*^{d} helper T cells from the chimera rather than residual syngeneic *H-2*^{k} T cells from the double chimera because it could be eliminated by anti-*H-2*^{d} antiserum plus complement.

In marked contrast, Sprent and von Boehmer (1976) found that a negative selection procedure virtually identical to that used by Heber-Katz and Wilson (1975) to eliminate alloreactive cells was not sufficient to allow cooperation with allogeneic B cells in a secondary antibody response to sheep erythrocytes in irradiated F_1 adoptive transfer recipients.

These apparently paradoxical results could all be reconciled by one hypothesis, that the genetic restriction was imposed by the process of priming the T cells—that is, by the macrophages presenting antigen at the time of initial exposure. Katz's (1973) T cells were all primed in a syngeneic environment. Heber-Katz and Wilson (1975) studied a primary response—that is, a response with T cells that had never been primed. Von Boehmer *et al.* (1975) and Waldmann *et al.* (1975) studied a secondary response, but the T cells were primed in the double-chimeric environment, in the presence of macrophages of both *H-2* types. In contrast, Sprent and von Boehmer used T cells which were primed in a syngeneic environment before negative selection to achieve tolerance. Thus, presumably the genetic restriction had already been imposed.

This unifying hypothesis was consistent with the demonstration by Erb and Feldmann (1975*b*) that only syngeneic macrophages (or ones which shared the *I-A* subregion with the T cells) could effectively prime helper T cells for cooperation with syngeneic B cells. However, the B cell populations (also containing macrophages) were syngeneic to the T cells. It was also consistent with the results of Pierce *et al.* (1976*a,b*), who used macrophages pretreated with antigen ("pulsed"), and washed, to present antigen. In a primary *in vitro* response, either syngeneic or allogeneic macrophages would work—that is, no genetic restriction was seen. In a secondary response, T cells cooperated with syngeneic B cells only when antigen was presented on the same type of macrophage as that used for priming. Thus, when antigen-pulsed *allogeneic* macrophages were used to prime the T cells, they cooperated with syngeneic B cells *only* in the presence of the same antigen-pulsed *allogeneic* macrophages, not antigen-pulsed syngeneic macrophages or even soluble antigen (which would be presented by syngeneic macrophages contaminating the B cells). In this system, only the T cells had to be primed to see genetic restriction, not the B cells (Pierce *et al.*, 1976*a*). However, it was subsequently shown that the restriction in this case was due to induction of antigen-specific suppressor T cells which blocked the primary response to GAT bound to macrophages other than those used for priming (Pierce and Kapp, 1978*a*). As discussed by these authors, it is not clear in the case of GAT whether (1) immunization with macrophage-bound GAT induces a genetically restricted "primed" helper T cell population and a population of GAT-specific suppressor T cells which blocks the *primary* response to GAT associated with any macrophage type but which cannot suppress already-primed helper T cells; or (2) the helper T cells are actually genetically unrestricted, but the

533

IMMUNE RESPONSE
GENES IN THE
REGULATION OF
MAMMALIAN
IMMUNITY

suppressor cells create the appearance of restriction by blocking the responses to GAT bound to any macrophage other than that used for priming. However, the later hypothesis would require the postulate of a mechanism by which suppressors are induced which act on the universe of all possible antigen-presenting cells *except* the one actually encountered in priming. The former hypothesis is more consistent with other studies discussed above and below.

This hypothesis that genetic restriction was imposed by initial contact with antigen on a particular macrophage was also strongly supported by Kappler and Marrack's (1976) finding that F_1 helper T cells, primed *in vivo* by injection of antigen-bearing parental macrophages, cooperated preferentially with B cells and macrophages of that parent, not the other parent. The restrictions mapped to *H-2* and in this case was *not* caused by suppression.

In an attempt to test this unifying hypothesis in their adoptive transfer system, Katz *et al.* (1976) did not see any effect of using T cells primed *in vivo* with KLH-bearing macrophages of allogeneic, compared with syngeneic, origin. They still cooperated equally well with syngeneic B cells, and not allogeneic B cells. However, the recipients to be primed were not at all tolerant to the antigen-bearing allogeneic macrophages with which they were injected. Thus, although it did not appear necessary in the case of Pierce *et al.* (1976*a,b*), one could argue in the other studies that tolerance was necessary in order to see cooperation across allogeneic barriers, even if tolerance alone were not sufficient. When Skidmore and Katz (1977) used naturally tolerant F_1 T cells and primed them in an adoptive transfer acutely irradiated parental recipient, they did see restriction to cooperation with B cells of that, not the opposite, parental *H-2* type. This result confirmed the role of priming in imposing a genetic restriction on cell cooperation in antibody production. However, it also resulted in the surprising finding that F_1 T cells primed in a parental strain recipient failed to cooperate with syngenic F_1 B cells, even though they cooperated with B cells of the strain used for priming. This unexpected restriction was shown by mixing experiments to be due to suppressor T cells.

This restricted cooperation of F_1 T cells with one parental strain or the other, due to priming (Kappler and Marrack, 1976; Skidmore and Katz, 1977), is the analogue of the F_1 experiments used to rule out the intimacy hypothesis for cytotoxicity and T cell–macrophage interactions (see Sections 8.1.1 and 8.2.1). The result was confirmed by using *in vitro* priming of F_1 T cells by antigen-pulsed macrophages of one parent or the other, in a study which was not complicated by suppression of the response to F_1 B cells (McDougal and Cort, 1978). In addition, the presence of two distinct populations of $(A \times B)$ F_1 T cells, one cooperating with parent A macrophages and B cells and the other with parent B macrophages and B cells, was demonstrated elegantly by several positive and negative selection procedures in both the mouse (Sprent, 1978*a–c*; Swierkosz *et al.*, 1978) and the guinea pig (Yamashita and Shevach, 1978).

This elimination of the intimacy hypothesis simultaneously excluded the like–like cellular interaction structure (CI) hypothesis in its original form, since there is ample evidence that Ia antigens are not allelically excluded—that is, that all F_1 Ia-positive lymphocytes bear Ia antigens of both parental haplotypes (Dickler and Sachs, 1974; Thomas *et al.*, 1977*b*; Klein, 1975). The presence of two separately restricted populations of T cells in the F_1 requires expression of different CI structures on different F_1 T cells (Skidmore and Katz, 1977) and so essentially converts the CI hypothesis into a dual recognition-type hypothesis in which the CI molecules would be indistinguishable from a second receptor for Ia on the antigen-specific T cells. This second receptor for Ia need not be, and probably would not be, encoded in the *I*-region of *H-2* as the original CI molecules were postulated to be. However, the clones bearing different Ia-specific receptors could be selected by the Ia antigens of the host environment. This concept, suggested by the chimeric experiments of Zinkernagel *et al.* (1978*a,b*), is not essentially different from the adaptive differentiation concept suggested by Katz (1977) to explain the chimera experiments described above (and see Section 8.3.3).

The role of priming in imposing at least one level of genetic restriction on T cell-B cell interaction seems fairly well established from these experiments. We have seen that in the case of cytotoxic responses to non-MHC antigens on cells, another level of genetic restriction occurs at the time of T cell maturation in the thymus, before antigen is ever encountered (see Section 8.1.3). It was therefore important to investigate the role of maturation environment on the genetic restrictions manifested by helper T cells.

8.3.3 The Role of Chimeric Environment on the Restriction Specificity of Helper T Cells

We have already seen in Section 8.3.2 that maturation in the F_1 environment of a chimera can expand the genetic restriction repertoire of homozygous parental helper T cells. A number of studies have confirmed the ability of A → (A × B) or A T cells selected from (A + B) → (A × B) chimeras to cooperate with parent B strain macrophages and B cells (Waldmann *et al.*, 1978; Kappler and Marrack, 1978; Sprent and von Boehmer, 1979; Erb *et al.*, 1979*b*; Singer *et al.*, 1979). [One apparent exception was reported by Erb *et al.* (1978).] In one study of *in vitro* responses, parent A T cells from (A + B) → (A × B) tetraparental bone marrow chimeras showed more easily demonstrated cooperativity with strain B macrophages and B cells than did T cells from single parent A → (A × B) chimeras (Erb *et al.*, 1979*b*), possibly relating to the presence of strain B macrophages and lymphoid cells in the former but not the latter environments. T cells from allophenic chimeras showed the least evidence of self-preference (Erb *et al.*, 1979*b*). In addition, Sprent and von Boehmer (1979) showed by positive and negative selection techniques that this expanded repertoire for cooperation was due to two distinct sets of parental T cells, one restricted to antigen plus self and the other restricted to antigen and B cells and macrophages of the opposite (allogeneic) parent. Thus, the explanation that the tolerant state in the A → (A × B) chimera merely removes mechanisms that would prevent a normally *unrestricted* clone of T cells from demonstrating its full potential for cooperation appears untenable. Rather, there are different clones with different restrictions, but any given clone is restricted in its ability to cooperate. However, it would still be theoretically possible for the tolerant state to have unmasked clones of helper T cells that were restricted to interact with antigen plus allogeneic macrophages and B cells, and that would have been present, but unable to function, even without maturation in the F_1 environment (thymus?).

This issue was addressed by Waldmann *et al.* (1978) who compared in parallel strain A T cells from A → (A × B) or (A + B) → (A × B) chimeras with strain A T cells from donors rendered neonatally tolerant to strain B. Whereas the strain A T cells from the chimeras could cooperate with B cells of either strain, and could be further restricted by priming in a secondary irradiated recipient of one strain or the other, the T cells from the neonatally tolerized mice could not cooperate with strain B macrophages and B cells even if they were primed in an F_1 secondary acutely irradiated recipient. Hence, tolerance alone was not sufficient to allow cooperation, a finding that is consistent with the negative selection experiments of Sprent and von Boehmer (1976). One might therefore conclude that the chimeric environment produces more than tolerance. Apparently it actually expands the repertoire, allowing maturation of T cell clones (in the thymus?) that would otherwise never have been seen in the peripheral lymphoid organs.

A further test of this hypothesis was the narrowing of restriction (contraction of the repertoire) which was seen in $F_1 \rightarrow$ parent chimeras (Sprent, 1978d; Waldmann et al., 1978; Kappler and Marrack, 1978; Erb et al., 1979b; Singer et al., 1979; Katz et al., 1978). This increased restriction was mapped to the K or I-A subregions of the H-2 complex of the chimeric recipient (Sprent, 1978d; Kappler and Marrack, 1978; Singer et al., 1979). This loss of the normal ability of (A × B) F_1 helper T cells to cooperate with parent B macrophages and B cells when the F_1 T cells matured in a strain A chimeric recipient held true despite the fact that the macrophages of the chimera were of F_1 origin, and even if the chimeric T cells were removed and primed to antigen in an acutely irradiated F_1 recipient (Sprent, 1978d). It also held for in vitro primary responses in which prior priming had never occurred (Singer et al., 1979; Erb et al., 1979b). Therefore the clones of T cells in the F_1 that would normally cooperate with parent B macrophages and B cells appeared to be absent from (never matured among?) the F_1 cells derived from the chimera. These results are analogous to those for cytotoxic T cells (Section 8.1.2) and suggest that the same processes are at work in both.

In view of these results, it was exciting to determine whether the restrictions were determined specifically by the thymus of the chimera, as had been found for cytotoxic T cells by Zinkernagel et al. (1978a) and Fink and Bevan (1978). Therefore, Waldmann et al. (1979a,b) produced thymic chimeras by lethally irradiating thymectomized F_1 mice and reconstituting them with syngeneic F_1 bone marrow (depleted of T cells) plus a thymus from one or the other parent or the F_1. They found that helper T cells from these chimeras, primed in the F_1 chimera, cooperated only with B cells and macrophages syngeneic to the donor of the thymus. Thus, the thymus appeared to be the critical organ determining the genetic restrictions on cooperation by helper T cells maturing in the chimera. Similar results were obtained by Hedrick and Watson (1979). In addition, another study by Katz et al., 1979a) found a preference for cooperation with B cells syngeneic to the donor of the thymus, but the preference was only partial (two-fold to four-fold in response). This same group also did not see an expansion of the repertoire of parent → F_1 chimeras (see Erb et al., 1978), although they did see a contraction of the repertoire of $F_1 \rightarrow$ parent chimeras, not only for T cells but in some cases possibly also for B cells (that is, a preference in which helper T cells these B cells would cooperate with) (Katz et al., 1978).

The concensus derived from many studies is that adaptive differentiation does occur, at least for helper T cells, so that the repertoire of clones available to cooperate with B cells and macrophages of different strains is determined by the K or I-A subregions of the H-2 complex of the thymus in which the T cells mature from stem cells, rather than the genotype of the T cells themselves. To the extent that Ir gene phenomena parallel or reflect genetic restriction phenomena (see Section 9), this conclusion is critical to the prediction of Ir gene behavior in chimeras (see Section 10) and the identity of Ir genes in general (see Section 12).

8.3.4 The Possible Role of the Macrophage in Apparent Genetic Restriction of T Cell– B Cell Interactions

One of the most controversial issues remaining in the field of genetic restrictions on T cell–B cell cooperation is whether or not there are, in fact, any restrictions on collaboration

535

IMMUNE RESPONSE
GENES IN THE
REGULATION OF
MAMMALIAN
IMMUNITY

between T cells and B cells at all, or whether the apparent restrictions reflect T cell–macrophage restrictions at all levels. We have already reviewed in Section 8.3.2 the convincing evidence that macrophages play a determining role in imposing genetic restrictions on helper T cells at the time of priming with antigen, just as they do in the case of T cell proliferative and delayed-type hypersensitivity responses (Section 8.2). The question, then, is whether these restrictions imposed by the thymus and during priming are subsequently reflected in true T cell–B cell restrictions during the mediation of help (for instance, the B cell may have to share Ia antigens with the macrophages used for priming), or whether the apparent T cell–B cell restrictions are due to a need for macrophages during the helper-effector phase of the process which are syngeneic with the macrophages used for priming. Most studies did not make this distinction, since the macrophages present were usually syngeneic with the B cells, which until very recently almost always consisted of a whole spleen population depleted only of T cells by treatment with antitheta serum plus complement.

The presence of large numbers of F_1 macrophages in the adoptive transfer irradiated F_1 recipient used in the test system of the original studies of Katz et al. (1973a) was taken as evidence that macrophages histocompatible with the helper T cells were not limiting during the actual mediation of help, and that therefore, the restriction was truly one between T cells and B cells. A similar argument could be made for the in vivo studies of Sprent and von Boehmer (1976), which involved adoptive transfer to an irradiated F_1 recipient. Moreover, $F_1 \rightarrow$ Parent A chimeric T cells, primed in an F_1 recipient and then tested by adoptive transfer into another F_1 recipient, failed to cooperate with B cells of parent B (Sprent, 1978d). In this last case, the T cells were genotypically F_1, primed in an F_1, and tested in the presence of F_1 macrophages, so it is hard to argue that F_1 macrophages were deficient at any stage.

In addition, a number of studies were designed to distinguish between a T cell–B cell restriction and a T cell–macrophage restriction. Sprent (1978a,b) used a positive selection procedure to separate subpopulations of primed F_1 T cells specific for antigen and restricted to interact with the B cells and/or macrophages of one parent. Four different approaches were then used to try to overcome this restriction by providing appropriate macrophages during the T cell–B cell interaction in the adoptive transfer F_1 recipient, which already had more F_1 macrophages than the parental macrophages in the B cell inoculum. The first was to increase the antigen dose to prevent parental macrophages accompanying the B cells from preempting the antigen. The second was to supplement the adoptive transfer mixture of cells with F_1 or parental macrophages (from peritoneal exudate or bone marrow). The third approach used B cells from thoracic duct lymph, which is very low in macrophages. In the absence of parental macrophages syngeneic to the B cells, only F_1 macrophages would be available to present antigen. Fourth, the F_1 recipients used as the hosts for the test response were irradiated and reconstituted with F_1 bone marrow six days earlier to allow newly differentiated F_1 macrophages to populate the lymphoid organs. Finally, the third and fourth approaches were combined to give every competitive advantage to the F_1 macrophages over the syngeneic ones contaminating the B cells. None of these attempts revealed any ability of the restricted F_1 T cells to cooperate with B cells of the other parent from that used for selection. Similarly, using a different type of positive selection system and an in vitro instead of in vivo test system, Swierkosz et al. (1978) obtained the same result—namely, adding extra macrophages of parent B did not unmask an ability of F_1 T cells, positively selected on antigen-pulsed parent B macrophages, to cooperate with parent A B cells and macrophages.

537

IMMUNE RESPONSE
GENES IN THE
REGULATION OF
MAMMALIAN
IMMUNITY

Likewise, in the guinea pig, positive selection of subpopulations of F_1 helper T cells by adherence to antigen-pulsed parental macrophages resulted in F_1 T cells which cooperated only with B cells syngeneic to the macrophages used for selection (Yamashita and Shevach, 1978). This genetic restriction was not overcome even by presenting antigen already bound to macrophages of either parent, in the absence of any other source of antigen. Thus (2 × 13) F_1 T cells selected on antigen-bearing strain 2 macrophages cooperated only with strain 2 B cells, not strain 13 B cells, regardless of whether antigen-bearing strain 2 or strain 13 macrophages were the source of antigen. On the one hand, the surprising ability of strain 13 antigen-bearing macrophages to present antigen to these selected strain 2-restricted F_1 T cells plus strain 2 B cells could be explained by carryover of antigen to the strain 2 macrophages from the B cell source, although several attempts to prevent such possible carryover did not change the results. On the other hand, such carryover would not explain the failure of strain 2 antigen-bearing macrophages to present antigen to the strain 2-restricted F_1 T cells for collaboration with strain 13 B cells if the only restrictions were between T cell and macrophage, not T cell and B cell. The presence of antigen already on strain 2 antigen-pulsed and washed macrophages should have given them a strong competitive edge over the strain 13 macrophages from the splenic B cell source.

All these results represent rather strong evidence that a restriction can be demonstrated in cooperation between T cells and B cells, not just T cells and macrophages. Moreover, because the selection of F_1 T cells by their ability to interact with parental macrophages bearing antigen appeared to determine the restriction or preference for cooperation with B cells syngeneic to the selecting macrophages, Sprent (1978a,b). Yamashita and Shevach (1978), Paul and Benacerraf (1977), and Benacerraf (1978) proposed a symmetrical relationship of B cells and macrophages relative to the pivotal T cell. They suggested that the T cell must recognize on the B cell the same combination of antigen plus Ia that it first encountered on the macrophage during priming in order for a helper effect to be manifested (Fig. 8). This idea is extremely appealing because of its symmetry, simplicity, and ability

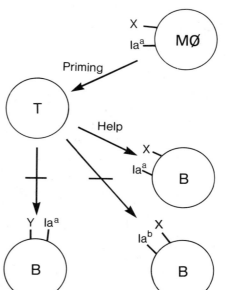

Figure 8. Symmetry between a macrophage involved in priming and a B cell receiving help in interaction with the pivotal T cell. The hypothesis suggests that the clone of T cells primed by exogenous antigen X presented on macrophage bearing Ia^a can help only those B cells bearing the same combination of antigen X plus Ia^a. B cells with antigen Y plus Ia^a or antigen X plus Ia^b are not helped. This restriction may be due to the existence of distinct clones of T cells with dual specificity as depicted in Fig. 7.

to tie together so many diverse findings, not the least of which is the similar expression of Ia antigens on B cells and macrophages, distinct from T cells.

However, there are also a number of studies which used similar approaches but obtained the opposite results. McDougal and Cort (1978) found that (A × B) F_1 T cells, primed *in vitro* with antigen-bearing strain B macrophages, could cooperate with strain A B cells if and only if supplementary strain B macrophages were added during the second culture. Thus, they found no restriction for the B cells, only one for the macrophages present during the mediation of T cell help. Similarly, Erb *et al.* (1978) found that F_1 T cells primed with antigen *in vitro* in the presence of macrophages of one parental sprain manifested a restriction for cooperation with B cells of that same parental strain, but that the failure to cooperate with B cells of the opposite parental strain could be overcome by the addition of macrophages syngeneic to those used for priming.

Finally, Singer *et al.* (1979) showed that in a primary *in vitro* response in which T cells, B cells, and macrophages were prepared separately, relatively free of either of the other cell types, F_1 T cells cooperated with B cells of either parental strain equally well in the presence of macrophages of either parental strain. Therefore, there appeared to be no requirement for T cell recognition of matching Ia antigenic determinants on B cells and macrophages. To see whether such a requirement would appear if the F_1 T cells had been primed to antigen *in vivo* (in the F_1), the same experiment was performed (at least in one direction) with primed F_1 T cells. The results were identical. Furthermore, these investigators found that the genetic restriction observed for (A × B) F_1 → A chimeric T cells (unprimed) to cooperate with parental strain A but not strain B B cells *in vitro* or *in vivo* could be overcome by the addition of strain A macrophages. Thus, this restriction induced by maturation in the chimera appeared to be a restriction for cooperation with macrophages but not with B cells. These results appeared incompatible with the theory, described above, which suggested symmetrical restriction for B cells and macrophages in cooperation with the same T cells.

These two groups of studies appear to yield mutually contradictory results, in experiments that seem to be directly comparable to one another. Since it is difficult to determine which type of result is the physiological rule and which the exception, the field remains extremely controversial. In the process of juxtaposing these several studies in this review, one systematic difference comes to mind which may distinguish studies yielding one result from those yielding the other. The studies in which the addition of appropriate macrophages could not overcome the genetic restriction on T cell–B cell collaboration all employed T cells primed with antigen *in vivo,* either homozygous T cells primed in a syngeneic environment or F_1 cells primed *in vivo* and selected for restriction to one or the other parental strain. The studies in which the restriction was overcome with added macrophages all used T cells either primed *in vitro* or completely unprimed, with one possible exception.* There may be a difference between the nature of priming *in vivo,* under phys-

---

*The single experiment which is hard to explain by the requirement for T-cell priming *in vivo* to see a restriction on cooperation with B cells is that in which Singer *et al.* (1979) used *in vivo*-primed (H-2^a × H-2^b)F_1 T cells to provide help *in vitro* for either H-2^a or H-2^b B cells. The ability of primed F_1 T cells to cooperate with B cells of either parental type is no surprise, since the F_1 T cells were not separated by positive or negative selection into two differently restricted subpopulations. What needs to be explained is why, when the response of macrophage-depleted T and B cells in culture did not occur without the addition of a source of macrophages, the addition of H-2^b macrophages allowed the F_1 T cells to cooperate with H-2^a B cells just as well as H-2^b B cells. On the one hand, this result could be explained by a hypothesis that priming leads to only T cell-B cell restriction and not T cell-macrophage restriction, since that is exactly what Yamashita and Shevach (1978)

539

IMMUNE RESPONSE
GENES IN THE
REGULATION OF
MAMMALIAN
IMMUNITY

iologic conditions, and *in vitro* in various culture situations. In addition, several studies described earlier (Heber-Hatz and Wilson, 1975; Pierce *et al.*, 1976*a,b*) showed that unprimed T cells may not be genetically restricted in the same way as are primed T cells. This observation could explain the finding that unprimed F_1 T cells did not require *H-2*-matched B cells and macrophages for an *in vitro* primary response (Singer *et al.*, 1979). The restriction which this last report showed could be overcome by adding appropriate macrophages was a restriction of the type induced by maturation of F_1 T cells in a parental chimeric recipient. Restriction of this type may be different from that induced by priming, as has also been suggested by Erb *et al.* (1979*b*). In the studies of Singer *et al.* (1979), the $F_1 \rightarrow$ parent chimeric T cells were unprimed, whether tested in an *in vitro* or *in vivo* primary response. In contrast, the $F_1 \rightarrow$ parent chimeric T cells which manifested the opposite behavior had been primed *in vivo* and were therefore tested in a secondary response (Sprent, 1978*d*). Even though the priming was accomplished in an irradiated F_1 recipient, in the presence of F_1 macrophages, only priming of clones restricted to recognize antigen in association with the parental strain Ia type of the chimera may have occurred, because of the restriction imposed by maturation in the chimeric host. Once primed, these restricted clones of F_1 T cells may now manifest an additional restriction for cooperation with B cells sharing the macrophage Ia type that had successfully primed them.

In summary, the field is still extremely controversial and diametrically opposite results abound. However, virtually, all the data may be reconciled with both the hypothesis that true genetic restrictions on the interaction between B cells and helper T cells do exist, and the appealing hypothesis that the combination of antigenic determinant and Ia antigen present on the macrophage in the first exposure of the T cell to antigen (priming) must be reiterated or mirrored on the B cell in order for that T cell to provide help. Two assumptions are required to reconcile the data. First, it is postulated that T cells primed *in vivo*, as opposed to unprimed T cells or T cells "primed" *in vitro*, are genetically restricted in their ability to cooperate with B cells. In fact, from the results of Yamashita and Shevach (1978), they may show only a B cell restriction and no longer need a *genetically-restricted* macrophage function to provide help, although this latter possibility is far less well established. Second, it is postulated that the narrowing of restriction imposed by maturation of F_1 T cells in the parental thymus of an $F_1 \rightarrow$ parent chimera represents a limitation of the repertoire of clones available to be primed by macrophages to those clones recognizing the Ia of one parental strain but not the other. However, this restriction is only a limitation on which macrophages can prime the T cells. Until the T cells are primed, they do *not* manifest a restriction on their ability to cooperate with B cells of either parent. Thus, the genetic restrictions imposed during maturation of T cells in the thymus and those imposed during *in vivo* priming with antigen are postulated to be restrictions of fundamentally different kinds. They restrict T cells for different phases of the induction and expression of help perhaps because they act on T cells at different stages of differentiation and/or activation (Fig. 9).

---

observed using primed F_1 T cells that had been separated into two restricted subpopulations. On the other hand, this experiment could be explained by the postulate that although removal of macrophages from the *H-2*[a] B cell population was sufficient to prevent a response in the absence of any added macrophages, enough *H-2*[a] macrophages were left to present antigen to *H-2*[a]-restricted F_1 T cells if any source of extra macrophages was added to provide a nonspecific "feeder" effect. This second possibility could be tested by repeating the same experiment using antigen-pulsed and washed *H-2*[b] macrophages to present antigen to *in vivo*-primed F_1 T cells and *H-2*[a] B cells. If either of these explanations holds, then all the other superficially conflicting results may be reconcilable.

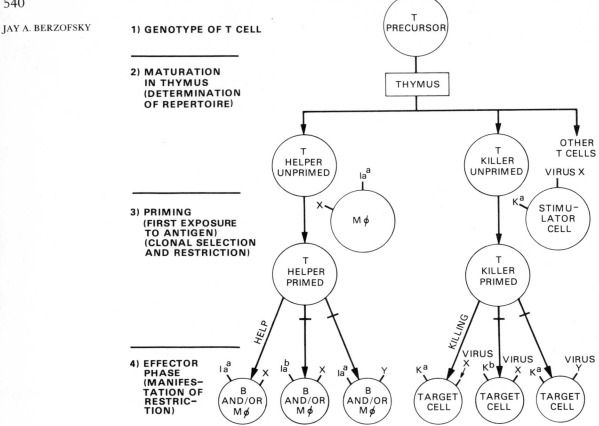

Figure 9. Four stages of induction and expression of genetic restriction of T cell interactions with other cells. Parallel stages exist for helper T cells (left) and cytotoxic T cells (right), except that the restriction elements for the former are Ia antigens and for the latter are H-2K/D antigens. (1) Some restriction is inherent in the T-cell genome, since (a × b)F$_1$ T-cell precursors raised in a parent *a* thymus do not recognize parent *b* as foreign. (2) Maturation in the thymus can expand or contract the repertoire of T cell clones recognizing different MHC antigens as "self." For instance, in a → (a × b)F$_1$ chimeras, the repertoire is expanded to recognize *b* as "self"; in (a × b)F$_1$ → a chimeras, the repertoire is contracted to recognize only *a* as "self," although a hole is left in the repertoire, since *b* is not recognized as foreign either, due to a higher level of restriction (1). This thymic maturation stage limits the choice of which antigen-presenting macrophages or stimulator cells can be recognized to further restrict the repertoire, probably by clonal selection, in stage (3). (3) Priming (initial exposure to antigen X) by a particular macrophage or stimulator cell then further restricts the available activated T cell clones to those which recognize that antigen (X) in association with the restriction element (Ia for helpers of K/D for killer T cell) on that macrophage or stimulator cell. This clonal selection is manifested by a restriction on which B cells and/or macrophages, or which target cells, can be recognized in the effector phase (stage 4) of the response.

9   Cellular Site of Expression: II. The Role of Macrophages in the Expression of Ir Genes

A massive accumulation of data has been building up over the last few years that indicates a striking parallelism between genetic restriction phenomena in systems not

541

IMMUNE RESPONSE
GENES IN THE
REGULATION OF
MAMMALIAN
IMMUNITY

involving known *Ir* gene control, and the behavior of responses controlled by *Ir* genes (for reviews, see Paul and Benacerraf, 1977; Thomas *et al.*, 1977*b*; Benacerraf, 1978; R. H. Schwartz *et al.*, 1978; Rosenthal, 1978; Benacerraf and Germain, 1978; Zinkernagel, 1978). In this and the next section, we shall examine the evidence that the two phenomena are different facets of the same underlying physiologic mechanism. These results will be explored in terms of the role of antigen-presenting cells (macrophages and their kin) in *Ir* gene expression and selection of immunogenic determinants in this section, and the role of the thymus and chimeric environments in the next section.

9.1 Macrophage Determination of Ir Phenotype

The first clear demonstration that the source of macrophages could be critical in the determination of *Ir* phenotype was in a T cell-proliferative response in the guinea pig (Shevach and Rosenthal, 1973). In a companion paper (Rosenthal and Shevach, 1973) discussed in Section 8.2, they had shown that T cells from immunized guinea pigs could be restimulated *in vitro* by antigen bound to syngeneic but not allogeneic macrophages, and that F_1 T cells could be restimulated by antigen bound to macrophages of either parent. In the subsequent study, Shevach and Rosenthal (1973) extended these experiments to the case of antigens under *Ir* gene control. Strain (2 × 13) F_1 T cells immune to a copolymer of glutamic acid and tyrosine (GT), to which strain 13 guinea pigs were responders and strain 2 guinea pigs nonresponders, could be restimulated by antigen bound to only F_1 or strain 13 macrophages, not strain 2 macrophages. The converse was true for the DNP conjugate of a glutamic acid-lysine copolymer (DNP-GL) to which strain 2 but not strain 13 animals responded. Thus, unlike antigens not under apparent *Ir* gene control, which could be successfully presented to F_1 T cells by macrophages of either parent (if the T cells had been primed in the F_1), antigens under *Ir* gene control could be presented by macrophages of only the responder parent. In addition, the presentation of an antigen under *Ir* gene control by (responder × nonresponder) F_1 macrophages to syngeneic F_1 T cells could be blocked by anti-Ia antisera against the responder parent, not the nonresponder parent (Shevach *et al.*, 1972; Shevach and Rosenthal, 1973). Thus, just as the MHC-linked antigens of the macrophage determined genetic restriction, the MHC-linked antigens of the macrophage determined *Ir* gene control. The parallelism was already apparent.

However, a concurrent study found no genetic influence of macrophage source on the *in vitro* antibody response to the terpolymer of glutamic acid, alanine, and tyrosine (GAT), under *Ir* gene control in mice (Kapp *et al.*, 1973). In view of the subsequent discovery of a critical role of macrophages in the *Ir* gene-controlled response to GAT by the same investigators (Pierce *et al.*, 1976*a,b*, 1977), it still remains unclear why no role of macrophages was seen in these earlier studies.*

In view of the conflicting results, and because of the lack of evidence for antigen-spe-

---

*However, several points may be relevant. First, the responses studied were primary *in vitro* responses (that is, the cells came from unimmunized animals). The same group did not see genetic restriction of antigen presentation for primary, as opposed to secondary, responses (Pierce *et al.*, 1976*b*, 1977). Thus even for an apparent exception, the parallelism between genetic restriction and *Ir* gene control holds. Second, the studies were done with allogeneic combinations of T cells and macrophages, rather than F_1 T cells and parental macrophages. Therefore, allogeneic effects might have accounted for the ability of low responder macrophages to present GAT to high-responder T and B cells. Third, a major role for active suppression in the *Ir* gene control of the response to GAT was subsequently found (see Section 5). Therefore, the failure of nonresponder T and B cells to respond to antigen presented by responder macrophages may have been due to suppression.

cific receptors on macrophages, the role of macrophages in *Ir* gene expression was not widely considered for several years. However, a series of similar studies, all using (high responder × low responder) F_1 T cells and parental B cells and/or macrophages, all firmly demonstrated the parallelism between genetic restriction of T cell–macrophage or T cell–B cell interaction and the apparent expression of *Ir* genes in macrophages or B cells, respectively. First, (high responder × low responder) F_1 carrier-immune helper T cells were found to help high-responder parental B cells (plus macrophages) but not low-responder parental B cells (plus macrophages) in the production of anti-hapten antibodies to a hapten-carrier conjugate (Katz *et al.*, 1973*c*). This result was interpreted as demonstrating either the expression of *Ir* genes in B cells or the influence of *Ir* genes in the genetic restriction of T cell–B cell interaction. It may turn out that these two interpretations are, in fact, mechanistically indistinguishable from one another.

Similarly, when both T and B cells were of (high responder × low responder) F_1 origin, the role of macrophages was studied by using GAT prebound to macrophages as either *in vivo* immunogen or *in vitro* stimulant (Pierce *et al.*, 1977). It was found that T cells and B cells from unimmunized F_1 mice made anti-GAT antibody in response to GAT on macrophages of the low-responder as well as of the high-responder type. However, T and B cells from F_1 mice immunized with GAT responded to GAT only on high responder, not low-responder, macrophages. F_1 mice immunized with GAT already bound to high responder macrophages gave results similar to those immunized with soluble GAT. In contrast, the restriction could be overcome by immunizing initially with GAT bound to low-responder parental macrophages. In this case, the F_1 T and B cells subsequently responded only to antigen on low responder macrophages, which must therefore bind antigen in a functional form. Thus, the genetic restrictions described earlier (Pierce *et al.*, 1976*b*; see Section 8.3) still applied in *Ir* gene-controlled systems, and two populations of (presumably T) cells existed in the F_1 which could potentially respond to antigen on macrophages of one parent or the other. The added effect of *Ir* gene control was that when soluble antigen was used to immunize, in the presence of F_1 macrophages bearing Ia antigens of both parental types, the clones of T cells specific for antigen plus high-responder Ia were preferentially "primed" (or expanded), instead of both subpopulations equally. Therefore, the *Ir* gene phenomenon appeared to be a perturbation on the genetic restriction phenomena described in Section 8. The failure to see even a primary response to GAT on low-responder macrophages, in spleen cells from F_1 animals immunized with soluble GAT, suggested a concomitant active suppression of this response, as subsequently demonstrated by Pierce and Kapp (1978*b*) and Germain and Benacerraf (1978*b*). In fact, immunization with GAT bound to F_1 macrophages induced helper T cells that would be restimulated by GAT bound to either low- or high-responder parental macrophages, but simultaneous injection of free GAT with GAT bound to F_1 macrophages also induced suppressor T cells which specifically blocked the response to GAT on low responder parental macrophages but not high-responder macrophages (Pierce and Kapp, 1978*b*). Since similar restrictions occur for other antigens the response to which is under *Ir* gene control, but for which suppression does not appear to play a predominant role, suppression of the response to antigen on low-responder macrophages may not be the primary mechanism of this *Ir* gene restriction. Rather, suppression may be manifested because a helper response is too weakly stimulated (see discussion of these studies in Section 5). This hypothesis is supported by the ability to induce suppressors even in high-responder mouse spleen cells by very high antigen concentrations or by normally immunogenic concentrations under conditions of relative macrophage depletion (Pierres and Germain, 1978; see also Ishizaka and Adachi, 1976; Feldmann and Kontiainen, 1976). Therefore, the difference between high and low responders for GAT appears to lie in the balance between, or relative ease of induction of, helper and suppressor T cell populations, not the absolute absence or presence of either cell type. The key question is why this balance is different for the T cell populations in the F_1 specific for the two parental *H-2* types when soluble GAT is given in the presence of F_1 macrophages displaying apparently equal amounts of Ia antigens of the two parental types.

A similar set of experiments was carried out for a pair of antigens of which the *Ir* gene control does not appear to involve a suppressive mechanism, (T,G)-A–L and (H,G)-A–L (Singer *et al.*, 1978). The elegant approach took advantage of the fact that B10 mice are high responders to (T,G)-A–L but low responders to (H,G)-A–L, while the reciprocal is true for B10.A mice. Thus, the experiments using (B10 $\times$ B10.A) F_1 T and B cells, depleted of macrophages and reconstituted with parental strain macrophages for a primary *in vitro* response to TNP-(T,G)-A–L or TNP-(H,G)-A–L, could be completely reciprocally internally controlled since each parental macrophage came from a responder to one antigen but a nonresponder to the other. In addition, the response to TNP-KLH, to which both strains respond, was measured as a further positive control. The result for each antigen was that only the respective high-responder macrophage, not the low-responder macrophage, could present that antigen. The gene determining the ability of the macrophages to successfully present these antigens mapped in the *K* or *I-A* subregions of *H-2*, indistinguishable by known recombinants from the *Ir-1* gene for (T,G)-A–L and (H,G)-A–L. These studies helped to establish that the macrophage played a key role in the expression of *Ir* genes. However, since the studies of both Pierce *et al.* (1977) and Singer *et al.* (1978) used F_1 T and B cells, they did not address the question of whether *Ir* genes were expressed in T cells or B cells as well.

Kappler and Marrack (1977) approached this issue for that portion of the response to TNP-sheep erythrocytes which cross-reacted with TNP-burro erythrocytes, which they showed to be under *Ir* gene control. In this case T cells from (high responder $\times$ low responder) F_1 mice immunized *in vivo* with sheep erythrocytes were cultured with parental B cells and macrophages and TNP-burro erythrocytes. The F_1 T cells provided only poor help for the low-responder B cells and macrophages. Importantly, the fact that this low responsiveness could not be overcome by the addition of high-responder macrophages suggested that the *Ir* gene "defect" was expressed at least in the B cells, although the experiments could not exclude expression in the other cell types as well. An even stronger study by the same group (Marrack and Kappler, 1978) employed the more widely studied antigen, (T,G)-A–L, in the same congenic strain combinations (B10 vs. B10.A, and C3H.SW vs. C3H) studied by other groups. These studies used parental macrophages pretreated ("pulsed") with antigen, and washed, as the sole source of antigen. In addition, in some experiments the B cell population was depleted of macrophages to eliminate antigen carryover from the pulsed macrophages. T cells from (B10 $\times$ B10.A) F_1 mice immunized *in vivo* with (T,G)-A–L provided help for high-responder B10, but not low-responder B10.A, B cells (depleted of macrophages) in a response to TNP-(T,G)-A–L, regardless of whether the antigen was presented on high-or low-responder macrophages. In addition, low-responder macrophages did not function with either type of B cell. Therefore, the conclusion was again reached that B cells, as well as macrophages, could express the *Ir* gene "defect."

In contrast, in a primary *in vitro* response to TNP-(T,G)-A–L, Hodes *et al.* (1979) found that both high-responder B10 and low-responder B10.A B cells produced equally high anti-TNP responses in the presence of B10, but not B10.A, macrophages, when help was provided by T cells from unimmunized (B10 $\times$ B10.A) F_1 mice. Thus, they concluded that macrophages but *not* B cells expressed the *Ir* gene defect. It can be seen that the

543

IMMUNE RESPONSE
GENES IN THE
REGULATION OF
MAMMALIAN
IMMUNITY

arguments for expression of *Ir* genes in B cells plus macrophages vs. macrophages alone exactly parallel the corresponding arguments for genetic restriction discussed in Section 8.3.4. On the one hand, this parallelism even of conflicting results strongly supports the notion that at least some *Ir* gene phenomena and genetic restriction phenomena are different facets of the same physiologic mechanism. On the other hand, it should be apparent that the same differences between *in vivo* priming and no or *in vitro* priming of T cells that distinguished the conflicting results for genetic restriction also apply to the corresponding experiments on *Ir* gene expression. It may be that helper T cells primed *in vivo* are genetically restricted in their cooperation with B cells, and therefore the *Ir* gene-dependent failure to prime that subpopulation of F_1 T cells that would cooperate preferentially with low-responder B cells results in an apparent *Ir* gene defect in B cells as well as in macrophages. The unprimed F_1 T cells may show the restriction only for macrophages (see Sections 8.3.4 and 12 for further details of this argument, which need not be repeated here).

Although the explanation based on priming may be the key distinction that explains the conflicting results, one additional factor applies in the case of *Ir* genes that did not apply for genetic restriction alone. The responses studied by Hodes *et al.* (1979) (as well as by Marrack and Kappler, 1978) were anti-TNP responses to TNP-(T,G)-A-L, so that only the help was (T,G)-A-L-specific, while the B cells were TNP-specific. Since the hapten TNP is not itself under *Ir* gene control, there is no reason to expect TNP-specific B cells to express a (T,G)-A-L *Ir* gene defect, unless this apparent "defect" reflects the genetic restriction specificity of the (T,G)-A-L-specific helper T cells. This caveat is particularly important since Howie and Feldmann (1978) found that a T cell helper factor for (T,G)-A-L could help (high-responder $\times$ low-responder) F_1 B cells to make antibody in the presence of high- but *not* low-responder macrophages if (T,G)-A-L was the antigen, but in the presence of macrophages of either type if DNP-(T,G)-A-L was the antigen. However, since Marrack and Kappler (1978) also studied hapten-specific B cells rather than (T,G)-A-L-specific B cells, this fact is not likely to be the only explanation for the failure of Hodes *et al.* (1979) to see an *Ir* gene restriction in B cell function as well. In fact, if the two different types of experiments (with and without *in vivo* priming, respectively) were to show the same opposing results even for (T,G)-A-L-specific B cells, then the similar results found for TNP-specific B cells would support the hypothesis that the *Ir* gene "defect" in B cells and macrophages is really a reflection of a limitation (or "defect") in the genetic restriction specificity of (T,G)-A-L-specific helper T cells.

The results of Howie and Feldmann (1978) mentioned above using (T,G)-A-L-specific helper factor to completely replace helper T cells (see Section 7) suggest an additional *Ir* gene function in the "presentation" of this helper factor to (high responder $\times$ low responder) F_1 B cells by high- but not low-responder parental macrophages. The alternative explanation would be that the factor itself, like the T cells from which it derived, was genetically restricted in its ability to interact with B cells and/or macrophages. The major weakness in this interpretation is that the factor works across complete allogeneic barriers, so that the only restriction would have to be the *Ir* gene restriction.

Perhaps the most definitive set of evidence for a common mechanism underlying MHC-genetic restriction of cell–cell interactions and *Ir* gene phenomena, especially on a molecular level, comes from studies of the antigen-specific T cell proliferative response in mice, where only T cells and macrophages are involved. The requirement for high-responder antigen-bearing macrophages to stimulate *in vivo*-primed (high responder $\times$ low responder) F_1 T cells has been shown for both GAT (Yano *et al.*, 1978) and GLPhe (R. H. Schwartz *et al.*, 1979a), but the key results regard the complementing *Ir* genes for GLPhe (see also Sections 1.3 and 4.2). Several approaches were used to demonstrate that both of the complementing *Ir* genes (α and β) must be present in the same cell, and that

545

IMMUNE RESPONSE
GENES IN THE
REGULATION OF
MAMMALIAN
IMMUNITY
·

at least one such cell requiring both genes is the antigen-presenting cell (macrophage) (Schwartz *et al.*, 1979a). First, tetraparental bone marrow radiation chimeras, made by reconstituting a lethally irradiated (low-responder $\alpha^+\beta^-$ × low-responder $\alpha^-\beta^+$) F_1 recipient with T cell-depleted bone marrow from both types of low-responder parents, failed to respond to GLPhe even though the two cell types in the chimera were mutually tolerant, and could respond to other antigens to which one or the other parent was a responder. The difference between these chimeras, which did not respond, and an F_1 hybrid which did respond, was the fact that the α and β genes were in different cells in the chimera, whereas they were both in the same cell in the F_1 hybrid. This result formally proved that the complementation between α and β *Ir* genes for GLPhe could not be between different classes of cells each possessing only one of the genes (see Munro and Taussig, 1975). Rather, the complementation was intracellular, and at least one cell type had to bear both genes. To determine which cell type required both genes, the familiar experiment using F_1 T cells and parental macrophages was performed. Neither the $\alpha^+\beta^-$ nor the $\alpha^-\beta^+$ low responder parental antigen-bearing macrophages could stimulate F_1 T cells, although they functioned for other antigens. In addition, a mixture of the two parental antigen-presenting cells did not stimulate. Only the F_1 hybrid or a recombinant strain antigen-presenting cell, containing both α and β genes, could present antigen. The reason this requirement for intracellular complementation between the α gene (in the *I-E* subregion) and the β gene (in the *I-A* subregion) in the antigen-presenting cell is so critical is that it fits exactly with the prediction of the theory that *Ir* gene phenomena are one facet of the genetic restriction mechanism that involves T cell recognition of antigen in association with surface Ia antigens on the antigen-presenting cell. Jones *et al.* (1978) and Cook *et al.* (1979c) had shown that in these same *H-2* haplotypes, one Ia antigen was made by complementation of an α chain encoded in the *I-E* subregion with a β chain encoded in the *I-A* subregion. Unless both genes were expressed in the same cell, the Ia molecule was not present on the cell surface (see Sections 1.3 and 4.2). This striking parallel between *Ir* gene complementation and Ia antigen complementation is unlikely to be fortuitous. If in fact it represents complementation of the same genes, the results strongly imply that one *Ir* gene product is the *Ia* antigen itself, and that the mechanism of action of these *Ir* genes is through genetic restrictions in the repertoire of clones of T cells which can recognize the particular antigen in association with different Ia molecules. This theory, which is thus given a firmer grounding at the molecular level, can still accommodate either altered-self or dual recognition-type mechanisms (see Section 8.1). The hypothesis that the Ia antigens on the macrophage alter or orient the antigen in a specific way would fit within the altered-self or neoantigenic determinant model; the hypothesis that the *Ir* gene defect is due to absence of a clone of T cells that can recognize that antigen in association with that particular Ia molecule would fit within either a dual recognition or altered-self model, depending on whether one or two receptors were required on the T cell and how closely physically associated the antigen must be with the Ia molecule. These ideas will be further explored in Sections 10 and 12. At this point, however, we can safely conclude that at least some *Ir* gene phenomena represent a facet of the same process that leads to MHC-linked genetic restrictions on cell interactions.

9.2   The Role of Macrophages in Determinant Selection by Ir Genes

Several studies using natural protein or polypeptide antigens, with unique, nonrepeating antigenic determinants, have extended the role of the macrophage to the determination of which antigenic determinants, under *Ir* gene control, can elicit a response. The first of these was the case of the T cell proliferative response to insulin in the guinea pig (Barcinski and Rosenthal, 1977; Rosenthal *et al.*, 1977; reviewed with additional data in Rosenthal, 1978). It was found that strain 2 guinea pigs responded to a determinant involving the A chain loop from residues A8 to A10, and therefore when immunized with pork insulin responded preferentially to pork insulin over sheep insulin, which differs in this determinant (see Section 4.1). In contrast, strain 13 guinea pigs immunized with pork insulin responded equally well to pork and sheep insulin, since the determinant in the B chain recognized by strain 13 animals does not differ in these two insulins. What was important for this discussion was that when T cells from strain (2×13) F_1 hybrid guinea pigs immunized with pork insulin were challenged with different insulins or free chains bound to parental macrophages, they responded to just those insulins or chains to which the donor of the macrophages would have responded. This result is exactly analogous to that of Shevach and Rosenthal (1973) except that instead of two different antigens such as DNP-GL and GT, one was dealing with two determinants on the same protein molecule. Thus, the antigen-presenting macrophage was critical in determining which determinants elicited a response from the F_1 T cells. Using the method of preventing DNA replication in one subpopulation of T cells by stimulating F_1 T cells with antigen on one parental strain macrophage, treating with BUdR and then light, and reculturing with the same or a different antigen-macrophage combination (as described in Section 8.2.1), it was shown that different populations of T cells in the F_1 were responding to the antigen on the different parental macrophages, and that these had the expected differences in antigen specificity. Thus, "determinant selection" by macrophages (as coined by Rosenthal *et al.*, 1977) was mediated by selective stimulation of the subpopulation of F_1 T cells specific for that determinant.

It was most likely that the responses to different antigenic determinants on insulin were controlled by different *Ir* genes, but because of the lack of intra-MHC recombinants in the guinea pig, these could not be separated. On the other hand, we had shown that different *Ir* genes, mapping in distinct subregions, *I-A* and *I-C*, of *H-2*, controlled the immune responses to different determinants on the single polypeptide chain antigen, sperm whale myoglobin, and that the antibody response and T cell proliferative response to a given determinant were controlled in parallel by the same apparent gene (Berzofsky, 1978*b*; Berzofsky *et al.*, 1979; see Section 4.1). In addition, we showed that the T cell proliferative response of macrophage-depleted lymph node T cells could be reconstituted with Kupffer cells, the tissue macrophage of the liver, which could be obtained relatively free of contaminating T and B lymphocytes (Richman *et al.*, 1979). The response to myoglobin was therefore the ideal situation in which to test whether "determinant selection" by macrophages was generalizable and if so, whether it was due to determinant-specific *Ir* genes functioning in the selection of T cell clones because of restrictions in macrophage–T cell interaction.

We found that macrophage-depleted lymph node T cells from myoglobin-immunized B6D2F$_1$ (low-responder $\times$ high-responder) F_1 mice responded to myoglobin *in vitro* only if reconstituted with F_1 or high-responder parental (*H-2*^{d}, DBA/2, or B10.D2) macrophages, but not if reconstituted with low-responder (*H-2*^{b}, B10) macrophages (Richman *et al.*, 1980). The failure of low-responder *H-2*^{b} macrophages to present myoglobin was antigen specific, since they presented ovalbumin to doubly immunized (myoglobin plus ovalbumin) F_1 T cells as well as did the congenic *H-2*^{d} macrophages. The novel feature was that we could now extend this familiar F_1 protocol to look at macrophages from recombinant strains of mice bearing only one or the other high-responder *Ir* genes, and could examine the proliferative response to fragments of myoglobin as well. Macrophages from B10.A (5R) mice, bearing high-responder I-C^d (Ir-Mb-2) but low-responder I-A^b could present only fragment (1–55) but not fragment (132–153) to the B6D2F$_1$ T cells immunized with whole myoglobin. Therefore, in addition, they could only partially reconstitute the *in vitro* response to whole myoglobin. In contrast, macrophages from D2.GD or B10.GD mice, bearing the high-responder I-A^d (Ir-Mb-1) but the low-responder I-C^b, presented both fragments. Thus, the same preparation of F$_1$ T cells responded exactly as would the strain from which the macrophages were derived (see Fig. 3). Therefore, the Ir genotype of the macrophage completely determined the Ir phenotype of the responding F$_1$ T cell population. Since the different Ir genes controlled the responses to different determinants, the overall effect was that the choice of macrophage used to present antigen determined the choice of antigenic determinants to which the T cells could respond. Hence, the phenomenon of "determinant selection" by macrophages appears to be due to the actions of different Ir genes for different determinants, operating in the selection of which T cell clones are stimulated. Consistent results were recently reported for the case of fibrinopeptide B by Thomas *et al.* (1979).

The mechanism of this selection is then open to all of the hypotheses which may explain the operation of Ir genes in T cell–macrophage interactions, which, as we have just seen in Section 9.1, is probably a facet of the genetic restrictions seen in T cell recognition of antigen in association with Ia molecules on the antigen-presenting cell. It is possible, as suggested by Rosenthal (1978), that the Ia molecules of high- and low-responder macrophages process or bind antigens differently from one another and thereby directly select which determinants are presented to T cells. As pointed out by Rosenthal (1978), this hypothesis avoids the need to postulate a dual specificity for T cells. Only the selected antigenic determinant need be recognized. Operationally, this theory behaves like a neoantigenic determinant or altered-self hypothesis. However, because we have seen parallel control of the antibody and T cell proliferative responses for the same region of the myoglobin molecule, and because we have no evidence for differences in processing of myoglobin by H-2^b versus H-2^d macrophages, we prefer the hypothesis that contains the following three elements: (1) that T cells have combinations of specificities for antigenic determinant and Ia molecule (whether via one receptor or two); (2) that fragment (132–153) of myoglobin is bound perfectly well to macrophages bearing I-A^b and I-C^d, but that no clone of T cells exists in the repertoire which has that combination of specificities [although there are clones specific for fragment (132–153) plus other Ia molecules, and other clones specific for I-C^d and other determinants of myoglobin] (Fig. 10); and (3) that the clone of helper T cells stimulated originally by a given determinant and Ia combination on the macrophage must recognize that same combination of antigenic determinant and Ia on the B cell in order to provide help (see also Figs. 8 and 9). The latter parallel control of T cell recognition of macrophage and B cell is consistent with the data and ideas of Paul and Benacerraf (1977), Sprent (1978a,b), Yamashita and Shevach (1978), and Benacerraf (1978). How such dual specificities might arise and what determines such "holes" in the repertoire of T cells will be discussed in Sections 10 and 12.

10 The Effect of T Cell Maturation in Chimeras on Ir Phenotype

In view of the striking parallels between Ir gene expression and genetic restrictions on the interaction of T cells with macrophages, B cells, stimulators, or targets described in

547

IMMUNE RESPONSE
GENES IN THE
REGULATION OF
MAMMALIAN
IMMUNITY

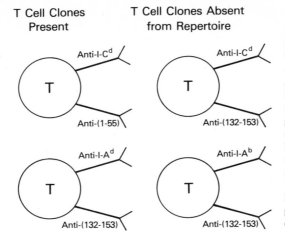

T Cell Clones Present · T Cell Clones Absent from Repertoire

Anti-I-C^d · Anti-(1-55) · T

Anti-I-C^d · Anti-(132-153) · T

Anti-I-A^d · Anti-(132-153) · T

Anti-I-A^b · Anti-(132-153) · T

Figure 10. Examples of some of the clones of helper T cells postulated to be present or absent in the (H-2^b × H-2^d)F$_1$ mouse immunized with whole sperm whale myoglobin (residues 1–153). Although dual specificity for antigenic determinant plus Ia is depicted by two receptors for simplicity, the hypothesis is compatible with a single receptor bearing dual specificity, or neoantigenic determinant specificity. Note that because of the two clones shown as missing from the repertoire, macrophages bearing only *I-A^b*- and *I-C^d*-encoded antigens have no T cells to which to present fragment (132–153). This hypothesis may explain the "selection" by *Ir* genes of which determinants on an antigen molecule are immunogenic in a given strain of mice.

Sections 8 and 9, it was exciting to ask whether the environment in which T cells matured from stem cells would alter their *Ir* phenotype in the way it could alter the genetic restrictions on their interactions with other cell types (see Sections 8.1.2 and 8.3.3). In the past year and a half, several studies have approached this question. The results have major implications for our understanding of *Ir* genes.

10.1 The Effect of the Chimeric Environment on Ir Gene Control of Cytotoxic Responses

10.1.1 Ir Genes Controlling Cytotoxic Responses: Diverse Mechanisms

Although we have not discussed them earlier in this chapter, a number of *Ir* gene-like phenomena have been described which appear to regulate the responses of cytolytic T cells against MHC-compatible targets bearing chemical, viral, or minor histocompatibility antigens. Many, but by no means all, of these genes map in the *H-2K* or *D* regions rather than the *I* region of *H-2* (Schmitt-Verhulst and Shearer, 1975, 1976; Gomard *et al.*, 1977; Simpson and Gordon, 1977; Hurme *et al.*, 1978; von Boehmer *et al.*, 1977; Billings *et al.*, 1978*a*; Zinkernagel *et al.*, 1978*d*; Doherty *et al.*, 1978; Pfizenmaier *et al.*, 1978). It is this fact that led to the proposal (Schmitt-Verhulst and Shearer, 1976; Zinkernagel, 1978) that *Ir* genes regulating a given type of response mapped in the same region as the class of histocompatibility antigen restricting that type of response because genetic restriction of T cell interaction with other cells was the process in which *Ir* genes operated.

However, not all putative *Ir* genes mapping in the *H-2K* or *D* regions behaved like classical *Ir* genes. In particular, an "*Ir* gene" mapping to *H-2K^k* was dominant, rather than recessive, in producing *low* responsiveness to vaccinia or Sendai virus or TNP associated with other *H-2* antigens on the same cell (Zingernagel *et al.*, 1978*d*; Levy and Shearer, 1979). This dominance suggested that the mechanism involved preemption of the response by antigen associated with *H-2K^k*. Mechanisms considered for this preemption include immunodominance (basically, preemption of the response by the most immunogenic antigen) and suppression. Immunodominance is consistent with most of the data, and is especially supported by the finding that ($H\text{-}2^k \times H\text{-}2^b$) F_1 cytotoxic T cells could be sensitized to $H\text{-}2D^b$ in $H\text{-}2^b$ irradiated recipients, but not in $H\text{-}2K^kD^b$ recombinant irradiated recipients (Zinkernagel *et al.*, 1978*d*). There are no data to support the hypothesis of suppression.

549

IMMUNE RESPONSE
GENES IN THE
REGULATION OF
MAMMALIAN
IMMUNITY

A potentially very exciting additional alternative was suggested by the recent finding of O'Neill and Blanden (1979) that the presence of $H\text{-}2K^k$ actually led to a fourfold to fivefold reduction in the amount of other $H\text{-}2K$ or D antigens expressed on the cell surface of F_1 hybrid cells, as measured by absorption of anti-$H\text{-}2$ antisera. This decrease was at least twofold greater than would be expected from a gene dosage effect alone, and correlated with differences in potential to stimulate or serve as targets in both alloantigen-specific and ectromelia virus-specific cytotoxic responses. Therefore, it was proposed that the effect of $H\text{-}2K^k$ might be to actually depress the amount of other $H\text{-}2$ antigens synthesized or at least expressed on the cell surface. This observation, if borne out, may have extremely far-reaching implications for the interaction of different $H\text{-}2$ genes at the level of RNA transcription, protein synthesis, or cell surface expression. It is analogous in some ways to the observation of Meruelo *et al.* (1978) and Meruelo (1979) that virus infection can lead to imbalanced expression of one $H\text{-}2$ antigen on the cell surface, correlating with genetic restriction. On the other hand, it cannot explain all the results, since Zinkernagel *et al.* (1978*d*) found no immunodominance of $H\text{-}2K^k$ for responses to lymphocytic choriomeningitis virus in the same experiments that showed striking effects for responses to vaccinia and Sendai viruses. Also, since this mechanism operates in the stimulator and target cells, not the responder cells, it does not explain the conversion of Ir phenotype of responder T cells in chimeras (see discussion of Zinkernagel *et al.*, 1978*e*, in Section 10.1.2).

Other Ir genes mapping to $H\text{-}2K$ or D, such as the low response to Sendai or vaccinia virus in association with $H\text{-}2D^k$, manifest recessiveness of low responsiveness (Zinkernagel *et al.*, 1978*d*) and so may be more analogous to classical I region Ir genes.

In addition, there are Ir genes for cytotoxic responses which map in the $I\text{-}A$ subregion of $H\text{-}2$, such as the requirement for $I\text{-}A^b$ to produce a cytotoxic response to the $H\text{-}Y$ male antigen (Simpson and Gordon, 1977; von Boehmer *et al.*, 1977 Hurme *et al.*, 1978). It has been suggested that these Ir genes are involved in the action of T helper cells required for the generation of a cytotoxic response. Thus, Ir genes which map in $H\text{-}2K$ or D are thought to reflect the genetic restriction specificity of the killer T cells themselves, while those that map in the I region are thought to reflect the genetic restriction specificity of helper T cells needed for the response (Zinkernagel, 1978; von Boehmer *et al.*, 1978). Independent evidence for the function of helper T cells in a cytotoxic response has recently been reported (Finberg *et al.*, 1979).

Finally, complementing Ir genes for the cytotoxic response to $H\text{-}Y$ antigen have been mapped to the $I\text{-}C$ subregion of haplotypes other than $H\text{-}2^b$ (Hurme *et al.*, 1978). The mechanism of their action is unknown.

10.1.2 The Role of the Chimeric Environment in Ir Gene-Controlled Cytotoxic Responses

Taking into consideration the several types of Ir genes for cytotoxic responses just described in Section 10.1.1, von Boehmer *et al.* (1978) made two types of parent (nonresponder) → F_1 chimeras. Cytotoxic responses can be generated to $H\text{-}Y$ antigen in association with H-$2K^k$, D^k, and D^b, but not K^b or D^d (Simpson and Gordon, 1977; von Boehmer *et al.*, 1977). However, an *Ir* gene mapping in *I*-A^b must also be present to obtain a response. Therefore, mice bearing $K^k I^k D^k$ and those bearing $K^b I^b D^d$ are both low responders, but for different reasons. Chimeras of the sort $K^k I^k D^k \rightarrow (K^k I^k D^k \times K^b I^b D^b)$ F$_1$ produced low-responder T cells; that is, the phenotype of the donor T cells was not altered. In contrast, T cells from $K^b I^b D^d \rightarrow (K^k I^k D^k \times K^b I^b D^b)$ F$_1$ chimeras killed male (*H-Y*-bearing) H-2^k or H-2^b targets (von Boehmer *et al.*, 1978). It was postulated that the cytotoxic T cells from the first type of chimera could recognize *H-Y* associated with H-$2K^k$, D^k, or D^b-encoded antigens from the recipient, and that helper T cells of the chimera could recognize I^b-encoded antigens present in the recipient thymus. However, these helper T cells could not help the cytotoxic cells which, being of donor origin, bore only I^k-, not I^b-encoded antigens. Similarly, the cytotoxic and helper T cells of the second type of chimera "learned" (that is, derived from precursors selected for specificity for) the same genetic restrictions in the recipient thymus, since the recipients were equivalent. However, in this case, the cytotoxic cells, of donor $K^b I^b D^d$ origin, bore I^b-encoded antigens recognized by the helper T cells and so could receive help.

A third type of chimera was made by reconstituting irradiated parental $K^k I^k D^k$ recipients with $(K^k I^k D^k \times K^b I^b D^b)$ F$_1$ bone marrow (von Boehmer *et al.*, 1978). F$_1$ T cells raised in this H-2^k chimeric recipient failed to kill H-2^k or H-2^b male (*H-Y$^+$*) targets, even though normal F$_1$ T cells could kill both. In this case, a similar explanation was postulated: *H-Y*-specific F$_1$ T cells differentiating in the H-2^k thymus were selected for clones recognizing $K^k I^k D^k$-encoded antigens only. Therefore, killer cells which could kill *H-Y$^+$*, K^k or D^k (but not D^b) target cells should be present. However, no helper cells specific for I^b could be generated, so no response could be seen.

If these explanations were true, then one would predict that a mixture of T cells from the first and third types of chimeras would respond to H-$2K^k$- and D^k- (but not D^b)-encoded antigens. T cells from the first type of chimera would provide helper cells specific for I^b-encoded antigens, even though their killer cells could not receive the help. T cells from the third type of chimera would supply killer cells specific for K^k- and D^k-encoded antigens, and these, bearing I^b-encoded antigens, would be able to receive help from the helper T cells provided by the first chimera. While this exact experiment has not been tried, a similar type of mixing experiment was attempted (von Boehmer and Haas, 1979). For this experiment, a fourth type of chimera, $(K^k I^k D^k \times K^b I^b D^b) \rightarrow K^b I^b D^d$ was prepared. This was found not to respond, presumably because it lacked killer cells specific for *H-Y* in association with K^k- D^k-, or D^b-encoded antigens, none of which were expressed in the recipient thymus. However, it should have had helper T cells specific for I^b-encoded antigens which were needed for the killer cells of the third type of chimera to respond. Therefore, if T cells from the third and fourth types of chimeras were mixed, both (H-2^k $\times$ H-2^b) F$_1$ in genotype but not responding for opposite reasons, the former should provide functional killers able to receive help, the latter should provide functional helpers, and the mixture should respond. In fact, a low but definite response was obtained. The successful prediction of complementation in the mixing experiment supports the hypothesis that the two types of chimeras failed to respond for different reasons, and in turn supports the hypothesis that the *Ir* phenotypes of populations of cytotoxic T cells restricted to K- and D-encoded antigens, and helper T cells restricted to *I*-encoded antigens, can be altered by the environment in which they mature.

Similar results were obtained by Zinkernagel *et al.* (1978*e*) in the case of vaccinia-specific cytotoxicity. T cells from chimeras in which clones specific for vaccinia plus K^k-

encoded antigens would not have been selected to mature were no longer influenced by the immunodominance of K^k, and so could respond to vaccinia plus D^b-encoded antigens. Also, $K^d D^d$ T cells maturing in a D^k environment could not kill D^k targets, just as $K^d D^k$ T cells could not. Again, the *Ir* phenotype was converted just as was the restriction specificity.

A third example of *Ir* gene conversion is more difficult to explain in terms of genetic restriction. The ability of cytotoxic T cells specific for TNP on syngeneic stimulators to kill TNP-modified allogeneic targets was found to be under *Ir* gene control (Billings *et al.*, 1978a). Responders bearing K^k I-A^k failed to effect cross-reactive killing, whereas H-2^b responders did. H-2^k cells which had matured in an (H-2^k $\times$ H-2^b) F_1 chimeric recipient manifested cross-reactive killing even when the stimulator cells used were TNP-modified H-2^k cells (Billings *et al.*, 1978b). Thus, they behaved just as do F_1 responder T cells which also manifested cross-reactive killing after stimulation with H-2^k stimulators. Therefore, the *Ir* phenotype of the donor H-2^k T cells was converted to that of the F_1 recipient, just as in the other cases above. The puzzling result is that induction of cross-reactive clones did not require stimulation with H-2^b stimulators. This contrasts with the findings for *Ir* gene-controlled antibody responses in chimeras (Section 10.2.1, especially Hodes *et al.*, 1979). If TNP-modified H-2^k stimulators elicit only clones which would have been also expanded in an H-2^k thymus, then normal H-2^k mice should show cross-reactive killing. Since no evidence for suppressive mechanisms could be found, the simplest explanation is that clones specific for H-2^b plus TNP, expanded in the F_1 thymus, cross-react with H-2^k plus TNP (since, after all, it is cross-reactive killing that is being studied), and that it is these which are stimulated by TNP-modified H-2^k stimulators and can then also kill TNP-bearing H-2^d targets in a second cross-reaction. While this explanation is consistent with the interpretations of the studies above, it is still not clear why such clones, which can be stimulated by TNP-modified H-2^k stimulator cells, are not generated in the H-2^k thymus. The simplest explanation would be that a neoantigenic determinant rather than dual recognition mechanism applies, so that these cells which cross-react with TNP-plus-H-2^k do not have a receptor for H-2^k alone. [Reasoning of a similar type regarding the crossreactive lysis of hapten-modified syngeneic targets by cytolytic T cells specific for allogeneic targets led Lemonnier *et al.* (1977) to prefer a neoantigenic determinant hypothesis.]

In any case, the parallel conversion of *Ir* gene phenotype and genetic restriction specificity in the chimeric environment strongly supports a common mechanism for these two traits. Equally importantly, it implies that *Ir* gene phenotype is not a property of the T cell's own genotype, but of the environment in which it matures. The nature of this environment and the way it might affect T cell repertoire are discussed further in the next section and Section 12.

10.2 The Effect of the Chimeric Environment on Ir Gene-Controlled Antibody Responses

10.2.1 Bone Marrow Chimeras

The conversion of *Ir* gene phenotype of helper T cells for an antibody response in a chimeric environment was first demonstrated by Kappler and Marrack (1978). Low-responder H-2^a T cells, from an H-$2^a \rightarrow (H$-$2^a \times H$-$2^b)$ F_1 chimera or selected with anti-H-2^b and complement from an (H-2^a + H-2^b) $\rightarrow$ (H-$2^a \times H$-2^b) F_1 chimera, could cooperate with H-2^b B cells and macrophages to produce an anti-TNP-(T,G)-A–L

551

IMMUNE RESPONSE
GENES IN THE
REGULATION OF
MAMMALIAN
IMMUNITY

response. However, although these *H-2*ᵃ chimeric T cells cooperated with syngeneic *H-2*ᵃ B cells and macrophages in responses to KLH and sheep erythrocytes, they still did not help *H-2*ᵃ B cells plus macrophages in a TNP-(T,G)-A–L response. Therefore, it appeared that the *Ir* phenotype of the T cell *population* was converted from low responder to high responder by virtue of the expansion of its genetic restriction repertoire. Only the clones which recognized *H-2*ᵇ B cells and macrophages could provide (T,G)-A–L-specific help (see Fig. 11).

On the one hand, this result could be interpreted as further evidence for expression of *Ir* genes in B cells and/or macrophages. On the other hand, it is possible that only clones restricted to cooperate with *H-2*ᵇ B cells and macrophages include (T,G)-A–L-specific clones. It is possible that no clones simultaneously specific for both *H-2*ᵃ and (T,G)-A–L could be generated (see Section 12).

Further results (Kappler and Marrack, 1978), still consistent with either explanation but further supporting the role of the chimeric environment in generation of *Ir* phenotype, were that (*H-2*ᵃ × *H-2*ᵇ) F₁ → *H-2*ᵃ (low-responder) chimeric T cells could not provide (T,G)-A–L-specific help, but they could not cooperate with *H-2*ᵇ B cells plus macrophages either. Therefore, the loss of high responsiveness paralleled the contraction of the genetic restriction repertoire (see Fig. 11). F₁ → *H-2*ᵇ T cells behaved as did high responder *H-2*ᵇ T cells in this response. The restrictions mapped to *H-2K* or *I-A*, just as does the *Ir* gene.

Similar results in (high responder × low responder) F₁ → parent chimeras were obtained by Hedrick and Watson (1979) in the *Ir* gene-controlled response to collagen.

Hodes *et al.* (1979) extended these results, in the case of (T,G)-A–L, to examine the three cell types involved in a primary *in vitro* antibody response. The chimeric environment was found to influence the responder phenotype of chimeric T cells, but not chimeric B cells or macrophages. As in their other studies of *in vitro* primary responses, this group observed genetic restriction of T cells only for macrophages, not B cells, so that low-

Genotype of T Cell Precursor	Genotype of Thymus in Which T Cells Mature	Helper Cell Response With B Cells and Macrophages of:	
		High Responder	Low Responder
Low responder	(High × Low responder) F₁ ⟶	+	−
High responder	(High × Low responder) F₁ ⟶	+	−
High responder × Low responder F₁	Low responder ⟶	−	−
	High responder ⟶	+	−

Figure 11. Effect of the thymic environment in chimeric recipients on the *Ir* phenotype of donor T cell precursors. This scheme summarizes the results described for (T,G)-A–L, collagen, and GLPhe in Sections 10.2.1 and 10.2.2. Apparent exceptions for (T,G)-A–L in *H-2*ᶠ mice and for lysozyme in *H-2*ᵇ mice are noted in the text.

responder $H-2^a$ B cells (normal or chimeric) could cooperate with F_1 or chimeric parent ($H-2^a$ or $H-2^b$) → F_1 T cells as long as $H-2^b$ macrophages were provided. Otherwise, the results were in agreement with those of Kappler and Marrack (1978) (see Fig. 11).

553

IMMUNE RESPONSE
GENES IN THE
REGULATION OF
MAMMALIAN
IMMUNITY

It should be pointed out that this issue of T cell–macrophage restriction only versus T cell–B cell restriction as well is critical to the earlier chimera studies of Bechtol and McDevitt (1976), Bechtol *et al.* (1974) and Press and McDevitt (1977) (see Section 3.1). A limitation of the restriction imposed by *Ir* genes to T cell–macrophage interactions would lead to a prediction of the results reported by Bechtol and McDevitt (1976) and Bechtol *et al.* (1974), since T cells of either parental origin in the allophenic chimera would be expected to cooperate with low-responder as well as high-responder B cells as long as high-responder macrophages were present in the chimera. Thus, anti-(T,G)-A–L antibodies of low-responder allotype would be produced. On the other hand, a T cell–B cell restriction imposed by *Ir* genes would result in anti-(T,G)-A–L production only by the high-responder B cells in the chimera, as reported by Press and McDevitt (1977). Since both studies employed *in vivo* immunization and the assay of serum antibody, we cannot attribute the differences to presence or absence of *in vivo* immunization. The most likely explanation lies in the different methods of constructing chimeras. Since different results can be obtained with radiation bone marrow chimeras when the method of elimination of T cells from the donor marrow or the dose of irradiation is varied (see Zinkernagel, 1978), it is not surprising that such differences might be found between radiation-bone marrow chimeras and allophenic tetraparental mice produced by fusing embryos. Which result is more "physiological" is still open to question.

Finally, two groups found exceptions to the above results that the *Ir* phenotype of T cells was determined by the environment during maturation, not the genotype of the T cells. First, Marrack and Kappler (1979) tested anti-(T,G)-A–L responses of chimeras produced from low-responder $H-2^f$ bone marrow reconstitution of ($H-2^b$ × $H-2^f$) F_1 irradiated recipients. It should be recalled that $H-2^f$ low responders were found to have a defect in helper T cell factor production to (T,G)-A–L (Munro and Taussig, 1975; Howie and Feldmann, 1977) but no defect in the ability of B cells to respond to the factor produced by other strains (Munro and Taussig, 1975), in contrast to $H-2^k$ and $H-2^a$ low-responders discussed above which had the opposite defect (see Sections 3.2.2 and 7.1). T cells derived from $H-2^f$ bone marrow, maturing in an ($H-2^b$ × $H-2^f$) recipient, were *not* converted to high responders, whether tested with $H-2^f$ or $H-2^b$ B cells and macrophages. Therefore, these chimera studies suggest that the $H-2^f$ *Ir* gene defect is expressed directly in T cells, and is not influenced by the environment in which they mature. This result would be consistent with the finding of T cell factor production by $H-2^a$ but not $H-2^f$ types of low responders, and would reaffirm the existence of two distinct classes of *Ir* gene "defects."*

*However, several puzzling points remain to be explained. First, the $H-2^s$ low-responder mice have been found to behave like $H-2^f$ low responders, with a defect in factor production by T cells but not factor acceptance by B cells and/or macrophages (Howie and Feldmann, 1977; Munro *et al.*, 1978). Nevertheless, the $H-2^s$ donor T cells, maturing in an ($H-2^b$ × $H-2^s$) recipient, were converted to high responders (M. Feldmann, personal communication). Therefore, either there are three distinct types of low responders, or there are conflicting results regarding the class of low responders which includes $H-2^f$ and $H-2^s$ strains. Second, the two classes of low responders behave in chimeras in a way *opposite* to what one might predict from their putative defects in factor production and acceptance. One simple view would be that strains such as $H-2^a$ whose T cells could produce factor in a syngeneic environment would have T cells that could already respond to (T,G)-A–L presented by syngeneic $H-2^a$ macrophages. The defect in factor acceptance would be reflected in $H-2^a$ B cells and/or macrophages or the ability of $H-2^a$ macrophages to present *factor* to B cells (Marrack and Kappler, 1978; Singer *et al.,* 1978; Howie and Feldmann, 1978). In this case, maturation in chimeras should not be necessary

The second exception to conversion of Ir gene phenotype of T cells in chimeras was recently reported by Hill and Frelinger (1979) in the case of the response to hen egg lysozyme. In this case, no chimeric combination allowed B10 (H-2^b) low-responder T cells to provide help for high- or low-responder B cells to lysozyme. Even though suppressive mechanisms exist in response to lysozyme (Sercarz *et al.*, 1978; Adorini *et al.*, 1979*a,b*; Araneo *et al.*, 1979), no suppression could be demonstrated that would explain the results. However, a suppressive mechanism specific for (affecting only) H-2^b T cells could not be excluded by the experiments since the H-2^b T cells produced no response which could be tested for suppression. If suppression is not the explanation, then this study provides further evidence that a second type of Ir gene mechanism exists, intrinsic to the helper T cell (see Section 12).

10.2.2 Thymus Chimeras

Because of the similarity between the effects of the chimeric environment on Ir gene function and T cell genetic restriction repertoire, it was of great interest to determine whether the thymus was the critical organ which exerted the former effects (on Ir genes) as had been shown for the latter effects (see Section 8). Two recent reports suggest that this is the case.

Hedrick and Watson (1979) transplanted high-responder H-2^b or low-responder H-2^d thymuses into congenitally athymic nude (H-$2^b \times H$-2^d) F_1 mice and tested the *in vivo* serum antibody response to collagen. Recipients of H-2^b thymus gave high responses comparable to those of H-2^b mice, whereas recipients of H-2^d thymus gave low responses comparable to those of H-2^d mice, even though both chimeras were immunologically competent in responses not controlled by known Ir genes.

A similar concurrent study was carried out for (T,G)-A–L (H. Waldmann, personal communication) and GLPhe (Waldmann *et al.*, 1979b), using adult-thymectomized lethally irradiated F_1 mice reconstituted with syngeneic F_1 bone marrow and grafted with an irradiated thymus from one parent or the other. In the case of (T,G)-A–L, the F_1 was (H-$2^k \times H$-2^d), and the response of T cells from the chimera paralleled that of the strain donating the thymus. Note that H-2^k, like H-2^a, low responders to (T,G)-A–L show a defect which is amenable to conversion in chimeras.

The results for GLPhe are potentially even more exciting. The F_1 recipients were a hybrid between two complementing low responder H-2 haplotypes, H-2^a and H-2^b. It will be recalled that the F_1 responds even though neither parent does (Dorf *et al.*, 1975) and that both genes must be expressed in the same antigen-presenting cell (macrophage) (R. H. Schwartz *et al.*, 1979a). Thymectomized and irradiated F_1 (responder) mice reconsti-

---

to induce H-2^a T cells to respond to (T,G)-A–L. On the other hand, if H-2^f T cells do not respond to (T,G)-A–L in a syngeneic environment (with H-2^f macrophages), expansion of their repertoire by maturation in an (H-$2^b \times H$-2^f) F_1 environment might result in expansion of clones which could respond to (T,G)-A–L on H-2^b macrophages, in contrast to the results of Marrack and Kappler (1979). Also, if H-2^f B cells and macrophages do not manifest the Ir gene defect (Munro and Taussig, 1975), one would have expected F_1 T cells to provide help for H-2^f B cells plus macrophages, also in contrast to the findings of Marrack and Kappler (1979). If these results are confirmed, they would imply that H-2^f low responders express both types of defects (a possible explanation for the difficulty in demonstrating complementation in (H-$2^{a \text{ or } k} \times H$-$2^f$)$F_1$ mice (Munro and Taussig, 1977; Rüde *et al.*, 1977; Deak *et al.*, 1978; Marrack and Kappler, 1978). They would also suggest that the Ir gene defect in H-2^f T cells involves more than just the genetic restriction repertoire induced during maturation (in the thymus). Since this conclusion would imply a totally different mechanism for one type of Ir gene, it is important to confirm (see Section 12).

tuted with F_1 (responder) bone marrow failed to respond when reconstituted with either parental type of thymus. Only an F_1 thymus graft resulted in an *in vivo* serum antibody response to GLPhe (Waldmann *et al.*, 1979*b*). This result is exciting because it suggests that the hybrid Ia molecule, present on the F_1 cells but not those of either parent (Jones *et al.*, 1978; Cook *et al.*, 1979*c*), is itself the crucial determinant which must be expressed in the thymus during maturation and selection of the T cell repertoire as well as in the macrophage which subsequently presents the antigen. Unfortunately, the critical control of grafting both parental thymuses into the same recipient was not done in this study, but these experiments are in progress (H. Waldmann, personal communication). One would predict, as in the case of the macrophage, that both *I* region genes must be expressed in the same thymic cell to achieve complementation.

That the T cell genome itself does not have to bear either complementing gene was shown elegantly by D. Longo and R. H. Schwartz (personal communication). They used B10.A (4R) recombinant T cells, which carry neither the α nor the β gene for GLPhe, which had matured in a responder ($I^a \times I^b$) F_1 chimera and were then immunized with GLPhe in a similar F_1, acutely irradiated, recipient with additional F_1 macrophages. Even though the T cells themselves carried neither complementing gene, they responded to GLPhe, presumably because appropriate clones were selected in the F_1 thymus of the chimera.

These results all imply that, even in the case of *Ir* genes whose expression appears to be determined by B cells and/or macrophages, the *T cell repertoire,* determined by the Ia antigens expressed in the thymus during T cell maturation, may be the primary site of *Ir* gene function. Once the T cell repertoire is fixed by this process, the later events are secondary. Thus, it may be suggested that at least some types of *Ir* genes are expressed in the T cell repertoire, if not the T cell genome, and that this expression is determined by the Ia antigens of the radioresistant portion of the thymus (see Figs. 9 and 11 and Section 12).

11 Ir Genes in Man and the Role of Ir Genes in Disease

This chapter would not be complete without at least a brief summary of the possible role *Ir* genes play in man and in particular in human disease. Since several more extensive reviews have been published (Mann, 1977; van Rood *et al.*, 1977; Bodmer, 1978; Mann and Murray, 1979; Snell *et al.*, 1976; Vladutiu and Rose, 1974), we shall concentrate on the general concepts rather than the experimental details.

It will be recalled that the human analogue of the mouse *H-2K* and *D* antigens are the *HLA-A, -B,* and *-C* antigens (see Section 1.3). In the mouse the serologically defined Ia antigens map in the same *I* region as the determinants that stimulate the mixed lymphocyte reaction although there is no evidence that they are identical. In the human, the latter are assigned to the *D* region, defined by mixed lymphocyte reactions, while the *DR* locus, defined by serologically detected determinants, is closely linked but appears to be distinguishable (Mann and Murray, 1979).

Since in man it is not ethical to immunize one person with the cells of another, one must depend on natural immunization, usually through pregnancy, to obtain alloantisera. Even the sera of multiparous women usually have relatively low titers compared to experimental alloantisera in other species, and the availability of sera with different specificities is beyond the experimenter's control. In addition, because genetic experiments are so limited, one must depend on secondary criteria for the specificity of antisera for Ia-like antigens such as reaction with B cells and monocytes but not T cells or nonlymphoid cells, or ability to block a mixed lymphocyte reaction, or, rarely, reaction with cell surface proteins of an appropriate molecular weight. Therefore B cell-specific alloantisera are often taken as synonymous with anti-Ia antisera.

One of the greatest limitations in studying human *Ir* genes and possibly related disease associations with *HLA* is the inability to do genetic breeding experiments (and as a corollary, the lack of inbred human populations). These difficulties, discussed further in the next Section (11.1) are so serious that it is doubtful whether much progress would have been made if it were not for man's overwhelming self-interest in his own nature and, in particular, human disease.

11.1 Linkage versus Linkage Disequilibrium

As just mentioned, the limitations on study of an outbred population not amenable to experimental breeding are tremendous. For this reason, most studies of human traits linked to *HLA* have depended on a greater-than-random association between the trait and a particular HLA allele, which requires the existence of *linkage disequilibrium,* rather than simple genetic linkage. Linkage disequilibrium refers to an association between particular alleles at distinct genetic loci greater (or less) than statistically expected if the alleles of these genes had reached a random assortment equilibrium in the population. This association may occur because one or both alleles have been introduced into the population too recently, relative to the genetic recombination frequency between the loci, to reach equilibrium, because some genetic mechanism suppresses recombination between them, because selective advantage is gained from their association, or sometimes because random drift has occurred in small populations. Linkage disequilibrium between alleles at two loci requires linkage between the loci, but the converse is not true.

To take advantage of this occasional occurrence of disequilibrium, the typical study compares the frequency of a particular *HLA* antigenic "specificity" in a particular disease with that in the general population. One striking example is that of the hereditary disease gluten-sensitive enteropathy or coeliac disease, in which more than 87% of patients but less than 22% of controls bore the antigen *HLA-B8* (Falchuk *et al.,* 1972). In such a situation, there is little doubt that there exists either a causative (or susceptibility) association between the B8 antigen itself and the disease, or a linkage disequilibrium between the gene(s) responsible for gluten-sensitive enteropathy and the gene for B8.

The important distinction to make is that the disease is not linked to *HLA-B8,* which is just one of many possible *alleles* at the *B* locus; rather, the *linkage disequilibrium* between the disease gene(s) and B8 implies a *linkage* to the *HLA-B* locus. When linkage disequilibrium can be clearly documented, as in this case, it implies genetic linkage. However, many pairs of linked genes do not manifest linkage disequilibrium between particular *alleles* at the two loci. Therefore, the absence of linkage disequilibrium in such a study does not imply the absence of genetic linkage between genes affecting a disease and *HLA*.

Many *HLA* types show much lower risk ratios for particular diseases than that just cited. In such cases where linkage disequilibrium is hard to prove statistically or is absent, one must resort to true linkage studies. With breeding studies impossible, one must rely on *pedigree* studies within single families. When there is true linkage, the particular *HLA* allele that is associated with a given disease can vary from family to family, but the *HLA*

557

IMMUNE RESPONSE
GENES IN THE
REGULATION OF
MAMMALIAN
IMMUNITY

locus which manifests linkage will be the same. In order for such studies to achieve statistical significance, the families must be large and preferably contain several generations of living members available for study. These requirements put great limitations on the opportunities to do such studies. However, when they can be properly carried out, they are much more convincing than the majority of studies that depend on linkage disequilibrium.

11.2   Possible Ir Genes in Man

While there are now many diseases that have been found to be associated with *HLA*, and some of these associations may be due to *Ir* genes (see Section 11.3.2), there are only a handful of traits linked to *HLA* which fit the definition of an *Ir* gene in that they control the level of an antigen-specific immune response.

One of the earliest and most widely studied human genes that fits this criterion is the *HLA*-linked gene (or genes) regulating ragweed allergic responses (Levine *et al.*, 1972; Marsh *et al.*, 1973, 1979; Blumenthal *et al.*, 1974). Not only is the trait affected clearly an immune response, but the linkage to *HLA* was demonstrated in a pedigree study of 57 members of a single large family (Blumenthal *et al.*, 1974), so that it does not depend solely on linkage disequilibrium. In addition, certain *HLA* antigens were positively correlated with high responsiveness and others negatively correlated with high responsiveness (Marsh *et al.*, 1979). Such a finding could be due to a single allele in positive or negative disequilibrium with different *HLA* alleles, or to different genes (or different alleles of the same gene) for high or low responsiveness each in a different linkage disequilibrium. However, genes which regulate Ig E levels further complicate the situation, and some problems which have been pointed out in these studies may reflect this multigenic control (Bias and Marsh, 1975).

A second type of possible human *Ir* gene was observed to regulate T cell-mediated cytotoxicity against autologous virus-infected cells. In humans, as in mice, cytotoxicity specific for cell-bound viral antigens (McMichael *et al.*, 1977), minor histocompatibility antigens such as the male (H-Y) antigen (Goulmy *et al.*, 1977), and hapten (Dickmeiss *et al.*, 1977) are restricted by the MHC to target cells which share at least one *HLA* antigen with the responder and autologous stimulator cells. In the case of influenza virus-specific cytotoxicity, poor responses were observed when the only shared antigen was *HLA-A2* (McMichael, 1978), and it was demonstrated that some, but not all, bearers of *HLA-A2* or -*B7* are low responders to influenza associated with these antigens (Biddison and Shaw, 1979). Since these individuals' virus-infected cells were efficiently lysed by high responders to *HLA-A2* or *B7*, the low responsiveness was not due to decreased expression of these antigens in the target cells from these individuals, but rather to regulatory *Ir* genes. In family studies, similar haplotype preferences for *HLA*-restricted influenza-specific cytotoxicity were observed, and preferences were shared by *HLA*-identical siblings, consistent with *HLA*-linked regulatory genes (Shaw and Biddison, 1979). In all, the *Ir* gene control of virus-specific T cell cytotoxicity in man appears to be very similar to that observed in the mouse (see Section 10.1.1).

Another human *Ir* gene(s) appears to control the *in vitro* cellular proliferative response to streptococcal antigens (Greenberg *et al.*, 1975). An association between responsiveness and *HLA* type was demonstrated. Such genes may play a role in susceptibility to the infectious disease and/or the risk of autoimmune sequelae (see below).

Because large numbers of human patients receive frequent injections of beef or pork insulin, and most develop some degree of insulin resistance while some also develop insulin allergy, it should be possible to search for correlations between *HLA* type and these immune responses to insulins. Preliminary results suggest that such correlations, which might be due to *Ir* genes, exist (C. R. Kahn, A. S. Rosenthal, and D. Mann, personal communication).

Finally, a human *Ir* gene was found to control *in vitro* proliferative responses to tetanus toxoid by T cells from immunized individuals (Sasazuki *et al.*, 1978a). Low responsiveness was found to be associated, in the Japanese population under study, with *HLA-B5* and more tightly with *HLA-DHO*, which is in linkage disequilibrium with *B5*. Surprisingly, low responsiveness was dominant, consistent with an *Is* rather than *Ir* gene. Similarly, in the *in vitro* proliferative response to vaccinia virus after primary immunization, there was a statistically significant increase in the frequency of *HLA-Cw3* among the lowest responders, possibly due to an *Ir* gene (de Vries *et al.*, 1977). Cutaneous hypersensitivity responses to other natural antigens can also show *HLA* associations (Buckley *et al.*, 1973). These results have implications for disease susceptibility as well as for the efficacy of prophylactic immunization. One additional fascinating example is that of leprosy, in which susceptibility to develop the tuberculoid form of the disease was association with *HLA* haplotype (van Rood *et al.*, 1977).

11.3 Ir Genes and Disease

11.3.1 Mouse Models

It is apparent that MHC-linked low responsiveness to bacterial antigens or to virus-infected autologous cells may result in greater host susceptibility to and morbidity or mortality from such infectious diseases. Since MHC-heterozygous individuals would have a lower probability of failure to respond to any given infectious organism, one type of "hybrid vigor" would result. Moreover, on a population level, the polymorphism of the MHC would be reinforced by such natural selective pressures of infectious disease.

Besides the obvious types of infectious diseases, susceptibility to tumor virus-induction of tumors would fall into this category. Well-documented examples include Friend leukemia virus (Lilly, 1968; Chesebro *et al.*, 1974; Chesebro and Wehrly, 1978), radiation leukemia virus (Meruelo *et al.*, 1977), and Rous sarcoma virus (Whitmore *et al.*, 1978).

However, there are at least two ways in which *Ir* high responsiveness would be to the detriment of the host. One case is that of infectious diseases in which the inflammatory or cytolytic host defense mechanisms cause more harm than the invading organism itself. An example is the case of lymphocytic choriomeningitis virus (Zinkernagel, 1978), in which genetic responder strains die rapidly, while nonresponders become carriers, but live relatively unharmed by their chronic infection.

The second situation in which high responsiveness is detrimental is the case of autoimmune diseases. Whether induced by viral (subacute sclerosing panencephalitis) or bacterial (rheumatic fever or acute glomerulonephritis) infection or of unknown etiology, the immune response against autologous proteins or tissues is central to the pathogenesis of these diseases. Experimental models in laboratory animals which manifest MHC linkage include autoimmune thyroiditis in the mouse (Tomazic *et al.*, 1974) and experimental allergic encephalitis (Bernard, 1976).

It is the last category, autoimmune diseases, that predominate among documented *HLA*-linked diseases in man. Because of ethical constraints on immunization of human subjects, it is often more difficult to demonstrate the role of an *Ir* gene *per se,* or even an autoimmune mechanism. However, in most human diseases linked to *HLA,* an autoimmune mechanism is at least strongly suspected. A partial list of examples is given in Table II, with the most clearly autoimmune listed first. It should be noted that in almost all cases, *HLA* linkage was recognized because of linkage disequilibrium with a particular *HLA* specificity. Searches for inheritance of these diseases linked to parental *HLA* types in family studies, where no linkage disequilibrium exists, are much more difficult because incomplete penetrance and multigenic influences, as well as environmental factors, are common in most of these diseases, and families are rarely large enough to demonstrate statistically significant correlations.

In most cases (for instance, gluten-sensitive enteropathy and multiple sclerosis) the association is (statistically) greater with the *HLA-D* locus antigen than with the *HLA-A* or *B* antigens in linkage disequilibrium with these. This observation is consistent with the

TABLE II. *Examples of HLA-Linked Diseases*

Disease	HLA specificities in linkage disequilibrium	References
Myasthenia gravis	A8	Kaakinen *et al.,* 1975; Möller *et al.,* 1976
Goodpasture's syndrome	DRw2	Rees *et al.,* 1978
Systemic lupus erythematosus	DRw2,3	Reinertsen *et al.,* 1978; Gibofsky *et al.,* 1978
Sjögren's (sicca) syndrome	B8, Dw3	Gershwin *et al.,* 1975; Chused *et al.,* 1977; Moutsopoulos *et al.,* 1978
Juvenile diabetes mellitus	B8, B15, DRw3, DYT	Singal and Blajchman, 1973; Nerup *et al.,* 1974; Thompsen *et al.,* 1975; Solow *et al.,* 1977; de Moerloose *et al.,* 1978; Cudworth and Festenstein, 1978 (review); Sasazuki *et al.,* 1978*b*
Addison's disease	B8	Thomsen *et al.,* 1975
Graves' disease	B8, Dw3, DRw3, DHO	Sasazuki *et al.,* 1978*b;* Farid *et al.,* 1979
Gluten-sensitive enteropathy (coeliac disease)	B8, Dw3	Falchuk *et al.,* 1972; Keuning *et al.,* 1976; Peña *et al.,* 1978; Mackintosh and Asquith, 1978 (review)
Rheumatoid arthritis	DRw4 (adult) Dw7,8 (juvenile)	Stastny, 1978; Gershwin *et al.,* 1978; Stastny and Fink, 1979
Ankylosing spondylitis	B27	Brewerton *et al.,* 1973; Schlosstein *et al.,* 1973; Caffrey and James, 1973; Sachs *et al.,* 1975; Möller and Olhagen, 1975; Sachs and Brewerton, 1978 (review)
Multiple sclerosis	A3, B7, Dw2	Bertrams *et al.,* 1972; Naito *et al.,* 1972; Jersild *et al.,* 1973; Winchester *et al.,* 1975; Terasaki *et al.,* 1976; Paty *et al.,* 1977; Batchelor *et al.,* 1978 (review)

identification of the *D* and *DR* loci in man with the *I* region of the mouse, and the suspected *Ir* gene mechanism involved. However, one clear exception is ankylosing spondylitis, strongly associated with *HLA-B27* (present in 85–95% of patients but only 6–10% of controls), but not with any known *D* locus antigen. It is possible in this case, which is less clearly autoimmune in pathogenesis than some of the other diseases listed, that another mechanism is responsible, although one cannot exclude the possibility of *Ir* genes more closely linked with *HLA-B* than *HLA-D* (comparable to *H-2K/D*-linked *Ir* genes for cytotoxic responses in the mouse). In other cases, a particular B cell alloantigen, not associated necessarily with a particular *DR* allele, shows a greater correlation with the disease than the *D* or *DR* allele in question (for instance, in gluten-sensitive enteropathy, multiple sclerosis, and Sjögren's syndrome). Since these B cell alloantigens are presumed to be Ia antigens, they may correspond to the actual *Ir* gene locus (see references in Table II and review by Mann and Murray, 1979). A particularly interesting example is that of Sjögren's syndrome (Moutsopoulos *et al.*, 1978; Mann and Murray, 1979). Patients with this disease fell into two groups: primary (isolated) sicca syndrome, in which DRw3 antigen was increased in frequency, and secondary sicca syndrome accompanying rheumatoid arthritis, in which DRw4 frequency was increased. However, patients from both groups were *invariably* positive when tested with the B cell alloantisera designated Ia172 and IaAGS. Therefore, these antisera may be detecting Ia antigens directly involved in the pathogenesis of the disease. As the genetic fine structure of genes coding for human Ia-like antigens becomes better defined, the significance of these disease correlations will be more completely understood.

11.4 Implications for Tissue Transplantation in Man

The studies of genetic restriction of immune cell interaction, discussed primarily in Section 8, are especially important for human bone marrow and thymus transplantation. Many of the genetic restriction phenomena described for cytotoxicity in mice have been found to occur in man as well (see Section 11.2). Other regulatory circuits in man have also been reported to manifest such restrictions (Engleman *et al.*, 1978). If the results found in thymus and bone marrow chimeras also apply to man, then the need for matching at all loci, especially the *HLA-D, DR* loci, becomes critical to the establishment of an immunocompetent recipient. As transplantation of bone marrow or thymus becomes more widely attempted in the treatment for severe immunodeficiency diseases, it will be crucial to keep these results in mind. Already maximal *HLA*-matching of bone marrow is attempted to prevent fatal graft-versus-host disease. However, as pointed out by Zinkernagel (1978), more thorough matching of thymus grafts may also be required if transplantation of thymus into totally athymic patients is to result in complete immunocompetence.

Finally, the results described in Section 6 on the specific induction of autoantiidiotypic antibodies to produce antigen-specific immunosuppression hold out great promise for organ transplantation in general. While some of the methods are difficult to extrapolate to an outbred population such as man, if they can be applied they will be enormously superior to the broad-spectrum immunosuppression, with all its accompanying risks, currently required.

In the preceding sections we have covered many subjects revolving around the theme of immune regulation via *Ir* genes. Regretably, many exciting aspects of the regulation of mammalian immunity had to be omitted or touched upon only fleetingly and peripherally. Now we must consider how the ideas we have reviewed fit into the broader issue of biological regulation and development addressed by this book.

Our story of immune response genes has had two focal points, which must, in some way, mirror one another: on the one hand, the receptor or receptors on the T cell which recognize antigen and other signals necessary to trigger an immune response; on the other hand, the cell surface molecules, encoded in the major histocompatibility complex, required on the antigen-bearing cells with which the T cells interact—stimulator and target cells for cytolytic T cell responses, macrophages and B cells for helper T cell responses.

Before we summarize these issues and the theories proposed to explain them, let us briefly review some of the other pathways of immune regulation which must interact with *Ir* genes and which were only touched upon in Sections 5–7.

We have seen that T cell receptors for antigen employ combining sites similar or identical to the combining sites of a subset of the antibodies with similar specificity. As such, T cell receptors may be subject to the same type of idiotype-specific regulation as are antibodies. Just as Jerne (1974) proposed a homeostatic network of idiotype-bearing antibodies and antiidiotypic antibodies which regulated them, so idiotype-bearing helper T cells may be regulated by suppressor T cells with complementary (antiidiotype) receptors [as demonstrated in the case of the antiazophenyl arsonate and antiphosphorylcholine antihapten responses by Owen and Nisonoff (1978) and Bottomly *et al.* (1978); evidence for a similar idiotype-specific suppressor T cell has been reported in a possibly T-independent anti-TNP antibody response by Bona and Paul (1979)]. Thus, two parallel feedback networks may be involved in regulation of the immune response, one among interacting antibodies, one among interacting cells (see Fig. 12). Yet since both are based on the same idiotypes shared by the antibody and cellular receptors, the two networks must interact at every step. The enormous complexity of such a composite multidimensional network of interacting cells and molecules boggles the mind and resists the ingenuity of the experimenter attempting to unravel it.

Another plane on which the network of interacting idiotypes on T cells and the network of interacting idiotypes on B cells and antibodies themselves intermingle is in the action of idiotype-specific helper T cells (see Fig. 12). Several groups have recently reported evidence for idiotype-specific helper T cells which must act in conjunction with the antigen-(carrier)-specific helper T cells in order to trigger the idiotype-bearing antigen-specific B cell (Woodland and Cantor, 1978; Hetzelberger and Eichmann, 1978; Eichmann *et al.,* 1978). The role of these in *Ir* gene-controlled responses was described for lysozyme in Section 5.2.

In addition to these regulatory networks based on idiotype, there are also complex antigen-specific regulatory circuits. At least two types of helper T cells have been demonstrated. In one study, two types were distinguished by differences in their longevity after thymectomy and in their sensitivity to antithymocyte antiserum (Feldman *et al.,* 1977; Erb and Feldmann, 1977). In another study, two types of helpers were distinguished by their

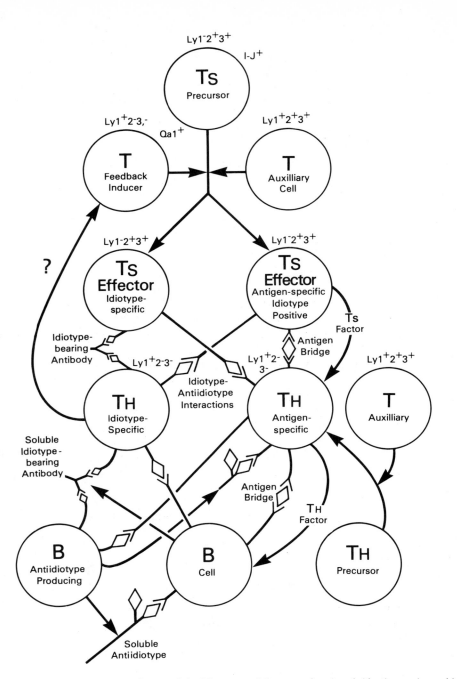

Figure 12. A partial illustration of some of the idiotype–antiidiotype and antigen-bridge interactions which have been suggested to be involved in the regulatory networks of the immune response. See text for references. This figure is intended to illustrate the complexity of the regulatory network as well as a sampling of the interactions postulated. Note the similarity to Fig. 3 for lysozyme, under *Ir* gene control. Note also the critical *omission* of macrophages and of Ia-specific T cell–macrophage and T cell–B cell interactions and restrictions on which *Ir* gene control of this network depends. See Figs. 6–11 for some of these.

adherence or nonadherence to nylon wool and by the presence or absence of *I-J* subregion-encoded Ia antigens on the cell surface (Tada *et al.*, 1978). In the latter study, the nylon-nonadherent Ia⁻ carrier-specific helper cell functioned only when hapten and carrier were on a single molecule, presumably because it recognized the hapten-specific B cell via an antigen bridge, similar to that described for lysozyme in section 5.2. The nylon-adherent Ia⁺ helper cell required only the presence of free carrier, and so may have been nonspecific in its final effector function even though its induction was antigen specific. Both cells are Ly1⁺2⁻3⁻ in cell surface antigen phenotype, and both (in both systems mentioned) are synergistic. In addition to these two Ly1⁺2⁻3⁻ antigen-specific helper cells and the Ly1⁺2⁻3⁻ idiotype-specific helper cells, there also appears to be an Ly1⁺2⁺3⁺ amplifier cell involved in augmenting the response (Tada, 1977; Tada *et al.*, 1977; Feldmann *et al.*, 1977) (see Fig. 12).

Maintaining the symmetry, in addition to antigen- and idiotype-specific suppressor T cells, there is an Ly1⁺2⁺3⁺ auxilliary T cell which is necessary for the Ly1⁻2⁺3⁺ suppressor precursor T cell to differentiate into the Ly1⁻2⁺3⁺ suppressor effector T cell (Tada, 1977). Moreover, to complicate matters with yet another feedback regulatory circuit, it now appears that an Ly1⁺2⁻3⁻ cell in the antigen-specific helper cell population (but distinguishable from the actual helper cell by another surface marker, Qa1) induces a feedback suppressor Ly1⁻2⁺3⁺ cell to dampen the response, but does so only in the presence of another auxilliary Ly1⁺2⁺3⁺ T cell (Eardley *et al.*, 1978; Cantor *et al.*, 1978; Eardley *et al.*, 1979; McDougal *et al.*, 1979) (see Section 5.2 and Fig. 12).

In the face of the overwhelming complexity of these interacting regulatory networks and feedback loops, the problems of *Ir* gene regulation of the immune response that we have wrestled with over the preceding eleven Sections seem simple by comparison. They operate in the midst of these other regulatory mechanisms, and sometimes employ them, as discussed in Section 5 on suppressive mechanisms in *Ir* gene control and Section 7 on antigen-specific soluble factors. It will be exciting to explore the reasons why certain types of antigenic determinants preferentially elicit suppression instead of help. Yet even in the cases such as the responses to GAT, GT, and hen lysozyme, where suppressor circuits and factors play a paramount role in the mediation of *Ir* gene control, there appears to be another, more fundamental mechanism in operation which determines low responsiveness in the first place. This mechanism, for the most part, appears to involve the genetically restricted T cell repertoire and the cell surface molecules these T cells recognize in addition to antigen.

Thus, we return to the dichotomy set forth at the beginning of this section: If, as the data of Sections 8, 9, and 10 strongly suggest, many if not all MHC-linked *Ir* genes operate by virtue of the MHC-linked genetic restrictions on the interactions of the pivotal T cell with the other cells involved in the response, is the *Ir* gene expressed in this T cell, or in the various cells with which it interacts?

There are two fundamentally distinct theories which cannot yet be distinquished by the available experimental evidence. Either of these hypotheses, in its essence, is compatible with the possibility that Ia-bearing genetically restricted macrophage factors play a critical role in these *Ir* gene phenomena (see Section 7.2). At one extreme is the theory that *Ir* genes, and perhaps all genetic restriction genes including those MHC genes governing cellular cytotoxicity, act solely in the antigen-presenting cell (and possibly B cell) or stimulator (and target cell), by chemically altering the exogenous antigen or at least determining its orientation (Rosenthal, 1978). According to this theory, the MHC encodes a number of enzymes, perhaps the Ia antigens themselves, which would be directly involved in "processing" or modifying antigens. Alternatively, the Ia antigens might have receptor function which allows them to bind certain amino acid sequences on certain classes of antigens and determine the orientation of the molecule seen by T cells (Benacerraf, 1978). In either case, the macrophage (or stimulator cell) MHC product would actively interact with antigen in such a way that the T cell would actually see a different exogenous antigen depending on which Ia antigen (or K/D antigen) was on the macrophage (or stimulator cell). Thus, both Ir gene control and genetic restriction phenomena could be explained completely on the basis of T cell clones with different specificity for different parts of or alterations of the original exogenous antigen. The cells need not have any specificity for Ia (or K/D) $per\ se$. Therefore, this theory would have the important advantage of requiring only a single T cell receptor (Rosenthal, 1978).

However, the theory also has two major weaknesses: (1) There is no direct evidence for any enzymatic or receptor activity of Ia or K/D-encoded antigens. While macrophages are widely believed to "process" antigen in some way, there is no evidence that this processing function is encoded within the MHC or is different among congenic strains differing only at H-2.* (2) The logical extension of a major attraction of this theory, that it requires only a T cell receptor recognizing antigen and not one recognizing MHC antigens $per\ se,$ is difficult to reconcile with the evidence that genetic restrictions and some Ir phenotypes are determined in chimeras during T cell maturation in the thymus, by the MHC antigens present on the radioresistant elements of the thymus, before any exposure to the exogenous antigen in question. If genetic restriction phenomena and some Ir gene phenomena are, in fact, two faces of the same underlying process, as they appear to be, then all genetic restrictions would be due to the selection of clones with different $antigen$ specificity. How such clones could be selected by the thymus MHC antigens in the absence of the pertinent exogenous antigens to be modified by them is hard to explain.

At the other extreme is the theory that Ia (and K/D-encoded) antigens are passive signals on macrophages (or B cells, and stimulator and target cells) which are recognized directly by receptors on T cells. Such receptors could be distinct from the T cell receptors for antigen [a dual receptor model (Janeway $et\ al.,$ 1976)], or they could reflect one part of a single T cell receptor which simultaneously recognized antigen and MHC antigen as a complex. Thus, each T cell clone would have a dual specificity, and could interact only with a cell having the proper combination of determinants (antigen and MHC) on its surface. The determination of genetic restriction phenomena in the thymus would then be the direct result of selection of clones with specificity for MHC antigens borne by the thymus. The observation of Ir gene phenomena would be due to the limitations on possible combinations of specificity for Ia (or K/D) and for exogenous antigen that could exist on a clone of T cells.† Why some combinations would be physiologically improbable or impos-

---

*On the other hand, it must be pointed out, as discussed in Section 8.1.1, that some evidence for direct interaction of viruses with H-2 K/D antigens has been demonstrated and correlates with the genetic restrictions observed on virus-specific T cell cytotoxicity (Bubbers and Lilly, 1977; Bubbers $et\ al.,$ 1978).

†Note that this hypothesis is consistent with the correlation between Ir gene specificity and T cell specificity noted in Section 6.2. On the other hand, exceptions to this generalization raise questions about this hypothesis placing Ir gene control in the T cell repertoire. For instance, the Ir gene for poly-L-lysine (PLL) and the copolymer of L-Glu, L-Lys (GL) could not be separated in a large number of outbred guinea pigs even though the two antigens showed virtually no cross-reactivity for T cell activation (Kantor $et\ al.,$ 1963). To reconcile this finding with the hypothesis, one would have to postulate that the gaps in the T cell repertoire for PLL plus self Ia and for GL plus self Ia happened to involve the same Ia molecule (restricting element), even though different T cell clones with different antigen specificity were involved. This possibility is not unreasonable,

sible (or disallowed), or, in other words, why certain gaps would exist in the dual specificity T cell repertoire, will be discussed shortly.

565

IMMUNE RESPONSE
GENES IN THE
REGULATION OF
MAMMALIAN
IMMUNITY

Before we can explore this issue more fully, it must be pointed out that for the cases in which this mechanism applies, *Ir* genes no longer exist in the sense originally envisioned of regulatory genes. In the most negative view, *Ir* gene control would be a mere epiphenomenon on the genetic restriction of cell interaction. From the more positive point of view, these *Ir* phenomena would represent the first glimpse of one of the most fundamental mechanisms by which interacting cells of an organism recognize each other. As such, this MHC-linked cell interaction mechanism would eclipse the complex idiotype interaction mechanisms in both its phylogenetic age and its breadth of applicability throughout the ontogeny and tissue diversity of the organism.

If no regulatory *Ir* gene exists in the original sense, where can one place the *Ir* gene "defects" observed. If the only relevant gene products which map in the MHC are the Ia (and K/D) antigens expressed on macrophages, B cells, stimulator and target cells, then one could argue that the *Ir* gene "defect" which one maps to the MHC must be expressed in these cells, not the responding T cell (Singer *et al.*, 1978). On the other hand, if there is no functional "defect" in these cells, but only in the T cell repertoire, then one would be tempted to place the *Ir* defect in the T cell repertoire, that is, in the population of available T cell clones. Since this repertoire does not appear to depend so much on the genes of the T cells themselves, but rather on the genes of the thymic epithelium (or other radioresistant thymic tissue) in the presence of which the T cells mature, the *Ir gene* would be expressed primarily in the thymus tissue. This *Ir* gene would act to establish the T cell repertoire (and gaps therein), so that subsequent expression of Ia on macrophages would be secondary: It would be necessary for the *Ir* "defect" to be observed, but the *Ir* "defect would be already established once the T cell repertoire was established in the thymus. Moreover, although the MHC genes of the T cell would not *encode* the T cell receptors, the MHC genes of the organism (in particular the thymus) would determine the *repertoire* of T cell receptors expressed in that organism and therefore determine gaps, or *Ir* gene defects, in the population of T cell clones in the organism. It is particularly important, in this regard, to determine by biochemical means whether antigen-specific T cells also have receptors for MHC antigens, or whether the specificity of T cell receptors is for a combined determinant encompasing antigen and Ia.

In addition to effects determined by the repertoire of T cells, there are also effects mediated by regulation of the expression of MHC antigens on the stimulator or target cell.

---

since the number of Ia molecules appears to be far lower than the number of potential antigens under *Ir* gene control. The corresponding postulate in the other hypothesis would be that the proposed receptor function of Ia molecules is of "broad" specificity.

Another case is that of delayed hypersensitivity to NP-GLPhe, in which the carrier (GLPhe) determines the *Ir* gene control even though the responding T cells are specific for the hapten (NP) (Weinberger *et al.*, 1979*c*). In that case, one could postulate the need for carrier-specific helper T cells to explain the findings. However since the responding T cells are hapten-specific, one must wonder why hapten-specific helper T cells either do not arise or are not adequate in this system. There is suggestive evidence in this system itself for placing the *Ir* gene control in a helper T cell. The *Ir* gene control observed operates only at the time of primary immunization. Once primed to NP on another carrier, even the low responders respond to NP-GLPhe, and do so in a genetically restricted fashion. This genetic restriction in the efferent limb of the response implies that the delayed hypersensitivity effector cells can recognize NP-GLPhe on a low-responder antigen-presenting cell. Therefore, the low-responder presenting cell (macrophage) can bind and effectively present NP-GLPhe to NP-specific T cells. The defect is therefore likely to be in the ability of GLPhe-specific helper T cells to recognize the GLPhe carrier moiety on these macrophages. Presumably, the helper T cell is required for priming, but not for the subsequent effector phase.

Evidence has been presented that viruses (Meruelo *et al.*, 1978; Meruelo, 1979), as well as other *H-2* antigens themselves (O'Neill and Blanden, 1979), can regulate the expression of *H-2* antigens on the cell surface and in turn modulate immune responsiveness. Whether this regulation occurs at the level of gene duplication, RNA transcription, protein synthesis, or incorporation into the cell membrane presents an exciting problem for the molecular biologist. However, these effects on the level of expression of MHC antigens must be distinguished from the postulated effects of MHC antigens on the antigenic determinants of exogenous antigens which are presented to the T cell, in the first theory above.

Several theories have been proposed to explain MHC-associated gaps in the T cell repertoire for exogenous antigen. One theory (Schwartz, 1978) does not require two T cell receptors. It proposes that the combination of exogenous antigen with a given Ia antigen mimics the combination of that Ia antigen with other endogenous, autologous, antigens. Since the animal is tolerant to its own endogenous molecules associated with Ia, it is tolerant to the mimicking combination. This tolerance theory has the advantage over an early tolerance theory invoking mimicry of Ia itself by the exogenous antigen alone (see Section 1) that it does not require that tolerance be dominant, merely codominant. It would therefore overcome the experimental objection that a (responder × nonresponder) F_1 hybrid is usually a responder. According to this theory of Schwartz (1978), the F_1 would still be tolerant to the antigen in association with the nonresponder parental Ia antigens, but it could respond to the antigen in association with the responder parental Ia antigens borne by the F_1 lymphoid cells. The net result would be a positive response.

Other hypotheses require two receptors on T cells, both related to V_H in their combining sites, but one specific for MHC antigens and the other for exogenous antigens. These receptors may be functionally equivalent and independent, with their net affinity equal to the product of the two affinities (Janeway *et al.*, 1976), or they may be functionally distinct from each other but joined in a single complex (Cohn and Epstein, 1978). The two hypotheses in which gaps in the dual specificity repertoire of T cells can be explained from the proposed origin of these receptors both require somatic diversification mechanisms. The first, proposed by Jerne (1971) and revised by von Boehmer *et al.* (1978) to encompass a dual receptor mechanism, is that all germ-line genes for T cell receptors code for binding sites specific for the MHC antigens of the species.* Under a negative selection mechanism in the thymus which produces self-tolerance, those clones of T cells with specificity for autologous MHC antigens are forced to mutate, or rather, cells which do not mutate are eliminated so that only mutant clones whose receptors no longer have a high affinity for self-MHC antigens survive and reach the periphery. These mutant clones form the pool of clones specific for all exogenous antigens. The clones specific for allogeneic MHC antigens remain unaltered and form a large fraction of the T cells found in the periphery [as observed by Binz and Wigzell (1975b)]. Since receptors specific for exogenous antigen had to derive by mutation or some other somatic process (such as recombination) from receptors for self-MHC antigens, clones with those receptors which could not easily be derived from receptors for self-MHC of a particular haplotype would be absent or infrequent in animals of that MHC haplotype. Therefore, gaps in the T cell repertoire, manifested as *Ir* gene defects linked to the MHC, would occur in animals of that haplotype. While the basic

---

*One problem with this theory is that the highly polymorphic MHC antigens of the species are changing by mutation. To maintain a germ-line gene repertoire encoding receptors for all MHC antigens of the species would require mutations in these genes to correspond to mutations in the MHC genes. However, since there is no obvious selective advantage for one individual to have receptors for a mutant MHC antigen arising in another individual, it is not clear how such a correspondence could be maintained.

567

IMMUNE RESPONSE
GENES IN THE
REGULATION OF
MAMMALIAN
IMMUNITY

elements were present in this original theory (Jerne, 1971), the discovery of genetic restriction phenomena in the interim required a revision to include two receptors (von Boehmer *et al.*, 1978). Under this proposal, two types of receptors would be required on each T cell, both initially specific for the same self-MHC antigen. Only one of these would be required to mutate to avoid negative selection in the thymus. Therefore, the cells with receptors for a *particular* self-MHC antigenic determinant would have receptors specific for exogenous antigens which were derived from the receptors for that self-MHC determinant. Gaps would exist not necessarily in an animal's total ability to recognize a particular exogenous antigen, but in its ability to recognize that antigen in *association* with a particular Ia (or K/D-encoded) antigen. For instance, we have discussed our own evidence in Section 9.2 (Richman *et al.*, 1980) that $H\text{-}2^d$ T cells from *in vivo*-immunized mice can recognize the fragment (132–153) of sperm whale myoglobin in association with $I\text{-}A^d$- but not $I\text{-}C^d$-encoded molecules although they can recognize other fragments, such as (1–55), in association with $I\text{-}C^d$-encoded Ia molecules. Therefore, clones exist that can recognize fragment (132–153) or $I\text{-}C^d$-encoded antigens, but no clone recognizes the combination of the two (see Fig. 10). These gaps in the repertoire would not necessarily be the same for different determinants on the same antigen molecule, as we have seen in the case of myoglobin.

A related, but in one respect opposite, somatic mutation theory was proposed by Langman (1978) and Cohn and Epstein (1978). According to this hypothesis, two distinct but coupled receptors (or binding sites) on T cells (anti-H and anti-X) recognize self-MHC antigens and exogenous antigens. Those clones with anti-H receptors specific for self-MHC antigens are *positively* selected to proliferate in the thymus, rather than deleted. The restrictions on possible anti-X binding sites associated with a particular anti-H (self) binding site is due to an exclusion rule. This rule is based on the notion that if both binding sites arise from the same V_H germ-line gene pool, a cell which is positively selected for its anti-H (self) receptor cannot use that same V_H germ-line gene as a starting point for somatic mutation to form the anti-X binding site. Therefore, anti-X specificities which can arise easily only from that germ-line gene do not occur in association with anti-H specificities for that particular self-MHC antigen. The net result is to produce similar gaps in the dual specificity T cell repertoire, but by a mechanism opposite to that of the previous theory.

If somatic mutation is not the primary mechanism for generation of receptor diversity, then the explanation of *Ir* gene defects by a dual receptor hypothesis requires some other mechanism for disallowing certain combinations of specificities. The lack of evidence for such a mechanism led Janeway *et al.* (1976) to postulate an *Ir* gene defect at the level of association between Ia and exogenous antigen on the surface of the macrophage (or other target cell). On the other hand, the lack of evidence for such a defect in Ia antigen association with exogenous antigen, and in fact, the evidence to the contrary (Pierce *et al.*, 1977), would lead one to search for a mechanism that would exclude certain combinations of specificities in the T cell repertoire. Such a mechanism may be at the level of clonal selection (such as a tolerance mechanism) or at the level of ontogenetic or even evolutionary generation of the receptor repertoire.

Any of the above mechanisms would allow *Ir* gene antigen specificity to reflect T cell receptor antigen specificity (Section 6.2.), even though T cell receptor combining sites are encoded by immunoglobulin V_H genes unlinked to the MHC (Section 6.1), while the *Ir* genes are MHC-linked.

As discussed in detail in Sections 8.3 and 9.1, it remains unresolved as to whether these *Ir* gene or Ia antigen restrictions apply in T cell–B cell interactions as they do in T cell–macrophage interactions. However, there is a large body of evidence that T cell–B cell interactions are also similarly restricted in at least some situations, especially after *in vivo* immunization. In addition, such restriction would have the aesthetic appeal that the same type of symmetry seen between the stimulator cell and target cell in cytotoxic T cell interactions would be seen between the stimulator (antigen presenting) macrophage and the target B cell with respect to the pivotal helper T cell in antibody responses. Such symmetrical restriction would also make it easier to explain the apparent *Ir* gene selection of B cells specific for particular antigenic determinants (Berzofsky *et al.*, 1977*b*; Berzofsky *et al.*, 1979).

It should be pointed out also, as discussed in Section 8.3.4 (see Fig. 9), that the genetic restrictions imposed at three successive stages may be fundamentally different types of restrictions because they act on T cells at different stages of differentiation and/or activation. First, a restriction imposed by the genome of the T cell itself prevents it from recognizing its own surface antigens as foreign, even though the clones which would recognize these surface antigens as "self" may be lost if maturation occurs in a foreign thymus. Second, the thymus in which the T cell precursor matures determines which cell surface antigens on macrophages or stimulator cells can be recognized (as "self") to produce a third level of restriction at the time of priming (first exposure to antigen). However, at this second stage, before priming, the population of T cells may not yet be restricted in its effector function. Thus, unprimed (A × B) F_1 helper T cells may be able to help parent A B cells in the presence of only parent B macrophages. (One could explain such a result in terms of clonal selection without postulating additional receptors, by postulating that the unprimed T cell population contains many genetically unrestricted clones. However, during priming with antigen, only the highly restricted clones would be expanded into a memory cell population.) Finally, the third stage of restriction occurs at the time of antigen priming by antigen-presenting macrophages (or stimulator cells in the case of cytotoxic responses). The primed or "secondary" T cell population is then limited to those clones which react in the effector phase only with B cells and macrophages (or target cells in the case of cytotoxic responses) which carry the same combination of antigen plus MHC (Ia or *K/D*-encoded) antigen as was present on the cell which did the priming (see Fig. 9). Clearly distinguishing these three stages of restriction may allow one to reconcile many apparent inconsistencies and controversies now in the literature, as I have tried to do for T cell–B cell vs. T cell–macrophage restrictions and *Ir* gene phenomena in Sections 8.3.4 and 9.1.

Finally, we must not ignore the evidence that another distinct type of *Ir* gene may exist in the genome of the T cell itself. This type of *Ir* gene is manifested in the failure of certain T cells to produce helper T cell factors specific for certain antigens (Munro and Taussig, 1975). That it is not solely a function of the genetic restriction specificity of these T cell populations is demonstrated by the failure to select high-responder clones when these low-responder T cells mature in the thymus of a (high-responder × low-responder) F_1 recipient (Marrack and Kappler, 1979; Hill and Frelinger, 1979). Since the antigen-specific T cell-derived helper factors bear Ia antigenic determinants (see Section 7), some part of these apparently T cell-derived molecules is by definition encoded within the MHC. Since they also specifically bind antigen, these factors are direct evidence for a complex, if not a single polypeptide chain, derived from the T cell and bearing both Ia determinants and an antigen-binding site. Unless the Ia portion of the factors actually derives from macrophages and not the T cells themselves (see Section 7), these findings imply that some part of the antigen-specific helper function of T cells involves molecules encoded in the MHC

569

IMMUNE RESPONSE
GENES IN THE
REGULATION OF
MAMMALIAN
IMMUNITY

of the T cell itself.* Whether the T cell receptor *per se* involves Ia as a constant region connected to V_H, or as a light chain associated with a V_HC immunoglobulinlike heavy chain (see Fig. 5), or does not directly include Ia at all, are exciting questions yet to be resolved by biochemical approaches. Nevertheless, even if no part of the T cell antigen receptor were found to be encoded in the *I*-region, the above evidence that T cell-derived *I* region-encoded molecules were involved at some stage in the antigen-specific "helper" mechanism would not be invalidated. Therefore, it is not unreasonable that genetic differences in the T cell *I* region could manifest themselves as antigen-specific *Ir* genes. It is probable that while most *Ir* genes studied appear to fall into the first category, acting by virtue of genetic restrictions on T cell interactions with macrophages and other cells, the existence of other types of *Ir* genes in the genome of the T cell itself will ultimately be confirmed.

It is clear that we have only scratched the surface of genetic regulatory mechanisms in the immune response. As we reach the limits of resolution of cellular immunology, the exciting questions posed will provide new and fertile ground for biochemical and molecular biological research. The reward may be not only a new understanding of the immune system itself, but also an insight into the fundamental genetic controls on cell interactions throughout the ontogeny of the organism.

ACKNOWLEDGMENTS

I thank Drs. Marc Feldmann, Jeffrey Frelinger, Sonoko Habu, Richard Hodes, C. Ronald Kahn, Ulrich Krawinkel, Peter Lonai, Dan Longo, Dean Mann, Bonnie Mathieson, Alan Rosenthal, Ronald H. Schwartz, Eli E. Sercarz, Alfred Singer, and Herman Waldmann for personnal communications and for permission to cite unpublished results, Dr. Ira Green for helpful suggestions on historical aspects, and Drs. Yoichi Kohno, Dean Mann, and Steven Shaw for suggestions and criticisms of the section on human *Ir* genes and HLA-associated diseases. I am grateful to Drs. Baruj Benacerraf, Robert Goldberger, Edward Max, David Sachs, Ronald Schwartz, Gene Shearer, Alfred Singer, Warren

---

*An alternative way to reconcile the suggested dual nature of helper T cell recognition (Section 8) with the presence of Ia determinants on T cell helper factors (Section 7.1) would be the hypothesis, suggested in Section 7.1, that the Ia moiety on these factors actually derives from the antigen-presenting macrophage and may be bound to the T cell receptor for antigen by the postulated second combining site associated with this receptor but specific for Ia. There is clear evidence for suppressor factors that the *I-J*-encoded Ia moiety derives from the T cell, but for helper factors it remains an open question. If this hypothesis were to be proven true, T cell helper factors could become direct evidence for an association between T cell receptors for exogenous antigen and those postulated to be present with specificity for MHC-encoded antigens. At the same time, the need to explain an association of T cell-derived Ia with the T cell receptor for antigen, as discussed above, would be obviated. Such an hypothesis would also explain why some low-responder strains [for example, H-2^f in the case of (T,G)-A–L)] fail to make a helper factor (because they do not have the right combination of specificities on any clone of T cells), but it would not explain why these are not converted to high responders in appropriate chimeras (Marrack and Kappler, 1979; Hill and Frelinger, 1979; see above and Section 10). It would also not explain, by itself, why genetically unrestricted helper factors, once made, fail to provide help for B cell plus macrophage populations of certain low-responder haplotypes [for example, H-2^k or H-2^a in the case of (T,G)-A–L)]. A more direct problem with the hypothesis is that the Ia moiety fails to dissociate from the antigen-binding site under conditions used to elute it from an affinity column (see Section 7.1). Thus the affinity of the putative T cell combining site for Ia would have to be considerably higher than that of anti-Ia antibodies, an unlikely situation.

Strober, Thomas Waldmann, and John Wunderlich for the major effort of critically read-ing the manuscript and for their many useful suggestions. My wife, Sharon, contributed valuable editorial assistance. Finally, I am grateful to my wife Sharon and sons Alexander and Marcus for enduring me during the 4 months of weekends from June to October 1979 during which this manuscript was written.

References

Ada, G. L., and Ey, P. L., 1975, Lymphocyte receptors for antigens, in: *The Antigens* (M. Sela, ed.), Vol. 3, pp. 189–269, Academic Press, New York.

Adorini, L., Miller, A., and Sercarz, E. E., 1979*a,* The fine specificity of regulatory T cells. I. Hen egg-white lysozyme-induced suppressor T cells in a genetically nonresponder mouse strain do not recognize a closely related immunogenic lysozyme, *J. Immunol.* **122**:871.

Adorini, L., Harvey, M. A., Miller, A., and Sercarz, E. E., 1979*b,* Fine specificity of regulatory T cells. II. Suppressor and Helper T cells are induced by different regions of hen egg-white lysozyme in a genetically nonresponder mouse strain, *J. Exp. Med.* **150**:293.

Adorini, L., Harvey, M., and Sercarz, E. E., 1979*c,* The fine specificity of regulatory T cells. IV. Idiotypic complementarity and antigen-bridging interactions in the anti-lysozyme response, *Eur. J. Immunol.* **9**:906.

Aguet, M., Andersson, L. C., Andersson, R., Wight, E., Binz, H., and Wigzell, H., 1978, Induction of specific immune unresponsiveness with purified mixed leukocyte culture-activated T lymphoblasts as autoimmu-nogen. II. An analysis of the effects measured at the cellular and serological levels, *J. Exp. Med.* **147**:50.

Ahmed, A., Scher, I., Sharrow, S. O., Smith, A. H., Paul, W. E., Sachs, D. H., and Sell, K. W., 1977, B-lymphocyte heterogeneity: Development and characterization of an alloantiserum which distinguishes B-lymphocyte differentiation alloantigens, *J. Exp. Med.* **145**:101.

Alkan, S. S., Nitecki, D. E., and Goodman, J. W., 1971, Antigen recognition and the immune response: The capacity of L-tyrosine-azobenzenearsonate to serve as a carrier for a macromolecular hapten, *J. Immunol.* **107**:353.

Alkan, S. S., Williams, E. B., Nitecki, D. E., and Goodman, J. W., 1972, Antigen recognition and the immune response. Humoral and cellular immune responses to small mono- and bifunctional antigen molecules, *J. Exp. Med.* **135**:1228.

Amsbaugh, D. F., Hansen, C. T., Prescott, B., Stashak, P. W., Barthold, D. R., and Baker, P. J., 1972, Genetic control of the antibody response to Type III pneumococcal polysaccharide in mice. I. Evidence that an X-linked gene plays a decisive role in determining responsiveness, *J. Exp. Med.* **136**:931.

Andersson, L. C., Binz, H., and Wigzell, H., 1976, Specific unresponsiveness to transplantation antigens induced by autoimmunisation with syngeneic, antigen-specific T lymphoblasts, *Nature (London)* **264**:778.

Andersson, L. C., Aguet, M., Wight, E., Andersson, R., Binz, H., and Wigzell, H., 1977, Induction of specific immune unresponsiveness using purified mixed leukocyte culture-activated T lymphoblasts as autoim-munogen. I. Demonstration of general validity as to species and histocompatibility barriers, *J. Exp. Med.* **146**:1124.

Andersson, L., Binz, H., and Wigzell, H., 1978, Idiotypic receptors on T lymphocytes with specificity for foreign MHC alloantigens, in: *Ir Genes and Ia Antigens* (H. O. McDevitt, ed.), pp. 597–606, Academic Press, New York.

Anfinsen, C. B., 1972, The formation and stabilization of protein structure, *Biochem. J.,* **128**:737.

Anfinsen, C. B., 1973, Principles that govern the folding of protein chains, *Science* **181**:223.

Anfinsen, C. B., Cuatrecasas, P., and Taniuchi, H., 1971, Staphylococcal nuclease, chemical properties and catalysis, in: *The Enzymes* (P. D. Boyer, ed.), Vol. 4, pp. 177–204, Academic Press, New York.

Araneo, B. A., Yowell, R. L., and Sercarz, E. E., 1979, *Ir* gene defects may reflect a regulatory imbalance. I. Helper T cell activity revealed in a strain whose lack of response is controlled by suppression, *J. Immunol.* **123**:961.

Armerding, D., and Katz, D. H., 1974, Activation of T and B lymphocytes in vitro. II. Biological and biochem-ical properties of an allogeneic effect factor (AEF) active in triggering specific B lymphocytes, *J. Exp. Med.* **140**:19.

Arquilla, E. R., and Finn, J., 1965, Genetic control of combining sites of insulin antibodies produced by guinea pigs, *J. Exp. Med.* **122**:771.

Bach, B. A., Greene, M. I., Benacerraf, B., and Nisonoff, A., 1979, Mechanisms of regulation of cell-mediated immunity. IV. Azobenzenearsonate-specific suppressor factor(s) bear cross-reactive idiotypic determinants the expression of which is linked to the heavy-chain allotype linkage group of genes, *J. Exp. Med.* **149**:1084.

Barcinski, M. A., and Rosenthal, A. S., 1977, Immune response gene control of determinant selection. I. Intramolecular mapping of the immunogenic sites on insulin recognized by guinea pig T and B cells, *J. Exp. Med.* **145**:726.

Batchelor, J. R., Compston, A., and McDonald, W. I., 1978, The significance of the association between HLA and multiple sclerosis, *Br. Med. Bull.* **34**:279.

Bechtol, K. B., and McDevitt, H. O., 1976, Antibody response of C3H ↔ (CKB × CWB) F₁ tetraparental mice to poly-L(Tyr, Glu)-poly-D,L-Ala–poly-L-Lys immunization, *J. Exp. Med.* **144**:123.

Bechtol, K. B., Freed, J. H., Herzenberg, L. A., and McDevitt, H. O., 1974, Genetic control of the antibody response to poly-L(Tyr, Glu)-poly-D,L-Ala–poly-L-Lys in C3H ↔ CWB tetraparental mice, *J. Exp. Med.* **140**:1660.

Benacerraf, B., 1975, *Immunogenetics and Immunodeficiency,* MTP, St. Leonard House, Lancaster, England.

Benacerraf, B., 1978, A hypothesis to relate the specificity of T lymphocytes and the activity of I region-specific Ir genes in macrophages and B lymphocytes, *J. Immunol.* **120**:1809.

Benacerraf, B., and Dorf, M. E., 1976, Genetic control of specific immune responses and immune suppressions by *I*-region genes, *Cold Spring Harbor Symp. Quant. Biol.* **41**:465.

Benacerraf, B., and Gell, P. G. H., 1959, Studies on hypersensitivity. I. Delayed and Arthus-type skin reactivity to protein conjugates in guinea pigs, *Immunology* **2**:53.

Benacerraf, B., and Germain, R. N., 1978, The immune response genes of the major histocompatibility complex, *Immunol. Rev.* **38**:71.

Benacerraf, B., and Katz, D. H., 1975, The histocompatibility-linked immune response genes, *Adv. Cancer Res.* **21**:121.

Benacerraf, B., and McDevitt, H. O., 1972, Histocompatibility-linked immune response genes, *Science* **175**:273.

Bennink, J. R., and Doherty, P. C., 1978, T-cell populations specifically depleted of alloreactive potential cannot be induced to lyse H-2-different virus-infected target cells, *J. Exp. Med.* **148**:128.

Ben-Sasson, S. Z., Paul, W. E., Shevach, E. M., and Green, I., 1975, In vitro selection and extended culture of antigen-specific T lymphocytes. II. Mechanisms of selection, *J. Immunol.* **115**:1723.

Ben-Sasson, S. Z., Lipscomb, M. F., Tucker, T. F., and Uhr, J. W., 1977, Specific binding of T lymphocytes to macrophages. II. Role of macrophage-associated antigen, *J. Immunol.* **119**:1493.

Ben-Sasson, S. Z., Lipscomb, M. F., Tucker, T. F., and Uhr, J. W., 1978, Specific binding of T lymphocytes to macrophages. III. Spontaneous dissociation of T cells from antigen-pulsed macrophages, *J. Immunol.* **120**:1902.

Bernard, C. C. A., 1976, Experimental autoimmune encephalomyelitis in mice. Genetic control of susceptibility, *J. Immunogenet.* **3**:263.

Bertrams, J., Kuwert, E., and Liedtke, U., 1972, HL-A antigens and multiple sclerosis, *Tissue Antigens,* **2**:405.

Berzofsky, J. A., 1978a, Genetic control of the immune response to mammalian myoglobins in mice. I. More than one *I*-region gene in *H-2* controls the antibody response, *J. Immunol.* **120**:360.

Berzofsky, J. A., 1978b, Genetic control of the antibody response to sperm whale myoglobin in mice, *Adv. Exp. Med. Biol.* **98**:225.

Berzofsky, J. A., and Schechter, A. N., 1980, On the concepts of crossreactivity and specificity in immunology, manuscript submitted for publication.

Berzofsky, J. A., Schechter, A. N., Shearer, G. M., and Sachs, D. H., 1977a, Genetic control of the immune response to staphylococcal nuclease. III. Time-course and correlation between the response to native nuclease and the response to its polypeptide fragments, *J. Exp. Med.* **145**:111.

Berzofsky, J. A., Schechter, A. N., Shearer, G. M., and Sachs, D. H., 1977b, Genetic control of the immune response to staphylococcal nuclease. IV. *H-2*-linked control of the relative proportions of antibodies produced to different determinants of native nuclease, *J. Exp. Med.* **145**:123.

Berzofsky, J. A., Pisetsky, D. S., Schwartz, R. H., Schechter, A. N., and Sachs, D. H., 1978, Genetic control of the immune response to staphylococcal nuclease in mice, *Adv. Exp. Med. Biol.* **98**:241.

Berzofsky, J. A., Richman, L. K., and Killion, D. J., 1979, Distinct *H-2*-linked *Ir* genes control both antibody and T cell responses to different determinants on the same antigen, myoglobin, *Proc. Natl. Acad. Sci. U.S.A.* **76**:4046.

Bevan, M. J., 1975a, Interaction antigens detected by cytotoxic T cells with the major histocompatibility complex as modifier, *Nature (London)* **256**:419.

571

IMMUNE RESPONSE
GENES IN THE
REGULATION OF
MAMMALIAN
IMMUNITY

Bevan, M. J., 1975b, The major histocompatibility complex determines susceptibility to cytotoxic T cells directed against minor histocompatibility antigens, *J. Exp. Med.* **142**:1349.

Bevan, M. J., 1977, In a radiation chimaera, host *H-2* antigens determine immune responsiveness of donor cytotoxic cells, *Nature (London)* **269**:417.

Bias, W. B., and Marsh, D. G., 1975, HL-A linked antigen E immune response genes: An unproved hypothesis, *Science* **188**:375.

Biddison, W. E., and Shaw, S., 1979, Differences in HLA antigen recognition by human influenza virus-immune cytotoxic T cells, *J. Immunol.* **122**:1705.

Billings, P., Burakoff, S. J., Dorf, M. E., and Benacerraf, B., 1978a, Genetic control of cytolytic T-lymphocyte responses. I. Ir gene control of the specificity of cytolytic T-lymphocyte responses to trinitrophenyl-modified syngeneic cells, *J. Exp. Med.* **148**:341.

Billings, P., Burakoff, S. J., Dorf, M. E., and Benacerraf, B., 1978b, Genetic control of cytolytic T-lymphocyte responses. II. The role of the host genotype in parental → F$_1$ radiation chimeras in the control of the specificity of cytolytic T-lymphocyte responses to trinitrophenyl-modified syngeneic cells, *J. Exp. Med.* **148**:352.

Binz, H., and Askonas, B. A., 1975, Inhibition of mixed leukocyte cultures by anti-idiotypic antibodies, *Eur. J. Immunol.* **5**:618.

Binz, H., and Wigzell, H., 1975a, Shared idiotypic determinants on B and T lymphocytes reactive against the same antigenic determinants. I. Demonstration of similar or identical idiotypes on IgG molecules and T-cell receptors with specificity for the same alloantigens, *J. Exp. Med.* **142**:197.

Binz, H., and Wigzell, H., 1975b, Shared idiotypic determinants on B and T lymphocytes reactive against the same antigenic determinants. II. Determination of frequency and characteristics of idiotypic T and B lymphocytes in normal rats using direct visualization, *J. Exp. Med.* **142**:1218.

Binz, H., and Wigzell, H., 1975c, Shared idiotypic determinants on B and T lymphocytes reactive against the same antigenic determinants. III. Physical fractionation of specific immunocompetent T lymphocytes by affinity chromatography using anti-idiotypic antibodies, *J. Exp. Med.* **142**:1231.

Binz, H., and Wigzell, 1975d, Shared idiotypic determinants on B and T lymphocytes reactive against the same antigenic determinants. IV. Isolation of two groups of naturally occurring, idiotypic molecules with specific antigen-binding activity in the serum and urine of normal rats, *Scand. J. Immunol.* **4**:591.

Binz, H., and Wigzell, H., 1976a, Successful induction of specific tolerance to transplantation antigens using autoimmunisation against the recipient's own natural antibodies, *Nature (London)* **262**:294.

Binz, H., and Wigzell, H., 1976b, Specific transplantation tolerance induced by autoimmunization against the individual's own, naturally occurring idiotypic, antigen-binding receptors, *J. Exp. Med.* **144**:1438.

Binz, H., and Wigzell, H., 1978, Induction of specific immune unresponsiveness with purified mixed leukocyte culture-activated T lymphoblasts as autoimmunogen. III. Proof for the existence of autoanti-idiotypic killer T cells and transfer of suppression to normal syngeneic recipients by T or B lymphocytes, *J. Exp. Med.* **147**:63.

Binz, H., Wigzell, H., and Bazin, H., 1976, T-cell idiotypes are linked to immunoglobulin heavy chain genes, *Nature (London)* **264**:639.

Binz, H., Frischknecht, H., Shen, F. W., and Wigzell, H., 1979, Idiotypic determinants on T-cell subpopulations, *J. Exp. Med.* **149**:910.

Black, S. J., Hämmerling, G. J., Berek, C., Rajewsky, K., and Eichmann, K., 1976, Idiotypic analysis of lymphocytes in vitro. I. Specificity and heterogeneity of B and T lymphocytes reactive with anti-idiotypic antibody, *J. Exp. Med.* **143**:846.

Blanden, R. V., and Andrew, M. E., 1979, Primary anti-viral cytotoxic T-cell responses in semiallogeneic chimeras are not absolutely restricted to host H-2 type, *J. Exp. Med.* **149**:535.

Blanden, R. V., Doherty, P. C., Dunlop, M. B. C., Gardner, I. D., Zinkernagel, R. M., and David, C. S., 1975, Genes required for cytotoxicity against virus-infected target cells in *K* and *D* regions of H-2 complex, *Nature (London)* **254**:269.

Blank, K. J., Freedman, H. A., and Lilly, F., 1976, T-lymphocyte response to Friend virus-induced tumor cell lines in mice of strains congenic at *H-2*, *Nature (London)* **260**:250.

Blank, K. J., Bubbers, J. E., and Lilly, F., 1977, *H-2*/viral protein interaction at the cell membrane as the basis for *H-2*-restricted T-lymphocyte immunity, in: *Immune System: Genetics and Regulation* (E. E. Sercarz, L. A. Herzenberg, and C. F. Fox, eds), pp. 607–614, Academic Press, New York.

Blomberg, B., Geckeler, W. R., and Weigert, M., 1972, Genetics of the immune response to dextran in mice, *Science* **177**:178.

Bluestein, H. G., Green, I., and Benacerraf, B., 1971, Specific immune response genes of the guinea pig. I. Dominant genetic control of immune responsiveness to copolymers of L-glutamic acid and L-alanine and L-glutamic acid and L-tyrosine, *J. Exp. Med.* **134**:458.

573

IMMUNE RESPONSE
GENES IN THE
REGULATION OF
MAMMALIAN
IMMUNITY·

Bluestein, H. G., Green, I., Maurer, P. H., and Benacerraf, B., 1972, Specific immune response genes of the guinea pig. V. Influence of the GA and GT immune response genes on the specificity of cellular and humoral immune responses to a terpolymer of L-glutamic acid, L-alanine, and L-tyrosine, *J. Exp. Med.* **135**:98.

Blumenthal, M. N., Amos, D. B., Noreen, H., Mendell, N. R., and Yunis, E. J., 1974, Genetic mapping of Ir locus in man: Linkage to second locus of HL-A, *Science* **184**:1301.

Bodmer, W. F. (ed.), 1978*a*, The HLA System, *Br. Med. Bull.* **34**(3):213.

Bodmer, J. G., 1978*b*, Ia antigens: Definition of HLA-DRw Specificities, *Br. Med. Bull.* **34**:233.

von Boehmer, H., and Haas, W., 1976, Cytotoxic T lymphocytes recognize allogeneic tolerated TNP-conjugated cells, *Nature (London)* **261**:141.

von Boehmer, H., and Haas, W., 1979, Distinct Ir genes for helper and killer cells in the cytotoxic response to H-Y antigen, *J. Exp. Med.* **150**:1134.

von Boehmer, H., Hudson, L., and Sprent, J., 1975, Collaboration of histoincompatible T and B lymphocytes using cells from tetraparental bone marrow chimeras, *J. Exp. Med.* **142**:989.

von Boehmer, H., Fathman, C. G., and Haas, W., 1977, H-2 gene complementation in cytotoxic T cell responses of female against male cells, *Eur. J. Immunol.* **7**:443.

von Boehmer, H., Haas, W., and Jerne, N. K., 1978, Major histocompatibility complex-linked immune-responsiveness is acquired by lymphocytes of low-responder mice differentiating in thymus of high-responder mice, *Proc. Natl. Acad. Sci. U.S.A.* **75**:2439.

Bona, C., and Paul, W. E., 1979, Cellular basis of regulation of expression of idiotype. I. T-suppressor cells specific for MOPC 460 idiotype regulate the expression of cells secreting anti-TNP antibodies bearing 460 idiotype, *J. Exp. Med.* **149**:592.

Bottomly, K., Mathieson, B. J., and Mosier, D. E., 1978, Anti-idiotype induced regulation of helper cell function for the response to phosphorylcholine in adult BALB/c mice, *J. Exp. Med.* **148**:1216.

Braendstrup, O., Werdelin, O., Shevach, E. M., and Rosenthal, A. S., 1979, Macrophage-lymphocyte clusters in the immune response to soluble protein antigen *in vitro*. VII. Genetically restricted and nonrestricted physical interactions, *J. Immunol.* **122**:1608.

Brewerton, D. A., Caffrey, M., Hart, F. D., James, D. C. O., Nicholls, A., and Sturrock, R. D., 1973, Ankylosing spondylitis and HL-A 27, *Lancet* **1973**(1):904.

Bubbers, J. E., and Lilly, F., 1977, Selective incorporation of *H-2* antigenic determinants into Friend virus particles, *Nature (London)* **266**:458.

Bubbers, J. E., Chen, S., and Lilly, F., 1978, Nonrandom inclusion of *H-2K* and *H-2D* antigens in Friend virus particles from mice of various strains, *J. Exp. Med.* **147**:340.

Buckley, C. E., III., Dorsey, F. C., Corley, R. B., Ralph, W. B., Woodbury, M. A., and Amos, D. B., 1973, HL-A-linked human immune-response genes, *Proc. Natl. Acad. Sci. U.S.A.* **70**:2157.

Burakoff, S. J., Germain, R. N., Dorf, M. E., and Benacerraf, B., 1976, Inhibition of cell-mediated cytolysis of trinitrophenyl-derivatized target cells by alloantisera directed to products of the K and D loci of the *H-2* complex, *Proc. Natl. Acad. Sci. U.S.A.* **73**:625.

Burnet, F. M., 1959, *The Clonal Selection Theory of Immunity,* Vanderbilt University Press, Nashville, Tennessee.

Bush, M. E., Alkan, S. S., Nitecki, D. E., and Goodman, J. W., 1972, Antigen recognition and the immune response. "Self-help" with symmetrical bifunctional antigen molecules, *J. Exp. Med.* **136**:1478.

Caffrey, M. F. P., and James, D. C. O., 1973, Human lymphocyte antigen association in ankylosing spondylitis, *Nature (London)* **242**:121.

Cantor, H., and Boyse, E. A., 1977, Lymphocytes as models for the study of mammalian cellular differentiation, *Immunol. Rev.* **33**:105.

Cantor, H., McVay-Boudreau, L., Hugenberger, J., Naidorf, K., Shen, F. W., and Gershon, R. K., 1978, Immunoregulatory circuits among T-cell sets. II. Physiologic role of feedback inhibition *in vivo:* Absence in NZB mice, *J. Exp. Med.* **147**:1116.

Carroll, M. C., and Capra, J. D., 1978, Studies on the murine Ss protein: Demonstration that the Ss protein is functionally the fourth component of complement, *Proc. Natl. Acad. Sci. U.S.A.* **75**:2424.

Cecka, J. M., Stratton, J. A., Miller, A., and Sercarz, E., 1976, Structural aspects of immune recognition of lysozymes. III. T cell specificity restriction and its consequences for antibody specificity, *Eur. J. Immunol.* **6**:639.

Chesebro, B., and Wehrly, K., 1978, *Rfv-1* and *Rfv-2,* two *H-2*-associated genes that influence recovery from Friend leukemia virus-induced splenomegaly, *J. Immunol.* **120**:1081.

Chesebro, B., Wehrly, K., and Stimpfling, J., 1974, Host genetic control of recovery from Friend leukemia virus-induced splenomegaly. Mapping of a gene within the major histocompatibility complex, *J. Exp. Med.* **140**:1457.

Cheung, N. K. V., Dorf, M. E., and Benacerraf, B., 1977, Genetic control of the primary humoral response to Glu^{56}Lys^{35}Phe^9, *Immunogenetics* **4**:163.

Chused, T. M., Kassan, S. S., Opelz, G., Moutsopouios, H. M., and Terasaki, P. I., 1977, Sjögren's syndrome associated with HLA-Dw3, *N. Engl. J. Med.* **296**:895.

Cohen, P. L., Scher, I., and Mosier, D. E., 1976, In vitro studies of the genetically determined unresponsiveness to thymus-independent antigens in CBA/N mice, *J. Immunol.* **116**:301.

Cohn, M., and Epstein, R., 1978, T-cell inhibition of humoral responsiveness. II. Theory on the role of restrictive recognition in immune regulation, *Cell. Immunol.* **39**:125.

Colombani, J., Colombani, M., Shreffler, D. C., and David, C. S., 1976, Separation of anti-Ia (I-region associated antigens) from anti-H-2 antibodies in complex sera, by absorption on blood platelets. Description of three new Ia specificities, *Tissue Antigens* **7**:74.

Cook, R. G., Siegelman, M. H., Capra, J. D., Uhr, J. W., and Vitetta, E. S., 1979a, Structural studies on the murine Ia alloantigens. IV. NH$_2$-terminal sequence analysis of allelic products of the I-A and I-E/C subregions, *J. Immunol.* **122**:2232.

Cook, R. G., Vitetta, E. S., Uhr, J. W., and Capra, J. D., 1979b, Structural studies on the murine Ia alloantigens. III. Tryptic peptide comparisons of allelic products of the *I-E/C* subregions, *Mol. Immunol.* **16**:29.

Cook, R. G., Vitetta, E. S., Uhr, J. W., and Capra, J. D., 1979c, Structural studies on the murine Ia alloantigens. V. Evidence that the structural gene for the *I-E/C* beta polypeptide is encoded within the *I-A* subregion, *J. Exp. Med.* **149**:981.

Cooper, M. G., 1972, Delayed-type hypersensitivity in the mouse. II. Transfer by thymus-derived (T) cells, *Scand. J. Immunol.* **1**:237.

Cooper, M. G., and Ada, G. L., 1972, Delayed-type hypersensitivity in the mouse. III. Inactivation of thymus-derived effector cells and their precursors, *Scand. J. Immunol.* **1**:247.

Corradin, G., and Chiller, J. M., 1979, Lymphocyte specificity to protein antigens. II. Fine specificity of T-cell activation with cytochrome *c* and derived peptides as antigenic probes, *J. Exp. Med.* **149**:436.

Cosenza, H., Leserman, L. D., and Rowley, D. A., 1971, The third cell type required for the immune response of spleen cells *in vitro, J. Immunol.* **107**:414.

Cotton, F. A., and Hazen, E. E., Jr., 1971, Staphylococcal nuclease X-ray structure, in: *The Enzymes* (P. D. Boyer, ed.), Vol 4, pp. 153–176, Academic Press, New York.

Cowing, C., Schwartz, B. D., and Dickler, H. B., 1978, Macrophage Ia antigens. I. Macrophage populations differ in their expression of Ia antigens, *J. Immunol.* **120**:378.

Cramer, M., Krawinkel, U., Melchers, I., Imanishi-Kari, T., Ben-Neriah, Y., Givol, D., and Rajewsky, K., 1979, Isolated hapten-binding receptors of sensitized lymphocytes. IV. Expression of immunoglobulin variable regions in (4-hydroxy-3-nitrophenyl) acetyl (NP)-specific receptors isolated from murine B and T lymphocytes, *Eur. J. Immunol.* **9**:332.

Cudworth, A. G., and Festenstein, H., 1978, HLA genetic heterogeneity in diabetes mellitus, *Br. Med. Bull.* **34**:285.

Cullen, S. E., Freed, J. H., and Nathenson, S. G., 1976, Structural and serological properties of murine Ia alloantigens, *Transplant. Rev.* **30**:236.

Cunningham, A. J., 1976, *The Generation of Antibody Diversity: A New Look,* Academic Press, New York.

David, C. S., 1976, Serologic and genetic aspects of murine Ia antigens, *Transplant. Rev.* **30**:299.

David, C. S., and Cullen, S. E., 1978, Murine Ia antigens: Identification and mapping of Ia.23 and further definition of the *I-E* subregion, *J. Immunol.* **120**:1659.

David, C. S., Shreffler, D. C., and Frelinger, J. A., 1973, New lymphocyte antigen system (*Lna*) controlled by the *Ir* region of the mouse *H-2* complex, *Proc. Natl. Acad. Sci. U.S.A.* **70**:2509.

David, C. S., Cullen, S. E., and Murphy, D. B., 1975, Serologic and biochemical studies of the Ia system of the mouse H-2 gene complex. Further evidence for an I-C subregion, *J. Immunol.* **114**:1205.

Davie, J. M., and Paul, W. E., 1970, Receptors on immunocompetent cells. I. Receptor specificity of cells participating in a cellular immune response, *Cell. Immunol.* **1**:404.

Davies, D. A. L., and Staines, N. A., 1976, A cardinal role for I-region antigens (Ia) in immunological enhancement, and the clinical implications, *Transplant. Rev.* **30**:18.

Deak, B. D., Meruelo, D., and McDevitt, H. O., 1978, Expression of a single major histocompatibility complex locus controls the immune response to poly-L-(Tyrosine, Glutamic acid)-poly-DL-Alanine—Poly-L-Lysine, *J. Exp. Med.* **147**:599.

Debré, P., Kapp, J. A., and Benacerraf, B., 1975a, Genetic control of specific immune suppression. I. Experimental conditions for the stimulation of suppressor cells by the copolymer L-glutamic acid^{50}—L-tyrosine^{50} (GT) in nonresponder BALB/c mice, *J. Exp. Med.* **142**:1436.

Debré, P., Kapp, J. A., Dorf, M. E., and Benacerraf, B., 1975*b*, Genetic control of specific immune suppression. II. H-2-linked dominant genetic control of immune suppression by the random copolymer L-glutamic acid^{50}-L-tyrosine^{50} (GT), *J. Exp. Med.* **142**:1447.

Debré, P., Waltenbaugh, C., Dorf, M., and Benacerraf, B., 1976*a*, Genetic control of specific immune suppression. III. Mapping of H-2 complex complementing genes controlling immune suppression by the random copolymer L-glutamic acid^{50}-L-tyrosine^{50} (GT), *J. Exp. Med.* **144**:272.

Debré, P., Waltenbaugh, C., Dorf, M., and Benacerraf, B., 1976*b*, Genetic control of specific immune suppression. IV. Responsiveness to the random copolymer L-glutamic acid^{50}-L-tyrosine^{50} induced in BALB/c mice by cyclophosphamide, *J. Exp. Med.* **144**:277.

Dickler, H. B., and Sachs, D. H., 1974, Evidence for identity or close association of the Fc receptor of B lymphocytes and alloantigens determined by the *Ir* region of the *H-2* complex, *J. Exp. Med.* **140**:779.

Dickler, H. B., Arbeit, R. D., Henkart, P. A., and Sachs, D. H., 1976, Association between Ia antigens and the Fc receptors of certain T lymphocytes, *J. Exp. Med.* **144**:282.

Dickmeiss, E., Soeberg, B., and Svejgaard, A., 1977, Human cell-mediated cytotoxicity against modified targets is restricted by HLA, *Nature (London)* **270**:526.

Doherty, P. C., and Bennink, J. R., 1979, Vaccinia-specific cytotoxic T-cell responses in the context of H-2 antigens not encountered in thymus may reflect aberrant recognition of a virus–H-2 complex, *J. Exp. Med.* **149**:150.

Doherty, P. C., Blanden, R. V., and Zinkernagel, R. M., 1976*a*, Specificity of virus-immune effector T cells for H-2K or H-2D compatible interactions: Implications for H-antigen diversity, *Transplant. Rev.* **29**:89.

Doherty, P. C., Götze, D., Trinchieri, G., and Zinkernagel, R. M., 1976*b*, Models for recognition of virally modified cells by immune thymus-derived lymphocytes, *Immunogenetics* **3**:517.

Doherty, P. C., Biddison, W. E., Bennink, J. R., and Knowles, B. B., 1978, Cytotoxic T-cell responses in mice infected with influenza and vaccinia viruses vary in magnitude with *H-2* genotype, *J. Exp. Med.* **148**:534.

Dorf, M. E., and Benacerraf, B., 1975, Complementation of *H-2*-linked *Ir* genes in the mouse, *Proc. Natl. Acad. Sci. U.S.A.* **72**:3671.

Dorf, M. E., and Stimpfling, J. H., 1977, Coupled complementation of immune response genes controlling responsiveness to the H-2.2 alloantigen, *J. Exp. Med.* **146**:571.

Dorf, M. E., Dunham, E. K., Johnson, J. P., and Benacerraf, B., 1974, Genetic control of the immune response: The effect of Non-*H-2* linked genes on antibody production, *J. Immunol.* **112**:1329.

Dorf, M. E., Stimpfling, J. H., and Benacerraf, B., 1975, Requirement for two *H-2* complex *Ir* genes for the immune response to the L-Glu, L-Lys, L-Phe terpolymer, *J. Exp. Med.* **141**:1459.

Dorf, M. E., Twigg, M. B., and Benacerraf, B., 1976, Genetic control of the immune response to the random linear terpolymer of L-glutamic acid, L-lysine and L-leucine (GLLeu) by complementing *Ir* genes, *Eur. J. Immunol.* **6**:552.

Dorf, M. E., Stimpfling, J. H., Cheung, N. K., and Benacerraf, B., 1978, Coupled complementation of *Ir* genes, in: *Ir Genes and Ia Antigens* (H. O. McDevitt, ed.), pp. 55–66, Academic Press, New York.

Dorf, M. E., Stimpfling, J. H., and Benacerraf, B., 1979, Gene dose effects in *Ir* gene-controlled systems, *J. Immunol.* **123**:269.

Dunham, E. K., Unanue, E. R., and Benacerraf, B., 1972, Antigen binding and capping by lymphocytes of genetic nonresponder mice, *J. Exp. Med.* **136**:403.

Eardley, D. D., and Sercarz, E. E., 1976, Modulation of help and suppression in a hapten-carrier system, *J. Immunol.* **116**:600.

Eardley, D. D., Hugenberger, J., McVay-Boudreau, L., Shen, F. W., Gershon, R. K., and Cantor, H., 1978, Immunoregulatory circuits among T-cell sets. I. T-helper cells induce other T-cell sets to exert feedback inhibition, *J. Exp. Med.* **147**:1106.

Eardley, D. D., Shen, F. W., Cantor, H., and Gershon, R. K., 1979, Genetic control of immunoregulatory circuits. Genes linked to the Ig locus govern communication between regulatory T-cell sets, *J. Exp. Med.* **150**:44.

Eichmann, K., 1972, Idiotypic identity of antibodies to streptococcal carbohydrate in inbred mice, *Eur. J. Immunol.* **2**:301.

Eichmann, K., 1974, Idiotype suppression. I. Influence of the dose and of the effector functions of anti-idiotypic antibody on the production of an idiotype, *Eur. J. Immunol.* **4**:296.

Eichmann, K., 1975, Idiotype suppression. II. Amplification of a suppressor T cell with anti-idiotypic activity, *Eur. J. Immunol.* **5**:511.

Eichmann, K., 1977, Heavy chain variable region idiotypes on helper T cells, in: *Immune System: Genetics and Regulation* (E. E. Sercarz, L. A. Herzenberg, and C. F. Fox, eds.), pp. 127–138, Academic Press, New York.

575

IMMUNE RESPONSE
GENES IN THE
REGULATION OF
MAMMALIAN
IMMUNITY

Eichmann, K., and Rajewsky, K., 1975, Induction of T and B cell immunity by anti-idiotypic antibody, *Eur. J. Immunol.* **5**:661.

Eichmann, K., Tung, A. S., and Nisonoff, A., 1974, Linkage and rearrangement of genes encoding mouse immunoglobulin heavy chains, *Nature (London)* **250**:509.

Eichmann, K., Falk, I., and Rajewsky, K., 1978, Recognition of idiotypes in lymphocyte interactions. II. Antigen-independent cooperation between T and B lymphocytes that possess similar and complementary idiotypes, *Eur. J. Immunol.* **8**:853.

Ellman, L., Green, I., Martin, W. J., and Benacerraf, B., 1970a, Linkage between the poly-L-lysine gene and the locus controlling the major histocompatibility antigens in strain 2 guinea pigs, *Proc. Natl. Acad. Sci. U.S.A.* **66**:322.

Ellman, L., Green, I., and Benacerraf, B., 1970b, Identification of the cell population responding to DNP-GL in lethally irradiated strain 13 chimeric guinea pigs reconstituted with strain 13 bone marrow and (2 × 13)F_1 lymph node and spleen cells, *Cell. Immunol.* **1**:445.

Ellner, J. J., Lipsky, P. E., and Rosenthal, A. S., 1977, Antigen handling by guinea pig macrophages: Further evidence for the sequestration of antigen relevant for activation of primed T lymphocytes, *J. Immunol.* **118**:2053.

Engleman, E. G., McMichael, A. J., Batey, M. E., and McDevitt, H. O., 1978, A suppressor T cell of the mixed lymphocyte reaction in man specific for the stimulating alloantigen. Evidence that identity at HLA-D between suppressor and responder is required for suppression, *J. Exp. Med.* **147**:137.

Ennis, F. A., Martin, W. J., Verbonitz, M. W., and Butchko, G. M., 1977, Specificity studies on cytotoxic thymus-derived lymphocytes reactive with influenza virus-infected cells: Evidence for dual recognition of H-2 and viral hemagglutinin antigens, *Proc. Natl. Acad. Sci. U.S.A.* **74**:3006.

Erb, P., and Feldmann, M., 1975a, Role of macrophages in *in vitro* induction of T-helper cells, *Nature (London)* **254**:352.

Erb, P., and Feldmann, M., 1975b, The role of macrophages in the generation of T-helper cells. II. The genetic control of the macrophage–T-cell interaction for helper cell induction with soluble antigens, *J. Exp. Med.* **142**:460.

Erb, P., and Feldmann, M., 1975c, The role of macrophages in the generation of T helper cells. III. Influence of macrophage-derived factors in helper cell induction, *Eur. J. Immunol.* **5**:759.

Erb, P., Feldmann, M., and Hogg, N., 1976, Role of macrophages in the generation of T helper cells. IV. Nature of genetically related factor derived from macrophages incubated with soluble antigens, *Eur. J. Immunol.* **6**:365.

Erb, P., Vogt, P., Meier, B., and Feldmann, M., 1977, The role of macrophages in the generation of T helper cells. V. Evidence for differential activation of short-lived T_1 and long-lived T_2 lymphocytes by macrophage factors GRF and NMF, *J. Immunol.* **119**:206.

Erb, P., Meier, B., Kraus, D., von Boehmer, H., and Feldmann, M., 1978, Nature of T cell-macrophage interaction in helper cell induction *in vitro*. I. Evidence for genetic restriction of T cell–macrophage interactions prior to T cell priming, *Eur. J. Immunol.* **8**:786.

Erb, P., Meier, B., and Feldmann, M., 1979a, Is genetically related macrophage factor (GRF) a soluble immune response (*Ir*) gene product? *J. Immunol.* **122**:1916.

Erb, P., Meier, B., Matsunaga, T., and Feldmann, M., 1979b, Nature of T-cell macrophage interaction in helper-cell induction in vitro. II. Two stages of T–helper-cell differentiation analyzed in irradiation and allophenic chimeras, *J. Exp. Med.* **149**:686.

Falchuk, Z. M., Rogentine, G. N., and Strober, W., 1972, Predominance of histocompatibility antigen HL-A8 in patients with gluten-sensitive enteropathy, *J. Clin. Invest.* **51**:160.

Farid, N. R., Sampson, L., Noel, E. P., Bernard, J. M., Mandeville, R., Larsen, B., Marshall, W. H., and Carter, N. D., 1979, A study of human leukocyte D locus related antigens in Graves' disease, *J. Clin. Invest.* **63**:108.

Fathman, C. G., and Nabholz, M., 1977, *In vitro* secondary mixed lymphocyte reaction (MLR). II. Interaction MLR determinants expressed by F_1 cells, *Eur. J. Immunol.* **7**:370.

Fathman, C. G., Cone, J. L., Sharrow, S. O., Tyrer, H., and Sachs, D. H., 1975, Ia alloantigen(s) detected on thymocytes by use of a fluorescence-activated cell sorter, *J. Immunol.* **115**:584.

Fathman, C. G., Watanabe, T., and Augustin, A., 1978, *In vitro* secondary MLR. III. Hybrid histocompatibility determinants, *J. Immunol.* **121**:259.

Feldmann, M., and Kontiainen, S., 1976, Suppressor cell induction *in vitro*. II. Cellular requirements of suppressor cell induction, *Eur. J. Immunol.* **6**:302.

Feldmann, M., Cone, R. E., and Marchalonis, J. J., 1973, Cell interactions in the immune response *in vitro*. VI. Mediation by T cell surface monomeric Ig M, *Cell. Immunol.* **9**:1.

577

IMMUNE RESPONSE
GENES IN THE
REGULATION OF
MAMMALIAN
IMMUNITY

Feldmann, M., Beverley, P. C. L., Woody, J., and McKenzie, I. F. C., 1977, T–T interactions in the induction of suppressor and helper T cells: Analysis of membrane phenotype of precursor and amplifier cells, *J. Exp. Med.* **145**:793.

Feldmann, M., Howie, S., and Kontiainen, S., 1979, Antigen-specific regulatory factors in the immune response, in: *Biology of the Lymphokines* (S. Cohen, E. Pick, and J. J. Oppenheim, eds.), pp. 391–419, Academic Press, New York.

Finberg, R., Greene, M. I., Benacerraf, B., and Burakoff, S. J., 1979, The cytolytic T lymphocyte response to trinitrophenyl-modified syngeneic cells. I. Evidence for antigen-specific helper T cells, *J. Immunol.* **123**:1205.

Fink, P. J., and Bevan, M. J., 1978, H-2 antigens of the thymus determine lymphocyte specificity, *J. Exp. Med.* **148**:766.

Foerster, J., Green, I., Lamelin, J.-P., and Benacerraf, B., 1969, Transfer of responsiveness to hapten conjugates of poly-L-lysine and of a copolymer of L-glutamic acid and L-lysine to lethally irradiated nonresponder guinea pigs by bone marrow or lymph node and spleen cells from responder guinea pigs, *J. Exp. Med.* **130**:1107.

Forman, J., 1975, On the role of the *H-2* histocompatibility complex in determining the specificity of cytotoxic effector cells sensitized against syngeneic trinitrophenyl-modified targets, *J. Exp. Med.* **142**:403.

Forman, J., Vitetta, E. S., Hart, D. A., and Klein, J., 1977*a*, Relationship between trinitrophenyl and H-2 antigens on trinitrophenyl-modified spleen cells. I. H-2 antigens on cells treated with trinitrobenzene sulfonic acid are derivatized, *J. Immunol.* **118**:797.

Forman, J., Vitetta, E. S., and Hart, D. A., 1977*b*, Relationship between trinitrophenyl and H-2 antigens on trinitrophenyl-modified spleen cells. II. Correlation between derivatization of H-2 antigens with trinitrophenyl and the ability of trinitrophenyl-modified cells to react functionally in the CML assay, *J. Immunol.* **118**:803.

Forman, J., Klein, J., and Streilein, J. W., 1977*c*, Spleen cells from animals neonatally tolerant to H-2K^k antigens recognize trinitrophenyl-modified H-2K^k spleen cells, *Immunogenetics* **5**:561.

Frelinger, J. A., Niederhuber, J. E., and Shreffler, D. C., 1976, Effects of anti-Ia sera on mitogenic responses. III. Mapping of the genes controlling the expression of Ia determinants on concanavalin A-reactive cells to the *I-J* subregion of the *H-2* gene complex, *J. Exp. Med.* **144**:1141.

Furie, B., Schechter, A. N., Sachs, D. H., and Anfinsen, C. B., 1975, An immunological approach to the conformational equilibrium of staphylococcal nuclease, *J. Mol. Biol.* **92**:497.

Gasser, D. L., 1969, Genetic control of the immune response in mice. I. Segregation data and localization to the fifth linkage group of a gene affecting antibody production, *J. Immunol.* **103**:66.

Gasser, D. L., and Shreffler, D. C., 1972, Involvement of *H-2* locus in a multigenically-determined immune response, *Nature (London), New Biol.* **235**:155.

Geczy, A. F., Geczy, C. L., and deWeck, A. L., 1976, Histocompatibility antigens and genetic control of the immune response in guinea pigs. III. Specific inhibition of antigen-induced lymphocyte proliferation by strain-specific anti-idiotypic antibodies, *J. Exp. Med.* **144**:226.

Geha, R. S., Milgrom, H., Broff, M., Alpert, S., Martin, S., and Yunis, E. J., 1979, Effect of anti-HLA antisera on macrophage-T-cell interaction, *Proc. Natl. Acad. Sci. U.S.A.* **76**:4038.

Gell, P. G. H., and Benacerraf, B., 1959, Studies on hypersensitivity. II. Delayed hypersensitivity to denatured proteins in guinea pigs, *Immunology* **2**:64.

Germain, R. N., and Benacerraf, B., 1978*a*, Antigen-specific T cell-mediated suppression III. Induction of antigen-specific suppressor T cells (T_{S2}) in L-Glutamic acid^{60}-L-alanine^{30}-L-tyrosine^{10} (GAT) responder mice by nonresponder-derived GAT-suppressor factor (GAT-T_sF), *J. Immunol.* **121**:608.

Germain, R. N., and Benacerraf, B., 1978*b*, The involvement of suppressor T cells in Ir gene regulation of secondary antibody responses of primed (responder × nonresponder)F$_1$ mice to macrophage-bound L-glutamic acid^{60}-L-alanine^{30}-L-tyrosine^{10}, *J. Exp. Med.* **148**:1324.

Germain, R. N., Thèze, J., Kapp, J. A., and Benacerraf, B., 1978*a*, Antigen-specific T-cell-mediated suppression. I. Induction of L-glutamic acid^{60}-L-alanine^{30}-L-tyrosine^{10} specific suppressor T cells in vitro requires both antigen-specific T-cell-suppressor factor and antigen, *J. Exp. Med.* **147**:123.

Germain, R. N., Thèze, J., Waltenbaugh, C., Dorf, M. E., and Benacerraf, B., 1978*b*, Antigen-specific T cell-mediated suppression. II. *In vitro* induction by I-J-coded L-glutamic acid^{50}-L-tyrosine^{50} (GT)-specific T cell suppressor factor (GT-T_sF) of suppressor T cells (T_{S2}) bearing distinct I-J determinants, *J. Immunol.* **121**:602.

Germain, R. N., Ju, S.-T., Kipps, T. J., Benacerraf, B., and Dorf, M. E., 1979, Shared idiotypic determinants on antibodies and T-cell-derived suppressor factor specific for the random terpolymer L-glutamic acid^{60}-L-alanine^{30}-L-tyrosine^{10}, *J. Exp. Med.* **149**:613.

Gershwin, M. E., Terasaki, P. I., Graw, R., and Chused, T. M., 1975, Increased frequency of HL-A8 in Sjogren's syndrome, *Tissue Antigens* **6**:342.

Gershwin, M. E., Terasaki, P. I., and Castles, J. J., 1978, B cell alloantigens in juvenile rheumatoid arthritis, *Tissue Antigens* **11**:71.

Gibofsky, A., Winchester, R. J., Patarroyo, M., Fotino, M., and Kunkel, H. G., 1978, Disease associations of the Ia-like human alloantigens. Contrasting patterns in rheumatoid arthritis and systemic lupus erythematosus, *J. Exp. Med.* **148**:1728.

Goding, J. W., White, E., and Marchalonis, J. J., 1975, Partial characterisation of Ia antigens on murine thymocytes, *Nature (London)* **257**:230.

Gomard, E., Duprez, V., Reme, T., Colombani, M. J., and Levy, J. P., 1977, Exclusive involvement of *H-2D^b* or *H-2K^d* product in the interaction between T-killer lymphocytes and syngeneic *H-2^b* or *H-2^d* viral lymphomas, *J. Exp. Med.* **146**:909.

Goodman, J. W., 1975, Antigenic determinants and antibody combining sites, in: *The Antigens* (M. Sela, ed.), Vol. 3, pp. 127–187, Academic Press, New York.

Gordon, R. D., Simpson, E., and Samelson, L. E., 1975, *In vitro* cell-mediated immune responses to the male specific (H-Y) antigen in mice, *J. Exp. Med.* **142**:1108.

Gordon, R. D., Mathieson, B. J., Samelson, L. E., Boyse, E. A., and Simpson, E., 1976, The effect of allogeneic presensitization on H-Y graft survival and in vitro cell-mediated responses to H-Y antigen, *J. Exp. Med.* **144**:810.

Goulmy, E., Termijtelen, A., Bradley, B. A., and van Rood, J. J., 1977, Y-antigen killing by T cells of women is restricted by HLA, *Nature (London)* **266**:544.

Green, I., Paul, W. E., and Benacerraf, B., 1966, The behavior of hapten-poly-L-lysine conjugates as complete antigens in genetic responder and as haptens in nonresponder guinea pigs, *J. Exp. Med.* **123**:859.

Greenberg, L. J., Gray, E. D., and Yunis, E. J., 1975, Association of HL-A5 and immune responsiveness in vitro to streptococcal antigens, *J. Exp. Med.* **141**:935.

Greene, M. I., Pierres, A., Dorf, M. E., and Benacerraf, B., 1977, The *I-J* subregion codes for determinants on suppressor factor(s) which limit the contact sensitivity response to picryl chloride, *J. Exp. Med.* **146**:293.

Greene, M. I., Bach, B. A., and Benacerraf, B., 1979, Mechanisms of regulation of cell-mediated immunity. III. The characterization of azobenzenearsonate-specific suppressor T-cell-derived-suppressor factors, *J. Exp. Med.* **149**:1069.

Grumet, F. C., 1972, Genetic control of the immune response: A selective defect in immunologic (IgG) memory in nonresponder mice, *J. Exp. Med.* **135**:110.

Hämmerling, G. J., 1976, Tissue distribution of Ia antigens and their expression on lymphocyte subpopulations, *Transpl. Rev.* **30**:64.

Hämmerling, G. J., and McDevitt, H. O., 1974, Antigen binding T and B lymphocytes. I. Differences in cellular specificity and influence of metabolic activity on interaction of antigen with T and B cells, *J. Immunol.* **112**:1726.

Hämmerling, G. J., Masuda, T., and McDevitt, H. O., 1973, Genetic control of the immune response: Frequency and characteristics of antigen-binding cells in high and low responder mice, *J. Exp. Med.* **137**:1180.

Hämmerling, G. J., Deak, B. D., Mauve, G., Hämmerling, U., and McDevitt, H. O., 1974, B lymphocyte alloantigens controlled by the *I* region of the major histocompatibility complex in mice, *Immunogenetics* **1**:68.

Hämmerling, G. J., Mauve, G., Goldberg, E., and McDevitt, H. O., 1975, Tissue distribution of Ia antigens: Ia on spermatozoa, macrophages, and epidermal cells, *Immunogenetics* **1**:428.

Hämmerling, G. J., Black, S. J., Berek, C., Eichmann, K., and Rajewsky, K., 1976, Idiotypic analysis of lymphocytes in vitro. II. Genetic control of T-helper cell responsiveness to anti-idiotypic antibody, *J. Exp. Med.* **143**:861.

Hanna, N., Ferraresi, R. W., and Leskowitz, S., 1973, In vitro correlates of hapten-specific delayed hypersensitivity, *Cell. Immunol.* **8**:155.

Hansen, T. H., and Levy, R. B., 1978, Alloantigens determined by a second *D* region locus elicit a strong *in vitro* cytotoxic response, *J. Immunol.* **120**:1836.

Hansen, T. H., and Shreffler, D. C., 1976, Characterization of a constitutive variant of the murine serum protein allotype, Slp, *J. Immunol.* **117**:1507.

Hansen, T. H., Cullen, S. E., and Sachs, D. H., 1977, Immunochemical evidence for an additional *H-2* region closely linked to *H-2D*, *J. Exp. Med.* **145**:438.

Hartmann, K., Dutton, R. W., McCarthy, M. M., and Mishell, R. I., 1970, Cell components in the immune response. II. Cell attachment separation of immune cells, *Cell. Immunol.* **1**:182.

Harvey, M. A., Adorini, L., Benjamin, C. D., Miller, A., and Sercarz, E. E., 1979, Idiotypy and antigenic specificity of T_h, T_s, and B cells induced by hen egg-white lysozyme, in: *T and B Lymphocytes: Recognition and Function* (F. L. Bach, B. Bonavida, E. S. Vitetta, and C. F. Fox, eds.), pp. 423–432, Academic Press, New York.

Hashim, G. A., Sharpe, R. D., Carvalho, E. F., and Stevens, L. E., 1976, Suppression and reversal of allergic encephalomyelitis in guinea pigs with a non-encephalitogenic analogue of the tryptophan region of the myelin basic protein, *J. Immunol.* **116**:126.

Haskill, J. S., Byrt, P., and Marbrook, J., 1970, In vitro and in vivo studies of the immune response to sheep erythrocytes using partially purified cell preparations, *J. Exp. Med.* **131**:57.

Hauptfeld, V., Klein, D., and Klein, J., 1973, Serological identification of an Ir-region product, *Science* **181**:167.

Hauptfeld, V., Hauptfeld, M., and Klein, J., 1974, Tissue distribution of I region-associated antigens in the mouse, *J. Immunol.* **113**:181.

Heber-Katz, E., and Wilson, D. B., 1975, Collaboration of allogeneic T and B lymphocytes in the primary antibody response to sheep erythrocytes in vitro, *J. Exp. Med.* **142**:928.

Hedrick, S. M., and Watson, J., 1979, Genetic control of the immune response to collagen. II. Antibody responses produced in fetal liver restored radiation chimeras and thymus reconstituted F_1 hybrid nude mice, *J. Exp. Med.* **150**:646.

Henkart, P. A., Schmitt-Verhulst, A.-M., and Shearer, G. M., 1977, Specificity of cytotoxic effector cells directed against trinitrobenzene sulfonate-modified syngeneic cells. Failure to recognize cell surface-bound trinitrophenyl dextran, *J. Exp. Med.* **146**:1068.

Herzenberg, L. A., Okumura, K., Cantor, H., Sato, V. L., Shen, F.-W., Boyse, E. A., and Herzenberg, L. A., 1976, T-cell regulation of antibody responses: Demonstration of allotype-specific helper T cells and their specific removal by suppressor T cells, *J. Exp. Med.* **144**:330.

Hess, M., 1976, Ia antigens: Isolation, chemical modification and structural characterization, *Transplant. Rev.* **30**:40.

Hetzelberger, D., and Eichmann, K., 1978, Recognition of idiotypes in lymphocyte interactions. I. Idiotypic selectivity in the cooperation between T and B lymphocytes, *Eur. J. Immunol.* **8**:846.

Hill, S. W., and Frelinger, J. A., 1979, Genetic control of the immune response in chimeric mice, submitted for publication.

Hill, S. W., and Sercarz, E. E., 1975, Fine specificity of the H-2 linked immune response gene for the gallinaceous lysozymes, *Eur. J. Immunol.* **5**:317.

Hodes, R. J., Hathcock, K. S., and Singer, A., 1979, Cellular and genetic control of antibody responses. VI. Expression of *Ir* gene function by *H-2ᵃ* accessory cells but not *H-2ᵃ* T or B cells in responses to TNP-(T,G)-A–L, *J. Immunol.* **123**:2823.

Howie, S., and Feldmann, M., 1977, In vitro studies on H-2-linked unresponsiveness to synthetic polypeptides. III. Production of an antigen-specific T helper cell factor to (T,G)-A–L, *Eur. J. Immunol.* **7**:417.

Howie, S., and Feldmann, M., 1978, Immune response (*Ir*) genes expressed at macrophage–B lymphocyte interactions, *Nature (London)* **273**:664.

Huber, B., Gershon, R. K., and Cantor, H., 1977, Identification of a B-cell surface structure involved in antigen-dependent triggering: Absence of this structure on B cells from CBA/N mutant mice, *J. Exp. Med.* **145**:10.

Hurme, M., Hetherington, G. M., Chandler, P. R., and Simpson, E., 1978, Cytotoxic T-cell responses to H-Y: Mapping of the Ir genes, *J. Exp. Med.* **147**:758.

Imanishi-Kari, T., Rajnavòlgyi, E., Takemori, T., Jack, R. S., and Rajewsky, K., 1979, The effect of light chain gene expression on the inheritance of an idiotype associated with primary anti-(4-hydroxy-3-nitrophenyl) acetyl (NP) antibodies, *Eur. J. Immunol.* **9**:324.

Isac, R., and Mozes, E., 1977, Antigen-specific T cell factors: A fine analysis of specificity, *J. Immunol.* **118**:584.

Ishizaka, K., and Adachi, T., 1976, Generation of specific helper cells and suppressor cells in vitro for the IgE and IgG antibody responses, *J. Immunol.* **117**:40.

Ishizaka, K., Kishimoto, T., Delespesse, G., and King, T. P., 1974, Immunogenic properties of modified antigen E. I. Presence of specific determinants for T cells in denatured antigen and polypeptide chains, *J. Immunol.* **113**:70.

Ishizaka, K., Okudaira, H., and King, T. P., 1975, Immunogenic properties of modified antigen E. II. Ability of urea-denatured antigen and α-polypeptide chain to prime T cells specific for antigen E, *J. Immunol.* **114**:110.

Janeway, C. A., Jr., and Paul, W. E., 1976, The specificity of cellular immune responses in guinea pigs. III. The precision of antigen recognition by T lymphocytes, *J. Exp. Med.* **144**:1641.

Janeway, C. A., Jr., Wigzell, H., and Binz, H., 1976, Two different V_H gene products make up the T-cell receptors, *Scand. J. Immunol.* **5**:994.

Janeway, C. A., Jr., Murphy, P. D., Kemp, J., and Wigzell, H., 1978, T cells specific for hapten-modified self

579

IMMUNE RESPONSE
GENES IN THE
REGULATION OF
MAMMALIAN
IMMUNITY

are precommitted for self major histocompatibility complex antigens before encounter with the hapten, *J. Exp. Med.* **147**:1065.

Jerne, N. K., 1971, The somatic generation of immune recognition, *Eur. J. Immunol.* **1**:1.

Jerne, N. K., 1974, Towards a network theory of the immune system. *Ann. Immunol. (Inst. Pasteur)* **125c**:373.

Jersild, C., Ammitzbøll, T., Clausen, J., and Fog, T., 1973, Association between HL-A antigens and measles antibody in multiple sclerosis, *Lancet* **1973**(1):151.

Jones, P. P., Murphy, D. B., and McDevitt, H. O., 1978, Two-gene control of the expression of a murine Ia antigen, *J. Exp. Med.* **148**:925.

Kaakinen, A., Pirskanen, R., and Tiilikainen, 1975, LD antigens associated with HL-A8 and myasthenia gravis, *Tissue Antigens* **6**:175.

Kabat, E. A., 1978, The structural basis of antibody complementarity, *Adv. Prot. Chem.* **32**:1.

Kabat, E. A., Wu, T. T., and Bilofsky, H., 1979, Evidence supporting somatic assembly of the DNA segments (minigenes), coding for the framework, and complementarity-determining segments of immunoglobulin variable regions, *J. Exp. Med.* **149**:1299.

Kano, S., Bloom, B. R., Shreffler, D. C., 1976, Blocking of MLC stimulation by anti-Ia sera: Studies with the virus plaque assay, *J. Immunol.* **117**:242.

Kantor, F. S., Ojeda, A., and Benacerraf, B., 1963, Studies on artificial antigens. I. Antigenicity of DNP-polylysine and DNP copolymer of lysine and glutamic acid in guinea pigs, *J. Exp. Med.* **117**:55.

Kapp, J. A., 1978, Immunosuppressive factors from lymphoid cells of nonresponder mice primed with L-glutamic acid^{60}L-alanine^{30}-L-Tyrosine^{10}. IV. Lack of strain restrictions among allogeneic, nonresponder donors and recipients, *J. Exp. Med.* **147**:997.

Kapp, J. A., Pierce, C. W., and Benacerraf, B., 1973, Genetic control of immune responses in vitro. II. Cellular requirements for the development of primary plaque-forming cell responses to the random terpolymer L-glutamic acid^{60}-L-alanine^{30}-L-tyrosine^{10} (GAT) by mouse spleen cells in vitro, *J. Exp. Med.* **138**:1121.

Kapp, J. A., Pierce, C. W., Schlossman, S., and Benacerraf, B., 1974, Genetic control of the immune response in vitro. V. Stimulation of suppressor T cells in nonresponder mice by the terpolymer L-glutamic acid^{60}-L-alanine^{30}-L-tyrosine^{10}(GAT), *J. Exp. Med.* **140**:648.

Kapp, J. A., Pierce, C. W., de la Croix, F., and Benacerraf, B., 1976, Immunosuppressive factor(s) extracted from lymphoid cells of nonresponder mice primed with L-glutamic acid^{60}-L-alanine^{30}-L-tyrosine^{10}(GAT). I. Activity and antigenic specificity, *J. Immunol.* **116**:305.

Kapp, J. A., Pierce, C. W., and Benacerraf, B., 1977, Immunosuppressive factor(s) extracted from lymphoid cells of nonresponder mice primed with L-glutamic acid^{60}-L-alanine^{30}-L-tyrosine^{10}(GAT). II. Cellular source and effect on responder and nonresponder mice, *J. Exp. Med.* **145**:828.

Kappler, J. W., and Marrack, P. C., 1976, Helper T cells recognized antigen and macrophage surface components simultaneously, *Nature (London)* **262**:797.

Kappler, J. W., and Marrack, P., 1977, The role of *H-2*-linked genes in helper T-cell function. I. In vitro expression in B cells of immune response genes controlling helper T-cell activity, *J. Exp. Med.* **146**:1748.

Kappler, J. W., and Marrack, P., 1978, The role of *H-2*-linked genes in helper T-cell function. IV. Importance of T-cell genotype and host environment in *I*-region and *Ir* gene expression, *J. Exp. Med.* **148**:1510.

Katz, D. H., 1976, The role of the histocompatibility gene complex in lymphocyte differentiation, *Transplant. Proc.* **8**:405.

Katz, D. H., 1977, *Lymphocyte Differentiation, Recognition, and Regulation,* Academic Press, New York.

Katz, D. H., and Benacerraf, B., 1975, The function and interrelationships of T-cell receptors, Ir genes and other histocompatibility gene products, *Transplant. Rev.* **22**:175.

Katz, D. H., and Unanue, E. R., 1973, Critical role of determinant presentation in the induction of specific responses in immunocompetent lymphocytes, *J. Exp. Med.* **137**:967.

Katz, D. H., Hamaoka, T., and Benacerraf, B., 1973*a*, Cell interactions between histoincompatible T and B lymphocytes. II. Failure of physiologic cooperative interactions between T and B lymphocytes from allogeneic donor strains in humoral response to hapten–protein conjugates, *J. Exp. Med.* **137**:1405.

Katz, D. H., Hamaoka, T., Dorf, M. E., and Benacerraf, B., 1973*b*, Cell interactions between histoincompatible T and B lymphocytes. The H-2 gene complex determines successful physiologic lymphocyte interactions, *Proc. Natl. Acad. Sci U.S.A.* **70**:2624.

Katz, D. H., Hamaoka, T., Dorf, M. E., Maurer, P. H., and Benacerraf, B., 1973*c*, Cell interactions between histoincompatible T and B lymphocytes. IV. Involvement of the immune response (Ir) gene in the control of lymphocyte interactions in responses controlled by the gene, *J. Exp. Med.* **138**:734.

Katz, D. H., Graves, M., Dorf, M. E., Dimuzio, H., and Benacerraf, B., 1975, Cell interactions between histoincompatible T and B lymphocytes. VII. Cooperative responses between lymphocytes are controlled by genes in the *I* region of the *H-2* complex, *J. Exp. Med.* **141**:263.

581

IMMUNE RESPONSE
GENES IN THE
REGULATION OF
MAMMALIAN
IMMUNITY

Katz, D. H., Dorf, M. E., and Benacerraf, B., 1976a, Control of T-lymphocyte and B-lymphocyte activation by two complementing *Ir*-GLϕ immune response genes, *J. Exp. Med.* **143**:906.

Katz, D. H., Chiorazzi, N., McDonald, J., and Katz, L. R., 1976b, Cell interactions between histoincompatible T and B lymphocytes. IX. The failure of histoincompatible cells is not due to suppression and cannot be circumvented by carrier-priming T cells with allogeneic macrophages, *J. Immunol.* **117**:1853.

Katz, D. H., Skidmore, B. J., Katz, L. R., and Bogowitz, C. A., 1978, Adaptive differentiation of murine lymphocytes. I. Both T and B lymphocytes differentiating in F$_1$ → parental chimeras manifest preferential cooperative activity for partner lymphocytes derived from the same parental type corresponding to the chimeric host, *J. Exp. Med.* **148**:727.

Katz, D. H., Katz, L. R., Bogowitz, C. A., and Skidmore, B. J., 1979a, Adaptive differentiation of murine lymphocytes. II. The thymic microenvironment does not restrict the cooperative partner cell preference of helper T cells differentiating in F$_1$ → F$_1$ thymic chimeras, *J. Exp. Med.* **149**:1360.

Keck, K., 1975, Ir-gene control of immunogenicity of insulin and A-chain loop as a carrier determinant, *Nature (London)* **254**:78.

Keck, K., 1977, Ir gene control of carrier recognition. III. Cooperative recognition of two or more carrier determinants on insulins of different species, *Eur. J. Immunol.* **7**:811.

Keuning, J. J., Peña, A. S., van Leeuwen, A., van Hooff, J. P., and van Rood, J. J., 1976, HLA-Dw3 Associated with coeliac disease, *Lancet* **1976**(1):506.

Kiefer, H., 1973, Binding and release of lymphocytes by hapten-derivatized nylon fibers, *Eur. J. Immunol.* **3**:181.

Kiefer, H., 1975, Separation of antigen-specific lymphocytes. A new general method of releasing cells bound to nylon mesh, *Eur. J. Immunol.* **5**:624.

Kindred, B., and Shreffler, D. C., 1972, H-2 dependence of co-operation between T and B cells in vivo, *J. Immunol.* **109**:940.

Kindt, T. J., 1975, Rabbit immunoglobulin allotypes: Structure, immunology, and genetics, *Adv. Immunol.* **21**:35.

Klein, J., 1975, *Biology of the Mouse Histocompatibility-2 Complex,* Springer-Verlag, New York.

Klein, J., and Hauptfeld, V., 1976, Ia antigens: Their serology, molecular relationships, and their role in allograft reactions, *Transplant. Rev.* **30**:83.

Kölsch, E., and Falkenberg, F. W., 1978, Genetic control of the immune response to phage fd. IV. Complementation between *H-2*-linked Ir genes, *J. Immunol.* **120**:2087.

Kontiainen, S., and Feldmann, M., 1977, Suppressor cell induction *in vitro*. III. Antigen-specific suppression by supernatants of suppressor cells, *Eur. J. Immunol.* **7**:310.

Kontiainen, S., and Feldmann, M., 1978, Suppressor-cell induction in vitro. IV. Target of antigen-specific suppressor factor and its genetic relationships, *J. Exp. Med.* **147**:110.

Kontiainen, S., Simpson, E., Bohrer, E., Beverley, P. C. L., Herzenberg, L. A., Fitzpatrick, W. C., Vogt, P., Torano, A., McKenzie, I. F. C., and Feldmann, M., 1978, T-cell lines producing antigen-specific suppressor factor, *Nature (London)* **274**:477.

Kontiainen, S., Howie, S., Maurer, P. H., and Feldmann, M., 1979, Suppressor cell induction *in vitro*. VI. Production of suppressor factors to synthetic polypeptides GAT and (T,G)-A–L from cells of responder and nonresponder mice, *J. Immunol.* **122**:253.

Koszinowski, U., and Ertl, H., 1975, Lysis mediated by T cells and restricted by H-2 antigen of target cells infected with vaccinia virus, *Nature (London)* **255**:552.

Koszinowski, U., and Thomssen, R., 1975, Target cell-dependent T cell-mediated lysis of vaccinia virus-infected cells, *Eur. J. Immunol.* **5**:245.

Krammer, P. H., 1978, Alloantigen receptors on activated T cells in mice. I. Binding of alloantigens and anti-idiotypic antibodies to the same receptors, *J. Exp. Med.* **147**:25.

Krammer, P. H., and Eichmann, K., 1977, T cell receptor idiotypes are controlled by genes in the heavy chain linkage group and the major histocompatibility complex, *Nature (London)* **270**:733.

Krawinkel, U., and Rajewsky, K., 1976, Specific enrichment of antigen-binding receptors from sensitized murine lymphocytes, *Eur. J. Immunol.* **6**:529.

Krawinkel, U., Cramer, M., Imanishi-Kari, T., Jack, R. S., Rajewsky, K., and Mäkelä, O., 1977a, Isolated hapten-binding receptors of sensitized lymphocytes. I. Receptors from nylon-wool enriched mouse T lymphocytes lack serological markers of immunoglobulin constant domains but express heavy chain variable portions, *Eur. J. Immunol.* **7**:566.

Krawinkel, U., Cramer, M., Mage, R. G., Kelus, A. A., and Rajewsky, K., 1977b, Isolated hapten-binding receptors of sensitized lymphocytes. II. Receptors from nylon wool-enriched rabbit T lymphocytes lack serological determinants of immunoglobulin constant domains but carry the *A* locus allotypic markers, *J. Exp. Med.* **146**:792.

Krawinkel, U., Cramer, M., Melchers, I., Imanishi-Kari, T., and Rajewsky, K., 1978, Isolated hapten-binding receptors of sensitized lymphocytes. III. Evidence for idiotypic restriction of T-cell receptors, *J. Exp. Med.* **147**:1341.

Krawinkel, U., Cramer, M., Kindred, B., and Rajewsky, K., 1979, Isolated hapten-binding receptors of sensitized lymphocytes. V. Cellular origin of receptor molecules, *J. Immunol.* **9**:815.

Landsteiner, K., 1921, Uber heterogenetisches antigen und hapten. XV. Mitteilung uber Antigene, *Biochem. Z.* **119**:294.

Landsteiner, K., and Simms, S., 1923, Production of heterogenetic antibodies with mixtures of the binding part of the antigen and protein, *J. Exp. Med.* **38**:127.

Langman, R. E., 1978, Cell-mediated immunity and the major histocompalibility complex, *Rev. Physiol. Biochem. Pharmacol.* **81**:1.

Lemonnier, F., Neauport-Sautes, C., Kourilsky, F. M., and Démant, P., 1975, Relationships between private and public H-2 specificities on the cell surface, *Immunogenetics* **2**:517.

Lemonnier, F., Burakoff, S. J., Germain, R. N., and Benacerraf, B., 1977, Cytolytic thymus-derived lymphocytes specific for allogeneic stimulator cells crossreact with chemically modified syngeneic cells, *Proc. Natl. Acad. Sci. U.S.A.* **74**:1229.

Leserman, L. D., Cosenza, H., and Roseman, J. M., 1972, Cell interactions in antibody formation in vitro. II. The interaction of the third cell and antigen, *J. Immunol.* **109**:587.

Levine, B. B., and Benacerraf, B., 1965, Genetic control in guinea pigs of immune response to conjugates of haptens and poly-L-lysine, *Science* **147**:517.

Levine, B. B., Ojeda, A., and Benacerraf, B., 1963*a*, Basis for the antigenicity of hapten-poly-L-lysine conjugates in random-bred guinea pigs, *Nature (London)* **200**:544.

Levine, B. B., Ojeda, A., and Benacerraf, B., 1963*b*, Studies on artificial antigens. III. The genetic control of the immune response to hapten-poly-L-lysine conjugates in guinea pigs, *J. Exp. Med.* **118**:953.

Levine, B. B., Stember, R. H., and Fotino, M., 1972, Ragweed hay fever: Genetic control and linkage to HL-A haplotypes, *Science* **178**:1201.

Levy, R. B., and Shearer, G. M., 1979, Regulation of T-cell-mediated lympholysis by the murine major histocompatibility complex. I. Preferential in vitro responses to trinitrophenyl-modified self K- and D-coded products in parental and F$_1$ hybrid mouse strains, *J. Exp. Med.* **149**:1379.

Levy, R. B., Shearer, G. M., and Hansen, T. H., 1978, Properties of *H-2L* locus products in allogeneic and *H-2* restricted, trinitrophenyl-specific cytotoxic responses, *J. Immunol.* **121**:2263.

Lichtenberg, L., Mozes, E., Shearer, G. M., and Sela, M., 1974, The role of thymus cells in the immune response to poly(Tyr, Glu)-poly DL Ala-polyLys as a function of the genetic constitution of the mouse strain, *Eur. J. Immunol.* **4**:430.

Lieberman, R., Paul, W. E., Humphrey, W., Jr., and Stimpfling, J. H., 1972, H-2-linked immune response (*Ir*) genes: Independent loci for *Ir-IgG* and *Ir-IgA* genes, *J. Exp. Med.* **136**:1231.

Lilly, F., 1968, The effect of histocompatibility-2 type on response to the Friend leukemia virus in mice, *J. Exp. Med.* **127**:465.

Lipscomb, M. F., Ben-Sasson, S. Z., and Uhr, J. W., 1977, Specific binding of T lymphocytes to macrophages. I. Kinetics of binding, *J. Immunol.* **118**:1748.

Lipsky, P. E., and Rosenthal, A. S., 1975, Macrophage–lymphocyte interaction. II. Antigen-mediated physical interactions between immune guinea pig lymph node lymphocytes and syngeneic macrophages, *J. Exp. Med.* **141**:138.

Lonai, P., and McDevitt, H. O., 1974, Genetic control of the immune response. In vitro stimulation of lymphocytes by (T,G)-A–L, (H,G)-A–L, and (Phe,G)-A–L, *J. Exp. Med.* **140**:977.

Lonai, P., Ben-Neriah, Y., Steinman, L., and Givol, D., 1978, Selective participation of immunoglobin V region and major histocompatibility complex products in antigen binding by T cells, *Eur. J. Immunol.* **8**:827.

Lozner, E. C., Sachs, D. H., Shearer, G. M., and Terry, W. D., 1974*a*, B-cell alloantigens determined by the H-2 linked Ir region are associated with mixed lymphocyte culture stimulation, *Science* **183**:757.

Lozner, E. C., Sachs, D. H., and Shearer, G. M., 1974*b*, Genetic control of the immune response to staphylococcal nuclease. I. *Ir*-Nase: Control of the antibody response to nuclease by the *Ir* region of the mouse *H-2* complex, *J. Exp. Med.* **139**:1204.

Luzzati, A. L., Taussig, M. J., Meo, T., and Pernis, B., 1976, Induction of an antibody response in cultures of human peripheral blood lymphocytes, *J. Exp. Med.* **144**:573.

Lyons, C. R., Tucker, T. F., and Uhr, J. W., 1979, Specific binding of T lymphocytes to macrophages. V. The role of Ia antigens on Mϕ in the binding, *J. Immunol.* **122**:1598.

McDevitt, H. O., 1968, Genetic control of the antibody response. III. Qualitative and quantitative characterization of the antibody response to (T, G)-A–L in CBA and C57 mice, *J. Immunol.* **100**:485.

McDevitt, H. O., and Benacerraf, B., 1969, Genetic control of specific immune responses, *Adv. Immunol.* **11**:31.

McDevitt, H. O., and Chinitz, A., 1969, Genetic control of the antibody response: Relationship between immune response and histocompatibility (H-2) type, *Science* **163**:1207.

McDevitt, H. O., and Sela, M., 1965, Genetic control of the antibody response. I. Demonstration of determinant-specific differences in response to synthetic polypeptide antigens in two strains of inbred mice, *J. Exp. Med.* **122**:517.

McDevitt, H. O., and Tyan, M. L., 1968, Genetic control of the antibody response in inbred mice: Transfer of response by spleen cells and linkage to the major histocompatibility (H-2) locus, *J. Exp. Med.* **128**:1.

McDevitt, H. O., Deak, B. D., Shreffler, D. C., Klein, J., Stimpfling, J. H., and Snell, G. D., 1972, Genetic control of the immune response. Mapping of the Ir-1 locus, *J. Exp. Med.* **135**:1259.

McDevitt, H. O., Delovitch, T. L., Press, J. L., and Murphy, D. B., 1976, Genetic and functional analysis of the Ia antigens: Their role in regulating the immune response, *Transplant. Rev.* **30**:197.

McDougal, J. S., and Cort, S. P., 1978, Generation of T helper cells *in vitro*. IV. F_1 T helper cells primed with antigen-pulsed parental macrophages are genetically restricted in their antigen-specific helper activity, *J. Immunol.* **120**:445.

McDougal, J. S., and Gordon, D. S., 1977a, Generation of T-helper cells in vitro. I. Cellular and antigenic requirements, *J. Exp. Med.* **145**:676.

McDougal, J. S., and Gordon D. S., 1977b, Generation of T-helper cells in vitro. II. Analysis of supernates derived from T-helper cell cultures, *J. Exp. Med.* **145**:693.

McDougal, J. S., Cort, S. P., and Gordon, D. S., 1977, Generation of T helper cells in vitro. III. Helper cell culture-derived factors are related to alloantigens coded for by the I region of the H-2 major histocompatibility complex, *J. Immunol.* **119**:1933.

McDougal, J. S., Shen, F. W., and Elster, P., 1979, Generation of T helper cells *in vitro*. V. Antigen-specific $Ly1^+$ T cells mediate the helper effect and induce feedback suppression, *J. Immunol.* **122**:437.

Mackintosh, P., and Asquith, P., 1978, HLA and Coeliac disease, *Br. Med. Bull.* **34**:291.

McMichael, A., 1978, HLA restriction of human cytotoxic T lymphocytes specific for influenza virus. Poor recognition of virus associated with HLA A2, *J. Exp. Med.* **148**:1458.

McMichael, A. J., Ting, A., Zweerink, H. J., and Askonas, B. A., 1977, HLA restriction of cell-mediated lysis of influenza virus-infected human cells, *Nature (London)* **270**:524.

Mage, R. G., Lieberman, R., Potter, M., and Terry, W. D., 1973, Immunoglobulin allotypes, in: *The Antigens* (M. Sela, ed.), Vol. 1, pp. 299–376, Academic Press, New York.

Mäkelä, O., and Karjalainen, K., 1977, Inherited immunoglobulin idiotypes of the mouse, *Immunol. Rev.* **34**:119.

Mann, D. L., 1977, Human lymphocyte alloantigens: Genetic control and relation to disease, *Clin. Haematol.* **6**:331.

Mann, D. L., and Murray, C., 1979, HLA alloantigens: Disease association and biological significance, *Sem. Haematol.* **16**:293.

Maoz, A., Feldmann, M., and Kontiainen, S., 1976, Enrichment of antigen-specific helper and suppressor T cells, *Nature (London)* **260**:324.

Maron, E., Scher, H. I., Mozes, E., Arnon, R., and Sela, M., 1973, Genetic control of immune response toward the loop region of lysozyme, *J. Immunol.* **111**:101.

Marrack, P., and Kappler, J. W., 1978, The role of H-2-linked genes in helper T-cell function. III. Expression of immune response genes for trinitrophenyl conjugates of poly-L(Tyr,Glu)-poly-D, L-Ala-poly-L-Lys in B cells and macrophages, *J. Exp. Med.* **147**:1596.

Marrack, P., and Kappler, J. W., 1979, The role of *H-2*-linked genes in helper T-cell function. VI. Expression of *Ir* genes by helper T cells, *J. Exp. Med.* **149**:780.

Marsh, D. G., Bias, W. B., Hsu, S. H., and Goodfriend, L., 1973, Association of the HL-A7 cross-reacting group with a specific reaginic antibody response in allergic man, *Science* **179**:691.

Marsh, D. G., Chase, G. A., Freidhoff, L. R., Meyers, D. A., and Bias, W. B., 1979, Association of HLA antigens and total serum immunoglobulin E level with allergic response and failure to respond to ragweed allergen Ra 3, *Proc. Natl. Acad. Sci. U.S.A.* **76**:2903.

Martin, W. J., Maurer, P. H., and Benacerraf, B., 1971, Genetic control of immune responsiveness to a glutamic acid, alanine, tyrosine copolymer in mice, *J. Immunol.* **107**:715.

Matzinger, P., and Mirkwood, G., 1978, In a fully H-2 incompatible chimera, T cells of donor origin can respond to minor histocompatibility antigens in association with either donor or host H-2 type, *J. Exp. Med.* **148**:84.

Maurer, P. H., Odstrchel, G., and Merryman, C. F., 1973, Genetic control of the immune response in guinea pigs to the known sequence polymer (Tyr-Glu-Ala-Gly)$_n$, *J. Immunol.* **111**:1018.

Maurer, P. H., Merryman, C. F., Lai, C.-H., and Ganfield, D. J., 1978, Dependence of immune responses of

583

IMMUNE RESPONSE
GENES IN THE
REGULATION OF
MAMMALIAN
IMMUNITY

"nonresponder" H-2^s mice on determinant concentration in poly (Glu^{60} Ala^{30} Tyr^{10}) and on complementation between nonresponder mice of the same H-2^p haplotype, *Immunochemistry* **15**:737.

Melchers, I., and Rajewsky, K., 1975, Specific control of responsiveness by two complementing Ir loci in the H-2 complex, *Eur. J. Immunol.* **5**:753.

Melchers, I., and Rajewsky, K., 1978, Functional complementation and polymorphism of H-2 linked immune response genes, in: *Ir Genes and Ia Antigens* (H. O. McDevitt, ed.), pp. 77–86, Academic Press, New York.

Meo, T., Vives, J., Miggiano, V., and Shreffler, D., 1973, A major role for the Ir-1 region of the mouse H-2 complex in the mixed leukocyte reaction, *Transplant. Proc.* **5**:377.

Meo, T., David, C. S., Rijnbeek, A. M., Nabholz, M., Miggiano, V. C., and Shreffler, D. C., 1975, Inhibition of mouse MLR by anti-Ia sera, *Transplant. Proc.* **7**(Suppl. 1):127.

Merryman, C. F., and Maurer, P. H., 1972, Genetic control of immune response to glutamic acid, alanine, tyrosine copolymers in mice. I. Association of responsiveness to H-2 genotype and specificity of the response, *J. Immunol.* **108**:135.

Meruelo, D., 1979, A role for elevated H-2 antigen expression in resistance to neoplasia caused by radiation-induced leukemia virus. Enhancement of effective tumor surveillance by killer lymphocytes, *J. Exp. Med.* **149**:898.

Meruelo, D., Lieberman, M., Ginzton, N., Deak, B., and McDevitt, H. O., 1977, Genetic control of radiation leukemia virus-induced tumorigenesis. I. Role of the major murine histocompatibility complex, *H-2*, *J. Exp. Med.* **146**:1079.

Meruelo, D., Nimelstein, S. H., Jones, P. P., Lieberman, M., and McDevitt, H. O., 1978, Increased synthesis and expression of H-2 antigens on thymocytes as a result of radiation leukemia virus infection: A possible mechanism for H-2 linked control of virus-induced neoplasia, *J. Exp. Med.* **147**:470.

Miller, J. F. A. P., Vadas, M. A., Whitelaw, A., and Gamble, J., 1975, H-2 gene complex restricts transfer of delayed-type hypersensitivity in mice, *Proc. Natl. Acad. Sci. U.S.A.* **72**:5095.

Miller, J. F. A. P., Vadas, M. A., Whitelaw, A., and Gamble, J., 1976, Role of major histocompatibility complex gene products in delayed-type hypersensitivity, *Proc. Natl. Acad. Sci. U.S.A.* **73**:2486.

Mishell, R. I., Dutton, R. W., and Raidt, D. J., 1970, Cell components in the immune response. I. Gradient separation of immune cells, *Cell. Immunol.* **1**:175.

Mitchell, G. F., Grumet, F. C., and McDevitt, H. O., 1972, Genetic control of the immune response: The effect of thymectomy on the primary and secondary antibody response of mice to poly-L(Tyr,Glu)-poly-D,L-Ala–poly-L-Lys, *J. Exp. Med.* **135**:126.

de Moerloose, P., Jeannet, M., Bally, C., Raffoux, C., Pointel, J.-P., and Sizonenko, P., 1978, HLA and DRw antigens in insulin-dependent diabetes, *Br. Med. J.* **1978**:823.

Möller, E., and Olhagen, B., 1975, Studies on the major histocompatibility system in patients with ankylosing spondylitis, *Tissue Antigens* **6**:237.

Möller, E., Hammarström, L., Smith, E., and Matell, G., 1976, HL-A8 and LD-8a in patients with myasthenia gravis, *Tissue Antigens* **7**:39.

Mond, J. J., Scher, I., Mosier, D. E., Blaese, M., and Paul, W. E., 1978, T-independent responses in B cell-defective CBA/N mice to Brucella abortus and to trinitrophenyl (TNP) conjugates of Brucella abortus, *Eur. J. Immunol.* **8**:459.

Moorhead, J. W., 1977*a*, Soluble factors in tolerance and contact sensitivity to 2,4-dinitrofluorobenzene in mice. I. Suppression of contact sensitivity by soluble suppressor factor released *in vitro* by lymph node cell populations containing specific suppressor cells, *J. Immunol.* **119**:315.

Moorhead, J. W., 1977*b*, Soluble factors in tolerance and contact sensitivity to DNFB in mice. II. Genetic requirements for suppression of contact sensitivity by soluble suppressor factor, *J. Immunol.* **119**:1773.

Mosier, D. E., 1967, A requirement for two cell types for antibody formation in vitro, *Science* **158**:1573.

Mosier, D. E., Mond, J. J., and Goldings, E. A., 1977, The ontogeny of thymic independent antibody responses *in vitro* in normal mice and mice with an X-linked B cell defect, *J. Immunol.* **119**:1874.

Moutsopoulos, H. M., Chused, T. M., Johnson, A. H., Knudsen, B., and Mann, D. L., 1978, B lymphocyte antigens in sicca syndrome, *Science,* **199**:1441.

Mozes, E., 1975, Expression of immune response (*Ir*) genes in T and B cells, *Immunogenetics* **2**:397.

Mozes, E., 1978, The nature and functions of specific immune response genes and their products, *Adv. Exp. Med. Biol.* **98**:429.

Mozes, E., and Fuchs, S., 1974, Linkage between immune response potential to DNA and X chromosome, *Nature (London)* **249**:167.

Mozes, E., and McDevitt, H. O., 1969, The effect of genetic control of immune response to synthetic polypeptides on the response to homologous DNP-polypeptide conjugates, *Immunochemistry* **6**:760.

Mozes, E., and Shearer, G. M., 1971, Contribution of bone marrow cells and lack of expression of thymocytes in genetic controls of immune responses for two immunopotent regions within poly-(Phe,Glu)-poly-Pro–poly-Lys in inbred mouse strains, *J. Exp. Med.* **134**:141.

585

IMMUNE RESPONSE
GENES IN THE
REGULATION OF
MAMMALIAN
IMMUNITY

Mozes, E., McDevitt, H. O., Jaton, J.-C., and Sela, M., 1969, The genetic control of antibody specificity, *J. Exp. Med.* **130**:1263.

Mozes, E., Shearer, G. M., and Sela, M., 1970, Cellular basis of the genetic control of immune responses to synthetic polypeptides. I. Differences in frequency of splenic precursor cells specific for a synthetic poly-peptide derived from multichain polyproline [(T,G)-Pro–L] in high and low responder inbred mouse strains, *J. Exp. Med.* **132**:613.

Muckerheide, A., Pesce, A. J., and Michael, J. G., 1977, Immunosuppressive properties of a peptic fragment of BSA, *J. Immunol.* **119**:1340.

Munro, A. J., and Taussig, M. J., 1975, Two genes in the major histocompatibility complex control immune response, *Nature (London)* **256**:103.

Munro, A. J., and Taussig, M. J., 1977, Complementation of immune response genes for (T,G)-A–L, *Nature (London)* **269**:355.

Munro, A. J., Taussig, M. J., Campbell, R., Williams, H., and Lawson, Y., 1974, Antigen-specific T-cell factor in cell cooperation: Physical properties and mapping in the left-hand (K) half of *H-2*, *J. Exp. Med.* **140**:1579.

Munro, A., Taussig, M., and Archer, J., 1978, I-region products and cell interactions: Contribution of non-H-2 genes to acceptor and factor for (T,G)-A–L, in: *Ir Genes and Ia Antigens* (H. O. McDevitt, ed.), pp. 487–491, Academic Press, New York.

Murphy, D. B., Herzenberg, L. A., Okumura, K., Herzenberg, L. A., and McDevitt, H. O., 1976, A new *I* subregion (*I-J*) marked by a locus (*Ia-4*) controlling surface determinants on suppressor T lymphocytes, *J. Exp. Med.* **144**:699.

Naito, S., Namerow, N., Mickey, M. R., and Terasaki, P. I., 1972, Multiple sclerosis: Association with HL-A3, *Tissue Antigens* **2**:1.

Nerup, J., Platz, P., Andersen, O. O., Christy, M., Lyngsøe, J., Poulsen, J. E., Ryder, L. P., Nielsen, L. S., Thomsen, M., and Svejgaard, A., 1974, HL-A antigens and diabetes mellitus, *Lancet* **1974**(2): 864.

Niederhuber, J. E., 1978, The role of I region gene products in macrophage–T lymphocyte interaction, *Immunol. Rev.* **40**:28.

Niederhuber, J. E., and Frelinger, J. A., 1976, Expression of Ia antigens on T and B cells and their relationship to immune-response functions, *Transplant. Rev.* **30**:101.

Neiderhuber, J. E., Allen, P., and Mayo, O., 1979, The expression of Ia antigenic determinants on macrophages required for the *in vitro* antibody response, *J. Immunol.* **122**:1342.

Okuda, K., and David, C. S., 1978, A new lymphocyte-activating determinant locus expressed on T cells, and mapping in *I-C* subregion, *J. Exp. Med.* **147**:1028.

O'Neill, H. C., and Blanden, R. V., 1979, Quantitative differences in the expression of parentally-derived H-2 antigens in F₁ hybrid mice affect T-cell responses, *J. Exp. Med.* **149**:724.

Ordal, J. C., and Grumet, F. C., 1972, Genetic control of the immune response: The effect of graft-versus-host reaction on the antibody response to poly-L(Tyr,Glu)-poly-D,L-Ala–poly-L-Lys in nonresponder mice, *J. Exp. Med.* **136**:1195.

Owen, F. L., and Nisonoff, A., 1978, Effect of idiotype-specific suppressor T cells on primary and secondary responses, *J. Exp. Med.* **148**:182.

Panfili, P. R., and Dutton, R. W., 1978, Alloantigen-induced T helper activity. I. Minimal genetic differences necessary to induce a positive allogeneic effect, *J. Immunol.* **120**:1897.

Parish, C. R., 1971, Immune response to chemically modified flagellin. II. Evidence for a fundamental rela-tionship between humoral and cell-mediated immunity, *J. Exp. Med.* **134**:21.

Paty, D. W., Cousin, H. K., Stiller, C. R., Boucher, D. W., Furesz, J., Warren, K. G., Marchuk, L., and Dossetor, J. B., 1977, HLA-D typing with an association of Dw2 and absent immune responses towards Herpes simplex (Type I) antigen in multiple sclerosis, *Transpl. Proc.* **IX**:1845.

Paul, W. E., 1970, Functional specificity of antigen-binding receptors of lymphocytes, *Transplant. Rev.* **5**:130.

Paul, W. E., and Benacerraf, B., 1977, Functional specificity of thymus-dependent lymphocytes, *Science* **195**:1293.

Paul, W. E., Shevach, E. M., Pickeral, S., Thomas, D. W., and Rosenthal, A. S., 1977, Independent popula-tions of primed F₁ guinea pig T lymphocytes respond to antigen-pulsed parental peritoneal exudate cells, *J. Exp. Med.* **145**:618.

Peña, A. S., Mann, D. L., Hague, N. E., Heck, J. A., van Leeuwen, A., van Rood, J. J., and Strober, W., 1978, Genetic basis of gluten-sensitive enteropathy, *Gastroenterology* **75**:230.

Pfizenmaier, K., Starzinski-Powitz, A., Rodt, H., Röllinghoff, M., and Wagner, H., 1976, Virus and trinitro-phenol hapten-specific T-cell-mediated cytotoxicity against H-2 incompatible target cells, *J. Exp. Med.* **143**:999.

Pfizenmaier, K., Trinchieri, G., Solter, D., and Knowles, B. B., 1978, Mapping of *H-2* genes associated with

T cell-mediated cytotoxic responses to SV40-tumour-associated specific antigens, *Nature (London)* **274**:691.

Pierce, C. W., and Kapp, J. A., 1978a, Antigen-specific suppressor T cell activity in genetically restricted immune spleen cells, *J. Exp. Med.* **148**:1271.

Pierce, C. W., and Kapp, J. A., 1978b, Suppressor T-cell activity in responder × nonresponder (C57BL/10 × DBA/1)F$_1$ spleen cells responsive to L-glutamic acid^{60}-L-alanine^{30}-L-tyrosine^{10}, *J. Exp. Med.* **148**:1282.

Pierce, C. W., Kapp, J. A., and Benacerraf, B., 1976a, Stimulation of antibody responses *in vitro* by antigen-bearing syngeneic and allogeneic macrophages, in: *The Role of Products of the Histocompatibility Gene Complex in Immune Responses* (D. H. Katz and B. Benacerraf, eds.), pp. 391–401, Academic Press, New York.

Pierce, C. W., Kapp, J. A., and Benacerraf, B., 1976b, Regulation by the *H-2* gene complex of macrophage-lymphoid cell interactions in secondary antibody responses *in vitro*, *J. Exp. Med.* **144**:371.

Pierce, C. W., Germain, R. N., Kapp, J. A., and Benacerraf, B., 1977, Secondary antibody responses *in vitro* to L-glutamic acid^{60}-L-alanine^{30}-L-tyrosine^{10} (GAT) by (responder × nonresponder)F$_1$ spleen cells stimulated by parental GAT-macrophages, *J. Exp. Med.* **146**:1827.

Pierres, M., and Germain, R. N., 1978, Antigen-specific T cell-mediated suppression. IV. Role of macrophages in generation of L-glutamic acid^{60}-L-alanine^{30}-L-tyrosine^{10} (GAT)-specific suppressor T cells in responder mouse strains, *J. Immunol.* **121**:1306.

Pisetsky, D. S., Berzofsky, J. A., and Sachs, D. H., 1978, Genetic control of the immune response to staphylococcal nuclease. VII. Role of non-*H-2*-linked genes in the control of the anti-nuclease antibody response, *J. Exp. Med.* **147**:396.

Prange, C. A., Fiedler, J., Nitecki, D. E., and Bellone, C. J., 1977, Inhibition of T-antigen-binding cells by idiotypic antisera, *J. Exp. Med.* **146**:766.

Press, J. L., and McDevitt, H. O., 1977, Allotype-specific analysis of anti-(Tyr,Glu)-Ala–Lys antibodies produced by Ir-1A high and low responder chimeric mice, *J. Exp. Med.* **146**:1815.

Ptak, W., Zembala, M., and Gershon, R. K., 1978, Intermediary role of macrophages in the passage of suppressor signals between T-cell subsets, *J. Exp. Med.* **148**:424.

Rajewsky, K., and Mohr, R., 1974, Specificity and heterogeneity of helper T cells in the response to serum albumins in mice, *Eur. J. Immunol.* **4**:111.

Ramseier, H., and Lindenmann, J., 1969, F$_1$ hybrid animals: Reactivity against recognition structures of parental strain lymphoid cells, *Pathol. Microbiol.* **34**:379.

Ramseier, H., and Lindenmann, J., 1971, Cellular receptors. Effect of antialloantiserum on the recognition of transplantation antigens, *J. Exp. Med.* **134**:1083.

Ramseier, H., and Lindenmann, J., 1972, Similarity of cellular recognition structures for histocompatibility antigens and of combining sites of corresponding alloantibodies, *Eur. J. Immunol.* **2**:109.

Ramseier, H., Aguet, M., and Lindenmann, J., 1977, Similarity of idiotypic determinants of T- and B-lymphocyte receptors for alloantigens, *Immunol. Rev.* **34**:50.

Ray, A., and Ben-Sasson, S., 1979, The separate roles of hapten and of carrier in the conformation of T cell-reactive determinants on ABA-protein conjugates, *J. Immunol.* **122**:1960.

Rees, A. J., Peters, D. K., Compston, D. A. S., and Batchelor, J. R., 1978, Strong association between HLA-DRw2 and antibody-mediated Goodpasture's syndrome, *Lancet* **1978**(1):966.

Rehn, T. G., Shearer, G. M., Koren, H. S., and Inman, J. K., 1976a, Cell-mediated lympholysis of N-(3-nitro-4-hydroxy-5-iodophenylacetyl)-β-alanylglycylglycyl modified autologous lymphocytes. Effector cell specificity to modified cell surface components controlled by *H-2K* and *H-2D* serological regions of the murine major histocompatibility complex, *J. Exp. Med.* **143**:127.

Rehn, T. G., Inman, J. K., and Shearer, G. M., 1976b, Cell-mediated lympholysis to *H-2*-matched target cells modified with a series of nitrophenyl compounds, *J. Exp. Med.* **144**:1134.

Reinertsen, J. L., Klippel, J. H., Johnson, A. H., Steinberg, A. D., Decker, J. L., and Mann, D. L., 1978, B-lymphocyte alloantigens associated with systemic lupus erythematosus, *N. Engl. J. Med.* **299**:515.

Reth, M., Imanishi-Kari, T., Jack, R. S., Cramer, M., Krawinkel, U., Hämmerling, G. J., and Rajewsky, K., 1977, The variable portion of T and B cell receptors for antigen: Binding sites for the hapten 4-hydroxy-3-nitro-phenacetyl in C57BL/6 mice, in: *Immune System: Genetics and Regulation* (E. E. Sercarz, L. A. Herzenberg, and C. F. Fox, eds.), pp. 139–149, Academic Press, New York.

Rich, S. S., and Rich, R. R., 1976, Regulatory mechanisms in cell-mediated immune responses. III. *I*-region control of suppressor cell interaction with responder cells in mixed lymphocyte reactions, *J. Exp. Med.* **143**:672.

Richman, L. K., Klingenstein, R. J., Richman, J. A., Strober, W., and Berzofsky, J. A., 1979, The murine Kupffer cell. I. Characterization of the cell serving accessory function in antigen-specific T cell proliferation, *J. Immunol.* **123**:2602.

587

IMMUNE RESPONSE
GENES IN THE
REGULATION OF
MAMMALIAN
IMMUNITY

Richman, L. K., Strober, W., and Berzofsky, J. A., 1980, Genetic control of the immune response to myoglobin. III. Determinant-specific, two *Ir* gene phenotype is regulated by the genotype of reconstituting Kupffer cells, *J. Immunol.* **124**:619.

van Rood, J. J., van Leeuwen, A., Termijtelen, A., and Keuning, J. J., 1976, B-cell antibodies, Ia-like determinants, and their relation to MLC determinants in man, *Transplant. Rev.* **30**:122.

van Rood, J. J., de Vries, R. R. P., and Munro, A., 1977, The biological meaning of transplantation antigens, in: *Progress in Immunology III* (T. E. Mandel, C. Cheers, C. S. Hosking, I. F. C. McKenzie, and G. J. V. Nossal, eds.), pp. 338–348, North-Holland, Amsterdam.

Roseman, J., 1969, X-ray resistant cell required for the induction of *in vitro* antibody formation, *Science* **165**:1125.

Rosenthal, A. S., 1978, Determinant selection and macrophage function in genetic control of the immune response, *Immunol. Rev.* **40**:136.

Rosenthal, A. S., and Shevach, E. M., 1973, Function of macrophages in antigen recognition by guinea pig T lymphocytes. I. Requirement for histocompatible macrophages and lymphocytes, *J. Exp. Med.* **138**:1194.

Rosenthal, A. S., Barcinski, M. A., and Blake, J. T., 1977, Determinant selection is a macrophage dependent immune response gene function, *Nature (London)* **267**:156.

Rosenwasser, L. J., Barcinski, M. A., Schwartz, R. H., and Rosenthal, A. S., 1979, Immune response gene control of determinant selection. II. Genetic control of the murine T lymphocyte proliferative response to insulin, *J. Immunol.* **123**:471.

Rotman, B., 1978, Genetic control of immunologic unresponsiveness to adjuvant-free solutions of β-D-galactosidase. I. Inheritance of the *Ir-Z1* and *ir-Z2* loci in mice, *J. Immunol.* **120**:1460.

Rubin, B., and Wigzell, H., 1973, Hapten-reactive helper lymphocytes, *Nature (London)* **242**:467.

Rüde, E., Günther, E., Meyer-Delius, M., and Liehl, E., 1977, Specificity of H-2-linked Ir gene control in mice: Recognition of defined sequence analogs of (T,G)-A–L, *Eur. J. Immunol.* **7**:520.

Sachs, D. H., 1976, The Ia antigens, in: *Contemporary Topics in Molecular Immunology* (H. N. Eisen and R. A. Reisfeld, eds.), Vol. 5, pp. 1–33, Plenum Press, New York.

Sachs, D. H., 1978, Evidence for an I-E subregion, in: *Ir Genes and Ia Antigens* (H. O. McDevitt, ed.), pp. 21–27, Academic Press, New York.

Sachs, D. H., and Cone, J. L., 1973, A mouse B-cell alloantigen determined by gene(s) linked to the major histocompatibility complex, *J. Exp. Med.* **138**:1289.

Sachs, D. H., Schechter, A. N., Eastlake, A., and Anfinsen, C. B., 1972, An immunologic approach to the conformational equilibria of polypeptides, *Proc. Natl. Acad. Sci. U.S.A.* **69**:3790.

Sachs, D. H., Berzofsky, J. A., Pisetsky, D. S., and Schwartz, R. H., 1978, Genetic control of the immune response to staphylococcal nuclease, *Springer Sem. Immunopathol.* **1**:51.

Sachs, J. A., and Brewerton, D. A., 1978, HLA, ankylosing spondylitis and rheumatoid arthritis, *Br. Med. Bull.* **34**:275.

Sachs, J. A., Sterioff, S., Robinette, M., Wolf, E., Curry, H. L. F., and Festenstein, H., 1975, Ankylosing spondylitis and the major histocompatibility system, *Tissue Antigens* **5**:120.

Sasazuki, T., Kohno, Y., Iwamoto, I., Tanimura, M., and Naito, S., 1978a, Association between an *HLA* haplotype and low responsiveness to tetanus toxoid in man, *Nature (London)* **272**:359.

Sasazuki, T., Kohno, Y., Iwamoto, I., Kosaka, K., Okimoto, K., Maruyama, H., Ishiba, S., Konishi, J., Takeda, Y., and Naito, A.,1978b, HLA and Graves' disease or diabetes mellitus in Japan, *N. Engl. J. Med.* **298**:630.

Scher, I., Frantz, M. M., and Steinberg, A. D., 1973, The genetics of the immune response to a synthetic double-stranded RNA in a mutant CBA mouse strain, *J. Immunol.* **110**:1396.

Schirrmacher, V., and Wigzell, H., 1972, Immune responses against native and chemically modified albumins in mice. I. Analysis of non-thymus-processed (B) and thymus-processed (T) cell responses against methylated bovine serum albumin, *J. Exp. Med.* **136**:1616.

Schirrmacher, V., and Wigzell, H., 1974, Immune responses against native and chemically modified albumins in mice. II. Effect of alteration of electric charge and conformation on the humoral antibody response and on helper T cell responses, *J. Immunol.* **113**:1635.

Schlossman, S. F., 1972, Antigen recognition: The specificity of T cells involved in the cellular immune response, *Transplant. Rev.* **10**:97.

Schlosstein, L., Terasaki, P. I., Bluestone, R., and Pearson, C. M., 1973, High association of an HL-A antigen, W27, with ankylosing spondylitis, *N. Engl. J. Med.* **288**:704.

Schmitt-Verhulst, A.-M., and Shearer, G. M., 1975, Bifunctional major histocompatibility-linked genetic regulation of cell-mediated lympholysis to trinitrophenyl-modified autologous lymphocytes, *J. Exp. Med.* **142**:914.

Schmitt-Verhulst, A.-M., and Shearer, G. M., 1976, Multiple *H-2*-linked immune response gene control of *H-2D*-associated T-cell-mediated lympholysis to trinitrophenyl-modified autologous cells: *Ir*-like genes mapping to the left of *I-A* and within the *I* region, *J. Exp. Med.* **144**:1701.

Schmitt-Verhulst, A.-M., Mozes, E., and Sela, M., 1974, Genetic control of the immune response to a thymus independent synthetic polypeptide, *Immunogenetics* **1**:357.

Schmitt-Verhulst, A.-M., Sachs, D. H., and Shearer, G. M., 1976, Cell-mediated lympholysis of trinitrophenyl-modified autologous lymphocytes. Conformation of genetic control of response to trinitrophenyl-modified H-2 antigens by the use of anti-H-2 and anti-Ia antibodies, *J. Exp. Med.* **143**:211.

Schmitt-Verhulst, A.-M., Pettinelli, C. B., Henkart, P. A., Lunney, J. K., and Shearer, G. M., 1978, *H-2-restricted cytotoxic effectors generated in vitro by the addition of trinitrophenyl-conjugated soluble proteins, *J. Exp. Med.* **147**:352.

Schroer, J. A., Inman, J. K., Thomas, J. W., and Rosenthal, A. S., 1979, H-2-linked *Ir* gene control of antibody responses to insulin. I. Anti-insulin plaque-forming cell primary responses, *J. Immunol.* **123**:670.

Schwartz, B. D., Kato, K., Cullen, S. E., and Nathenson, S. G., 1973, *H-2* histocompatibility alloantigens. Some biochemical properties of the molecules solubilized by NP-40 detergent, *Biochemistry* **12**:2157.

Schwartz, B. D., Paul, W. E., and Shevach, E. M., 1976, Guinea-pig Ia antigens: Functional significance and chemical characterization, *Transplant. Rev.* **30**:174.

Schwartz, M., Waltenbaugh, C., Dorf, M., Cesla, R., Sela, M., and Benacerraf, B., 1976, Determinants of antigenic molecules responsible for genetically controlled regulation of immune responses, *Proc. Natl. Acad. Sci. U.S.A.* **73**:2862.

Schwartz, M., Lifshitz, R., Givol, D., Mozes, E., and Haimovich, J., 1978, Cross-reactive idiotypic determinants on murine anti-(T,G)-A-L antibodies, *J. Immunol.* **121**:421.

Schwartz, R. H., 1978, A clonal deletion model for Ir gene control of the immune response, *Scand. J. Immunol.* **7**:3.

Schwartz, R. H., and Paul, W. E., 1976, T-lymphocyte-enriched murine peritoneal exudate cells. II. Genetic control of antigen-induced T-lymphocyte proliferation, *J. Exp. Med.* **143**:529.

Schwartz, R. H., Dorf, M. E., Benacerraf, B., and Paul, W. E., 1976a, The requirement for two complementing *Ir-Glφ* immune response genes in the T-lymphocyte proliferative response to poly(Glu^{53}Lys^{36}Phe^{11}), *J. Exp. Med.* **143**:897.

Schwartz, R. H., Fathman, C. G., and Sachs, D. H., 1976b, Inhibition of stimulation in murine mixed lymphocyte cultures with an alloantiserum directed against a shared Ia determinant, *J. Immunol.* **116**:929.

Schwartz, R. H., David, C. S., Sachs, D. H., and Paul, W. E., 1976c, T-lymphocyte-enriched murine peritoneal exudate cells. III. Inhibition of antigen-induced T-lymphocyte proliferation with anti-Ia antisera, *J. Immunol.* **117**:531.

Schwartz, R. H., Dickler, H. B., Sachs, D. H., and Schwartz, B. D., 1976d, Studies of Ia antigens on murine peritoneal macrophages, *Scand. J. Immunol.* **5**:731.

Schwartz, R. H., David C. S., Dorf, M. E., Benacerraf, B., and Paul, W. E., 1978a, Inhibition of dual *Ir* gene-controlled T-lymphocyte proliferative response to poly (Glu^{56}Lys^{35}Phe^9)$_n$ with anti-Ia antisera directed against products of either *I-A* or *I-C* subregion, *Proc. Natl. Acad. Sci. U.S.A.* **75**:2387.

Schwartz, R. H., Berzofsky, J. A., Horton, C. L., Schechter, A. N., and Sachs, D. H., 1978b, Genetic control of the T lymphocyte proliferative response to staphylococcal nuclease: Evidence for multiple MHC-linked *Ir* gene control, *J. Immunol.* **120**:1741.

Schwartz, R. H., Yano, A., and Paul, W. E., 1978c, Interaction between antigen-presenting cells and primed lymphocytes: An assessment of Ir gene expression in the antigen-presenting cell, *Immunol. Rev.* **40**:153.

Schwartz, R. H., Solinger, A. M., Ultee, M., and Margoliash, E., 1978d, Genetic control of the T-lymphocyte proliferative response to cytochrome c, *Adv. Exp. Med. Biol.* **98**:371.

Schwartz, R. H., Yano, A., Stimpfling, J. H., and Paul, W. E., 1979a, Gene complementation in the T-lymphocyte proliferative response to poly (Glu^{55} Lys^{36}Phe^9)$_n$. A demonstration that both immune response gene products must be expressed in the same antigen-presenting cell, *J. Exp. Med.* **149**:40.

Schwartz, R. H., Merryman, C. F., and Maurer, P. H., 1979b, Gene complementation in the T lymphocyte proliferative response to poly (Glu^{57}Lys^{38}Tyr^5): Evidence for effects of polymer handling and gene dosage, *J. Immunol.* **123**:272.

Schwartz, R. H., Solinger, A. M., Ultee, M. E., Margoliash, E., Yano, A., Stimpfling, J. H., Chen, C., Merryman, C. F., Maurer, P. H., and Paul, W. E., 1979c, Ir gene complementation in the murine T-lymphocyte proliferative response, in: *T and B Lymphocytes: Recognition and Function* (F. L. Bach, B. Bonavida, E. S. Vitetta, and C. F. Fox, eds.), pp. 261–275, Academic Press, New York.

Seidman, J. G., Max, E. E., and Leder, P., 1979, A κ-immunoglobulin gene is formed by site-specific recombination without further somatic mutation, *Nature (London)* **280**:370.

Senyk, G., Williams, E. B., Nitecki, D. E., and Goodman, J. W., 1971, The functional dissection of an antigen molecule: Specificity of humoral and cellular immune responses to glucagon, *J. Exp. Med.* **133**:1294.

Sercarz, E. E., Yowell, R. L., Turkin, D., Miller, A., Araneo, B. A., and Adorini, L., 1978, Different functional specificity repertoires for suppressor and helper T cells, *Immunol. Rev.* **39**:108.

Shaw, S., and Biddison, W. E., 1979, HLA-linked genetic control of the specificity of human cytotoxic T-cell responses to influenza virus, *J. Exp. Med.* **149**:565.

Shearer, G. M., 1974, Cell-mediated cytotoxicity to trinitrophenyl-modified syngeneic lymphocytes, *Eur. J. Immunol.* **4**:527.

Shearer, G. M., Mozes, E., and Sela, M., 1971, Cellular basis of the genetic control of immune response to synthetic polypeptides. II. Frequency of immunocompetent precursors specific for two distinct regions within (Phe,G)-Pro–L, a synthetic polypeptide derived from multichain polyproline, in inbred mouse strains, *J. Exp. Med.* **133**:216.

Shearer, G. M., Mozes, E., and Sela, M., 1972, Contribution of different cell types to the genetic control of immune responses as a function of the chemical nature of the polymeric sides chains (poly-L-Prolyl and poly-DL-Alanyl) of synthetic immunogens, *J. Exp. Med.* **135**:1009.

Shearer, G. M., Rehn, T. G., and Garbarino, C. A., 1975, Cell-mediated lympholysis of trinitrophenyl-modified autologous lymphocytes. Effector cell specificity to modified cell surface components controlled by the *H-2K* and *H-2D* serological regions of the murine major histocompatibility complex, *J. Exp. Med.* **141**:1348.

Shearer, G. M., Rehn, T. G., and Schmitt-Verhulst, A.-M., 1976, Role of the murine major histocompatibility complex in the specificity of *in vitro* T-cell-mediated lympholysis against chemically-modified autologous lymphocytes, *Transplant. Rev.* **29**:222.

Shearer, G. M., Schmitt-Verhulst, A. M., and Rehn, T. G., 1977, Significance of the major histocompatibility complex as assessed by T-cell-mediated lympholysis involving syngeneic stimulating cells, in: *Contemporary Topics in Immunobiology,* Vol. 7 (O. Stutman, ed.), pp. 221–243, Plenum Press, New York.

Shevach, E. M., and Rosenthal, A. S., 1973, Function of macrophages in antigen recognition by guinea pig T lymphocytes. II. Role of the macrophage in the regulation of genetic control of the immune response, *J. Exp. Med.* **138**:1213.

Shevach, E. M., Paul, W. E., and Green, I., 1972, Histocompatibility-linked immune response gene function in guinea pigs. Specific inhibition of antigen-induced lymphocyte proliferation by alloantisera, *J. Exp. Med.* **136**:1207.

Shiozawa, C., Singh, B., Rubenstein, S., and Diener, E., 1977, Molecular control of B cell triggering by antigen-specific T cell-derived helper factor, *J. Immunol.* **118**:2199.

Shreffler, D. C., and David, C. S., 1975, The *H-2* major histocompatibility complex and the I immune response region: Genetic variation, function, and organization, *Adv. Immunol.* **20**:125.

Shreffler, D., David, C., Götze, D., Klein, J., McDevitt, H., and Sachs, D., 1974, Genetic nomenclature for new lymphocyte antigens controlled by the *I* region of the *H-2* complex, *Immunogenetics* **1**:189.

Silver, D. M., McKenzie, I. F. C., and Winn, H. J., 1972, Variations in the responses of C57BL/10 J and A/ J mice to sheep red blood cells. I. Serological characterization and genetic analysis, *J. Exp. Med.* **136**:1063.

Simpson, E., and Gordon, R. D., 1977, Responsiveness to HY antigen: Ir gene complementation and target cell specificity, *Immunol. Rev.* **35**:59.

Singal, D. P., and Blajchman, M. A., 1973, Histocompatibility (HL-A) antigens, lymphocytotoxic antibodies and tissue antibodies in patients with diabetes mellitus, *Diabetes* **22**:429.

Singer, A., Dickler, H. B., and Hodes, R. J., 1977, Cellular and genetic control of antibody responses *in vitro.* II. Ir gene control of primary Ig M responses to trinitrophenyl conjugates of Poly-L-(Tyr,Glu)-poly-D,L-Ala-poly-L-Lys and poly-L-(His,Glu)-poly-D,L-Ala-poly-L-Lys, *J. Exp. Med.* **146**:1096.

Singer, A., Cowing, C., Hathcock, K. S., Dickler, H. B., and Hodes, R. J., 1978, Cellular and genetic control of antibody responses in vitro. III. Immune response gene regulation of accessory cell function, *J. Exp. Med.* **147**:1611.

Singer, A., Hathcock, K. S., and Hodes, R. J., 1979, Cellular and genetic control of antibody responses. V. Helper T-cell recognition of H-2 determinants on accessory cells but not B cells, *J. Exp. Med.* **149**:1208.

Sjöberg, O., Andersson, J., and Möller, G., 1972, Requirement for adherent cells in the primary and secondary immune response *in vitro, Eur. J. Immunol.* **2**:123.

Skidmore, B. J., and Katz, D. H., 1977, Haplotype preference in lymphocyte differentiation. I. Development of haplotype-specific helper and suppressor activities in F_1 hybrid-activated T cell populations, *J. Immunol.* **119**:694.

Smith, S. M., Ness, D. B., Talcott, J. A., and Grumet, F. C., 1977, Genetic control of Ig M responses to (T,G)-A–L, *H-2* and *Ig-1* linkage, *Immunogenetics* **4**:221.

Snell, G. D., Dausset, J., and Nathenson, S., 1976, *Histocompatibility,* Academic Press, New York.

Solinger, A. M., Ultee, M. E., Margoliash, E., and Schwartz, R. H., 1979, The T-lymphocyte response to cytochrome *c*. I. Demonstration of a T-cell heteroclitic proliferative response and identification of a topographic antigenic determinant on pigeon cytochrome *c* whose immune recognition requires two complementing major histocompatibility complex-linked immune response genes, *J. Exp. Med.* **150**:830.

589

IMMUNE RESPONSE
GENES IN THE
REGULATION OF
MAMMALIAN
IMMUNITY

Solow, H., Hidalgo, R., Blajchman, M., and Singal, D. P., 1977, HLA-A, B, C and B-lymphocyte alloantigens in insulin-dependent diabetes, *Transplant. Proc.* **IX**:1859.

Sprent, J., 1978a, Restricted helper function of F_1 hybrid T cells positively selected to heterologous erythrocytes in irradiated parental strain mice. I. Failure to collaborate with B cells of the opposite parental strain not associated with active suppression, *J. Exp. Med.* **147**:1142.

Sprent, J., 1978b, Restricted helper function of F_1 hybrid T cells positively selected to heterologous erythrocytes in irradiated parental strain mice. II. Evidence for restrictions involving helper cell induction and T–B collaboration, both mapping to the K-end of the *H-2* complex, *J. Exp. Med.* **147**:1159.

Sprent. J., 1978c, Two subgroups of T helper cells in F_1 hybrid mice revealed by negative selection to heterologous erythrocytes in vivo, *J. Immunol.* **121**:1691.

Sprent, J., 1978d, Restricted helper function of F_1 → parent bone marrow chimeras controlled by *K*-end of *H-2* complex, *J. Exp. Med.* **147**:1838.

Sprent, J., and von Boehmer, H., 1976, Helper function of T cells depleted of alloantigen-reactive lymphocytes by filtration through irradiated F_1 hybrid recipients. I. Failure to collaborate with allogeneic B cells in a secondary response to sheep erythrocytes measured in vivo, *J. Exp. Med.* **144**:617.

Sprent, J., and von Boehmer, H., 1979, T-helper function of parent → F_1 chimeras. Presence of a separate T-cell subgroup able to stimulate allogeneic B cells but not syngeneic B cells, *J. Exp. Med.* **149**:387.

Stastny, P., 1978, Association of the B-cell alloantigen DRw4 with rheumatoid arthritis, *N. Engl. J. Med.* **298**:869.

Stastny, P., and Fink, C. W., 1979, Different HLA-D associations in adult and juvenile rheumatoid arthritis, *J. Clin. Invest.* **63**:124.

Stingl, G., Katz, S. I., Shevach, E. M., Wolff-Schreiner, E., and Green, I., 1978, Detection of Ia antigens on Langerhans cells in guinea pig skin, *J. Immunol.* **120**:570.

Stout, R. D., Murphy, D. B., McDevitt, H. O., and Herzenberg, L. A., 1977, The Fc receptor on thymus-derived lymphocytes. IV. Inhibition of binding of antigen-antibody complexes to Fc receptor-positive T cells by anti-Ia sera, *J. Exp. Med.* **145**:187.

Swanborg, R. H., 1975, Antigen-induced inhibition of experimental allergic encephalomyelitis. III. Localization of an inhibitory site distinct from the major encephalitogenic determinant of myelin basic protein, *J. Immunol.* **114**:191.

Swierkosz, J. E., and Swanborg, R. H., 1975, Suppressor cell control of unresponsiveness to experimental allergic encephalomyelitis, *J. Immunol.* **115**:631.

Swierkosz, J. E., Rock, K., Marrack, P., and Kappler, J. W., 1978, The role of *H-2*-linked genes in helper T-cell function. II. Isolation on antigen-pulsed macrophages of two separate populations of F_1 helper T cells each specific for antigen and one set of parental *H-2* products, *J. Exp. Med.* **147**:554.

Swierkosz, J. E., Marrack, P., and Kappler, J. W., 1979, The role of H-2-linked genes in helper T cell function. V. *I*-region control of helper T cell interaction with antigen-presenting macrophages, *J. Immunol.* **123**:654.

Szenberg, A., Marchalonis, J. J., and Warner, N. L., 1977, Direct demonstration of murine thymus-dependent cell surface endogenous immunoglobulin, *Proc. Natl. Acad. Sci. U.S.A.* **74**:2113.

Tada, T., 1977, Regulation of the antibody response by T cell products determined by different *I* subregions, in: *Immune System: Genetics and Regulation* (E. E. Sercarz, L. A. Herzenberg, and C. F. Fox, eds.), pp. 345–361, Academic Press, New York.

Tada, T., and Okumura, K., 1979, The role of antigen-specific T cell factors in the immune response, *Adv. Immunol.* **28**:1.

Tada, T., Taniguchi, M., and Takemori, T., 1975, Properties of primed suppressor T cells and their products, *Transplant. Rev.* **26**:106.

Tada, T., Taniguchi, M., and David, C. S., 1976, Properties of the antigen-specific suppressive T-cell factor in the regulation of antibody response of the mouse. IV. Special subregion assignment of the gene(s) that codes for the suppressive T-cell factor in the *H-2* histocompatibility complex, *J. Exp. Med.* **144**:713.

Tada, T., Taniguchi, M., and Okumura, K., 1977, Regulation of antibody response by antigen-specific T cell factors bearing *I* region determinants, in: *Progress in Immunology III* (T. E. Mandel, C. Cheers, C. S. Hosking, I. F. C. McKenzie, and G. J. V. Nossal, eds.), pp. 369–377, North-Holland, Amsterdam.

Tada, T., Takemori, T., Okumura, K., Nonaka, M., Tokuhisa, T., 1978, Two distinct types of helper T cells involved in the secondary antibody response: Independent and synergistic effects of Ia^- and Ia^+ helper T cells, *J. Exp. Med.* **147**:446.

Takemori, T., and Tada, T., 1975, Properties of antigen-specific suppressive T-cell factor in the regulation of antibody response of the mouse. I. In vivo activity and immunochemical characterization, *J. Exp. Med.* **142**:1241.

591

IMMUNE RESPONSE
GENES IN THE
REGULATION OF
MAMMALIAN
IMMUNITY

Taniguchi, M., and Miller, J. F. A. P., 1978, Specific suppressive factors produced by hybridomas derived from the fusion of enriched suppressor T cells and a T lymphoma cell line, *J. Exp. Med.* **148**:373.

Taniguchi, M., Hayakawa, K., and Tada, T., 1976*a*, Properties of antigen-specific suppressive T cell factor in the regulation of antibody response of the mouse. II. In vitro activity and evidence for the I region gene product, *J. Immunol.* **116**:542.

Taniguchi, M., Tada, T., and Tokuhisa, T., 1976*b*, Properties of the antigen-specific suppressive T-cell factor in the regulation of antibody response of the mouse. III. Dual gene control of the T-cell-mediated suppression of the antibody response, *J. Exp. Med.* **144**:20.

Taniguchi, M., Saito, T., and Tada, T., 1979, Antigen-specific suppressive factor produced by a transplantable I-J bearing T-cell hybridoma, *Nature (London)* **278**:555.

Taussig, M. J., 1974, T cell factor which can replace T cells *in vivo*, *Nature (London)* **248**:234.

Taussig, M. J., and Holliman, A., 1979, Structure of an antigen-specific suppressor factor produced by a hybrid T-cell line, *Nature (London)* **277**:308.

Taussig, M. J., and Munro, A. J., 1974, Removal of specific cooperative T cell factor by anti-H-2 but not by anti-Ig sera, *Nature (London)* **251**:63.

Taussig, M. J., Mozes, E., and Isac, R., 1974, Antigen-specific thymus cell factors in the genetic control of the immune response to poly (tyrosyl, glutamyl)-poly-D, L-alanyl–poly-lysyl, *J. Exp. Med.* **140**:301.

Taussig, M. J., Munro, A. J., Campbell, R., David, C. S., and Staines, N. A., 1975, Antigen-specific T-cell factor in cell cooperation. Mapping within the *I* region of the *H-2* complex and ability to cooperate across allogeneic barriers, *J. Exp. Med.* **142**:694.

Taussig, M. J., Corvalán, J. R. F., Binns, R. M., and Holliman, A., 1979, Production of an H-2-related suppressor factor by a hybrid T-cell line, *Nature (London)* **277**:305.

Taylor, R. B., and Iverson, G. M., 1971, Hapten competition and the nature of cell-cooperation in the antibody response, *Proc. R. Soc. London, Ser. B* **176**:393.

Taylor, R. B., and Wortis, H. H., 1968, Thymus dependence of antibody response: Variation with dose of antigen and class of antibody, *Nature (London)* **220**:927.

Terasaki, P. I., Park, M. S., Opelz, G., and Ting, A., 1976, Multiple sclerosis and high incidence of a B lymphocyte antigen, *Science* **193**:1245.

Thèze, J., Kapp, J., and Benacerraf, B., 1977*a*, Immunosuppressive factor(s) extracted from lymphoid cells of nonresponder mice primed with L-glutamic acid^{60}-L-alanine^{30}-L-tyrosine^{10} (GAT). III. Immunochemical properties of the GAT-specific suppressive factor, *J. Exp. Med.* **145**:839.

Thèze, J., Waltenbaugh, C., Dorf, M. E., and Benacerraf, B., 1977*b*, Immunosuppressive factor(s) specific for L-glutamic acid^{50}-L-tyrosine^{50} (GT). II. Presence of *I-J* determinants on the GT-suppressive factor, *J. Exp. Med.* **146**:287.

Thèze, J., Waltenbaugh, C., Germain, R. N., and Benacerraf, B., 1977*c*, Immunosuppressive factor(s) specific for L-glutamic acid^{50}-L-tyrosine^{50}. IV. *In vitro* activity and immunochemical properties, *Eur. J. Immunol.* **7**:705.

Thomas, D. W., and Shevach, E. M., 1976, Nature of the antigenic complex recognized by T lymphocytes. I. Analysis with an in vitro primary response to soluble protein antigens, *J. Exp. Med.* **144**:1263.

Thomas, D. W., and Shevach, E. M., 1977*a*, Nature of the antigenic complex recognized by T lymphocytes. II. T-cell activation by direct modification of macrophage histocompatibility antigens, *J. Exp. Med.* **145**:907.

Thomas, D. W., and Shevach, E. M., 1977*b*, Nature of the antigenic complex recognized by T lymphocytes: Specific sensitization by antigens associated with allogeneic macrophages, *Proc. Natl. Acad. Sci. U.S.A.* **74**:2104.

Thomas, D. W., Yamashita, U., and Shevach, E. M., 1977*a*, Nature of the antigenic complex recognized by T lymphocytes. IV. Inhibition of antigen-specific T cell proliferation by antibodies to stimulator macrophage Ia antigens, *J. Immunol.* **119**:223.

Thomas, D. W., Yamashita, U., and Shevach, E. M., 1977*b*, The role of Ia antigens in T cell activation, *Immunol. Rev.* **35**:97.

Thomas, D. W., Clement, L., and Shevach, E. M., 1978, T lymphocyte stimulation by hapten-conjugated macrophages: a model system for the study of immunocompetent cell interactions, *Immunol. Rev.* **40**:182.

Thomas, D. W., Meltz, S. K., and Wilner, G. D., 1979, Nature of the T lymphocyte recognition of macrophage-associated antigens. II. Macrophage determination of guinea pig T cell responses to human fibrinopeptide B, *J. Immunol.* **123**:1299.

Thompson, K., Harris, M., Benjamini, E., Mitchell, G., and Noble, M., 1972, Cellular and humoral immunity: A distinction in antigenic recognition, *Nature (London), New Biol.* **238**:20.

Thomsen, M., Platz, P., Andersen, O. O., Christy, M., Lyngsøe, J., Nerup, J., Rasmussen, K., Ryder, L. P.,

Nielsen, L. S., and Svejgaard, A., 1975, MLC typing in juvenile diabetes mellitus and idiopathic Addison's disease, *Transplant. Rev.* **22**:125.

Tomazic, V., Rose, N. R., and Shreffler, D. C., 1974, Autoimmune thyroiditis. IV. Localization of genetic control of the immune response, *J. Immunol.* **112**:965.

Turkin, D., and Sercarz, E. E., 1977, Key antigenic determinants in regulation of the immune response, *Proc. Natl. Acad. Sci. U.S.A.* **74**:3984.

Tyan, M. L., McDevitt, H. O., and Herzenberg, L. A., 1969, Genetic control of the antibody response to a synthetic polypeptide: Transfer of response with spleen cells or lymphoid precursors, *Transplant. Proc.* **1**:548.

Unanue, E. R., and Cerottini, J.-C., 1970, The immunogenicity of antigen bound to the plasma membrane of macrophages, *J. Exp. Med.* **131**:711.

Unanue, E. R., Dorf, M. E., David, C. S., and Benacerraf, B., 1974, The presence of I-region-association antigens on B cells in molecules distinct from immunoglobulin and H-2K and H-2D, *Proc. Natl. Acad. Sci. U.S.A.* **71**:5014.

Urba, W. J., and Hildemann, W. H., 1978, *H-2*-linked recessive *Ir* gene regulation of high antibody responsiveness to TNP hapten conjugated to autogenous albumin, *Immunogenetics* **6**:433.

Vladutiu, A. O., and Rose, N. R., 1974, HL-A antigens: association with disease, *Immunogenetics* **1**:305.

de Vries, R. R. P., Kreeftenberg, H. G., Loggen, H. G., and van Rood, J. J., 1977, In vitro responsiveness to vaccinia virus and HLA, *N. Engl. J. Med.* **297**:692.

Waldmann, H., Pope, H., and Munro, A. J., 1975, Cooperation across the histocompatibility barrier, *Nature (London)* **258**:728.

Waldmann, H., Pope, H., and Munro, A. J., 1976, Cooperation across the histocompatibility barrier: $H-2^d$ T cells primed to antigen in an $H-2^d$ environment can cooperate with $H-2^k$ B cells, *J. Exp. Med.* **144**:1707.

Waldmann, H., Pope, H., Brent, L., and Bighouse, K., 1978, Influence of the major histocompatibility complex on lymphocyte interactions in antibody formation, *Nature (London)* **274**:166.

Waldmann, H., Pope, H., Bottles, C., and Davies, A. J. S., 1979a, The influence of thymus on the development of MHC restrictions exhibited by T-helper cells, *Nature (London)* **277**:137.

Waldmann, H., Munro, A., and Maurer, P. H., 1979b, The role of MHC gene products in the development of the T-cell repertoire, in: *Recent Trends in the Immunobiology of Bone Marrow Transplantation* (S. Thierfelder, H. Rodt and H. J. Kolb, eds.), Vol. 25 (supplement to BLUT), Springer-Verlag, New York.

Waldmann, T. A., Strober, W., and Blaese, R. M., 1980, T and B cell immunodeficiency diseases, in: *Clinical Immunology* (C. Parker, ed.), pp. 314–375, W. B. Saunders, Philadelphia.

Waldron, J. A., Jr., Horn, R. G., and Rosenthal, A. S., 1973, Antigen-induced proliferation of guinea pig lymphocytes in vitro: obligatory role of macrophages in the recognition of antigen by immune T-lymphocytes, *J. Immunol.* **111**:58.

Waltenbaugh, C., 1979, Specific and nonspecific suppressor T-cell factors, in: *Biology of the Lymphokines* (S. Cohen, E. Pick, and J. J. Oppenheim, eds.), pp. 421–442, Academic Press, New York.

Waltenbaugh, C., Debré, P., and Benacerraf, B., 1976, Analysis of the cross-reactive immune suppression induced by the random copolymers L-glutamic acid^{50}-L-tyrosine^{50} (GT), L-glutamic acid^{60}-L-alanine^{40} (GA), and L-glutamic acid^{60}-L-alanine^{30}-L-tyrosine^{10} (GAT), *J. Immunol.* **117**:1603.

Waltenbaugh, C., Debré, P., Thèze, J., and Benacerraf, B., 1977a, Immunosuppressive factor(s) specific for L-glutamic acid^{50}-L-tyrosine^{50} (GT). I. Production, characterization, and lack of H-2 restriction for activity in recipient strain, *J. Immunol.* **118**:2073.

Waltenbaugh, C., Thèze, J., Kapp, J. A., and Benacerraf, B., 1977b, Immunosuppressive factor(s) specific for L-glutamic acid^{50}-L-tyrosine^{50}(GT). III. Generation of suppressor T cells by a suppressive extract derived from GT-primed lymphoid cells, *J. Exp. Med.* **146**:970.

Waltenbaugh, C., Dessein, A., and Benacerraf, B., 1979, Characterization of the primary IgM response to GAT and GT: conditions required for the detection of IgM antibodies, *J. Immunol.* **122**:27.

Warner, C. M., McIvor, J. L., Maurer, P. H., and Merryman, C. F., 1977, The immune response of allophenic mice to the synthetic polymer L-glutamic acid, L-lysine, L-phenylalanine. II. Lack of gene complementation in two nonresponder strains, *J. Exp. Med.* **145**:766.

Warner, C. M., Berntson, T. J., Eakley, L., McIvor, J. L., and Newton, R. C., 1978, The immune response of allophenic mice to 2,4-dinitrophenyl (DNP)-bovine gamma globulin. I. Allotype analysis of anti-DNP antibody, *J. Exp. Med.* **147**:1849.

Warner, N. L., 1974, Membrane immunoglobulins and antigen receptors on B and T lymphocytes, *Adv. Immunol.* **19**:67.

Watson, J., and Riblet, R., 1974, Genetic control of responses to bacterial lipopolysaccharides in mice. I. Evi-

593

IMMUNE RESPONSE
GENES IN THE
REGULATION OF
MAMMALIAN
IMMUNITY

dence for a single gene that influences mitogenic and immunogenic responses to lipopolysaccharide, *J. Exp. Med.* **140**:1147.

Weinberger, J. Z., Greene, M. I., Benacerraf, B., and Dorf, M. E., 1979*a*, Hapten-specific T-cell responses to 4-hydroxy-3-nitrophenyl acetyl. I. Genetic control of delayed-type hypersensitivity by V_H and I-A-region genes, *J. Exp. Med.* **149**:1336.

Weinberger, J. Z., Germain, R. N., Ju, S.-T., Greene, M. I., Benacerraf, B., and Dorf, M. E., 1979*b*, Hapten-specific T-cell responses to 4-hydroxy-3nitrophenyl acetyl. II. Demonstration of idiotypic determinants on suppressor T cells, *J. Exp. Med.* **150**:761.

Weinberger, J. Z., Benacerraf, B., and Dorf, M. E., 1979*c*, Ir gene controlled carrier effects in the induction and elicitation of hapten-specific delayed-type hypersensitivity responses, *J. Exp. Med.* **150**:1255.

Wernet, P., 1976, Human Ia-type alloantigens: Methods of detection, aspects of chemistry and biology, markers for disease states, *Transplant. Rev.* **30**:271.

Whitmore, A. C., Babcock, G. F., and Haughton, G., 1978, Genetic control of susceptibility of mice to Rous sarcoma virus tumorigenesis. II. Segregation analysis of strain A.SW-associated resistance to primary tumor induction, *J. Immunol.* **121**:213.

Wilson, D. B., Lindahl, K. F., Wilson, D. H., and Sprent, J., 1977, The generation of killer cells to trinitrophenyl-modified allogeneic targets by lymphocyte populations negatively selected to strong alloantigens, *J. Exp. Med.* **146**:361.

Winchester, R. J., Ebers, G., Fu, S. M., Espinosa, L., Zabriskie, J., and Kunkel, H. G., 1975, B-cell alloantigen Ag 7a in multiple sclerosis, *Lancet* **1975**(2):814.

Woodland, R., and Cantor, H., 1978, Idiotype-specific T helper cells are required to induce idiotype-positive B memory cells to secrete antibody, *Eur. J. Immunol.* **8**:600.

Yamamoto, H., Hamaoka, T., Yoshizawa, M., Kuroki, M., and Kitagawa, M., 1977, Regulatory functions of hapten-reactive helper and suppressor T lymphocytes. I. Detection and characterization of hapten-reactive suppressor T-cell activity in mice immunized with hapten-isologous protein conjugate, *J. Exp. Med.* **146**:74.

Yamashita, U., and Shevach, E. M., 1978, The histocompatibility restrictions on macrophage T-helper cell interaction determine the histocompatibility restrictions on T-helper cell B-cell interaction, *J. Exp. Med.* **148**:1171.

Yano, A., Schwartz, R. H., and Paul, W. E., 1977, Antigen presentation in the murine T-lymphocyte proliferative response. I. Requirement for genetic identity at the major histocompatibility complex, *J. Exp. Med.* **146**:828.

Yano, A., Schwartz, R. H., and Paul, W. E., 1978, Antigen presentation in the murine T-lynphocyte proliferative response. II. Ir-GAT-controlled T lymphocyte responses require antigen-presenting cells from a high responder donor, *Eur. J. Immunol.* **8**:344.

Yowell, R. L., Araneo, B. A., Miller, A., and Sercarz, E. E., 1979, Amputation of a suppressor determinant on lysozyme reveals underlying T-cell reactivity to other determinants, *Nature (London)* **279**:70.

Zeicher, M., Mozes, E., and Lonai, P., 1977, Lymphocyte alloantigens associated with X-chromosome-linked immune response genes, *Proc. Natl. Acad. Sci. U.S.A.* **74**:721.

Zeiger, A. R., and Maurer, P. H., 1976, Genetic control of the immune response in guinea pigs to the known sequence polymer (Tyr-Ala-Glu-Gly)$_n$, *J. Immunol.* **117**:708.

Zembala, M., and Asherson, G. L., 1974, T cell suppression of contact sensitivity in the mouse. II. The role of soluble suppressor factor and its interactions with macrophages, *Eur. J. Immunol.* **4**:799.

Zinkernagel, R. M., 1976, Virus-specific T-cell-mediated cytotoxicity across the H-2 barrier to virus-altered alloantigen, *Nature (London)* **261**:139.

Zinkernagel, R. M., 1978, Thymus and lymphohemopoietic cells: their role in T cell maturation in selection of T cells' H-2-restriction specificity and in H-2 linked Ir gene control, *Immunol. Rev.* **42**:224.

Zinkernagel, R. M., and Doherty, P. C., 1974*a*, Restriction of *in vitro* T cell-mediated cytotoxicity in lymphocytic choriomeningitis within a syngeneic or semiallogeneic system, *Nature (London)* **248**:701.

Zinkernagel, R. M., and Doherty, P. C., 1974*b*, Immunological surveillance against altered self components by sensitized T lymphocytes in lymphocytic choriomeningitis, *Nature (London)* **251**:547.

Zinkernagel, R. M., and Doherty, P. C., 1975, *H-2* compatibility requirement for T-cell-mediated lysis of target cells infected with lymphocytic choriomeningitis virus, *J. Exp. Med.* **141**:1427.

Zinkernagel, R. M., Callahan, G. N., Althage, A., Cooper, S., Klein, P. A., and Klein, J., 1978*a*, On the thymus in the differentiation of "H-2 self-recognition" by T cells: Evidence for dual recognition? *J. Exp. Med.* **147**:882.

Zinkernagel, R. M., Callahan, G. N., Althage, A., Cooper, S., Streilein, J. W., and Klein, J., 1978*b*, The lymphoreticular system in triggering virus plus self-specific cytotoxic T cells: Evidence for T help, *J. Exp. Med.* **147**:897.

Zinkernagel, R. M., Callahan, G. N., Klein, J., and Dennert, G., 1978*c*, Cytotoxic T cells learn specificity for self H-2 during differentiation in the thymus, *Nature (London)* **271**:251.

Zinkernagel, R. M., Althage, A., Cooper, S., Kreeb, G., Klein, P. A., Sefton, B., Flaherty, L., Stimpfling, J., Shreffler, D., and Klein, J., 1978*d*, *Ir*-genes in *H-2* regulate generation of anti-viral cytotoxic T cells. Mapping to *K* or *D* and dominance of unresponsiveness, *J. Exp. Med.* **148**:592.

Zinkernagel, R. M., Althage, A., Cooper, S., Callahan, G., and Klein, J., 1978*e*, In irradiation chimeras, K or D regions of the chimeric host, not of the donor lymphocytes, determine immune responsiveness of antiviral cytotoxic T cells, *J. Exp. Med.* **148**:805.

Index

Pattern regulation (*cont.*)
  from first cleavage to midblastula, 197–200
  in gastrula, 292
  and homeostasis, 134
  importance of egg geometry for, 259
  inhibitors of cell division and, 256–257
  in mid- and late blastula, 288
  from midblastula to gastrula, 229–233
      in neurula, 292
  in oocyte and unfertilized egg, 153–154
  in oogenesis, 276–277
  partitioning mechanisms, 276
  and postfertilization regulation, 180–181
  and prevention of twinning, 198
  by regenerating adult, 134
  self-adjusting, 134
  and self-organization, 233
  and separated blastomeres, 194, 257–258
  and single yolk mass, 277
  surface, 199
  in transplantation, 254
  and UV irradiation, 180
  and vegetal cytoplasm arrangements, 197
  of vegetal hemisphere, 197
  and yolk mass coherence, 181
Penicillium chrysogenum, trans-inhibition in, 415
Peptide bonds
  cis–trans isomer, 47, 48
  distortions from planarity, 47
  proteolytic cleavage of, 48
Peptide isomers, 47–48
Periplasmic binding proteins, activity measurement
      of, 421
Periplasmic proteins, 423–424
  N-terminal sequences of, 423
Peritrichous bacteria, defined, 396
Phage(s)
  double-stranded DNA, 102, 116
  filamentous, 102
  giant, 120
  lambda, 119, 120, 126–127
  φ80, 115
  P2, 120
  P22, 117, 119–120, 121, 126–127
  single-stranded, 102
  T3, 119
  T4, 119, 126–127
Phage assembly pathways, central feature of, 102
Phase variation, in flagellar synthesis, 393–394
Phosphorylase, covalent conversion of, 26–28
Phosphorylation–dephosphorylation regulation,
      24–28
Phosphotransferase system
  and enzyme synthesis regulation, 416
  and secondary components in, 403
Phototaxis, defined, 378
Physarum, genetic morphogenesis studies, 102

Physiological competition hypothesis, 221
Physiological regulation, role of limited proteolysis
      in, 422
Pigment segregation, mechanisms, 274–275
Pinocytosis, 272
  of oocyte plasma membrane, 146
Pinocytotic vesicles
  within oocyte, 272
  and unit membrane formation, 272
Plasma membrane of full-grown oocyte, 145–146
Platelet activation, in vegetal hemisphere, 208
Pleurodeles
  cortical granules in, 146
  oocyte microvilli, 146
Point mutations, in single-domain proteins, 71
Point symmetry
  defined, 80
  and subunit aggregation, 80
Polar granules, role in germ-line determination, 270
Polymerization
  rate of, 114
  tail tube, 114
Polynucleotides, functional versatility of sequences
      of, 395
Polypeptide chains
  extended, 64
  loops, 64
  turns, 63
Polyspermy, 149, 150, 157
  block to, 281
  fast block to, 158
Positional information, mechanism for self-
      organization, 246
Postfertilization wave(s), 166, 281
  and gray crescent formation, 162
  and surface changes, 162–163
Posttranscriptional modification, and methyl-
      accepting chemotaxis proteins, 404
Pre-B cells, 445–448
  detection of, 446
  general properties of, 448
Precursor cells, and dynamic determination, 249
Precursor shell(s)
  dissociation of scaffolding from, 121–122
  electron microscopy study of, 117
  enlargement of, 122
  with/without scaffolding protein, 118
Pregastrular movement, 190
Pregastrulation, and gene expression, 213
Pregastrulation movements, 289
Premelanosomes, origins of, 274
Presumptive primordial germ cells, 191
Previtellogenic oocyte, asymmetry of, 270
Previtellogenic period
  blood vessel invasion, 269
  centriolar–nuclear axis formation in, 268
  DNA increase in, 268–269